COPING WITH CHAOS

WILEY SERIES IN NONLINEAR SCIENCE

Series Editors: **ALI H. NAYFEH, Virginia Tech**
ARUN V. HOLDEN, University of Leeds

Abdullaev	THEORY OF SOLITONS IN INHOMOGENOUS MEDIA
Nayfeh	METHOD OF NORMAL FORMS
Nayfeh & Balachandran	APPLIED NONLINEAR DYNAMICS
Nayfeh & Pai	LINEAR AND NONLINEAR STRUCTURAL MECHANICS
Ott, Sauer & Yorke	COPING WITH CHAOS
Rozhdestvensky	MATCHED ASYMPTOTICS OF LIFTING FLOWS

COPING WITH CHAOS

Analysis of Chaotic Data and The Exploitation of Chaotic Systems

Edited by

Edward Ott
University of Maryland

Tim Sauer
George Mason University

James A. Yorke
University of Maryland

A Wiley-Interscience Publication
John Wiley & Sons, Inc.
New York / Chichester / Brisbane / Toronto / Singapore

This text is printed on acid-free paper.

Library of Congress Cataloging in Publication Data:

Coping with chaos : analysis of chaotic data and the exploitation of chaotic systems / [edited by] Edward Ott, Tim Sauer, James A. Yorke.
p. cm.
Includes bibliographical references.
ISBN 0–471–02556–9 (alk. paper)
1. Chaotic behavior in systems. 2. Numerical calculations.
I. Ott, Edward. II. Sauer, Tim. III. Yorke, James A.
Q172.5.C45C67 1994
003'.7--dc20 93-49071
CIP

Printed in the United States of America

10 9 8 7 6 5 4 3 2 1

Contents

Preface

The existence of chaotic dynamics has been known for a long time among mathematicians, starting with Poincaré's work at the turn of the century and continuing with the subsequent pioneering studies of Birkhoff, Cartwright and Littlewood, Levinson, Smale, and Kolmogorov and his students. At first, knowledge of this work remained largely confined to the mathematical community. Starting in the mid-1970s, and stimulated by the availability of digital computers, this situation rapidly changed, as the broad impact and occurrence of chaos in the sciences and engineering began to be widely recognized. It has been demonstrated that chaos is relevant to problems in fields as diverse as ecology, chemistry, fluid mechanics, solid state devices, biology, and celestial mechanics.

More recently, the study of chaotic dynamics has entered a new phase. In addition to the original pursuits of demonstrating chaos in a wide range of situations, and studying the properties of chaotic dynamics, many researchers are interested in the general set of problems that we call **coping with chaos**. By this term we mean utilizing the basic knowledge of the theory of chaos either to analyze chaotic experimental time series data, or else to use the presence of chaos to achieve some practical goal. In most of these cases, the equations underlying the dynamics either are not known, or are known in principle but are too complex to solve. How can the researcher most efficiently exploit available measurements when the equations are missing or not helpful?

Remarkable progress has been made on this group of interrelated problems during the last fifteen years. This progress has been spread over a variety of disciplines, consistent with the unifying conceptual power of the paradigms of chaos. With this volume, we have attempted to provide a single place in which to gather some of the fundamental articles that have shaped this pursuit up to the present time, and to give the reader a point of entry to the rapidly expanding scientific literature.

The collection we have assembled is a selection from a diversity of topics and viewpoints. Specific choices were dependent upon our personal tastes and biases. Space restrictions and our wish to keep the book affordable constrained us to include fewer articles than we had originally intended. We hope that the bibliography at the end of the book will aid the reader in developing a more comprehensive picture.

Part I consists of five chapters of background material which will illuminate the reprinted articles that follow. Chapter 1 will familiarize the reader with the basic concepts of chaos, and Chapters 2–4 introduce the important topics of dimension, symbolic dynamics, Lyapunov exponents, and entropy. Chapter 5 is a survey of fundamental concepts on attractor reconstruction from time series data.

The reprints that follow are organized under two separate parts. Part II deals with methods of analyzing experimental time series data from chaotic systems, including embedding techniques, calculation of dimension and Lyapunov exponents, and determination of periodic orbits and symbolic dynamics. Part III deals with studies in which the unique attributes of chaos are utilized for a practical purpose, including prediction of chaotic time series, noise filtering of chaotic data, control of chaotic systems, and the use of chaotic signals for communication. The book concludes with a research bibliography of articles directed toward coping with chaos.

Edward Ott
College Park, MD

Tim Sauer
Fairfax, VA

James A. Yorke
College Park, MD

COPING WITH CHAOS

PART I

Background

Chaos is the score on which reality is written.

— Henry Miller

CHAPTER 1

Introduction to Chaos

The purpose of this chapter is to provide an elementary introduction to the subject of chaos. The concepts discussed here will provide a basis for the discussions in the subsequent chapters and the reprints that follow.

1.1 Dynamical systems

A *dynamical system* may be thought of as any set of equations giving the time evolution of the state of a system from a knowledge of its previous history. Examples are Maxwell's equations, the Navier-Stokes equations, and Newton's equations for the motion of a particle with suitably specified forces. A common setting is a system of k first-order autonomous ordinary differential equations

$$\dot{\mathbf{x}} = \mathbf{F}(\mathbf{x}), \tag{1.1}$$

where $\mathbf{x} = (x^{(1)}, x^{(2)}, \ldots, x^{(k)})$ denotes k state components, considered as a vector in k-dimensional phase space, $\mathbf{F}(\mathbf{x}) = (F^{(1)}(\mathbf{x}), F^{(2)}(\mathbf{x}), \ldots, F^{(k)}(\mathbf{x}))$ is a k-dimensional vector function of $\mathbf{x}$, and $\dot{\mathbf{x}}$ denotes the time derivative $d\mathbf{x}/dt$. Under general conditions on the form of the equations, the existence and uniqueness properties of solutions hold. Then Equation (1.1) is a dynamical system, because given any initial state $\mathbf{x}(0)$ we can, in principle, solve Equation (1.1) to unambiguously determine the *trajectory* (or *orbit*) $\mathbf{x}(t)$ for all subsequent time. In this system, the time variable t is continuous.

It is also important to consider dynamical systems where time is a discrete variable. Let n be an integer-valued (discrete) time variable. Then another example of a dynamical system is a k-dimensional *map*

$$\mathbf{x}_{n+1} = \mathbf{G}(\mathbf{x}_n), \tag{1.2}$$

where $\mathbf{x}$ is again a k-vector and $\mathbf{G}(\mathbf{x})$ is a k-dimensional vector function of $\mathbf{x}$. Equation (1.2) is a dynamical system because given an initial condition $\mathbf{x}_0$, the equation yields $\mathbf{x}_1$ via $\mathbf{x}_1 = \mathbf{G}(\mathbf{x}_0)$, which then yields $\mathbf{x}_2$ via $\mathbf{x}_2 = \mathbf{G}(\mathbf{x}_1)$, and so on, thus generating the discrete time trajectory $\mathbf{x}_0, \mathbf{x}_1, \mathbf{x}_2, \mathbf{x}_3, \ldots$

Maps arise naturally in many applications. In addition, even when the natural statement of a problem is in continuous time, it is often possible and

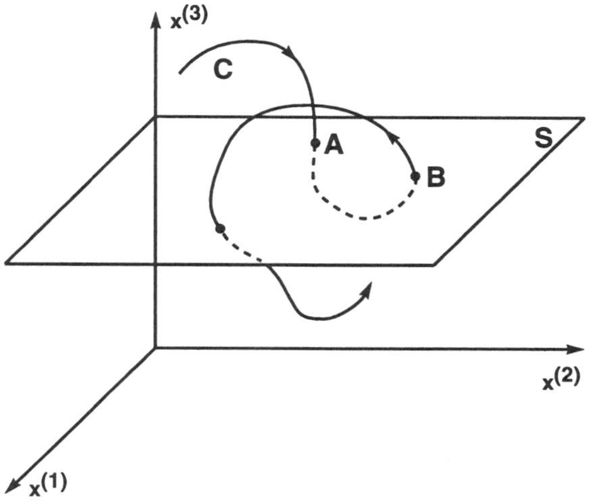

Figure 1.1: *The Poincaré surface of section technique.*

profitable to reduce the problem to a map. This can be done by the Poincaré surface of section technique illustrated in Figure 1.1 for the case of $k = 3$ first-order autonomous ordinary differential equations. The *state-space*, or *phase-space*, of the system is coordinatized by the three rectangular variables $x^{(1)}$, $x^{(2)}$, and $x^{(3)}$. Thus a point in this space corresponds to a state of the system, and, as the system evolves with time, the point $\mathbf{x}(t) = (x^{(1)}, x^{(2)}, x^{(3)})$ moves through the space tracing out a curve (labeled C in Figure 1.1) representing the system trajectory. Now say that we choose a surface in this space, which we call the *surface of section*. In Figure 1.1 this surface, labeled S, was chosen as the plane $x^{(3)} =$ constant, but we emphasize that the choice is a matter of convenience. Every time the trajectory C pierces S, as at points A and B in Figure 1.1, we record the point of piercing on the surface S. For this case, we can label these points by the coordinates $(x^{(1)}, x^{(2)})$. Let A represent the nth piercing of the surface of section and B the $(n+1)$th piercing. We specify A by the two-component vector $\mathbf{y}_n = (x_n^{(1)}, x_n^{(2)})$, where $x_n^{(i)} = x^{(i)}(t_n)$, and t_n is the value of the continuous time variable t at the instant of the nth piercing of S.

Given A (i.e., given $\mathbf{y}_n$), Equation (1.1) can in principle be solved for $\mathbf{x}(t)$ for $t > t_n$ by considering A as an initial condition, and $\mathbf{x}(t)$ can be followed until the next surface of section piercing at B. Thus A uniquely determines B. We conclude that there exists a two-dimensional map

$$\mathbf{y}_{n+1} = \mathbf{G}(\mathbf{y}_n), \tag{1.3}$$

and this map could in principle be iterated to find all subsequent piercings of S. In general, the Poincaré surface of section technique reduces a k-dimensional, continuous-time dynamical system to a $(k-1)$-dimensional map.

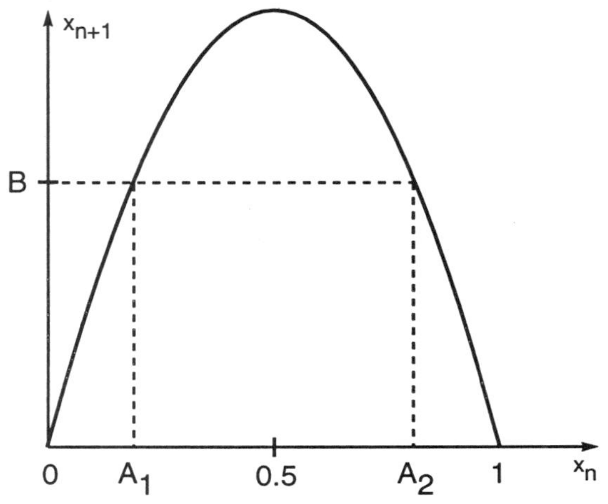

Figure 1.2: *Noninvertibility of the logistic map.*

It is also important to distinguish between maps that are *noninvertible* and maps that are *invertible*. The map $\mathbf{G}$ in Equation (1.2) is invertible if, given $\mathbf{x}_{n+1}$, we can solve Equation (1.2) to determine $\mathbf{x}_n$ uniquely. If this is so, we can define an inverse function $\mathbf{G}^{-1}$ such that

$$\mathbf{x}_n = \mathbf{G}^{-1}(\mathbf{x}_{n+1}). \tag{1.4}$$

As an example of a noninvertible case consider the one-dimensional map

$$x_{n+1} = r x_n (1 - x_n), \tag{1.5}$$

where r is a parameter. This is called the logistic map. As shown in Figure 1.2, for x_{n+1} =B there are two possible values of x_n (namely, A_1 and A_2) from which B could have come. Thus given a value of x, we cannot deduce the orbit which led up to it. As an example of an invertible map, we consider the Hénon map (Hénon, 1976), which is two-dimensional:

$$\begin{aligned} x_{n+1}^{(1)} &= \alpha - (x_n^{(1)})^2 + \beta x_n^{(2)} \\ x_{n+1}^{(2)} &= x_n^{(1)}. \end{aligned} \tag{1.6}$$

Given $\mathbf{x}_{n+1} = (x_{n+1}^{(1)}, x_{n+1}^{(2)})$, Equation (1.6) can be solved for $\mathbf{x}_n$:

$$\begin{aligned} x_n^{(1)} &= x_{n+1}^{(2)} \\ x_n^{(2)} &= \frac{1}{\beta}(x_{n+1}^{(1)} - \alpha + (x_{n+1}^{(2)})^2). \end{aligned} \tag{1.7}$$

Hence Equation (1.6) is an invertible map, provided that $\beta \neq 0$.

A map obtained from a system of the general form of Equation (1.1) via a Poincaré surface of section is necessarily invertible, because given B in Figure 1.1 we can integrate the ordinary differential equations backward in time to find A. Even though this is so, it is often useful to use a noninvertible map to approximate an invertible Poincaré surface of section (or other) map. For example, if $\beta \neq 0$, Equation (1.6) is always invertible. But, if β is very small, the evolution of $x^{(1)}$ is approximated by the noninvertible one-dimensional map

$$x_{n+1}^{(1)} = \alpha - (x_n^{(1)})^2. \tag{1.8}$$

The advantage of this approximation is that now we are dealing with a one-dimensional map rather than a two-dimensional map.

1.2 Attractors

Phase space volumes are conserved in Hamiltonian systems, such as those that arise from the Newtonian mechanics of particles moving without friction. This fact, called Liouville's Theorem, says that if all points in a subset of phase space with a positive finite volume are evolved forward over some time interval, then the resulting set has the same volume as the initial set. We call such a system *conservative*.

In this book we shall mainly be interested in *nonconservative* systems, by which we mean systems whose evolutions do not preserve phase space volumes (and cannot be made to do so by a smooth change of variables). In particular, we shall be most interested in systems which contract volumes in all (or at least some parts) of phase space.

For the case of a map, the factor by which the k-dimensional volume of a differential element at point $\mathbf{x}$ changes upon one iterate of the map $\mathbf{G}$ is given by the magnitude of the Jacobian determinant of $\mathbf{G}$,

$$J = |\det[\mathbf{DG}(\mathbf{x})]|.$$

For example, for the two-dimensional map given by Equation (1.6),

$$\mathbf{DG}(\mathbf{x}) = \begin{pmatrix} \dfrac{\partial x_{n+1}^{(1)}}{\partial x_n^{(1)}} & \dfrac{\partial x_{n+1}^{(1)}}{\partial x_n^{(2)}} \\ \dfrac{\partial x_{n+1}^{(2)}}{\partial x_n^{(1)}} & \dfrac{\partial x_{n+1}^{(2)}}{\partial x_n^{(2)}} \end{pmatrix} = \begin{pmatrix} -2x_n^{(2)} & \beta \\ 1 & 0 \end{pmatrix},$$

and $J = |\beta|$. Thus if $|\beta| < 1$, any area is contracted on each iterate by the factor $|\beta|$.

For the case of the system given by Equation (1.1), the rate of change of the differential volume $dV(t)$ following an orbit $\mathbf{x}(t)$ is given by the divergence of the flow

$$\frac{d}{dt}\left[dV(t)\right] = (\nabla \cdot \mathbf{F})|_{\mathbf{x}=\mathbf{x}(t)} \left[dV(t)\right],$$

where

$$\nabla \cdot \mathbf{F} \equiv \frac{\partial F^{(1)}(\mathbf{x})}{\partial x^{(1)}} + \ldots + \frac{\partial F^{(k)}(\mathbf{x})}{\partial x^{(k)}}.$$

For example, consider the Lorenz equations:

$$\begin{aligned} \frac{dx^{(1)}}{dt} &= -\sigma x^{(1)} + \sigma x^{(2)} \\ \frac{dx^{(2)}}{dt} &= -x^{(1)}x^{(3)} + rx^{(1)} - x^{(2)} \\ \frac{dx^{(3)}}{dt} &= x^{(1)}x^{(2)} - bx^{(3)}. \end{aligned} \tag{1.9}$$

These equations were first obtained as a crude model of convective fluid flow. The divergence of the Lorenz equations is $\nabla \cdot \mathbf{F} = -(\sigma + 1 + b)$, independent of $\mathbf{x}$. Thus, an initial phase space volume $V(0)$ shrinks with time as $V(t) = V(0)\exp[-(\sigma + 1 + b)t]$.

Systems with phase space contraction, as in Equations (1.6) and (1.9), are commonly characterized by the presence of *attractors*. In particular, if one considers an appropriate region of phase space, then it will often be the case that initial conditions in that region eventually converge with increasing time to some bounded subset of the region called the attractor. The set of initial conditions that converge to a particular attractor is called the *basin of attraction* for the attractor.

As an example of an attractor, consider the damped harmonic oscillator

$$\ddot{x} + a\dot{x} + bx = 0.$$

The evolution of a trajectory $(x^{(1)}, x^{(2)}) = (x, \dot{x})$ is shown in Figure 1.3**a**. Here we assume $a > 0$, which corresponds to a friction term that removes energy

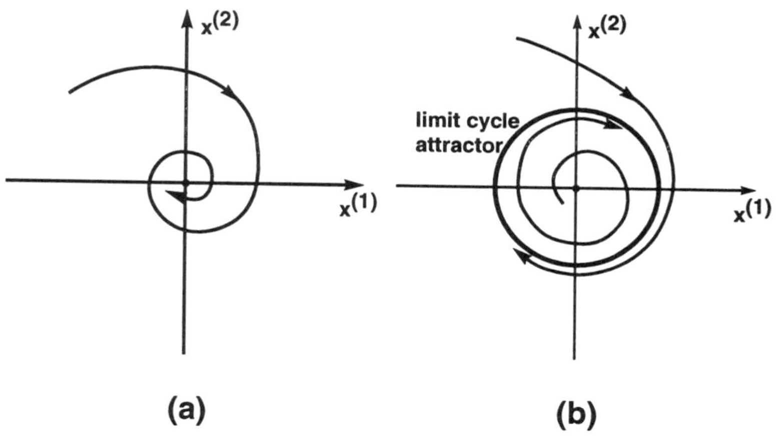

Figure 1.3: *(a) Attractor for the damped harmonic oscillator. (b) Limit cycle attractor.*

from the system. The attractor is the point at rest (in this case the origin). For a damped periodically driven oscillator, the asymptotic motion is typically a limit cycle, tracing out an attractor which is a closed curve in the phase space (Figure 1.3**b**).

In the two examples of Figure 1.3, the attractors were a point (Figure 1.3**a**), which is a set of dimension zero, and a closed curve (Figure 1.3**b**), which is a set of dimension 1. For many other cases the attracting set can be much more irregular, and in fact, can have a dimension that is not an integer. Such sets have been called *fractals*, and when they are attractors they are called strange attractors. The existence of a strange attractor in a physical context was first shown by (Lorenz, 1963) for Equation (1.9).

As an example of a strange attractor, consider the Hénon map (Equation 1.6). Figure 1.4**a** shows the result of plotting 10^4 consecutive points obtained by iterating Equation (1.6) with parameters $\alpha = 1.4$ and $\beta = 0.3$ (with initial transient behavior deleted). The result is essentially a picture of the chaotic attractor.

Figures 1.4**b** and 1.4**c** show successive enlargements of the small rectangles in the preceding figure. Figure 1.4**b** looks like a number of straight parallel lines. However, the magnification of the rectangle, shown in Figure 1.4**c**, shows still more lines. In fact, examination of any region of the attractor at greater and greater magnification will reveal similar pictures. Thus the attractor has repeated structure on arbitrarily small length scales. This is a characteristic of a fractal object.

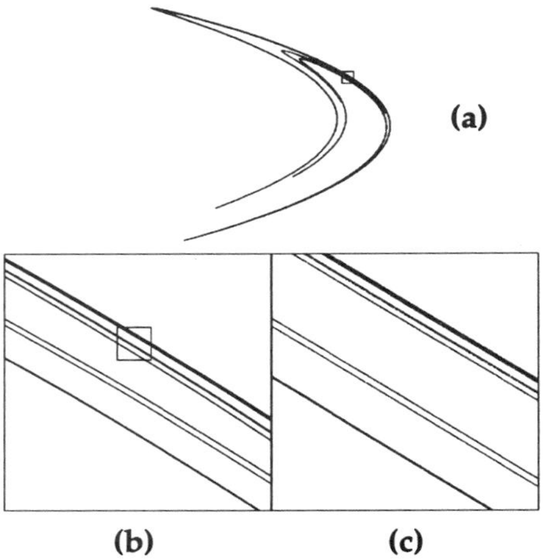

Figure 1.4: *The Hénon chaotic attractor. (a) Full set. (b) Enlargement of region defined by the rectangle in (a). (c) Enlargement of region defined by the rectangle in (b).*

Strange attractors, such as the one in Figure 1.4, can be characterized by a spectrum of dimension values, and this will be discussed in the next chapter. For the purposes of this introductory chapter, we shall restrict attention to one convenient and simple definition of the dimension of a set. Namely, we consider the *box-counting dimension* (alternatively called the *capacity dimension*), defined to be

$$D_0 = \lim_{\epsilon \to 0} \frac{\ln M(\epsilon)}{\ln(1/\epsilon)}, \tag{1.10}$$

where we imagine the attracting set in the k-dimensional phase space to be covered by k-dimensional cubes of edge length ϵ. In dimension $k = 2$ the "cube" is a square, while in dimension $k = 1$ the cube is an interval of length ϵ. Here $M(\epsilon)$ is the minimum number of such cubes needed to cover the set. In practice, one can use a grid formed by marking off intervals of equal length along the axes of a rectangular k-dimensional coordinate system, and it suffices to select the $M(\epsilon)$ cubes covering the set from this grid.

For a point attractor (Figure 1.3**a**), $M(\epsilon) = 1$ independent of ϵ, and Equation (1.10) yields $D_0 = 0$, as it should. For a limit cycle attractor, as in Figure 1.3**b**, we have that $M(\epsilon) \approx l/\epsilon$, where l is the length of the closed curve in the figure; hence, for this case Equation (1.10) gives a dimension $D_0 = 1$, again as it should. Numerical analysis of the Hénon attractor shown in Figure 1.4, however, yields a number between 1 and 2 ($D_0 \approx 1.3$). Hence this attractor is a strange attractor, that is, it is fractal.

Equation (1.10) implies the scaling

$$M(\epsilon) \sim \epsilon^{-D_0} \tag{1.11}$$

for ϵ small. In nature, fractal phenomena will not fit Equation (1.10) in the sense of a mathematical limit as $\epsilon \to 0$. Rather, they will satisfy the scaling relation Equation (1.11) over some (preferably wide) range of length scales which is relevant to the experiment.

The middle-third Cantor set is perhaps the simplest fractal, and can be analyzed completely. Its construction is illustrated in Figure 1.5. The set is formed by starting with the closed line interval from 0 to 1, dividing it in thirds, discarding the open middle third $(\frac{1}{3}, \frac{2}{3})$, then dividing the two remaining thirds into thirds (each of length $\frac{1}{9}$), discarding the two open middle intervals, and so on ad infinitum. The Cantor set is the closed set of points left in the limit of this repeated process.

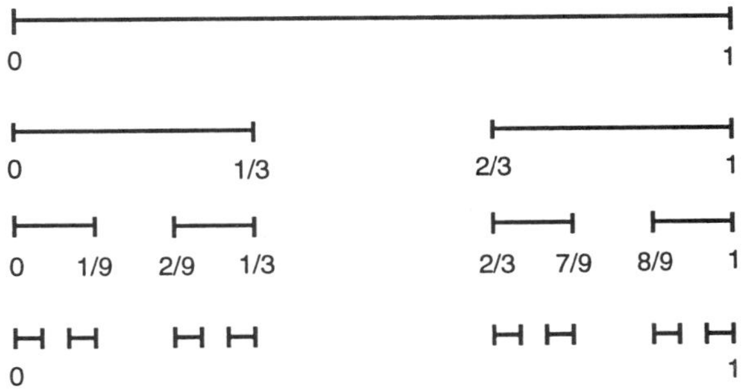

Figure 1.5: *Construction of the middle-third Cantor set.*

It is particularly easy to calculate the box-counting dimension of the middle-third Cantor set. Set $\epsilon = (\frac{1}{3})^n$ with n an integer. Then we see that $M(\epsilon) = 2^n$ and Equation (1.10) (in which $\epsilon \to 0$ corresponds to $n \to \infty$) yields $D_0 = \frac{\ln 2}{\ln 3}$, a number between 0 and 1, and hence the set is a fractal.

1.3 Chaos

Before giving a definition of chaotic motion on a strange attractor we give an example. In particular, we consider the experiment of Moon and Holmes (1979) illustrated in Figure 1.6. When this apparatus is at rest the steel beam has two stable equilibrium positions, one with the beam bending toward the magnet on the right, the other with the beam bending toward the magnet on the left. In the experiment the horizontal position of the apparatus was varied sinusoidally with time. The resulting acceleration caused the position of the beam to vary in an erratically appearing manner, as shown in Figure 1.7**a**.

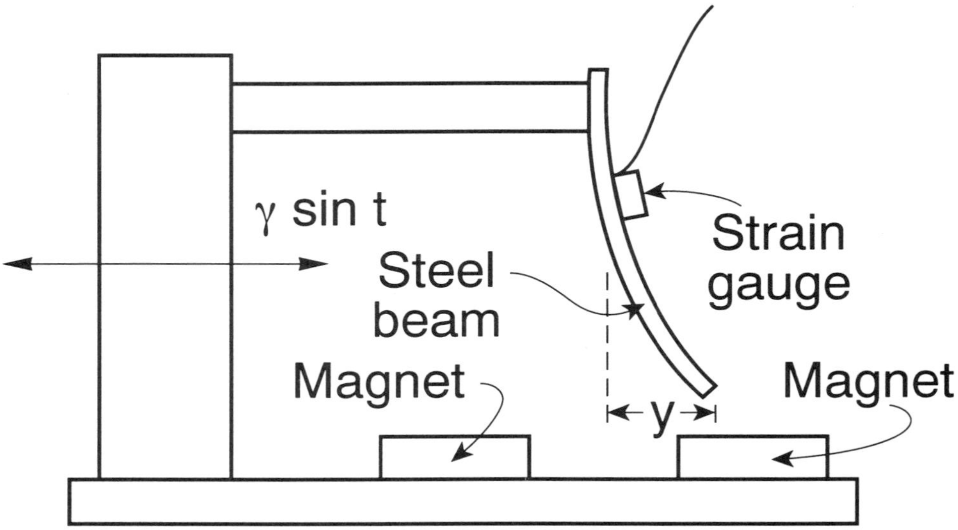

Figure 1.6: *The apparatus of Moon and Holmes.*

Although the apparatus appears to be very simple, one might attempt to explain the complex appearance of the signal by complexities in the physical experiment, such as higher-order vibrational modes on the beam, small inaccuracies in the sinusoidal shaking, and so on. To show that one need not invoke such complications, Moon and Homes numerically solved a simple model of their experiment, the sinusoidally forced Duffing equation:

$$\ddot{y} + \nu\dot{y} + (y^3 - y) = g \sin t. \tag{1.12}$$

Figure 1.7**b** shows the result of this calculation. We see that the same kind of complex motion is present in the solution of the theoretical, noiseless Equation (1.12) as in the laboratory experiment.

The system motion illustrated in Figure 1.7 is typical of the erratic, random-appearing evolution that is a hallmark of chaotic dynamics. Such behavior is extremely common, and as illustrated by Equation (1.12), can be present in even relatively simple dynamical systems. A natural question is whether the complexity of chaotic dynamics only occurs if the system complexity is great enough.

The answer to this question is as follows. If we are dealing with a system of first-order autonomous ordinary differential equations [as in Equation (1.1)], chaos is only possible if the system dimensionality is three or greater: $k \geq 3$.

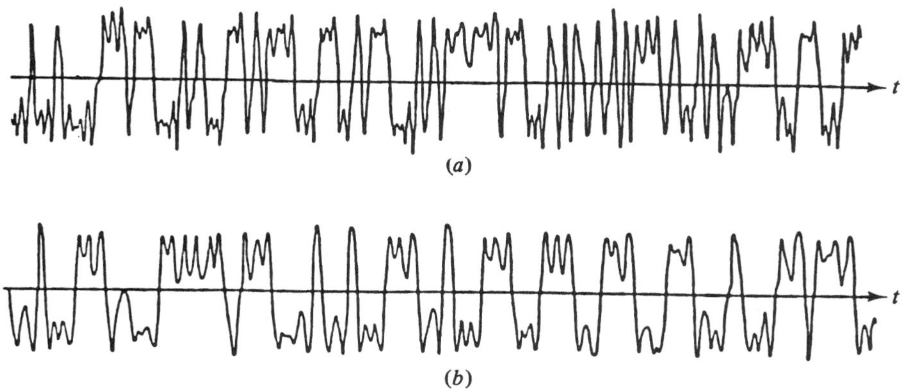

Figure 1.7: *(a) Experimental strain gauge signal. (b) y versus t from the numerical solution of Equation (1.12).*

Thus, if we are presented with a problem involving only two coupled first order autonomous ordinary differential equations, we can immediately conclude that the solutions will not be chaotic.

If the system is an invertible map, we require $k \geq 2$ for chaotic solutions to be possible. This is in accord with the result that a Poincaré surface of section for Equation (1.1) results in a map whose dimensionality is one lower than the number of equations in Equation (1.1).

Finally, if we are dealing with a noninvertible map, then chaos is not ruled out even in the one-dimensional case. Thus, for example, the one-dimensional map, Equation (1.5), has chaotic solutions for many values of the parameter r.

Referring to Equation (1.12), we may ask how this single, second-order nonautonomous equation fits into the discussion above. To answer this we recast the problem in the form of Equation (1.1) by introducing $x^{(1)} = \dot{y}$, $x^{(2)} = y$, and $x^{(3)} = t$, in terms of which we obtain

$$\begin{aligned} \dot{x}^{(1)} &= g \sin x^{(3)} - \nu x^{(1)} - (x^{(2)})^3 + x^{(2)} \\ \dot{x}^{(2)} &= x^{(1)} \\ \dot{x}^{(3)} &= 1. \end{aligned} \tag{1.13}$$

Thus $k = 3$ and chaos is not ruled out for Equation (1.12).

So far we have only spoken of chaos as yielding time series that appear complex. This is a qualitative statement, and one would like a more precise definition of chaos. We say that motion on an attractor is chaotic if it displays *sensitive dependence on initial conditions*, by which we mean the following. Say we have an orbit whose long-term limit fills out the attractor (as, for example, the orbit of the Hénon map shown in Figure 1.4). Now consider the initial condition for this orbit $\mathbf{x}(0)$ and a second initial condition $\mathbf{x}'(0)$ which is displaced from

$\mathbf{x}(0)$ by an infinitesimal distance $\delta\mathbf{x}(0)$,

$$\mathbf{x}'(0) = \mathbf{x}(0) + \delta\mathbf{x}(0).$$

Examine the orbit $\mathbf{x}(t)$ generated by the initial condition $\mathbf{x}(0)$ and the orbit $\mathbf{x}'(t)$ generated by the initial condition $\mathbf{x}'(0)$. If, for a typical orientation of the differential vector $\delta\mathbf{x}(0)$, the displacement $\delta\mathbf{x}(t) = \mathbf{x}'(t) - \mathbf{x}(t)$ grows exponentially with time, then we say there is sensitive dependence on initial conditions and the attractor is chaotic. More precisely, chaos implies that there is an orientation of $\delta\mathbf{x}(0)$ such that

$$h \equiv \lim_{t\to\infty} \frac{1}{t} \ln\left(\frac{||\delta\mathbf{x}(t)||}{||\delta\mathbf{x}(0)||}\right) > 0.$$

The quantity h is called a Lyapunov exponent. See Chapter 4 for an in-depth discussion of Lyapunov exponents.

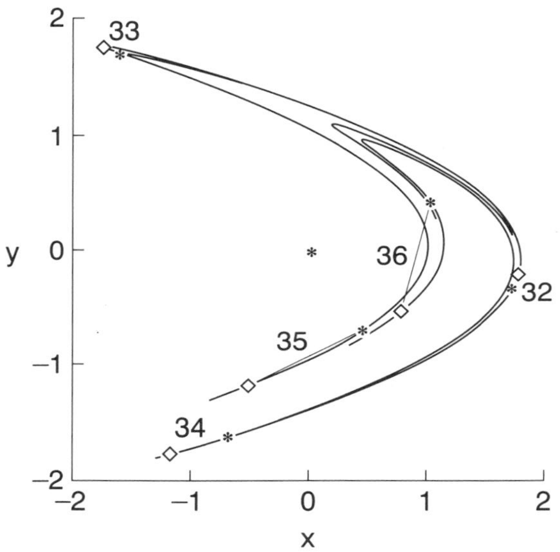

Figure 1.8: *Evolution of one initial condition of the Hénon map which results from two separate computations.*

The existence of chaos means that as time goes on, a small random error grows rapidly (i.e., exponentially) and can eventually completely alter the solution. As an illustration, Figure 1.8 shows the attractor for the Hénon map with parameters $\alpha = 1.4$ and $\beta = 0.3$. Superimposed on this figure we compare trajectories produced by two different computations, both starting from the same initial condition $(x_0^{(1)}, x_0^{(2)}) = (0, 0)$. A straight line joins the corresponding orbit points for the two versions, for iterates 32–36.

The two computations are done identically except that one uses single precision and the other uses double precision; thus the computations differ by $\sim 10^{-14}$ on each iterate. In spite of this extremely small error, the two orbits have separated by a distance comparable to the attractor size by the 36th iterate. Thus we cannot meaningfully calculate the orbit for the Hénon map using 10^{-14} computer roundoff, for more than a handful of iterates. This is the result of the exponential sensitivity inherent in chaos: given an initial state for the above Hénon map, it becomes essentially impossible to predict where the state will reside 40 iterates hence.

It is interesting to note in this connection that the first work on exponential sensitivity in an application was that of Lorenz (1963), who was interested in whether there might be a fundamental limitation to long-term weather prediction. It seems likely that weather involves chaotic atmospheric motion. If so, exponentially sensitive dependence on initial conditions imposes such a limitation. On the other hand, short-term prediction is not ruled out. Indeed, as we shall see later in Chapter 10, the knowledge that a time series originates from some deterministic chaotic process (rather than a stochastic process) opens up new possibilities for short-term prediction of future behavior from past behavior.

References

M. HÉNON, A two-dimensional mapping with a strange attractor. Comm. Math. Phys. **50**, 69 (1976).

E. LORENZ, Deterministic nonperiodic flow. J. Atmos. Sci. **20**, 130 (1963).

F.C. MOON, P.J. HOLMES, A magnetoelastic strange attractor. J. Sound. Vib. **65**, 285 (1979).

CHAPTER 2

Dimension

2.1 Natural measure

In our definition of the box-counting dimension we covered the attractor in k-dimensional phase space with k-dimensional cubes from an ϵ-grid. A typical orbit on the chaotic attractor will bounce around on the attractor, visiting various covering cubes in turn. For a given initial condition $\mathbf{x}_0$, we can define the fraction of time that the orbit spends in a given cube C_i as

$$\mu(C_i, \mathbf{x}_0) = \lim_{T \to \infty} \frac{\eta(C_i, \mathbf{x}_0, T)}{T}, \tag{2.1}$$

where $\eta(C_i, \mathbf{x}_0, T)$ is the amount of time that the orbit starting at $\mathbf{x}_0$ at time $t = 0$ spends in cube i in the time interval $0 \leq t \leq T$. If $\mu(C_i, \mathbf{x}_0)$ is the same for all initial conditions $\mathbf{x}_0$ in the basin of attraction, with the possible exception of a set of $\mathbf{x}_0$ whose total Lebesgue measure (i.e., total k-dimensional volume) is zero, then we drop the $\mathbf{x}_0$ argument of $\mu(C_i, \mathbf{x}_0)$ and say that $\mu(C_i)$ is the *natural measure* of the cube C_i. In the absence of any detailed knowledge of the orbit, we can think of $\mu(C_i)$ as the *probability* that an orbit point falls in cube i.

In general, when some property holds for all points in some set, except for a set of measure zero, then we say that the property holds for *almost every* point with respect to the measure. For example, assuming the existence of a natural measure, we can say that $\mu(C_i)$ is the same for almost every point in the basin of attraction with respect to Lebesgue measure. Thus, if we were to pick a point in the basin of the attractor randomly with a uniform probability density per unit k-dimensional volume, then the probability of picking an $\mathbf{x}_0$ for which $\mu(C_i, \mathbf{x}_0)$ was not the natural measure $\mu(C_i)$ would be zero. An initial condition $\mathbf{x}_0$ for which $\mu(C_i, \mathbf{x}_0) = \mu(C_i)$ for all cubes C_i is called *typical* with respect to Lebesgue measure in the basin.

More generally, we can consider the natural measure of any set S in the phase space (not necessarily a cube), and we denote this $\mu(S)$. The natural measure μ is a *probability measure*, because $\mu(S) \geq 0$ for any S, and $\mu(S) = 1$ when S is the entire phase space. By definition, a measure must also satisfy the condition of countable additivity; namely, if $\{S_i\}$ is any countable collection of disjoint subsets of the phase space, then

$$\mu(\bigcup_i S_i) = \sum_i \mu(S_i).$$

Given a map **G**, we say μ is an *invariant measure* for the map **G** if for any set S,

$$\mu(S) = \mu\left[\mathbf{G}^{-1}(S)\right],$$

where $\mathbf{G}^{-1}(S)$ is the set of all points that map to S under **G**. [Thus $\mathbf{G}^{-1}(S)$ for a set S is defined even when **G** is noninvertible in the sense of Section 1.1.] We say that an invariant probability measure μ is *ergodic* if μ cannot be decomposed into two other invariant probability measures

$$\mu = a\mu_1 + (1-a)\mu_2$$

with $a \neq 0, 1$ and $\mu_1 \neq \mu_2$. The condition of ergodicity implies that the time average of a smooth phase space function over a typical orbit can be replaced by the phase space average over the measure.

2.2 The spectrum of D_q dimensions

The definition of the box-counting dimension was given in Equation (1.10). As ϵ decreases, the rate of increase in the number of covering cubes $M(\epsilon)$ yields the dimension D_0, according to the formula $M(\epsilon) \sim \epsilon^{-D_0}$. It turns out that this is too crude a characterization for some purposes. In particular, it is usually the case that for small ϵ, one can form a collection consisting of most of the $M(\epsilon)$ covering cubes, such that this collection contains only a very small fraction of the natural measure (Farmer et al., 1983). Conversely, only a small fraction of the $M(\epsilon)$ cubes is needed to cover most of the natural measure. Furthermore, this situation becomes more extreme as ϵ is decreased. That is, the fraction of cubes needed to cover a given amount of the natural measure (say 90%) decreases strongly as ϵ is made smaller. The box-counting dimension is insensitive to this phenomenon, since it counts all covering cubes equally in its definition. The count of covering cubes $M(\epsilon)$ is simply the number of cubes for which $\mu(C_i) \neq 0$. It is desirable to have a sharper characterization that is capable of weighting higher probability cubes more strongly.

Such a definition was introduced by Grassberger (1983) and Hentschel and Procaccia (1983). They define a "spectrum" of dimensions as

$$D_q = \frac{1}{q-1} \lim_{\epsilon \to 0} \frac{\ln I(q,\epsilon)}{\ln \epsilon}, \tag{2.2}$$

where $I(q, \epsilon)$ is

$$I(q, \epsilon) = \sum_{i=1}^{M(\epsilon)} [\mu(C_i)]^q , \tag{2.3}$$

and q is a continuous index, $-\infty < q < \infty$. It can be shown that D_q is nonincreasing as a function of q: $D_{q_1} \leq D_{q_2}$ if $q_1 > q_2$. From Equation (2.3) we see that q specifies the degree to which higher probability cubes are more strongly weighted; the larger q is, the more strongly are the higher probability cubes weighted in the sum for $I(q, \epsilon)$. For $q = 0$, Equations (2.2) and (2.3) yield the box-counting dimension as in Equation (1.10).

For the case where all the cubes have approximately the same measure $\mu(C_i) \approx 1/M(\epsilon)$, Equation (2.3) yields $I(q, \epsilon) \approx M(\epsilon)^{1-q}$. Thus $D_q = D_0$ for all q. As we have already said, however, for typical chaotic attractors, $\mu(C_i)$ varies wildly with i, and as a consequence, D_q will be found to vary with q. A measure for which D_q varies with q is called a *multifractal* measure.

Letting $q \to 1$ and applying L'Hopital's rule, Equation (2.2) becomes

$$D_1 = \lim_{\epsilon \to 0} \left[\frac{\sum_i \mu(C_i) \ln \mu(C_i)}{\ln \epsilon} \right] , \tag{2.4}$$

which is called the *information dimension* . The properties of the information dimension are described in the Section 2.3.

Another case of particular interest is the case $q = 2$,

$$D_2 = \lim_{\epsilon \to 0} \left[\frac{\ln \sum_i \mu^2(C_i)}{\ln \epsilon} \right] , \tag{2.5}$$

which is called the *correlation dimension* . The correlation dimension is interesting because it is relatively easy to determine from experimental data. The basic idea (Grassberger and Procaccia, 1983) is as follows. Say we are given data consisting of a set of n orbit points $\mathbf{z}_1, \mathbf{z}_2, \ldots, \mathbf{z}_n$ on the attractor, where n is large and the orbit has been sampled at some fixed time interval δt. We wish to use this data to estimate the quantity $I(2, \epsilon)$ for the natural measure of the attractor for small ϵ. Grassberger and Procaccia argue that the *correlation integral*

$$\hat{C}(\epsilon, n) = \frac{1}{n^2} \sum_{j=1}^{n} \sum_{i=1}^{n} U(\epsilon - |\mathbf{z}_i - \mathbf{z}_j|) \tag{2.6}$$

provides such an estimate, where $U(\cdot)$ is the unit step function: $U(x) = 1$ if x is positive, and is zero otherwise. The summation over i and j in Equation (2.6) gives the number of pairs of the n points of the orbit that are separated by less than ϵ. More precisely, letting $n \to \infty$, we define

$$\hat{C}(\epsilon) = \lim_{n \to \infty} \hat{C}(\epsilon, n). \tag{2.7}$$

Assuming that $\hat{C}(\epsilon)$ gives a good approximation of $I(2, \epsilon)$, we could then find the correlation dimension as

$$D_2 = \lim_{\epsilon \to 0} \frac{\ln \hat{C}(\epsilon)}{\ln \epsilon}, \tag{2.8}$$

where it is assumed that the given orbit is typical in the sense that it generates the natural measure of the attractor. For finite n, as in the case of an actual experiment, one plots a graph of $\ln \hat{C}(\epsilon, n)$ versus $\ln \epsilon$ in the range $\epsilon_1 > \epsilon > \epsilon_2$, where ϵ_1 is of the order of the size of the attractor in phase space, and ϵ_2 is of the order of the smallest spacing $|\mathbf{z}_i - \mathbf{z}_j|$. With luck, this data can be fit reasonably well by a straight line, whose slope then provides an estimate of D_2.

To see the correspondence between $\hat{C}(\epsilon)$ and $I(2, \epsilon)$, we write $\hat{C}(\epsilon)$ as

$$\hat{C}(\epsilon) = \lim_{n \to \infty} \frac{1}{n} \sum_{j=1}^{n} \left[\frac{1}{n} \sum_{i=1}^{n} U(\epsilon - |\mathbf{z}_i - \mathbf{z}_j|) \right].$$

For large n the term in the square bracket tends to the natural measure in the ball $B_\epsilon(\mathbf{z}_j)$ of radius ϵ centered at $\mathbf{z} = \mathbf{z}_j$. Thus

$$\hat{C}(\epsilon) = \lim_{n \to \infty} \frac{1}{n} \sum_{j=1}^{n} \mu \left[B_\epsilon(\mathbf{z}_j) \right],$$

which is the average of the quantity $\mu[B_\epsilon(\mathbf{z}_j)]$ over the orbit; it is the average of the function of $\mathbf{z}$ given by $\mu[B_\epsilon(\mathbf{z})]$ over the natural measure. On the other hand, if we let $C_\epsilon(\mathbf{z})$ denote the cube from the ϵ-grid in which $\mathbf{z}$ lies, then we can rewrite $I(2, \epsilon)$ as

$$I(2, \epsilon) = \lim_{n \to \infty} \frac{1}{n} \sum_{j} \mu \left[C_\epsilon(\mathbf{z}_j) \right].$$

Thus the expressions for $\hat{C}(\epsilon)$ and $I(2, \epsilon)$ involve similar averages, the difference between the two being that one involves a ball of radius ϵ [namely, $B_\epsilon(\mathbf{z})$], while the other involves a cube of edge length ϵ [namely, $C_\epsilon(\mathbf{z})$]. This distinction is not expected to make a difference in the ϵ-scaling, so that $I(2, \epsilon) \sim \hat{C}(\epsilon)$, and Equation (2.8) follows.

2.3 Properties of the information dimension

The information dimension D_1 is special in that it can be thought of as the dimension of the "core set," which is the part of the attractor that contains most of the natural measure. To see what is meant by this, let θ denote a number in the range $0 < \theta \leq 1$, and let $M(\epsilon, \theta)$ be the minimum number of ϵ-grid cubes needed to cover a fraction θ of the natural measure. Now define the *θ-box-counting dimension* as

$$D_0(\theta) = \lim_{\epsilon \to 0} \frac{\ln M(\epsilon, \theta)}{\ln(1/\epsilon)}. \tag{2.9}$$

For $\theta = 1$, this reduces to the usual box-counting dimension, $D_0(1) \equiv D_0$. The more interesting situation is where θ is less than 1. In that case, there is evidence (see Farmer et al., 1983) that for typical chaotic attractors, $D_0(\theta)$ is constant with θ and is equal to the information dimension:

$$D_0(\theta) = D_1 \text{ for } 0 < \theta < 1.$$

This is what we mean by saying that D_1 is the box-counting dimension of the core set.

Another fundamental property of D_1 has to do with the local scaling of the natural measure in a small ball. Again let $B_\epsilon(\mathbf{x})$ denote the k-dimensional ball of radius ϵ centered at the point $\mathbf{x}$. Then we define the *pointwise dimension* of the measure μ at the point $\mathbf{x}$ as

$$\alpha(\mathbf{x}) = \lim_{\epsilon \to 0} \frac{\ln \mu(B_\epsilon(\mathbf{x}))}{\ln \epsilon}. \tag{2.10}$$

Thus $\mu\left[B_\epsilon(\mathbf{x})\right] \sim \epsilon^{\alpha(\mathbf{x})}$. The pointwise dimension $\alpha(\mathbf{x})$ is also called the *singularity exponent* at $\mathbf{x}$ by some authors. It can be shown that $\alpha(\mathbf{x})$ takes on the same value for almost every $\mathbf{x}$ with respect to the natural measure on the attractor, and that this common value is the information dimension. That is, if we form the set of all $\mathbf{x}$ for which $\alpha(\mathbf{x}) \neq D_1$, then the natural measure of this set is zero. If $\mathbf{x}$ is some point on a *typical* orbit on the attractor, then $\alpha(\mathbf{x}) = D_1$. For usual chaotic attractors, however, we emphasize that there are points in the zero natural measure set for which $\alpha(\mathbf{x}) \neq D_1$, and these points are important also. In fact, they are responsible for the difference between D_0 and D_1 and for the variation of D_q with q for multifractal measures. This will be the subject of further discussion in Section 2.5.

2.4 An example: The generalized baker's map

A simple two-dimensional map can be shown to exhibit all the peculiar properties of typical natural measures. In the generalized baker's map (Farmer et al., 1983; Ott, 1993), (x_{n+1}, y_{n+1}) is defined in terms of (x_n, y_n) by

$$x_{n+1} = \begin{cases} \lambda_a x_n & \text{for } y_n < \gamma \\ (1 - \lambda_b) + \lambda_b x_n & \text{for } y_n \geq \gamma \end{cases} \tag{2.11}$$

$$y_{n+1} = \begin{cases} y_n/\gamma & \text{for } y_n < \gamma \\ (y_n - \gamma)/\delta & \text{for } y_n \geq \gamma \end{cases} \tag{2.12}$$

where $\gamma + \delta = 1$ and $\lambda_a + \lambda_b \leq 1$. The action of this map is illustrated in Figure 2.1. The unit square in the xy-plane is mapped to a pair of vertical strips contained in the original unit square.

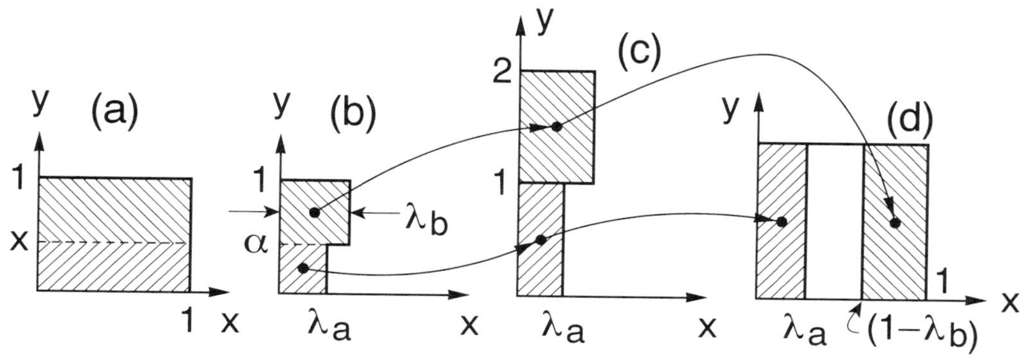

Figure 2.1: *Action of the generalized baker's map on the unit square. (a)→ (b): Horizontal compression. (b) → (c): Vertical elongation. (c) → (d): Placing the λ_a-width strip back in the unit square.*

If we apply the map again, the two strips in Figure 2.1 are mapped to four narrower strips, as shown in Figure 2.2. After n applications there will be 2^n strips of still smaller widths. In the limit as $n \to \infty$ the attractor is a Cantor set of vertical lines, a strange attractor.

We now wish to obtain the dimension spectrum D_q for this attractor. Since the natural measure is uniform in y we can express D_q as

$$D_q = 1 + \hat{D}_q, \tag{2.13}$$

where $\hat{D}_q$ is the dimension of the measure projected to the x-axis. Thus $\hat{D}_q$ is given by Equation (2.2) with $I(q, \epsilon)$ of Equation (2.3) replaced by

$$\hat{I}(q, \epsilon) = \sum_{i=1}^{\hat{M}(\epsilon)} \hat{\mu}^q(\hat{C}_i), \tag{2.14}$$

where we divide the interval $0 \leq x \leq 1$ into intervals of length ϵ and $\hat{\mu}(\hat{C}_i)$ denotes the projected natural measure in the ith ϵ–interval $\hat{C}_i$. We now express $\hat{I}(q, \epsilon)$ as

$$\hat{I}(q, \epsilon) = \hat{I}_a(q, \epsilon) + \hat{I}_b(q, \epsilon), \tag{2.15}$$

where $\hat{I}_a$ denotes that part of $\hat{I}$ that arises from ϵ-intervals containing points in $0 \leq x \leq \lambda_a$ and $\hat{I}_b$ arises from ϵ-intervals containing points in $(1 - \lambda_b) \leq x \leq 1$.

From the construction of Figure 2.1, we see that the fraction of the attractor measure in the strip $0 \leq x \leq \lambda_a$ is γ, while that in the strip $(1 - \lambda_b) \leq x \leq 1$ is $(1 - \gamma) \equiv \delta$. Furthermore, if we were to view the strip $0 \leq x \leq \lambda_a$ after magnification of the x-scale by a factor $1/\lambda_a$, then an exact replica of the entire attractor would be reproduced, and similarly for the strip $(1 - \lambda_b) \leq x \leq 1$ under a horizontal magnification of $1/\lambda_b$. This self-similarity yields

$$\begin{aligned} \hat{I}_a(q, \epsilon) &= \gamma^q \hat{I}(q, \epsilon/\lambda_a) \\ \hat{I}_b(q, \epsilon) &= \delta^q \hat{I}(q, \epsilon/\lambda_b). \end{aligned} \tag{2.16}$$

From Equation (2.15)and Equation (2.16) it follows that

$$\hat{I}(q, \epsilon) = \gamma^q \hat{I}(q, \epsilon/\lambda_a) + \delta^q \hat{I}(q, \epsilon/\lambda_b). \tag{2.17}$$

From Equation (2.2) we assume that $\hat{I}(q, \epsilon)$ takes the form $\hat{I}(q, \epsilon) \sim K\epsilon^{(q-1)\hat{D}_q}$ asymptotically for small ϵ.

Putting this in Equation (2.16) yields a transcendental equation for D_q,

$$\gamma^q \lambda_a^{(1-q)(D_q-1)} + \delta^q \lambda_b^{(1-q)(D_q-1)} = 1. \tag{2.18}$$

For $q \neq 1$, this equation can be solved (by numerical means, if necessary) for D_q; if $q = 1$, we can recover the information dimension D_1 by expanding Equation (2.18) to first order in $(q - 1)$,

$$D_1 = 1 + \frac{\gamma \ln(1/\gamma) + \delta \ln(1/\delta)}{\gamma \ln(1/\lambda_a) + \delta \ln(1/\lambda_b)}. \tag{2.19}$$

In the case $\lambda_a = \lambda_b$, Equation (2.18) can be solved explicitly to give

$$D_q = 1 + \frac{1}{q-1} \frac{\ln(\gamma^q + \delta^q)}{\ln \lambda_a}.$$

The important point is that D_q varies with q. Thus, for example, the box-counting dimension D_0 of the entire attractor typically exceeds D_1, which gives the dimension of the core region of the attractor. The only case where D_q from Equation (2.18) is constant with q is when *both* the stretching and the compression are uniform, $\lambda_a = \lambda_b$, and $\gamma = \delta = 1/2$, in which case $D_q = 1 + (\ln 2)/(\ln \lambda_a^{-1})$, independent of q.

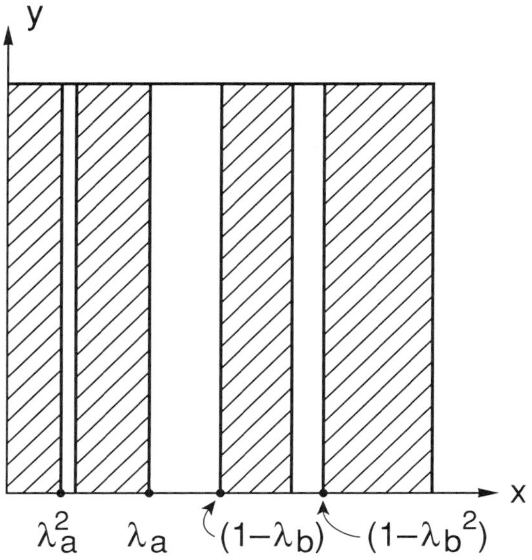

Figure 2.2: *The image of the square under two applications of the generalized baker's map is the four shaded strips shown.*

2.5 The singularity spectrum $f(\alpha)$

To each cube C_i of the ϵ grid we can further associate a singularity index $\alpha_i = \ln \mu(C_i)/\ln \epsilon$, so that

$$\mu(C_i) \equiv \epsilon^{\alpha_i}. \tag{2.20}$$

We count all those cubes for which α_i falls in a small range α to $\alpha + \delta\alpha$. For small enough ϵ there will be many cubes in this range. As an idealization, we replace $\delta\alpha$ by a differential $d\alpha$. We then assume that the number of cubes $dN(\alpha)$ in the range α to $\alpha + d\alpha$ is of the form

$$dN(\alpha) = \rho(\alpha)\epsilon^{-f(\alpha)}d\alpha. \tag{2.21}$$

We can motivate this form as follows. For every point $\mathbf{x}$ on the attractor we calculate its pointwise dimension $\alpha(\mathbf{x})$. We then form the set of all points with a particular value $\alpha(\mathbf{x}) = \alpha$, and we denote the box-counting dimension of this set by $\tilde{f}(\alpha)$. If we interpret α_i in Equation (2.20) as the pointwise dimension at the point in the center of the cube, then it is natural to interpret $f(\alpha)$ in Equation (2.21) as $\tilde{f}(\alpha)$. This has, in fact, been shown rigorously to be correct for certain cases (e.g., for hyperbolic attractors).

To relate the quantity $f(\alpha)$ to D_q, we follow Halsey et al. (1986) and use

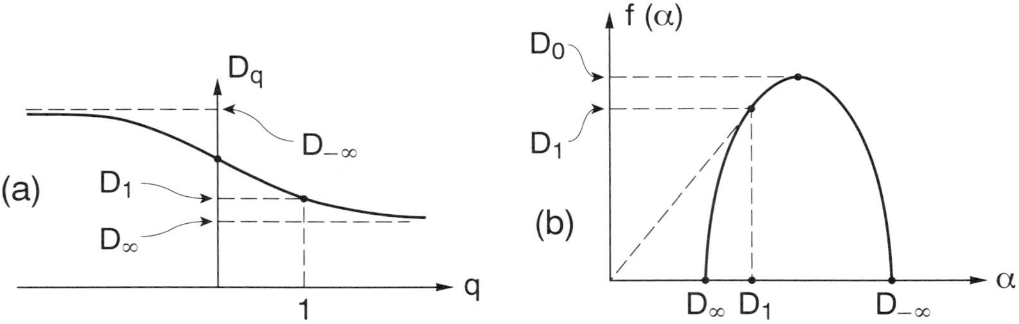

Figure 2.3: *(a) D_q versus q and (b) $f(\alpha)$ versus α for the generalized baker's map. In cases of nonhyperbolic attractors, D_q can have discontinuities in its derivative (not present for the generalized baker's map). Figure (b) also illustrates some general properties of $f(\alpha)$: (1) the maximum value of $f(\alpha)$ is D_0; (2) $f(D_1) = D_1$; and (3) the line joining the origin to the point on the $f(\alpha)$ curve where $\alpha = D_1$ is tangent to the curve.*

Equation (2.21) to express $I(q, \epsilon)$ in Equation (2.3) as

$$\begin{aligned} I(q,\epsilon) &= \int d\alpha' \rho(\alpha') \epsilon^{-f(\alpha')} \epsilon^{q\alpha'} \qquad (2.22) \\ &= \int d\alpha' \rho(\alpha') \exp\{\ln(1/\epsilon)[f(\alpha') - q\alpha']\}. \end{aligned}$$

For ϵ small, $\ln(1/\epsilon)$ is large, and the main contribution to the integral comes from the neighborhood of the maximum of the function $f(\alpha') - q\alpha'$.

For the case where $f(\alpha')$ is a smooth function of α', the maximum of this function occurs when $\alpha' = \alpha(q)$, where

$$f'(\alpha(q)) = q, \qquad (2.23)$$

provided $f''(\alpha(q)) < 0$. Here $f'(\alpha) = df/d\alpha$ and $f''(\alpha) = d^2f/d\alpha^2$. Equation (2.22) then yields

$$I(q,\epsilon) \sim \exp\{\ln(1/\epsilon)[f(\alpha(q)) - q\alpha(q)]\},$$

which when inserted in Equation (2.2) gives

$$D_q = \frac{1}{q-1}[q\alpha(q) - f(\alpha(q))]. \qquad (2.24)$$

Given $f(\alpha)$ we thus obtain the D_q spectrum via Equations (2.23) and (2.24). To proceed in the opposite direction, multiply Equation (2.24) by $q - 1$, differentiate with respect to q, and use Equation (2.23). This gives

$$\alpha(q) = \tau'(q) \qquad (2.25)$$

where $\tau(q) \equiv (q-1)D_q$. Equation (2.24) then yields

$$f(\alpha(q)) = q\tau'(q) - \tau(q). \tag{2.26}$$

For a given value of q, Equations (2.25) and (2.26) yield the corresponding values of f and α, thus parametrically specifying the function $f(\alpha)$. Figure 2.3 shows D_q versus q and $f(\alpha)$ versus α for the generalized baker's map. The $f(\alpha)$ spectrum is a useful alternative characterization of multifractal measures and has been directly determined in a number of physical experiments.

References

J.D. FARMER, E. OTT, J.A. YORKE, The dimension of chaotic attractors. Physica D7, 153 (1983).

P. GRASSBERGER, Generalized dimensions of strange attractors. Phys. Lett. A **97**, 227 (1983).

P. GRASSBERGER, I. PROCACCIA, Measuring the strangeness of strange attractors. Physica D **9**, 189 (1983).

T.C. HALSEY, M.H. JENSEN, L.P. KADANOFF, I. PROCACCIA, B.I. SHRAIMAN, Fractal measures and their singularities: the characterization of strange sets. Physical Rev. A **33**, 1141 (1986).

H.G.E. HENTSCHEL, I. PROCACCIA, The infinite number of generalized dimensions of fractals and strange attractors. Physica D **8**, 435 (1983).

E. OTT, *Chaos in Dynamical Systems*. Cambridge University Press, New York (1993).

CHAPTER 3

Symbolic Dynamics

Chaotic attractors are typified by sensitive dependence on initial conditions, fractal structure, and the presence of unstable periodic orbits located densely throughout the attractor. The existence of infinitely many periodic points is a reflection of the complexity of chaotic dynamics. On the other hand, the availability of periodic orbits throughout the phase space explored by the attractor can be turned to advantage in some applications. In particular, methods for controlling chaos (see Chapter 12) depend on this property of chaotic attractors.

Information about the periodic orbits of a dynamical system can often be captured in terms of sequences of symbols. These symbols code the movement of the orbit around the attractor.

3.1 Itineraries

Consider the quadratic map $f(x) = 4x(1-x)$ on the unit interval. Denote the subintervals $L = [0, 1/2]$ and $R = [1/2, 1]$. Then to a typical orbit of f one can assign a sequence of symbols, called the *itinerary* of the orbit, as follows. The first symbol is L or R according to the location of the initial condition x on the right- or left-hand side of 1/2. Likewise, the ith symbol is L or R depending on the location of $f^{i-1}(x)$. For example, the initial condition $x_0 = 1/3$ begets the orbit $\{\frac{1}{3}, \frac{8}{9}, \frac{32}{81}, \ldots\}$, whose itinerary begins $LRL\ldots$. For the initial condition $x_0 = \frac{1}{4}$, the orbit is $\{\frac{1}{4}, \frac{3}{4}, \frac{3}{4}, \ldots\}$, which terminates in the fixed point $x = \frac{3}{4}$. The itinerary for this orbit is L followed by an infinite string of R's, which we abbreviate $L\overline{R}$.

It is informative to know the subintervals corresponding to all initial points whose itineraries begin with a specified sequence of symbols. For example, the set of initial conditions whose itinerary begins with LRL forms an interval, shown in Figure 3.1. The subintervals in Figure 3.1 give information about the future behavior of the initial conditions lying in them. The subinterval marked LRL consists of orbits that start out in the interval $L = [0, 1/2]$, whose first iterate lands in $R = [1/2, 1]$, and whose second iterate lands in $[0, 1/2]$. For example, $x = 1/3$ lies in LRL. On the other hand, $x = 1/4$ lies in LRR because its first and second iterates are in R.

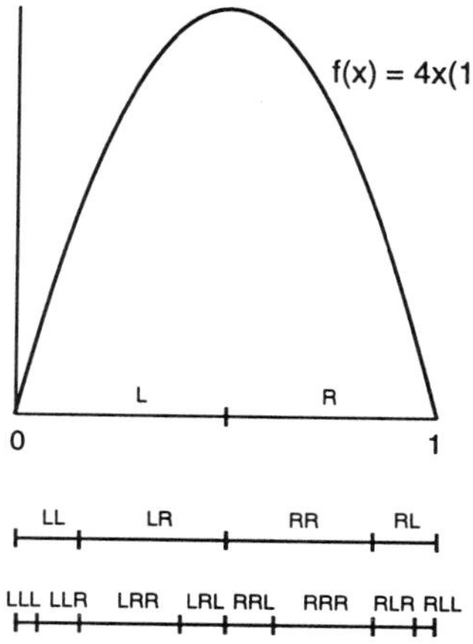

Figure 3.1: *Schematic itineraries for* $G(x) = 4x(1-x)$.

You may see some patterns in the arrangement of the subintervals of Figure 3.1. It turns out that the rule for dividing an interval into its two subintervals is the following. Count the number of R's in the sequence. If this number is even, then the sequence for the left subinterval is the one ending in L and that for the right subinterval ends in R. If odd, the two are interchanged. For example, the interval in Figure 3.1 labeled LR is divided into LRR on the left and LRL on the right.

A graphical way of specifying the possible itineraries for the logistic map is shown in Figure 3.2. We call this the *transition graph* for the subintervals L and R. For every path through this graph with directed edges, there exists an orbit with an itinerary satisfying the sequence of symbols determined by the path. Since the graph is fully connected, all possible sequences of L and R are possible.

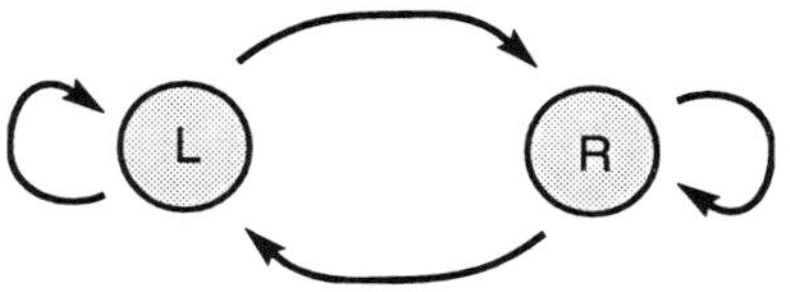

Figure 3.2: *Transition graph for logistic map.*

This information about possible itineraries is sometimes expressed in another completely equivalent way, with a *transition matrix*. If the number of subintervals is K, the transition matrix is a $K \times K$ matrix of zeros and ones. The ijth entry is 1 if subinterval i can be followed by subinterval j in a symbol sequence, and zero otherwise. The transition matrix for the quadratic map is

$$\begin{pmatrix} 1 & 1 \\ 1 & 1 \end{pmatrix}.$$

The subintervals shown in Figure 3.1 decrease to zero as a function of the length of the defining symbol sequence. Therefore, an infinite symbol sequence that is allowed by the transition graph (or, equivalently, the transition matrix), corresponds to a single point in the interval.

3.2 Symbolic dynamics

The orbits of f, except for the point $x = 1/2$ and its preimages, form a *conjugacy* with the space of symbol sequences. Let **S** denote the set of all infinite sequences of form

$$S_0 S_1 S_2 S_3 \ldots$$

where each S_i is either L or R. Denote by H the map which identifies an orbit with its corresponding symbol sequence. Then the fact that H is one-to-one allows a map s to be defined on the symbol space **S**, where $s = HfH^{-1}$. According to the diagram in Figure 3.3, the dynamics of f on R^1 are mirrored by a dynamical system s on the symbol space **S**.

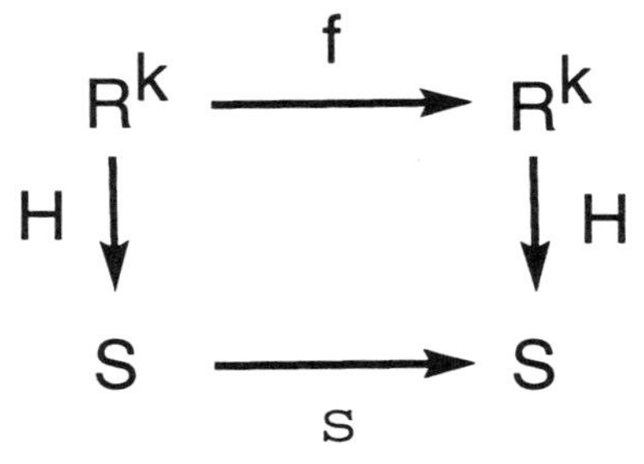

Figure 3.3: *The map f and the shift map s on the symbol space* **S** *are conjugate maps.*

The map s on **S** is called the *shift map*, and simply moves each symbol in the sequence one step to the left. The leftmost symbol is dropped. It is now clear that periodic orbits of f correspond to periodic symbol sequences in **S**. For example, the orbit $\{\frac{3}{4}, \frac{3}{4}, \ldots\}$, which represents the fixed point $x = \frac{3}{4}$, has itinerary $RRR\ldots$. On the other hand, since $LRLRLR\ldots$ is an allowable symbol sequence for f, it belongs to the symbol space **S** and must correspond to some period two orbit of f (one can check that the orbit of f is $\{(5-\sqrt{5})/8, (5+\sqrt{5})/8\}$).

The concept of itineraries make it easy to explain, for the example of the quadratic map, what we mean by sensitive dependence on initial conditions. If we specify the first $n+1$ symbols of the itinerary, we have 2^{n+1} choices. Most of the 2^{n+1} subintervals are rather small for large n; in fact the lengths go to zero, as we have said.

Consider any one of these small subintervals for a large value of n, corresponding to some sequence of symbols $S_0 \cdots S_n$. This subinterval has in turn subintervals corresponding to both $S_0 \cdots S_n LL$ and $S_0 \cdots S_n RL$. If we choose a point from each, we have a pair of initial conditions that lie extremely close (as close as desired, by taking n large), but which map $n+2$ iterates later to subinterval LL and RL, respectively. As seen in Figure 3.1, the LL and RL subintervals lie over 1/2 unit apart.

Therefore the quadratic map exhibits sensitive dependence on initial conditions. Throughout most of the unit interval one can find pairs of initial conditions, arbitrarily close together, so that their orbits separate by at least 1/2 unit after a sufficient number of iterates.

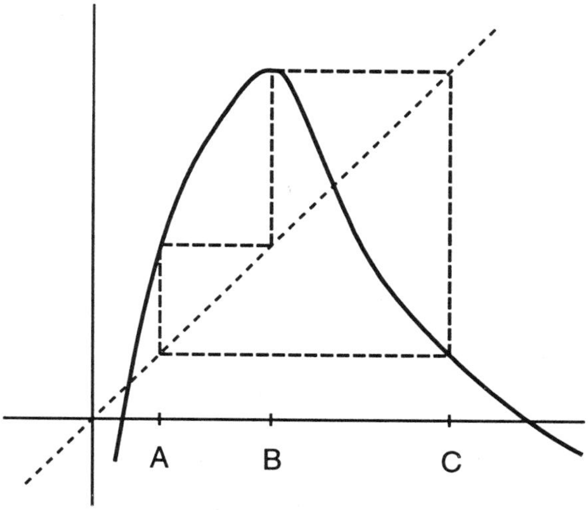

Figure 3.4: *A map with a period three orbit.*

For a second example, in Figure 3.4 we show a map that has a period three orbit. As in the quadratic case, divide the subintervals L and R at the point B

corresponding to the maximum of the function.

The existence of the period three orbit in Figure 3.4 guarantees that $f[A, B] = [B, C]$, and that $f[B, C] = [A, C]$. Let the symbol L represent the interval $[A, B]$, and R represent $[B, C]$. It follows that in the itinerary of an orbit, L must be followed by R, although R can be followed by either L or R. For this example, not all sequences of the symbols L and R correspond to trajectories of f. Some of the possible itinerary subintervals are shown schematically in Figure 3.5. Notice the absence of LL and any sequence containing LL. The corresponding transition graph is shown in Figure 3.6. The transition matrix is

$$\begin{pmatrix} 0 & 1 \\ 1 & 1 \end{pmatrix}.$$

Using the symbol sequences of this map, we can understand the fact that period three implies chaos. As for the logistic map, the length of the subintervals becomes as small as desired for large n. If we use one of the subintervals, corresponding to $S_0 \cdots S_n$, then the subintervals $S_0 \cdots S_n LR$ and $S_0 \cdots S_n RL$ map a nonzero distance apart (at least the length of the RR subinterval), after $n + 2$ iterates, after beginning quite close together.

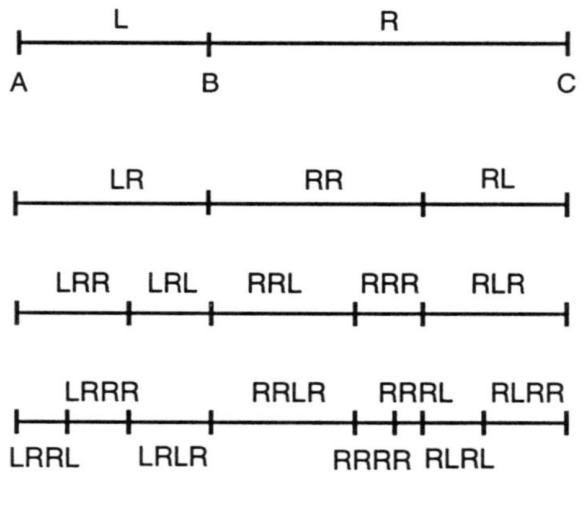

Figure 3.5: *Schematic itineraries for period three map.*

For both of these examples, it can be seen from the corresponding symbol space **S** that the number of periodic orbits increases exponentially with the period. In fact, the rate of exponential increase is given by the topological entropy, which we investigate in Chapter 4.

We emphasize that the existence of complex orbit structure need not result in a chaotic attractor. In particular, for the map in Figure 3.4, the period three orbit is stable and hence an attractor (because the period three orbit goes through the maximum of the map function, at which point the slope of the map function is zero). Furthermore, it is possible that this is the only attractor of the map. In this case all the other infinite number of orbits satisfying all possible symbol sequences (consistent with the transition graph in Figure 3.6) are nonattracting. That is, *typical* initial conditions (with respect to Lebesgue measure) do not tend to them, but rather tend to the period three orbit. In this case we still say there is chaos, but it is nonattracting. The book of Collet and Eckmann (1980) is a good source for further information on the symbolic dynamics of one-dimensional maps.

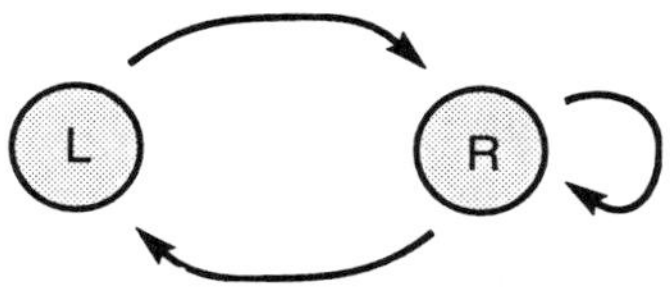

Figure 3.6: *Transition graph for map with period three orbit.*

The generalized baker's map of the previous chapter is a two-dimensional map that can be analyzed using its itineraries. Referring to Figure 2.1, we will assign the symbol L to the left vertical strip, and R to the right vertical strip. Notice that in this case, the set of orbits that share an itinerary $S_0S_1S_2S_3\ldots$ is a vertical line segment. This is in contrast to the one-dimensional cases discussed above, for which an infinite symbol sequence defined a unique trajectory.

In order to construct a symbol space $\mathbf{S}$ for this map, it is necessary to consider bi-infinite sequences; that is, sequences of form

$$\ldots S_{-2}S_{-1}S_0S_1S_2\ldots$$

The index i runs from $-\infty$ to $+\infty$, and S_i is L (respectively, R) if $f^i(x)$ lies in the left (respectively, right) vertical strip. The set of points with symbol sequence $\ldots S_{-2}S_{-1}S_0$ consists of a horizontal line segment, and so each bi-infinite sequence $\ldots S_{-2}S_{-1}S_0S_1S_2\ldots$ defines a unique point in the unit square. There are no restrictions on the ordering of the symbols L and R; that is, the transition graph of Figure 3.2 applies to the bi-infinite sequences.

The symbolic dynamics of the Hénon map is a good deal more complicated. For the quadratic map, we achieved a good symbol set by dividing the unit interval at the critical point $x = 1/2$ to make a binary partition. Similar reasoning justifies the choices we made for partitioning phase space into two sets for the period three map and for the generalized baker's map. For the Hénon map, this principle dictates that a partition for symbolic dynamics might be generated by a curve in the plane passing through all "primary tangencies" between stable and unstable manifolds. This partition into two sets, resulting in itineraries with two symbols, was suggested by Grassberger and Kantz (1985). Further study of this partition by Cvitanovich et al. (1988) led to rules for distinguishing possible symbol sequences from forbidden symbol sequences for attractor trajectories, called pruning rules. We refer the interested reader to Grassberger et al. (1989) for a more complete description and further theoretical work. Chapter 9 of this book deals with methods of analyzing experimental data to reconstruct the symbolic dynamics of the underlying attractor.

References

P. COLLET, J.-P. ECKMANN, *Iterated Maps on the Interval as Dynamical Systems.* Birkhauser, Basel (1980).

P. CVITANOVIC, G.H. GUNARATNE, I. PROCACCIA. Phys. Rev. A **38**, 1503 (1988).

P. GRASSBERGER, H. KANTZ, Generating partitions for the dissipative Hénon map. Phys. Lett. A **113**, 235 (1985).

P. GRASSBERGER, H. KANTZ, U. MÖNIG, On the symbolic dynamics of the Hénon map. J. Phys. A **22**, 5217 (1989).

CHAPTER 4

Lyapunov Exponents and Entropy

As we have seen, two key aspects of chaos are the stretching of infinitesimal displacements and the existence of complex orbit structure in the form of a vast variety of possible orbits. The stretching property is closely related to sensitive dependence on initial conditions, which was discussed in the first chapter. The existence of complex orbit structure is exemplified in the exponential increase, as a function of itinerary length, of the number of distinct symbol sequences that represent orbits, as described in the preceding chapter. It is important to have *quantitative* ways of characterizing these two aspects.

A quantitative characterization of stretching properties is provided by the Lyapunov exponents, while a characterization of complex orbit structure is provided by entropy. It should be emphasized that the two aspects of stretching (i.e., sensitive dependence on initial conditions) and complex orbit structure are not independent. Rather, they may be thought of as two sides of the same "chaos coin."

4.1 Lyapunov exponents

Assume that $\mathbf{x}_n$ denotes a k-dimensional vector, and that we are interested in a dynamical system specified by a map,

$$\mathbf{x}_{n+1} = \mathbf{G}(\mathbf{x}_n). \tag{4.1}$$

Consider an orbit displaced from the original orbit, say $\mathbf{x}_n \to \mathbf{x}_n + \delta\mathbf{x}_n$, where $\delta\mathbf{x}_n$ is an infinitesimal vector. We can arrive at an equation describing the evolution of $\delta\mathbf{x}_n$ by differentiating Equation (4.1):

$$\delta\mathbf{x}_{n+1} = \mathbf{DG}(\mathbf{x}_n)\delta\mathbf{x}_n \tag{4.2}$$

where $\mathbf{DG}(\mathbf{x})$ denotes the $k \times k$ Jacobian matrix of partial derivatives of $\mathbf{G}(\mathbf{x})$ with respect to the k components of $\mathbf{x}$. Let $\mathbf{y}_n = \delta\mathbf{x}_n/|\delta\mathbf{x}_0|$; the vector $\mathbf{y}_n$ is called a *tangent vector* and the space in which $\mathbf{y}_n$ lies is the *tangent space*. The evolution equation for $\mathbf{y}_n$ is

$$\mathbf{y}_{n+1} = \mathbf{DG}(\mathbf{x}_n)\mathbf{y}_n. \tag{4.3}$$

Clearly the evolution of the tangent vector $\mathbf{y}_n$ depends on both the orbit $\{\mathbf{x}_n\}$, which is determined by the initial condition $\mathbf{x}_0$, and on the initial orientation of the unit tangent vector $\mathbf{y}_0$. We are interested in the exponential rate at which the magnitude of $\mathbf{y}$ grows or shrinks per iterate of the map. Define

$$h(\mathbf{x}_0, \mathbf{y}_0) = \lim_{n\to\infty} h(\mathbf{x}_0, \mathbf{y}_0, n) \tag{4.4}$$

where

$$h(\mathbf{x}_0, \mathbf{y}_0, n) = \frac{1}{n} \ln |\mathbf{y}_n|. \tag{4.5}$$

We call $h(\mathbf{x}_0, \mathbf{y}_0)$ a *Lyapunov exponent* and $h(\mathbf{x}_0, \mathbf{y}_0, n)$ a *finite-time Lyapunov exponent*.

Equivalently, we can speak of the *Lyapunov number* L, which is given in terms of the Lyapunov exponent h by $L = e^h$. The Lyapunov number represents an average factor by which the magnitude of the infinitesimal vector displacement is multiplied on each iterate.

Combining Equations (4.3) and (4.4) we have

$$\begin{aligned} h(\mathbf{x}_0, \mathbf{y}_0, n) &= \frac{1}{n} \ln |\mathbf{DG}^n(\mathbf{x}_0)\mathbf{y}_0| \\ &= \frac{1}{n} \ln |\mathbf{DG}(\mathbf{x}_{n-1}) \cdots \mathbf{DG}(\mathbf{x}_0)\mathbf{y}_0| \end{aligned} \tag{4.6}$$

where $\mathbf{G}^n(\mathbf{x}_0)$ denotes the nth iterate of $\mathbf{G}$,

$$\mathbf{G}^n(\mathbf{x}) \equiv \underbrace{\mathbf{G}(\mathbf{G}(\mathbf{G}(\ldots(\mathbf{G}}_{n}(\mathbf{x}))\ldots))),$$

and $\mathbf{DG}^n$ is the Jacobian matrix of the map $\mathbf{G}^n$. Applying the $k \times k$ matrix $\mathbf{DG}^n(\mathbf{x}_0)$ to all unit tangent vectors $\mathbf{y}_0$ results in an ellipsoid, whose k principal radii are the finite-time Lyapunov numbers $L_i(\mathbf{x}_0, n)$, for $i = 1, 2, \ldots, k$. The principal directions of the ellipsoid are the k perpendicular eigenvectors of the real symmetric matrix $[\mathbf{DG}^n(\mathbf{x}_0)] \cdot [\mathbf{DG}^n(\mathbf{x}_0)]^T$. The principal radii of the ellipsoid are the square roots of the k eigenvalues, which are sometimes called the *singular values* of $\mathbf{DG}^n(\mathbf{x}_0)$. Taking the $n \to \infty$ limit in Equation (4.4), we can expect there to be k possible values of the Lyapunov exponent depending on the initial orientation of the vector $\mathbf{y}_0$. We denote these values $h_i(\mathbf{x}_0)$, and assign the index i according to the value of the corresponding exponent:

$$h_1(\mathbf{x}_0) \geq h_2(\mathbf{x}_0) \geq \ldots \geq h_k(\mathbf{x}_0). \tag{4.7}$$

Osledec's multiplicative ergodic theorem (Osledec, 1968) guarantees that if the orbit from $\mathbf{x}_0$ generates an ergodic probability measure, then the limit in

Equation (4.4) exists and the values of the $h_i(\mathbf{x}_0)$ are the same for almost every $\mathbf{x}_0$ with respect to the ergodic measure. Here, as in Chapter 2, we are primarily concerned with orbits on a chaotic attractor, and we assume that there exists an ergodic natural measure on the attractor. In particular, for almost every $\mathbf{x}_0$ with respect to Lebesgue measure in the basin of the attractor, the orbit from $\mathbf{x}_0$ generates the natural measure. Thus, for such $\mathbf{x}_0$, Osledec's theorem tells us that the set of k numbers $\{h_i(\mathbf{x}_0)\}$ will be the same for all $\mathbf{x}_0$ in the basin except for a possible set of Lebesgue measure zero. Alternatively, the k numbers $h_i(\mathbf{x}_0)$ will be the same for all $\mathbf{x}_0$ on the attractor except for a possible set of natural measure zero. In such circumstances it is common practice to drop the $\mathbf{x}_0$ dependence of the Lyapunov exponent, $h_i(\mathbf{x}_0) \to h_i$, with the understanding that h_i is that value of $h_i(\mathbf{x}_0)$ obtaining for almost every $\mathbf{x}_0$.

In general, we define a chaotic attractor by the condition that there is net average stretching for at least one orientation of $\mathbf{y}_0$, and almost every $\mathbf{x}_0$:

$$h_1 > 0. \tag{4.8}$$

This condition formalizes the notion of sensitive dependence on initial conditions. In general, the h_i with $i \geq 2$ for a chaotic attractor can be positive or negative (or in rare cases zero). If

$$h_1 \geq h_2 \geq \ldots \geq h_j > 0 > h_{j+1} \geq h_{j+2} \geq \ldots \geq h_k$$

it is sometimes (rather loosely) said that there are j stretching directions and $k - j$ contracting directions.

To summarize our description of Lyapunov exponents, we can imagine an infinitesimal k-dimensional ball in phase space that is mapped forward in time by n iterates. The result will be an infinitesimal ellipsoid with k principal radii. The ratio of these radii to the initial radius will then be of the order of e^{nh_i}. This is illustrated in Figure 4.1 for the case of a two-dimensional map ($k = 2$).

In the preceding, we have been discussing Lyapunov exponents for a map. A parallel discussion goes through for the case of continuous time dynamical systems. For a k-dimensional system of first-order ordinary differential equations

$$\dot{\mathbf{x}} = \mathbf{F}(\mathbf{x}) \tag{4.9}$$

we consider the infinitesimally displaced orbit $\mathbf{x}(t) + \delta\mathbf{x}(t)$ and the tangent vector $\mathbf{y}(t) = \delta\mathbf{x}(t)/|\delta\mathbf{x}(0)|$, in terms of which we have

$$\dot{\mathbf{y}} = \mathbf{DF}(\mathbf{x}(t))\mathbf{y}. \tag{4.10}$$

The Lyapunov exponents are then given by

$$h(\mathbf{x}(0), \mathbf{y}(0)) = \lim_{t\to\infty} \frac{1}{t} \ln |\mathbf{y}(t)|, \tag{4.11}$$

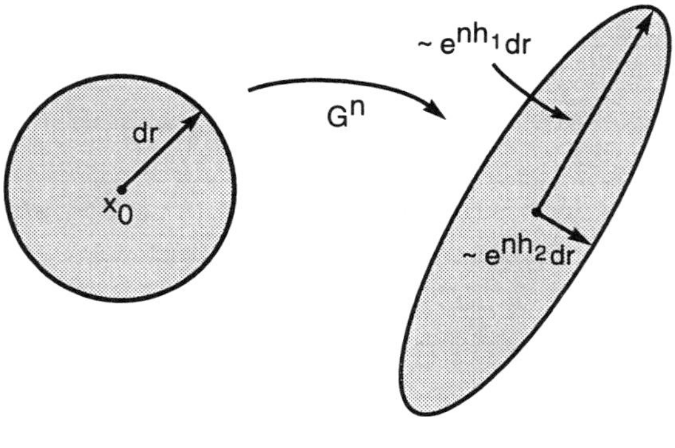

Figure 4.1: *Evolution of an initial infinitesimal ball after n iterates of the map* $(k = 2)$.

yielding k exponents for typical $\mathbf{x}(0)$ in the attractor basin as in Equation (4.7). (For a continuous time system one of the exponents will be zero, corresponding to the initial location of $\mathbf{x}(0)$ on the chaotic attractor with initial orientation of $\mathbf{y}_0$ tangent to the orbit through $\mathbf{x}_0$.)

4.2 Baker's map example

As an example we can calculate the Lyapunov exponents for the generalized baker's map, Equations (2.11-2.12) and Figure 2.1. The Jacobian matrix of the generalized baker's map is

$$\mathbf{DG}(\mathbf{x}) = \begin{bmatrix} \left\{ \begin{array}{l} \lambda_a \text{ if } y < \gamma \\ \lambda_b \text{ if } y > \gamma \end{array} \right\} & 0 \\ 0 & \left\{ \begin{array}{l} \gamma^{-1} \text{ if } y < \gamma \\ \delta^{-1} \text{ if } y > \gamma \end{array} \right\} \end{bmatrix}. \tag{4.12}$$

Thus we have

$$\mathbf{DG}^n(\mathbf{x}_0) = \begin{bmatrix} \lambda_a^{n_1}\lambda_b^{n_2} & 0 \\ 0 & \gamma^{-n_1}\delta^{-n_2} \end{bmatrix} \tag{4.13}$$

where n_1 and n_2 are the number of times the orbit $\mathbf{x}_1, \mathbf{x}_2, \ldots, \mathbf{x}_n$ fell in the regions $y < \gamma$ and $y > \gamma$, respectively; $n_1 + n_2 = n$. Since $\mathbf{DG}^n(\mathbf{x}_0)$ is diagonal, the principal directions of the ellipsoid of tangent vectors are aligned with the x and y axes, and the finite-time Lyapunov exponents are

$$\begin{aligned} h_1(\mathbf{x}_0, n) &= \frac{n_1}{n}\ln\gamma^{-1} + \frac{n_2}{n}\ln\delta^{-1} \\ h_2(\mathbf{x}_0, n) &= \frac{n_1}{n}\ln\lambda_a + \frac{n_2}{n}\ln\lambda_b. \end{aligned}$$

For $\mathbf{x}_0$ chosen as a typical point with respect to Lebesgue measure in the unit square, $0 \le x \le 1, 0 \le y \le 1$, the limits

$$\lim_{n\to\infty} \frac{n_1}{n} \text{ and } \lim_{n\to\infty} \frac{n_2}{n}$$

are just the natural measures of the attractor in the regions $y < \gamma$ and $y > \gamma$, respectively. Since the natural measure is uniform in y for $0 \le y \le 1$ we have

$$\lim_{n\to\infty} \frac{n_1}{n} = \gamma \text{ and } \lim_{n\to\infty} \frac{n_2}{n} = \delta.$$

Thus the Lyapunov exponents for the generalized baker's map are

$$\begin{aligned} h_1 &= \gamma \ln \gamma^{-1} + \delta \ln \delta^{-1} \\ h_2 &= \gamma \ln \lambda_a + \delta \ln \lambda_b. \end{aligned} \qquad (4.14)$$

There will usually be a zero measure set of points on the attractor which do not share the Lyapunov exponents of the attractor shown in Equation (4.14). An example of a point in this zero measure set for the generalized baker's map is the fixed point $\mathbf{x}_0 = (x, y) = (0, 0)$. For this initial condition, $n_1/n = 1$ and $n_2/n = 0$, yielding $h_1 = \ln \gamma^{-1}$ and $h_2 = \ln \lambda_a$, which are different. We note, however, that an arbitrarily small perturbation of the initial condition causes the ensuing orbit to move exponentially away from $(0, 0)$, producing an orbit with exponents given by (4.14).

In the example above we have been able to compute the Lyapunov exponents simply and analytically. Unfortunately, this is not possible for most systems encountered in practice. Thus the question arises as to how Lyapunov exponents can be calculated numerically from a known dynamical system. In the next section we describe an algorithm for determining Lyapunov exponents from a system whose equations are known. When this approach is combined with attractor reconstruction techniques, it can be used to extract Lyapunov exponents from time series, as illustrated in the reprints of Chapter 8.

4.3 Numerical calculation of Lyapunov exponents

A technique for numerically calculating the h_i (Benettin et al., 1980) proceeds sequentially, first determining h_1, then h_2, and so on. To find h_1, pick an arbitrarily oriented $\mathbf{y}_0$. Thus $\mathbf{y}_0$ will typically consist of a mixture of components in all the principal directions of $\mathbf{DG}^n$. For increasing n, the magnitudes of the various components of the initial $\mathbf{y}_0$ grow roughly as e^{nh_i}, for $i = 1, 2, \ldots, k$. Since h_1 is the largest exponent it eventually dominates. Thus a plot of $\ln |\mathbf{y}_n|$ for an arbitrarily chosen orientation of $\mathbf{y}_0$ invariably grows approximately linearly with n at the average rate h_1. Thus we can numerically estimate h_1 as the slope of the best fit straight line to a numerical plot of $\ln|\mathbf{y}_n|$ versus n.

To find h_2 we arbitrarily choose two tangent vectors $\mathbf{y}_0^{(1)}$ and $\mathbf{y}_0^{(2)}$. These vectors define a parallelogram of area A_0. Iterating the two vectors forward n steps, $\mathbf{y}_n^{(1)}$ and $\mathbf{y}_n^{(2)}$ define a parallelogram of area denoted A_n. Since h_1 and h_2 are the two largest exponents, the parallelogram becomes aligned with the plane corresponding to the directions of these two strongest stretchings (or if $h_2 < 0$, weakest contraction). Thus we expect that $A_n \sim e^{n(h_1+h_2)}$ for large n. Hence we can obtain an estimate of h_2 by fitting numerically obtained data for $\ln A_n$ versus n by a straight line, calculating its slope, and then subtracting the previously determined value of h_1. The calculation of A_n is slightly tricky due to the fact that $\mathbf{y}_n^{(1)}$ and $\mathbf{y}_n^{(2)}$ tend to become more and more parallel (in the direction of strongest stretching) as n increases, making the determination of the area susceptible to error. This problem can easily be avoided by replacing the nearly parallel vectors with a new pair

$$(\mathbf{y}_n^{(1)}, \mathbf{y}_n^{(2)}) \to (\mathbf{y}_n^{(1)*}, \mathbf{y}_n^{(2)*})$$

such that the new pair lie in the plane determined by the old pair, are perpendicular to each other (i.e., $\mathbf{y}_n^{(1)*} \cdot \mathbf{y}_n^{(2)*} = 0$), and define a square of the same area A_n as the parallelogram defined by $\mathbf{y}_n^{(1)}$ and $\mathbf{y}_n^{(2)}$. The renormalization leaves the magnitude of A_n, and its subsequent evolution, unaffected.

This idea generalizes to the computation of all k Lyapunov exponents. To find h_3 one begins with three initially chosen tangent vectors, and determines the evolution of the three-dimensional volume subtended by the vectors. The growth of the natural log of this volume gives $h_1 + h_2 + h_3$, from which h_3 can be extracted, and similarly for $h_4, \ldots, h_k$. (For a more or less equivalent but more efficient version of this method, put in the context of experimental time series, see the reprinted paper by Eckmann et al. (1986) in Chapter 8.)

4.4 Lyapunov dimension

It was conjectured by Kaplan and Yorke (1979) that there is a relationship between the Lyapunov exponents and the fractal dimension of a typical chaotic attractor. Let p denote the largest integer such that

$$\sum_{i=1}^{p} h_i \geq 0 \tag{4.15}$$

(recall the ordering in Equation (4.7)). Define the *Lyapunov dimension* D_L by

$$D_L = p + \frac{1}{|h_{p+1}|} \sum_{i=1}^{p} h_i. \tag{4.16}$$

Recall from Chapter 2 that the information dimension is $D_1 = \lim_{q \to 1} D_q$. The Kaplan-Yorke conjecture (Kaplan and Yorke, 1979) is that for typical attractors, the information dimension of the attractor is equal to the Lyapunov dimension:

$$D_1 = D_L. \tag{4.17}$$

In the case of a two-dimensional map with $h_1 > 0 > h_2$ and $h_1 + h_2 < 0$ (e.g., the Hénon map and the generalized baker's map), Equation (4.16) yields

$$D_L = 1 + \frac{h_1}{|h_2|}. \tag{4.18}$$

Since we have explicitly calculated the information dimension and the Lyapunov exponents for the generalized baker's map, the conjecture can be checked for that map. Inserting the Lyapunov exponents from Equation (4.14) for the generalized baker's map into the Lyapunov dimension formula (Equation (4.18)) yields

$$D_1 = 1 + \frac{\gamma \ln(1/\gamma) + \delta \ln(1/\delta)}{\gamma \ln(1/\lambda_a) + \delta \ln(1/\lambda_b)}. \tag{4.19}$$

which agrees with the value for D_1 obtained in Chapter 2 (see Equation (2.19)).

4.5 Entropy

In his formulation of the study of information coding, transmission, and decoding, Shannon introduced a notion of the degree of uncertainty in being able to predict the output of a probabilistic event. Say an experiment has r possible outcomes whose probabilities are $p_1, p_2, \ldots, p_r$. Then the *Shannon entropy* is

$$H_s = \sum_{i=1}^{r} p_i \ln \frac{1}{p_i}. \tag{4.20}$$

For example, in the case where one of the p_i's is one and the others all zero, Equation (4.20) gives $H_s = 0$ (we define $p \ln(1/p) \equiv 0$ for $p = 0$), corresponding to the fact that there is no uncertainty in the outcome of the experiment: we know with certainty that the event whose p_i is one occurs. Uncertainty is at a maximum when all outcomes are equally probable. In that case $p_1 = p_2 = \ldots = p_r = 1/r$, and the entropy H_s assumes its largest possible value of $\ln r$. Notice that this H_s increases with r, corresponding to the increased uncertainty of having more possible equal probability events.

Kolmogorov (1958) applied Shannon's information characterization, Equation (4.20), to ergodic theory. Let μ be an ergodic invariant probability measure for a map G. (For example, the case of most interest to us is that where μ is the natural measure on a chaotic attractor.) Let R be a bounded region of phase space containing the measure μ, and imagine that we partition R into a finite number of subregions, $R = R_1 \cup \ldots \cup R_r$. Then we can form an entropy function for the partition $\{R_i\}$,

$$H(\{R_i\}) = \sum_{i=1}^{r} \mu(R_i) \ln[\mu(R_i)]^{-1}, \tag{4.21}$$

which gives the average information gain when one is told that the orbit lies in one of the elements R_i of the partition (in the absence of any other prior information).

Next Kolmogorov considers the preimages $\mathbf{G}^{-1}(R_i)$ of the R_i and examines the r^2 intersections

$$R_i \bigcap \mathbf{G}^{-1}(R_j)$$

for each pair $(i, j), 1 \le i \le r, 1 \le j \le r$. Collecting all such *nonempty* intersections we have a new partition $\{R_i^{(2)}\}$, where $1 \le i \le r_2$, and r_2 is the number of nonempty intersections. Next, form a third-stage partition $\{R_i^{(3)}\}$ from the r_3 nonempty intersections

$$R_i \bigcap \mathbf{G}^{-1}(R_j) \bigcap \mathbf{G}^{-2}(R_k)$$

for $i, j, k = 1, 2, \ldots, r$. Higher-stage partitions are similarly formed. Consider the quantity

$$h(\mu, \{R_i\}) = \lim_{n\to\infty} \frac{1}{n} H(\{R_i^{(n)}\}). \tag{4.22}$$

Since

$$\lim_{n\to\infty} \frac{1}{n} H(\{R_i^{(n)}\}) = \lim_{n\to\infty} [H(\{R_i^{(n+1)}\}) - H(\{R_i^{(n)}\})] \tag{4.23}$$

we can regard $h(\mu, \{R_i\})$ as the average information gained by going from the partition at stage n to the finer partition at stage $n+1$, for large n. The *metric entropy* of the measure μ (also called the *Kolmogorov entropy* or the *Kolmogorov-Sinai entropy*) is by definition the supremum of $h(\mu, \{R_i\})$ over all possible initial partitions:

$$h(\mu) \equiv \sup_{\{R_i\}} h(\mu, \{R_i\}). \tag{4.24}$$

There are many definitions of entropy, which serve various purposes. We have defined the metric entropy for invariant measures. The topological entropy of a map $\mathbf{G}$ is a characterization of the complexity of the map dynamics, independent of any invariant measure or measures that the map might admit. Let r_n denote the number of nonempty elements $\{R_i^{(n)}\}$ in the nth stage partition. Then the *topological entropy*,

$$h_T \equiv \sup_{\{R_i\}} \lim_{n \to \infty} \frac{1}{n} \ln r_n \tag{4.25}$$

gives the exponential rate of increase of r_n with n for large n. It can be shown that

$$h_T \geq h(\mu). \tag{4.26}$$

The metric and topological entropies provide different ways of characterizing chaos. We say that the dynamics on an invariant set with invariant measure μ is chaotic for almost every initial condition with respect to the measure μ, if $h(\mu) > 0$. Again we emphasize the case where μ is the natural measure on a chaotic attractor. On the other hand, we say that the overall dynamics of the map $\mathbf{G}$ admits chaotic orbits if $h_T > 0$. In the latter case, the map might only have nonchaotic attractors (e.g., attracting periodic orbits), but there may exist a *nonattracting* chaotic invariant set somewhere in the phase space. An example is the nonattracting chaotic set accompanying the attracting period three orbit of the one-dimensional map shown in Figure 3.4 of Chapter 3.

Grassberger and Procaccia (1983) give a generalization of the entropy concept which includes both the metric and topological entropies. Specifically they replace the Shannon entropy of Equation (4.20) by Renyi's q-order entropy

$$H_q = \frac{1}{1-q} \ln \sum_{i=1}^{p} p_i^q \tag{4.27}$$

where, as before, $p_1, p_2, \ldots, p_r$ are the probabilities of r possible events (compare with Equations (2.2) and (2.3)). Proceeding as before, but using Equation (4.27) in place of Equation (4.20), we obtain the q-order entropy

$$h_q(\mu) \equiv \sup_{\{R_i\}} \lim_{n \to \infty} \frac{1}{n} H_q(\{R_i^{(n)}\}) \tag{4.28}$$

where $H_q(\{R_i^{(n)}\}) = \frac{1}{1-q} \ln \sum_i [\mu(R_i^{(n)})]^q$. Note that, if we take the limit $q \to 1$ in Equation (4.27), we obtain the Shannon entropy, while if we set $q = 0$ it gives $\ln r$. Thus we see that

$$\lim_{q \to 1} h_q(\mu) = h(\mu) \tag{4.29}$$

and

$$h_{q=0}(\mu) = h_T. \tag{4.30}$$

References

G. BENETTIN, L. GALGANI, A. GIORGILLI, AND J.-M. STRELCYN, Lyapunov characteristic exponents for smooth systems and for Hamiltonian system: A method for computing all of them. Part 2: Numerical Application, Meccanica **15**, 21 (1980).

P. GRASSBERGER, I. PROCACCIA, Measuring the strangeness of strange attractors. Physica D **9**, 189 (1983).

J.L. KAPLAN, J.A. YORKE, Chaotic behavior of multidimensional difference equations, in *Functional Differential Equations and Approximation of Fixed Points*, edited by H.-O. Peitgen and H.-O. Walter, Lecture Notes in Mathematics **730** (Springer, Berlin, 1979), 204.

A.N. KOLMOGOROV, A new metric invariant of transitive dynamical systems and automorphisms in Lebesgue spaces. Dokl. Acad. Nauk. SSSR **119**, 861 (1958).

V.I. OSLEDEC, A multiplicative ergodic theorem: Lyapunov characteristic numbers for dynamical systems. Trudy Mosk. Obsch. **19**, 179 (1968) [Trans. Mosc. Math. Soc. **19**, 197 (1968).]

CHAPTER 5

The Theory of Embedding

The concepts of chaos have been applied increasingly often to the description of the time evolution of systems in engineering and the physical and biological sciences. As a result, much light has been shed on the question of distinguishing deterministic processes from stochastic processes.

This distinction is important because the prediction of the future behavior of a system is possible only to the extent that the system is deterministic, meaning that information from the system's past unambiguously determines its future states. Then the possibility exists of predicting the future with a deterministic model, and of exploiting these predictions to characterize, monitor, or control the system.

For a purely stochastic system, on the other hand, predictions are by definition statistical in nature. The predictability of a stochastic signal is limited to mean values, higher moments, linear autocorrelation, and other statistical quantities. From our point of view, the distinction between deterministic and stochastic is more practical than philosophical. The real question is whether the *model* used to describe given data should be deterministic or stochastic. For a deterministic system with many degrees of freedom, a stochastic model may be as useful as any other.

The impact of the discovery of chaos lies in the realization that nonlinear systems with few degrees of freedom, while deterministic in principle, can create output signals that look complex, and mimic stochastic signals from the point of view of conventional time series analysis. The reason for this is that trajectories that have nearly identical initial conditions will separate from one another at an exponentially fast rate. This exponential separation causes chaotic systems in the laboratory to exhibit much of the same medium to long-term behavior as stochastic systems. The key fact is that short-term prediction is not ruled out for chaotic systems, if there are a reasonably low number of active degrees of freedom. It is the ability to predict a short time ahead that is the basis of the new techniques disseminated in this book. These techniques allow the signals measured from real-world systems to be examined for deterministic origins, and possible exploitation using a deterministic model.

Before the recognition of chaos, a complex signal was commonly assumed to be the output of a complicated system with a large number of active degrees of freedom. If there are hundreds of active independent modes in the system, while technically the system is deterministic, there will be no way in practice to make use of this fact. In such a case, there is no alternative to modeling the signal as the output of a stochastic source. It is therefore important when analyzing a complex signal to determine whether it results from low-dimensional chaos. Methods for making this determination include measurements of attractor dimension, Lyapunov exponents, and periodic orbits as described in the reprints of Part II of this book, as well as more direct methods such as time series prediction error, discussed in Part III.

5.1 Measurements and state representation

In a laboratory setting, it is seldom the case that all relevant dynamical variables can be measured. Imagine a mechanical system consisting of gears, levers and other components. The system state might be given by specifying the positions of these components. However, in a practical case, it is likely that only a limited number of the positions of the components can be measured. How can we proceed to study the dynamics in such a situation?

As another example, consider a system consisting of a fluid in a rectangular container undergoing thermal convective motion. The fluid is governed by the Navier-Stokes equations, which are a system of coupled partial differential equations. To describe the motion, one can expand the fluid velocity in a Fourier series, relative to the three spatial coordinates. Assuming convergence of the Fourier series, the infinite set of Fourier coefficients satisfy a set of first-order coupled nonlinear ordinary differential equations. We can now view the system state as given by the Fourier coefficients; that is, the system is modeled by an infinite-dimensional dynamical system. There is no hope of measuring the full system state – it consists of the infinite number of Fourier coefficients. Again, we would like to know how to proceed in analyzing the dynamics when only a limited number of measurements (generally *not* measurements of Fourier coefficients) are available.

A key element in resolving this general class of problems is provided by embedding theory. In typical situations, points on the dynamical attractor in the full system phase space have a one-to-one correspondence with measurements of a limited number of variables. This is a powerful fact. By definition, a point in the phase space carries complete information about the current system state. If the equations defining the system dynamics are not explicitly known, this phase space is not directly accessible to the observer. *A one-to-one correspondence means that the phase space state can be identified by measurements.*

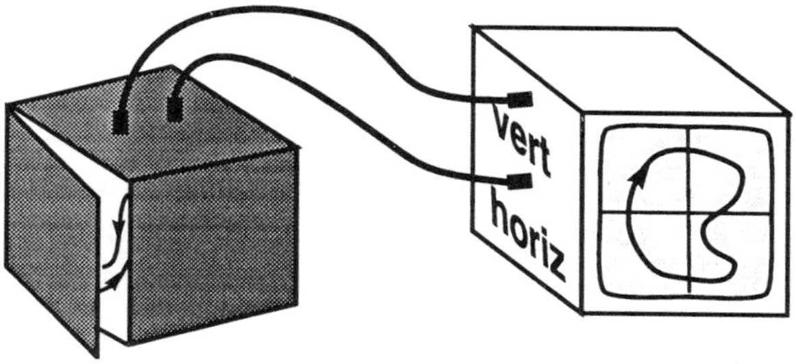

Figure 5.1: *The dynamical system is represented by the black box on the left. Two simultaneous independent measurements are plotted against one another on the display screen at right. Each state* $\mathbf{x}$ *of the dynamical system gives rise to a dot* $\mathbf{F}(\mathbf{x})$ *on the screen. The goal is for the reconstruction map* $\mathbf{F}$ *to be one-to-one – meaning that each state* $\mathbf{x}$ *corresponds to a different* $\mathbf{F}(\mathbf{x})$.

Assume we can simultaneously measure m variables $(y_1(t), \ldots, y_m(t))$, which we denote by the vector $\mathbf{y}(t)$. For example, in the case of the convecting fluid, $y_1(t), \ldots, y_m(t)$ might be measurements of the magnitude of the fluid velocity at m different points in the fluid. The m-dimensional vector $\mathbf{y}$ can be viewed as a function of the system state $\mathbf{x}(t)$ in the full system phase space:

$$\mathbf{y} = \mathbf{F}(\mathbf{x}) = (f_1(\mathbf{x}), \ldots, f_m(\mathbf{x})). \tag{5.1}$$

(In the case of the fluid, the m values of the velocity magnitudes are uniquely determined by a knowledge of the infinite set of Fourier coefficients.) We call the function $\mathbf{F}$ the *measurement function*, and we call the m-dimensional vector space in which the vectors $\mathbf{y}$ lie the *reconstruction space*.

We have grouped the measurements as a vector-valued function $\mathbf{F}$ of $\mathbf{x}$. The fact that $\mathbf{F}$ is a function is a consequence of the definition of state – information about the system is determined uniquely by the state, so each measurement is a well-defined function of $\mathbf{x}$.

As we discuss in detail below, as long as m is taken sufficiently large, the measurement function $\mathbf{F}$ generically defines a one-to-one correspondence between the attractor states in the full phase space and m-vectors $\mathbf{y}$. See Figure 5.1. By one-to-one, we mean that for a given $\mathbf{y}$ there is a *unique* $\mathbf{x}$ on the attractor such that $\mathbf{y} = \mathbf{F}(\mathbf{x})$. When there is a one-to-one correspondence, each vector formed of m measurements is a proxy for a single system state, and the fact that the entire system information is determined by a state $\mathbf{x}$ is transferred to be true as well for the measurement vector $\mathbf{F}(\mathbf{x})$. In order for this to be so, it turns out that it is enough to take m larger than twice the box-counting dimension of the phase space attractor.

The one-to-one property is useful because the state of a deterministic dynamical system, and thus its future evolution, is completely specified by a point in the full phase space. Suppose that when the system is in a given state $\mathbf{x}$ one observes the vector $\mathbf{F}(\mathbf{x})$ in the reconstruction space, and that this is followed one second later by a particular event. If $\mathbf{F}$ is one-to-one, each appearance of the measurements represented by $\mathbf{F}(\mathbf{x})$ will be followed one second later by the same event. This is because there is a one-to-one correspondence between the attractor states in phase space and their image vectors in reconstruction space. Thus there is predictive power in measurements $\mathbf{y}$ that are matched to the system state $\mathbf{x}$ in a one-to-one manner.

5.2 Topological embedding

Two different types of embedding are relevant to attractor reconstruction. The first type is topological embedding, which is nothing more than a one-to-one continuous correspondence between the vectors. We will start our description using this basic form of embedding. The second type is differentiable embedding, which means that the differential structure of the attractor is preserved, including such quantities as Lyapunov exponents. Differentiable embeddings are harder to achieve, especially when fractal structure exists. We discuss differentiable embeddings in Section 5.4.

We begin by considering a k-dimensional Euclidean space R^k. Points in this space are specified by giving k real coordinate values $x_1, \ldots, x_k$, which we represent as the components of a vector $\mathbf{x}$. Let $\mathbf{F}$ be a continuous function from R^k to R^m,

$$\mathbf{y} = \mathbf{F}(\mathbf{x})$$

where $\mathbf{y}$ is in R^m and $\mathbf{x}$ is in R^k. If A is a subset of R^k, we denote by $\mathbf{F}(A)$ the set of all points $\mathbf{y} = \mathbf{F}(\mathbf{x})$ generated as $\mathbf{x}$ ranges over A, and we call $\mathbf{F}(A)$ the *image* of A in R^m. We say that $\mathbf{F}$ is *one-to-one* on A if given any $\mathbf{y}$ in the image $\mathbf{F}(A)$, there is one and only one $\mathbf{x}$ in A such that $\mathbf{y} = \mathbf{F}(\mathbf{x})$. In other words, if $\mathbf{x}_1$ and $\mathbf{x}_2$ are both in A, then $\mathbf{F}(\mathbf{x}_1) = \mathbf{F}(\mathbf{x}_2)$ necessarily implies that $\mathbf{x}_1 = \mathbf{x}_2$. If $\mathbf{F}$ is one-to-one, then the inverse map $\mathbf{F}^{-1}$ can be defined.

In a typical experimental situation, the set A we are interested in is an attractor, which is a compact (closed and bounded) subset of R^k which is invariant under the dynamical system. The goal is to use measurements to construct the function $\mathbf{F}$ so that $\mathbf{F}(A)$ is a copy of A that can be analyzed. A finite time series of measurements will produce a finite set of the points that make up $\mathbf{F}(A)$. If enough points are present, we can hope to discern some of the properties of $\mathbf{F}(A)$, and consequently, of A.

For a continuous one-to-one map of a compact set, the inverse map $\mathbf{F}^{-1}$ will be continuous as a map on $\mathbf{F}(A)$. A one-to-one map on A that is continuous

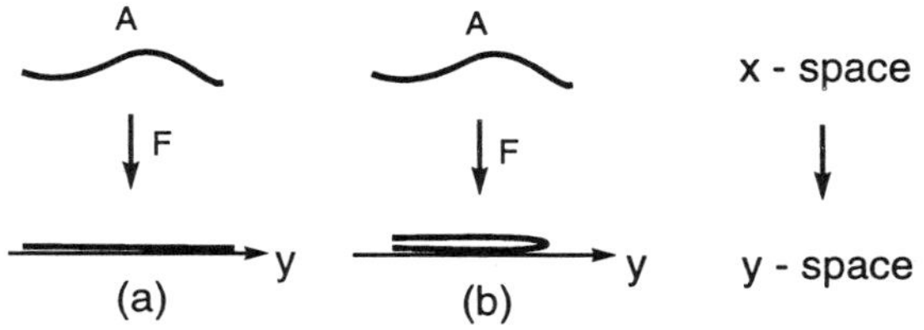

Figure 5.2: *Embeddings and non-embeddings of a curve segment. The image of A is a topological embedding in (a). In (b), neither* **F** *nor small perturbations of it are embeddings.*

and has a continuous inverse is called a *topological embedding* of A. Our first goal is to find a condition on the set A and the function **F** that makes it virtually certain that **F** is a topological embedding of A.

For example, assume that A is a finite length segment of a curve in R^k. Let **F** be a function from R^k to R^1, the real line. Depending on **F**, the set A may or may not be topologically embedded in the real line. Figure 5.2**a** shows an embedding. In Figure 5.2**b**, the function **F** (which may be visualized as taking the curve, folding it, and then placing the folded curve on the real line) fails to embed A. For the latter function, pairs of points which are far apart on A are brought together on $\mathbf{F}(A)$, violating the one-to-one condition. One can also notice that the function in Figure 5.2**b** cannot be easily fixed up to be an embedding. Small perturbations in **F** may change the details of the overlap, but it would take a significant overhaul to remove the overlap entirely.

The situation is similar if we consider the image of a curve segment under functions **F** from R^k to R^2. Although it is possible to conceive of functions **F** that are topological embeddings of the curve segment A, as in Figure 5.3**a**, there are still others, exemplified by Figure 5.3**b**, that are not embeddings, and furthermore cannot be changed to be an embedding by a small perturbation. In Figure 5.3**b** the function **F** fails to be one-to-one at the self-intersection point of $\mathbf{F}(A)$, denoted by a dot in the figure.

For functions from R^k to R^3, however, the situation is significantly different: virtually all functions applied to a curve segment A are topological embeddings. It is still possible to imagine a function **F** that fails to be an embedding, as in Figure 5.4**b**. However, in three dimensions, there clearly exists a small perturbation of **F** that pulls apart the self-intersection of $\mathbf{F}(A)$ at the point P, as in Figure 5.4**a**. That is, given any function **F**, either **F** is an embedding of the curve segment A or there exists a small perturbation of **F** that is an embedding.

This is a general property of smooth functions **F** from R^k to R^m. Compact

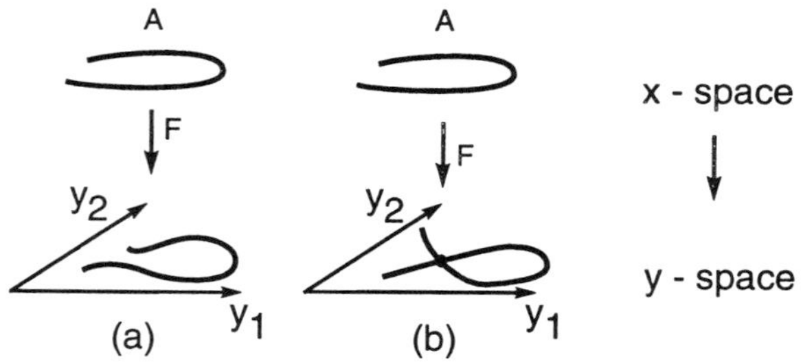

Figure 5.3: *Mapping a curve segment to two-dimensional space. The image of A is a topological embedding in (a). In (b), neither* **F** *nor small perturbations of it are embeddings.*

submanifolds of dimension d within R^k will be topologically embedded in R^m, except for relatively unlikely exceptional **F**, as long as $m > 2d$. This fact was proved in the stronger sense of differentiable embedding in Whitney (1936). In the case of Figure 5.4, $m = 3$ and $d = 1$. Figure **5.4b** illustrates an exceptional case for **F**, in that the image of the curve is not an embedding. But almost every perturbation of **F** (all those that don't cause the self-intersection P to persist) results in a topological embedding.

This fact, that for any given map **F** of a curve segment to three-dimensional space, almost every perturbation is "good" in the sense that it is a topological embedding of A, is obvious from Figure 5.4. In order to handle cases that are not quite so clear-cut (e.g., when A is fractal), it is important to pin down more precisely what we mean by "almost every" in this situation.

A property of functions is called *generic in the C^n topology*[1] if for every function $\mathbf{F}(\mathbf{x})$ that does not have the property, there exists a perturbation $\delta\mathbf{F}(\mathbf{x})$ such that the magnitude of $\delta\mathbf{F}(\mathbf{x})$ and the magnitude of the derivatives of $\delta\mathbf{F}(\mathbf{x})$ up to order n (i.e., $|\delta\mathbf{F}|$ and $|\partial^m \delta\mathbf{F}_j / \partial x_{i_1} \partial x_{i_2} \cdots \partial x_{i_m}|$ for $1 \leq m \leq n$) are arbitrarily small and such that $\mathbf{F} + \delta\mathbf{F}$ does have the property. That is, for any ϵ, no matter how small, we can find a $\delta\mathbf{F}$ whose Taylor series coefficients of order n or less are all less than ϵ, and such that $\mathbf{F} + \delta\mathbf{F}$ has the property.

As D. Ruelle (1989) has written, "It is good to know whether a property is generic or not, but it should be understood that *generic* does not imply

[1] In this book, our usage of the term "generic" makes it equivalent to "dense." Generic is sometimes used in the following more restrictive sense: a property is generic if the set of functions which possess the property is a *residual* set, which is a set that is a countable intersection of dense open subsets of the function space. In particular, this implies the set is dense, which corresponds to the usage in this book.

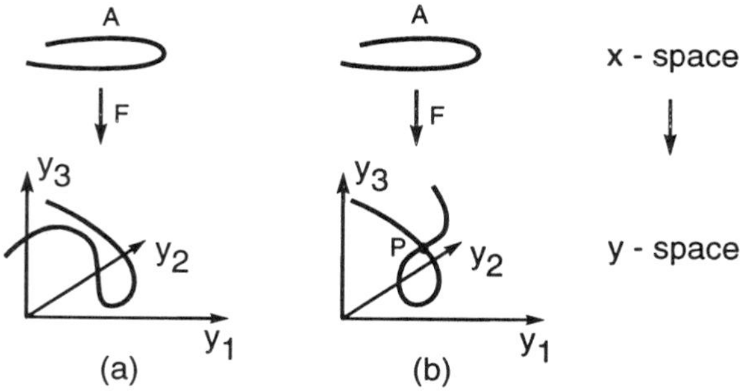

Figure 5.4: *Mapping of a curve segment to three dimensions.* **F** *is an embedding in (a). It is not an embedding in (b), although small perturbations of* **F** *yield an embedding.*

usually true." For example, generic subsets of real numbers can have arbitrarily small probability (Lebesgue measure). Recently, a stronger characterization was introduced, called prevalence. A property is called *prevalent among* C^n *functions* if whenever **F** does not have the property, not only do arbitrarily small C^n perturbations have the property, but small perturbations have the property with probability one. For a precise definition and discussion of "probability one" in this context, see Sauer et al. (1991). If a property is prevalent for functions **F**, we will often say that *almost every* function **F** has the property, or that **F** has the property *with probability one.*

To keep matters as simple as possible, we will always consider C^1 perturbations in our discussions. When we use the terms "generic" and "prevalent" we will mean "among C^1 functions."

For functions from R^k to R^m, the property that **F** is a topological embedding of a d-dimensional compact submanifold A is a prevalent (and therefore generic) property of C^1 functions, if $m > 2d$. In fact, a very simple set of first-order perturbations suffices. If we consider the finite-dimensional space of $m \times k$ matrices **L**, then no matter what **F** does to the submanifold A, the function $\mathbf{F}(\mathbf{x}) + \mathbf{L}\mathbf{x}$ *is* a topological embedding of A for almost every **L** in the sense of probability. (The set of exceptional matrices **L** has mk-dimensional Lebesgue measure zero.)

In the discussion so far, we have considered A to be a compact smooth submanifold of the k-dimensional phase space. In many applications, A will be the attractor of a dynamical system, and may not be a manifold, or even have integer dimension. Somewhat surprisingly, if the box-counting dimension is used, the requirement for embedding a fractal set is the same as for manifolds

– namely, that the number of measurements m is greater than twice the box-counting dimension of the set A. That leads to the following general statement.

A. Topological Embedding: Simultaneous Measurements

Assume that A is a compact subset of R^k of box-counting dimension D_0. If $m > 2D_0$, then almost every C^1 function $\mathbf{F} = (f_1, \ldots, f_m)$ from R^k to R^m is a topological embedding of A into R^m.

The intuitive reason for the condition $m > 2D_0$ can be seen by considering generic intersections of smooth surfaces in m-dimensional Euclidean space R^m. Two sets of dimensions d_1 and d_2 in R^m may or may not intersect. If they do intersect, and the intersection is generic, then they will meet in a surface of dimension

$$d_I = d_1 + d_2 - m. \tag{5.2}$$

(If this number is negative, generic intersections do not occur.) If the surfaces lie in special (i.e., nongeneric) position relative to one another, the intersection may be special and have a different dimension. See Figure 5.5 for some examples.

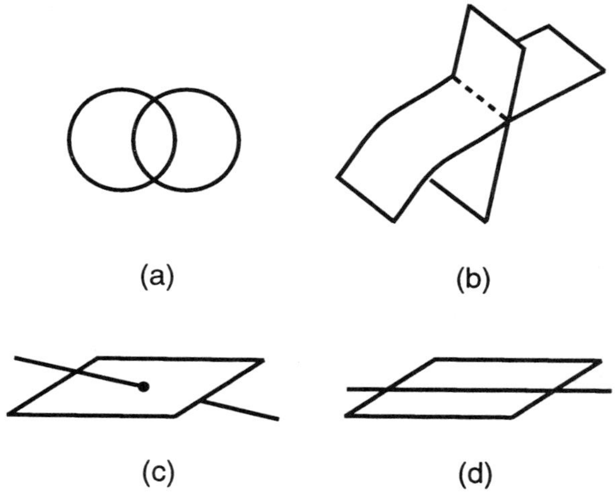

Figure 5.5: *(a) Generic intersection:* $d_1 = d_2 = 1, m = 2$. *(b) Generic intersection:* $d_1 = d_2 = 2, m = 3$. *(c) Generic intersection:* $d_1 = 2, d_2 = 1, m = 3$. *(d) Nongeneric intersection:* $d_1 = 2, d_2 = 1, m = 3$, *but* $d_I = 1$.

For example, as shown in Figure 5.5, two circles in the plane, if they intersect, will generically meet in a set of dimension $1 + 1 - 2 = 0$ – a finite set of points

– unless the circles lie on top of one another. In the latter nongeneric case, almost every smooth perturbation of the position of the circles results in a zero-dimensional intersection. (The exceptional cases, which form a probability zero set of perturbations, are those that perturb the circles in lockstep, so that they remain atop one another.) For circles in three-dimensional space, since $1+1-3<0$, generic intersections do not occur. Almost every perturbation of intersecting circles in R^3 breaks the intersection.

The requirement $m > 2d$ can now be thought of as the necessary condition for the image $\mathbf{F}(A)$ not to intersect *itself*. For a d-dimensional set mapped into R^m, in order to make sure that we can perturb away unlucky intersections of one part of $\mathbf{F}(A)$ with another (perhaps originally distant) part of itself, we must require $d+d-m<0$, or $m>2d$.

For fractal sets, if d is taken to be the box-counting dimension $d = D_0$, the requirement $m > 2d$ for avoiding self-intersection still holds, although it is harder to draw pictures. This follows from the fact that two generically intersecting fractal sets in R^m, of box-counting dimensions d_1 and d_2, will not intersect if $m > d_1 + d_2$. The following argument shows why this is so.

Let S_1 and S_2 be compact sets of box-counting dimensions d_1 and d_2, respectively, lying in R^m. We will consider perturbing one of the sets by linear transformations of R^m and looking to see whether the perturbed set still intersects the other set. Our argument will work for small and large perturbations equally well. For simplicity, we will allow perturbations chosen from the set M of $m \times m$ matrices whose entries are bounded by 1 in absolute value. However, the choice of 1 is arbitrary and can be replaced by any nonzero number in what follows.

To perturb a point $\mathbf{y}$ by a matrix $\mathbf{Q}$ from M means replacing $\mathbf{y}$ by $\mathbf{y}+\mathbf{Q}\mathbf{y} = (\mathbf{I}+\mathbf{Q})\mathbf{y}$, where $\mathbf{I}$ is the identity map. The matrices from M move $\mathbf{y}_a$ with equal probability in all directions. For any pair of points $\mathbf{y}_a$ and $\mathbf{y}_b$, the probability that the particular $\mathbf{Q}$ chosen from the set of matrices M causes the perturbed point $(\mathbf{I}+\mathbf{Q})\mathbf{y}_a$ and $\mathbf{y}_b$ to lie within ϵ of one another is (at most) on the order of ϵ^m. The reason is that this probability is the same as the probability that for a given $\mathbf{Q}$, the point $(\mathbf{I}+\mathbf{Q})\mathbf{y}_a$ lies in the ϵ-ball centered at $\mathbf{y}_b$, which has volume proportional to ϵ^m. If the points $\mathbf{y}_a$ and $\mathbf{y}_b$ are distant from one another, then the probability will be zero. Even if they are near, the probability is limited to ϵ^m.

Similarly, if B_1 and B_2 are ϵ-boxes lying near S_1 and S_2, respectively, the probability that $(\mathbf{I}+\mathbf{Q})(B_1)$ and B_2 intersect is on the order of ϵ^m. The set S_1 (respectively, S_2) can be covered by essentially ϵ^{-d_1} (respectively, ϵ^{-d_2}) boxes of size ϵ, so the number of pairs of boxes is proportional to $\epsilon^{-(d_1+d_2)}$. Putting everything together, the probability that no distinct pair of boxes collide in the intersection $S_2 \cap (\mathbf{I}+\mathbf{Q})S_1$ is proportional to $\epsilon^{-(d_1+d_2)}\epsilon^m = \epsilon^{m-d_1-d_2}$. If

$m > d_1 + d_2$, the probability of choosing a perturbation that has nonempty intersection is negligible for small ϵ.

We conclude that sets of box-counting dimensions d_1 and d_2 will generically miss one another in R^m if $m > d_1 + d_2$. (In fact, the probabilistic nature of the argument we gave means that they will miss one another with probability one.) As in the case where the set was nonfractal (a smooth surface), this allows the images of different parts of a set of box-counting dimension D_0 to miss one another if $m > 2D_0$, verifying that almost every choice of independent simultaneous measurements will lead to a topological embedding.

5.3 Delay coordinates

In the previous section we described how the phase space of possible states for a dynamical attractor can be reconstructed using m simultaneous measurements which are not in special position with regard to one another. Although it seems impossible to specify in exact terms what "not in special position" means in this context, we found that in theory almost every measurement function $\mathbf{F}$ in Equation (5.1) has this property. In fact, we found that small linear perturbations are sufficient to destroy the special position in any given $\mathbf{F}$, at least in theory. The number of simultaneous measurements required is $m > 2D_0$, where D_0 is the box-counting dimension of the attractor.

Now assume our ability to make independent measurements is limited. In the most limited case, we might be in the position of having available only the measurement of a single scalar as a function of time $y(t)$. In a fluid experiment, the measured quantity $y(t)$ might be the magnitude of the fluid velocity measured at a single position in the fluid volume. Since the measurement depends only on the system state, we can represent such a situation by $y(t) = f(\mathbf{x}(t))$, where f is the single measurement function, evaluated when the system is in state $\mathbf{x}(t)$. We assign to $\mathbf{x}(t)$ the *delay coordinate vector*

$$\mathbf{H}(\mathbf{x}(t)) = (y(t-\tau), \ldots, y(t-m\tau)) = (f(\mathbf{x}(t-\tau)), \ldots, f(\mathbf{x}(t-m\tau))). \quad (5.3)$$

Note that for an invertible dynamical system (e.g., the set of ordinary differential equations (1.1)), given $\mathbf{x}(t)$, we can view the state $\mathbf{x}(t-\tau)$ at any previous time $t-\tau$ as being a function of $\mathbf{x}(t)$. This is true because we can start at $\mathbf{x}(t)$ and use the dynamical system to follow the trajectory backward in time to the instant $t-\tau$. Hence $\mathbf{x}(t)$ uniquely determines $\mathbf{x}(t-\tau)$. To emphasize this, we define $h_1(\mathbf{x}(t)) = f(\mathbf{x}(t-\tau)), \ldots, h_m(\mathbf{x}(t)) = f(\mathbf{x}(t-m\tau))$. Then, if we write $\mathbf{y}(t)$ for $(y(t-\tau), \ldots, y(t-m\tau))$, we can express Equation (5.3) as

$$\mathbf{y} = \mathbf{H}(\mathbf{x}), \quad (5.4)$$

where $\mathbf{H}(\mathbf{x}) = (h_1(\mathbf{x}), \ldots, h_m(\mathbf{x}))$.

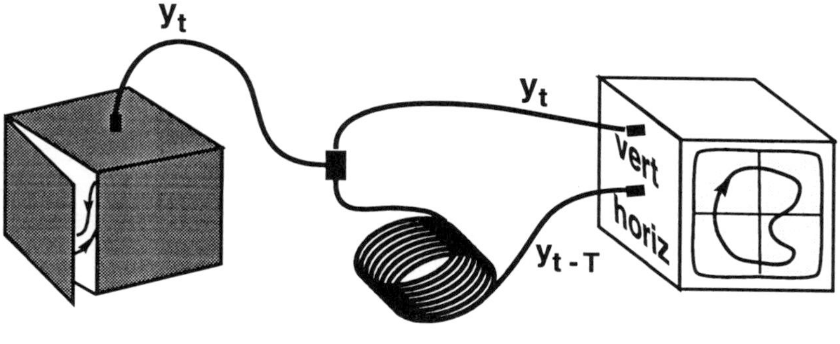

Figure 5.6: *The single measurement of the dynamical system on the left is plotted against a time-delayed version of itself on the display screen on the right. The delay is caused by differing lengths of transmission lines from the black box to the display. Each state* **x** *of the dynamical system gives rise to a dot* **F**(**x**) *on the screen. As in the simultaneous, independent measurement case, the goal is for the reconstruction map* **F** *to be one-to-one.*

The delay coordinate function **H** can be viewed as a special choice of the measurement function **F** in Equation (5.1). For this special choice, the requirement that the measurements not lie in special position is brought into question, since the components of the delay coordinate vector in Equation (5.3) are constrained – they are simply time-delayed versions of the same measurement function f. This is relevant when considering small perturbations of **H**. Small perturbations in the measuring process are introduced by way of the scalar measurement function f, and influence coordinates of the delay coordinate function **H** in an interdependent way. Although it was determined in the simultaneous measurement case that almost every perturbation of $h_1, \ldots, h_m$ leads to a one-to-one correspondence, these independent perturbations (which have access to all coordinates of phase space) may not be achievable by perturbing the single measurement function f.

A simple example will illustrate this point. Suppose the set A contains a single periodic orbit (a topological circle) whose period is equal to the delay time τ. This turns out to be a bad choice of τ, since each delay coordinate vector from the periodic orbit will have the form $(h(\mathbf{x}), \ldots, h(\mathbf{x}))$ for some **x**, and lie along the straight line $y_1 = \ldots = y_m$ in R^m. But a circle cannot be continuously mapped to a line without points overlapping, violating the one-to-one property. Notice that this problem will afflict all measurement functions h, so that perturbing h will not help.

In this case, we cannot perturb our way out of the problem by making the measurement function more generic; the problem is built-in. Although this case is a particularly bad one because of a poor choice of τ, it shows us that the reasoning for the simultaneous measurement case does not extend to delay coordinates, since it gives an obviously wrong conclusion in this case. This problem can be avoided, for example, by perturbing the time delay τ (if indeed that is possible in the experimental setting). In any case, a little extra analysis beyond the geometric arguments we made for the simultaneous measurements case needs to be done. This analysis was begun by Takens (1981) and was extended in Sauer et al., (1991). The result can be stated as follows.

B. Topological Embedding: Delay Coordinates

> **Assume that a continuous time dynamical system has a compact invariant set A (e.g., A may be a chaotic attractor) of box-counting dimension D_0, and let $m > 2D_0$. Let τ be the time delay. Assume that A contains only a finite number of equilibria (i.e., fixed states of the dynamical system), a finite number of periodic orbits of period $p\tau$ for $3 \le p \le m$, and that there are no periodic orbits of period τ or 2τ. Then, with probability one, a choice of the measurement function h yields a delay-coordinate function $\mathbf{H}$ which is one-to-one from A to $\mathbf{H}(A)$.**

The one-to-one property is guaranteed to fail not only when the sampling rate is equal to the frequency of a periodic orbit, as discussed above, but also when the sampling rate is twice the frequency of a periodic orbit, that is, when A contains a periodic orbit of minimum period 2τ. (Surprisingly, it is fine to sample at three times the frequency of a periodic orbit – see below.) To see that this is so, define the function $\eta(\mathbf{x}) = h(\mathbf{x}) - h(\phi_{-\tau}(\mathbf{x}))$ on the periodic orbit, where ϕ_t denotes the action of the dynamics over time t. The function η is either identically zero or it is nonzero for some $\mathbf{x}$ on the periodic orbit, in which case it has the opposite sign at the image point $\phi_{-\tau}(\mathbf{x})$, and changes sign on the periodic orbit. In any case, $\eta(\mathbf{x})$ has a root $\mathbf{x}_0$. Since the period is 2τ, we have $h(\mathbf{x}_0) = h(\phi_{-\tau}(\mathbf{x}_0)) = h(\phi_{-2\tau}(\mathbf{x}_0)) = \ldots$. Then the delay coordinates map $\mathbf{x}_0$ and $\phi_{-\tau}(\mathbf{x}_0)$ to the same point in R^m. Since the minimum period is 2τ, the points $\mathbf{x}_0$ and $\phi_{-\tau}(\mathbf{x}_0)$ are distinct, so $\mathbf{H}$ is not one-to-one for any observation function h. (Again, the problem might be eliminated by changing τ.)

On the other hand, no such problem arises from a single periodic orbit of period 3τ, or any period not equal to τ or 2τ. In fact, the restrictions on the periodic orbits can be weakened to the following: for $p \le m$, the box-counting dimension of the set of points which are periodic of period p must be less than $p/2$. We include these conditions on the periodic orbits for the sake of completeness. For practical cases, barring an unfortunate choice of τ, the important condition for embedding is $m > 2D_0$.

Unfortunate choices of τ that are not ruled out by the theory are those that are unnaturally small or large in comparison to the time constants of the system. Such values of τ will cause the correlation between successive measurements to be excessively large or small, causing the effectiveness of the reconstruction to degrade in real-world cases, where noise is present.

The choice of optimal time delay τ for unfolding the reconstructed attractor is an important and largely unresolved problem. A commonly used rule of thumb is to set τ to be the time lag required for the autocorrelation function to become negative, or alternatively, the time lag required for the autocorrelation function to decrease by a factor of e (from its value at time lag 0). Another approach, that of Fraser and Swinney (1986), incorporates the concept of mutual information, borrowed from Shannon's information theory, which provides a measure of the general independence of two variables. They suggest choosing the delay time which produces the first local minimum of the mutual information of the observed quantity and its delayed value. Other aspects of the problem have been clarified by Liebert and Schuster (1988) and Gibson et al. (1992).

Finally, although we have discussed using simultaneous measurements and delay coordinates separately in statements A and B, there is no theoretical restriction against mixing the two. Analogous theorems can be proved which allow one-to-one reconstruction of the attractor with m_1 delay coordinates and m_2 independent simultaneous coordinates, as long as $m_1 + m_2 > 2D_0$.

5.4 Differentiable embeddings

Assume that A is a compact smooth d-dimensional submanifold of R^k. A circle is an example of a smooth 1-manifold; a sphere and torus are examples of 2-manifolds. A smooth d-manifold has a well-defined d-dimensional tangent space at each point. If $\mathbf{F}$ is a smooth function from one manifold to another, then the Jacobian derivative $\mathbf{DF}$ maps tangent vectors to tangent vectors. More precisely, for each point $\mathbf{x}$ on A, the map $\mathbf{DF}(\mathbf{x})$ is a linear map from the tangent space at $\mathbf{x}$ to the tangent space at $\mathbf{F}(\mathbf{x})$. If for all $\mathbf{x}$ in A, no nonzero tangent vectors map to zero under $\mathbf{DF}(\mathbf{x})$, then $\mathbf{F}$ is called an *immersion*.

Figure 5.7**a** shows an immersion of the circle A. In Figure 5.7**b**, although $\mathbf{F}$ is one-to-one on A, the function $\mathbf{F}$ fails to be an immersion at the pinch point. Figure 5.7**c** shows an immersion which is not a topological embedding.

A smooth function $\mathbf{F}$ on a smooth manifold is called a *differentiable embedding* if $\mathbf{F}$ and $\mathbf{F}^{-1}$ are one-to-one immersions. In particular, a differentiable embedding is automatically a topological embedding. In addition, the tangent spaces

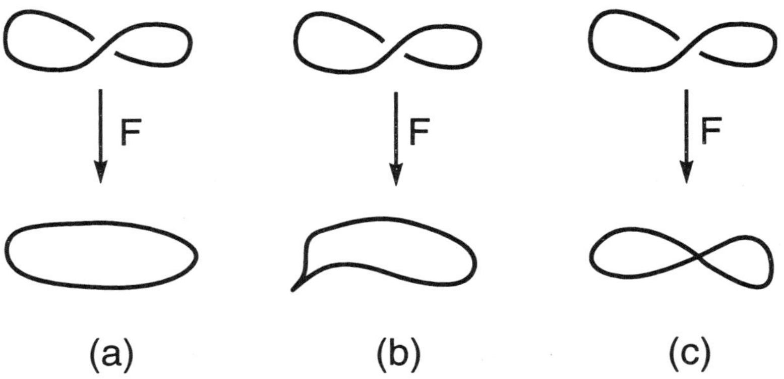

Figure 5.7: *(a) The functions* $\mathbf{F}$ *and* $\mathbf{F}^{-1}$ *are one-to-one immersions, so* $\mathbf{F}$ *is a differentiable embedding of the topological circle* A. *(b)* $\mathbf{F}$ *is a topological embedding of* A *but is not an immersion. (c)* $\mathbf{F}$ *is an immersion but is not one-to-one.*

of A and $\mathbf{F}(A)$ are isomorphic. In particular, the image $\mathbf{F}(A)$ is also a smooth manifold of the same dimension as A.

Whitney (1936) proved that if A is a smooth d-manifold in R^k, and if $m > 2d$, then a typical map from R^k to R^m is a differentiable embedding when restricted to A. Takens (1981) proved a result in this context for delay coordinate functions. The following is a version of Takens' theorem from Sauer et al. (1991).

C. Differentiable Embedding: Delay Coordinates

Assume that a continuous time dynamical system has a compact invariant smooth manifold A of dimension d, and let $m > 2d$. Let τ be the time delay. Assume that A contains only a finite number of equilibria, no periodic orbits of periods τ or 2τ, and only a finite number of periodic orbits of period $p\tau$ for $3 \leq p \leq m$. Assume that the Jacobians of the return maps of those periodic orbits have distinct eigenvalues. Then, with probability one, a choice of the measurement function h yields a delay coordinate function H which is a differentiable embedding from A to $\mathbf{H}(A)$.

This version differs from Takens' original statement in two ways: instead of assuming the genericity of the dynamical system, the requirements on the system (in particular, the relation of periodic orbits to the time delay) are made explicit; and generic is replaced by probability one.

From the point of view of extracting information from an observed dynamical system, there are some advantages to having a differentiable, as compared with a topological, embedding. These advantages stem from the fact that metric properties are preserved by the reconstruction.

First, because a differentiable embedding is a C^1-diffeomorphism from the phase space attractor to the reconstructed attractor, there is a uniform upper bound on the stretching done by both $\mathbf{H}$ and $\mathbf{H}^{-1}$. (This is in contrast to the topological embedding case. One typically knows that $\mathbf{H}$ is continuously differentiable, giving an upper bound on the stretching of $\mathbf{H}$ on a compact attractor. However, in general, $\mathbf{H}^{-1}$ can be extremely nondifferentiable, even though $\mathbf{H}$ is one-to-one.) Functions with an upper bound on stretching (meaning there is a uniform constant C such that $|\mathbf{H}(\mathbf{x}_1) - \mathbf{H}(\mathbf{x}_2)| \leq C|\mathbf{x}_1 - \mathbf{x}_2|$) are called *Lipschitz;* if $\mathbf{H}$ and $\mathbf{H}^{-1}$ are Lipschitz, then $\mathbf{H}$ is called *bi-Lipschitz*. Under virtually any reasonable definition of dimension which measures metric information (including Hausdorff, information, box-counting, and correlation dimensions), the dimension of a set is unchanged under a bi-Lipschitz map. Therefore, the hypotheses of statement C above imply that the fractal dimensions of every closed subset of A can be calculated in principle by delay coordinate embedding.

Of course, in order to investigate the dimension of an arbitrary set S, one first needs to find a d-dimensional differentiable manifold A that contains S and which satisfies those hypotheses, and to use at least $2d$ delay coordinates. In many cases, no natural manifold structure exists for an attractor A. In those cases, it is necessary to rely on a separate theorem (see Ding et al. (1993) in Chapter 7), which says that, in general, the correlation dimension of the reconstructed attractor $\mathbf{H}(A)$ in R^m equals the correlation dimension of A, as long as the number of delay coordinates m is greater than the correlation dimension of A. This fact is independent of whether $\mathbf{H}$ is one-to-one.

A second advantage of a differentiable embedding (should one exist) is that all Lyapunov exponents of the phase space attractor are reproduced in the reconstruction. If ϕ_t denotes the dynamical flow on the phase space manifold A, then $\psi_t \equiv \mathbf{H}\phi_t\mathbf{H}^{-1}$ defines the flow in the reconstructed manifold. Since $\mathbf{H}$ and $\mathbf{H}^{-1}$ are differentiable, the chain rule $\mathbf{D}\psi_t = \mathbf{DHD}\phi_t\mathbf{DH}^{-1}$ shows that $\mathbf{D}\psi_t$ and $\mathbf{D}\phi_t$ are similar matrices. As a result, the d Lyapunov exponents of the dynamics on A will be reproduced on the manifold $\mathbf{H}(A)$. In practice, the reconstructed manifold will be embedded as a submanifold of R^m, so there will be $m-d$ irrelevant dimensions. Thus numerical procedures using experimental data produce $m - d$ "spurious" Lyapunov exponents not associated with the dynamics on A, which are artifacts of the embedding process. Distinguishing the spurious from the true Lyapunov exponents is one of the goals of the reprints in Chapter 8.

5.5 Embedding of filtered data

Time series data acquired from an experiment is often filtered. In some instances, the filtering is beyond the control of the experimentalist; for example, when a filter is intrinsic in the data collection apparatus. In other cases, filtering can assist in the separation of underlying signal from unwanted noise. However, caution must be exercised, since dynamical properties of interest may be changed under filtering.

Assume that the measured time series is denoted by $\{x_i\}$. Then

$$z_{i+1} = \sum_{j=0}^{L} \alpha^j x_{i-j} = x_i + \alpha x_{i-1} + \ldots + \alpha^L x_{i-L} \tag{5.5}$$

where $|\alpha| < 1$, is a simple example of a filtered series. If L is finite, this is called a finite impulse response (FIR) filter because the hypothetical input sequence $x_i = \{1, 0, 0, \ldots\}$ leads to a finite sequence of nonzero z_i.

Replacing Equation (5.5) with an infinite sum leads to an infinite impulse response (IIR) filter. Setting $L = \infty$, one checks using Equation (5.5) that

$$z_{i+1} = \alpha z_i + x_i, \tag{5.6}$$

which is a more compact definition of z_i. This is a discrete-time version of an RC low-pass filter.

What is the effect of using the filtered time series $\{z_i\}$ in the place of the original $\{x_i\}$ in calculations which depend on delay coordinates? The theoretical answers are that FIR filters will not change dynamical quantities, while IIR filters may change them.

Of course, the practical use of this theoretical information must be tempered with the knowledge that an IIR filter is a limiting case of an FIR filter. The more the FIR filter looks like an IIR filter, the larger the potential that numerical effects of the latter will be felt, because of the presence of noise in the data and insufficiency of the data. Understanding this uneasy dichotomy is important for numerical calculations, and we will pursue it in a detailed example below.

First we show that an IIR filter can change the dimension of the attractor. This was discovered by Badii and Politi (1986); see in particular Badii et al. (1988) and Mitschke et al. (1988). They point out that under an IIR filter, the reconstructed attractor can have a different dimension than the original attractor that generated the unfiltered data.

This can easily be verified for Lyapunov dimension. As defined in Chapter 4, the Lyapunov dimension of an attractor is the number $D_L = k + (h_1 + \ldots + h_k)/|h_{k+1}|$, where the Lyapunov exponents of the attractor are $h_1 \geq h_2 \geq \ldots$, and where k is determined by $h_1 + \ldots + h_k \geq 0 > h_1 + \ldots + h_{k+1}$. Consider the example of the Hénon map:

$$\begin{aligned} x_{i+1} &= A - x_i^2 + By_i \\ y_{i+1} &= x_i. \end{aligned} \tag{5.7}$$

If we consider the measured signal to be the x-coordinate, putting the signal through the IIR filter of Equation (5.6) is equivalent to augmenting the dynamical system with a new variable z:

$$\begin{aligned} x_{i+1} &= A - x_i^2 + By_i \\ y_{i+1} &= x_i \\ z_{i+1} &= \alpha z_i + x_i \end{aligned} \tag{5.8}$$

Here we consider z to be the output of the IIR filter.

For the standard parameter values $A = 1.4$ and $B = 0.3$, the Lyapunov exponents of system (5.7) are approximately $h_1 = 0.4$ and $h_2 = -1.6$, which amounts to a Lyapunov dimension of $D_L \approx 1.25$. System (5.8) has the same two plus a new Lyapunov exponent of $\ln|\alpha|$. From the formula for Lyapunov dimension it is obvious that for $\alpha > e^{h_2} \approx 0.2$, the Lyapunov dimension will rise. If we set $\alpha = 0.9$, for example, $D_L \approx 2.2$.

Thus Lyapunov dimension is changed by application of this IIR filter. The same effect shows up in calculations of the correlation dimension. As we will see below in Figure 5.8, the correlation dimension of the attractors of systems (5.7) and (5.8) with $\alpha = 0.9$ are approximately 1.2 and 2.1, respectively.

For FIR filters, on the other hand, mathematical theorems exist which say that the delay coordinate properties are unchanged by the filter. First of all, the one-to-one property holds. That is, using as coordinates the output of a full-rank FIR filter whose inputs are delay coordinates collected from a generic measurement function will give a one-to-one representation of the attractor, as long as the number of filtered delay coordinates used is greater than twice the box-counting dimension of the attractor. Secondly, the correlation dimension D_2 of the attractor reconstructed with the filtered data is the same as the correlation dimension of the original attractor, again assuming that the number of FIR-filtered delay coordinates used is $m > D_2$.

These facts fit in the following more general context. Let $\mathbf{M}$ be an $m \times w$ matrix of rank m, and define the *filtered delay coordinate vector* $\mathbf{b}$ by

$$\mathbf{b} = \mathbf{M}[x_t, x_{t-\tau}, \ldots, x_{t-(w-1)\tau}]^T. \tag{5.9}$$

Note that the reconstruction vector $\mathbf{b}$ is an m-dimensional vector. For the FIR filter of Equation (5.5), the matrix is

$$\mathbf{M} = \begin{pmatrix} 1 & \alpha & \cdots & \cdots & \alpha^L & & & \\ & 1 & \alpha & \cdots & \cdots & \alpha^L & & \\ & & \ddots & \ddots & & & \ddots & \\ & & & 1 & \alpha & \cdots & \cdots & \alpha^L \end{pmatrix}. \tag{5.10}$$

More general matrices M can be used, as long as they are of full rank.

For the filtered delay coordinates of Equation (5.9), it is shown in Sauer et al. (1991) that the number of coordinates needed to insure a one-to-one correspondence is $m > 2D_0$. For the special case of a smooth manifold, where D_0 is an integer, this was proved independently in Broomhead et al. (1992) by a modification of Takens' original proof.

For the dimension preservation question, first notice that unlike the IIR case, none of the commonly used dimensions can increase under an FIR filter. This again follows because a finite matrix such as Equation (5.10) can cause at most finite stretching (it has a finite matrix norm in the Euclidean metric), and so the representation map shown in Equation (5.9) is Lipschitz.

Can dimension decrease under an FIR filter? Again the answer is no. The conventional wisdom is that applying an FIR filter is like a change of coordinates. A change of coordinates, and its inverse, are Lipschitz, and should not change dimension. This argument is not entirely relevant to the present question because most useful FIR filters do not have inverses because they are not represented by square matrices. In particular, they have a nontrivial nullspace, compressing some directions to zero (examine the $m \times (m + L)$ matrix of Equation (5.10) for a representative example). Even if an inverse map existed, it could not be Lipschitz. Although the map cannot be bi-Lipschitz, it can be proved that a filtered delay coordinate map, with a generic measurement function, preserves correlation dimension. The number m of filtered delay coordinates needed in Equation (5.9) so that the reconstructed attractor preserves correlation dimension is $m > D_2$. Note that the number of filtered delay coordinates needed is independent of w, the length of the filter.

The main theoretical conclusion is that FIR filters preserve the one-to-one property and do not change the correlation dimension. This means that whatever distortions occur in the relationship between points, they do not change the infinitesimal structure of the attractor.

However, with data measured in the laboratory, the infinitesimal structure of the attractor may or may not be visible above the noise level of the experiment. In particular, consider the effect of noise on determining the correlation dimension D_2. FIR filters are capable of taking an input signal for which the correlation dimension scaling region is visible, and producing an output for which it is obscured by noise, even though the theoretical fact is that the dimension does not change.

Figure 5.8 shows the effect of filtering on a time series corresponding to the x-coordinate of the Hénon map (5.7). Each part of the figure is a correlation dimension estimate plotted as a function of ϵ. Figure 5.6**a** is a dimension calculation for the unfiltered data, and the succeeding parts of the figure show dimension calculations for increasing filter length L.

The solid lines in Figure 5.8 show correlation dimension estimates for $m = 3$, an embedding in three-dimensional space; the dashed lines represent $m = 10$. Each correlation dimension estimate plotted for a distance ϵ is the slope of the least squares line (on a log-log plot) along the correlation integral $\hat{C}(\epsilon)$ (see Section 2.2) from $\epsilon/2$ to 2ϵ. Thus the point plotted for $\epsilon = 2^{-8}$ is an approximation of the slope of $\hat{C}(\epsilon)$ over the range $2^{-9} < \epsilon < 2^{-7}$. The length of time series used for this calculation was 10^5.

Figures 5.8**b–e** use increasing values of L, the FIR filter length, with α fixed at 0.9. Remember that we know the correct answer is $D_2 = 1.2$ for all cases of finite L. Figure 5.8**b**, for $L = 5$, shows that the $m = 3$ embedding correctly calculates the dimension on the scale shown, but that the scaling region for $m = 10$ has already moved toward the left. In Figure 5.8**c**, for $L = 10$, the scaling region for $m = 10$ has moved completely off the page (toward the left). For larger values of L, shown in Figures 5.6**d** and **e**, the range of ϵ shown gives the wrong answer; the scaling region for the correct $D_2 = 1.2$ exists for smaller ϵ than is shown.

As L increases, the scaling region moves toward smaller ϵ, until it is no longer visible at the scale of this study. Figure 5.8**f** corresponds to $L = \infty$, so that an IIR filter is being applied. In summary, the correlation dimension D_2 remains near 1.2 for all FIR filters, but the scaling region tends to zero as $L \to \infty$. In the limit of the IIR filter, $D_2 \approx 2.1$. The point is that when L is beyond 10, the scaling region of the true correlation dimension exists only for ϵ smaller than the scale of this exercise, so that for all practical purposes, the correct D_2 cannot be measured. To the extent to which an FIR filter resembles an IIR filter up to the ambient noise level, the correct scaling region may become invisible, and the incorrect dimension could be inferred.

A similar effect is shown to occur when bleaching a time series acquired from a chaotic system, as discussed by Theiler and Eubank (1993). The object of bleaching is to fit the best linear predictor model to the data, and then subtract it from the time series and analyze the residuals. Producing the residuals in this way can be viewed as applying a particular FIR filter to the data. Although the correlation dimension does not change, the attractor undergoes distortion on small scales. As linear models of increasing order are subtracted from the data, the effective scaling region for correlation dimension calculation retreats

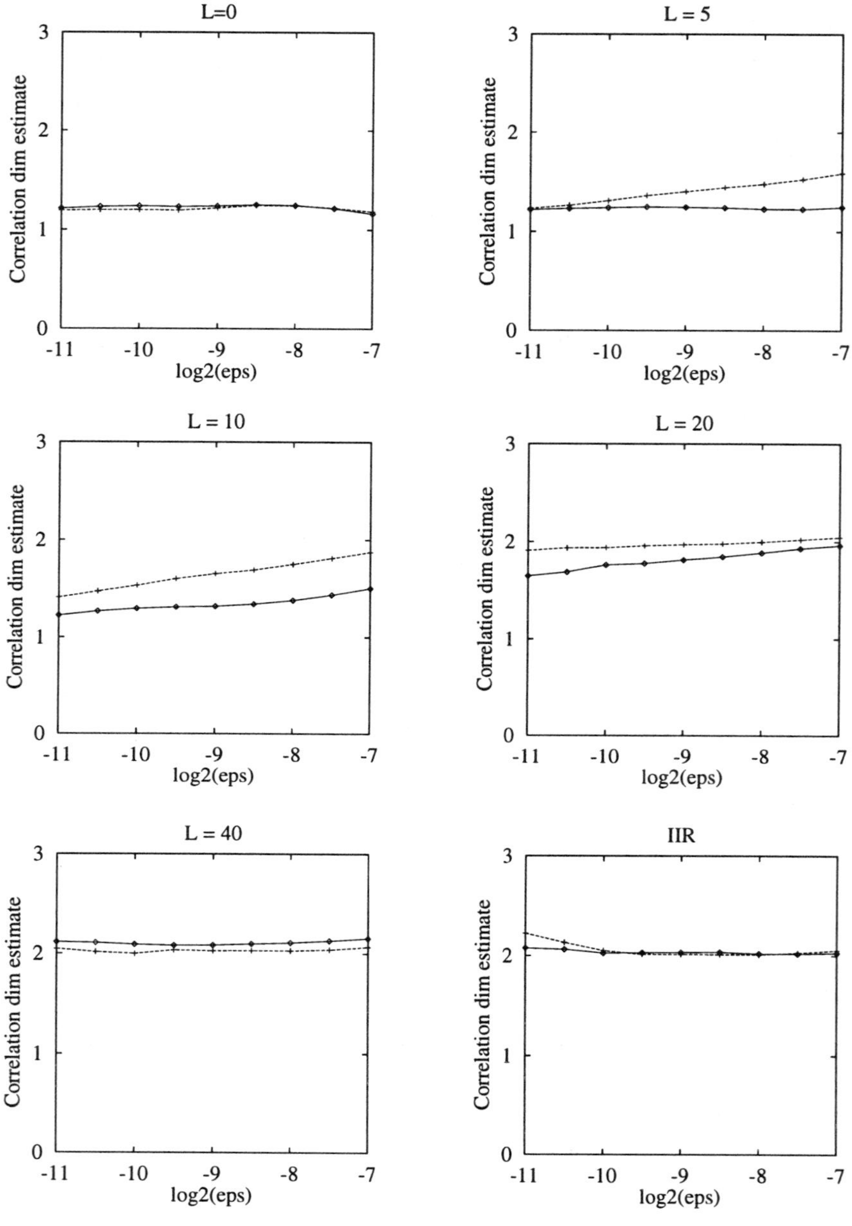

Figure 5.8: *Estimates for correlation dimension as a function of pairwise distance* ϵ*, for (a)* x*-coordinate of the Hénon map (5.7); (b) same as (a), but output of FIR filter with* $L = 5$*; (c) L=10; (d) L=20; (e) L=40; (f) L=*∞*, system (5.8). Solid curve: embedding dimension* $m = 3$*. Dashed curve:* $m = 10$*.*

toward infinitesimal lengths. Thus bleaching can be detrimental to dimension calculations and other forms of data analysis.

Finally, we emphasize that *every* full rank filter **M**, as in Equation (5.9), preserves correlation dimension, as long as the measurement function h of the dynamical system is generic. An example which at first glance seems to contradict this is the dynamical system (5.8), where we momentarily ignore the fact that it is the augmented Hénon map. Assume $\alpha = 0.9$ as before. Although the attractor of system (5.8) has correlation dimension $D_2 \approx 2.1$, if the FIR filter

$$w_i = z_i - \alpha z_{i-1} \tag{5.11}$$

is applied to the output z_i, the resulting signal is $w_i = z_i - \alpha z_{i-1} = x_{i-1}$, which has $D_2 \approx 1.2$ (because it is the x-coordinate of the Hénon map). The FIR filter of Equation (5.11) seems to have reduced the correlation dimension, which would run counter to the theorem described above. This is mentioned by Broomhead et al. (1992).

The resolution of this contradiction is that the measurement function h, which we have taken to be the z-coordinate of Equation (5.8), is nongeneric. The fault does not lie in the choice of FIR filter of Equation (5.11) – any full rank filter preserves dimension, for almost every choice of measurement function. The problem lies instead with the failure of the measurement function of system (5.8) to be generic. One can check that if z is replaced by any generic nonlinear combination of the variables x, y, z, the correct correlation dimension of approximately 2.1 will be calculated for the filtered output w_i of the system (5.8). (This in turn focuses attention on an important fact about the output of an IIR filter, which is that the filter output itself, in this case z, is a special choice, and not necessarily a generic measurement function.) In fact, if the dimension calculation is repeated with the measurement function $h = z^2$ (i.e., replacing the occurences of z in Equation (5.11)) with z^2, the correct dimension of $D_2 = 2.1$ is again obtained for the filtered signal.

Finite impulse response filters can be devised that are more sophisticated than Equation (5.5). The reprint by Broomhead and King (1986) in Chapter 6 advocates using for **M** the matrix of right singular vectors which result from the singular value decomposition (SVD) of the autocovariance matrix of the series. This transforms the original delay coordinates to principal component coordinates. Since the right singular vectors form an orthogonal set, the matrix **M** will be a full rank matrix, so the theory we have referred to guarantees that the correlation dimension can be calculated as long as the number of components kept is greater than the dimension. This filter is used in the reprint of Albano et al. (1988) in Chapter 7 to improve dimension measurements of chaotic attractors.

In summary, the application of a finite impulse response (FIR) filter to delay coordinates does not change (in theory) the number of independent coordinates needed to represent an attractor. For a one-to-one representation, $m > 2D_0$ coordinates are sufficient, where m is the output dimension of the filter. To calculate correlation dimension, $m > D_2$ is sufficient.

These are theoretical results and are therefore only a starting point. For a fixed amount of finite resolution data, practical issues may dominate. In particular, as shown in Figure 5.8, an FIR filter may obscure the scaling region in the observable range, increasing the difficulty of calculating dimension. Infinite impulse response (IIR) filters may change the dimension, even in theory. On the other hand, there are FIR filters which can improve the quality of the embedding provided by delay coordinates.

References

R. BADII, A. POLITI, On the fractal dimension of filtered chaotic signals. In *Dimensions and Entropies in Chaotic Systems*, ed. G. Mayer-Kress (Springer, Berlin) (1986).

R. BADII, G. BROGGI, B. DERIGHETTI, M. RAVAIN, S. CILIBERTO, A. POLITI, M.A. RUBIO, Dimension increase in filtered signals. Phys. Rev. Lett. **60**, 979 (1988).

D.S. BROOMHEAD, J.P. HUKE, M.R. MULDOON, Linear filters and nonlinear systems. J. Roy. Stat. Soc. B**54**, 373 (1992).

A. FRASER, H. SWINNEY, Independent coordinates for strange attractors from mutual information. Phys. Rev. A **33**, 1134 (1986).

J. GIBSON, J.D. FARMER, M. CASDAGLI, S. EUBANK, An analytic approach to practical state space reconstruction. Physica D **57**, 1 (1992).

W. LIEBERT, H. SCHUSTER, Proper choice of the time delay for the analysis of chaotic time series. Phys. Lett. A **142**, 107 (1988).

F. MITSCHKE, M. MOLLER, W. LANGE, Measuring filtered chaotic signals. Phys. Rev. A **41**, 1169 (1990).

D. RUELLE, *Elements of Differentiable Dynamics and Bifurcation Theory*. Academic Press, New York (1989).

T. SAUER, J.A. YORKE, M. CASDAGLI, Embedology. J. Stat. Phys. **65**, 579 (1991).

F. TAKENS, Detecting strange attractors in turbulence. Lecture Notes in Math. **898**, Springer-Verlag (1981).

J. THEILER, S. EUBANK, Don't bleach chaotic data. Chaos **4** (1993).

H. WHITNEY, Differentiable manifolds. Ann. Math. **37**, 645 (1936).

PART II

Analysis of Data from Chaotic Systems

Given for one instant an intelligence which could comprehend all the forces by which nature is animated and the respective positions of the beings which compose it, if moreover this intelligence were vast enough to submit these data to analysis, it would embrace in the same formula both the movements of the largest bodies in the universe and those of the lightest atom; to it nothing would be uncertain, and the future as the past would be present to its eyes.

— P. Laplace

Analogies prove nothing, that is quite true, but they can make one feel more at home.

— S. Freud

CHAPTER 6

The Practice of Embedding

N. Packard, J. Crutchfield, D. Farmer, R. Shaw
Geometry from a time series.
Phys. Rev. Lett. **45**, 712 (1980).

D.S. Broomhead, G.P. King
Extracting qualitative dynamics from experimental data.
Physica D **20**, 217-236 (1986).

M. Kennel, R. Brown, H. Abarbanel
Determining embedding dimension for phase-space reconstruction using a geometrical construction.
Phys. Rev. A **45**, 3403-3411 (1992).

D. Kaplan, L. Glass
Direct test for determinism in a time series.
Phys. Rev. Lett. **68**, 427-430 (1992).

Editors' Notes

The idea of reconstructing chaotic attractors from experimental data using delay coordinates entered common usage around 1980. By this time, delay plots of scientific data had begun to be displayed; for examples, see Glass and Mackey (1979), where the technique arose naturally to display trajectories from a delay differential equation, and Roux et al. (1980), who used it for the analysis of experimental data. The influential paper of **Packard et al.** (1980)[1] attributes the idea of delay coordinates to a communication with D. Ruelle.

In a conference proceedings published in 1981, F. Takens gave a mathematical proof of the legitimacy of delay coordinates. He showed that if the dynamical system and the observed quantity are generic, then the delay coordinate map from a d-dimensional smooth compact manifold to $2d+1$-dimensional

[1]Papers that are reprinted in this chapter are identified with boldface lettering in the Editors' Notes.

reconstruction space is a diffeomorphism. Even more useful would have been a result which tied the reconstruction dimension to the dimension of the attractor, rather than the dimension of the phase space containing the attractor, which could presumably be much larger. R. Mañé published a paper in the same conference proceedings claiming that a generic projection of a set of Hausdorff dimension D_H to a Euclidean space of dimension greater than D_H was one-to-one on the set. This turns out to be false. [See the counterexamples in Sauer et al. (1991). The first to note the problem was Mañé, who circulated a correction shortly after publication. No condition solely on the Hausdorff dimension guarantees that a generic projection is one-to-one.] Although the result is not correct, it is correct in spirit, and became very influential, along with the paper of Takens.

This gap was later filled by the paper of Sauer et al. (1991), which directly ties the reconstruction dimension to the dimension (box-counting instead of Hausdorff) of the attractor, and makes explicit the genericity assumptions on the dynamics. Under these assumptions, they show that a measured quantity generically leads to a one-to-one delay coordinate map as long as the reconstruction dimension is greater than twice the box-counting dimension of the attractor. Rather than including the rather technical reprints of Takens (1981), Mañé (1981), and Sauer et al. (1991) in this volume, we have instead summarized the relevant results in Chapter 5.

The idea of filtering the delay coordinates was first put forward by **Broomhead and King** (1986). They suggest that a great practical advantage can be gained by replacing raw delay coordinates with the dominant modes recovered from the singular value decomposition.

Although there is a definitive result that using greater than $2D_0$ delay coordinates results in an embedding of the attractor, this fact seems less than helpful at the time of data collection, when the attractor dimension is not known. In fact, approximating the dimension requires embedding the data in some way. A common approach to this problem is to embed the data in a sufficiently high dimension m such that the fractal dimension (usually the correlation dimension D_2) can be measured. The number of delay coordinates necessary for this step (in theory) is the smallest integer greater than D_2 (see Ding et al. (1993), reprinted in Chapter 7). Then, assuming $D_2 \approx D_0$, a reasonable embedding dimension can be found.

Kennel et al. (1992) consider the problem of finding the lowest possible number of delay coordinates to achieve an embedding of the attractor. (We emphasize that although $m > 2D_0$ *guarantees* an embedding, in specific cases it is common that fewer delay coordinates may be sufficient.) The authors describe a practical method which relies on the fact that choosing too low a value for m results in points that are far apart in the original phase space being moved close together in the reconstruction space. This is called the problem of "false nearest neighbors," and is diagnosed by charting the change in the number of nearest neighbors as m increases.

One of the motivations of embedding techniques for attractor reconstruction is to distinguish a deterministic process from a stochastic process. Traditional signal processing relies heavily on spectral analysis of the signal, which is sufficient if the system producing the signal is linear, and produces a signal characterized by a finite number of frequencies. Chaotic attractors, on the other hand, ordinarily produce time series which have a continuous frequency power spectrum instead of a discrete set of frequencies. Because of this, it is not possible to distinguish a chaotic attractor from a stochastic noise process on the grounds of power spectrum.

Kaplan and Glass (1992) analyze the tangent directions of evolution of the embedded delay vectors. The absence of determinism would imply that the tangents in a small box in reconstruction space are uncorrelated. In that case, if the N tangent vectors within a small box are added (using vector addition), the expected length of the sum vector is of the order of $N^{-1/2}$. This gives a null hypothesis with which to compare possibly deterministic signals.

Finally, we mention that although most of the papers in this section deal with delay coordinates, the theory explained in Chapter 5 works equally well for simultaneous independent coordinates, or even for a mixture of simultaneous and delay coordinates (see Sauer et al. (1991)). The paper of Guckenheimer and Buzyna (1986) in Chapter 7 is an example of the use of simultaneous coordinates to measure dimension.

Further reading

There has been much discussion of the quality of embeddings in general, and the particular problem of detecting low-dimensional chaos using time series. The following papers form a small sample of the work that illuminates these issues. Many more can be found in the Bibliography at the end of this book.

BUZUG, T., PFISTER, G., Optimal delay time and embedding dimension for delay-time coordinates by analysis of the global static and local dynamical behavior of strange attractors. Phys. Rev. A **45**, 7073 (1992).

CASDAGLI, M., Chaos and deterministic versus stochastic nonlinear modeling. J. Roy. Stat. Soc. B **54**, 303 (1991).

CASDAGLI, M., EUBANK, S., FARMER, J.D., GIBSON, J., State space reconstruction in the presence of noise. Physica D **51**, 52 (1991).

Geometry from a Time Series

N. H. Packard, J. P. Crutchfield, J. D. Farmer, and R. S. Shaw
Dynamical Systems Collective, Physics Department, University of California, Santa Cruz, California 95064
(Received 13 November 1979)

It is shown how the existence of low-dimensional chaotic dynamical systems describing turbulent fluid flow might be determined experimentally. Techniques are outlined for reconstructing phase-space pictures from the observation of a single coordinate of any dissipative dynamical system, and for determining the dimensionality of the system's attractor. These techniques are applied to a well-known simple three-dimensional chaotic dynamical system.

PACS numbers: 47.25.-c

Lorenz originally demonstrated that very simple low-dimensional systems could display "chaotic" or "turbulent" behavior.[1] Attractors which display such behavior were termed "strange attractors" by Ruelle and Takens,[2] who then went on to conjecture that these strange attractors are the cause of turbulent behavior in fluid flow. The experiments of Gollub and Swinney have strengthened the conjecture,[3] but the question still remains: How can we discern the nature of the strange attractor underlying turbulence from observing the actual fluid flow?

Data obtained by experimentalists examining turbulent fluid flow often take the form of a "time series," which is to say, a series of values sampled at regular intervals. We address here the problem of using such a time series to reconstruct a finite-dimensional phase-space picture of the sampled system's time evolution. From this picture we can then obtain the asymptotic properties of the system, such as the positive Liapunov characteristic exponents, which are a measure of how chaotic the system is,[4-6] and topological characteristics such as the attractor's topological dimension. We illustrate these reconstruction methods by applying them to a time series obtained from sampling one coordinate of a three-dimensional chaotic dynamical system first studied by Rossler,[7] and then comparing the resulting values of the Liapunov exponents to those obtained by a different method.

The dynamical system of interest is a set of three ordinary differential equations:

$$\begin{aligned} \dot{x} &= -(y+z), \\ \dot{y} &= x + 0.2y, \\ \dot{z} &= 0.4 + xz - 5.7z. \end{aligned} \tag{1}$$

These equations have a chaotic attractor which is illustrated in Fig. 1, which was obtained from an analog computer simulation.

The heuristic idea behind the reconstruction method is that to specify the state of a three-dimensional system at any given time, the measurement of *any* three independent quantities should be sufficient, where "independent" is not yet formally defined, but will become operationally defined. We conjecture that any such sets of three independent quantities which uniquely and smoothly label the states of the attractor are diffeomorphically equivalent. The three quantities typically used are the values of each state-space

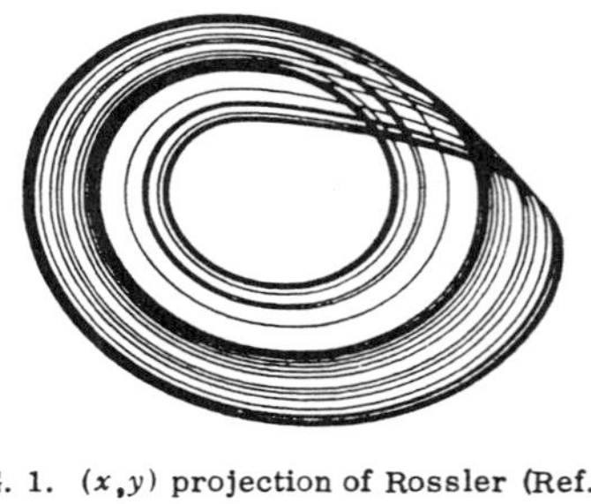

FIG. 1. (x,y) projection of Rossler (Ref. 7).

coordinate, $x(t)$, $y(t)$, and $z(t)$. We have found that beginning with a time series obtained by sampling a single coordinate of Eq. (1), one can obtain a variety of three independent quantities which appear to yield a faithful phase-space representation of the dynamics in the original x, y, z space. One possible set of three such quantities is the value of the coordinate with its values at two previous times,[8] e.g., $x(t)$, $x(t-\tau)$, and $x(t-2\tau)$. Another set obtained by making the time delays small, and taking differences is $x(t)$, $\dot{x}(t)$, and $\ddot{x}(t)$. Figure 2 shows a reconstruction of the $(x, \dot{x})$ picture from the time series taken from sampling the x coordinate of Eq. (1). Comparison of Figs. 1 and 2 certainly indicates that topological characteristics and geometrical form of the attractor remain intact when viewed in the $(x, \dot{x})$ coordinates. For an experimentalist observing some chaotic phenomenon, such as turbulent fluid flow, the construction of phase-space coordinates might not be as simple as the case illustrated above. In many cases the experimentalist has no *a priori* knowledge of how many dimensions a dynamical description would require, nor the quantities appropriate to the construction of such a description. So far there is no universally applicable method of phase-space construction, though the nature of the phenomenon might suggest possible alternatives. In a study of fluid turbulence, for example, the experimentalist might try using the velocity of the fluid in different directions, at different points in space, and at different times.

After having obtained a phase-space picture like that shown in Fig. 2, if the attractor is of sufficiently simple topology, one can use methods which have been previously developed[1,4] to construct a one-dimensional return map, and then from the return map one can obtain the positive characteristic exponent of the attractor. Roughly speaking, the procedure consists of making a cut along the attractor, coordinatizing it with the unit interval (0,1), and accumulating a return map by observing where successive passes of the trajectory through the cut occur. The result is a return map of the form $x(n+1) = f(x(n))$, and the positive characteristic exponent is found by computing

$$\lambda = \lim_{N\to\infty} \frac{1}{N} \sum_{i=1}^{N} \ln \left| \frac{df}{dx} \right|_{x_i}$$

or alternately, by computing

$$\lambda = \int_0^1 P(x') \ln \left| \frac{df}{dx} \right|_{x'} dx'$$

if one knows (or has accumulated empirically) the equilibrium probability distribution $P(x)$. See Shaw[4] for more complete discussions of the computation of the characteristic exponents with use of this method.

Equations (1) are sufficiently simple that one can explicitly obtain a new set of three differential equations describing the dynamics of the state space comprised of a coordinate along with its first and second derivatives. Table I contains a comparison of the characteristic exponents for the original system [Eq. (1)], the transformed $(y, \dot{y}, \ddot{y})$ system, and the $(x, \dot{x})$ reconstructed return map, and shows very good agreement. The first two entries were obtained using the method of neighboring trajectories,[5,9] and the third entry was obtained using the return map method outlined above. The former method requires explicit knowledge of the dynamical equations, while the latter method depends on the dynamical system's attractor having sufficiently simple topology.

When trying to apply these reconstruction techniques to actual turbulence data, one of the first

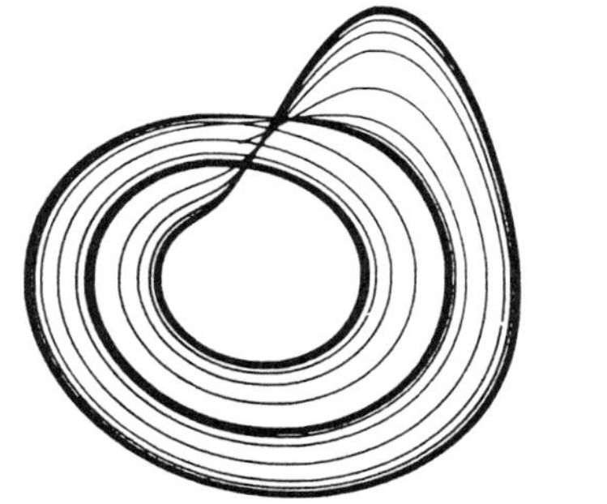

FIG. 2. $(x, \dot{x})$ reconstruction from the time series.

TABLE I. Comparison of characteristic exponents from original (x, y, z) system, transformed $(y, \dot{y}, \ddot{y})$ system, and construction of return map from $(x, \dot{x}, \ddot{x})$ system.

	Characteristic exponent value
(x, y, z) system [Eq. (1)]	0.0677 ± 0.0005
$(y, \dot{y}, \ddot{y})$ system [Eq. (2)]	0.0680 ± 0.0005
$(x, \dot{x})$ return map reconstruction	0.0677 ± 0.0001

questions will be exactly what dimension the system's attractor is. Note that the topological dimension of the attractor is directly related to the number of nonnegative characteristic exponents (see Bennetin, Galgani, and Strelcyn,[5] Shimada and Nagashima,[9] and Crutchfield[10]). A spectrum of all negative characteristic exponents implies a pointlike zero-dimensional attractor; one zero characteristic exponent with all others negative implies a one dimensional attractor; one positive and one zero characteristic exponent corresponds to the observation of folded-sheet-like structures making up the attractor; two positive characteristic exponents correspond to volumelike structures; and so on. The case of two zero characteristic exponents corresponds to a two-torus (two dimensional), but this should be distinguishable from a sheetlike chaotic attractor by the observation of two sharp incommensurate frequencies in the power spectrum. The dimension referred to above is the topological dimension of the attractor; we must realize that the Cantor-set structure typical of these objects implies a nonintegral fractal dimension[11] which can be expressed in terms of the characteristic exponents.[12] However, at any finite degree of resolution the observed topological dimension will be some integer value, though nonintegral dimension might be obtained by varying the resolution of the observation to see scaling in the structure of the attractor.

We now outline a procedure for determining the dimension of a smooth dynamical system. We begin with the idea that the "dimension" of a system being observed corresponds to the number of independent quantities needed to specify the state of the system at any given instant. Thus the observed dimension of an attractor is the number of independent quantities needed to specify a point on the attractor. For an attractor in an n-dimension phase space, we can discover the number of independent quantities needed for such a specification by slicing the attractor with $(n-1)$-dimension hypersheets defined by one coordinate being constant. The topological dimension of the attractor corresponds to the minimum number of sheets which, when intersected with each other and the attractor, will yield a countable number of points.[11]

If one chooses as phase-space coordinates the value of some variable along with time-delayed values of the same variable, this slicing of phase space can be accomplished by constructing conditional probability distributions. We define the kth-order conditional probability distribution of a coordinate x, $P(x \mid x_1, x_2, \ldots; \tau)$, as the probability of observing the value x given that x_1 was observed time τ before, x_2 was observed time 2τ before, and so on. If we take τ to be small, the k conditions are equivalent to specification of the value of x at some time along with the value of all its derivatives up to order $k-1$. In fact, we must have $\tau \ll I/\Lambda$, where I is the degree of accuracy with which one can specify a state, and where Λ is the sum of all the positive-characteristic exponents, otherwise the information generating properties of the flow would randomize the samples with respect to each other.[4] In practice, it is easy to choose τ one or two orders of magnitude smaller than I/Λ. The dimensionality of the attractor is the number of conditions needed to yield an extremely sharp conditional probability distribution, in which case the system is determined by the conditions. These conditional probability distributions have been accumulated for the system given by Eq. (1), illustrated by the sequence in Fig. 3. We observe that the second-order conditional probability distribution is extremely sharp, implying that the attractor is two dimensional (sheets), which is indeed the observed structure. Other methods for determining the dimension of attractors will be reported elsewhere.[13]

The presence of observational noise in an experiment would be manifested in the increased width of the sharpest peaks obtainable in the sequence of probability distributions. For a noise level of δ, the width of the nth (sharp) probability distribution should be $\sim n\delta$. Thus for high-dimensional attractors, low-noise data is of paramount importance.

We have outlined techniques for reconstructing a phase-space picture from observing a single coordinate of any dynamical system. For systems which have only one positive-characteristic exponent along with sufficiently simple topology, we can obtain its value. We have also outlined a procedure for determining the dimensionality of an attractor from the observation of a single coordinate. All these techniques should be directly applicable to time series obtained from observing turbulence, as well as any other physical system, to construct a finite-dimensional phase-space picture of the system's attractor, provided such a low-dimensional structure exists. These ideas have recently been utilized by Roux *et al.*[14] to construct a phase-space picture of the chaotic attractor underlying chemical turbulence in the Belousof-Zhabotinsky reaction.

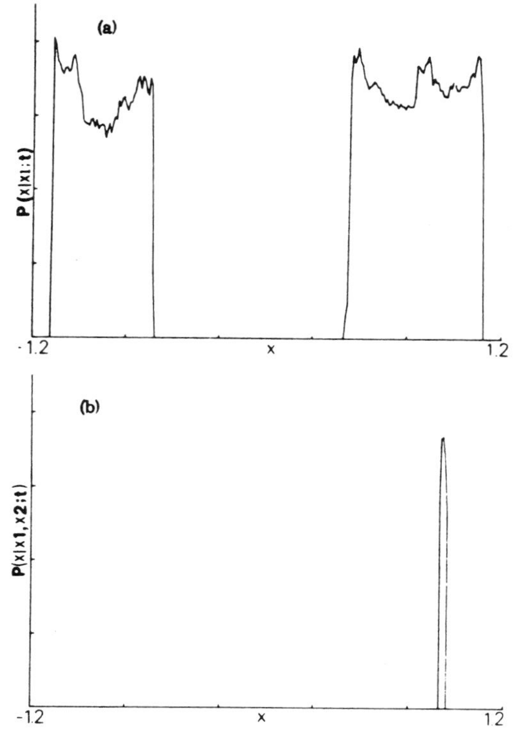

FIG. 3. Conditional-probability-distribution sequence for Eq. (1): (a) x vs $P(x|x_1;t)$; (b) x vs $P(x|x_1,x_2;t)$; where $x_1 = 0$, $x_2 = 0.495$, and $t = 0.2$ time units.

We have benefitted from many stimulating conversations with R. Abraham, W. Burke, J. Guckenheimer, and T. Jacobson. We also thank F. Bridges for the use of his microcomputer. This work was supported in part by the National Science Foundation under Grant No. 44315 0-21299 and in part by the John and Fanny Hertz Foundation.

---

[1]E. N. Lorenz, J. Atmos. Sci. 20, 130 (1963).

[2]D. Ruelle and E. Takens, Commun. Math. Phys. 50, 69 (1976).

[3]J. P. Gollub and H. L. Swinney, Phys. Rev. Lett. 35, 927 (1975).

[4]R. S. Shaw, Ph.D. thesis, University of California at Santa Cruz, 1978 (to be published).

[5]G. Bennentin, L. Galgani, and J. M. Strelcyn, Phys. Rev. A 14, 2338 (1976).

[6]Ya. B. Piesin, Dokl. Akad. Nauk SSSR 226, 196 (1976) [Sov. Math. Dokl. 17, 196 (1976)].

[7]O. E. Rossler, Phys. Lett. 57A, 397 (1976).

[8]D. Ruelle, private communication.

[9]Shimada and Nagashima, Prog. Theor. Phys. 61, 1605 (1979).

[10]J. P. Crutchfield, Senior thesis, University of California at Santa Cruz, 1979 (unpublished).

[11]B. Mandelbrot, *Fractals: Form, Chance, and Dimension* (Freeman, San Francisco, 1977).

[12]H. Mori, Prog. Theor. Phys. 63, 1044 (1980).

[13]H. Froehling, J. P. Crutchfield, J. D. Framer, N. H. Packard, R. S. Shaw, and L. Wennerberg, "On Determining the Dimension of Chaotic Flows" (to be published).

[14]J. C. Roux, A. Rossi, S. Bachelart, and C. Vidal, Phys. Lett. 77A, 391 (1980).

EXTRACTING QUALITATIVE DYNAMICS FROM EXPERIMENTAL DATA

D.S. BROOMHEAD and Gregory P. KING*
Royal Signals and Radar Establishment, St Andrews Road, Great Malvern, Worcestershire, WR14 3PS, UK

Received 1 July 1985
Revised manuscript received 25 November 1985

We consider the notion of qualitative information and the practicalities of extracting it from experimental data. Our approach, based on a theorem of Takens, draws on ideas from the generalized theory of information known as singular system analysis due to Bertero, Pike and co-workers. We illustrate our technique with numerical data from the chaotic regime of the Lorenz model.

1. Introduction

In this paper we consider the notion of *qualitative* information† and how it may be extracted from experimental time series. That this type of information might be recovered from a time series was first suggested by Packard et al. [1]. These authors suggested that a phase portrait, equivalent in some sense to that of the underlying dynamical system, could be reconstructed from time derivatives formed from the data. Another method of phase portrait reconstruction was suggested independently by Takens [2]. This method, which we shall discuss below, is known as the "method of delays". In his paper, Takens provided both approaches with a firm theoretical foundation. As general experimental tools, however, they remain ill-defined since they do not take account of problems associated with the process of measurement.

In this paper we develop another method which draws on Takens' proof and on ideas from the generalized theory of information, known as singular system analysis, recently developed by Bertero, Pike and co-workers [3]. By casting the problem in an information theoretic context, a framework is established which allows us to address the problems associated with the noisy, finite precision, sampled data produced by an experimental measurement. We are further able to resolve in a self-consistent fashion the well-known ambiguities inherent in the application of the method of delays: the need for an ad hoc lag time and the choice of dimension for the space in which the data are plotted.

In section 2 of this paper we introduce some of the relevant language of dynamical systems theory, the definition of qualitative dynamics and the concept of equivalence relations, discuss Whitney's embedding theorem, and review the method of delays. Section 3 is primarily concerned with the singular system analysis and contains the main results of this paper. An example of the methodology introduced in this section is applied to a time series, obtained from the Lorenz model, in section 4. Certain details of the implementation of the methodology are relegated to an appendix. A brief conclusion is given in section 5.

*Also at: Imperial College of Science and Technology, Department of Mathematics, Huxley Building, Queens Gate, London, SW7 2BZ, UK.

†By "qualitative information" we mean that knowledge which may be obtained from a qualitative analysis of a dynamical system.

2. Dynamical systems and the method of delays

2.1. *Dynamical systems theory*

Consider a dynamical system formally as:

$$\frac{\mathrm{d}y}{\mathrm{d}t} = \boldsymbol{F}(y), \tag{2.1}$$

where each $y = (y_1, y_2, \ldots)$ represents a state of the system and may be thought of as a point in a suitably defined space – which we shall call *phase space*, S. The dimensionality of S, since it controls the number of possible states, will be associated with the number of a priori degrees of freedom of the system. The vector field, $\boldsymbol{F}(y)$, is in general a non-linear operator acting on points in S. Under well-known conditions on $\boldsymbol{F}(y)$ (i.e., $\boldsymbol{F}(y)$ is locally Lipschitz) equation (2.1) defines an initial value problem in the sense that a unique solution curve passes through each point y in the phase space. Formally we may write the solution at time t given an initial value y_0 as $y(t) = \varphi_t y_0$. φ_t represents a one parameter family of maps of the phase space into itself. We can conceive of solutions to all possible initial value problems for the system by writing them collectively as $\varphi_t S$. This may be thought of as a flow of points in the phase space.

Initially the dimension of the set $\varphi_t S$ will be that of S itself. As the system evolves, however, it is generally the case for so-called dissipative systems that the flow contracts onto sets of lower dimension. These are called *attractors*. For the purposes of the present work it will be assumed that the attractor of interest exists within a smooth manifold which we call M. This, too, will generally have a dimension less than that of S. On the attractor the system has fewer degrees of freedom and consequently requires less information to specify its state. Physically this corresponds to self-organization and is common in systems driven far from thermodynamic equilibrium. For example, in the Belousov–Zhabotinski chemistry experiment [4] there are about 30 chemical species participating in the reaction – thus dim $S = 30$. However, in the parameter regime where there exist oscillations with a single fundamental frequency, the attractor is a limit cycle in S and hence has a dimension of 1. A more extreme example is found in fluid dynamic experiments such as the Couette–Taylor system [5] where S is in fact a function space since equation (2.1) is actually a partial differential equation – thus dim $S = \infty$. However, when the system control parameters are adjusted so as to stabilize time-independent Taylor-vortex flow, the attractor is a fixed point in S and hence has a dimension of 0.

2.2. *Qualitative dynamics*

The complete solution of eq. (2.1) is equivalent to a complete knowledge of the family φ_t. However, in most problems of interest obtaining this knowledge is not a practical proposition. It was Poincaré who observed that a great deal of qualitative information about the dynamics could nevertheless be obtained [6]. A qualitative study of eq.

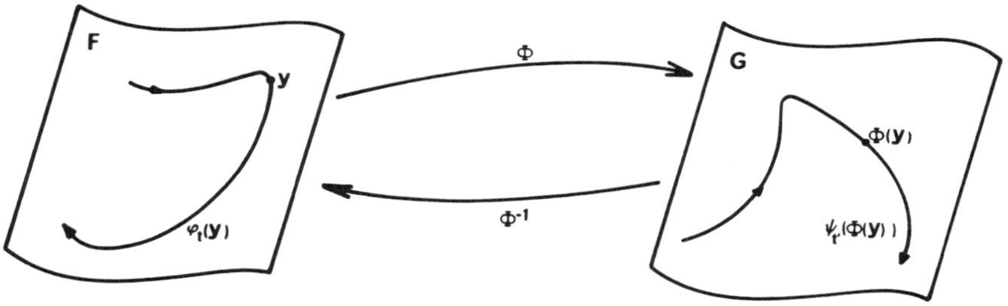

Fig. 1. Schematic representation of an equivalence relation between two vector fields $\boldsymbol{F}$ and $\boldsymbol{G}$. The invertible map, Φ, takes orbits $\varphi_t(y)$ of $\boldsymbol{F}$ into orbits $\psi_{t'}(\Phi(y))$ of $\boldsymbol{G}$.

(2.1) results in a geometric description of its orbits. This will be referred to as the *phase portrait*. In order to give meaning to the idea of qualitative information and to be able to compare and classify the phase portraits of different systems, an equivalence relation between differential equations must be introduced [7, 8].

In general, two C^r vector fields, $\boldsymbol{F}$ and $\boldsymbol{G}$, are said to be C^k equivalent ($k \leq r$) if there exists a C^k diffeomorphism, Φ, which takes orbits $\varphi_t(y)$ of $\boldsymbol{F}$ to orbits $\psi_{t'}(\Phi(y))$ of $\boldsymbol{G}$ in such a way as to preserve their orientation. Intuitively, one can think of this as meaning that Φ is an invertible, possibly non-linear, change of coordinates, which, though distorting the flow, will do so smoothly and will not confuse the order in which the points on the trajectory are visited. When $k = 0$ Φ is a homeomorphism – that is, continuous and one-to-one in both directions. This is known as *topological*, or C^0, *equivalence*. If, in addition, one has the above smoothness conditions on Φ ($k \geq 1$), then one has the stronger *differentiable equivalence*. These ideas are represented pictorially in fig. 1.

The following are useful consequences of topological equivalence:

(1) $y \in M$ is a singularity of $\boldsymbol{F}$ iff $\Phi(y)$ is a singularity of $\boldsymbol{G}$.
(2) The orbit of y for the vector field $\boldsymbol{F}$, is closed iff the orbit of $\Phi(y)$ for $\boldsymbol{G}$ is closed.
(3) The image of the ω-limit set of the orbit of y for $\boldsymbol{F}$ under Φ is the ω-limit set of the orbit of $\Phi(y)$ for $\boldsymbol{G}$ and similarly for the α-limit set.

This means that important topological objects defined by the flow are preserved by the equivalence relation. Furthermore, topological equivalence preserves the stability properties of fixed points, but does not, however, distinguish between nodes, improper nodes, and foci. For this the stronger condition of differentiable equivalence is required. It follows, therefore, that the classification of solutions to differential equations into qualitatively distinct types may be made on the basis of their topological or differentiable equivalence. Thus members of the same equivalence class will be said to have the same qualitative dynamics.

2.3. *Embeddings of manifolds*

Dynamical systems in phase spaces of widely differing dimensions may belong to the same equivalence class provided that their asymptotic dynamics are confined to attracting manifolds of the same dimensionality. This forms the basis of a technique which introduces the ideas of qualitative dynamics into the experimental domain. As has been observed, it is a common phenomenon in physical systems that self-organization gives rise to system evolution on low dimensional manifolds. This leads to the possibility that physically disparate systems may give rise to qualitatively equivalent dynamics, and, moreover, that all members of a given equivalence class may be represented by a canonical model equation. The apparent problem with pursuing this idea is that, in general, it is not at all clear what needs to be measured in order to relate to the dynamics on the underlying attractor. For example, in the Couette–Taylor flow experiment laser-doppler velocimetry enables the measurement of the evolution of one component of the velocity at a point in the fluid – apparently a very low-level description of an evolution in a function space. If it were possible to abstract in some way the low-dimensional manifold from its high-dimensional phase space, this difficulty might be avoided. Such an abstraction process may be achieved by an embedding of the manifold in a lower dimensional space.

An *embedding* is a smooth map, say Φ, from the manifold M to a space U such that its image $\Phi(M) \subset U$ is a smooth submanifold of U and that Φ is a diffeomorphism between M and $\Phi(M)$. In other words, the embedding of M in U is a "realization" of M as a submanifold within U. In particular, the fact that the embedding gives a diffeomorphism between two manifolds means that we have an important prerequisite with which to set up a differentiable equivalence relation. A general existence theorem for embeddings in Euclidean

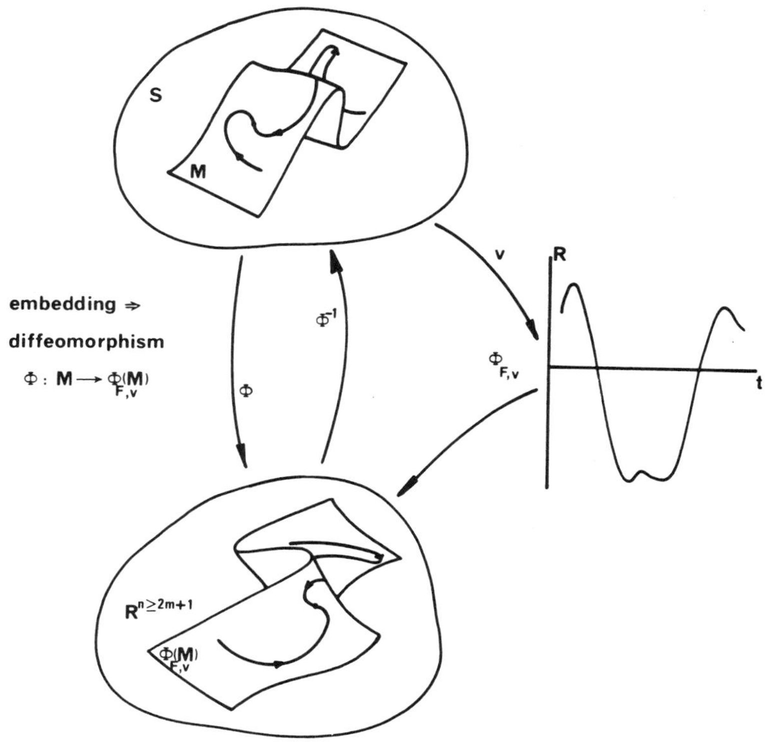

Fig. 2. A schematic representation of the method of delays. The asymptotic dynamics of an experimental system is assumed to correspond to an evolution on an m-dimensional submanifold of the state space S. A sequence of real-valued measurements, v, is used to construct a map $\Phi_{F,v}$ into a Euclidean n-space, $\mathbf{R}^n$. The image of M, $\Phi_{F,v}(M)$, is a submanifold of $\mathbf{R}^n$ under the hypotheses of Takens' theorem. Moreover, the evolution on M is C^1-equivalent to that on $\Phi_{F,v}(M)$.

spaces was given by Whitney [9] who proved that a smooth (C^2) m-dimensional manifold (which is compact and Hausdorff) may be embedded in $\mathbb{R}^{2m+1}$. This theorem is the basis of reconstruction techniques for phase portraits from time series measurements proposed by Packard et al. [1] and by Takens [2].

The present paper is concerned with systems for which the underlying dynamics may be associated with a flow corresponding to a physical process continuous in time. The relevant theorem for flows was proved by Takens and is the basis for the work to be described. In the present notation his theorem (theorem 2) states:

Let M be a compact manifold of dimension m. For pairs $(\boldsymbol{F}, v)$, $\boldsymbol{F}$ a smooth (i.e. C^2) vectorfield and v a smooth function on M, it is a generic property that $\Phi_{F,v}(y)$: $M \to \mathbb{R}^{2m+1}$, defined by

$$\Phi_{F,v}(y) = (v(y), v(\varphi_1(y)), \dots, v(\varphi_{2m}(y)))^{\mathrm{T}}$$

is an embedding, where φ_t is the flow of $\boldsymbol{F}$.

Here $v(y)$ corresponds to the value of a measurement made on the system in a state given by $y \in M$.

The conceptual framework discussed here is illustrated in fig. 2, where it is shown how the above theorem provides an explicit construction of an embedding implied by Whitney's theorem. In practice it is necessary to relate the above to a time series of measurements made on the system:

$$v_1, v_2, \ldots, v_i, v_{i+1}, \ldots,$$

where $v_i \equiv v(\varphi_i(y))$. Clearly here we are dealing with a sampled time series for which the sampling interval need not correspond to the unspecified and arbitrary interval implied by the time one map, φ_1, utilized in the theorem. We shall call the practical implementation of this theorem the *method of delays*. The details of the method will now be discussed.

2:4. *Method of delays*

At this stage it is convenient to introduce some vocabulary. The space which contains the image of $\Phi_{F,v}$ will be called the *embedding space* and its dimension the *embedding dimension*. We will denote the embedding dimension by n to emphasize the fact that it will not, in general, equal $2m + 1$ since the dimension of M is not known a priori. Nevertheless, it is supposed that $n \geq 2m + 1$ to satisfy the Whitney embedding theorem.

In applying the method of delays a useful concept is an "(n, J)-window" which makes visible n elements of the time series. When $J = 1$ the elements are consecutive, and when $J > 1$ there is an interval of J sample times between each visible element. We shall refer to an $(n, 1)$-window as an *n-window*. At any stage the elements visible in the (n, J)-window constitute the components of a vector in the embedding space, $\mathbb{R}^n$. As the time series is advanced step-wise through the window, a sequence of vectors in the embedding space is generated. These form a discrete trajectory. To represent this we use the notation

$$x_i = \Phi_{F,v}(\varphi_i(y))$$
$$= (v_i, v_{i+J}, \ldots, v_{i+(n-1)J})^T.$$

(a) A 5-window:

$$v_1 \; v_2 \; \cdots \; v_{i-1} \; \boxed{v_i \; | \; v_{i+1} \; | \; v_{i+2} \; | \; v_{i+3} \; | \; v_{i+4}} \; v_{i+5} \; \cdots$$

$$x_i = (v_i, v_{i+1}, v_{i+2}, v_{i+3}, v_{i+4})^T$$

(b) A (5,3)-window:

$$v_1 \; v_2 \; \cdots \; \boxed{v_i} \; \cdots \; \boxed{v_{i+3}} \; \cdots \; \boxed{v_{i+6}} \; \cdots \; \boxed{v_{i+9}} \; \cdots \; \boxed{v_{i+12}} \; \cdots$$

$$x_i = (v_i, v_{i+3}, v_{i+6}, v_{i+9}, v_{i+12})^T$$

(c) The trajectory matrix for a 5-window:

$$X = N^{-\frac{1}{2}} \begin{bmatrix} x_1^T \\ x_2^T \\ \cdot \\ \cdot \\ \cdot \end{bmatrix} = N^{-\frac{1}{2}} \begin{bmatrix} v_1 & v_2 & v_3 & v_4 & v_5 \\ v_2 & v_3 & v_4 & v_5 & v_6 \\ & & \cdot & & \\ & & \cdot & & \\ & & \cdot & & \end{bmatrix}$$

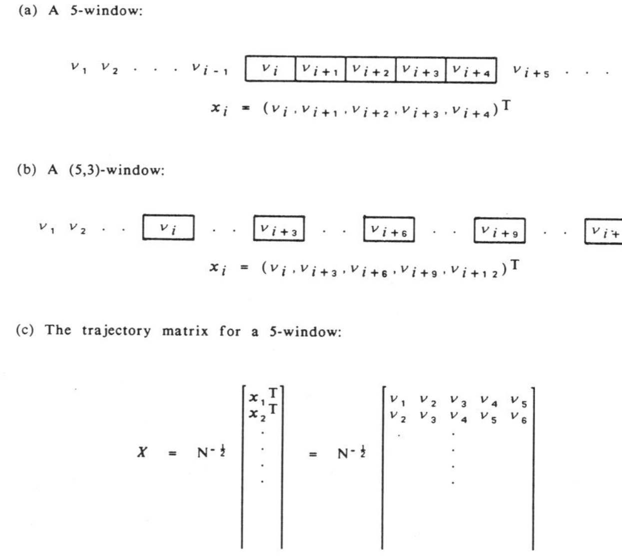

Fig. 3. Illustrations of the use of the method of delays: (a) The construction of vectors in $\mathbb{R}^5$ from a sequence of measurements of v using a lag-time of one sample-time. (b) As in (a), but with a lag-time of three sample-times. (c) The construction of the trajectory matrix, X, from vectors obtained by application of a 5-window (see (a)) to a time series.

The construction of vectors using an (n, J)-window is illustrated in fig. 3.

There are several difficulties in applying the method of delays in its present form. These can be traced to the fact that Takens' theorem makes no direct contact with the process of measurement. In particular, there are several time scales that are unspecified. The most obvious are the sampling time, τ_s, the "lag time", $\tau_L = J\tau_s$, and the window length, $\tau_w = n\tau_L$. In practice, short sampling times are employed to produce good approximations to smooth trajectories. However, this has the unfortunate effect of creating highly correlated samples within an n-window – thus causing the trajectory to lie close to the diagonal in the embedding space. To avoid this, the more general (n, J)-windows, with τ_L large enough to introduce a degree of statistical independence between the components, are used.

Takens' theorem appears to provide information about the choice of embedding dimension, stating that $n \geq 2m + 1$. This, however, is of little practical relevance since $m = \dim M$ is not generally known a priori. The approach taken in published work has been to increase n systematically, until trajectories no longer appear to intersect [10]. This is at best a rather subjective criterion, becoming rapidly unworkable in higher dimensions or in the presence of noise. A further difficulty is that plotting data in this way introduces an artificial symmetry into the phase portrait. A consequence of this is that time-averaged moments of trajectories projected onto the coordinate axes become independent of coordinate for long time series. More confusing, from the point of view of interpreting the multi-dimensional structure of attractors, is the fact that many projections onto orthogonal planes are identical. For example, the projection onto the (i, j)-plane is the same as that onto all the $(i+k, j+k)$-planes for all k such that $i+k, j+k \leq n$. Indeed, as one increases the embedding dimension, in each new embedding space the artificial symmetry is increased.

This undesirable property results from choosing a basis for the embedding space in an arbitrary manner. Intuitively one could hope for a choice of basis such that as n increases beyond $2m+1$ the attractor will be found, with invariant geometry, confined to a subspace of fixed dimension. In the next section we show how an analysis of the information content of the time series can be used to derive such a basis. This will have the additional advantage of being able to deal with experimental noise in a systematic fashion, thereby remedying a deficiency in the implementation of the method of delays as described above.

3. A statistical approach to the method of delays

In this section we develop a *singular system* approach to the method of delays which deals with the ambiguities and limitations described in the previous section. A consequence of this analysis will be that we eliminate the need to introduce statistical independence through the use of a general (n, J)-window. Thus, we set $J = 1$ and dispense with arbitrary lag times.

3.1. *The singular system*

The application of an n-window to a time series of N_T data points results in a sequence of $N = N_T - (n-1)$ vectors, $\{\boldsymbol{x}_i \in \mathbb{R}^n | i = 1, 2, \ldots, N\}$ in the embedding space. Such a sequence can be used to construct a *trajectory matrix*, $\boldsymbol{X}$, which contains the complete record of patterns which have occurred within the window:

$$\boldsymbol{X} = N^{-1/2} \begin{bmatrix} \boldsymbol{x}_1^{\mathrm{T}} \\ \boldsymbol{x}_2^{\mathrm{T}} \\ \vdots \\ \boldsymbol{x}_N^{\mathrm{T}} \end{bmatrix}, \tag{3.1}$$

where $N^{-1/2}$ has been introduced as a convenient normalization. The trajectory matrix and its transpose may be thought of as linear maps between the spaces $\mathbb{R}^n$ and $\mathbb{R}^N$. The embedding space, $\mathbb{R}^n$, is the space of all n-element patterns and is the natural object of interest. A similar interpretation of $\mathbf{R}^N$ is clearly possible. Indeed, the columns of $\boldsymbol{X}$ bear the same relationship to one another as do the rows. However, for the present purposes we shall use $\mathbf{R}^N$ in a simple way which utilizes the obvious property that the standard basis vectors $\{\boldsymbol{e}_i \in \mathbf{R}^N\}$, can be used as an indexing system for points on the trajectory in $\mathbf{R}^n$ (where $\boldsymbol{e}_i$ is the ith column of the $N \times N$ unit matrix). That is, formally we can extract the ith vector, $\boldsymbol{x}_i$, from the trajectory matrix by operating from the left with $N^{1/2}\boldsymbol{e}_i^{\mathrm{T}}$:

$$\boldsymbol{x}_i^{\mathrm{T}} = N^{1/2}\boldsymbol{e}_i^{\mathrm{T}}\boldsymbol{X}. \tag{3.2}$$

It follows directly that operating with a general vector, $\boldsymbol{w}^{\mathrm{T}} = N^{1/2}\sum_{i=1}^{N} w_i \boldsymbol{e}_i^{\mathrm{T}}$, will produce a linear combination of vectors on the trajectory in $\mathbf{R}^n$. Then $\boldsymbol{w}^{\mathrm{T}}\boldsymbol{X}$ is the mean position in the embedding space relative to a measure induced on the sampled attractor by the choice of $\boldsymbol{w}$. Alternatively, it corresponds to a weighted time-average over the trajectory. We note that in this context the concepts of space- and time-average are interchangeable: Just as the standard basis of $\mathbf{R}^N$ indexes the points in the embedding space so also does it represent the labelling of a time-sequence.

The triple $(\boldsymbol{X}, \mathbf{R}^N, \mathbf{R}^n)$ can be analysed using a singular system as considered in the generalized theory of information developed by Bertero, Pike, and their co-workers [3]. In the following we shall develop a singular system analysis based on the method of delays.

3.2. *Independence and orthogonality*

A central concept in distinguishing chaotic systems from stochastic, Brownian-like systems is that of *number of degrees of freedom*. Unfortunately, there are numerous definitions of number of degrees of freedom depending on the context in which it arises. For example, in signal processing it is the number of modes used to describe a signal which contain significant power. In dynamical systems it is usually used to specify the dimensionality of an attracting manifold within the phase space, the dimensionality of the attractor itself or even the dimension of the whole phase space. We shall show that the signal processing definition when related to a dynamical system interpretation becomes the dimensionality of the subspace in which the embedded manifold is to be found. This has shortcomings since, unlike $m = \dim M$, it is not an invariant of the embedding process. However, it will be shown to give a reasonable upper bound to m.

We begin, therefore, by calculating the dimensionality of the subspace which contains the embedded manifold. To do this we need to know the number of linearly independent vectors that can be constructed from the trajectory in the embedding space by forming linear combinations of the $\boldsymbol{x}_i$. It has already been established that vectors in $\mathbb{R}^N$ give rise to such linear combinations when they act on the trajectory matrix. Consider the set of vectors $\{\boldsymbol{s}_i \in \mathbb{R}^N\}$, which we assume give a set of linearly independent vectors in $\mathbb{R}^n$ by their action on $\boldsymbol{X}$. We shall also assume, with no loss of generality, that the latter have been orthonormalized. Hence, in general, they constitute part of a complete orthonormal basis, $\{\boldsymbol{c}_i | i = 1, \ldots, n\}$, for the embedding space. By construction, the following relationship holds:

$$\boldsymbol{s}_i^{\mathrm{T}}\boldsymbol{X} = \sigma_i \boldsymbol{c}_i^{\mathrm{T}}, \tag{3.3}$$

where the $\{\sigma_i\}$ are a set of real constants which will be used to fix the normalization of both sets of vectors.

The orthonormality of the $\{\boldsymbol{c}_i\}$ imposes the following condition:

$$\boldsymbol{s}_i^{\mathrm{T}}\boldsymbol{X}\boldsymbol{X}^{\mathrm{T}}\boldsymbol{s}_j = \sigma_i \sigma_j \delta_{ij}, \tag{3.4}$$

where δ_{ij} is the Kronecker delta. The $N \times N$ matrix $\boldsymbol{\Theta} = \boldsymbol{X}\boldsymbol{X}^{\mathrm{T}}$ is real symmetric, and hence its eigenvectors form a complete orthonormal basis for $\mathbf{R}^N$. In particular, the above equation is solved by the eigenvectors of $\boldsymbol{\Theta}$:

$$\boldsymbol{\Theta}\boldsymbol{s}_i = \sigma_i^2 \boldsymbol{s}_i \tag{3.5}$$

provided the $\{\sigma_i^2\}$ are interpreted as the corresponding eigenvalues. It should be noted that real matrices of the form $\boldsymbol{XX}^{\mathrm{T}}$ are non-negative definite. Therefore, the $\{\sigma_i\}$ are real constants which may themselves be taken to be non-negative without loss of generality.

Eq. (3.1) may be used to show that $\boldsymbol{\Theta}$ is actually an array of scalar products of all pairs of points on the trajectory in the embedding space:

$$\boldsymbol{\Theta} = N^{-1}\begin{bmatrix} \boldsymbol{x}_1^{\mathrm{T}}\boldsymbol{x}_1 & \boldsymbol{x}_1^{\mathrm{T}}\boldsymbol{x}_2 \cdots \boldsymbol{x}_1^{\mathrm{T}}\boldsymbol{x}_N \\ \boldsymbol{x}_2^{\mathrm{T}}\boldsymbol{x}_1 & \boldsymbol{x}_2^{\mathrm{T}}\boldsymbol{x}_2 \cdots \boldsymbol{x}_2^{\mathrm{T}}\boldsymbol{x}_N \\ \vdots & \vdots \qquad \vdots \end{bmatrix}, \tag{3.6}$$

and for this reason it will be called the *structure matrix* of the trajectory. Equally, it may be interpreted as being composed of the correlations between all pairs of patterns to have appeared in the n-window. The considerable redundancy in specifying the correlations between all pairs of patterns results in $\boldsymbol{\Theta}$ having low rank. This is borne out by eq. (3.3) which shows that there are, at most, n of the $\{\sigma_i\}$ which are non-zero. Because of this the difficulty of diagonalizing $\boldsymbol{\Theta}$ when N is large can be avoided.

To explore this point further, we look for an inverse relationship to eq. (3.3) whereby an expression for the vector $\boldsymbol{s}_i$ which yields a particular $\boldsymbol{c}_i$ is obtained. This is:

$$\boldsymbol{Xc}_i = \sigma_i \boldsymbol{s}_i \tag{3.7}$$

which, when $\sigma_i \neq 0$, may be derived by taking the transpose of eq. (3.3), operating from the left with $\boldsymbol{X}$, and using eq. (3.5). The $\{\boldsymbol{s}_i\}$ have been shown to be an orthogonal set, therefore, the following eigenvalue equation can be derived analogously to eq. (3.5):

$$\boldsymbol{\Xi c}_i = \sigma_i^2 \boldsymbol{c}_i. \tag{3.8}$$

Here $\boldsymbol{\Xi} = \boldsymbol{X}^{\mathrm{T}}\boldsymbol{X}$ is a real, symmetric $n \times n$ matrix which may be written as:

$$\boldsymbol{\Xi} = \frac{1}{N}\sum_{i=1}^{N} \boldsymbol{x}_i \boldsymbol{x}_i^{\mathrm{T}} \tag{3.9}$$

using the definition of the trajectory matrix. This is the time-average of the dyadic product $\boldsymbol{x}_i\boldsymbol{x}_i^{\mathrm{T}}$. Thus $\boldsymbol{\Xi}$ is the *covariance matrix* of the components of the $\{\boldsymbol{x}_i\}$, averaged over the entire trajectory; that is, the time-averaged correlation between all pairs of elements in the n-window. Expressed in terms of the original time series, this has the form:

$$\boldsymbol{\Xi} = \frac{1}{N} \times \begin{bmatrix} \sum_{i=1}^{N} v_i v_i & \sum_{i=1}^{N} v_i v_{i+1} & \cdots & \sum_{i=1}^{N} v_i v_{i+n-1} \\ \vdots & \vdots & & \vdots \\ \sum_{i=1}^{N} v_{i+n-1} v_i & \sum_{i=1}^{N} v_{i+n-1} v_{i+1} & \cdots & \sum_{i=1}^{N} v_{i+n-1} v_{i+n-1} \end{bmatrix}$$

Equation (3.8) is far more tractable than eq. (3.5) since it is implicit to the approach that the embedding dimension is small.

Returning to the question of the rank of $\boldsymbol{\Theta}$, we note that the derivation of eq. (3.8) from eq. (3.5) shows that the non-zero eigenvalues of the structure matrix equal the non-zero eigenvalues of the covariance matrix. That is, rank $\boldsymbol{\Theta}$ = rank $\boldsymbol{\Xi} = n' \le n$. One is thus led to the observation that $\mathbb{R}^N$ can be decomposed into a subspace of dimension n' and its orthogonal complement. The n'-dimensional subspace is spanned by a set $\{\boldsymbol{s}_i | i = 1, \ldots, n'\}$ which is such that each corresponding average over the $\boldsymbol{x}_i$ gives rise uniquely to a basis vector $\boldsymbol{c}_i \in \mathbb{R}^n$ according to eq. (3.3). The complementary subspace, spanned by the set $\{\boldsymbol{s}_i | i = n' + 1, \ldots, N\}$, is the kernal of $\boldsymbol{X}$ mapping onto the origin of the embedding space through eq. (3.3).

The complete set $\{\boldsymbol{s}_i\}$ is a basis for the construction of all possible averages over the trajectory. The significance of the decomposition is that only the averages associated with the n'-dimensional subspace give a non-trivial vector in $\mathbb{R}^n$. There are at most n' linearly independent vectors in the embedding space that may be constructed from the trajectory. Therefore, it might be supposed that the number n', the rank of $\boldsymbol{\Xi}$, is the dimensionality of the subspace containing the embedded manifold. However, account must be taken of the fact that $\boldsymbol{X}$ has contributions from sources of experimental noise. This point is addressed in the next section.

3.3. *Singular value decomposition and noise*

To study the effect of noise it is necessary to discuss the eigenvectors and their associated spectrum. Consider the orthogonal $n \times n$ matrix $\boldsymbol{C}$ which has columns consisting of the vectors $\{\boldsymbol{c}_i\}$, $\boldsymbol{C} = (\boldsymbol{c}_1, \boldsymbol{c}_2, \ldots, \boldsymbol{c}_n)$, and the diagonal matrix $\boldsymbol{\Sigma} = \mathrm{diag}(\sigma_1, \sigma_2, \ldots, \sigma_n)$, where the ordering $\sigma_1 \geq \sigma_2 \geq \cdots \geq \sigma_n \geq 0$ is assumed. Using this notation, eq. (3.8) is:

$$\boldsymbol{\Xi C} = \boldsymbol{C \Sigma}^2. \tag{3.10}$$

Using the definition of $\boldsymbol{\Xi}$ it follows that:

$$(\boldsymbol{XC})^{\mathrm{T}}(\boldsymbol{XC}) = \boldsymbol{\Sigma}^2. \tag{3.11}$$

The matrix $\boldsymbol{XC}$ is the trajectory matrix projected onto the basis $\{\boldsymbol{c}_i\}$. This result expresses the fact that in the basis $\{\boldsymbol{c}_i\}$ the components of the trajectory are uncorrelated since the $\{\boldsymbol{c}_i\}$ are obtained from the diagonalization of the covariance matrix. It was in anticipation of this result that we omitted consideration of general (n, J)-windows. This result also shows that each σ_i^2 is the mean square projection of the trajectory onto the corresponding $\boldsymbol{c}_i$. Therefore, the spectrum $\{\sigma_i^2\}$ has information about the extent to which the trajectory explores the embedding space. One may think of the trajectory as exploring on average, an n-dimensional ellipsoid. The $\{\boldsymbol{c}_i\}$ then give the directions and the $\{\sigma_i\}$ the lengths of the principal axes of the ellipsoid. In the previous section the trajectory was found to be confined to a subspace of dimension equal to the rank of $\boldsymbol{\Xi}$. In a noisy environment this needs qualification since the presence of noise will tend to smear out the deterministic behaviour, and, in the directions associated with small or vanishing σ_i, the noise will dominate. Therefore, the rank of $\boldsymbol{\Xi}$ is an upper bound to the dimensionality of the subspace explored by the deterministic component of the trajectory.

The above formalism, as is well known from linear algebra, can be used to express the singular value decomposition of a singular, linear map [11]. The significance of this in the context of information theory has been discussed recently in a different application [3]. Here, we are interested in the singular value decomposition of the trajectory matrix:

$$\boldsymbol{X} = \boldsymbol{S\Sigma C}^{\mathrm{T}}, \tag{3.12}$$

where $\boldsymbol{S}$ is the $N \times n$ matrix of eigenvectors of Θ, n' of which have non-zero eigenvalues. The vectors of $\boldsymbol{C}$ and $\boldsymbol{S}$ will henceforth be referred to as the *singular vectors* of $\boldsymbol{X}$, while the elements of the diagonal matrix, $\boldsymbol{\Sigma}$, will be called the associated *singular values*.

To exemplify the effect of noise consider the following simple model of a time series with a noise component ξ_j: $v_j = \bar{v}_j + \xi_j$. Here, and below, an overbar indicates a quantity associated with the deterministic component. For white noise uncorrelated with $\bar{v}_j$ the autocorrelation function of the time series can be written $\langle v_0 v_j \rangle = \langle \bar{v}_0 \bar{v}_j \rangle + \langle \xi^2 \rangle \delta_{0j}$, where the angle brackets refer to a time average. Generally, in the limit of an infinite time series of statistically stationary data, the covariance matrix is expected to have the Toeplitz structure: $\Xi_{ij} = g([i-j]\tau_s)$, where $g(\tau)$ is the autocorrelation function of the continuous time series. In this limit the covariance matrix for the above time series may be decomposed into two parts: $\boldsymbol{\Xi} = \bar{\boldsymbol{\Xi}} + \langle \xi^2 \rangle \mathbb{1}_n$, where $\mathbb{1}_n$ is the $n \times n$ unit matrix. In this case, the singular values of $\boldsymbol{X}$ are shifted uniformly:

$$\sigma_i^2 = \bar{\sigma}_i^2 + \langle \xi^2 \rangle, \quad i = 1, 2, \ldots, n, \tag{3.13}$$

where $\bar{\sigma}_i^2$ is an eigenvalue of $\bar{\boldsymbol{\Xi}}$. Thus the noise causes *all* the singular values of the trajectory matrix to be non-zero. Hence, the trajectory appears to explore all dimensions of the embedding space. We need, therefore, a method whereby these two types of contribution may be distinguished.

In the simple case of white noise, the existence of a non-zero constant background or noise floor is a noteable characteristic which can be used to distinguish the deterministic component. More generally, the independent measurement of a time series consisting only of the experimental noise will enable the calculation of its root mean square projections onto the $\{c_i\}$. By comparing these with the corresponding singular values for the time series containing the deterministic signal we can define a signal-to-noise ratio which may be associated with each singular vector. The technique will allow the identification of those singular values which are noise dominated even when the noise is not white. It is not always possible to measure the noise separately and for this reason there is interest in the signal processing community in developing so-called *cross-validation* techniques whereby the internal consistency of the data itself may be used to estimate the noise level [12].

Given that a suitable method has been used to partition the singular value spectrum, we move on to consider the implied partitioning of the embedding space. Consider the matrices $\boldsymbol{p}^{(i)}$: $\boldsymbol{p}^{(i)}_{jk} = \delta_{ij}\delta_{jk}$, which are representations of projection operators onto the basis functions $\{c_i\}$. These may be used to construct projection operators onto the corresponding subspaces of the embedding space:

$$\boldsymbol{Q} = \sum_{\sigma_i \approx \text{noise}} \boldsymbol{p}^{(i)}, \tag{3.14}$$

$$\boldsymbol{P} = \sum_{\sigma_i > \text{noise}} \boldsymbol{p}^{(i)}. \tag{3.15}$$

Inserting the identity $\boldsymbol{P} + \boldsymbol{Q} = \mathbb{1}_n$ into eq. (3.12) gives the following:

$$\boldsymbol{X} = \bar{\boldsymbol{X}} + \Delta \boldsymbol{X}, \tag{3.16}$$

where

$$\bar{\boldsymbol{X}} = \boldsymbol{S P \Sigma C}^{\mathrm{T}} \tag{3.17}$$

is the deterministic part of the trajectory matrix, and

$$\Delta \boldsymbol{X} = \boldsymbol{S Q \Sigma C}^{\mathrm{T}} \tag{3.18}$$

is the noise-dominated part. This separation allows the rejection of a portion of the inherent noise of the experiment. In a signal processing context this is known as rejection of *out-of-band* noise. Here it amounts to oversampling the data in order to average the noise over more points in the window.

From now on our attention will shift to the reduced trajectory matrix $\bar{\boldsymbol{X}}$ since this contains all the information about the deterministic trajectory that we can sensibly extract from the experiment. A useful form for $\bar{\boldsymbol{X}}$ may be obtained by substituting eq. (3.7) into eq. (3.17):

$$\bar{\boldsymbol{X}} = \sum_{\sigma_i > \text{noise}} (\boldsymbol{X} \boldsymbol{c}_i) \boldsymbol{c}_i^{\mathrm{T}}. \tag{3.19}$$

This expression uses only quantities obtainable from the diagonalization of the covariance matrix. In this form there is an obvious interpretation of $\bar{\boldsymbol{X}}$ which relates directly to the methodology of plotting phase portraits. Here $\boldsymbol{c}_i^{\mathrm{T}}$ represents a coordinate axis of the embedding space referred to the standard basis, while $(\boldsymbol{X}\boldsymbol{c}_i)$ is a column vector containing a time series of the ith component, in the basis $\{\boldsymbol{c}_i\}$, of the vectors in the trajectory.

Clearly $\bar{\boldsymbol{X}}$ is an $N \times n$ matrix, however, by construction it consists of a trajectory confined to the deterministic subspace of $\mathbb{R}^n$ having dimension $d \leq n$ (where d is the number of the $\{\sigma_i\}$ above the noise floor). It is straightforward to define vectors restricted to this subspace. We shall distinguish these by a square bracket notation. Thus, $\bar{\boldsymbol{X}}$ restricted to the deterministic subspace is an $N \times d$ rectangular matrix the jth row of which is $[\boldsymbol{x}_j^{\mathrm{T}}\boldsymbol{c}_1, \boldsymbol{x}_j^{\mathrm{T}}\boldsymbol{c}_2, \ldots, \boldsymbol{x}_j^{\mathrm{T}}\boldsymbol{c}_d]$. In the case that the experimental noise is not white it may be necessary to relabel the $\{\boldsymbol{c}_i\}$ such that the first d vectors span the deterministic subspace.

3.4. *On the choice of time scales*

In the following paragraphs we outline an approach to the choice of sampling time, τ_s, and the window length, τ_w, so that the formalism developed above can be applied to time series data. It is more usual in the method of delays to consider the choice of embedding dimension rather than the window length. However, the window length has particular significance in the singular system analysis since it determines the *form* of the singular spectrum.

Before we can consider the effect of the window length on the form of the singular spectrum, we must first establish a sampling criterion. Observe that for fixed τ_w increasing n by decreasing τ_s generates additional singular vectors capable of describing more rapid variations within the window. The corresponding singular values will, as we have said earlier, represent the "power" in the time series which corresponds to such variations. On physical grounds one can assume the existence of an inner time scale. Over times less than this inner scale the data does not vary significantly to within the precision of the measurement. In this case decreasing τ_s further results in additional singular vectors with singular values in the noise floor defined by the precision of the experiment. At this point we say the spectrum has converged. Therefore, a natural choice for τ_s for a given τ_w is that necessary to achieve convergence of the singular spectrum.

Having determined a sampling time that ensures a convergent spectrum, we now consider how τ_w affects the form of the spectrum. Clearly, as τ_w is decreased we approach the inner scale for which there is no measureable variation of the data within the window. The singular spectrum is thus reduced to a triviality. On the other hand as τ_w is increased the information to be represented within the window increases. This has the corresponding effect of increasing the number of significant singular values. Indeed in the limit $\tau_w \to \infty$ it can be shown that our method becomes a discrete Fourier transform [13]. Clearly, it is important to have a criterion for the choice of τ_w.

Takens, in his proof of the theorem given in section 2, required on generic grounds the exclusion of data with interger periods less than τ_w. It can be shown that for realistic measurements the generic argument is too weak. However, it is sufficient, for band-limited data, to replace this with a constraint on τ_w:

$$\tau_w \le \tau^*, \tag{3.20}$$

where $\tau^* = 2\pi/\omega^*$ and ω^* is the band-limiting frequency. By band-limited data we mean here that the Fourier spectrum of the time series contains no frequencies with significant power greater than a cutoff frequency known as the band-limit. Furthermore, there is an obvious lower bound on τ_w:

$$\tau_w \ge (2m+1)\tau_s. \tag{3.21}$$

Since m is unknown, however, the only consistent a priori estimate is $\tau_w = \tau^*$.

The above arguments provide a self-consistent approach to the choice of a sampling time and window length which will satisfy the postulates of Takens' theorem. In our limited experience the use of $\tau_w = \tau^*$ has proved satisfactory. However, it must be emphasized that the derivation of this estimate for τ_w used a sufficient condition which in some circumstances may prove to be too strong. A complete answer to this problem must circumvent a fundamental limitation of Takens' theorem: the implicit assumption of data with infinite precision. It is clear from the above analysis that it is necessary to specify two quantities – the embedding dimension and a time scale (τ_w or τ_s). It is equally clear that Takens' theorem makes no mention of a time scale. This is due to an assumption that successive measurements contain new information whatever the time interval between them. For finite precision measurements this is manifestly untrue. Thus it is insufficient merely to require $2m+1$ measurements to specify an embedding – a time scale is also required. It should be emphasized that this problem exists for all data analysis techniques that rely on the construction of an embedding (e.g., entropy and dimension calculations using time series data). Theoretically the problem is still open; in practice the effects of sampling and window length on the results should be investigated.

3.5. *Concluding comments*

The derivation of a coordinate system by diagonalizing a covariance matrix is known as the Karhunen–Loeve method [13] and is widely known in the signal processing and pattern recognition fields. One feature of a basis obtained in this way is that it produces an optimum compression of information. By this we mean that in order to distinguish points in a set of interest (in this case the trajectory in $\mathbb{R}^n$) to within a given accuracy the Karhunen–Loeve basis requires the fewest components to be specified. Thus, for a fixed embedding dimension the error produced by projecting onto the first ν basis vectors is, when averaged over the set, minimized if the first ν singular vectors are used. This suggests a systematic sequence of coarse grainings – the small scale limit of which is the grain size set by the noise as measured by the noise floor of the singular value spectrum. In contrast the standard basis for $\mathbb{R}^n$ – the one implicit in a naive implementation of the method of delays – gives the worst possible information compression since for long times the rms projections of the trajectory onto the basis vectors are all equal.

An additional consequence of deriving a basis by diagonalization of the covariance matrix is that orthogonality in the embedding space is related to the statistical properties of the time series. An outcome of this is that a trajectory, when plotted against the $\{c_i\}$, will appear coherent to the extent that it deviates from a gaussian random process. This is an important point since it is known that deterministic chaos is non-gaussian. The possibility of using higher multi-point correlation properties of the time series to distinguish chaos from stochasticity has not to our knowledge been systematically investigated [14], although these concepts do underlie attempts to define dimensions and entropies for attracting sets [15].

4. Application to the Lorenz model

In this section we illustrate the singular system approach with data obtained from the Lorenz model [16]. Applications to experimental data will appear elsewhere [17].

The Lorenz model is defined by the equations

$$\frac{\mathrm{d}X}{\mathrm{d}t} = \sigma(Y - X),$$

$$\frac{\mathrm{d}Y}{\mathrm{d}t} = rX - Y - XZ,$$

$$\frac{\mathrm{d}Z}{\mathrm{d}t} = -bZ + XY.$$

In this work the equations were solved with a fourth-order Runge–Kutta routine using a step size of either 0.003 or 0.009. When the parameters take on the values $\sigma = 10$, $b = 8/3$, and $r = 28$, the system produces turbulent dynamics. The time-evolution is organized by two unstable foci and an intervening saddle point. Fig. 4 shows a projection of the phase-space orbit of the system onto the XY-plane. The study of the continuous system can be replaced by the study of a discrete dynamical system known as the Poincaré map. This is obtained by recording the intersections of the trajectory with a surface of section oriented transverse to the flow. The surface of section shown in fig. 5a

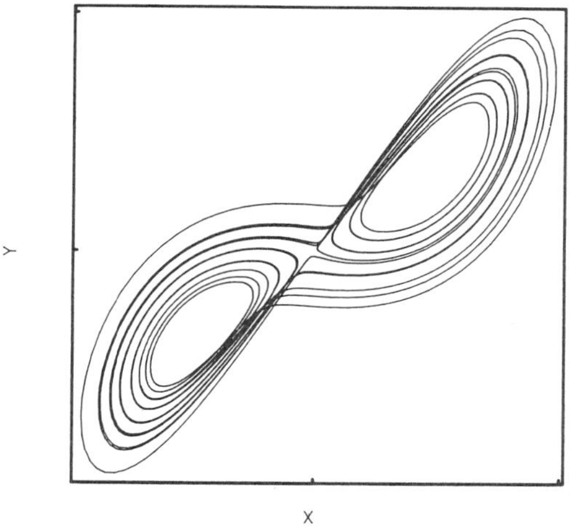

Fig. 4. The attractor of the full Lorenz model ($\sigma = 10$, $b = 8/3$, $r = 28$) projected onto the XY-plane.

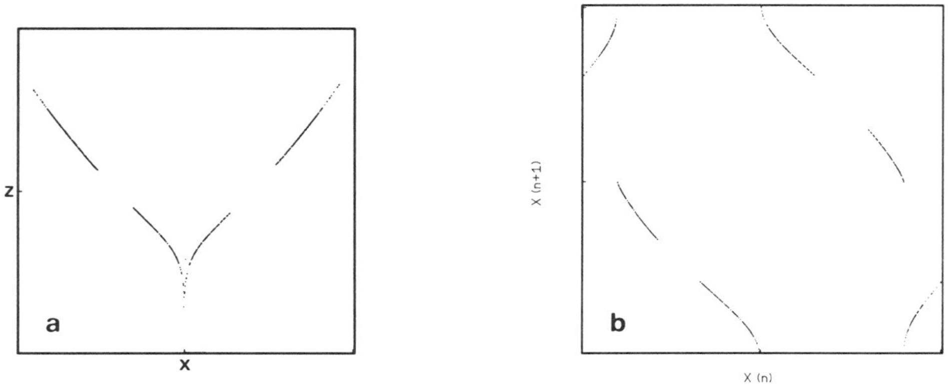

Fig. 5. (a) The intersection of the attractor of the full Lorenz model (see fig. 4) with a planar surface of section containing the Z-axis and passing through the two unstable foci. (b) The corresponding first-return map using the X-coordinates of the points of intersection.

was chosen to contain the two unstable foci and to be transverse to their unstable manifolds (it in fact contains the Z-axis). An approximately one-dimensional first-return map constructed from the Poincaré map is shown in fig. 5b. This map has been calculated to within a suitable approximation [14] and forms a connection between the full system and analytic theory.

We now turn our attention to the time series of the X variable shown in fig. 6a. Using the approach outlined in section 3.4, we choose a window length by estimating the band-limiting frequency of the power spectrum of $X(t)$ shown in fig. 6b. This gives a window length of approximately a tenth of the period of oscillation about the unstable foci [18]. In Lorenz units we have $\tau_w = 0.063$.

Let us now consider the choice of sampling time. Fig. 7a illustrates the effect of changing the sampling time on the singular value spectrum while maintaining the window length at the above value. This was done by using a time series, sampled with a step size of 0.003, and generating more sparsely

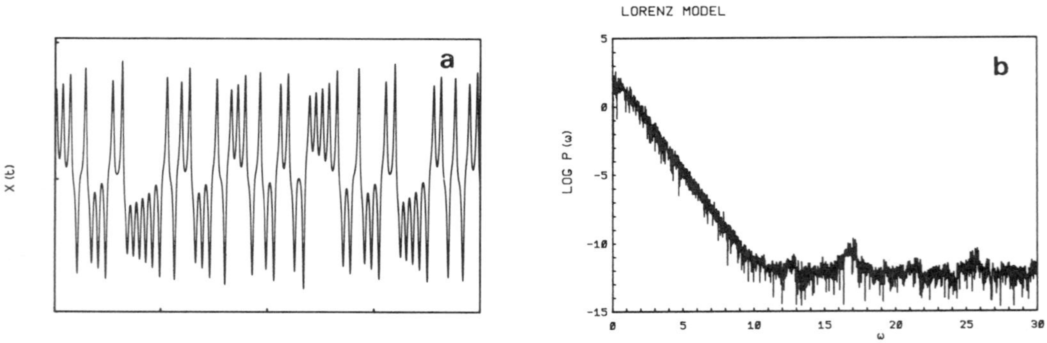

Fig. 6. (a) A sample of the time series $X(t)$ corresponding to motion on the attractor of the full Lorenz model (see fig. 4). (b) The corresponding power spectrum constructed from $X(t)$. Frequencies are scaled with the frequency associated with the unstable foci [18].

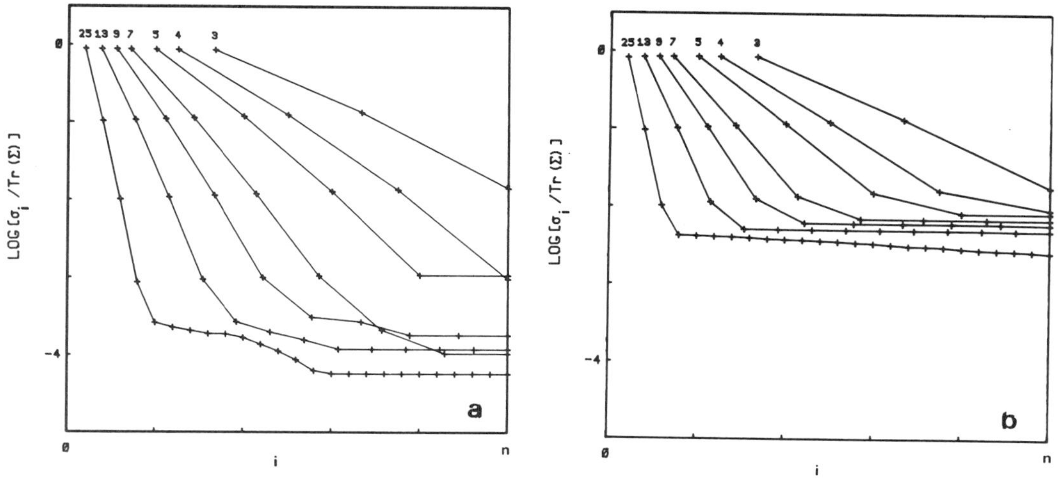

Fig. 7. (a) The normalized singular spectra $\{\log_{10}(\sigma_i/\Sigma\sigma_i)\}$ obtained from $X(t)$ data of the Lorenz model (see fig. 6a). The figure illustrates the effect of changing the sampling time, τ_s, for fixed window length, $\tau_w = n\tau_s$. Each curve involves the diagonalization of an $n \times n$ symmetric matrix. The matrix elements are obtained from the autocorrelation function of the time series which was calculated using $\sim 2 \times 10^4$ data points. The noise floors are a result of numerical noise in the calculations. (b) As with (a), the data were, however, passed through a simulated 6-bit A/D converter. Quantization error dominates the noise in this case.

sampled time series from it. The example shown in fig. 7a began with 25 samples within τ_w†. Omitting every other data point gave a time series sampled at half the rate and having 13 samples in the window. Similarly, taking points at $\frac{1}{3}$, $\frac{1}{4}$, $\frac{1}{6}$, $\frac{1}{8}$, and $\frac{1}{12}$ the original rate gave time series with 9, 7, 5, 4, and 3 samples in the window. For sampling rates yielding 7 or more points in the window, the spectrum has two distinct parts – one part which can be associated with the noise floor, and the other part which can be associated with the deterministic components of the data. The distinguishing features of the noise floor are its magnitude and flatness. The magnitude represents rounding errors within the computer as measured by the magnitude of spurious negative eigenvalues generated by the diagonalization routine. This should be contrasted with fig. 7b for which the data were passed through a simulated 6-bit analog-to-digital converter. In this case the noise floor is higher since it is dominated by quantization noise.

It is clear from figs. 7a and b that the form of the singular spectrum is insensitive to the range of sampling times used. In particular, the effect of decreasing τ_s is essentially to increase the number of singular values in the noise floor. On the basis of this analysis we choose to work with data sampled at intervals $\tau_s = 0.009$ which implies an embedding dimension of 7. This is largely a matter of computational ease since our data is very clean. With noisy data it is often better to increase the sampling rate in order to be able to average over more data points within the window.

The singular value spectrum shown in fig. 8 results from the above choices of τ_s and τ_w. It is clear from figs. 7a and 8 that the important dynamics will be confined to a 4-dimensional subspace of the embedding space. In the absence of any prior knowledge about the system it would be necessary to consider the dynamics in this 4-

†Note that this implies $\tau_w = 0.075$. This choice was made for convenience in producing the figure and does not have a significant effect on the spectra obtained, cf. figs. 7a and 8.

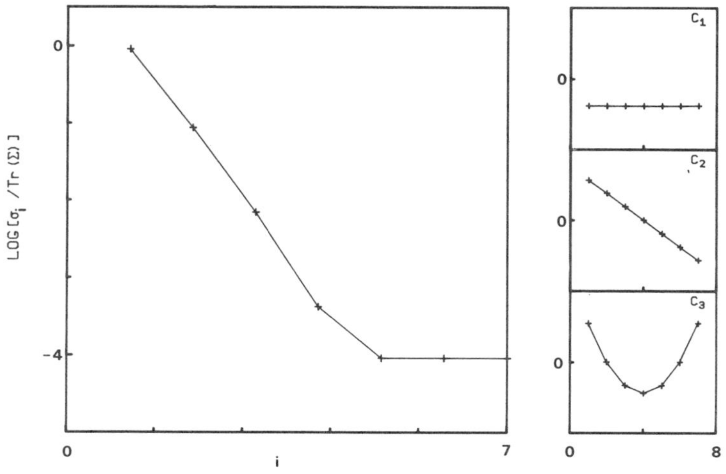

Fig. 8. The singular spectrum and the first three singular vectors obtained from the unquantized $X(t)$ Lorenz data using $\tau_s = 0.009$ and $\tau_w = 7\tau_s$. (The ordinates of each singular vector plot cover the range $[-0.5, +0.5]$).

dimensional space. However, since we know that the Lorenz attractor can be embedded in a 3-dimensional Euclidean space, we shall only consider the projection onto the subspace spanned by the first three singular vectors. These vectors, which are shown in fig. 8, have a form reminiscent of orthogonal polynomials. In fig. 9 we show the projections of the trajectory onto the planes spanned by $(\boldsymbol{c}_1, \boldsymbol{c}_2)$, $(\boldsymbol{c}_1, \boldsymbol{c}_3)$, and $(\boldsymbol{c}_2, \boldsymbol{c}_3)$. The details of this have been given in section 3.3. We interpret the polynomial appearance of the $\{\boldsymbol{c}_i\}$ to mean that the components of $\boldsymbol{c}_i^{\mathrm{T}}\boldsymbol{x}_j$ are averaged "time derivatives" of the time series. This recalls the approach of Packard et al. [1]. An important difference, however, is that, unlike the sequences of components generated here, the time derivatives of the time series are not statistically independent. Moreover, time derivatives estimated by finite differences are sensitive to noise since no averaging process is involved.

The projections shown in fig. 9 do not exhibit the spurious symmetries inherent in the basic method of delays and readily suggest the 3-dimensional form of the reconstructed attractor. Indeed the use of a suitable orthogonal transform enables one to generate stereo pairs as shown in fig. 10. We believe these to be a valuable tool for the development of geometric intuition about attractors extracted from experimental data [19].

Figures 9 and 10 demonstrate the clear qualitative relationship between the Lorenz attractor and its reconstruction from the $X(t)$ time series. The major features which are obviously preserved are the number of fixed points, their stability properties, and the disposition of the flow about them. The extent of the relationship can be further demonstrated by constructing a surface of section in the embedding space. The surface chosen in fig. 11a is analogous to that used in fig. 5a – viz., it contains the two unstable foci and is transverse to their unstable manifolds (and in fact contains the c_3-axis). The pseudo one-dimensional first-return map is shown in fig. 11b. We note that the return map of fig. 11b is indistinguishable from the return map in fig. 5b obtained from the full Lorenz model. While this result is gratifying, it is unexpected since in general one expects the two maps to be related by a diffeomorphism.

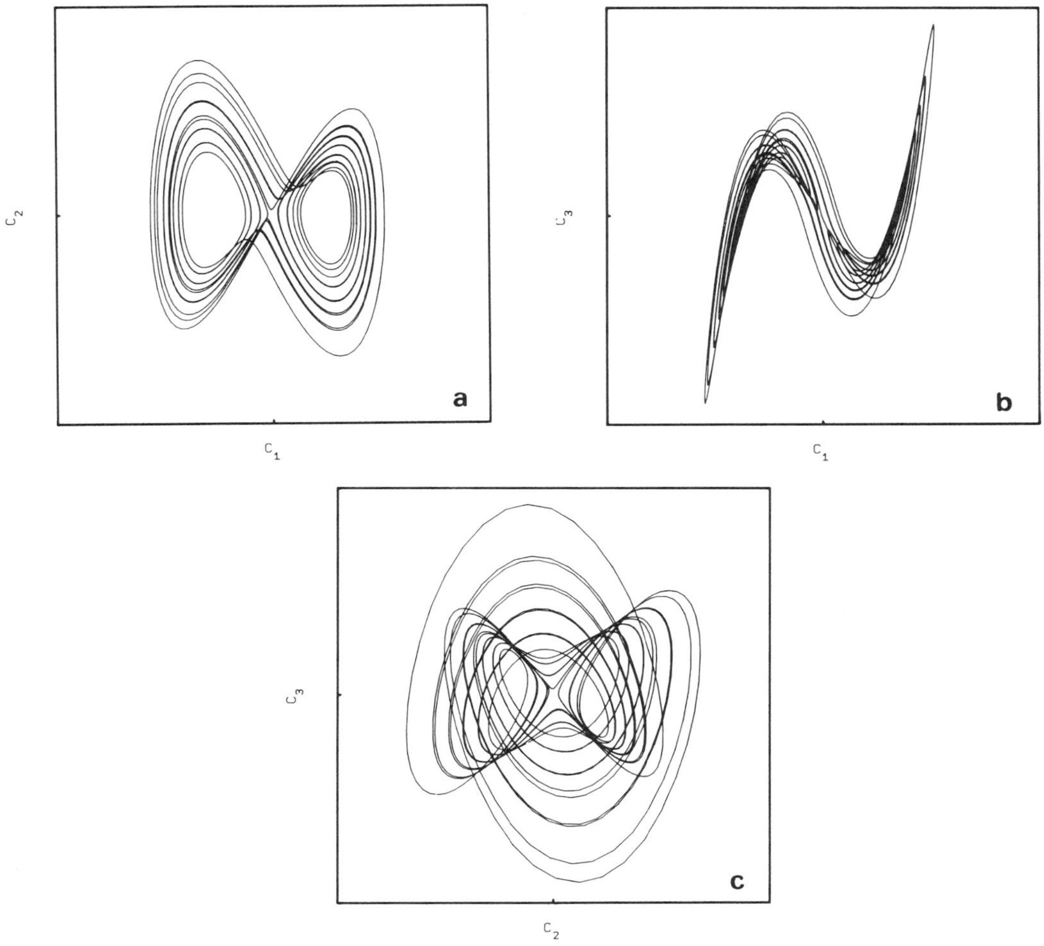

Fig. 9. Plots of the trajectory, $\boldsymbol{X}$, projected onto the three mutually orthogonal planes spanned by the singular vectors $\{\boldsymbol{c}_1, \boldsymbol{c}_2, \boldsymbol{c}_3\}$ shown in fig. 8. The ith point on the $(\boldsymbol{c}_j, \boldsymbol{c}_k)$-plane is given by $(\boldsymbol{c}_j^{\mathrm{T}}\boldsymbol{x}_i, \boldsymbol{c}_k^{\mathrm{T}}\boldsymbol{x}_i)$.

The closeness of the two maps can be understood by considering the form of the singular vectors $\boldsymbol{c}_1$ and $\boldsymbol{c}_2$ (see fig. 8). Recall that the vector $\boldsymbol{x}$ is constructed from a sequence of (seven) consecutive samples of the time series $X(t)$. Therefore, the component $\boldsymbol{c}_1^{\mathrm{T}}\boldsymbol{x}$ corresponds to the average of $X(t)$ in the window, and the component $\boldsymbol{c}_2^{\mathrm{T}}\boldsymbol{x}$ corresponds to an averaged central difference approximation to $\mathrm{d}X(t)/\mathrm{d}t$. The surface of section used in fig. 11b is the plane $\boldsymbol{c}_2^{\mathrm{T}}\boldsymbol{x} = 0$ and the coordinate used for the return map is $\boldsymbol{c}_1^{\mathrm{T}}\boldsymbol{x}$. Thus the surface of section picks out turning points in $X(t)$ and the return map is constructed from the window-averaged value of $X(t)$. On the other hand, by inspection of fig. 4 one can see that the surface of section chosen for the full Lorenz model will intersect the trajectory near its turning points in the XY-plane. Since the coordinate used for the

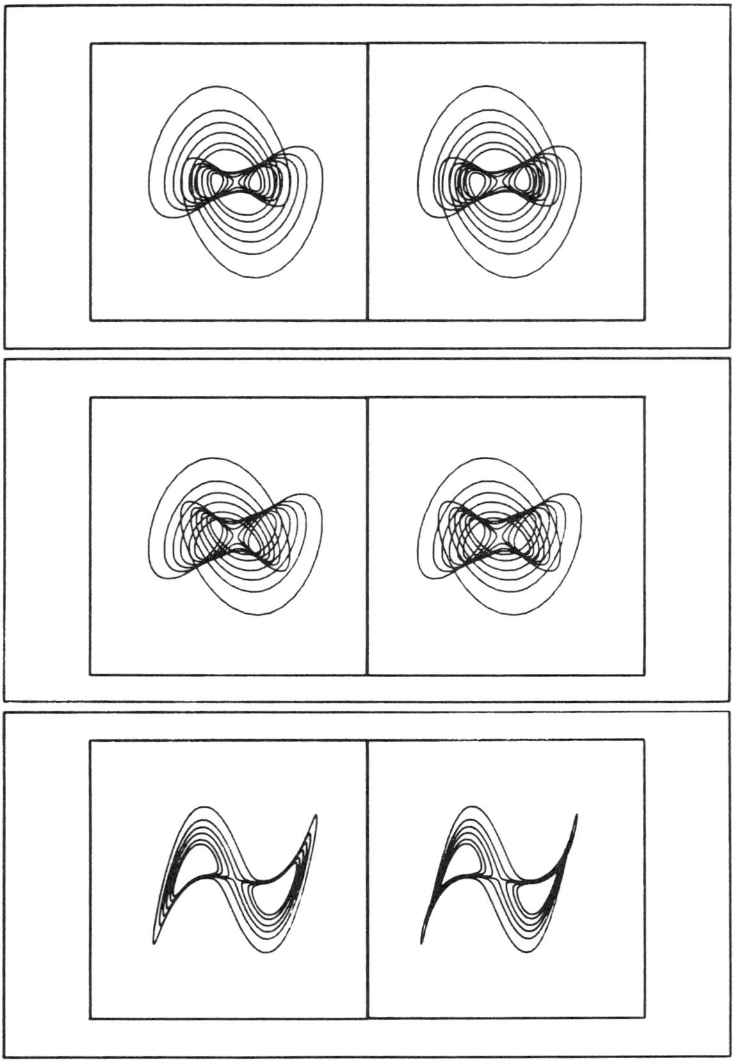

Fig. 10. Stereo pairs of $\boldsymbol{X}$ in the subspace spanned by $\{c_1, c_2, c_3\}$ corresponding to the trajectory shown in fig. 9.

return map in fig. 5b was chosen to be $X(t)$, this map is close to that shown in fig. 11b. Thus the apparent congruence of the two maps is a feature of the simple example we have taken and should not be expected in a more general setting.

Finally, we illustrate the advantages of the singular system approach in extracting qualitative dynamics from data corrupted by noise. For this we use $X(t)$ data obtained with $\tau_s = 0.009$ and quantized to 6-bits of precision. Fig. 12a shows the

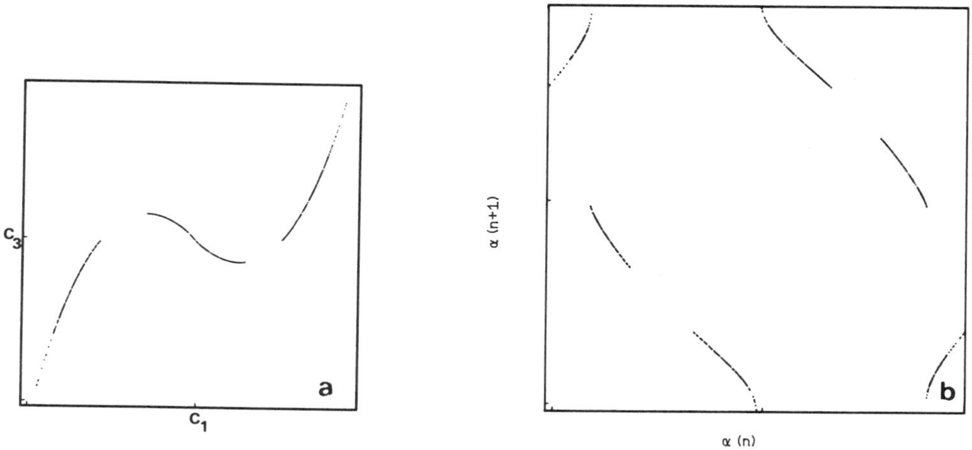

Fig. 11. (a) The intersection of the attractor reconstructed from $X(t)$ data with a planar surface of section containing the c_3-axis and passing through the two unstable foci. (b) The corresponding first-return map using the c_1-coordinate of the points of intersection. ($\alpha = c_1^T x$ where x is a point on the surface of section.)

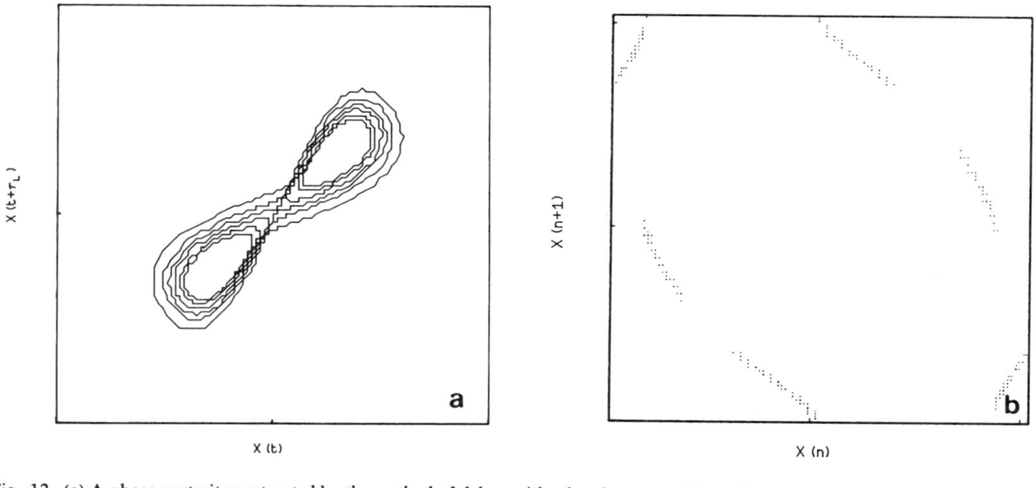

Fig. 12. (a) A phase portrait constructed by the method of delays with a lag time, $\tau_L = 10\tau_s$, using $X(t)$ data obtained with $\tau_s = 0.009$ and quantized to 6-bits of precision. (b) The corresponding first-return map obtained using a surface of section perpendicular to the $(X(t), X(t+\tau_L))$-plane and containing the two unstable foci.

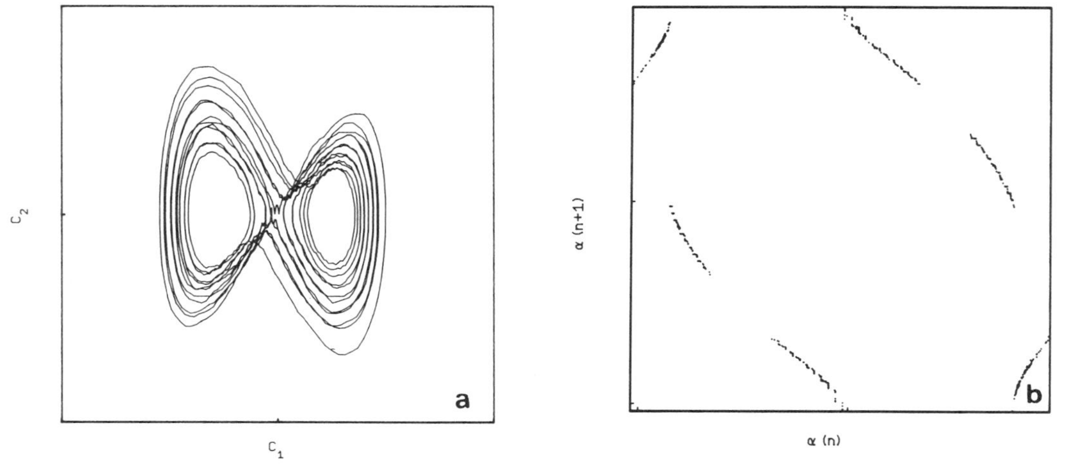

Fig. 13. (a) A phase portrait constructed by the singular system method with a window length $\tau_w = 13\tau_s$ using $X(t)$ data obtained with $\tau_s = 0.009$ and quantized to 6-bits of precision. Compare with fig. 12a. (b) The corresponding first-return map obtained using a surface of section perpendicular to the (c_1, c_2)-plane and containing the two unstable foci. Compare with fig. 12b.

phase portrait constructed by the method of delays with $\tau_L = 10\tau_s$, and fig. 12b shows the corresponding first-return map (which was obtained on a surface of section analogous to those used previously). The effect of the noise is obvious. These figures should be compared with figs. 13a and b which have been obtained from the same data using the singular system method. Following the prescription described in section 3.4, we require a window length larger than that obtained previously. This is due to the increased level of noise which results in a smaller value of ω^*. For the present data the prescription suggests a window length of $13\tau_s$. The singular spectrum for this window length and noise level has only three singular values above the noise floor, but the forms of their corresponding singular vectors are unchanged from those shown in fig. 8. The improved quality of the phase portrait and first-return map is due to the fact that the singular system approach generates averages of data within the window. In contrast the method of delays uses raw (unaveraged) data.

5. Conclusions

Dynamical systems theory has provided a language and a point of view for the study of nonlinear phenomena in physical systems. However, it is a geometric theory and hence requires that experimental results be analysed for geometric information. The work reported here and developed elsewhere [20] is intended ultimately to provide the experimentalist with statistical tools for qualitative analysis.

Acknowledgements

We gratefully acknowledge useful discussions with E.R. Pike, E. Jakeman, R. Jones, M. Johnson, I. Clarke, and J. Mather.

Appendix

In this appendix we give some practical details and timings for the implementation of the methodology presented in this paper.

As in section 3, consider a discrete time series of an observable, v, of length $N_T = N + (n-1)$:

$v_1, v_2, \ldots, v_{N_T}$

The (k, l)th element of the $n \times n$ covariance matrix, Ξ, is:

$$\Xi_{kl} = \frac{1}{N} \sum_{i=1}^{N} v_{i+k-1} v_{i+l-1}. \tag{A.1}$$

It is easy to show that:

$$\Xi_{k+1,l+1} = \Xi_{kl} + \frac{1}{N} \{ v_{N+k} v_{N+l} - v_k v_l \}. \tag{A.2}$$

Therefore, the time involved in calculating Ξ is essentially the time required to calculate the first row of this matrix. A typical time for the calculation of a 25×25 covariance matrix for a data set with $N_T = 32{,}768$ on a 16-bit laboratory minicomputer possessing a floating point hardware unit is about 3 minutes. The diagonalization of a matrix of this size takes only about 10 seconds.

The amount of data required for the calculation of Ξ should follow the usual rules for ensuring the convergence of the autocorrelation function of a stationary process. Since the purpose at this stage is to extract the second-order statistics from the time series for the calculation of the Karhunen–Loeve basis, the requirement on the amount of data is far less formidable than would be needed for the calculation of higher-order quantities such as Lyapunov exponents, the dimension of the manifold, etc.

References

[1] N.H. Packard, J.P. Crutchfield, J.D. Farmer and R.S. Shaw, "Geometry from a time series", Phys. Rev. Lett. 45 (1980) 712.

[2] F. Takens, "Detecting strange attractors in turbulence", Lecture Notes in Mathematics, D.A. Rand and L.-S. Young, eds. (Springer, Berlin, 1981) p. 366.

[3] (a) M. Bertero and E.R. Pike, "Resolution in diffraction-limited imaging, a singular value analysis. I: The case of coherent illumination", Opt. Acta. 29 (1982) p. 727.
(b) E.R. Pike, J.G. McWhirter, M. Bertero and C. de Mol, "Generalised information theory for inverse problems in signal processing", IEE Proceedings, 131, pt. F, No. 6 (Oct. 1984) p. 660. Also see the references therein.

[4] D. Edelson, R.J. Field, R.M. Noyes, "Mechanistic details of the Belousov–Zhabotinskii oscillations", Int. J. Chem. Kinet. 7 (1975) 417.

[5] R.C. Di Prima and H.L. Swinney, "Instabilities and transition in flow between concentric rotating cylinders", in: Hydrodynamic Instabilities and the Transition to Turbulence, H.L. Swinney and J.P. Gollub, eds. (Springer, Berlin, 1981) p. 139.

[6] H. Poincaré, "Mémoire sur les courbes définies par les équations différentielles I–VI", Oeuvre I. Gauthier-Villar: Paris (1880–1890).
H. Poincaré, "Sur les équations de la dynamique et le Problème de trois corps", Acta Math. 13 (1890) 1–270.
H. Poincaré, "Les Méthodes Nouvelles de la Mécanique Céleste", 3 vols. (Gauthier-Villars, Paris 1899).

[7] J. Guckenheimer and P. Holmes, Nonlinear Oscillations, Dynamical Systems, and Bifurcations of Vector Fields (Springer, Berlin, 1983).

[8] J. Palis Jr. and W. de Melo, Geometric Theory of Dynamical Systems: An Introduction (Springer, New York, 1982).

[9] (a) H. Whitney, "Differentiable Manifolds", Ann. Math. 37 (1936) 645.
(b) M.W. Hirsch, Differential Topology (Springer, New York, 1976).
(c) for a good introduction to embeddings:
D.R.J. Chillingworth, Differential Topology with a View to Applications, Research Notes in Mathematics 9 (Pitman, London 1976).

[10] J.C. Roux, R.H. Simoyi and H.L. Swinney, "Observation of a strange attractor", Physica 8D (1983) 257–266.

[11] G.W. Stewart, Introduction to Matrix Computations (Academic Press, New York, 1973).

[12] J.G. McWhirter, private communication.

[13] P.A. Devijver and J. Kittler, Pattern Recognition: A Statistical Approach (Prentice-Hall, New York, 1982).

[14] D.S. Broomhead, J.N. Elgin, E. Jakeman, S. Sarkar, S.C. Hawkins and P. Drazin, "Statistical properties in the chaotic regime of the Maxwell–Bloch equations", Optics Comms. 50 (1984) 56.

[15] P. Grassberger and I. Procaccia, "Dimensions and entropies of strange attractors from a fluctuating dynamics approach", Physica 13D (1984) 34.

[16] E.N. Lorenz, "Deterministic nonperiodic flow", J. Atmospheric Sci. 20 (1963) 130.

[17] D.S. Broomhead, R. Jones and G.P. King, "Qualitative dynamics of an electronic oscillator", in preparation.

[18] G. Rowlands, "An approximate analytic solution of the Lorenz equations", J. Phys. A16 (1983) 585.

[19] O. Rössler, "Different types of chaos in two simple differential equations", Z. Naturforsch 31a (1976) 1661.

[20] D.S. Broomhead and G.P. King, "On the qualitative analysis of experimental dynamical systems", to appear in: Nonlinear Phenomena and Chaos, S. Sarkar, ed. (Adam Hilger, Bristol, 1986).

Determining embedding dimension for phase-space reconstruction using a geometrical construction

Matthew B. Kennel

Institute for Nonlinear Science and Department of Physics, University of California, San Diego, La Jolla, California 92093-0402

Reggie Brown

Institute for Nonlinear Science, University of California, San Diego, La Jolla, California 92093-0402

Henry D. I. Abarbanel

Institute for Nonlinear Science, Department of Physics,
and Marine Physical Laboratory, Scripps Institution of Oceanography,
University of California, San Diego, Mail Code R-002, La Jolla, California 92093-0402

(Received 24 April 1991)

We examine the issue of determining an acceptable minimum embedding dimension by looking at the behavior of near neighbors under changes in the embedding dimension from $d \rightarrow d+1$. When the number of nearest neighbors arising through projection is zero in dimension d_E, the attractor has been unfolded in this dimension. The precise determination of d_E is clouded by "noise," and we examine the manner in which noise changes the determination of d_E. Our criterion also indicates the error one makes by choosing an embedding dimension smaller than d_E. This knowledge may be useful in the practical analysis of observed time series.

PACS number(s): 05.45.+b, 02.40.+m

I. INTRODUCTION

It has become quite familiar in the analysis of observed time series from nonlinear systems to make a time-delay reconstruction of a phase space in which to view the dynamics. This is accomplished by utilizing time-delayed versions of an observed scalar quantity: $x(t_0+n\Delta t)=x(n)$ as coordinates for the phase space [1,2]. From the set of observations, multivariate vectors in d-dimensional space

$$\mathbf{y}(n)=(x(n),x(n+T),\ldots,x(n+(d-1)T)) \quad (1)$$

are used to trace out the orbit of the system. Time evolution of the y's is given by $\mathbf{y}(n)\rightarrow\mathbf{y}(n+1)$. In practice, the natural questions of what time delay T and what embedding dimension d to use in this reconstruction have had a variety of answers. In this paper we provide a clean, direct answer to the question: What is the appropriate value of d to use as the embedding dimension? We do so by directly addressing the topological issue raised by the embedding process. Our procedure identifies the number of "false nearest neighbors," points that appear to be nearest neighbors because the embedding space is too small, of every point on the attractor associated with the orbit $\mathbf{y}(n)$, $n=1,2,\ldots,N$. When the number of false nearest neighbors drops to zero, we have unfolded or embedded the attractor in $\mathbb{R}^d$, a d-dimensional Euclidian space.

The observations, $x(n)$, are a projection of the multivariate state space of the system onto the one-dimensional axis of the $x(n)$'s. The purpose of time-delay (or any other [3]) embedding is to unfold the projection back to a multivariate state space that is representative of the original system. The general topological result of Mañé and Takens [4,5] states that when the attractor has dimension d_A, all self-crossings of the orbit (which is the attractor) will be eliminated when one chooses $d>2d_A$ [6]. These self-crossings of the orbit are a result of the projection, and the embedding process seeks to undo that. The Mañé and Takens result is only a sufficient condition as can be noted by recalling that the familiar Lorenz [7] attractor, $d_A=2.06$, can be embedded by the time-delay method in $d=3$. This is in contrast to the theorem, which only informs us that $d=5$ will surely do the job. The open question is, given a scalar time series, what is the appropriate value for the minimum embedding dimension d_E? From the point of view of the mathematics of the embedding process it does not matter whether one uses the minimum embedding dimension d_E or any $d \geq d_E$, since once the attractor is unfolded, the theorem's work is done. For a physicist the story is quite different. Working in any dimension larger than the minimum required by the data leads to excessive computation when investigating any subsequent question (Lyapunov exponents, prediction, etc.) one wishes to ask. It also enhances the problem of contamination by round-off or instrumental error since this "noise" will populate and dominate the additional $d-d_E$ dimensions of the embedding space where no dynamics is operating.

The usual method of choosing the minimum embedding dimension d_E is to compute some invariant on the attractor. By increasing the embedding dimension used for the computation one notes when the value of the invariant stops changing. Since these invariants are geometric properties of the attractor, they become independent of d for $d \geq d_E$ (i.e., after the geometry is unfolded). The problem with this approach is that it is often very data intensive and is certainly subjective. Furthermore, the analysis does not indicate the penalty one pays for choosing too low an embedding dimension. We have already discussed the penalty for choosing too large an embedding dimension.

In this paper we present calculations based on the idea that in the passage from dimension d to dimension $d+1$ one can differentiate between points on the orbit $\mathbf{y}(n)$ that are "true" neighbors and points on the orbit $\mathbf{y}(n)$ which are "false" neighbors. A false neighbor is a point in the data set that is a neighbor solely because we are viewing the orbit (the attractor) in too small an embedding space ($d < d_E$). When we have achieved a large enough embedding space ($d \geq d_E$), all neighbors of every orbit point in the multivariate phase space will be true neighbors. A simple example of this behavior is found in the Hénon map [8] of the plane to itself. In Fig. 1 we show the attractor for the Hénon map as a projection onto a $d=1$ dimensional phase space (the x axis) as well as in a $d=d_E=2$ dimensional embedding space. The points **A** and **B** appear to be neighbors in the projection onto the x axis. However, they are neighbors solely because we are viewing the orbit of the attractor in too small an embedding space. When viewed in a $d=2$ dimensional embedding space they are no longer neighbors. Points **A** and **B** are examples of false neighbors. In contrast points **A** and **C** are true neighbors. This follows because they are neighbors in $d=1$ dimension (x-axis) as well as $d=d_E=2$ and all higher embedding dimensions. We will return to and fully develop this point in Sec. II.

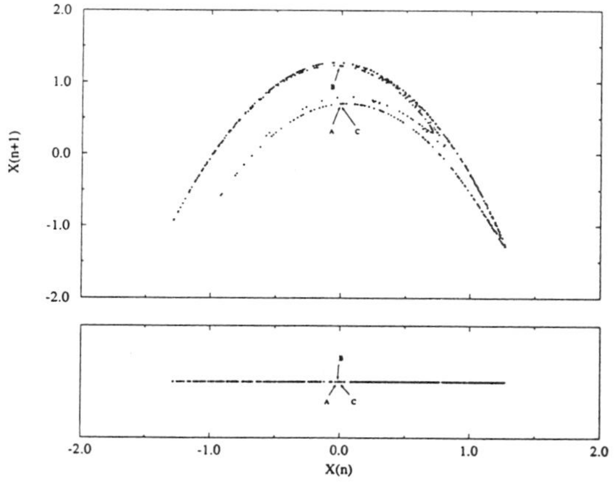

FIG. 1. The R^1 and R^2 embeddings of the x coordinate of the Hénon map of the plane. It is known that for this map $d_E=2$. The points **A** and **B** are false neighbors while the points **A** and **C** are true neighbors.

In the process of writing this paper we found an article by Liebert, Pawelzik, and Schuster (LPS) [9] which reports on the same basic idea but implements it in quite a different fashion. We find rather distinct results from these authors. We have several comments on their observations and conclusions. Their method appears to be somewhat more time consuming, but only by a constant factor. Furthermore, their method does not yield one of the desirable features of the present work. Namely, we provide an estimate of the error encountered in using too small an embedding dimension.

We also found an older paper by Bumeliene, Lasiene, Pyragas, and Cenys [10] (BLPC) which cites an even earlier paper by Pyragas and Cenys [11] that has the essential geometric idea of seeking false neighbors, and again implements it in another fashion. We will comment on these papers as well.

The question of what time delay T to use in the embedding is logically independent of the topological question of how large the space must be to eliminate false neighbors. The determination of the time lag T requires information which is independent of the topological arguments of Mañé and Takens. This can be understood since their argument works *in principle* for any lag T, and so must be independent of considerations about T. The issue of the time lag is dynamical, not geometric. For the purpose of determining T we use the information theoretic techniques of Fraser and Swinney [12,13].

II. THE METHOD OF FALSE NEIGHBORS

One of the important features of an attractor is that it is often a compact object in phase space. Hence, points of an orbit on the attractor acquire neighbors in this phase space. The utility of these neighbors, among other things, is that they allow the information on how phase-space neighborhoods evolve to be used to generate equations for the prediction of the time evolution of new points on or near the attractor [14]. They also allow accurate computations of the Lyapunov exponents of the system [15,16].

In an embedding dimension that is too small to unfold the attractor, not all points that lie close to one another will be neighbors because of the dynamics. Some will actually be far from each other and simply appear as neighbors because the geometric structure of the attractor has been projected down onto a smaller space (cf. Sec. I). If we are in d dimensions and we denote the rth nearest neighbor of $\mathbf{y}(n)$ by $\mathbf{y}^{(r)}(n)$, then from Eq. (1), the square of the Euclidian distance between the point $\mathbf{y}(n)$ and this neighbor is

$$R_d^2(n,r) = \sum_{k=0}^{d-1} [x(n+kT) - x^{(r)}(n+kT)]^2 \, . \tag{2}$$

In going from dimension d to dimension $d+1$ by time-delay embedding we add a $(d+1)$th coordinate onto each of the vectors $\mathbf{y}(n)$. This new coordinate is just $x(n+Td)$. We now ask what is the Euclidean distance, *as measured in dimension* $d+1$, between $\mathbf{y}(n)$ and the same rth neighbor as determined in dimension d? *After* the addition of the new $(d+1)$th coordinate the distance between $\mathbf{y}(n)$ and the same rth nearest neighbor we determine in d dimensions is

$$R_{d+1}^2(n,r)=R_d^2(n,r)+[x(n+dT)-x^{(r)}(n+dT)]^2 \ . \tag{3}$$

A natural criterion for catching embedding errors is that the increase in distance between $\mathbf{y}(n)$ and $\mathbf{y}^{(r)}(n)$ is large when going from dimension d to dimension $d+1$. The increase in distance can be stated quite simply from Eqs. (2) and (3). We state this criterion by designating as a false neighbor any neighbor for which

$$\left[\frac{R_{d+1}^2(n,r)-R_d^2(n,r)}{R_d^2(n,r)}\right]^{1/2} = \frac{|x(n+Td)-x^{(r)}(n+Td)|}{R_d(n,r)} > R_{\text{tol}} \ , \tag{4}$$

where R_{tol} is some threshold. We will investigate the sensitivity of our criterion to R_{tol} in our numerical work below, and we will find that for $R_{\text{tol}} \geq 10$ the false neighbors are clearly identified. It is sufficient to consider only nearest neighbors ($r=1$) and interrogate every point on the orbit ($n=1,2,\ldots,N$) to establish how many of the N nearest neighbors are false. We record the results of the computations as the proportion of all orbit points which have a false nearest neighbor.

Before we report on these computations we remark that this criterion, by itself, is not sufficient for determining a proper embedding dimension. To illustrate this we note that when we used this criterion to examine the embedding dimension for white "noise," the criterion erroneously reported that this noise could be embedded in a quite small dimensional space. (By "noise" we mean very-high-dimensional attractors associated with computerized random number generators.) The problem turns out to be that even though $\mathbf{y}^{(1)}(n)$ is the nearest neighbor to $\mathbf{y}(n)$, it is not necessarily *close* to $\mathbf{y}(n)$. Indeed, as we moved up in embedding dimension, with finite data from a noise signal, the actual values of $R_d(n)\equiv R_d(n,r=1)$ were comparable with the size of the attractor R_A. Thus the nearest neighbor to $\mathbf{y}(n)$ is not close to $\mathbf{y}(n)$. This behavior follows from the fact that trying to uniformly populate an object in d dimensions with a fixed number of points means that the points must move further and further apart as d increases. We note that as we increased the number of data points in the noise signal the embedding dimension (that dimension where the number of false neighbors drops to nearly zero) systematically increased. In the limit of an infinite amount of data we would find that the embedding dimension d also diverged to infinity. This is in contrast to the low-dimensional attractor common to many dynamical systems. For a low-dimensional dynamical system increasing the number of data points on the attractor will not change d_E.

However, in practical settings the number of data points is often not terribly large. We have implemented the following criterion to handle the issue of limited data set size: If the nearest neighbor to $\mathbf{y}(n)$ is not close $[R_d(n)\approx R_A]$ and it is a false neighbor, then the distance $R_{d+1}(n)$ resulting from adding on a $(d+1)$th components to the data vectors will be $R_{d+1}(n)\approx 2R_A$. That is, even distant but nearest neighbors will be stretched to the extremities of the attractor when they are unfolded from each other, if they are false nearest neighbors.

We write this second criterion as

$$\frac{R_{d+1}(n)}{R_A} > A_{\text{tol}} \ . \tag{5}$$

In our work we advocate using this pair of criteria jointly by declaring a nearest neighbor (as seen in dimension d) as false if either test fails.

As a measure of R_A we chose the value

$$R_A^2=\frac{1}{N}\sum_{n=1}^{N}[x(n)-\bar{x}]^2 \ ,$$

where

$$\bar{x}=\frac{1}{N}\sum_{n-1}^{N}x(n) \ .$$

Other choices for R_A, such as the absolute deviation about $\bar{x}$, did not change our results. The reader may choose whatever estimate of the attractor size which appeals to her or him.

In Fig. 2 we show the result of applying the first criterion alone, of applying the second criterion alone, and

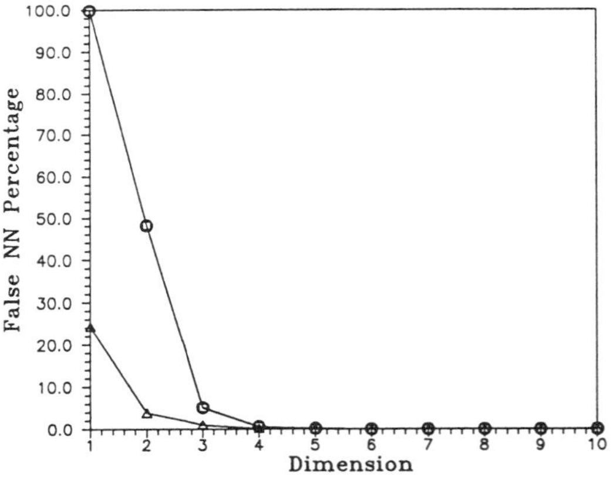

FIG. 2. The percentage of false nearest neighbors for 24 000 data points from the Lorenz-II equations. The data were output at $\Delta t=0.05$ during the integration. A time lag $T=11\Delta t=0.55$, which is the location of the first minimum in the average mutual information for this system, was used in forming the time-delayed vectors. Three different criteria are compared for detecting false nearest neighbors. First is the change in distance of the nearest neighbors in d dimensions when the component $x(n+Td)$ is added to the vectors, Eq. (4). These points are marked with squares. Second is the criterion which compares R_{d+1} to the size of the attractor R_A, Eq. (5). These points are marked with triangles. The third criterion applies both of these to the data. These points are marked with circles. In this last case a point which fails either test is declared false.

then applying them jointly when a data set from the three-dimensional model of Lorenz [17]

$$\dot{x} = -y^2 - z^2 - a(x - F) ,$$
$$\dot{y} = xy - bxz - y + G , \quad (6)$$
$$\dot{z} = bxy + xz - z ,$$

is used to generate the data. We use the values $a = 0.25$, $b = 4.0$, $F = 8.0$, and $G = 1.0$ where Lorenz points out irregular behavior is encountered. The attractor has a dimension, d_A, slightly greater than 2.5. We produced the data used in this figure with a variable order Adams integrator with output at $\Delta t = 0.05$. The first minimum of the average mutual information occurs at $T \approx 0.85 = 17\Delta t$. This is the time delay T, we used in reconstructing the phase-space vectors $\mathbf{y}(n)$ from samples of $x(n) = x(t_0 + n\Delta t)$. In Fig. 2 a total of 25 000 data points were used. It is clear that the joint criterion [Eqs. (4) and (5)] as well as each of the individual criteria [Eqs. (4) or (5)] mentioned above yield an embedding dimension of $d_E = 6$ for this attractor. In this case the result is actually the same as the $d > 2d_A$ sufficient bound of Mañé and Takens. For this computation we used the values $R_{\text{tol}} = 15.0$ and $A_{\text{tol}} = 2.0$. We will report on the dependence of the method on the number of data points N and on the value of R_{tol} in Sec. III.

The results are quite different for noise. In Fig. 3 we show a similar comparison of false-nearest-neighbor criteria for noise uniformly distributed in $x(n)$ between -1 and 1. We used a time delay $T = 1$, and again $N = 25\,000$. Here we see that the first criterion, Eq. (4), fails to indicate the need for a high embedding dimension. However, the second criterion, Eq. (5), yields the expected answer.

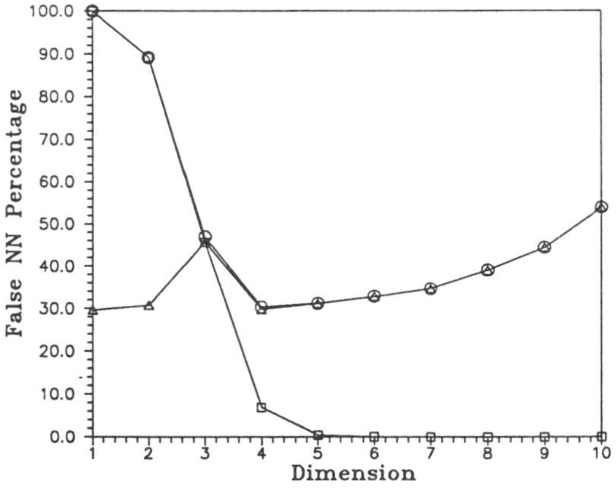

FIG. 3. The same as Fig. 2 but the data come from a random set of numbers uniformly distributed in the interval $[-1.0, 1.0]$. The comparison between Fig. 2 and this figure is quite striking and points out clearly the difference between a low-dimensional chaotic signal and noise (a high-dimensional chaotic signal).

Of course, the joint criterion also works and, as one would expect tracks the second criterion as d increases. This striking difference between low-dimensional chaos and high-dimensional chaos was seen in all examples we tried. Henceforth, we quote only the result of applying our criteria jointly: *a nearest neighbor which fails either test is declared false.*

It is important to note that in each case we are able to determine the quality of the embedding because the percentage of false neighbors is reported. For example, in the case of the Lorenz 84 attractor noted above, if we chose to use an embedding dimension of $d = 5$ rather than $d = 6$, there would have been 0.18% false neighbors remaining. It is likely that this is an acceptably small number for many purposes. A physicist might well chose to accept this error to make more efficient any further computations performed on the data from this system. Actually the error in choosing only $d = 4$ is not substantial since the percentage of false neighbors is only 0.59%. One may even interpret these neighborhoods as isolated instances of very large local Lyapunov exponents. A very high rate of instability growth may cause R_{d+1}/R_d to be very large, even if the embedding is correct. In other investigations [18], we have found that for short-time intervals, the distribution in the finite-time Lyapunov exponents may be very broad. The implication for the present situation is that there may be a few neighborhoods, whose largest local Lyapunov exponent falls in the high extreme of the distribution, for which the false-neighbor criterion (and any other similar type of measure) will fail due to inherent dynamical reasons. Carefully disentangling false neighbors from high local exponents is still an unresolved issue, but the problem does not appear to be terribly severe.

III. NUMERICAL RESULTS FOR SEVERAL EXAMPLES

A. Clean data and "noise"

We have implemented the method just described on a variety of simple models. Let us begin with the Lorenz-II model previously discussed. Using the same time delay, $T = 17\Delta t$, we examined the dependence of our method on the tolerance R_{tol} and the length of the data set N. In Fig. 4 we show the false-neighbor percentage in $d = 1$ as a function of these two variables. It is clear that except for very small amounts of data $N \approx 100$ and large values of R_{tol}, the method of false neighbors would not indicate $d = 1$ as a good embedding dimension. The same dependence, false-nearest-neighbor percentage as a function of R_{tol} and N, is shown in Fig. 5 for $d = 5$. It is clear that when the number of data points N is sufficient to fill out an attractor in $d = 5$ dimensions, and the tolerance level is $R_{\text{tol}} \geq 15$, we can with confidence select $d = 5$ as the embedding dimension. As noted above one could just as well choose $d = 6$, where the percentage of false nearest neighbors drops to exactly zero. However, the error associated with $d = 5$ is so small (approximately 0.18%) one might as well use $d = 5$. We have examined this system for $d = 1, 2, \ldots, 10$.

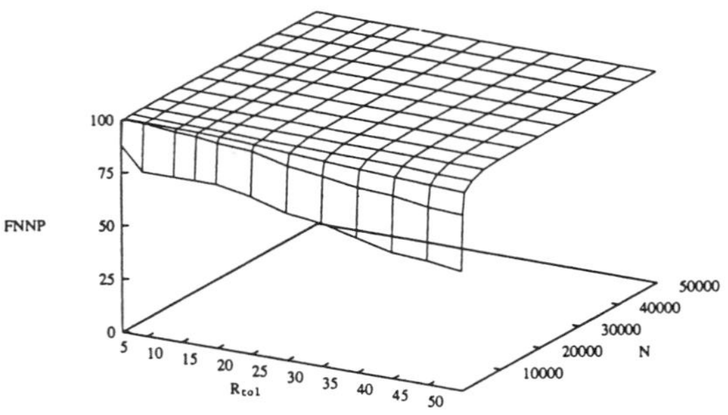

FIG. 4. Data from the Lorenz-II system in $d=1$. The false-nearest-neighbor percentage is evaluated as a function of R_{tol} and the length of the data set N. The time lag in forming all multivariate vectors is $T=0.85$. $d=1$ is clearly a good choice for embedding this system.

The message here is clear. For a fixed amount of data one cannot always explore the finest details of the attractor. So some very close near neighbors might still appear false. With large amounts of data, this appearance is removed, and the trueness of the neighbors is exposed. When tackling the analysis of a data set from an unknown source, a bit of judgment is called for. We see no significant harm in suggesting that whatever dimension yields a percentage of false nearest neighbors below 1% should work fine as an effective embedding dimension.

Returning to uniform noise we show in Figs. 6–8 the dependence on N and R_{tol} of the false-nearest-neighbor percentage in $d=1$, $d=4$, and $d=8$. The percentage falls to a plateau in each case and never indicates that a low dimension would qualify as a good embedding. The false-nearest-neighbor percentage stays about 25% even as both N and R_{tol} grow.

In Figs. 9–12 we show our results for the Ikeda [19,20] map of the plane to itself,

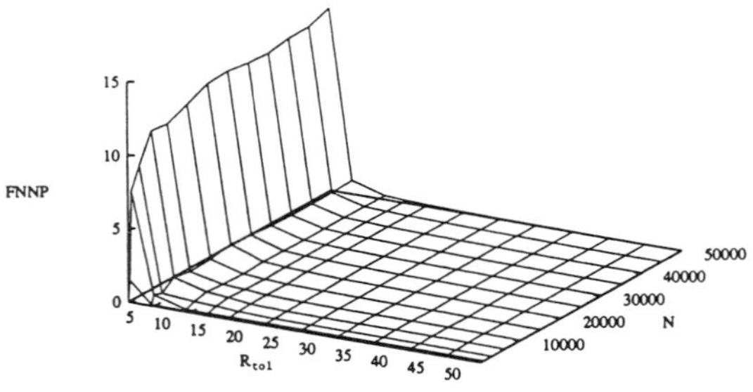

FIG. 5. Data from the Lorenz-II system in $d=5$. The false-nearest-neighbor percentage is evaluated as a function of R_{tol} and the length of the data set N. The time lag in forming all multivariate vectors is $T=0.85$. $d=5$ is clearly a good choice for embedding this system.

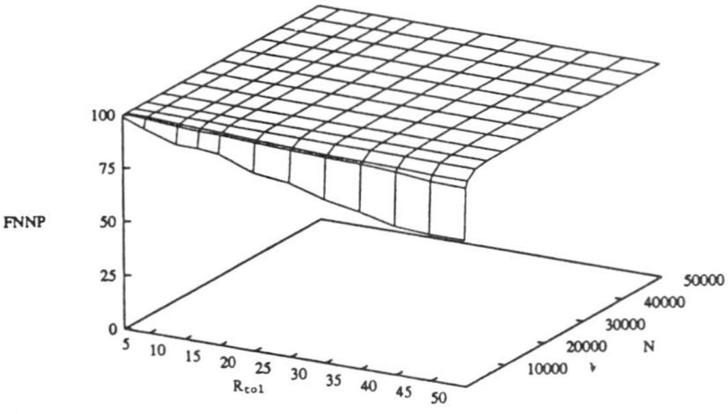

FIG. 6. Data from a uniform noise distribution $[-1,1]$ in $d=1$. The false-nearest-neighbor percentage is evaluated as a function of R_{tol} and the length of the data set N.

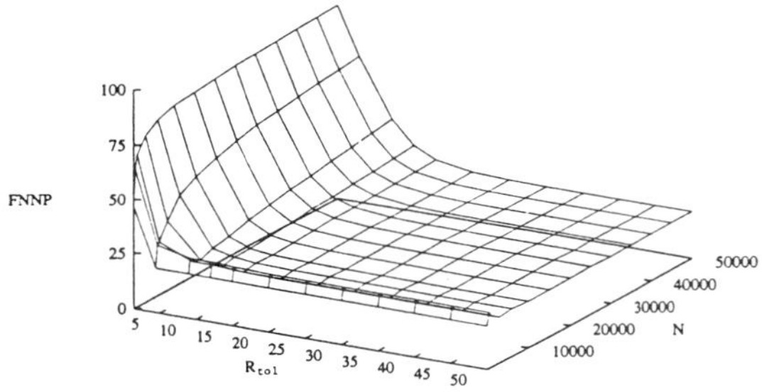

FIG. 7. Data from a uniform noise distribution $[-1,1]$ in $d=4$. The false-nearest-neighbor percentage is evaluated as a function of R_{tol} and the length of the data set N.

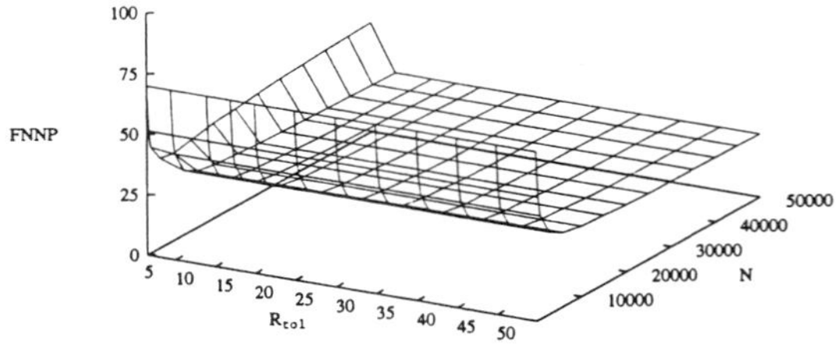

FIG. 8. Data from a uniform noise distribution $[-1,1]$ in $d=8$. The false-nearest-neighbor percentage is evaluated as a function of R_{tol} and the length of the data set N.

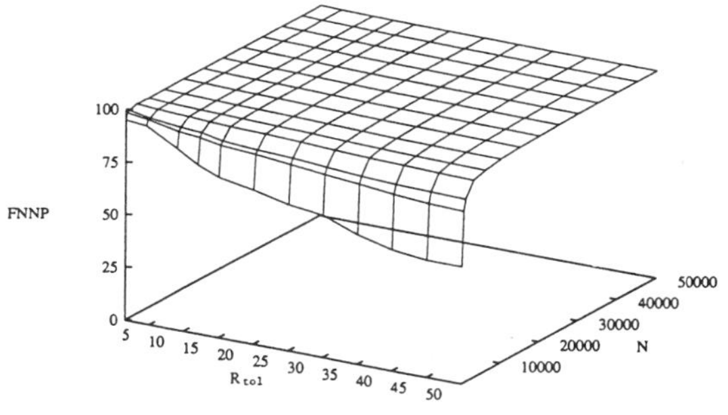

FIG. 9. Data from the two-dimensional Ikeda map in $d=1$. The false-nearest-neighbor percentage is evaluated as a function of R_{tol} and the length of the data set N.

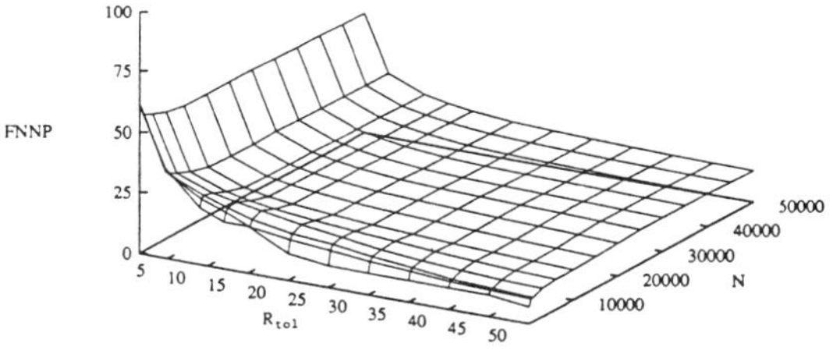

FIG. 10. Data from the two-dimensional Ikeda map in $d=2$. The false-nearest-neighbor percentage is evaluated as a function of R_{tol} and the length of the data set N.

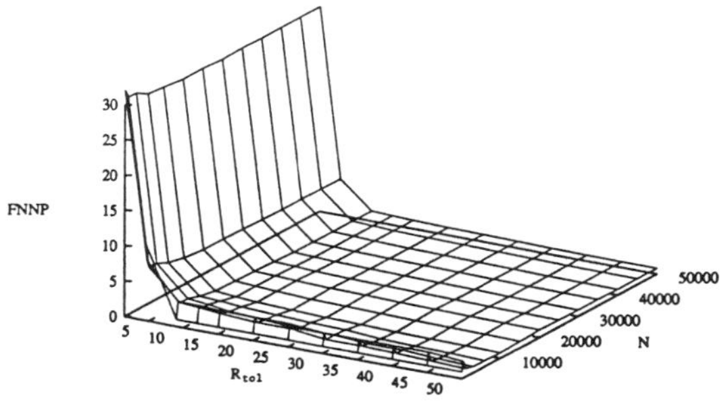

FIG. 11. Data from the two-dimensional Ikeda map in $d=3$. The false-nearest-neighbor percentage is evaluated as a function of R_{tol} and the length of the data set N.

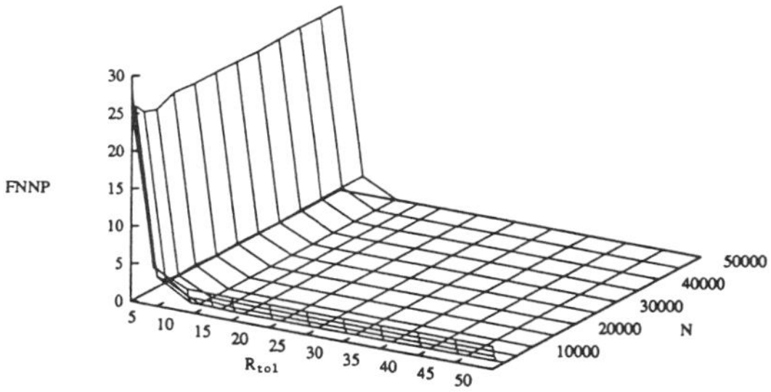

FIG. 12. Data from the two-dimensional Ikeda map in $d=4$. The false-nearest-neighbor percentage is evaluated as a function of R_{tol} and the length of the data set N. $d=4$ is clearly a good choice for embedding this system with the use of time delay coordinates. This is consistent with the observations in Ref. [15].

$$z(n+1)=p+Bz(n)\exp[i\kappa-i\alpha/(1+|z(n)|^2)]\ , \qquad (7)$$

where $p=1.0$, $B=0.9$, $\kappa=0.4$, $\alpha=6.0$, and $z(n)$ is complex. The dimension of the attractor associated with this map is $d_A\approx 1.8$. It is known from other considerations [15] that an embedding dimension of $d=4$ is required to unfold the attractor by time-delay embedding. This property shows up quite clearly in the figures where results from $d=1$, 2, 3, and 4 are shown.

We also applied our methods to the determination of a minimum embedding dimension for the Lorenz-I [7] model, the Hénon map [8], the Rössler attractor [21] and the Mackey-Glass differential-delay equation [22]. The results are just as striking as those just shown, and in the latter two cases are consistent with the minimum embedding dimensions found by LPS. We forbear from drowning the reader in data, since the point is quite clear. Perhaps we should add, however, that in going through the data set and determining which points are near neighbors of the point $\mathbf{y}(n)$ we use the sorting method of a "k-d" tree to reduce the computation time from $O(n^2)$ to $O(N\log_{10}[N])$ [23]. For this reason each of the calculations required takes minutes on a SUN Sparcstation 2 workstation.

B. Realistic Data

Data are never delivered to the user in the pristine, no-noise, state of our examples above. Time series obtained from real experimental apparatus always contains both signal and noise. The results shown for "pure noise" certainly indicate that the effective embedding dimension will degrade as the signal-to-noise ratio decreases from plus infinity. We have examined the way in which noise degrades the results one obtains for our set of example systems. The basic trend we expect, and see, is that as the signal to noise ratio is lowered, the effective embedding dimension rises.

In Fig. 13 we show the results of applying our joint false-nearest-neighbor criterion to 25 000 points on the Lorentz-I attractor. The Lorenz-I equations are [7]

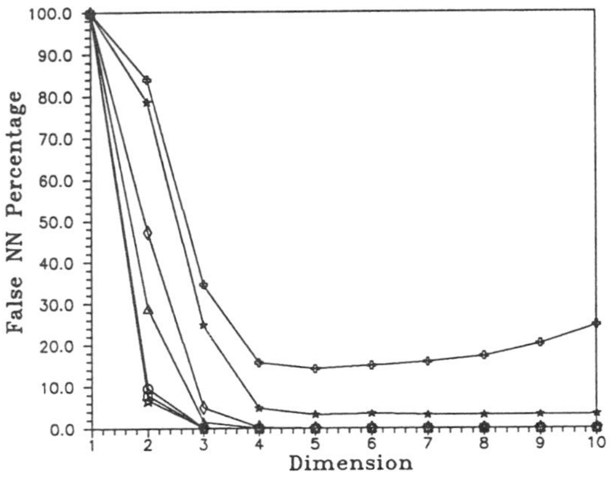

FIG. 13. Data from the Lorenz-I model which have been contaminated with uniformly distributed noise in the interval $[-L,L]$. 41 000 points from the attractor are used here. The percentage of false nearest neighbors is shown for $L/R_A=0.0$, 0.005, 0.01, 0.05, 0.1, 0.5, and 1.0. For this system $R_A \approx 12.7$. Even with $L/R_A=10\%$, the error in choosing $d=4$ as the embedding dimension is only 0.27%. The false-nearest-neighbor method is rather robust against noise contamination. The result for $L/R_A=0$ is marked with an open star; for $L/R_A=0.005$, it is marked with a square; for $L/R_A=0.01$, with a circle; for $L/R_A=0.05$, with a triangle; for $L/R_A=0.1$, with a diamond; for $L/R_A=0.5$, with a filled star; and for $L/R_A=1.0$, with an open cross.

$$\dot{x}=\sigma(y-x) ,$$

$$\dot{y}=-xz+rx-y ,$$

$$\dot{z}=xy-bz ,$$

and we use parameter values $\sigma=16$, $b=4$, and $r=45.92$. The data were generated by a variable order Adams code with output at $\Delta t=0.02$. For this system the first minimum of the average mutual information is at $T\approx 0.1=5\Delta t$, and we used tolerance values of $R_{\text{tol}}=15.0$ and $A_{\text{tol}}=2.0$. The orbits have been contaminated with noise uniformly distributed in the interval $[-L,L]$ with $L/R_A=0,0$, 0.005, 0.01, 0.05, 0.1, and 0.5. The last noise level corresponds to a signal-to-noise ratio $[20\log_{10}(R_A/L)]$ of about 6 dB in power; namely, a very low signal-to-noise ratio. The increase in embedding dimension for this attractor in the presence of noise is really quite slow, and only for the last, very contaminated, case does it fail. (When the signal-to-noise ratio falls to 0 dB, the false-nearest-neighbor criterion fails, and we find the Lorenz-I signal contaminated at that level to be indistinguishable from noise.) In the case where $L/R_A=10\%$, and $d=3$, we have 7.8% false nearest neighbors. For $d=4$ this number drops to 2.5% and plateaus for higher embedding dimensions.

Figure 13 illustrates our approach to embedding dimensions. The practical issue, to say it again, is what embedding dimension should one use to capture the features of the data observed. We would surely choose $d=4$ as an effective embedding dimension and interpret the plateau at 2.5% as a noise effect. Indeed, one of the first tasks one would perform with noise observed data is to apply one of the several methods of noise reduction, or separation, that have been recently suggested [24–28]. The techniques we have presented will give an accurate starting point by choosing an embedding dimension in which to perform the noise reduction. After separating the signal from the contamination, one can easily go back to the false-nearest-neighbor method to determine another (probably smaller) embedding dimension to use in analyzing the signal. Actually since two of the methods require either knowing the dynamics [24–26] or observing a clean trajectory of the dynamics [27], it is only in the case where no knowledge of the "clean" system is required [28] that these remarks might come into play. In the probabilistic method [27] where only a clean trajectory has been observed, the present false-nearest-neighbor technique will give a direct, useful answer about the embedding dimension in which to clean subsequent noise data.

IV. CONCLUSIONS AND COMMENTS

We begin this section by commenting on the relationship between this paper and that of LPS [9] and the paper of BLPC [10]. It should be mentioned that the general ideas behind all of these methods have been around for some time, for example, see Fig. 3 in Ref. [29], and in fact are just generalizations to the realm of nonlinear dynamics of questions dealt with in state-space formulations of linear control systems.

Both the LPS paper and this one provide an implementation of the same basic idea. We each use time-delay coordinates and attribute the disappearance of false neighbors as an indication of a minimum embedding dimension for the data. From this point forward our approaches are different. The LPS technique uses small neighborhoods whereas we use individual neighbors. They then compute two distances, $D_{d+1}(n;r,d)$ and $D_{d+1}(n;r,d+1)$. $D_{d+1}(n;r,d)$ is the distance between $\mathbf{y}(n)$ and its rth neighbor. For this case the calculation that determines which is a near neighbor is performed in d dimensions, while the calculation of the distance between neighbors is performed in $(d+1)$ dimensions. $D_{d+1}(n;r,d+1)$ is also the distance between $\mathbf{y}(n)$ and its rth neighbor. However, for this case both the calculation that determines which is a near neighbor *and* the calculation of the distance to that near neighbor are performed in $(d+1)$ dimensions. The ratio $D_{d+1}(n;r,d)/D_{d+1}(n;r,d+1)\equiv Q_1$ defines their quantity Q_1.

LPS then go on to define Q_2 via $Q_2 \equiv D_d(n;r,d)/D_d(n;r,d+1)$. In this case the calculations for distances between neighbors are performed in d dimensions. The calculation used to identify the rth nearest neighbor is performed in d dimensions for the numerator and $(d+1)$ dimensions for the denominator. LPS then examine the geometric mean of the product Q_1Q_2 over a neighborhood (in their reported case the first ten nearest neighbors). Our first criterion, Eq. (4), can be seen as the ratio of part of their Q_1 to part of their Q_2. LPS's statistic, called W, is the logarithm of the arithmetical average over neighborhoods of the individual geometric means.

For the purposes of finding a minimum embedding dimension our method and that of LPS use the same underlying principle; however, the statistic and interpretation in our case are simpler. The proportion of false neighbors is on an absolute scale, always bounded between 0 and 1, whereas with LPS it is not immediately clear what constitutes a sufficiently small statistic unless the convergence is sharp. If one sees a plateau of 0.1% false neighbors with increasing embedding dimension, in contrast to perhaps a plateau at 10%, one could be rather confident that a good reconstruction of a clean chaotic attractor has been accomplished. LPS did not examine the effect of noise on their algorithm, but in a brief comparision we found that on pure uniform noise, their statistic continued to decrease with larger embedding dimension, whereas ours plateaued at a comparatively high level. This is not surprising as their statistic uses the same information as ours, but without the additional criterion that we found necessary to guard against pure noise. For moderate noise levels (10%) added to the Lorenz-II attractor, the LPS statistic did not have a definite plateau with increasing embedding dimension as did the false-neighbors criterion, and so we found the identification of minimum embedding dimension easier to make in the latter case.

LPS also use their statistic to choose the time lag for embedding. They define a new quantity $\tilde{W}=W/T$ and search for a minimum in $\tilde{W}$ as a function of T. We admit to not fully understanding the motivation behind the division by T. The W quantity alone gives the general behavior that we intuitively expect, and also observe with the false-neighbors statistic. For large T, the proportion of false neighbors and also the W statistic increase, because the attractor looks more like noise, as the elements of the state-space vector become more decorrelated due to the positive Lyapunov exponent. For sufficiently small time lags, the attractor eventually collapses onto a one-dimensional object, and so the estimated embedding dimension will be spuriously low. The division by T appears to be a way of ameliorating the latter problem, if one then ignores new minima in $\tilde{W}$ created at large T by the division.

In contrast to the case for embedding dimension, the theorems do not define a "correct" time delay to use, rather one must choose a condition that simply gives reasonable results given a finite amount of data at a certain sampling rate, and the LPS criterion is another one to do so. Our method, as it is attuned specifically to the gross errors created by an improper embedding, appears in practice to be less sensitive to time delay than LPS's method. This fact may be either a virtue or a vice, depending on one's outlook.

Our example of the Ikeda map also demonstrates that the choice of embedding dimension does very little to eliminate the problem of spurious Lyapunov exponents as LPS suggest it might. With time delay embedding dimensions larger than that of the "true" dynamics, one must still identify, by dynamical means, the Lyapunov exponents which describe the system and not the artifacts of time-delay embedding by Euclidian coordinates [15].

The BLPC paper is closer in spirit to ours than the LPS work. BLPC identify a quantity $\sigma_d(\epsilon)$ which is the average over the attractor of the value of $[x(n+dT)-x(m+dT)]^2$ for all pairs of points (labeled by n and m) with $\mathbf{y}(n)$ and $\mathbf{y}(m)$ having all components within a distance ϵ of each other divided by the number of such points on the orbit. This is essentially the average over the attractor of what we have called $R^2_{d+1}(nr)-R^2_d(n,r)$ for all neighbors within a sphere of radius $\epsilon\sqrt{d}$ of the data point $\mathbf{y}(n)$. They then display $\sigma_d(\epsilon)$ for various dimensions d with various selections of the neighborhood size ϵ and the noise level as added to the equations of evolution of the system. When there is a sharp break in the curve of this quantity versus dimension, they argue this dimension should be chosen as the embedding dimension. For dimensions above this break, the value of $\sigma_d(\epsilon)$ should be of order ϵ^2. They also state that this minimum embedding dimension d_E is related to the fractal dimension of the attractor d_A by $d_E=1+[\text{integer part of }(d_A)]$.

In general idea this is quite close to what we have done. BLPC must choose a neighborhood size, of course, and as far as we can see from their work they did not systematically investigate the effect this size has on their results. They also always work with 4096 data points, and we suspect this is likely to be sufficient for the data they investigated. We must choose a value for the tolerance levels we use in our criteria for false nearest neighbors, but we have demonstrated the independence of this choice over a very wide range of values. By choosing a neighborhood size, apparently in a fashion unrelated to the size of the attractor, BLPC in effect determine how many neighbors will be included as they move around the attractor. This number will vary as one moves about the attractor since the neighborhood size is fixed and the density on the attractor is inhomogeneous. One can see from the data of BLPC that the value of $\sigma_d(\epsilon)$ does not reach down to ϵ^2 above d_E since they are, in some fashion, counting points with true neighbors within a ball of radius $\epsilon\sqrt{d}$ more than once. This is not severe really, but the estimate of when one has reached d_E could be spoofed by this when there is noise in the data. Also we comment that the example of the Ikeda map we present here shows that their suggested connection between d_E and d_A is not correct in general. Choosing neighborhood sizes and asking when points stop moving out of those neighborhoods, and this is the essence of the BLPC algorithm, entails additional work and computing cost over our procedure using nearest neighbors only. This is because points are likely to appear in many neighborhoods, especially when the neighborhood size is not small, as is the case in several of the BLPC examples.

Finally the simplicity of the computations required for our implementation of the false-neighbor idea coupled with the ability to estimate (quantitatively and directly) the error implied by various choices of embedding dimensions should allow the use of our procedures in a large variety of settings. Indeed, we have successfully applied this method to experimental data [30] from several sources, and we have been able to determine, with no trouble and quite efficiently, minimum embedding dimensions even in the presence of contamination by noise.

ACKNOWLEDGMENTS

We thank the members of INLS and the members of the UCSD–MIT–Lockheed-Sanders Nonlinear Signal Processing Group for numerous discussions on this subject; the comments and assistance of Steve Isabelle, Doug Mook, Cory Myers, Al Oppenheim, and J. J. Sidorowich were especially helpful. The data on the Mackey-Glass system were provided to us by Sidorowich. This work was supported in part under the DARPA-University Research Initiative, URI Contract No. N00014-86-K-0758, in part by the U.S. Department of Energy, Office of Basic Energy Sciences, Division of Engineering and Geosciences, under Contract No. DE-FG03-90ER14138.

---

[1] J.-P. Eckmann and D. Ruelle, Rev. Mod Phys. **57**, 617 (1985).

[2] T. S. Parker and L. O. Chua, *Practical Numerical Algorithms for Chaotic Systems* (Springer-Verlag, New York, 1990).

[3] P. S. Landa and M. Rosenblum, Physica D **48**, 232 (1991).

[4] R. Mañé, in *Dynamical Systems and Turbulence, Warwick, 1980,* edited by D. Rand and L. S. Young, Lecture Notes in Mathematics 898 (Springer, Berlin, 1981), p. 366.

[5] F. Takens, in *Dynamical Systems and Turbulence, Warwick, 1980,* edited by D. Rand and L. S. Young, Lecture Notes in Mathematics 898 (Springer, Berlin, 1981), p. 366.

[6] A recent revisit to the theorem of Mañé and Takens is found in Tim Sauer, J. A. Yorke, and M. Casdagli (unpublished).

[7] E. N. Lorenz, J. Atmos. Sci. **20**, 130 (1963). We refer to this as the Lorenz-I model.

[8] M. Hénon, Commun. Math. Phys. **50**, 69 (1976).

[9] W. Liebert, K. Pawelzik, and H. G. Schuster, Europhys. Lett. **14**, 521 (1991).

[10] S. Bumeliene, G. Lasiene, K. Pyragas, and A. Cenys, Litov. Fiz. Sbornik **28**, 569 (1988).

[11] K. Pyragas and A. Cenys, Litov. Fiz. Sbornik **27**, 437 (1987).

[12] A. M. Fraser and H. L. Swinney, Phys. Rev. A **33**, 1134 (1986); A. M. Fraser, IEEE Trans. Info. Theory **35**, 245 (1989); A. M. Fraser, Physica D **34**, 391 (1989).

[13] W. Liebert and H. G. Schuster, Phys. Lett. A **142**, 107 (1989). This paper evaluates the average mutual information function of the previous reference by an efficient numerical algorithm.

[14] H. D. I. Abarbanel, R. Brown, and J. B. Kadtke, Phys. Rev. A **138**, 1782 (1990); M. C. Casdagli, Physica D **35**, 335 (1989); J. D. Farmer and J. J. Sidorowich, Phys. Rev. Lett. **59**, 845 (1987).

[15] R. Brown, P. Bryant, and H. D. I. Abarbanel, Phys. Rev. A **43**, 2787 (1991); Phys. Rev. Lett. **65**, 1523 (1990). Also see the review article by H. D. I. Abarbanel, R. Brown, and M. B. Kennel, Int. J. Mod. Phys. B **5**, 1347 (1991).

[16] J.-P. Eckmann, S. O. Kamphorst, D. Ruelle, and S. Ciliberto, Phys. Rev. A **34**, 4971 (1986).

[17] E. N. Lorenz, Tellus **36A**, (1984). We refer to this as the Lorenz-II model.

[18] H. D. I. Abarbanel, R. Brown, and M. B. Kennel, J. Nonlinear Sci. **1**, 175 (1991).

[19] K. Ikeda, Opt. Commun. **30**, 257 (1979).

[20] S. M. Hammel, C. K. R. T. Jones, and J. V. Moloney, J. Opt. Soc. Am B **2**, 552 (1985).

[21] O. E. Rössler, Phys. Lett. A **57**, 397 (1976).

[22] M. Mackey and L. Glass, Science **197**, 287 (1977).

[23] J. H. Friedman, J. L. Bentley, and R. A. Finkel, ACM Trans. Math. Software **3**, 209 (1977).

[24] S. M. Hammel, Phys. Lett. A **148**, 421 (1990).

[25] J. D. Farmer and J. J. Sidorowich, Physica D **47**, 373 (1991).

[26] H. D. I. Abarbanel, R. Brown, S. M. Hammel, P.-F. Marteau, and J. J. Sidorowich (unpublished).

[27] P.-F. Marteau and H. D. I. Abarbanel, J. Nonlinear Sci. **1**, 313 (1991).

[28] E. J. Kostelich and J. A. Yorke, Physica D **41**, 183 (1990).

[28] E. J. Kostelich and J. A. Yorke, Physica D **41**, 183 (1990).

[29] N. Packard *et al.*, Phys. Rev. Lett. **45**, 9 (1980).

[30] We are grateful to T. Carroll and L. Pecora from the U.S Naval Research Laboratory for providing us with their data. The results of this analysis using the tool of false nearest neighbors and other analysis will appear in the near future.

Direct Test for Determinism in a Time Series

Daniel T. Kaplan and Leon Glass
Department of Physiology, McGill University, 3655 Drummond Street, Montreal, Quebec, Canada H3G 1Y6
(Received 8 August 1991)

A direct test for deterministic dynamics can be established by measurement of average directional vectors in a coarse-grained d-dimensional embedding of a time series. Theoretical analysis of the statistical properties of a random time series using the same embedding technique is possible by consideration of classical results concerning random walks in d dimensions. Examples are given to show the clear differences between deterministic dynamics, such as may be generated by chaotic systems, and stochastic dynamics.

PACS numbers: 05.45.+b, 05.40.+j, 87.10.+e

Complex time series are ubiquitous in nature and in man-made systems, and a variety of measures have been proposed to characterize them. For example, power spectra [1] are particularly suitable for analysis of linear systems, where their interpretation is often transparent, whereas the dimension [2,3] and Lyapunov number [4] have been used to study geometrical and temporal properties of chaotic dynamics. However, none of these measures can be readily applied directly to determine if the dynamics are generated by a deterministic, rather than a stochastic, process. Here we propose a novel method to characterize a time series that is directed towards the analysis of whether the time series is generated by a deterministic system. The method is based on the observation that the tangent to the trajectory generated by a deterministic system [5] is a function of position in phase space, and therefore all the tangents to the trajectory in a given region of phase space will have similar orientations.

The method is illustrated in Fig. 1(a). The left-hand panel shows the x component of the deterministic chaotic Lorenz attractor [6] embedded in a two-dimensional phase space with the abscissa given by $x(t)$ and the ordinate by $x(t-\tau)$. This phase plane is coarse grained into a 16×16 grid, and each pass k of the trajectory through a box j of phase space generates a vector of unit length, called the *trajectory vector* $v_{k,j}$, whose direction is determined by the vector between the point where the trajectory enters the box and the point where it leaves the box, that is, the average direction of that pass of the trajectory in the box. The resultant vector from the vector addition of all the passes through each box, normalized by the number of passes n_j through the box, is $V_j=\sum_k v_{k,j}/n_j$. This gives a coarse caricature of the dynamics. In regions of the phase space where the vectors are well aligned in the 2D embedding, the resultant vector is almost of unit length. In the regions of phase space where trajectories cross, the length of the resultant vector is reduced. Now consider the results of performing the same procedure in a 3D phase space with coordinates of $x(t)$, $x(t-\tau)$, $x(t-2\tau)$. The phase space is coarse grained into a $16\times16\times16$ grid, and the resultant vector average V_j is determined and plotted as a projection onto the $x(t)$-$x(t-\tau)$ plane. Now the crossings of the trajectories are resolved and the geometry of the flow on the attractor is well approximated. In contrast, in Fig. 1(b), we consider embedding a synthesized random signal [7] that has an autorcorrelation function $\Psi(T)$ identical to that of the Lorenz attractor. Now there is a spaghetti mess that is not resolved in the 2D or 3D embeddings.

A statistical characterization of the dynamics can be generated by first determining the resultant vector V_j in each box j in phase space. The average value of this sum for a d-dimensional embedding of the dynamics is denoted $\bar{L}_n^d=\langle|V_j|\rangle_{n_j=n}$ where the angular brackets denote an average over all boxes containing n passes of the trajectory. Figure 2 shows $\bar{L}_n^d$ for the data displayed in Fig. 1. The strong correlations between the directions of the vectors in each box for the Lorenz attractor are reflected in the high values of $\bar{L}_n^d$ that are close to the maximal value of 1, for the 3D embedding. In contrast, for the randomized signal, $\bar{L}_n^d$ decreases approximately as $n^{-1/2}$.

In general, we wish to embed the signal in d dimen-

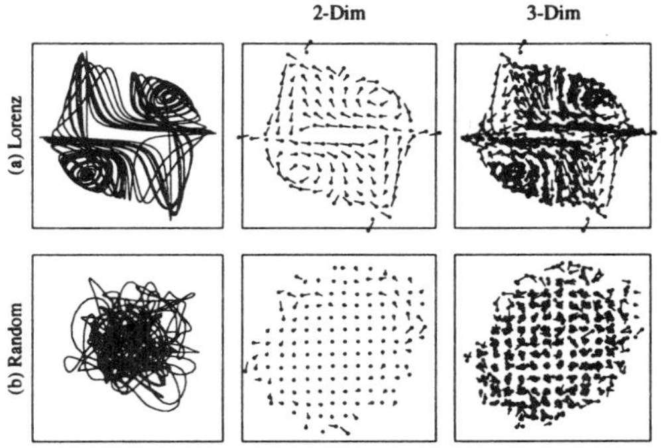

FIG. 1. An embedding plot $x(t-\tau)$ vs $x(t)$ and coarse-grained flow averages for the x component of Lorenz equations and for a random signal with the identical power spectrum. Embedding lag $\tau=0.75$. The length of the arrows shows the alignment of the trajectory vectors in the corresponding box, and the direction shows the direction of mean flow. Scales: (a) $[-20,20]$ and (b) $[-30,30]$. The signals shown in the trajectory plots cover 50 time units; the other plots cover 655 time units.

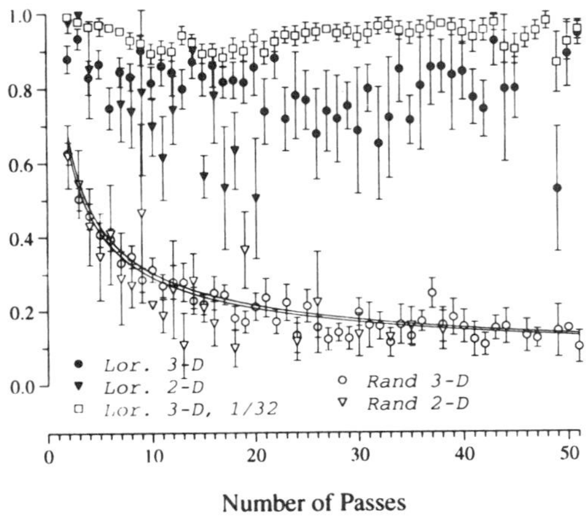

FIG. 2. $\bar{L}_n^d$ vs n for $d=2,3$ for cases shown in Fig. 1 for a signal length of 655 time units and gridding of 16^d. Error bars show the standard error in the estimate of the mean. The lines show the theoretical values computed from Eq. (1) for random flights in 2D and 3D. The open squares represent a 3D embedding with grid 32^3. $\tau=0.75$.

sions and to develop statistical criteria to distinguish deterministic dynamics from dynamics that involve stochastic processes. To carry out the statistical analysis, we consider random walks in d dimensions [8-11]. A random walk consists of n steps of unit length in d dimensions where the angle from each step to the next is chosen randomly. In this Letter we consider the average displacement per step, $\bar{R}_n^d$, which, for large n, is [11]

$$\bar{R}_n^d=\left(\frac{2}{nd}\right)^{1/2}\frac{\Gamma((d+1)/2)}{\Gamma(d/2)}, \tag{1}$$

where Γ is the gamma function. Thus, $\bar{R}_n^d=c_d n^{-1/2}$, where c_d equals $\pi^{1/2}/2$, $4/(6\pi)^{1/2}$, $3\pi^{1/2}/32^{1/2}$ for $d=2,3,4$, respectively, and $\lim_{d\to\infty}c_d=1$. Although Eq. (1) is valid only for the asymptotic limit when n is large, it agrees with the analytical values for $n=2$ to within about 3% [12]. In Fig. 2, the solid curves show the theoretical values of Eq. (1) for $d=2$ (lower curve) and $d=3$ (upper curve), in close agreement with the randomized time series.

To analyze a time series it is necessary to set three parameters: the time lag τ of the embedding, the number of dimensions d of the embedding, and the edge length ϵ of a box. For a system with an m-dimensional attractor, a necessary condition to embed the attractor is $d\geq m$, although hints of deterministic structure may be seen at smaller d. As ϵ decreases, longer data sets must be used in order to generate multiple passes through individual boxes of phase space. For high-dimensional attractors, the necessity of long data sets and the memory size of computers provide practical limitations on the utility of the method.

Choosing the value of the time lag in the embedding is a subtle issue that has been considered in the related question of the determination of the dimension of attractors. In this earlier work, suggestions for the time lag include a fraction of the first zero of the autocorrelation function [13] and the minimum of the mutual information function [14]. In the present case, there is an interplay between the autocorrelation function $\Psi(T)$ and the mean direction in each box. For Gaussian random processes [7] the structure of the coarse-grained embedded vector field can be approximated in terms of $\Psi(T)$. We consider the covariance matrix [15] of the $2d$-dimensional vector

$$\xi=[x(t),x(t-\tau),\ldots,x(t-(d-1)\tau),\quad \Delta x(t),\Delta x(t-\tau),\ldots,\Delta x(t-(d-1)\tau)],$$

where $\Delta x(t)=x(t+b)-x(t)$ and $b>0$ is the time scale for passage through a box. The first d components of ξ describe position in the phase space; the second d components describe the direction of trajectories through that position. The covariance matrix is written $\langle\xi^T\xi\rangle$, where the angular brackets denote the average over time, and can be written in terms of $\Psi(T)$. The off-diagonal elements

$$\langle\Delta x(t-j\tau)x(t-k\tau)\rangle=\Psi((j-k)\tau-b)-\Psi((j-k)\tau), \tag{2}$$

$$\langle\Delta x(t-j\tau)\Delta x(t-k\tau)\rangle=2\Psi((j-k)\tau)-\Psi((j-k)\tau-b)-\Psi((j-k)\tau+b) \tag{3}$$

correspond to the first and second finite-difference derivatives of $\Psi(T)$ evaluated at multiples of τ. When they are zero, the directional elements of ξ are independent of each other and the positional elements of ξ. Although $\langle\xi^T\xi\rangle$ provides an accurate description of the statistical dependence between the components of ξ for a Gaussian random process, chaotic and other nonlinear systems have higher-order correlations that are not included in this formulation.

In order to show how the structure of the vector field changes with τ, it is convenient to construct a single number that summarizes the set of $\bar{L}_n^d$. One way of doing this is to construct a weighted average of V_j over all the occupied boxes

$$\bar{\Lambda}=\left\langle\frac{(V_j)^2-(R_{n_j}^d)^2}{1-(R_{n_j}^d)^2}\right\rangle. \tag{4}$$

For a deterministic system $\bar{\Lambda}=1$, while for a random walk $\bar{\Lambda}=0$.

In the Lorenz system (Fig. 3), $\bar{\Lambda}$ falls off slowly with τ—this is due to sensitive dependence on initial conditions destroying the deterministic connection between the

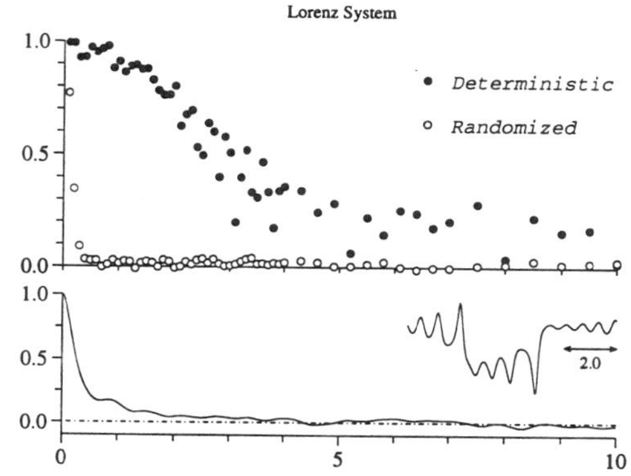

FIG. 3. $\bar{\Lambda}$ vs τ for a gridding of 32^3 for the signals of Fig. 1; $\Psi(\tau)$.

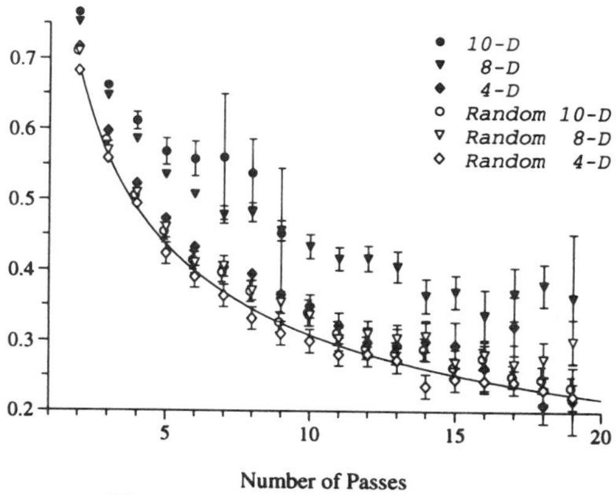

FIG. 4. $\bar{L}_n^d$ vs n for $d=4,8,10$ for $x(t)$ from Eq. (5), and for a random signal with the same $\Psi(T)$. $\tau=75$, near the first zero crossing of $\Psi(T)$. A gridding of 4^d was employed for $d=8,10$ and of 16^4 for $d=4$.

elements of ξ for large τ. For the randomized signal, $\bar{\Lambda}$ falls off similarly to $\Psi(T)$ [16].

We now consider the application of this technique to a high-dimensional system. Previous work on delay-differential equations has shown that, for long time delays, attractors of high dimension are found in the delay equation

$$\frac{dx}{dt}=\frac{ax(t-\delta)}{1+x(t-\delta)^c}-bx(t) \tag{5}$$

that has been proposed as a model for nonlinear feedback control in physiology [17]. For $a=0.2$, $b=0.1$, $c=10$, and $\delta=100$, Eq. (5) has an attractor whose estimated dimension is ~ 7.5 [3].

Figure 4 shows $\bar{L}_n^d$ for $x(t)$ and for a random time series with the same $\Psi(T)$, each of length 1.32×10^5 time units. The box edge length ϵ needs to be large to ensure that many boxes are traversed more than once by the trajectory. The coarse resolution causes $\bar{L}_n^d<1$ for the deterministic system, but even so, $\bar{L}_n^{10}>\bar{L}_n^8>\bar{L}_n^4$, indicating that the higher-dimensional embeddings are untangling the trajectory. In contrast, for the random time series, $\bar{L}_n^{10}\approx\bar{L}_n^8$. At the low resolution of coarse graining used for $d=8,10$, almost all the boxes are at the extremes of the trajectory cloud, where there is a directional bias towards the center even for the random signal. This causes the slight difference between $\bar{R}_n^d$ and the random signal's $\bar{L}_n^d$. Nonetheless, the $\bar{L}_n^d$ for the deterministic system are clearly above those for the randomized time series. This remains true when random noise is added to $x(t)$, even for signal-to-noise ratios as poor as 20 dB (for $d=10$). The statistic $\bar{\Lambda}$ shows a complex dependence on τ due to the delays contained in Eq. (5), but $\bar{\Lambda}$ for the deterministic system stays above $\bar{\Lambda}$ for the random system for τ as large as 200 [18].

The method sketched out in this Letter is directed at establishing whether a time series is generated by a deterministic system. Although in principle $\bar{L}_n^d=1$ in a deterministic system, this is observed only as $\epsilon\to 0$. This limit on ϵ requires infinitely long data sets and would be defeated by even small amounts of measurement noise. In a practical situation with finite ϵ, two comparisons can be made: $\bar{L}_n^d$ can be compared to $\bar{R}_n^d$ to establish whether there is any evidence for a deterministic mechanism, and the values of $\bar{L}_n^d$ can be compared to those generated from $\langle\xi^T\xi\rangle$ to establish whether the determinism indicated by $\bar{L}_n^d$ goes beyond that which would be found in a randomly forced linear dynamical system. For the linear system, appropriate choice of τ and ϵ causes $\bar{L}_n^d\to\bar{R}_n^d$, so the two comparisons become equivalent. The fact that the method provides intrinsic estimates of the error in its statistic $\bar{L}_n^d$, coupled with its ability to analyze high-dimensional systems, should make the method useful in the analysis of complex dynamics from diverse sources.

We thank Hiroyuki Ito, Steven Skates, and Michael Marder for useful discussions. This research is supported by grants from the Natural Sciences and Engineering Research Council of Canada, the Heart and Stroke Foundation of Quebec, and les Fonds des Recherches en Santé du Québec. D.T.K. is a Postdoctoral Fellow of the North American Society of Pacing and Electrophysiology.

[1] E. O. Brigham, *The Fast Fourier Transform* (Prentice Hall, Englewood Cliffs, NJ, 1974).
[2] J. D. Farmer, Physica (Amsterdam) **4D**, 366 (1982).
[3] P. Grassberger and I. Procaccia, Phys. Rev. Lett. **50**, 346 (1983); Physica (Amsterdam) **9D**, 189 (1983).
[4] A. Wolf, J. B. Swift, H. L. Swinney, and J. A. Vastano, Physica (Amsterdam) **16D**, 285 (1985).
[5] F. Takens, in *Dynamical Systems and Turbulence,*

Warwick, 1980, edited by D. A. Rand and L. S. Young (Springer-Verlag, Berlin, 1981), p. 366.

[6] E. N. Lorenz, J. Atmos. Sci. **20**, 130 (1963). Parameters: $r=28$, $b=\frac{8}{3}$, $\sigma=10$.

[7] The randomized time series are generated by randomizing the phases of the discrete Fourier transform of the original signal. The resulting signals are examples of Gaussian random processes equivalent to passing Gaussian random noise through a linear filter. D. T. Kaplan and R. J. Cohen, Circ. Res. **67**, 886 (1990); J. Theiler, B. Galdrikian, A. Longtin, S. Eubank, and J. D. Farmer (to be published).

[8] Lord Rayleigh, Philos. Mag. **37**, 321 (1919).

[9] S. Chandrasekhar, Rev. Mod. Phys. **15**, 1–89 (1943).

[10] K. V. Mardia, *Statistics of Directional Data* (Academic, London, 1972); K. V. Mardia, J. Appl. Stat. **15**, 115 (1988).

[11] W. Feller, *An Introduction to Probability Theory and Its Applications, Volume II* (Wiley, New York, 1966). This result follows from the density function on p. 255 and is in accord with results in Refs. [8,9] for lower dimensions.

[12] Computations not shown give $\bar{R}_2^d = 2^{d-2}\Gamma(d/2)\Gamma(d/2)/\pi^{1/2}\Gamma(d-\frac{1}{2})$. This equals $2/\pi$, $2/3$, $32/15\pi$ for $d=2,3,4$, respectively. $\bar{R}_2^d$ approaches $1/\sqrt{2}$ for $d\rightarrow\infty$ in accord with earlier calculations: J. M. Hammersley, Ann. Math. Stat. **21**, 447 (1950); R. D. Lord, Ann. Math. Stat. **25**, 794 (1954).

[13] A. I. Mees, P. E. Rapp, and L. S. Jennings, Phys. Rev. A **36**, 340 (1987); A. M. Albano, J. Muench, and C. Schwartz, Phys. Rev. A **38**, 3017 (1988).

[14] A. M. Fraser and H. L. Swinney, Phys. Rev. A **33**, 1134 (1986).

[15] Feller, Ref. [11], pp. 80–87.

[16] The dependence of $\bar{\Lambda}$ on τ is more complex in other situations. For example, dynamical systems such as the Rossler system or a lightly damped harmonic oscillator have $\Psi(T)$ that oscillate with an envelope that falls off only slowly to zero. Random signals with such $\Psi(T)$ show $\bar{L}_n^d \rightarrow \bar{R}_n^d$ when $\Psi(\tau)\rightarrow 0$ and when τ is near a maximum or minimum of $\Psi(T)$, particularly for small ϵ.

[17] M. C. Mackey and L. Glass, Science **197**, 287 (1977); L. Glass and M. C. Mackey, Ann. N.Y. Acad. Sci. **316**, 214 (1979).

[18] All calculations reported here were performed on a SUN SparcStation IPC with 8 Mb RAM. Details on the sorting algorithms that enable analysis of high-dimensional systems will be presented elsewhere. A typical run with a time series of 250000 points and 2^{20} boxes takes ~ 100 CPU seconds.

CHAPTER 7

Dimension Calculations

A.M. Albano, J. Muench, C. Schwartz, A.I. Mees, P.E. Rapp
Singular-value decomposition and the Grassberger-Procaccia algorithm
Phys. Rev. A **38**, 3017 (1988).

J.-P. Eckmann, D. Ruelle
Fundamental limitations for estimating dimensions and Lyapunov exponents in dynamical systems
Physica D **56**, 185 (1992).

M. Ding, C. Grebogi, E. Ott, T. Sauer, J.A. Yorke
Plateau onset for correlation dimension: When does it occur?
Phys. Rev. Lett. **70**, 3872 (1993).

J. Theiler, S. Eubank, A. Longtin, B. Galdrakian, J.D. Farmer
Testing for nonlinearity in time series: The method of surrogate data
Physica D **58**, 77 (1992).

A. Brandstater, H.L. Swinney
Strange attractors in weakly turbulent Couette-Taylor flow
Phys. Rev. A **35**, 2207 (1987).

J. Guckenheimer, G. Buzyna
Dimension measurements for geostrophic turbulence
Phys. Rev. Lett. **51**, 1438 (1983).

Editors' Notes

Several authors have written on the calculation of dimension from experimental data. Much of the research has centered on reliable computation of the correlation dimension as suggested by Grassberger and Procaccia (1983).

Albano et al. (1988) present results on how the singular-value decomposition (cf. Broomhead and King (1986) in Chapter 6) can be used for the measurement of correlation dimension.

One of the problems encountered in experimental measurement of dimension by embedding is that there often is only a limited amount of available data. Blind use of the Grassberger-Procaccia algorithm for finding correlation dimension can lead to spurious results that are an artifact of the finiteness of the available data series. The question of when the data series is long enough is addressed in the paper of **Eckmann and Ruelle** (1992) and in Essex and Nerenberg (1991). **Ding et al.** (1993) also address the question of the effect of limited data, but from the point of view of how it affects the behavior of the dimension measurement as a function of the dimensionality of the delay coordinate reconstruction space.

Theiler et al. (1992) present a useful empirical method for validating dimension measurements by comparison with the result of applying the same dimension measurement algorithm to surrogate data. By "surrogate data" the authors mean data than is not deterministic but has been generated so that it has certain chosen statistical properties (e.g., the frequency power spectrum) in common with the original experimental data. This method can be used as a possible means to validate not only dimension measurements but also any of the usual invariants computed from data.

An excellent example of the determination of dimension in a physical system is provided by the papers of **Brandstater and Swinney** (1987), who provide a careful and instructive analysis of Couette-Taylor flow data. Another example is the paper of **Guckenheimer and Buzyna** (1983), also on a fluid system. This latter paper uses an embedding space consisting of several *simultaneous* measurements of the fluid temperature at different spatial positions. This is in contrast with much of the subsequent experimental work which uses delay coordinates from a single measured time series. It is our feeling that simultaneous measurements will often give superior results, and should be used, if available.

Further reading

An excellent general introduction to the issues involved in dimension calculation is:

THEILER, J., Estimating fractal dimensions. J. Opt. Soc. Am. **A7**, 1055 (1990).

Singular-value decomposition and the Grassberger-Procaccia algorithm

A. M. Albano, J. Muench,* and C. Schwartz†
Department of Physics, Bryn Mawr College, Bryn Mawr, Pennsylvania 19010

A. I. Mees
Department of Mathematics, University of Western Australia, Nedlands, Perth, Western Australia, Australia 6009

P. E. Rapp
Department of Physiology and Biochemistry, The Medical College of Pennsylvania, Philadelphia, Pennsylvania 19129
(Received 14 January 1988)

A singular-value decomposition leads to a set of statistically independent variables which are used in the Grassberger-Procaccia algorithm to calculate the correlation dimension of an attractor from a scalar time series. This combination alleviates some of the difficulties associated with each technique when used alone, and can significantly reduce the computational cost of estimating correlation dimensions from a time series.

I. INTRODUCTION

The correlation dimension[1,2] has become the most widely used measure of chaotic behavior.[3] It has been used in the analysis of hydrodynamic experiments,[4,5] laser systems,[6] neutron-star luminosities,[7] neuronal[8] and electroencephalographic[9] signals, business cycles,[10] etc.

One of the reasons for this popularity is the relative ease with which it can be calculated from a scalar time series. The calculation typically starts by reconstructing the system's trajectory in an "embedding space" using the method of "lags" or "time delays."[11,12] The dimension of the reconstructed trajectory is then calculated using, more often than not, an algorithm due to Grassberger and Procaccia,[13] although there are also widely used ones due to Termonia and Alexandrovitch[14] and Badii and Politi.[15] It has been known for some time, however, that while it may be easy to calculate an apparently precise value for a dimension, getting a reliably *accurate* one is often quite something else.[16,17]

Broomhead and King[18] have suggested that singular-value decompositions[19] may provide an alternative way of estimating dimensions. This procedure identifies orthogonal directions in the embedding space which may be ordered according to the magnitude of the variance of the trajectory's projection on them. The ordering is done using the singular values of the embedding. The number of these directions "visited" by the reconstructed trajectory, and indicated by large singular values, is an estimate of the dimension of the smallest space that contains the trajectory. Mees, Rapp, and Jennings,[20] however, point out that the number of large singular values may depend on the details of the embedding and the accuracy of the data as much as they do on the dynamics of the system. Mere counting of large singular values may, therefore, not give a reliable estimate of dimension. However, in subsequent papers Broomhead, Jones, and King[21] and Mees and Rapp[22] have shown that it is possible to estimate the dimension of a state-space manifold by using an appropriately modified procedure.

In this paper, we show that combining singular-value decompositions and the Grassberger-Procaccia algorithm can alleviate some of the ambiguities associated with each technique when used alone. A singular-value decomposition leads to a statistically independent set of variables spanning the embedding space. While the numerical examples shown here are all done with the Grassberger-Procaccia algorithm, singular-value decomposition will apply equally well to other methods of estimating dimension.

In Sec. II we briefly review the embedding procedure and the Grassberger-Procaccia algorithm, incorporating details that will be needed in a subsequent analysis as well as some refinements in the implementation which have been introduced since the original formulation of the algorithm. In Sec. III we review singular-value decompositions and discuss the question of statistical independence. We also establish a connection between the covariance matrix of the embedding and the autocorrelation function of the time series which can serve as a basis for setting limits on the "window length" of the embedding—i.e., the time spanned by each embedding vector.

Section IV presents some computational evidence that for a given time series, the singular-value spectrum depends principally on the window length, corroborating results obtained earlier by Caputo, Malraison, and Atten,[23] in connection with correlation dimensions. We conclude with some suggestions for an interactive procedure combining singular-value decompositions and the Grassberger-Procaccia algorithm.

II. EMBEDDING AND THE GRASSBERGER-PROCACCIA ALGORITHM

A. Embedding

Let a continuous signal $v(t)$ be measured at equal "sampling intervals," T_s, to yield a time series,

$$[v(k) \mid k=1,2,\ldots,N_T]\ , \qquad (2.1)$$

where $v(k)$ is an abbreviation for $v(kT_s)$. We presume $v(t)$ to be one of n state variables which completely describe a dynamical system, the trajectory of which lies on a d-dimensional ($d \le n$) attractor X in the system's phase space. Generalizations to multivariable time series are relatively straightforward. For simplicity of presentation they will not be described here. If the system's temporal evolution is chaotic, X is a strange attractor characterized by a noninteger dimension. For the applications in which we are interested, n is unknown and $v(t)$ is the only measured quantity.

Packard *et al.*[11] and Takens[12] have shown that starting from the time series (2.1), one may "embed" or reconstruct the trajectory in an M-dimensional "embedding space" by means of the vectors,

$$\begin{aligned}
&\mathbf{y}(1)=(v(1),\, v(1+L),\ldots,\, v(1+(M-1)L))\ ,\\
&\mathbf{y}(2)=(v(1+J),\, v(1+J+L),\ldots,\, v(1+J+(M-1)L))\ ,\\
&\quad \ldots,\\
&\mathbf{y}(p)=(v(1+(p-1)J),\, v(1+(p-1)J+L),\ldots,\, v(1+(p-1)J+(M-1)L))\ ,\\
&\quad \ldots\ .
\end{aligned} \qquad (2.2)$$

Here, L is the "lag," or the number of sampling intervals between successive components of an embedding vector, and J is the number of sampling intervals between the first components of successive vectors.[24] The time $(M-1)L$, spanned by each embedding vector, is the "window length" of the embedding.[18] L is introduced to allow for the fact that, in an experiment, the sampling interval is often set without accurate prior knowledge of the time scales intrinsic to the system being studied. Determining an appropriate value of L then becomes part of the analysis. We return to this point in Secs. III and IV. The number J (or a set of J's) describe how the time series is sampled to create a set of embedding vectors with a computationally manageable size from a possibly large data set.[25]

Theorems due to Takens[12] and Mañé[26] state that if the embedding dimension M and the dimension n of a manifold containing the attractor satisfy the inequality ("Takens criterion")

$$M \ge 2n+1\ , \qquad (2.3)$$

then given the assumption of an infinite amount of noise-free data one has a proper embedding except when the system has special symmetries. In particular, the dimension of the embedded or reconstructed attractor is the same as that of the system's phase space attractor.[26] In practice these requirements are not met. However, experimental[6] and computational[17] studies with progressively increased data sets suggest that good approximations to the dimension can often be obtained with fairly small amounts of data. For example, it is possible to estimate the dimension of the Hénon attractor to within 6% of its literature value with 500 data points.[6,17]

B. Grassberger-Procaccia

In an M-dimensional embedding, the "correlation integral," $C_M(r)$ is defined as the fraction of the distances between embedding vectors that do not exceed r:

$$C_M(r)=(1/N_p)\sum_{\{i,j\}} \Theta(r-|\mathbf{y}(i)-\mathbf{y}(j)|)\ , \qquad (2.4)$$

where $\Theta(\cdot)$ is the Heaviside unit-step function. To avoid artificial correlates due to measurements being taken at nearly the same time,[27,28] the sum should be taken only over embedding vectors which are not too closely spaced in time. That is, a number K is chosen and the sum is taken over those i's and j's for which $|i-j|>K$. In the following, we will take $|\mathbf{y}(i)-\mathbf{y}(j)|$ to be the Euclidean distance between $\mathbf{y}(i)$ and $\mathbf{y}(j)$.[29] N_p is the number of distances used in the sum. Grassberger and Procaccia[13] show that the correlation dimension D_2 is given by

$$D_2=\lim_{M\to\infty}\lim_{r\to 0} D_2(M;r)\ , \qquad (2.5)$$

where $D_2(M;r)$ is the slope of the log-log plot of $C_M(r)$ versus r:

$$D_2(M;r)=d[\log_e C_M(r)]/d[\log_e(r)]\ . \qquad (2.6)$$

In calculations involving actual experimental data, neither of the limits in Eq. (2.5) can be taken. Small values of r are blurred by noise and by limitations on experimental accuracy, while large values of M are precluded by practical limitations on data set sizes and computing times. In practice, for a given value of M, one looks for a "plateau" in the plot of $D_2(M;r)$ versus $\log_e(r)$—that is, a range, r_L to r_U (the "scaling region"), over which $D_2(M;r)$ has a constant value $d(r_L,r_U)$, say, to within some tolerance, $\pm\Delta d$. If this plateau is common to a number of embedding dimensions exceeding $[2d(r_L,r_U)+1]$, then it is taken to be the correlation dimension for the range of lengths (r_L,r_U).[16]

Although seemingly quite simple, this prescription needs to be applied with some care. Figure 1 shows the slope $D_2(M;r)$ versus $\log_e(r)$ for the Lorenz attractor,

$$(dx/dt,\, dy/dt,\, dz/dt) = (-10(x-y),\, x(28-z)-y,\, xy-(\tfrac{8}{3})z)\ . \qquad (2.7)$$

The calculations used 1000 embedding vectors formed from a data set of 10 000 values of x with $T_S=0.01$ and $L=5$. For each embedding dimension, a value of J was chosen so as to pick embedding vectors uniformly from the entire data set. The plots are for embedding dimensions of (*a*) 5, (*b*) 10, (*c*) 15, (*d*) 20, and (*e*) 25. The

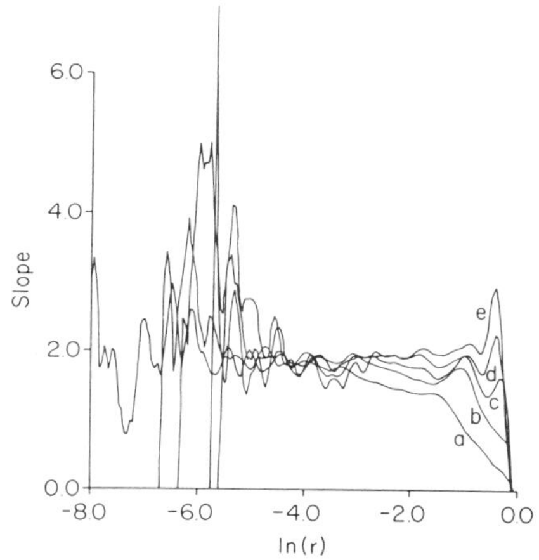

FIG. 1. Plot of slope $=d[\log_e C_M(r)]/d \log_e(r)$ for data generated by the Lorenz equations with parameter values $T_S=0.01$, $L=5$. The five curves correspond to (*a*) $M=5$, (*b*) $M=10$, (*c*) $M=15$, (*d*) $M=20$, (*e*) $M=25$.

length scale r is normalized to *the largest interpoint distance in each embedding.* This is done in order to use a length scale derived from the size of the attractor. With this normalization, the correlation integral $C_M(r)$ [Eq. (2.4)] reaches its maximum value ("saturates") at the same value of r making it easier to compare plateaus for different embedding dimensions. [This normalization, however, makes the plots of $\log_e C_M(r)$ versus $\log_e r$ inappropriate for calculating the entropy.] In previous work (see, e.g., Refs. 6 and 9), such comparisons were facilitated by plotting $D_2(M;r)$ versus $\log_e C_M(r)$.

For small r's [$\log_e(r) < -5$] as well as for large r's [$\log_e(r) > -2$], the slopes generally increase with increasing embedding dimension. The behavior at small r is characteristic of noise which has infinite dimension.[6,13,23,26] At small distances, one has experimental noise, either inherent in the system or contributed by measuring instruments, as well as "computational noise," arising from round-off errors. At large r's, the increase in the slope is induced in part by using a relatively small number of embedding vectors in a high-dimensional space.[23] It becomes more pronounced and extends to smaller values of r as M is increased. Thus in the process of increasing the embedding dimension to check the convergence of the calculation, one causes the scaling region to shrink, making it difficult to ascertain where the convergence does occur. Any procedure that can legitimately reduce the required size of the embedding dimension will therefore be helpful. The singular value decomposition does this.

III. SINGULAR-VALUE DECOMPOSITION AND AUTOCORRELATION

A. Singular-value decomposition

It is well known[19] that any $N \times M$ matrix $\underline{A}$, with $N \geq M$, may be expressed as

$$\underline{A} = \underline{VSU}^T , \tag{3.1}$$

where $\underline{U}^T$ means the transpose of $\underline{U}$. $\underline{S}$ is an $M \times M$ diagonal matrix,

$$S_{i,j} = \delta_{i,j} s(i), \quad i=1,2,\ldots,M \tag{3.2}$$

$\underline{V}$ is an $N \times M$ matrix with orthogonal columns,

$$(\underline{V}^T\underline{V})_{i,j} = \delta_{i,j} , \tag{3.3}$$

and U is an $M \times M$ orthogonal matrix,

$$(\underline{U}^T\underline{U})_{i,j} = (\underline{UU}^T)_{i,j} = \delta_{i,j} . \tag{3.4}$$

The elements $s(i)$ of the diagonal matrix $\underline{S}$ are known as the "singular values" of A.

B. Covariance and statistical independence

To relate the embedding process to singular-value decompositions, we follow Broomhead and King[18] and consider the case when $\underline{A}$ is the "trajectory matrix"—that is, up to a normalization, the matrix whose rows are the M-dimensional embedding vectors formed from the time series (2.1):

$$\underline{A} = \frac{1}{\sqrt{N}} \begin{bmatrix} \mathbf{y}(1) \\ \mathbf{y}(2) \\ \vdots \\ \mathbf{y}(N) \end{bmatrix} . \tag{3.5}$$

The embedding defines a set of points in an M-dimensional space which may be described by a multivariate distribution whose variables are the M components of the embedding vectors, Eq. (2.2). We may rewrite these as

$$\mathbf{y}(k) = (y_1(k), y_2(k), \ldots, y_i(k), \ldots, y_M(k)) .$$

From the definitions of the trajectory matrix $\underline{A}$ [Eq. (3.5)] and of the covariance matrix $\mu_{i,j}$ of this multivariate distribution[30] we have

$$\mu_{i,j} = (1/N)\Sigma_k y_i(k) y_j(k) = (\underline{A}^T\underline{A})_{i,j} . \tag{3.6}$$

If each column of $\underline{A}$ has zero mean $\Sigma_k y_i(k)=0$ (which we presume hereafter), the off-diagonal elements of the covariance matrix are the (unnormalized) "correlation coefficients" of the distribution. The correlation coefficients measure the statistical dependence of the variables y_i; $i=1,2,\ldots,M$, on each other, or the redundancy of the information which they contain. Variables with vanishing correlation coefficients are, by definition, statistically independent.[30]

The transformation

$$\underline{A} \rightarrow \underline{A}' = \underline{AU} , \tag{3.7}$$

or

$$y_i \rightarrow y_i' = \Sigma_j y_j U_{ji} , \tag{3.8}$$

diagonalizes the covariance matrix:[18,20]

$$\underline{U}^T(\underline{A}^T\underline{A})\underline{U} = \underline{S}^2 . \tag{3.9}$$

The squares of the singular values $[s(i)]^2$ are the eigenvalues of $\underline{A}^T\underline{A}$, while the columns of $\underline{U}$ are its eigenvectors.

Since $\underline{U}$ is orthogonal, its eigenvectors form an orthonormal basis for the embedding space. The directions of the eigenvectors are called the "principal axes." In addition, the Euclidean distances $|\mathbf{y}(i)-\mathbf{y}(j)|$ used in the calculation of the correlation integral Eq. (2.4) are invariant under this transformation. The correlation dimension itself, therefore, is also invariant. $\underline{A}'$ is called the "matrix of principal components" of $\underline{A}$, y_i'' is the ith "principal component" of y', or its projection along the ith principal axis.

It also follows from the above that the eigenvalue $[s(i)]^2$ is the variance of the ith principal component. If, for some j, $[s(j)]^2=0$, then the reconstructed trajectory does not "visit" principal axis j. In the absence of noise, this means that the rank of the covariance matrix, which is equal to the number of its nonzero eigenvalues, is the dimension of the smallest subspace of the embedding space that contains the reconstructed trajectory.[18] Noise prevents any eigenvalue from vanishing, so that the dimension is estimated by counting the number of "large" eigenvalues.[18,20]

Although a singular-value decomposition is mathematically equivalent to a diagonalization of the covariance matrix, the former turns out to be more robust numerically. In this work, we use a version of the code developed by Mees, Rapp, and Jennings[20] implementing the Golub-Reinsch algorithm.[31]

C. Autocorrelation

Using Eqs. (2.2) and (3.5), we may write the k,i element of the trajectory matrix as

$$A_{k,i}=(\mathbf{y}(k))_i=v(1+(i-1)J+(k-1)L)\ , \qquad (3.10)$$

so that the (i,j)th element of the covariance matrix $\underline{A}^T\underline{A}$ is

$$(\underline{A}^T\underline{A})_{i,j}=(1/N)\sum_{p_i} v(p_i)v(p_i+(j-i))\ , \qquad (3.11)$$

where $p_i=1+(k-1)J+(i-1)L$. In the limit that the number N of embedding vectors is large, $(\underline{A}^T\underline{A})_{i,j}$ becomes proportional to the value at $t=(j-i)L$ of the signal's autocorrelation function,

$$g(t)=[\Sigma_k v(kL)v(kL+t)]/\{\Sigma_k[v(kL)]^2\}\ . \qquad (3.12)$$

That is,

$$\lim_{N\to\infty}(\underline{A}^T\underline{A})_{i,j}=\sigma^2 g[(j-i)LT_S]\ , \qquad (3.13)$$

where σ^2 is the variance of the time series.

This intimate relationship between the covariance matrix and the autocorrelation function helps to illuminate the critical importance of the window length $(M-1)L$ in the embedding and in subsequent calculations. By definition, $g(0)=1$, and for decorrelated signals, $g(t)\to 0$ for $t \gg t_c$, where t_c is a time interval characteristic of the decay of the autocorrelation. Thus, for these signals, in the limit of large windows, where $(M-1)L \gg t_c$, the covariance matrix approaches the unit matrix, which has M degenerate unit eigenvalues. This means that regardless of the choice of M, the rank of the matrix is always M. If we took the rank of the covariance matrix to be an estimate of the trajectory's dimension, we would be led to conclude that we are observing noise. On the other hand, in the limit of very small windows $(M-1)L \ll t_c$, all the elements of the covariance matrix approach 1. In the extreme case when all elements are equal to 1, the matrix has one nonzero eigenvalue equal to M and $(M-1)$ zero eigenvalues corresponding to the case when the trajectory is projected onto a line segment on the space's main diagonal, $y_1=y_2=\cdots=y_M$.

These effects are evident in Fig. 2 which shows the normalized eigenvalues for 16-dimensional embeddings of the Lorenz attractor with lags $L=1,2,\ldots,9$. The graphs with the larger eigenvalues correspond to larger windows. It does not seem possible to deduce from these graphs that these eigenvalues pertain to an attractor with a correlation dimension that is slightly greater than 2.

IV. GRASSBERGER AND PROCACCIA IN ROTATED EMBEDDING SPACE

A. Testing convergence

The process of checking the convergence of a Grassberger-Procaccia calculation in embedding spaces of relatively high dimensions can obscure the very convergence that is being tested. Counting the number of "large" singular values of a trajectory matrix does not give a reliable estimate of the dimension of the embedded trajectory. However, since the principal components of the embedding vectors form a statistically independent set of variables and since the relative contributions of these variables to the distances used in the calculation of the correlation dimension are directly measured by the eigenvalues, there are obvious advantages to combining these two techniques.

The combination proceeds as follows: For a given embedding, a singular-value decomposition is performed, yielding the matrix of singular values S, and the orthogonal matrix U of Eq. (3.1). The trajectory matrix is rotated [Eq. (3.7)] to get the matrix of principal components which is then used in a Grassberger-Procaccia calculation.

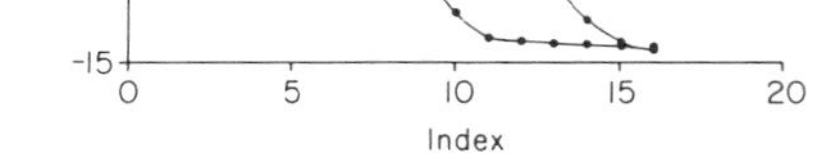

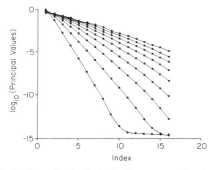

FIG. 2. Normalized principal-value spectrum for data generated from the Lorenz equations. The diagram shows normalized principal values $\log_{10}(s_j^2/\Sigma s_i^2)$ as a function of index j. The nine spectra correspond, from bottom to top, to lag$=1,2,3,\ldots,9$. The embedding dimension is 16 in each case. One thousand vectors were used in each calculation.

The convergence of the calculation may be checked by performing the calculations in a series of subspaces of the embedding space spanned by principal axes with the largest singular values. The procedure is considered to have converged if the calculations yield similar results for a few of the largest subspaces. Actually, this check need not always be done explicitly. A few calculations are usually enough to establish how large a singular value must be in order for its principal components to introduce a measurable difference in the dimension calculation.

To illustrate this we show in Fig. 3 some results for the Lorenz attractor Eq. (2.7). To compare the effects of principal components characterized by singular values that span several orders of magnitude, we have chosen a nine-dimensional embedding with lag, $L=1$. This gives a window length that would be too small for a "good" dimension calculation (see Sec. V) but does give a range of eigenvalues spanning 15 orders of magnitude. The figure consists of four superimposed graphs of slope versus $\log_e(r)$ obtained by using principal components corresponding to the largest 2, 3, 4, and 9 eigenvalues, that is, 2, 3, 4, and 9 columns of the principal component matrix $\underline{AU}$ were used in the calculations. The *normalized eigenvalues* $[s(i)]^2/\{\sum_{j=1,M}[s(j)]^2\}$ are 0.98, 1.62×10^{-2}, 1.69×10^{-4}, 1.48×10^{-6}, 1.65×10^{-8}, 7.46×10^{-11}, 4.35×10^{-13}, 9.25×10^{-15}, and 4.37×10^{-15}. In this case, inclusion of principal components with normalized singular values less than 10^{-4} has a negligible effect on the results of the calculation. This was found to be true for other embeddings of the Lorenz attractor as well as for the Rössler attractor [Eq. (4.1)] and a 3-torus [Eq. (4.2)]. In the case of Fig. 3, the calculations clearly agree for $\log_e(r) > -3$ when only the two largest principal components are used.

Again we explicitly note that this diagram does not demonstrate a successful calculation of the dimension of the Lorenz attractor. Because $L=1$, there is no plateau in the derivative of the correlation integral that can be used to estimate the dimension. The issue of an appropriate lag is addressed presently. The purpose of this calculation is to demonstrate that, given an appropriate singular-value spectrum, only a few columns of the principal component matrix are required to estimate accurately the correlation integral. Using the Golub-Reinsch algorithm, singular-value decompositions can be rapidly computed. For example, the double precision decomposition of a 2000×10 matrix can be performed in approximately 84 sec on an 8-MHz personal computer with a floating point coprocessor. Singular-value decomposition is not a major computational cost because the computational costs of dimension-estimation procedures are mainly in the calculation of the correlation integral. A correlation integral calculation using 1000 ten-dimensional vectors takes some 15 min on the same computer. Because the time required to calculate these integrals increases rapidly with embedding dimension, the results summarized in Fig. 3 are of considerable practical significance.

A comparison of Figs. 1 and 3 displays quite vividly the advantage of implementing the algorithm in terms of the principal components. One can check for convergence without introducing noiselike behavior at large distances.

B. Window length

Results obtained by Broomhead and King[18] and by Mees, Rapp, and Jennings[20] suggest that for a given time series, the normalized singular values may depend only on the window length $(M-1)L$, and not separately on the embedding dimension M, or the lag L. Figure 4 shows a semilog plot of the normalized eigenvalues versus the index for a number of embeddings with window lengths that are nearly equal. The values of embedding

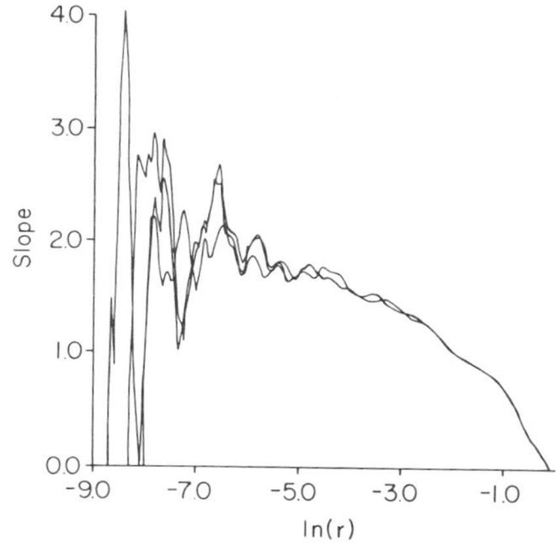

FIG. 3. Plot of slope vs $\log_e(r)$ for data generated by the Lorenz equations with parameter values $T_s=0.01$ and $L=1$. Matrix A was formed by 1000 vectors embedded in a nine-dimensional space. Correlation integrals and the corresponding derivatives shown here were calculated using the first 2, 3, 4, and 9 columns of the rotated principal component matrix $A\cdot U$.

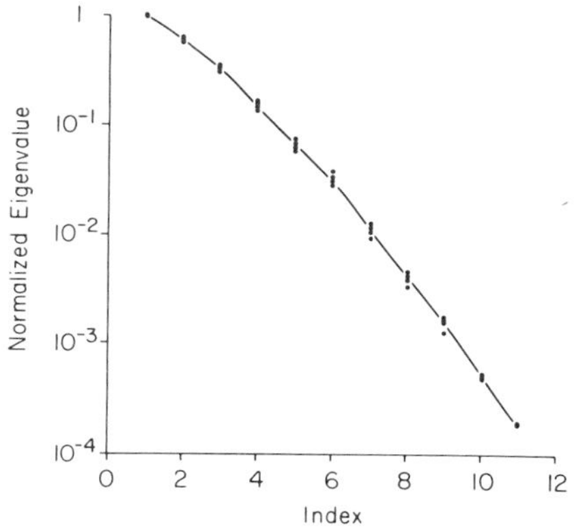

FIG. 4. Normalized principal values $\log_{10}(s_j^2/\Sigma s_i^2)$ plotted as a function of index j for data generated by the Lorenz equations. In each case, matrix A contained 1000 rows. The values of embedding dimension M and lag L are $(M,L)=(5,24)$, (6,19), (7,16), (8,14), (9,12), (10,11), and (11,10). The solid line connects the eigenvalues for the 11-dimensional embedding.

dimension and lag (M,L) used were (5,24), (6,19), (7,16), (8,14), (9,12), (10,11), and (11,10). The eigenvalues are seen to fall essentially on a single curve. The solid lines connect eigenvalues for the 11-dimensional embedding with a lag of 10.

This dependence on window length rather than separately on embedding dimension and lag is also displayed by results of correlation dimension calculations. On the basis of extensive calculations on the Lorenz attractor, Caputo *et al.* suggest that the local slope obeys a "universal law" that depends only on the window length.[23] Figure 5 shows plots of the slope of $\log_e[C_M(r)]$ versus $\log_e(r)$ for the Lorenz attractor, Eq. (2.7), for $(M,L)=(5,24)$, (8,14), (9,12), and (11,10). The thick, dark curve is a superposition of results from the three largest subspaces of the 8-, 9-, and 11-dimensional embeddings, while the light curve is from the 5-dimensional embedding using all five components. Note that embedding dimension 5 does not satisfy Taken's criterion (2.3) for a proper embedding. Nevertheless, in the scaling region $-3<\log_e(r)<-1$, the results for five dimensions do not differ from the rest by more than 10%. Of course many n-dimensional manifolds (and attractors contained in them) can be embedded in fewer than $2n+1$ dimensions, and indeed what we have here is an embedding of $\mathbb{R}^3$ in $\mathbb{R}^5$. Figures 1 and 5 show the relative effects of varying the embedding dimension for fixed lag and of varying both lag and embedding dimension, while keeping the window length approximately constant. It could be argued that the results in Fig. 5 might reflect a behavior unique to the Lorenz equations. Additional computations on other systems suggest that this is not the case.

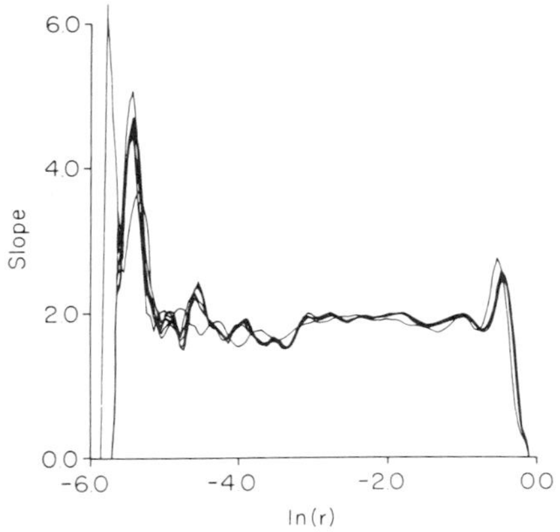

FIG. 5. Plot of slope vs $\log_e(r)$ for data generated by the Lorenz equations. In each case, the trajectory matrix A had 1000 rows. The curves correspond to embedding dimension M and lag L of $(M,L)=(5,24)$, (8,14), (9,12), and (11,10). The thick dark curve results from the superposition of three calculations each for $M=8$, 9, and 11. In each embedding dimension M, the first $M-2$, $M-1$, and M columns, respectively, of the rotated matrix $A\cdot U$ were used in the dimension calculation. The thin curve corresponds to the five-dimensional embedding in which all five columns were used in the dimension calculation.

Figure 6 is similar to Fig. 5, but for the Rössler attractor,

$$(dx/dt,\, dy/dt,\, dz/dt)$$
$$=(-y-z,\, x+0.2y,\, 0.4+xz-5.7z)\,, \quad (4.1)$$

for $(M,L)=(6,13)$, (7,11), (9,8). Figure 7 shows results for the 3-torus,

$$x(k)=\sin(\omega_0 k)+\sin(2^{3/2}\omega_0 k)+\sin(3^{3/2}\omega_0 k)\,, \quad (4.2)$$

with $\omega_0=6.0\times10^{-3}$ and for $(M,L)=(7,20)$, (9,15), (10,13), and (11,12). All cases show remarkable stability as both lag and embedding dimension are changed while the window length remains approximately constant.

Upon reflection, the dependence of the correlation integral on window length rather than on lag or embedding dimension should have been expected. Similar window lengths compare similar segments of a trajectory. Changes of lag within the window simply give different discrete approximations of the correlation integral.

V. WHAT IS A GOOD WINDOW?

In Sec. III B we saw that the window length used in the embedding crucially affects the singular-value spectrum. It also affects the outcome of attempts to estimate the attractor's dimension. The importance of choosing an appropriate window was pointed out by Froehling *et al.*[32] who proposed using 10% of the "folding time" which was defined as "the average time between "folding" of adjacent sheets of the attractor.[32] Unfortunately this criterion is difficult to implement in a systematic way for all attractors. Aspects of the importance of window

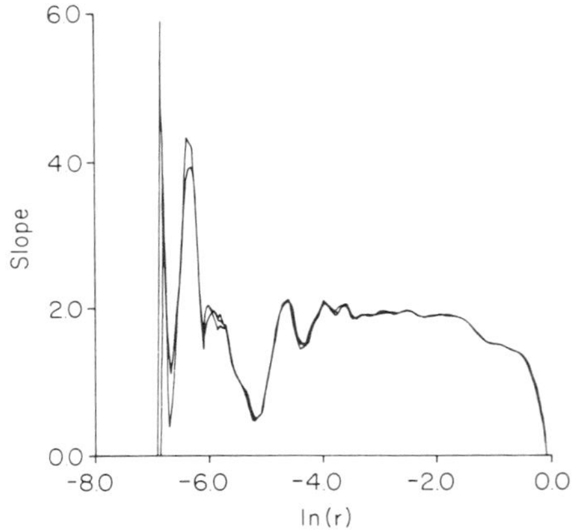

FIG. 6. Plot of slope vs $\log_e(r)$ for data generated by the Rössler equations ($T_S=0.05$). In each case matrix A consisted of 1000 rows. The curves correspond to embedding dimension M and lag L of $(M,L)=(6,13)$, (7,11), and (9,8).

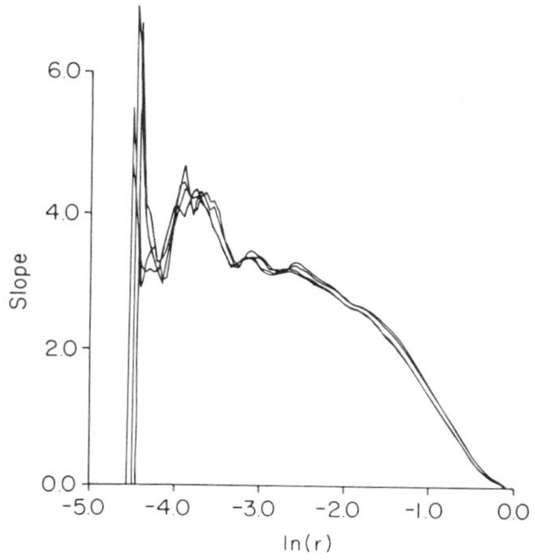

FIG. 7. Plot of slope vs $\log_e(r)$ for the 3-torus Eq. (4.2). In each case matrix A consisted of 1000 rows. The curves correspond to embedding dimension M and lag L of $(M,L)=(7,20)$, (9,15), (10,13), and (11,12).

value to dimension estimates using the Grassberger-Procaccia algorithm are shown in Figs. 1 and 3. The various graphs in Fig. 1 are results of calculations using different window lengths. There is so much variability between graphs that one cannot tell if the calculations are converging to a limit. Figure 3, on the other hand, shows that the calculations agree, but it does not display a scaling region, and so does not yield a value for the correlation dimension. Problems such as these have been known for some time and there have been a number of efforts to define an optimum window.[17]

The first issue to address is why there should be an optimal window. It is clear that there is a lower bound to the window length. If the window is too short, noise dominates. Up to a point, we want the largest window possible. The question is, why should we have to stop? Why not use giant windows? The answer to this question is not to be found in the Grassberger-Procaccia algorithm, but rather in the Takens's theorem[12] that produces the theoretical foundations for the calculation. Takens's theorem states that

$$(v(J),\ v(J+L),\ v(J+2L),\ldots,\ v(J+(M-1)L))$$

is an embedding. However, if the system decorrelates, as it will in chaotic cases, we must ensure that, as required by the theorem, all of these points are on the same trajectory by ensuring that $(M-1)L$ is not too big. Data generated by numerical integration or in noisy experiments do not present a trajectory but a perturbation of a trajectory. This puts an upper bound on an acceptable value of $(M-1)L$. Effective window length is thus bounded from below and from above. The present object of our investigation is to locate the optimal window within this range.

The relationship between the covariance matrix and the autocorrelation function of the signal [Eq. (3.13)] suggests that a time-scale characteristic of the autocorrelation function is an appropriate time scale to use. In previous work we have used the first zero of the autocorrelation function,[33] but subsequent calculations on data from simulations as well as from experiments show that the correlation time, which is defined as the time required for the autocorrelation function to decrease to $1/e$ of its original value, is more robust. For band-limited data, Broomhead and King use the inverse of the band-limiting frequency.[18] While straightforward in principle, numerical estimation of the band-limiting frequency poses some practical problems.

In slightly different terms but in similar context, Caputo, Malraison, and Atten[23] use a number of the order of the first return time, while Fraser and Swinney[34] use the first minimum of the mutual information function. We have performed a series of calculations of mutual information for these three example systems that are analogous to those presented here of the autocorrelation function. In summary, the mutual information function did not prove to be a reliable indicator of optimal window length. Additionally, significant technical problems are associated with numerical estimation of mutual information that are very nearly as complex as those encountered in dimension calculations themselves. A systematic account of these results is given in Martinerie *et al.*[35]

To test if an "optimum" window length does exist and, if it exists, whether it is related to a time scale characteristic of the correlation function, we performed a series of calculations using a variety of windows. The analysis was performed on the data used for Figs. 5–7. These were the Lorenz attractor [Eq. (2.7)], the Rössler attractor [Eq. (4.1)], and a 3-torus [Eq. (4.2)]. The results of the calculations are shown in Figs. 8–10, respectively, and are summarized in Table I.

For our present purposes, the optimum window is

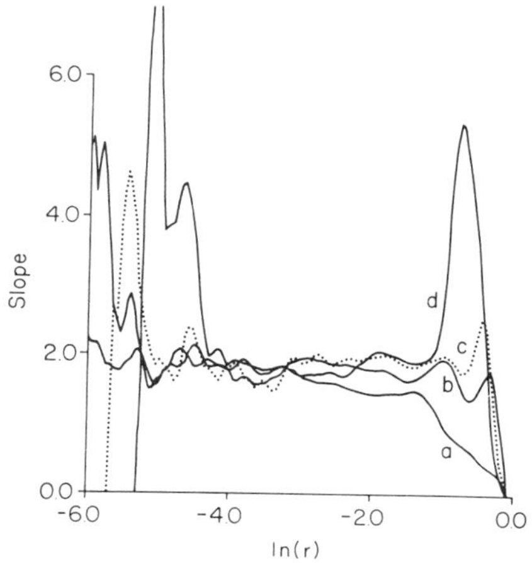

FIG. 8. Plot of slope vs $\log_e(r)$ for data generated by the Lorenz equations. In each case matrix A consisted of 1000 rows. The curves correspond to embedding dimension M and lag L of (a) $(M,L)=(5,6)$, (b) (8,10), (c) (11,10), and (d) (9,30).

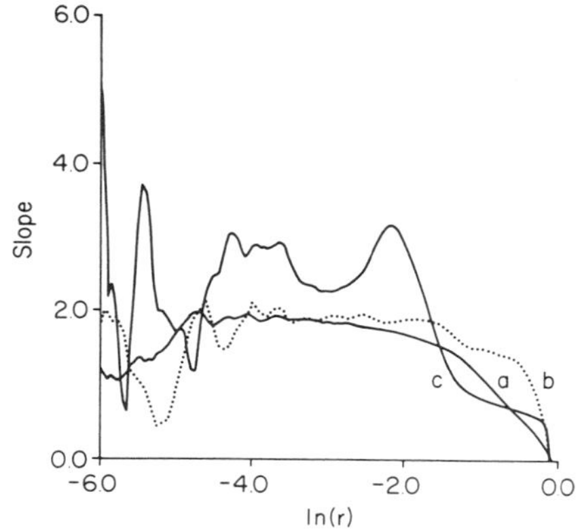

FIG. 9. Plot of slope vs $\log_e(r)$ for data generated by the Rössler equations. In each case matrix A consisted of 1000 rows. The curves correspond to embedding dimension M and lag L of (a) $(M,L)=(9,3)$, (b) $(7,11)$, (c) $(9,12)$.

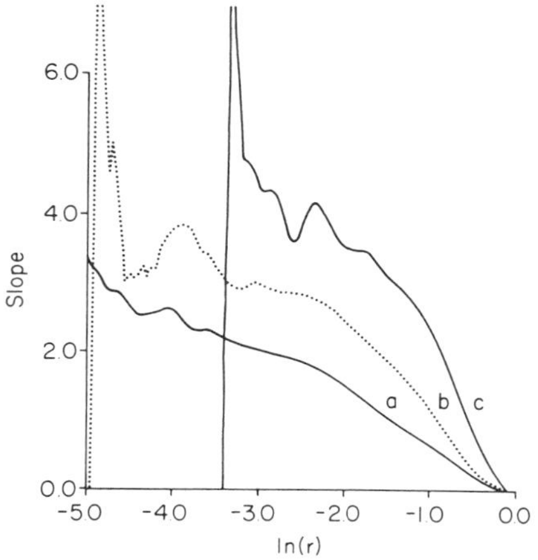

FIG. 10. Plot of slope vs $\log_e(r)$ for the 3-torus. In each case matrix A consisted of 1000 rows. The curves correspond to embedding dimension M and lag L of (a) $(M,L)=(9,5)$, (b) $(11,12)$, (c) $(10,20)$.

merely that which gives the broadest plateau at the attractor's known correlation dimension. For each window, the time series is embedded in an M-dimensional space that satisfies Takens's criterion (2.3). A singular-value decomposition is performed on the trajectory matrix, and the transformation Eq. (3.7) is done. This is followed by a Grassberger-Procaccia calculation in a subspace of the embedding space spanned by principal axes with normalized eigenvalues exceeding 1.0×10^{-4}.

Figure 8 shows results for the Lorenz attractor using a data set with a correlation time of $30T_S$. The four graphs are labeled according to the window length used in the calculation: (a) $24T_S$, (b) $70T_S$, (c) $100T_S$, (d) $240T_S$. Curves (a) and (b) corresponding to the smaller window lengths show low values of the slope for large values of r $[\log_e(r)>-3.5]$ and oscillate as the small, noise-dominated length scales are reached. On the other hand, curve (d), with a window length that is eight times the correlation time, shows an anomalously large peak at large values of r $[\log_e(r)>-1.0]$, while becoming as erratic as the curves for small windows at the small length scales. Curve (c), with a window length three times the correlation time, displays a plateau in the region $-3.0<\log_e(r)<-0.6$, at a slope of 2.0 ± 0.1.

Figure 9 shows similar results for the Rössler attractor, for a data set with a correlation time of $23T_S$. Curve (a) is for a window of $24T_S$, (b) for $66T_S$, and (c) for $96T_S$. Figure 10 is for a 3-torus with a correlation time of $62\tau_S$, curves (a), (b), and (c) corresponding, respectively, to window lengths of $40T_S$, $120T_S$, and $180T_S$.

The results of comparisons of window length and the autocorrelation function are summarized in Table I. In all three cases, windows of the order of the correlation time or smaller result in small or nonexistent plateaus with values of the slope that are smaller than the attractor's dimension except in noise-dominated regions at small distance scales. On the other hand, windows of the order of four times the correlation time or greater result in similarly small or nonexistent plateaus with values of the slope that are larger than the attractor's dimension. In all three cases, the broadest plateaus, at values within 10% of the attractor's dimension, are obtained for windows that are between 1.6 and 3.5 times the correlation time. In these numerical experiments, $T_{\min}$ tracks T_{window} slightly better than T_{AC}. However, as previously mentioned, numerical difficulties are frequently encountered in estimating $T_{\min}$. T_{AC} estimates are much more robust against variations in size and quality of the data

TABLE I. Comparison of window length with the autocorrelation function. T_{AC} is the autocorrelation time, T_{zero} is the time of the first zero crossing of the autocorrelation function, and $T_{\min}$ is the time of the first local minimum of the autocorrelation function. T_{window} is the window length of those tested that results in the broadest plateau at the attractor's known dimension. All times are reported in units of sample interval T_S.

	T_{AC}	T_{zero}	$T_{\min}$	T_{window}	T_{window}/T_{AC}
Lorenz	30	122	72	100	3.33
Rössler	23	32	60	66	2.87
3-torus	62	90	122	100	1.61

set. For this reason, the present discussion focuses on T_{AC}.

These results indicate that though the correlation time is an imperfect predictor of the optimal window, the rate at which a signal decorrelates is certainly a determinant of optimal window. That this should be so was anticipated by our previously outlined examination of Takens's theorem. The Lyapunov exponents, particularly the system's largest Lyapunov exponent, provide a quantitative measure of the average exponential divergence of nearby trajectories in the phase space of a dynamical system.[36] We hypothesize that if the product (window$\cdot L_{max}$), where L_{max} is the largest Lyapunov exponent, is too large, the signal decorrelates and the scaling region is lost. The smaller the maximum Lyapunov exponent, the larger we can take the window. Tests of this hypothesis using a numerical procedure for estimating Lyapunov exponents constructed by Wolf *et al.*[37] are now underway.

While these admittedly anecdotal results confirm the overall dependence of the results of the calculations on the window length, they do not give a precise prescription that would allow determination of an "optimum" in terms of the signal's correlation time. They do, however, suggest a relatively narrow range over which the window length may be varied to maximize the plateau.

We propose the following interactive procedure for combining a singular-value decomposition and the Grassberger-Procaccia algorithm.

- Choose an embedding dimension M and a lag L so that the window $(M-1)L$ is a few times larger than the correlation time.
- Perform a singular-value decomposition and rotate the embedding space using the matrix $\underline{U}$ of eigenvectors of the covariance matrix.
- Perform Grassberger-Procaccia calculations in subspaces of the rotated space spanned by principal axes with eigenvalues exceeding 10^{-4}. If this results in a dimension that does not satisfy Takens' criterion, increase M until it does. (In practice, 10^{-4} is not invariably an appropriate criterion. However, it cuts off dimensions that contribute no more than 0.01% to the variance of the multivariate data. If the quality of the data used and the precision of the calculations warrant it, more or less stringent criteria may be used.)
- Vary the window length to maximize the plateau.

VI. FURTHER PROBLEMS AND POSSIBILITIES

Establishing r_L and r_U, the lower and upper bounds of the derivative plateau, presents computational problems that are not readily resolved. In calculations using noisy experimental data, the plateau boundaries may not be obvious. Numerical experimentation demonstrates that even small variations in r_L and r_U can have significant effects on the dimension estimate. Caswell and Yorke[16] have suggested including explicit values of r_L and r_U in dimension calculations; $D_c\,(r_L,r_U)$ would be reported. In many experimental investigations, the absolute value of dimension is not of interest, rather the change in dimension in response to changes in experimentally controlled parameters (Reynolds number, temperature, cognitive activity) may be of interest. In this case, the importance of systematic errors associated with plateau boundary estimates can be reduced by reporting the ratio of dimensions estimated with common values of r_L and r_U. In some cases no single number can accurately approximate the derivative of the correlation dimension. In these instances, producing the Slope versus $\log_e(r)$ plot is perhaps the only legitimate means of presenting the results of these calculations.

The results presented here have addressed the question of window length. The question of epoch length has not been considered. If N_T data points are collected at a uniform sample interval T_s, then the epoch length, the duration of the record, is given by $T_E=T_s(N_T-1)$. Selection of an appropriate epoch length is a much more difficult problem than selecting a window because its resolution turns on questions of signal stationarity. This problem has already been encountered in classical signal analysis, and the generalization of a classical technique used in the characterization of nonstationary signals may be applicable here. The Wigner transform[38] estimates the instantaneous energy for a given time and frequency. Others have shown[7,39] that it is possible to generalize the concept of dimension to construct a continuous dimension spectrum. We have proposed[17] that it should be possible to construct an analog to the Wigner transform that can be applied successfully to the dimension spectra of nonstationary signals.

ACKNOWLEDGMENTS

We thank N. B. Abraham for his numerous illuminating comments and suggestions. The implementation of the Golub-Reinsch singular-value decomposition algorithm we used is due to L. S. Jennings, to whom our profuse thanks are extended. This work was supported in part by Contract No. 22919-PH with the U.S. Army Research Office (A.M.A.), Grant No. NS19716 from the U.S. National Institutes of Health (P.E.R.), and a Helena Rubinstein Foundation Grant (C.S.). A.I.M. thanks King's College, University of Cambridge for support and hospitality.

*Present address: Department of Oceanography, University of Washington, Seattle, WA 98109.

†Present address: Bell Communications Research, 331 Newman Springs Road, Red Bank, NJ 07701-7020.

1. H. G. E. Hentschel and I. Procaccia, Phys. Rev. Lett. **50**, 346 (1983).
2. J. D. Farmer, E. Ott, and J. A. Yorke, Physica **7D**, 153 (1983).
3. See, e.g., *Dimensions and Entropies in Chaotic Systems*, edited by G. Mayer-Kress (Springer, Berlin, 1986).
4. B. Malraison, P. Atten, P. Bergé, and M. Dubois, J. Phys. **50**,

897 (1983).

[5]A. Brandstäter, J. Swift, H. L. Swinney, A. Wolf, J. D. Farmer, E. Jen, and P. J. Crutchfield, Phys. Rev. Lett. **51**, 1442 (1983).

[6]See, e.g., A. M. Albano, J. Aboundadi, T. H. Chyba, C. E. Searle, S. Yong, R. S. Gioggia, and N. B. Abraham, J. Opt. Soc. Am. B **2**, 102 (1985); G. P. Puccioni, A. Poggi, W. Gadomski, F. T. Arecchi, and J. R. Tredicce, Phys. Rev. A **34**, 2073 (1986); N. B. Abraham, A. M. Albano, T. H. Chyba, L. M. Hoffer, M. F. H. Tarroja, S. P. Adams, and R. S. Gioggia, in *Instabilities and Chaos in Quantum Optics,* edited by F. T. Arecchi and R. G. Harrison (Springer, Berlin, 1987).

[7]W. Voges, H. Atmanspacher, and H. Scheingraber, Astrophys. J. **320**, 794 (1987); H. Atmanspacher, H. Scheingraber, and W. Voges, Phys. Rev. A **37**, 1314 (1988).

[8]P. E. Rapp, I. D. Zimmerman, A. M. Albano, G. C. de Guzman, and N. N. Greenbaun, Phys. Lett. **110A**, 335 (1985); G. Mpitsos, in *Dynamic Patterns of Complex Systems,* edited by J. A. S. Kelso, A. J. Mandell, and M. F. Schlesinger (World Scientific, Singapore, 1988).

[9]A. M. Albano, N. B. Abraham, G. C. de Guzman, M. F. H. Tarroja, D. K. Bandy, R. S. Gioggia, P. E. Rapp, I. D. Zimmerman, N. N. Greenbaun, and T. R. Bashore, in Ref. 3, p. 231; P. E. Rapp, I. D. Zimmerman, A. M. Albano, G. C. de Guman, N. N. Greenbaun, and T. R. Bashore, in *Nonlinear Oscillations in Chemistry and Biology,* edited by H. G. Othmer (Springer, Berlin, 1987).

[10]W. A. Brock and C. L. Sayres (unpublished).

[11]N. H. Packard, J. P. Crutchfield, J. D. Farmer, and R. S. Shaw, Phys. Rev. Lett. **45**, 712 (1980).

[12]F. Takens, in *Dynamical Systems of Turbulence,* edited by D. A. Rand and L.-S. Young (Springer, Berlin, 1981).

[13]P. Grassberger and I. Procaccia, Physica **9D**, 189 (1983); Phys. Rev. Lett. **50**, 349 (1983); Phys. Rev. A **28**, 2591 (1983); Physica **13D**, 34 (1984).

[14]Y. Termonia and Z. Alexandrovitch, Phys. Rev. Lett. **45**, 1265 (1983).

[15]R. Badii and A. Politi, J. Stat. Phys. **40**, 725 (1985).

[16]W. E. Caswell and J. A. Yorke, in Ref. 3, p. 123.

[17]A. M. Albano, A. I. Mees, G. C. de Guzman, and P. E. Rapp, in *Chaotic Biological Systems,* edited by A. V. Holden (Pergamon, Oxford, 1987).

[18]D. S. Broomhead and G. P. King, Physica **20D**, 217 (1986); in *Nonlinear Phenomena and Chaos,* edited by S. Sarkar (Hilger, Bristol, 1986).

[19]G. H. Golub and C. F. van Loan, *Matrix Computations* (Johns Hopkins University Press, Baltimore, 1983).

[20]A. I. Mees, P. E. Rapp, and L. S. Jennings, Phys. Rev. A **36**, 340 (1987).

[21]D. S. Broomhead, R. Jones, and G. P. King (unpublished).

[22]A. I. Mees and P. E. Rapp (unpublished).

[23]J. G. Caputo, B. Malraison, and P. Atten, in Ref. 3, p. 180.

[24]The terminology used here is in the spirit of a system of nomenclature discussed at the Conference on the Brain as a Dynamical System, Pecos River Ranch, New Mexico, 1987 (unpublished). It was hoped that a more or less uniform terminology would make it possible, at least for those who use nonlinear dynamics in the study of electrical signals from the brain, to communicate with each other the details of their work with minimal ambiguity. This nomenclature is still being discussed.

[25]In most of the calculations presented here, J is chosen so as to get 10^3 equally spaced embedding vectors from data sets consisting of between 5×10^3 and 10×10^3 points, the limit on the number of embedding vectors being dictated mainly by the fact that most of the calculations were done on microcomputers.

[26]J.-P. Eckmann and D. Ruelle, Rev. Mod. Phys. **57**, 617 (1985).

[27]J. Theiler, Phys. Rev. A **34**, 2427 (1986).

[28]P. Grassberger, Nature **323**, 609 (1986).

[29]Subsequent analysis will require an orthogonal transformation of the embedding space. The Euclidean distance is chosen here to make the dimension calculation invariant under this transformation.

[30]See, e.g., M. Schwartz, *Information, Transmission, Modulation, and Noise* (McGraw-Hill, New York, 1970), p. 375.

[31]G. B. Golub and C. Reinsch, Numer. Math. **14**, 403 (1970).

[32]H. Froehling, J. P. Crutchfield, D. Farmer, and N. H. Packard, Physica **3D**, 605 (1981).

[33]A. M. Albano, L. Smilowitz, P. E. Rapp, G. C. de Guzman, and T. R. Bashore, in *The Physics of Phase Space,* edited by Y. S. Kim and W. W. Zachary (Springer, Berlin, 1987).

[34]A. M. Frazer and H. L. Swinney, Phys. Rev. A **33**, 1135 (1986).

[35]J. Martinerie, A. M. Albano, A. I. Mees, T. R. Bashore, and P. E. Rapp (unpublished).

[36]V. I. Osledec, Trans. Moscow Math. Soc. **19**, 197 (1968); Ya. B. Pesin, Russ. Math. Surveys **2**, 55 (1977).

[37]A. Wolf, J. B. Swift, H. L. Swinney, and J. A. Vastano, Physica **16D**, 285 (1985); related investigations have also been presented by I. Shimada and T. Nagashima, Prog. Theor. Phys. **61**, 1605 (1979); G. Benettin, L. Galgani, A. Giorgilli, and J.-M. Strelcyn, Meccanica **15**, 9 (1980); **15**, 21 (1980).

[38]E. Wigner, Phys. Rev. **40**, 749 (1932); T. A. C. M. Claasen and W. F. G. Mecklenbrauker, Phillips J. Res. **35**, 217 (1980); **35**, 276 (1980); **35**, 372 (1980).

[39]T. C. Halsey *et al.*, Phys. Rev. A **33**, 1141 (1986).

Fundamental limitations for estimating dimensions and Lyapunov exponents in dynamical systems

J.-P. Eckmann[1] and D. Ruelle
I.H.E.S., 91440 Bures-sur-Yvette, France

Received 2 October 1991
Revised manuscript received 3 December 1991
Accepted 14 January 1992
Communicated by P.E. Rapp

We show that values of the correlation dimension estimated over a decade from the Grassberger–Procaccia algorithm cannot exceed the value $2\log_{10} N$ if N is the number of points in the time series. When this bound is saturated it is thus not legitimate to conclude that low dimensional dynamics is present. The estimation of Lyapunov exponents is also discussed.

The purpose of this note is to question the validity of a number of recently published estimates of dimensions of attractors which are based on rather short time series. The values obtained are like 6 or 7, and we shall argue that they are probably a reflection of the small number of data points rather than of the dimension of a hypothetical attractor. Our conclusions go in the same direction as those of Grassberger [1] discussing work of Nicolis and Nicolis [2], and Procaccia [3] discussing work of Tsonis and Elsner [4]. Our analysis is however more precise, and somewhat more optimistic than that of Procaccia (we believe dimension estimates twice as large as those allowed by ref. [3]). See also the more pessimistic bounds of Smith [5].

While it is obvious that a short time series of low precision must lead to spurious results, we wish to argue that – even with good precision data – wrong (too low) dimensions will be obtained. A similar analysis will apply to estimates of Lyapunov exponents.

Let (u_i) be a (scalar) time series with $i = 1, \ldots, N$ (the choice of sampling time unit will be discussed below). Using an embedding dimension m, we first reconstruct a trajectory in $\mathbb{R}^m$, with $x_n = (u_n, u_{n+1}, \ldots, u_{n+m-1})$. (This method, advocated by one of us (DR), was first documented in ref. [6].) Then, according to the Grassberger–Procaccia algorithm (GP) [7], we count the number $\mathcal{N}(r)$ of pairs of points with mutual distance $\leq r$. Note now that $\mathcal{N}(r)$ varies from 0 to $\frac{1}{2}(N-m)(N-m+1) \approx \frac{1}{2}N^2$. The algorithm next calls for plotting $\log \mathcal{N}(r)$ versus $\log r$. For small r, the slope of this plot is an estimate of the correlation dimension d [7]. (For larger r, the plot is not expected to be linear.) Thus, the method assumes

$$\mathcal{N}(r) \approx \text{const.} \times r^d, \tag{1}$$

and, if D is the diameter of the reconstructed

[1]Permanent address: Département de Physique Théorique, Université de Genève, 1211 Geneva 4, Switzerland.

From *Physica D* **56** (1992). Reprinted with permission from Elsevier Science Publishers.

attractor, we should have

$$\eta(D) \approx \tfrac{1}{2}N^2 \tag{2}$$

so that

$$\mathscr{N}(r) \approx \frac{N^2}{2}\left(\frac{r}{D}\right)^d. \tag{3}$$

The determination of the slope of $\log \mathscr{N}(r)$ requires using several values of r, and these should be "small" compared to D. But, obviously, we also need $\mathscr{N}(r)$ large with respect to 1, for statistical reasons. This forces

$$\tfrac{1}{2}N^2\left(\frac{r}{D}\right)^d \gg 1 \quad \text{and} \quad \frac{r}{D} = \rho \ll 1. \tag{4}$$

Taking logarithms, we find the requirement

$$2\log N > d\log(1/\rho). \tag{5}$$

From this it is clear that the GP-algorithm will *not produce dimensions larger than*

$$d_{\max} = \frac{2\log N}{\log(1/\rho)}. \tag{6}$$

Using decimal logarithms, and $\rho = 0.1$, we see that if $N = 1000$, then $d \leq 6$, and if $N = 100000$, then $d \leq 10$. Values of ρ larger than 0.1 might be adequate but, since the method is interesting mainly in very nonlinear situations, this would have to be justified. Thus, if the GP method yields a dimension of 6 for $N = 1000$ points, the result is probably worthless.

In case (u_i) is obtained by discretizing a continuous time signal, we have $N = T/\tau$ where T is the total recording time and τ the sampling time. One can of course try to make N large in (6) by taking τ small, but an easy geometric argument shows that τ should not be so small that consecutive points x_n, x_{n+1} on a reconstructed orbit are closer than the typical distance of points x_n, x_p which are close on the attractor, but for which $|n-p|$ is large.

When can then a dimension estimate be considered reliable? First of all, the GP plots should be displayed and their linearity at small $\log r$ should be verified, as well as equality of slopes for different embedding dimensions. Next, the estimated dimension should be well below the quantity (6), obtained by using an honest value of N (not one artificially boosted by interpolation). A trick introduced by Scheinkman and Le Baron [8] may be of use to check the value of $d_{\max}$ in (6): they perform the GP algorithm both on the original time series (u_i) and on a "scrambled" series obtained by randomly permuting the u_i. The "scrambled" dimension is expected to be approximately equal to $d_{\max}$ and should be well above the "true" dimension.

We now briefly discuss the estimation of Lyapunov exponents. The situation is here somewhat worse than for the dimension. Any method to determine a Lyapunov exponent from an experimental time series requires that near a sequence of points x_n one finds other points x_{n+k} (for some k) so that the rate of divergence of orbits can be estimated. The number of points in a ball of radius r around a point x is

$$\mathscr{N}'(r) \approx \text{const.}' \times r^d, \tag{1$'$}$$

with

$$\mathscr{N}'(D) \approx N \tag{2$'$}$$

so that

$$\mathscr{N}'(r) \approx N\left(\frac{r}{D}\right)^d. \tag{3$'$}$$

(Strictly speaking, the information dimension rather than the correlation dimension should be used here, but the difference is not expected to be significant for present purposes.) As before, this forces

$$N\left(\frac{r}{D}\right)^d \gg 1 \quad \text{and} \quad \rho = \frac{r}{D} \ll 1 \tag{4'}$$

so that we must have

$$\log N > d \log(1/\rho) .$$

This says that the number of points N needed to estimate Lyapunov exponents is about the *square* of that needed to estimate the dimension.

The conclusion of what we have said above is obvious, but worth repeating: to extract useful dynamical information from time series (dimensions, Lyapunov exponents, etc.), *long* time series of high quality are necessary. We hope that the challenge of providing more such time series can be met.

References

[1] P. Grassberger, Nature 323 (1986) 609–612; 326 (1987) 524.

[2] C. Nicolis and G. Nicolis, Nature 326 (1987) 523.

[3] I. Procaccia, Nature 333 (1988) 498–499.

[4] A.A. Tsonis and J.B. Elsner, Nature 333 (1988) 545–547.

[5] L. Smith, Phys. Lett. A 133 (1988) 283.

[6] N.H. Packard, J.P. Crutchfield, J.D. Farmer and R.S. Shaw, Phys. Rev. Lett. 45 (1980) 712–716.

[7] P. Grassberger and I. Procaccia, Phys. Rev. Lett. 58 (1987) 2387–2389.

[8] J.A. Scheinkman and B. Le Baron, Nonlinear dynamics and stock returns, J. Business, to appear.

Plateau Onset for Correlation Dimension: When Does it Occur?

Mingzhou Ding,[(1)] Celso Grebogi,[(2),(3)] Edward Ott,[(2),(4)] Tim Sauer,[(5)] and James A. Yorke[(3)]

[(1)]*Center for Complex Systems and Department of Mathematics, Florida Atlantic University, Boca Raton, Florida 33431*

[(2)]*Laboratory for Plasma Research, University of Maryland, College Park, Maryland 20742*

[(3)]*Department of Mathematics and Institute for Physical Science and Technology, University of Maryland, College Park, Maryland 20742*

[(4)]*Departments of Physics and of Electrical Engineering, University of Maryland, College Park, Maryland 20742*

[(5)]*Department of Mathematical Sciences, George Mason University, Fairfax, Virginia 22030*

(Received 22 February 1993)

Chaotic experimental systems are often investigated using delay coordinates. Estimated values of the correlation dimension in delay coordinate space typically increase with the number of delays and eventually reach a plateau (on which the dimension estimate is relatively constant) whose value is commonly taken as an estimate of the correlation dimension D_2 of the underlying chaotic attractor. We report a rigorous result which implies that, for long enough data sets, the plateau begins when the number of delay coordinates first exceeds D_2. Numerical experiments are presented. We also discuss how lack of sufficient data can produce results that seem to be inconsistent with the theoretical prediction.

PACS numbers: 05.45.+b

The estimation of the correlation dimension [1] of a presumed chaotic time series has been widely used by scientists to assess the nature of a variety of experimental as well as model systems, ranging from simple circuits to chemical reactions to the human brain. It is also known that many factors, such as noise and a lack of data, can hinder the successful application of the dimension extraction algorithm. In this paper, we address two issues related to the understanding of these difficulties, namely, what happens in an ideal situation (i.e., long data string with low noise) and what could be expected when the data set is small. In particular, we focus on the character of the dependence of the estimated correlation dimension on the dimension of the delay coordinate reconstruction space.

Consider an n-dimensional dynamical system that exhibits a chaotic attractor. A correlation integral $C(\epsilon)$ [1] is defined to be the probability that a pair of points chosen randomly on the attractor with respect to the natural measure ρ is separated by a distance less than ϵ on the attractor. The correlation dimension D_2 [1] of the attractor is then defined as $D_2 = \lim_{\epsilon\to 0}\log C(\epsilon)/\log\epsilon$. Assume that we measure and record a trajectory of finite duration L on the attractor at N equally spaced discrete times, $\{\mathbf{x}_i\}_{i=1}^{N}$, where $\mathbf{x}_i \in \mathbf{R}^n$. The correlation integral $C(\epsilon)$ is then approximated by

$$C(N,\epsilon) = \frac{2}{N(N-1)}\sum_{j=1}^{N}\sum_{i=j+1}^{N}\Theta(\epsilon - |\mathbf{x}_i - \mathbf{x}_j|)\,, \quad (1)$$

where $\Theta(x) = 1$ for $x > 0$ and $\Theta(x) = 0$ for $x \le 0$. In the limit $L, N \to \infty$, $C(N,\epsilon) \to C(\epsilon)$.

The dynamical information of a chaotic experimental system is often contained in a time series, $\{y_i = y(t_i)\}_{i=1}^{N}$, obtained by measuring a single scalar function $y = h(\mathbf{x})$ where $\mathbf{x} \in \mathbf{R}^n$ is the original phase space variable. From $\{y_i\}_{i=1}^{N}$ one reconstructs an m-dimensional vector $\mathbf{y}_i$ using the delay coordinates [2,3]

$$\mathbf{y}_i = \{y(t_i), y(t_i - T), \ldots, y(t_i - (m-1)T)\}\,, \quad (2)$$

where $T > 0$ is the delay time and m is the dimension of the reconstruction space. We call the mapping from $\{\mathbf{x}_i\}$ in $\mathbf{R}^n$ to $\{\mathbf{y}_i\}$ in $\mathbf{R}^m$ the "delay coordinate map." Results in Ref. [4] show that, for typical $T > 0$ and $m > 2D_0$, this delay coordinate map is one to one. Here D_0 is the box-counting dimension of the original chaotic attractor.

Our main focus is to estimate correlation dimension from a time series using delay coordinates [Eq. (2)]. As a point of departure for subsequent discussions, we first report a theorem [5,6] which shows that, for estimating the correlation dimension, $m \ge D_2$ suffices. We emphasize that this result holds true irrespective of whether the delay coordinate map is one to one or not. This is contrary to the commonly accepted notion that an embedding (one to one and differentiable) is needed for dimension estimation, leading to the false surmise that m needs to be at least $2D_2+1$ to guarantee a correct dimension estimation (see [7] for further discussion).

Consider an n-dimensional map $G: \mathbf{R}^n \to \mathbf{R}^n$. Let A be an attractor of G in $\mathbf{R}^n$ with a natural probability measure ρ. For a function $h: \mathbf{R}^n \to \mathbf{R}$, define a delay coordinate map $F_h: \mathbf{R}^n \to \mathbf{R}^m$ as

$$F_h(\mathbf{x}) = [h(\mathbf{x}), h(G^{-1}(x)), \ldots, h(G^{-(m-1)}(\mathbf{x}))]\,.$$

The projected image of the attractor A under F_h has an induced natural probability measure $F_h(\rho)$ in $\mathbf{R}^m$. Furthermore, assume that G has only a finite number of periodic points of period less than or equal to m in A. The following result then applies.

Theorem.—If $D_2(\rho) \le m$, then for almost every h, $D_2(F_h(\rho)) = D_2(\rho)$.

The theorem says that the correlation dimension is preserved under the delay coordinate map with $m \ge D_2(\rho)$. Similar results hold for flows generated by ordinary differential equations. "Almost every" in the statement is understood in the sense of prevalence defined in Ref. [4]; roughly speaking, we can regard this "almost every" as meaning that the functions h that do not give the stated result are very scarce and are not expected to occur in practice. The above dimension preservation result also holds for almost all general projection maps meeting the condition in the theorem. To illustrate, consider a closed curve with a uniform measure in $\mathbf{R}^3$. The dimension of this curve is 1. The projected image of this curve onto the plane still has a dimension of 1 but is generally self-intersecting. Thus the map is not one to one but preserves dimension information. One can further project the image to $\mathbf{R}^1$ and obtain an interval, which again has a dimension of 1 but bears little resemblance to the original curve in $\mathbf{R}^3$.

In applications D_2 is commonly extracted from a time series as follows (see Refs. [8–11] for reviews). First, an m-dimensional trajectory is constructed using Eq. (2). Then, the correlation integral $C_m(N,\epsilon)$ is computed according to Eq. (1), where m indicates the dimensionality of the reconstruction space. From the curve $\log C_m(N,\epsilon)$ vs $\log\epsilon$ one then locates a linear scaling region for small ϵ and estimates the slope of the curve over the linear region. This slope, denoted $\bar{D}_2^{(m)}$, is then taken as an estimate of the correlation dimension $D_2^{(m)}$ of the projection of the attractor to the m-dimensional reconstruction space. If these estimates $\bar{D}_2^{(m)}$, plotted as a function of m, appear to reach a plateau for a range of large enough m values, then we denote the plateaued value $\bar{D}_2$ and take $\bar{D}_2$ an estimate of the true correlation dimension D_2 for the system. From the theorem it is clear that the onset of this plateau should ideally start at $m = \text{Ceil}(D_2)$, where $\text{Ceil}(D_2)$, standing for ceiling of D_2, denotes the smallest integer greater than or equal to D_2.

Our original interest in the current problem was motivated by published reports (see Refs. [12–17] for a sample) where $\bar{D}_2^{(m)}$ plateaus at m values that are considerably greater than $\bar{D}_2$. A particular concern is that, when this happens, what does it imply regarding the correctness of the assertion that $\bar{D}_2$ is an estimate of the true correlation dimension D_2 of the underlying chaotic process? In an attempt to answer this question we have obtained new results on the systematic behavior of the correlation integrals. Based on these behaviors we are able to explain how factors such as a lack of sufficient data can produce results, resembling those seen in the experimental reports cited above, which seem to be inconsistent with the theorem. Furthermore, we find that even in cases where the plateau onset of $\bar{D}_2^{(m)}$ occurs at m values considerably greater than $\text{Ceil}(\bar{D}_2)$, there are situations where the plateaued $\bar{D}_2$ is a good estimate of the true correlation dimension D_2. See Refs. [18–25] for other relevant works addressing the issue of short data sets and noise.

To study the numerical aspects of dimension estimation we use chaotic time series generated by the Mackey-Glass equation [26] $dy(t)/dt = ay(t-\tau)/\{1+[y(t-\tau)]^c\} - by(t)$, where we fix $a = 0.2$, $b = 0.1$, $c = 10.0$, and $\tau = 100.0$, and take as the initial condition $y(t) = 0.5$ for $t \in [-\tau, 0]$. The numerical integration of this equation is done by the following iterative scheme [1]:

$$y(t+\delta t) = \frac{2-b\delta t}{2+b\delta t}\,y(t) + \frac{\delta t}{2+b\delta t}\left\{\frac{ay(t-\tau)}{1+[y(t-\tau)]^{10}} + \frac{ay(t-\tau+\delta t)}{1+[y(t-\tau+\delta t)]^{10}}\right\}, \tag{3}$$

where δt is the integration step size. We choose $\delta t = 0.1$. Equation (3) is then a 1000-dimensional map, which, aside from being an approximation to the original equation, is itself a dynamical system. The time series, generated with a sampling time $t_s = 10.0$, are normalized to the unit interval so that the reconstructed attractor lies in the unit hypercube in the reconstruction space. The norm we use to calculate distances in Eq. (1) is the max-norm in which the distance between two points is the largest of all the component differences. To reconstruct the attractor, we follow Eq. (2) and take the delay time to be $T = 20.0$ The dimension of the reconstruction space is varied from $m = 2$ to $m = 25$.

The first time series, used to illustrate the theorem, consists of 50000 points. For each reconstructed attractor at a given m we calculate the correlation integral $C_m(N,\epsilon)$ according to Eq. (1). In Fig. 1 we display $\log_2 C_m(N,\epsilon)$ vs $\log_2\epsilon$ for $m = 2$–8, 11, 15, 19, 23. For each m we identify a scaling region for small ϵ and fit a straight line through the region. The open circles in Fig. 2 show the values of $\bar{D}_2^{(m)}$ so estimated as a function of m. For $m \le 7$, $\bar{D}_2^{(m)} \approx m$. For $m \ge 8$, $\bar{D}_2^{(m)}$ plateaus at $\bar{D}_2$ which has a value of about 7.1. Identifying $\bar{D}_2$ with the true correlation dimension D_2 of the underlying attractor, this result is consistent with the prediction by the theorem that the onset of the plateau occurs at $m = \text{Ceil}(D_2)$.

The second time series, used to illustrate the effect

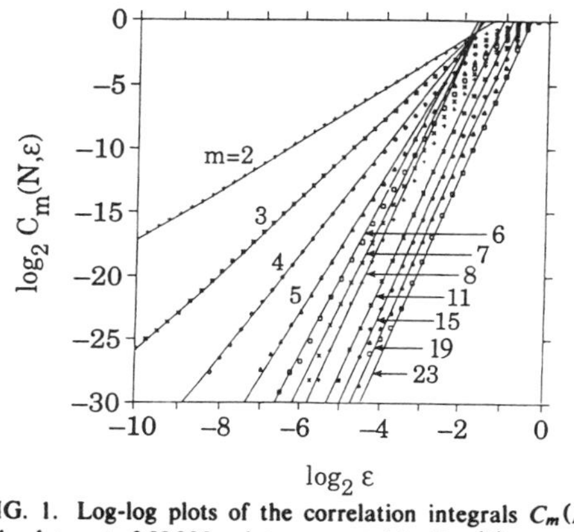

FIG. 1. Log-log plots of the correlation integrals $C_m(N,\epsilon)$ for the data set of 50000 points generated by Eq. (3).

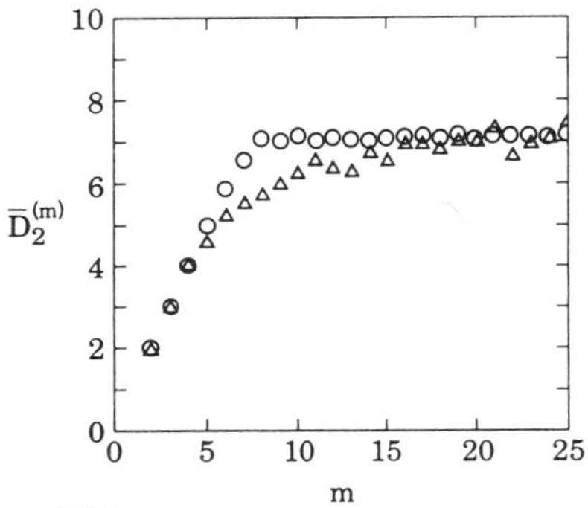

FIG. 2. $\bar{D}_2^{(m)}$ vs m plotted as open circles for the long data set ($N=50000$ and Fig. 1), $\bar{D}_2^{(m)}$ vs m plotted as triangles for the short data set ($N=2000$ and Fig. 3).

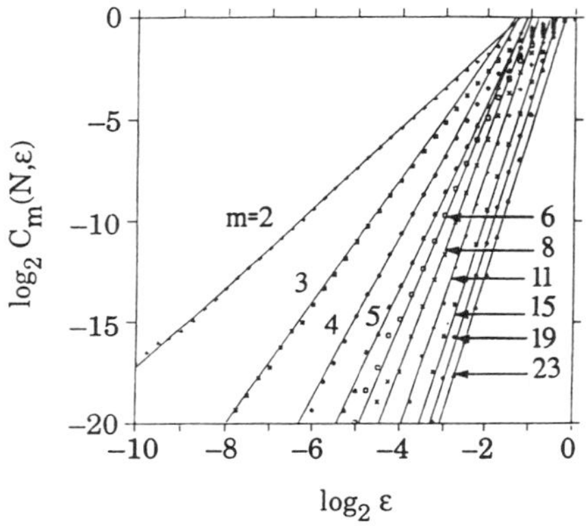

FIG. 3. $\log_2 C_m(N,\epsilon)$ vs $\log_2\epsilon$ for the data set of 2000 points.

due to a lack of data, consists of 2000 points. The $\log_2 C_m(N,\epsilon)$ vs $\log_2\epsilon$ curves are shown in Fig. 3 for $m=2$-6, 8, 11,15, 19, 23. The values of $\bar{D}_2^{(m)}$ in this case are plotted using triangles as a function of m in Fig. 2. This function attains an approximate plateau which begins at $m=16$ and extends beyond $m=25$. The slope averaged over the plateau is about 7.05 which is consistent with the value of 7.1 obtained using the long data set ($N=50000$) plotted as open circles in Fig. 2. But the D_2 estimates for the short data set fall systematically under that for the long data set for $5\leq m\leq 13$. Thus the plateau does not begin until m is substantially larger than Ceil($\bar{D}_2$). This behavior has also been seen in many experimental studies. In what follows we explain the origin of this apparent inconsistency by exploring the systematic behavior of correlation integrals.

Figure 4 is a schematic diagram of a set of correlation integrals for $m=2$ to $m=13$. A dashed line is fit through the scaling region for each m. For $m\leq 5$, $\bar{D}_2^{(m)}\approx m$. For $m\geq 6$, $\bar{D}_2^{(m)}$ plateaus at $\bar{D}_2\approx 5.7$. This value is an estimate of the true D_2 for the system. This figure exhibits several features that are typical of correlation integrals for chaotic systems. The first feature we note is that the horizontal distance between $\log C_m(N,\epsilon)$ and $\log C_{m+1}(N,\epsilon)$ for $m\geq 6$ in the scaling regions is roughly a constant. This constant is predicted, for large m and small ϵ, to be $\Delta h=\Delta v/D_2$, where $\Delta v=TK_2$ with K_2 the correlation entropy [27] and T the delay time in Eq. (2). Two other significant features exhibited by Fig. 4 are as follows. For $m\leq 9$, $\log_2 C_m(N,\epsilon)$ increases with a gradually diminishing slope; while for $m\geq 11$, after exiting the linear region, the log-log plots in Fig. 4 first increase with a slope that is steeper than that in the linear scaling region and then level off to meet the point (0,0). These two different trends give rise to an uneven distribution in the extent of the scaling regions for different m with the most extended scaling region occurring at $m=10$.

In Refs. [8,28] arguments are presented to show that the trend observed for relatively small m is due to an "edge effect" resulting from the finite extent to the reconstructed attractor. Ding *et al.* [5] show that the steeper slope observed for relatively large m is caused by foldings occurring on the original attractor. This can be illustrated analytically [5] for the tent map [29]. For $m=1$, the correlation integral for the reconstructed tent map attractor is $C_1(\epsilon)=\epsilon(2-\epsilon)$. For $m=2$, $C_2(\epsilon)$ is written as $C_2(\epsilon)=C_1(\epsilon/2)+R(\epsilon)$. The first term arises because a pair of points y_j and y_l in the time series satisfying $|y_j-y_l|<\epsilon/2$ give rise to a pair of two-dimensional points, $\mathbf{y}_{j+1}=\{y_{j+1},y_j\}$ and $\mathbf{y}_{l+1}=\{y_{l+1},y_l\}$, satisfying $|\mathbf{y}_{j+1}-\mathbf{y}_{l+1}|<\epsilon$. The folding of the tent map at $y=\frac{1}{2}$ leads to situations in which $|y_j-y_l|>\epsilon/2$, but $|y_{j+1}-y_{l+1}|<\epsilon$. Thus the folding in the attractor underlies the correction term $R(\epsilon)$ which is calculated [5] to be $R(\epsilon)=\epsilon^2/2$ for $0\leq\epsilon<\frac{2}{3}$ and $R(\epsilon)=3\epsilon-7\epsilon^2/4-1$ for $\frac{2}{3}\leq\epsilon\leq 1$. For $0\leq\epsilon<\frac{2}{3}$, $d\log_2 C_2(\epsilon)/d\log_2\epsilon=1+\epsilon/(2+\epsilon)$. This derivative is 1 when $\epsilon=0$ ($D_2=1$ for the attractor) and increases due to the term $\epsilon/(2+\epsilon)$ whose presence reflects the influence of $R(\epsilon)$ which, in turn, is caused by the folding on the attractor.

From Eq. (1), the range of $C_m(N,\epsilon)$ is $\log_2 2/N^2$

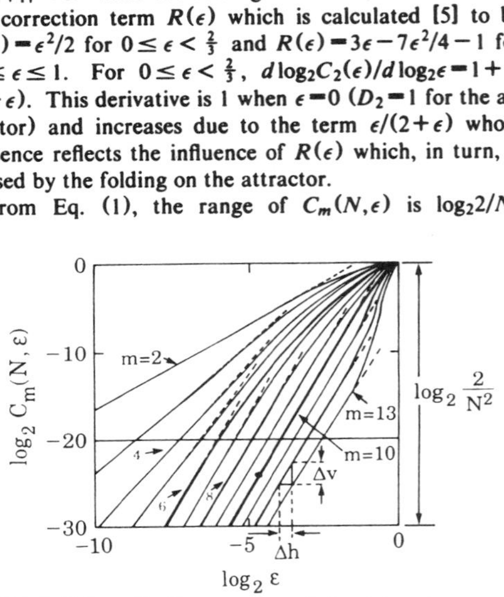

FIG. 4. Schematic diagram of correlation integrals.

$\leq \log_2 C_m(N,\epsilon) \leq 0$. Imagine a time series of $N = 2000$ points generated by the system underlying Fig. 4. The plots of $\log_2 C_m(N,\epsilon)$ vs $\log_2 \epsilon$ for this data set roughly correspond to the portion of Fig. 4 above the horizontal line drawn at $\log_2 C_m(N,\epsilon) = \log_2[2/(2000)^2] \approx -20$. Since the upper boundary points of the scaling regions for $m = 6$ and 7 are under this horizontal line, the correct dimension is not obtained for $m = 6$ and 7. In fact, if we fit a straight line to an *apparent* linear region above $\log_2 C_m(N,\epsilon) = -20$ for $m = 6$ we obtain a slope which is markedly smaller than the actual dimension. However, since the upper boundary points of the scaling regions for $m \geq 8$ are above the horizontal line, we can still expect to obtain the correct estimate of $D_2 = 5.7$ for $m \geq 8$. Thus the plateau onset is delayed due to a lack of data.

The same consideration applies to the short data set generated by Eq. (3). In particular, imagine that we restrict our attention to the region $\log_2 C_m(N,\epsilon) > -20$ in Fig. 1 and fit a line through an *apparent* linear range for the $m = 8$ data in this region. The slope of this straight line is about 5.9, which is roughly the same as that of 5.8 estimated using 2000 points. Thus, by knowing the correlation integrals for a large data set, we can roughly predict the outcome of a dimension measurement based on a smaller subset of this data.

We remark that if one extends the range of m values beyond what is shown in Fig. 2, at large enough m, $\bar{D}_2^{(m)}$ will start to deviate from the plateau behavior and increase monotonically with m. This is caused by the finite length of the data set and can be understood from the systematic behavior of correlation integrals seen in Fig. 4. A lack of sufficient data will not only delay the plateau onset, but also make the deviation from the plateau behavior occur at smaller values of m, thus shortening the plateau length from both sides. This can again be understood with reference to Fig. 4.

This work was supported by the U.S. DOE (Office of Scientific Computing, Office of Energy Research) and the NSF (Divisions of Mathematical and Physical Sciences). M.D.'s research is also supported in part by a grant from the National Institute of Mental Health.

---

[1] P. Grassberger and I. Procaccia, Phys. Rev. Lett. **50**, 346 (1983); Physica (Amsterdam) **9D**, 189 (1983).

[2] J.-P. Eckmann and D. Ruelle, Rev. Mod. Phys. **57**, 617 (1985).

[3] F. Takens, in *Dynamical Systems and Turbulence*, edited by D. Rand and L. S. Young (Springer-Verlag, Berlin, 1981), p. 366; R. Mañe, *ibid.*, p. 230.

[4] T. Sauer, J. A. Yorke, and M. Casdagli, J. Stat. Phys. **65**, 579 (1991).

[5] M. Ding, C. Grebogi, E. Ott, T. Sauer, and J. A. Yorke, Physica (Amsterdam) D (to be published).

[6] T. Sauer and J. A. Yorke (to be published).

[7] To get a rough idea of how widespread the misconception about the relevance of embedding to dimension estimation is, we conducted a literature search using the Science Citation Index for the years from 1987 to 1992 by looking for papers citing both Ref. [1] and Ref. [3]. We found 183 such papers. We then randomly selected a sample of 22 of these papers for closer examination. The following is what we found. Among the 22 papers there are 15 of them that calculate correlation dimension from time series. Five of these 15 papers make explicit connections between $2D+1$ and dimension estimation. The rest of these papers ignore this issue entirely. Based on this information we estimate that, during the period from 1987 to 1992, there are at least 42 papers (probably many more) where the authors implicitly or explicitly assumed that a one-to-one embedding is needed for dimension calculation. In addition, among the papers we researched for this work, only [8] and [9] imply that $m \geq D_2$ is sufficient for estimating D_2, although no justification is given.

[8] J. Theiler, J. Opt. Soc. Am. A **7**, 1055 (1990).

[9] P. Grassberger, in *Chaos*, edited by A. V. Holden (Manchester Univ. Press, Manchester, 1986).

[10] P. Grassberger, T. Schreiber, and C. Schaffrath, Int. J. Bifurcation Chaos **1**, 521 (1991).

[11] *Dimensions and Entropies in Chaotic Systems*, edited by G. Mayer-Kress (Springer-Verlag, Berlin, 1986).

[12] J. Holzfuss and G. Mayer-Kress, in *Dimensions and Entropies in Chaotic Systems* (Ref. [11]).

[13] A. Branstater and H. L. Swinney, Phys. Rev. A **35**, 2207 (1987).

[14] U. Hübner, W. Klische, N. B. Abraham, and C. O. Weiss, in *Measures of Complexity and Chaos*, edited by N. B. Abraham *et al.* (Plenum, New York, 1990).

[15] K. R. Sreenivasan, in *Dimensions and Entropies in Chaotic Systems* (Ref. [11]).

[16] A. A. Tsonis and J. B. Elsner, Nature (London) **333**, 545 (1988).

[17] D. V. Vassiliadis, A. S. Sharma, T. T. Eastman, and K. Papadopoulos, Geophys. Res. Lett. **17**, 1841 (1990).

[18] A. Ben-Mizrachi, I. Procaccia, and P. Grassberger, Phys. Rev. A **29**, 975 (1984).

[19] J. G. Caputo, in *Measures of Complexity and Chaos*, edited by N. B. Abraham *et al.* (Plenum, New York, 1989).

[20] J. G. Caputo, B. Malraison, and P. Atten, in *Dimensions and Entropies in Chaotic Systems* (Ref. [11]).

[21] J.-P. Eckmann and D. Ruelle, Physica (Amsterdam) **56D**, 185 (1992).

[22] N. A. Gershenfeld, Physica (Amsterdam) **55D**, 155 (1992).

[23] J. W. Havstad and C. L. Ehlers, Phys. Rev. A **39**, 845 (1989).

[24] M. Möller, W. Lange, F. Mischke, N. B. Abraham, and U. Hübner, Phys. Lett. A **138**, 176 (1989).

[25] A. R. Osborne and A. Provenzale, Physica (Amsterdam) **35D**, 357 (1989).

[26] M. C. Mackey and L. Glass, Science **197**, 287 (1977).

[27] P. Grassberger and I. Procaccia, Phys. Rev. A **28**, 2591 (1983); Physica (Amsterdam) **13D**, 34 (1984).

[28] L. A. Smith, Phys. Lett. A **133**, 283 (1988).

[29] The tent map is defined as $y_{n+1} = T(y_n)$, where $T(y) = 2y$ if $0 \leq y < 1/2$ and $T(y) = 2 - 2y$ if $1/2 \leq y \leq 1$.

Testing for nonlinearity in time series: the method of surrogate data

James Theiler[a,b], Stephen Eubank[a,b,c,d], André Longtin[a,b], Bryan Galdrikian[a,b] and J. Doyne Farmer[a,c,d]

[a]*Theoretical Division, Los Alamos National Laboratory, Los Alamos, NM 87545, USA*
[b]*Center for Nonlinear Studies, Los Alamos National Laboratory, Los Alamos, NM 87545, USA*
[c]*Santa Fe Institute, 1660 Old Pecos Trail, Santa Fe, NM 87501, USA*
[d]*Prediction Company, 234 Griffin Street, Santa Fe, NM 87501, USA*

Received 4 October 1991
Revised manuscript received 11 February 1992
Accepted 3 March 1992

We describe a statistical approach for identifying nonlinearity in time series. The method first specifies some linear process as a null hypothesis, then generates surrogate data sets which are consistent with this null hypothesis, and finally computes a discriminating statistic for the original and for each of the surrogate data sets. If the value computed for the original data is significantly different than the ensemble of values computed for the surrogate data, then the null hypothesis is rejected and nonlinearity is detected. We discuss various null hypotheses and discriminating statistics. The method is demonstrated for numerical data generated by known chaotic systems, and applied to a number of experimental time series which arise in the measurement of superfluids, brain waves, and sunspots; we evaluate the statistical significance of the evidence for nonlinear structure in each case, and illustrate aspects of the data which this approach identifies.

1. Introduction

The inverse problem for a nonlinear system is to determine the underlying dynamical process in the practical situation where all that is available is a time series of data. Algorithms have been developed which can in principle make this distinction, but they are notoriously unreliable, and usually involve considerable human judgement. Particularly for experimental data sets, which are often short and noisy, simple autocorrelation can fool dimension and Lyapunov exponent estimators into signalling chaos where there is none. Most authors agree that the methods contain many pitfalls, but it is not always easy to avoid them. While some data sets very cleanly exhibit low-dimensional chaos, there are many cases where the evidence is sketchy and difficult to evaluate. Indeed, it is possible for one author to claim evidence for chaos, and for another to argue that the data is consistent with a simpler explanation [1–4].

The real complication arises because low-dimensional chaos and uncorrelated noise are not the only available alternatives. The erratic fluctuations that are observed in an experimental time series owe their dynamical variation to a mix of various influences: chaos, nonchaotic but still nonlinear determinism, linear correlations, and noise, both in the dynamics and in the measuring apparatus. While we are motivated by the prospect of ultimately disentangling these influences, we take as a more modest goal the detection of nonlinear structure in a stationary time series. (We will not attempt to characterize non-stationary time series – see refs. [5–9] for a

From *Physica D* **58** (1992). Reprinted with permission from Elsevier Science Publishers.

discussion of some of the problems arising in the estimation of nonlinear statistics from non-stationary data.)

Positive identification of chaos is difficult; the usual way to detect low-dimensional behavior is to estimate the dimension and then see if this value is small. With a finite time series of noisy data, the dimension estimated by the algorithm will at best be approximate and often, outright wrong. One can guard against this by attempting to identify the various sources of error (both systematic and statistical), and then putting error bars on the estimate (see, for example, refs. [10–18]). But this can be problematic for non-linear algorithms like dimension estimators: first, assignment of error bars requires some model of the underlying process, and that is exactly what is not known; further, even if the underlying process were known, the computation of an error bar may be analytically difficult if not intractable.

The goal of detecting nonlinearity is considerably easier than that of positively identifying chaotic dynamics. Our approach is to specify a well-defined underlying linear process or null hypothesis, and to determine the distribution of the quantity we are interested in (dimension, say) for an ensemble of surrogate data sets which are just different realizations of the hypothesized linear stochastic process. Then, rather than estimate error bars on the dimension of the original data, we put error bars on the value given by the surrogates. This can be done reliably because we have a model for the underlying dynamics (the null hypothesis itself), and because we have many realizations of the null hypothesis, we can estimate the error bar numerically (from the standard deviation of all estimated dimensions of the surrogate data sets) and avoid the issue of analytical tractibility altogether.

While this article elaborates on preliminary work described in an earlier publication [19], our aim is to make this exposition self-contained. In section 2, we express the problem of detecting nonlinearity in terms of statistical hypothesis testing. We introduce a measure of significance, develop various null hypotheses and discriminating statistics, and describe algorithms for generating surrogate data. Section 3 demonstrates the technique for several computer-generated examples under a variety of conditions: large and small data sets, high and low-dimensional attractors, and various levels of observational and dynamical noise. In section 4, we illustrate the application of the method to several real data sets, including fluid convection, electroencephalograms (EEG), and sunspots. With real data, there is always room for human judgment, and we argue that besides formally rejecting a null hypothesis, the method of surrogate data can also be useful in an informal way, providing a benchmark, or control experiment, against which the actual data can be compared.

2. Statistical hypothesis testing

The formal application of the method of surrogate data is expressed in the language of statistical hypothesis testing. This involves two ingredients: a null hypothesis against which observations are tested, and a discriminating statistic. The null hypothesis is a potential explanation that we seek to show is inadequate for explaining the data; and the discriminating statistic is a *number* which quantifies some aspect of the time series. If this number is different for the observed data than would be expected under the null hypothesis, then the null hypothesis can be rejected.

It is possible in some cases to derive analytically the distribution of a given statistic under a given null hypothesis, and this approach is the basis of many existing tests for nonlinearity (e.g., see refs. [20–26]). In the method of surrogate data, this distribution is estimated by direct Monte Carlo simulation. An ensemble of surrogate data sets are generated which share given properties of the observed time series (such as mean, variance, and Fourier spectrum) but are otherwise random as specified by the null hypothesis. For each surrogate data set, the discriminating statistic is computed, and from this ensemble of statistics, the distribution is approximated.

While this approach can be computationally intensive, it avoids the analytical derivations which can be difficult if not impossible. This leads to increased flexibility in the choice of null hypotheses and discriminating statistics; in particular, the hypothesis and statistic can be chosen independently of each other. The method of surrogate data is basically an application of the "bootstrap" method of modern statistics. These methods have by now achieved widespread popularity for reasons that are well described in Efron's 1979 manifesto [27]. A more recent reference, which applies the bootstrap in a context very similar to ours is by Tsay [28].

2.1. Computing significance

Let Q_D denote the statistic computed for the original time series, and Q_{H_i} for the ith surrogate generated under the null hypothesis. Let μ_H and σ_H denote the (sample) mean and standard deviation of the distribution of Q_H.

If multiple realizations are available for the observational data, then it may be possible to compare the two distributions (observed data and surrogate) directly, using for instance the Kolmogorov–Smirnov or Mann–Whitney test, which compare the full distributions, or possibly a Student-t test which only compares their means. For the present purposes, however, we consider that only one experimental data set is available[#1], and we use a kind of t test.

[#1] Of course, it is always possible to create several realizations out of that single set by chopping up the data; we have not tried this approach, but just as the convergence of numerical algorithms like correlation dimension and Lyapunov exponent estimation are compromised by shortened data sets, so we suspect will be their power to reject a null hypothesis. This is only a suspicion, however; it would be worthwhile to compare the relative power of several short data sets versus that of one long data set.

We define our measure of "significance" by the difference between the original and the mean surrogate value of the statistic, divided by the standard deviation of the surrogate values:

$$\mathcal{S} \equiv \frac{|Q_D - \mu_H|}{\sigma_H} . \tag{1}$$

The significance is properly a dimensionless quantity, but it is natural to call the units of $\mathcal{S}$ "sigmas". If the distribution of the statistic is gaussian (and numerical experiments indicate that this is often a reasonable approximation), then the p-value is given by $p = \mathrm{erfc}(\mathcal{S}/\sqrt{2})$; this is the probability of observing a significance $\mathcal{S}$ or larger if the null hypothesis is true.

A more robust way to define significance would be directly in terms of p-values with rank statistics. For example, if the observed time series has a statistic which is in the lower one precentile of all the surrogate statistics (and at least a hundred surrogates would be needed to make this determination), then a (two-sided) p-value of $p = 0.02$ could be quoted. We have used eq. (1) for the investigations reported here because the computational effort in that case is not as severe.

2.1.1. Estimating error bars on significance

Our plots of significance include error bars; these are meant only as a rough guide and are computed assuming that the statistics are distributed as a gaussian.

We write the error bar on $\mathcal{S}$ as $\Delta\mathcal{S}$, and it is computed by standard propagation of errors methodology. Here

$$\begin{aligned}\left(\frac{\Delta\mathcal{S}}{\mathcal{S}}\right)^2 &= \left(\frac{\Delta|\mu_H - \mu_D|}{|\mu_H - \mu_D|}\right)^2 + \left(\frac{\Delta\sigma_H}{\sigma_H}\right)^2 \\ &= \frac{(\Delta\mu_H)^2 + (\Delta\mu_D)^2}{(\mu_H - \mu_D)^2} + \left(\frac{\Delta\sigma_H}{\sigma_H}\right)^2 . \end{aligned} \tag{2}$$

Now the error of the sample mean based on N observations is given by $(\Delta\mu)^2 = \sigma^2/N$, and the error of the sample standard deviation is

$(\Delta\sigma)^2 = \sigma^2/2N$, so we can write

$$\left(\frac{\Delta\mathcal{S}}{\mathcal{S}}\right)^2 = \frac{\sigma_H^2/N_H + \sigma_D^2/N_D}{(\mu_H - \mu_D)^2} + \frac{1}{2N_H}. \tag{3}$$

The absolute error bar is then given by

$$\Delta\mathcal{S} = \sqrt{(1 + \tfrac{1}{2}\mathcal{S}^2)/N_H + (\sigma_D/\sigma_H)^2/N_D}. \tag{4}$$

When only a single realization of the time series is available, we take $\sigma_D = 0$ and ignore the second term in the above equation. This reports the error bar on the significance of the specific realization.

In our numerical experiments, we use several realizations of the time series under question. However, the significance we report is not based on the collective evidence of the several, but is the average significance of each realization taken individually. The error bar in that case describes the expected error of our estimate of this average. Note that this differs from the error reported for a single realization.

2.2. Toward a hierarchy of null hypotheses

The null hypothesis defines the nature of the candidate process which may or may not adequately explain the data. Our null hypotheses usually specify that certain properties of the original data are preserved – such as mean and variance – but that there is no further structure in the time series. The surrogate data is then generated to mimic these preserved features but to otherwise be random. There is some latitude in choosing which features ought to be preserved: certainly mean and variance, and possibly also the Fourier power spectrum. If the raw data is discretized to integer values, then the surrogate data should be similarly discretized.

Ultimately we envision a hierarchy of null hypotheses against which time series might be compared. Beginning with the simplest hypotheses, and increasing in generality, the following sections outline some of the possibilities that we have considered.

2.2.1. Temporally independent data

The first (and easiest) question to answer about a time series is whether there is evidence for any dynamics at all. The null hypothesis in this case is that the observed data is fully described by independent and identically distributed (IID) random variables. If the distribution is further assumed to be gaussian, then surrogate data can be readily generated from a standard pseudorandom number generator, normalized to the mean and variance of the original data.

To test the hypothesis of IID noise *with arbitrary amplitude distribution* in an analysis of stock market returns, Schienkman and LeBaron [29] generated surrogate data by shuffling the time-order of the original time series. The surrogate data is obviously guaranteed to have the same amplitude distribution as the original data, but any temporal correlations that may have been in the data are destroyed. Breeden and Packard also used a shuffling process along with a sophisticated nonlinear predictor to prove that there was some dynamical structure to a time series of quasar data which were sampled nonuniformly in time [30].

2.2.2. Ornstein–Uhlenbeck process

A very simple case of non-IID noise is given by the Ornstein–Uhlenbeck process [31]. A discrete sampling of this process yields a model of the form

$$x_t = a_0 + a_1 x_{t-1} + \sigma e_t, \tag{5}$$

where e_t is uncorrelated gaussian noise of unit variance. The coefficients a_0, a_1, and σ collectively determine the mean, variance, and autocorrelation time of the time series. In fact, the autocorrelation function is exponential in this case,

$$A(\tau) \equiv \frac{\langle x_t x_{t-\tau}\rangle - \langle x_t\rangle^2}{\langle x_t^2\rangle - \langle x_t\rangle^2} = e^{-\lambda|\tau|},$$

where $\langle\ \rangle$ denotes an average over time t, and $\lambda = -\log a_1$.

To make surrogate data sets, the mean μ, variance v, and first autocorrelation $A(1)$ are estimated from the original time series; from these the coefficients are fit: $a_1 = A(1)$, $a_0 = \mu(1 - a_1)$, and $\sigma^2 = v(1 - a_1^2)$. Finally, one generates the surrogate data by iterating eq. (5), using a pseudorandom number generator for the unit variance gaussian e_t.

2.2.3. Linearly autocorrelated gaussian noise

We can generalize the above null hypothesis by extending eq. (5) to arbitrary order. This leads to the hypothesis that is generally associated with linearity. We emphasize that we are discussing linear *gaussian* processes here (see Tong [26, pp. 13, 14] for a brief description of some of the surprising properties of linear nongaussian processes); Section 2.2.4 describes one approach toward a nongaussian null hypothesis. The model is described by fitting coefficients a_k and σ to a process

$$x_t = a_0 + \sum_{k=1}^{q} a_k x_{t-k} + \sigma e_t \,, \tag{7}$$

which mimics the original time series in terms of mean, variance, and the autocorrelation function for delays of $\tau = 1, \ldots, q$. This is an autoregressive (AR) model; a more general model includes a moving average (MA) of time delayed noise terms as well, and the combination is called an ARMA model. For large enough q, the models are essentially equivalent. The null hypothesis in this case is that all the structure in the time series is given by the autocorrelation function, or equivalently, by the Fourier power spectrum.

One algorithm for generating surrogate data under this null hypothesis is again to iterate eq. (7), where the coefficients have been fit to the original data. We describe an alternative algorithm in section 2.4.1 which involves randomizing the phases of a Fourier transform. (To our knowledge, this algorithm was first suggested in this context by Osborne et al. [5], and independently in refs. [15, 32].) The alternative algorithm generates surrogate data which by construction has the same Fourier spectrum as the original data. While the two algorithms are essentially equivalent, we use the Fourier transform method because it is numerically stabler. If the values of the coefficients in eq. (7) are mis-estimated slightly, it is possible that iterating the equation will lead to a time series which diverges to infinity; this is particularly problematic if the raw time series is nearly periodic or highly sampled continuous data.

We remark that this is the null hypothesis that is associated with residual-based tests for nonlinearity. For instance, see refs. [22–24, 33, 34]. In these tests, a model of the form of eq. (7) is fit to the data, and the residuals

$$\epsilon_t = x_t - \left(a_0 + \sum_{k=1}^{q} a_k x_{t-k}\right) \tag{8}$$

are tested against a null of temporally independent noise. In ref. [19], we argue that it is usually preferable to use the method of surrogate data on the raw data directly, rather than working with residuals.

2.2.4. Static nonlinear transform of linear gaussian noise

One way to generalize the above null hypothesis to cases where the data is nongaussian is to suppose that although the dynamics is linear, the observation function may be nonlinear. In particular, we hypothesize that there exists an "underlying" time series $\{y_t\}$, consistent with the null hypothesis of linear gaussian noise, and an observed time series $\{x_t\}$ given by

$$x_t = h(y_t) \,. \tag{9}$$

Since x_t depends only on the current value of y_t and not on derivatives or past values, the filter $h(\,)$ is said to be "static" or "instantaneous". To permit the generation of surrogate data, we must further assume (as part of the null hypothesis) that the observation function $h(\,)$ is effectively invertible.

In section 2.4.3, an algorithm for generating surrogate data corresponding to this null hypothesis is described. Its effect is to shuffle the time-order of the data but in such a way as to preserve the linear correlations of the underlying time series $y_t = h^{-1}(x_t)$. One advantage of shuffling over, for example, a smooth fit to the function $h(\,)$, is that any discretization that was present in the original data will be reflected in the surrogate data.

Note that time series in this class are strictly speaking nonlinear, but that the nonlinearity is not in the dynamics. Most conventional tests for nonlinearity would (correctly) conclude that the time series is nonlinear, but would not indicate whether the nonlinearity was in the dynamics or in the amplitude distribution. By using surrogate data tailored to this specific null hypothesis, it becomes possible to make such fine distinctions about the nature of the dynamics.

2.2.5. *More general null hypotheses*

Eventually, we would like to extend this list to consider more general cases. A natural next step is a null hypothesis that the dynamics is a noisy limit cycle. Such time series cannot be described by a linear process, even if viewed through a static nonlinear transform. Yet it is often of great interest, particularly in systems driven by seasonal cycles, to determine the nature of the inter-seasonal variation.

There is another large class of nonlinear stochastic processes which are not predictable even in the mean; among these are the conditional heteroscedastic models (for which the variance is conditioned on the past, but not the mean) in favor among economists. While there is definite nonlinear structure in these time series, it is not manifested in enhanced predictability by nonlinear models. (For instance, it may be possible to predict the magnitude $|x_t|$ from past values of x, but not the sign.)

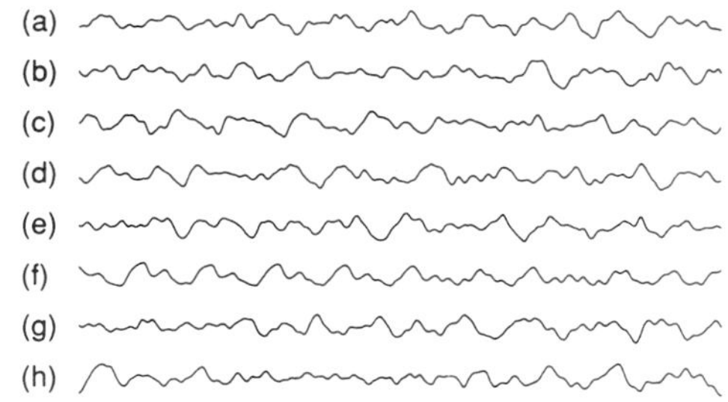

Fig. 1. Shown is a time series from the Mackey–Glass equation with $\tau = 30$, which is known to be low-dimensional and chaotic, and seven surrogate time series generated by the WFT algorithm. It is often not obvious by eye which is the actual data set and which are the surrogates. In this case it is series (f) which is the real one.

2.3. *Battery of discriminating statistics*

The method of surrogate data can in principle be used with virtually any discriminating statistic. Formally, all that is required to reject a null hypothesis is that the statistic have a different distribution for the data than for the surrogates. However, the method is more useful if the statistic actually provides a good estimate of a physically interesting quantity; in that case, one may not only formally reject a null hypothesis, but also informally characterize the nature of the nonlinearity.

Since we were motivated by the possibility that the underlying dynamics may be chaotic, our original choices for discriminating statistics were the correlation dimension, Lyapunov exponent, and forecasting error. Ideally, dimension counts degrees of freedom, Lyapunov exponent quantifies the sensitivity to initial conditions, and forecasting error tests for determinism. One of the ultimate aims in this project is to understand the conditions in which one or the other of these methods will be more effective.

We should remark that a danger in using a battery of statistics is that one of them, by chance, will show up as significant. This effect can be formally accounted for by keeping strict count of the number of tests used, and increasing the threshold of significance accordingly. The formal approach tends to be more conservative than necessary, since the tests are not really independent of each other, but it is still a recommended practice to maintain a reasonably high threshold of significance.

2.3.1. *Correlation dimension*, ν

Dimension is an exponent which characterizes the scaling of some bulk measure with linear size. A number of algorithms are available [17, 35] for estimating the dimension of an underlying strange attractor from a time series; we chose a box-assisted variation [36] (see Grassberger [37] for an elegant alternative) of the Grassberger–Procaccia–Takens algorithm [38–40] to compute a correlation integral, and the best estimator of Takens [12] for the dimension itself. To compute a dimension, it is necessary to choose some range of sizes over which the scaling is to be estimated. The Takens estimator requires only an upper cutoff size; we used one-half of the rms variation in the time series for this value. (See Ellner [41] for an estimator that takes both an upper and a lower cutoff.)

We will concede that this choice is a bit arbitrary; one might prefer a more sophisticated algorithm for choosing a good scaling range. L. Smith (personal communication) has suggested choosing the range "by eye" for the raw data and then keeping this range for the surrogates. From the point of view of the formal test, it does not really matter, but if we are to ask for insight as well as a rejected null, then it is important to use a good dimension estimator. In the $N \to \infty$ limit, the estimator we describe will not converge to the actual precise dimension of the attractor, but we note that it will converge fairly rapidly to a number which is often reasonably close to actual dimension (of course, one can always contrive counterexamples); in particular, it will properly indicate low-dimensionality when it sees it. While we do not claim that this is the optimal dimension estimator, we believe that it is a useful one.

2.3.2. *Forecasting error*, ϵ

A system is deterministic if its future can be predicted. A natural statistic in this case is some average of the forecasting errors obtained from nonlinear modeling. The method we use entails first splitting the time series into a fitting set of length N_f, and a testing set of length N_t, with $N_f + N_t = N$, the length of the time series; then fitting a local linear model [42] to the fitting set, locality given by the number of neighbors k; and finally, using this model to forecast the values in the testing set, and comparing them with the actual values.

The prediction error $e_t = x_t - \hat{x}_t$ is the difference between the actual value of x and the predicted value, $\hat{x}$; we define our discriminating statistic as the log median absolute prediction error.

Several modeling parameters must be chosen, including the partitioning of the data set into fitting (N_f) and testing (N_t) segments, the number of steps ahead to predict (T), and number of neighbors (k) used in the local linear fit. We arbitrarily chose to divide the fitting and testing sets equally, with $N_f = N_t = \frac{1}{2}N$, and to predict one step ahead, so $T = 1$. For oversampled continuous data, a larger T would be more appropriate. The choice of k is also important. For the results in this article, we set $k = 2m$, which is twice the minimum number needed for a fit, but we note that this is often not optimal. Indeed, Casdagli [43, 44] has advocated sweeping the parameter k in a local linear forecaster as an exploratory method to look for nonlinearity in the first place.

2.3.3. *Estimated Lyapunov exponent*, λ

Following standard practices [45–47], we compute Lyapunov exponents by multiplying Jacobian matrices along a trajectory, with the matrices computed by local linear fits, and we use QR decomposition to maintain orthogonality.

We have found that numerical estimation of even the largest Lyapunov exponent can be problematic in the presence of noise. Indeed, for our surrogate data sets, for which the linear dynamics is contracting, we often obtain positive Lyapunov exponents. This indicates that our Lyapunov exponent estimator (which, as we have described, is fairly standard) is seriously flawed, something we might not have noticed had we not tested with linear stochastic time series. We are aware of at least one group whose Lyapunov exponent estimator explicitly considers the effects of noise [48–50]. While our estimator is arguably still useful as a statistic which formally distinguishes original data from surrogate data, it would be better to use a discriminating statistic which correctly quantifies some feature of the dynamics. For that reason, we have avoided using the Lyapunov exponent estimator in this article.

2.3.4. *Other discriminating statistics*

We have found that using the correlation integral ($C(r)$ for some value of r) directly as a discriminating statistic generally provides a more powerful discrimination than the estimated dimension itself, but of course it is less useful as an informal tool. L. Smith (personal communication) has suggested using a statistic which characterizes the linearity of a $\log C(r)$ versus $\log r$ curve. We have also considered but not implemented two-sided forecasting – predicting x_t from the past *and* future values: $x_{t-1}, \ldots$; $x_{t+1}, \ldots$, instead of the usual forecasting which uses only the past (this was inspired by the simple noise reduction technique suggested by Schreiber and Grassberger [51]). In our forecasting, we are careful to distinguish the "training" set from the "testing" set, so that the forecasting statistic is an out-of-sample error; but the in-sample fitting error may also suffice as a discriminating statistic. We have found that the BDS test [33], which was designed to test for any temporal correlation at all – linear or nonlinear, can readily be extended to test other null hypotheses; we use the same statistic, but we do not pre-whiten the data, and instead of relying on an analytical derivation of the distribution function, we use surrogate data. Higher and cross moments provide another class of discriminating statistic; in fact, many of these are the basis of traditional tests for nonlinearity in a time series (e.g., see refs. [22–24]). We have found that a simple skewed difference statistic, defined by $Q = \langle (x_{t+m} - x_t)^3 \rangle / \langle (x_{t+m} - x_t)^2 \rangle$, is both rapidly computable and often quite powerful. Informally, this statistic indicates the asymmetry between rise and fall times in the time series. The most direct example we know is due to Brock, Lakonishok, and LeBaron [52], who used technical trading rules as discriminating statistics for financial data; here there is no difficulty interpreting the informal meaning of the statistic: it is how much money you should have made using that rule in that market.

2.4. *Algorithms for generating surrogate data*

In this section, we describe algorithms we use for generating surrogate data. The first two are consistent with the hypothesis of linearly correlated noise described in section 2.2.3, and the third adjusts for the possibility of a static nonlinear transform as discussed in section 2.2.4.

2.4.1. *Unwindowed Fourier transform (FT) algorithm*

This algorithm is based on the null hypothesis that the data come from a linear gaussian process. The surrogate data are constructed to have the same Fourier spectra as the raw data. The algorithm is described in more detail in ref. [19], but we briefly note the main features. First, the Fourier transform is computed for positive and negative frequencies $f = 0, 1/N, 2/N, \ldots, 1/2$, and without the benefit of windowing. Although windowing is generally recommended when it is the power spectrum which is of ultimate interest [53], we originally chose not to use windowing because what we wanted was for the real and surrogate data to have the same power spectrum; we were not concerned with the spectrum, *per se*. The Fourier transform has a complex amplitude at each frequency; to randomize the phases, we multiply each complex amplitude by $e^{i\phi}$, where ϕ is independently chosen for each frequency from the interval $[0, 2\pi]$. In order for the inverse Fourier transform to be real (no imaginary components), we must symmetrize the phases, so that $\phi(f) = -\phi(-f)$. Finally, the inverse Fourier transform is the surrogate data.

One limitation of this algorithm is that it does not reproduce "pure" frequencies very well. What happens is that nearby frequencies in Fourier space are "contaminated" and then because their phases are randomized, they end up "beating" against each other and producing spurious low-frequency effects. (We are grateful to S. Ellner for pointing this out to us.) This may not be too surprising since it is difficult to make a linear stochastic process with a long coherence time. Put another way, the time series should not only be much larger than the dominant periodicities but also much longer than the coherence time of any given frequency if one is to try and model it with a linear stochastic process.

A second problem, which is most evident for highly sampled continuous data, is that spurious high frequencies can be introduced. This can be understood as an artifact of the Fourier transform which assumes the time series is periodic with period N. This means that there is a jump-discontinuity from the last to the first point. We recommend tailoring the length N of the data set so that $x[0] \approx x[N-1]$. This should not be a problem if the time series is stationary and much longer than its dominant frequency. We have done this for the experimental results in this article.

2.4.2. *Windowed Fourier transform (WFT) algorithm*

The problem of spurious high frequencies can also be addressed by windowing the data before taking the Fourier transform. The time series is multiplied by a function $w(t) = \sin(\pi t/N)$ which vanishes at the endpoints $t = 0$ and $t = N$. This suppresses the jump discontinuity from the last to the first point, and seems to effectively get rid of the high frequency effect. However, it also introduces a spurious low-frequency from the power spectrum of $w(t)$ itself. We have done experiments where we simply set the magnitude of the offending frequency ($f = 1/N$) to zero; this seems to work well for stationary time series, but if there is significant power at that frequency in the original data, it too will be suppressed.

2.4.3. *Amplitude adjusted Fourier transform (AAFT) algorithm*

The algorithm in this section generates surrogate data sets associated with the null hypothesis in section 2.2.4, that the observed time series is a monotonic nonlinear transformation of a linear gaussian process. The idea is to first rescale the values in the original time series so they are gaussian. Then the FT or WFT algorithm can be used to make surrogate time series which have the same Fourier spectrum as the rescaled data. Finally, the gaussian surrogate is then rescaled back to have the amplitude distribution as the original time series.

Denote the original time series by $x[t]$, with $t = 0, \ldots, N-1$. The first step is to make a gaussian time series $y[t]$, where each element is generated independently from a gaussian pseudorandom number generator. Next, we re-order the time sequence of the gaussian time series so that the ranks of both time series agree; that is, if $x[t]$ is the nth smallest of all the x's, then $y[t]$ will be the nth smallest of all the y's. Therefore, the re-ordered $y[t]$ is a time series which "follows" the original time series $x[t]$ and which has a gaussian amplitude distribution. Using the FT or WFT algorithm, a surrogate, call it $y'[t]$, of the gaussian time series can be created. If the original time series $x[t]$ is time re-ordered so that it follows $y'[t]$ in the sense that the ranks agree, then the time-re-ordered time series provides a surrogate of the original time series which matches its amplitude distribution. Further, the "underlying" time series ($y[t]$ and $y'[t]$) are gaussian and have the same Fourier power spectrum.

3. Experiments with numerical time series

To properly gauge the utility of the surrogate data approach will eventually require many tests with data from both numerical and laboratory experiments. In this section we illustrate several aspects of the method with data whose underlying dynamics is known. In the next section, we consider several examples with real data.

3.1. Linear gaussian data

First, we note that a time series which actually is generated by a linear process should by construction give a negative result (that is, the null hypothesis should *not* be rejected). We checked this by generating some time series with two simple linear processes, a moving average

$$x_t = e_t + ae_{t-1} \tag{10}$$

and an autoregressive

$$x_t = ax_{t-1} + e_t\,. \tag{11}$$

We used an embedding dimension $m = 3$ and computed correlation dimension from $N = 4096$ points. The "correct" dimension for both processes is $\nu = m = 3$, though as fig. 2 shows, the estimates were always biased low. The bias increases for data which are more highly autocorrelated ($|a|$ larger) but the point we wish to make is that the bias is the same for the original data and for the surrogates. The null hypothesis is not rejected.

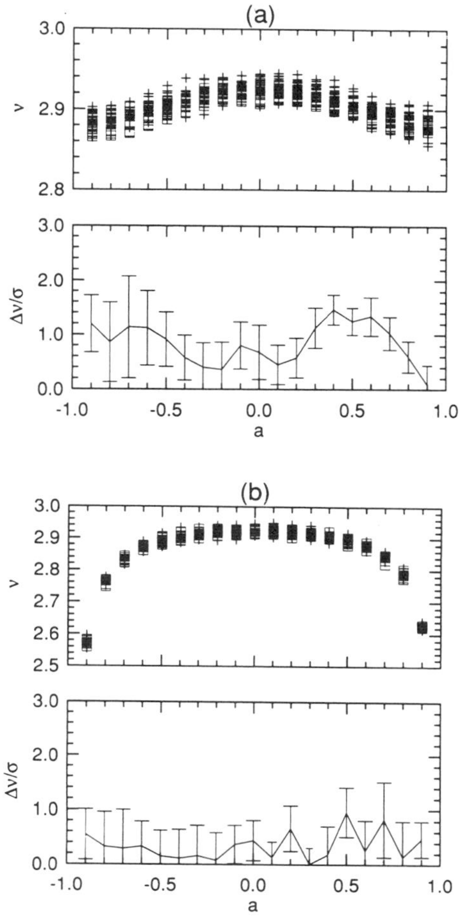

Fig. 2. Significance of evidence for nonlinearity for linear gaussian time series generated by (a) a moving average process, and (b) an autoregressive process. The coefficient in each case is a. The estimated dimension is shown for five realizations of the linear process (□) and thirty realizations of surrogate data (+). Note that the dimension does not distinguish the original from the surrogate data. The value we obtain for significance is shown in the lower panels and in neither case is significant.

3.2. Variation with number of data points and complexity of attractor

Using the FT algorithm, we showed in ref. [19] that increasing the number of points in a time series increases the significance with which nonlinearity can be detected in a time series that is known to be chaotic; and increasing the complexity of the chaotic time series decreases the ability to distinguish from linearity. This basic point is illustrated again here using the AAFT algorithm (see fig. 3); while the significance is not as large using this more general null hypothesis, the qualitative behavior is the same. Time series are generated by summing n independent trajectories of the Hénon map [54]; such time series will have a dimension $n\nu$ where $\nu \approx 1.2$ is the dimension of a single Hénon trajectory. For the largest data sets, with $N = 8192$ points, our dimension estimator obtained correlation dimensions of 1.215 ± 0.008, 2.279 ± 0.014, 3.48 ± 0.02, and 4.81 ± 0.06 using embedding dimensions $m = 3$, 4, 5, and 6, for $n = 1$, 2, 3, and 4, respectively.

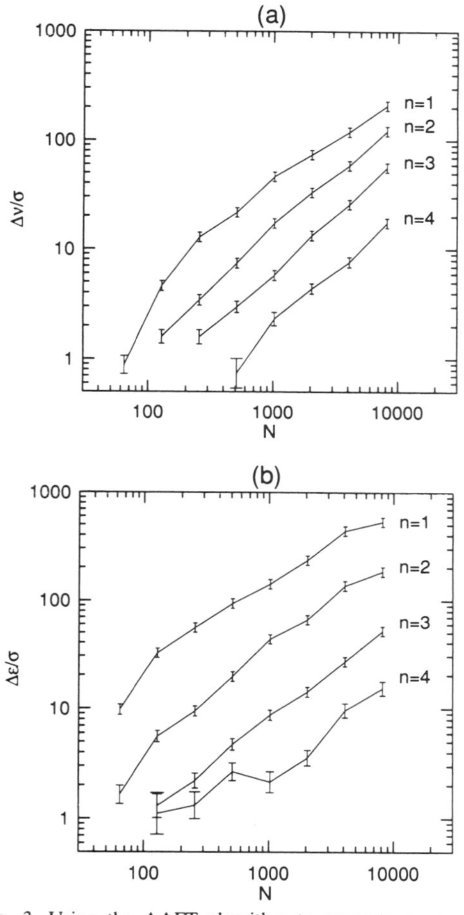

Fig. 3. Using the AAFT algorithm to generate surrogate data, the significance as a function of the number N of data points is computed for time series obtained by summing n independent trajectories of the Hénon map. For both (a) correlation dimension and (b) forecasting error, the significance increases with number of data points and decreases with the complexity of the system.

3.3. Effect of observational and dynamical noise

To test whether nonlinear determinism can be detected even when it is mixed with noise, we added both dynamical (η) and observational (ε) noise to the cosine map: $y_t = \lambda \cos(\pi y_{t-1}) + \eta_t$; $x_t = y_t + \varepsilon_t$. We chose a value $\lambda = 2.8$ which is in the chaotic regime when the external noise is zero. (The cosine map was used instead of the Hénon map because it does not "blow up" in the presence of too much dynamical noise.) In fig. 4,

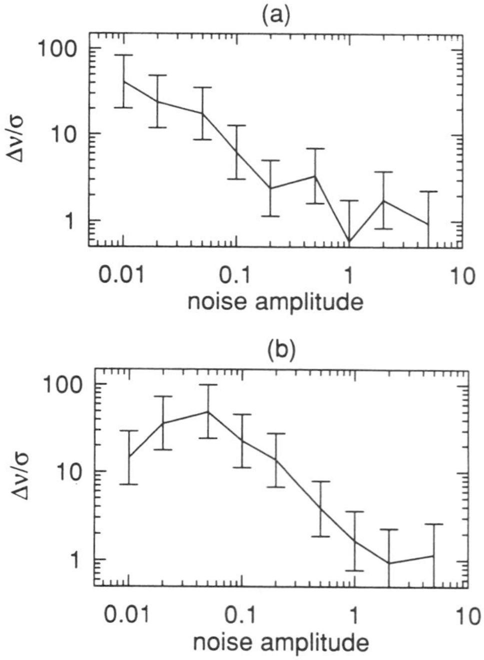

Fig. 4. Effect of noise on significance for a short time series of $N = 512$ points, derived from the cosine map with $\lambda = 2.8$: (a) observational noise; (b) dynamical noise.

we plot significance as a function of noise level for both dynamical and observational noise. As expected, significance decreases with increasing noise level, though we remark that the nonlinearity is still observable even with considerable noise. In the absence of noise, the rms amplitude of the signal is 0.36; thus we are able to detect significant nonlinearity even with a signal to noise ratio of one, using a time series of length $N = 512$. We also note that the decrease in significance with increased dynamical noise is not always monotonic; low levels of dynamical noise can make the nonlinearity more evident.

3.4. *Continuous data*

In most experiments, data is better described as a flow than a map. Although there is a formal equivalence, data which arise from processes that are continuous in time are often sampled at a much faster rate than is characteristic of the underlying dynamics. For these data sets, the effects of autocorelation can be quite large, and the importance of testing against a null hypothesis that includes autocorrelation becomes paramount.

We illustrate this point with numerical experiments on data obtained from the Mackey–Glass differential delay equation [55]

$$\frac{\mathrm{d}x}{\mathrm{d}t} = -bx(t) + \frac{ax(t-\tau)}{1 + [x(t-\tau)]^{10}}, \tag{12}$$

with $a = 0.2$, $b = 0.1$, and $\tau = 30$. Grassberger and Procaccia [39] compute a correlation dimension of $\nu \approx 3.0$ for these parameters.

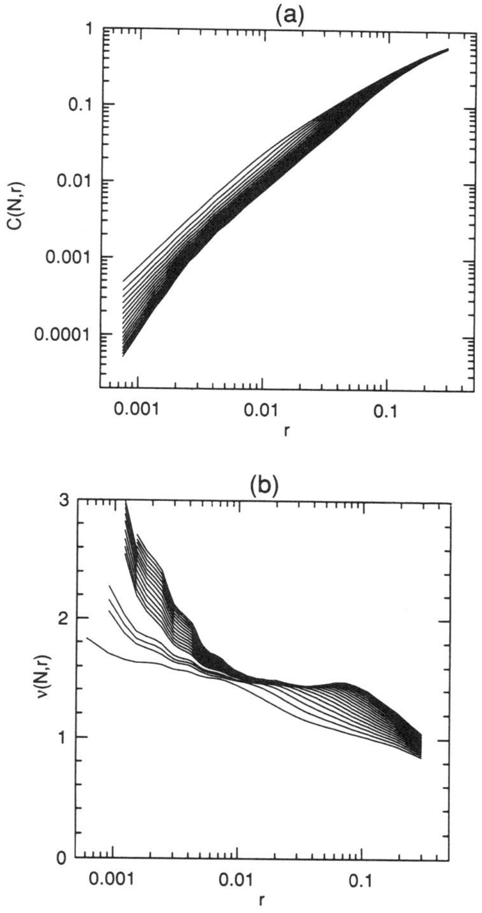

Fig. 5. (a) Correlation integral $C(N, r)$ for $N = 4096$ points and embedding dimensions $m = 3, \ldots, 19$ from oversampled Mackey–Glass data. (b) Estimated correlation dimension ν according to Takens estimator as a function of cutoff r.

3.4.1. *A poor embedding*

We oversample the data ($\Delta t = 0.1$) and use a deliberately poor embedding strategy – straight time-delay coordinates with a lag time of one sample period. We estimate correlation dimension with $N = 4096$ points and compute distances between all pairs of distinct vectors (despite the advice in refs. [2, 56]). Fig. 5 shows the correlation integral and estimated dimension as a function of the upper cutoff value R. There is about a decade of roughly constant slope, which might be taken to indicate convergence to a low correlation dimension.

For this example, the dimension statistic was computed as the Takens best estimator [12] at an upper cutoff of $R = 0.02$. (For comparison, the

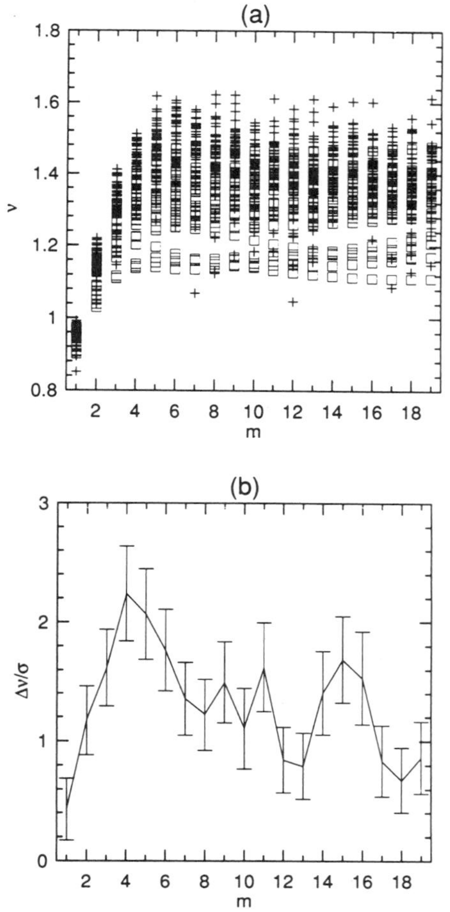

Fig. 6. (a) Estimated correlation dimension versus embedding dimension for oversampled ($\Delta t = 0.1$) Mackey–Glass data (□) and for surrogates generated using the WFT algorithm (+). (b) Significance of nonlinearity in no case exceeds three sigmas.

RMS value for this time series is $R_{\text{rms}} = 0.25$.) Fig. 6a shows an apparent convergence of the estimated dimension as a function of embedding dimension. A naive interpretation of this figure is that the time series arises from a low-dimensional strange attractor. However, as fig. 6a shows, the surrogate data also converge to a low dimension; the convergence is evidently an artifact of the autocorrelation. Indeed, fig. 6b shows that the dimension statistic in this case does not even provide evidence for nonlinearity.

3.4.2. A better embedding

From the same Mackey–Glass process, we recompute correlation dimension and the signifi-

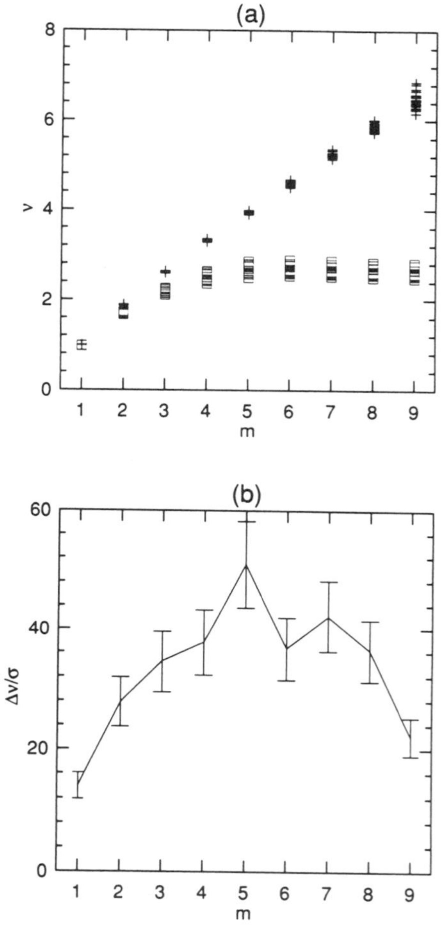

Fig. 7. Same as previous figure, except that a better embedding and a better algorithm were used for estimating the dimension. Not only is the evidence for nonlinearity extremely significant in this case, but it is also evident that the process is low-dimensional.

cance of evidence for nonlinearity using a better (though probably still not optimal) choice of embedding. We sample at a much lower rate, $\Delta t = 3.0$, and again use straight delay coordinates with lag time of one sample period. We estimate the correlation dimension as described in section 2.3.1 with $N = 4096$ points, and we avoid pairs of points which are closer together in time than one hundred sample periods. In fig. 7, we see that the evidence for nonlinearity is extremely significant. Indeed, we also see positive evidence of low-dimensional behavior (the estimated dimension ν converges with m) which we know is not an artifact of autocorrelation.

4. Examples with real data

We report some results on experimental time series from several sources. These results should be taken as illustrative, and not necessarily typical of the class which they represent. In particular, we have not yet attempted to "normalize" our findings with others that have previously appeared in the literature.

4.1. *Rayleigh–Benard convection*

Data from a mixture of ^{3}He and superfluid ^{4}He in a Rayleigh–Benard convection cell [57] provides an example where the evidence for nonlinear structure is extremely significant. The significance as obtained with the dimension and forecasting statistics from a time series of $N = 2048$ points are shown in fig. 8. Further, the dimension statistic indicates that the flow is in fact low-dimensional; while the measured dimension of 3.8 may be due to an artifact of some kind, we are at least assured that it is not an artifact of autocorrelation or of nongaussian amplitude distribution. Farmer and Sidorowich [42] used this data to demonstrate the enhanced predictability using nonlinear rather than linear predictors.

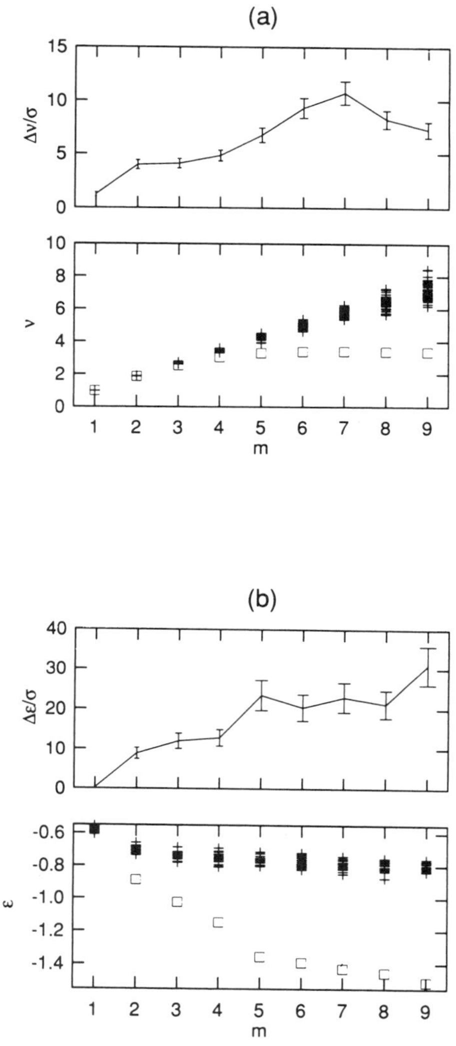

Fig. 8. Data from a fluid convection experiment exhibits very significant nonlinear structure, using (a) dimension, and (b) forecasting error. The top panel in these figures show the significance, measured in "sigmas", and the bottom panel shows the values of the statistics, with squares (□) for the original data and pluses (+) for the AAFT-generated surrogates. Both panels plot these statistics against the embedding dimension m. Not only is the evidence for nonlinear structure statistically significant, but the estimated dimension of about $\nu = 3.8$ suggests that the underlying dynamics is in fact low-dimensional chaos.

4.2. The human electroencaphalogram (EEG)

The electroencephalogram (EEG) is to the brain what the electrocardiogram (EKG) is to the heart. It has become a widely used tool for the monitoring of electrical brain activity, and its potential for diagnosis is still being explored. A number of researchers have applied the methods that were developed for the analysis of chaotic time series to EEG time series. While it was hoped that the characterization of deterministic structure in EEG would eventually lead to insights about the workings of the brain, the shorter term goal was to use the nonlinear properties of the time series as a diagnostic tool [58, 59].

Although we feel a more systematic survey is in order, we have not examined any EEG data which gives positive evidence of low-dimensional chaos. However, we have found examples where nonlinear structure was evident. We present here two cases, one positive and one negative. The two time series are from the same individual, eyes closed and resting; one is from a probe at the left occipital (O1), and the other from the left central (C3) part of the skull. The sampling rate is 150 Hz, and $N = 2048$ time samples are taken. The two time series are not necessarily contemporaneous. Using the dimension statistic, the first data set shows no significant evidence for nonlinearity, but the second data set exhibits about eight sigmas. Even in the significant case, we do not see any evidence that the time series is in fact low-dimensional (the correlation dimension ν does not converge with increasing embedding dimension m). We are formally able to reject the null hypothesis that the data arise from a linear stochastic process, but by comparing the surrogate data to the real data, we see no reason to expect that the "significant" data arises from a low-dimensional chaotic attractor.

4.3. The sunspot cycle

Our final example is the well known and much studied eleven year sunspot cycle [44, 60–66].

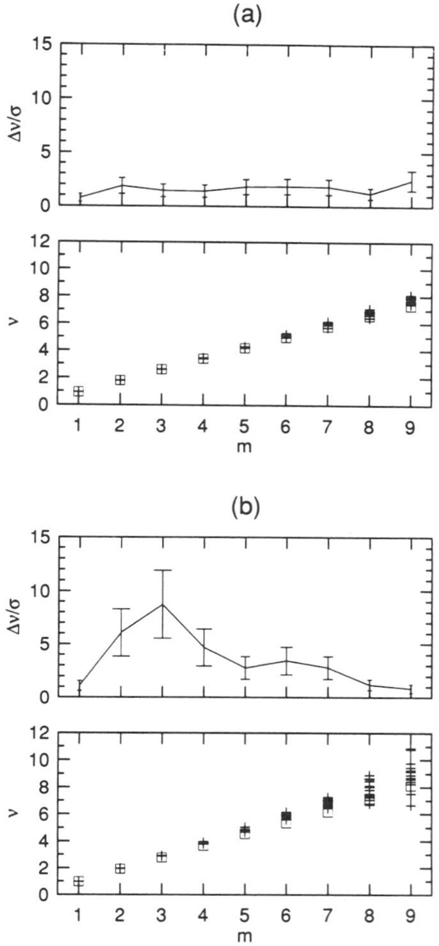

Fig. 9. Data from two electroencephalogram (EEG) time series. Using the dimension statistic, the first (a) shows no nonlinear structure, while the second (b) exhibits significant nonlinear structure at the eight sigma level. The evidence for low-dimensional chaos, however, is weak, since the estimated dimension increases almost as rapidly with embedding dimension for the original time series as it does for the surrogates.

First, we used the FT algorithm for generating surrogate data, but we were careful to use a length of time series ($N = 287$) for which the first and last data point both corresponded to minima; this avoids introducing the spurious high

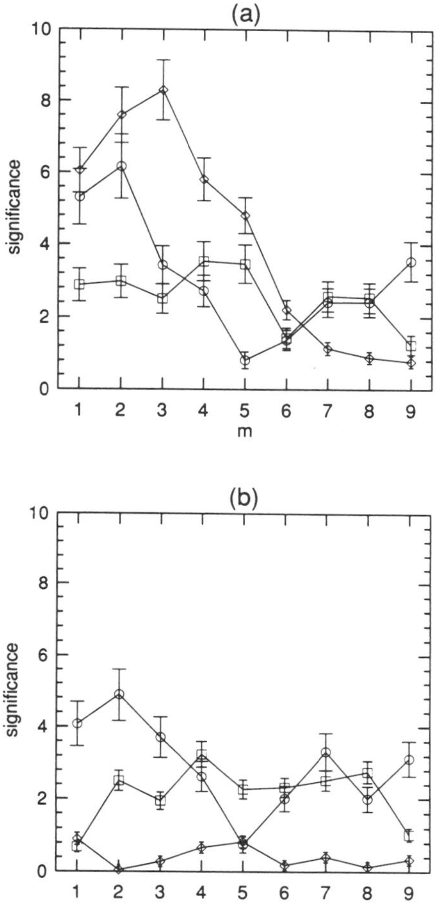

Fig. 10. Significance of nonlinearity in the annual sunspot series; (a) against the null hypothesis of linear gaussian noise (surrogates generated by FT algorithm), and (b) the null hypothesis of amplitude corrected linear gaussian noise (surrogates generated by AAFT algorithm). For both plots, the discriminating statistics are estimated dimension (◇), log median prediction error (□), and the skew statistic described in the text (○).

frequencies that we discussed in section 2.4.1. As fig. 10a shows, it is possible to quite confidently reject the null hypothesis of linear gaussian noise; this is in agreement with the numerous authors who obtained better agreement using nonlinear models instead of linear models. However, when we expand the null hypothesis to include a static nonlinear observation of an underlying linear gaussian process, the evidence for dynamical nonlinear structure is less dramatic. Using the dimension statistic, there is no significance; the prediction statistic gives that the evidence is just significant; the cubed difference statistic $Q = \langle (x_{t+m} - x_t)^3 \rangle / \langle (x_{t+m} - x_t)^2 \rangle$, which is a measure of the time irreversibility of the data, provides a more significant rejection of the hypothesis of static nonlinear filter of an underlying linear process.

5. Comparison to other work

Numerous authors have carefully compared their dimension estimates for real data against similar estimates for white noise. A few have extended this informal control to other forms of correlated noise. Grassberger [2] showed that a reported dimension for climate data could be reproduced with data from an Ornstein–Uhlenbeck process. Osborne et al. [5], criticized the Grassberger–Procaccia algorithm on the basis that the low dimension it gave to nonstationary data on ocean currents it also gave to data generated by randomizing the phases of the Fourier transform. Kaplan and Cohen [32] argued that fibrillation was not usefully described as chaotic, again since randomly phased data gave similar dimensions. Smith [67] has used the FT algorithm to generate surrogates which are used to assess the predictability of geophysical time series. Weiss [62], described a comparison of the sunspot time series against a particular stochastic model. Brock et al. [52] used technical trading rules to distinguish stock market data from surrogates generated by several stochastic models. And Ellner [68] showed that a variety of "plausible alternatives" might adequately explain measles and chickenpox data, despite earlier claims of chaos.

Brock and coworkers in particular [33, 52,

69–71], and the economics community in general [29, 34, 72, 73], have been extremely active in the development of statistical tools for time series analysis. While the choice of null hypotheses for financial time series tends to be different than for more physical time series (autocorrelation plays a lesser role, for example), the overall methodologies are quite similar. Classical statisticians [20–25, 28] have long considered tests for nonlinearity, and are becoming increasingly aware of low-dimensional chaos (just as physicists are becoming increasingly aware of the importance of the statistical approach); we cite Tong [26] as *the* review which most neatly and comprehensively ties these two fields together.

6. Conclusion

In this article, we have described an approach for evaluating the statistical significance of evidence for nonlinearity in a stationary time series. The test properly fails to find nonlinear structure in linear stochastic systems, and correctly identifies nonlinearity in several well-known examples of low-dimensional chaotic time series, even when contaminated with dynamical and observational noise. We illustrated the method with several experimental data sets, and confirmed the evidence for nonlinear structure in some systems, while failing to see such structure in other time series.

Acknowledgements

We are pleased to acknowledge Martin Casdagli, Xiangdong He, Willian Brock, Blake LeBaron, Steve Ellner, Peter Grassberger, Mark Berge, Tony Begg and Bette Korber for useful discussions. We are especially grateful to Lenny Smith for illuminating disagreements and a careful reading of the manuscript. We also thank Bob Ecke for providing convection data, and Paul Nuñez and Arden Nelson for providing EEG data. This work was partially supported by the National Institute for Mental Health under grant 1-R01-MH47184-01, and performed under the auspices of the Department of Energy.

References

[1] C. Nicolis and G. Nicolis, Nature 311 (1984) 529.
[2] P. Grassberger, Nature 323 (1986) 609.
[3] C. Nicolis and G. Nicolis, Nature 326 (1987) 523.
[4] P. Grassberger, Nature 326 (1987) 524.
[5] A.R. Osborne, A.D. Kirwin, A. Provenzale and L. Bergamasco, Physica D 23 (1986) 75.
[6] G. Mayer-Kress, in: Directions In Chaos, Vol. I, ed. B.-L. Hao (World Scientific, Singapore, 1988), pp. 122–147.
[7] A.R. Osborne and A. Provenzale, Physica D 35 (1989) 357.
[8] A. Provenzale, A.R. Osborne R. Soj, Physica D 47 (1991) 361.
[9] J. Theiler, Phys. Lett. A 155 (1991) 480.
[10] J.B. Ramsey and J.-J. Yuan, Phys. Lett. A 134 (1989) 287.
[11] J. Theiler, Phys. Rev. A 41 (1990) 3038.
[12] F. Takens, in: Dynamical Systems and Bifurcations (Groningen, 1984), eds. B.L.J. Braaksma, H.W. Broer and F. Takens, Lecture Notes in Mathematics, Vol. 1125 (Springer, Berlin, 1985), pp. 99–106.
[13] W.E. Caswell and J.A. Yorke, in: Dimensions and Entropies in Chaotic Systems – Quantification of Complex Behavior, Springer Series in Synergetics, Vol. 32 (Springer, Berlin, 1986), pp. 123–136.
[14] J. Holzfuss and G. Mayer-Kress, in: Dimensions and Entropies in Chaotic Systems – Quantification of Complex Behavior, Springer Series in Synergetics, Vol. 32 (Springer, Berlin, 1986), pp. 114–122.
[15] J. Theiler, Quantifying Chaos: Practical Estimation of the Correlation Dimension, Ph.D. thesis (Caltech, 1988).
[16] M. Möller, W. Lange, F. Mitschke, N.B. Abraham and U. Hübner, Phys. Lett. A 138 (1989) 176.
[17] J. Theiler, J. Opt. Soc. Am. A 7 (1990) 1055.
[18] R.L. Smith, in: Nonlinear Modeling and Forecasting, eds. M. Casdagli and S. Eubank, of SFI Studies in the Sciences of Complexity, Vol. XII (Addison-Wesley, Reading, MA, 1992) pp. 115–136.
[19] J. Theiler, B. Galdrikian, A. Longtin, S. Eubank and J.D. Farmer, in: Nonlinear Modeling and Forecasting, eds. M. Casdagli and S. Eubank, SFI Studies in the Sciences of Complexity, Vol. XII (Addison-Wesley, Reading, MA, 1992) pp. 163–188.
[20] T. Subba Rao and M.M. Gabr, J. Time Series Anal. 1 (1980) 145.
[21] M.J. Hinich, J. Time Series Anal. 3 (1982) 169.
[22] A.I. McLeod and W.K. Li, J. Time Series Anal. 4 (1983) 269.

[23] D.M. Keenan, Biometrika 72 (1985) 39.
[24] R.S. Tsay, Biometrika 73 (1986) 461.
[25] R.S. Tsay, Stat. Sin. 1 (1991) 431.
[26] H. Tong, Non-linear Time Series: A Dynamical System Approach (Clarendon Press, Oxford, 1990).
[27] B. Efron, SIAM Rev. 21 (1979) 460.
[28] R.S. Tsay, Appl. Stat. 41 (1992) 1.
[29] J.A. Scheinkman and B. LeBaron, J. Business 62 (1989) 311.
[30] J.L. Breeden and N.H. Packard, Nonlinear analysis of data sampled nonuniformly in time, Physica D 58 (1992) 273, these Proceedings.
[31] G.E. Uhlenbeck and L.S. Ornstein, Phys. Rev. 36 (1930) 823; reprinted in: Noise and Stochastic Processes, ed. N. Wax (Dover, New York, 1954).
[32] D.T. Kaplan and R.J. Cohen, Circulation Res. 67 (1990) 886.
[33] W.A. Brock, W.D. Dechert and J. Scheinkman, A test for independence based on the correlation dimension, Social Systems Research Institute, University of Wisconsin at Madison, technical report 8702 (1986).
[34] T.-H. Lee, H. White and C.W.J. Granger, Testing for neglected nonlinearity in time series models: A comparison of neural network methods and alternative tests, J. Econometrics, to appear.
[35] E.J. Kostelich and H.L. Swinney, in: Chaos and Related Natural Phenomena, eds. I. Procaccia and M. Shapiro (Plenum, New York, 1987), p. 141.
[36] J. Theiler, Phys. Rev. A 36 (1987) 4456.
[37] P. Grassberger, Phys. Lett. A 148 (1990) 63.
[38] P. Grassberger and I. Procaccia, Phys. Rev. Lett. 50 (1983) 346.
[39] P. Grassberger and I. Procaccia, Physica D 9 (1983) 189.
[40] F. Takens, Invariants related to dimension and entropy, in: Atas do 13° Colóqkio Brasiliero de Matemática (1983).
[41] S. Ellner, Phys. Lett. A 133 (1988) 128.
[42] J.D. Farmer and J.J. Sidorowich, in: Evolution, Learning and Cognition, ed. Y.C. Lee (World Scientific, Singapore, 1988), pp. 277–330.
[43] M. Casdagli, in: Modeling Complex Phenomena, eds. L. Lam and V. Naroditsky (Springer, New York, 1992), p. 131.
[44] M. Casdagli, Chaos and deterministic versus stochastic nonlinear modeling, J. R. Stat. Soc. B 54 (1992) 303.
[45] M. Sano and Y. Sawada, Phys. Rev. Lett. 55 (1985) 1082.
[46] J.-P. Eckmann and D. Ruelle, Rev. Mod. Phys. 57 (1985) 617.
[47] J.-P. Eckmann, S.O. Kamphorst, D. Ruelle and S. Ciliberto, Phys. Rev. A 34 (1986) 4971.
[48] S. Ellner, A.R. Gallant, D. McCaffrey and D. Nychka, Phys. Lett. A 153 (1991) 357.
[49] D. McCaffrey, S. Ellner, A.R. Gallant and D. Nychka, Estimating the Lyapunov exponent of a chaotic system with nonparametric regression, J. Am. Stat. Assoc., to appear.
[50] D. Nychka, S. Ellner, D. McCaffrey and A.R. Gallant, J. R. Stat. Soc. B 54 (1992) 399.
[51] T. Schreiber and P. Grassberger, Phys. Lett. A 160 (1991) 411.
[52] W.A. Brock, J. Lakonishok and B. LeBaron, Simple technical trading rules and the stochastic properties of stock returns, J. Finance, to appear.
[53] R.B. Blackman and J.W. Tukey, The Measurement of Power Spectra (Dover, New York, 1959).
[54] M. Hénon, Commun. Math. Phys. 50 (1976) 69.
[55] M.C. Mackey and L. Glass, Science 197 (1977) 287.
[56] J. Theiler, Phys. Rev. A 34 (1986) 2427.
[57] H. Haucke and R. Ecke, Physica D 25 (1987) 307.
[58] P.E. Rapp, I.D. Zimmerman, A.M. Albano, G.C. deGuzman, N.N. Greenbaum and T.R. Bashore, in: Nonlinear Oscillations in Biology and Chemistry, ed. H.G. Othmer (Springer, Berlin, 1986), pp. 175–205.
[59] P.E. Rapp, T.R. Bashore, J.M. Martinerie, A.M. Albano, I.D. Zimmerman and A.I. Mees, Brain Topography 2 (1989) 99.
[60] G.U. Yule, Philos. Trans. R. Soc. London A 226 (1927) 267.
[61] H. Tong and K.S. Lim, J. R. Stat. Soc. B 42 (1980) 245.
[62] N.O. Weiss, Phil. Trans. R. Soc. London A 330 (1990) 617.
[63] J. Kurths and A.A. Ruzmaikin, Solar Phys. 126 (1990) 407.
[64] A. Weigend, B. Huberman and D. Rummelhart, Intern. J. Neural Systems 1 (1990) 193.
[65] M. Mundt, W.B. Maguire II and R.B. P. Chase, J. Geophys. Res. 96 (1991) 1705.
[66] A. Weigend, B.A. Huberman and D.E. Rummelhart, Predicting sunspots and exchange rates with connectionist networks, in: Nonlinear Modeling and Forecasting, eds. M. Casdagli and S. Eubank, SFI Studies in the Sciences of Complexity, Vol. XII (Addison-Wesley, Reading, MA, 1992), pp. 397–434.
[67] L. Smith, Identification and prediction of deterministic dynamical systems, this volume.
[68] S. Ellner, Detecting low-dimensional chaos in population dynamics data: a critical review, in: Chaos and Insect Ecology, eds. J.A. Logan and F.P. Hain (University of Virginia Press, Blacksburg, VA, 1991), pp. 65–92.
[69] W.A. Brock and C.L. Sayers, J. Monetary Econ. 22 (1988) 71.
[70] W.A. Brock and W.D. Dechert, Statistical inference theory for measures of complexity in chaos theory and nonlinear science, in: Measures of Complexity and Chaos, eds. N. Abraham et al. (Plenum, New York, 1989), pp. 79–98.
[71] W.A. Brock and S.M. Potter, in: Nonlinear Modeling and Forecasting, eds. M. Casdagli and S. Eubank, SFI Studies in the Sciences of Complexity, Vol. XII (Addison-Wesley, Reading, MA, 1992), pp. 137–162.
[72] D.A. Hsieh, J. Business 62 (1989) 339.
[73] D.A. Hsieh, J. Finance 46 (1991) 1839.

Strange attractors in weakly turbulent Couette-Taylor flow

A. Brandstater and Harry L. Swinney
Department of Physics and the Center for Nonlinear Dynamics, The University of Texas, Austin, Texas 78712-9990
(Received 7 August 1986)

We have conducted an experiment on the transition from quasiperiodic to weakly turbulent flow of a fluid contained between concentric cylinders with the inner cylinder rotating and the outer cylinder at rest. Power spectra, phase-space portraits, and circle maps obtained from velocity time-series data indicate that the nonperiodic behavior which we have observed is deterministic, that is, it is described by strange attractors. We discuss various problems that arise in computing the dimension of strange attractors constructed from experimental data and show that these problems impose severe requirements on the quantity and accuracy of data necessary for determining dimensions greater than about 5. In the present experiment the attractor dimension increases from 2 at the onset of turbulence to about 4 at a Reynolds number 50% above the onset of turbulence.

I. INTRODUCTION

The determination of the fractal dimension d of strange attractors has become a standard diagnostic tool in the analysis of dynamical systems.[1–5] The dimension, roughly speaking, measures the number of degrees of freedom that are relevant to the dynamics. Most work on dimension has concerned maps, such as the Hénon map, or systems given by a few coupled ordinary differential equations, such as the Lorenz and Rössler models. For any chaotic system described by differential equations, d must be greater than 2, but d could be much much larger for a system described by *partial* differential equations, such as the Navier-Stokes equations. Indeed, the best rigorous estimates of the dimension of attractors for turbulent flow yield quite large numbers for the attractor dimension (say 10^8).[6–8]

Recent experiments on the transition to chaos in the Couette-Taylor system,[9] convection,[10] and a differentially heated annulus[11] have shown that the chaotic (strange) attractors for these systems have surprisingly low dimension. However, efforts to compute dimension from experimental data have revealed a number of problems that do not arise in the analysis of model systems where essentially unlimited amounts of data of arbitrary accuracy can easily be generated.[3,5] Using our Couette-Taylor data as an example, we will examine in this paper some of the problems that arise in determining dimension from laboratory data.

After a brief discussion of the experimental methods in Sec. II, we will show in Sec. III that photographs, velocity power spectra, phase-space portraits, and circle maps alone provide evidence that the observed nonperiodic behavior is *deterministic* and *low dimensional*, that is, *chaotic*. In Sec. IV we present the dimension calculations, emphasizing the limitations and possible pitfalls in such calculations. Section V is a discussion and summary.

II. EXPERIMENTAL METHODS

The Reynolds number for the Couette-Taylor system can be defined as $R = a\Omega(b-a)/\nu$, where a and b are, respectively, the radii of the inner and outer cylinders, Ω is the angular velocity of the inner cylinder, and ν is the kinematic viscosity. In our system we have $a = 5.205$ cm and $b = 5.947$ cm, which gives a radius ratio of 0.875. At this radius ratio the critical Reynolds number for the onset of Taylor vortex flow in an infinite system is $R_c = 118.4$.[12] The upper and lower boundaries of the annulus are Teflon rings that are fastened to the outer cylinder, and the ratio of the fluid height to the gap between the cylinders is 20.0. The frequency of rotation of the inner cylinder is locked to a crystal reference oscillator.

The working fluid is orange oil (Cargille No. 19604), chosen because it has the same index of refraction at the laser wavelength (488 nm) as the borosilicate glass outer cylinder. The fluid kinematic viscosity is 0.0109 cm^2/s at 25 °C. In the laser Doppler velocimetry measurements the fluid is seeded with 0.22-μm diam spherical titanium dioxide particles, and the flow visualization studies are made using a dilute suspension of small platelet particles (Kalliroscope AQ1000, Ref. 13). The cylinder system is immersed in a bath that is controlled in temperature to 0.1 °C. The bath has the same refractive index as the glass cylinder and the working fluid, and the windows of the bath are parallel glass plates; therefore, the optics are all planar for the laser Doppler velocimetry measurements.

The radial component of the velocity is measured (at points midway in the gap between the inner and outer cylinders) by the laser Doppler crossed-beam technique. A 256-channel pulse correlator is used to determine the correlation function of the pulse train from the photomultiplier; each pulse corresponds to the detection of a single photon. In the Reynolds-number range of interest the signal is sufficiently strong so that a correlation function (oscillating at the Doppler shift frequency) can be obtained in a time that is short compared to the time required for the velocity to change significantly—if needed, as many as 100 velocity points can be obtained per characteristic oscillation period. Thus the Doppler shifts extracted from the correlation functions at successive time intervals yield essentially the instantaneous velocity.[14,15] The velocity data files each contain 32 768 points.

The spatial resolution of the laser Doppler measurements is set by the size of the scattering volume. This sample volume is approximately an ellipsoid of revolution of length 100 μm in the azimuthal direction and 30 μm in the axial and radial directions; for comparison, the gap between the cylinders is 7.42 mm. The concentration of the 0.22-μm diam spherical seed particles is adjusted so that the scattering volume contains about one particle on the average.

III. DETERMINISTIC NONPERIODIC FLOW (CHAOS)

A. Photographs, time series, and spectra

Previous experiments have shown that the Couette-Taylor system with a radius ratio of 0.875 and aspect ratio around 20 passes through the following sequence of well-defined states as the Reynolds number is increased:[16–18] the basic (Couette) flow; Taylor vortex flow, which is time independent; wavy vortex flow, which is periodic in time; modulated wavy vortex flow, which is quasiperiodic; and, finally, chaotic (weakly turbulent) flow. In this paper we are concerned with the transition from modulated wavy vortex flow to chaos. Figure 1 shows photographs of the flow for a sequence of Reynolds numbers ranging from $R/R_c = 10.6$, which is in the modulated wavy vortex flow regime, to $R/R_c = 26.3$, which is well beyond the onset of chaos ($R/R_c = 11.7 \pm 0.2$).[19]

The photographs show that some small-scale spatial structure is discernible below the onset of chaos, but the small-scale structure in successive vortex pairs appears to be the same. Above the onset of chaos there is an increasing amount of small-scale structure, and it is no longer the same in different vortex pairs. Nevertheless, a high degree of spatial order persists above the onset of chaos; therefore, we prefer to call this flow chaotic rather than turbulent, but, following current usage, we also sometimes call this flow "weakly turbulent." The increase in the number of degrees of freedom with increasing Reynolds number, qualitatively clear from the photographs, will be quantified by the dimension calculations given in Sec. IV.

Velocity time series and power spectra are shown for a sequence of Reynolds numbers in Figs. 2 and 3, respectively. In Figs. 2, 3(a), and 3(b) the upper two graphs are for modulated wavy vortex flow, while the lower two graphs in each figure are for chaotic flow. In modulated wavy vortex flow there are two traveling azimuthal waves[17,18] on the Taylor vortices, the first with rotation frequency $\omega_1/m_1\Omega = 0.34$ and the second with rotation frequency $\omega_2/m_2\Omega = 0.44$; in the present experiment the azimuthal wave numbers for both waves were the same, $m_1 = m_2 = 4$. The values 0.34 and 0.44 for the wave frequencies are approximate; actually the ratio ω_2/ω_1 increases slowly but in a strictly monotone fashion with increasing Reynolds number in the vicinity of the onset of chaos—*no* frequency locking is observed.[20]

The velocity power spectra for modulated wavy vortex flow contain two fundamental frequency components and their combinations, as Fig. 3 illustrates with data at $R/R_c = 10.4$ and 11.4. The background level in these two spectra is instrumental noise rather than fluid noise; the instrumental noise is white and independent of Reynolds number [see Fig. 3(b)]. The onset of nonperiodic flow is marked by a rise in the background noise above the instrumental noise level, as can be seen in the spectra at $R/R_c = 12.0$ and 15.5. However, even at $R/R_c = 15.5$ most of the spectral energy is still in the components at ω_1 and ω_2 (and their combinations), and the flow is still highly ordered, as can be seen in the photographs in Fig. 1. At larger Reynolds numbers the amplitudes of the components at ω_1 and ω_2 decrease with increasing Reynolds number, and these components disappear at $R/R_c = 22.0$ and 20.0, respectively.[9,16,21]

B. Spectral evidence for deterministic behavior

Broadband noise in power spectra can arise from stochastic or deterministic processes, but the decay in the spectral power at large ω is different for the two cases.[22,23] For a process governed by an nth-order *stochastic* differential equation, Sigeti and Horsthemke[23] have shown that the spectral power at high frequencies follows a power law, $P(\omega) \propto \omega^{-2n}$. In contrast, the assumption that the solution of *deterministic* equations is infinitely differentiable leads to the conclusion that the spectrum falls off exponentially, or at least faster than ω^{-s} for arbitrarily large s. Greenside *et al.*[22] examined numerically a stochastic differential equation model and a deterministic model, and they found a power-law decay for the first case (with n equal to the order of the equation, 2) and an exponential decay for the second case, which was the Lorenz model.

Greenside *et al.* analyzed data obtained by Ahlers and co-workers[24] for weakly turbulent convection and found power-law behaviors, ω^{-4} in some cases and ω^{-2} in other cases. Another study of turbulent convection, by Atten *et al.*,[25] showed exponential decay at high frequencies.

Figure 4 shows two graphs of our velocity power spectra obtained at $R/R_c = 15.2$, one on a semilogarithmic scale and the other on a log-log scale. The exponential decay given by the linear behavior on the semilogarithmic plot indicates that the noise in our data arises from a deterministic rather than a stochastic process.

C. Phase space portraits and Poincaré sections

We have constructed multidimensional phase-space portraits from the velocity time series $\{V(t_k);\ t_k = k\,\Delta t,\ k = 1, \ldots, 32\,768\}$ by the method of time delays:[26,27] trajectories in an m-dimensional space pass through the points $\{V(t_k),\ V(t_k+\tau), \ldots, V(t_k+(m-1)\tau)\}$, where m is the embedding dimension and τ is a time delay.

For an infinite amount of noise-free data the choice of time delay τ is arbitrary,[27] but for real laboratory data a good choice of τ is essential for geometrical as well as numerical analysis of a phase portrait. As $\tau \to 0$, the trajectory approaches the identity line $V(t) = V(t+p\tau)$ for all positive integers $p < m$. There are other values of τ for which the data points are concentrated in a small region of phase space, so that the local structure of the attractor

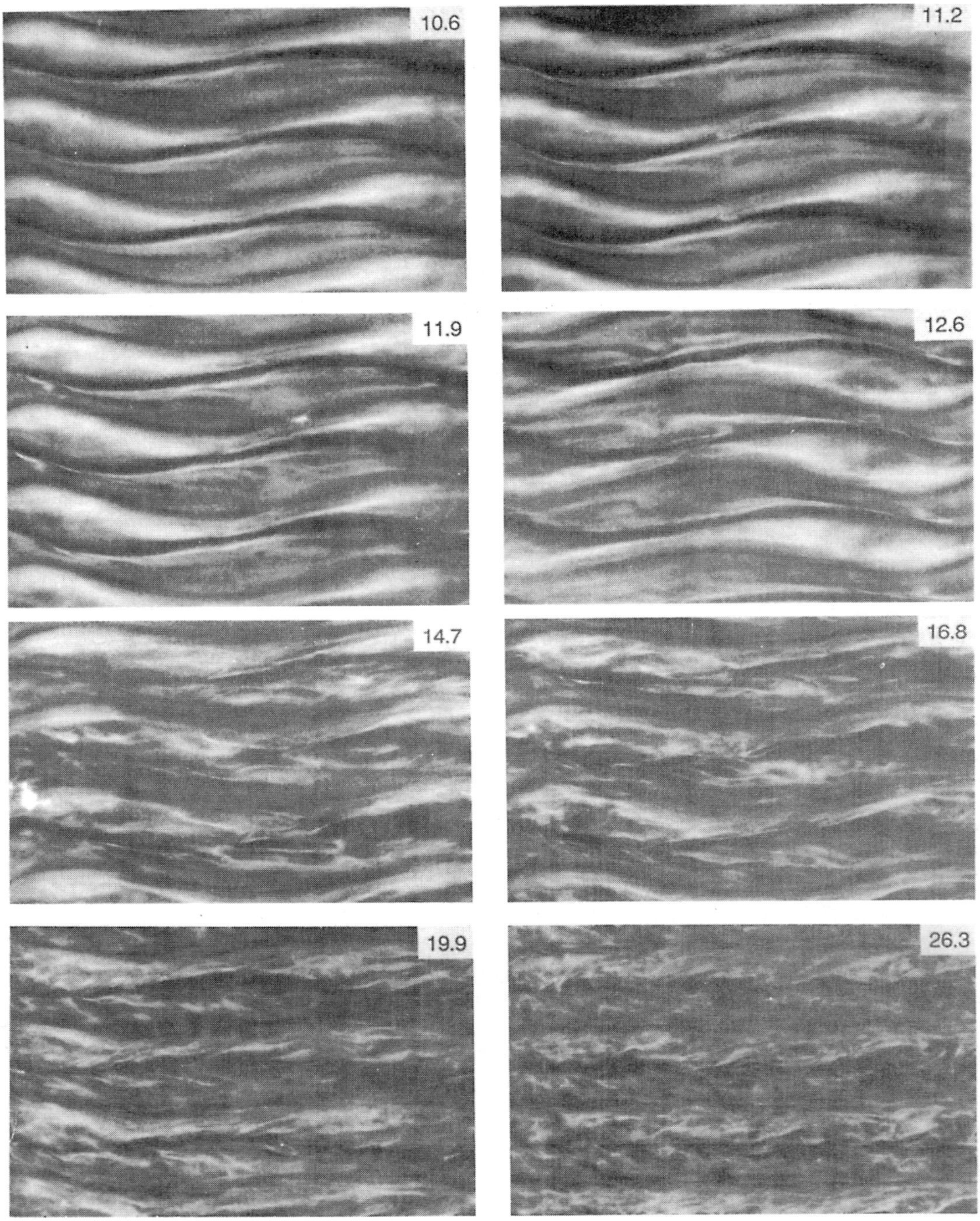

FIG. 1. Photographs of the flow as a function of Reynolds number in the region of transition from modulated wavy vortex flow to chaotic flow. The values of R/R_c are given on the photographs. The first two pictures are of modulated wavy vortex flow, while the remainder are in the chaotic regime. The flow patterns were rendered visible using a suspension of small platelets (Ref. 13).

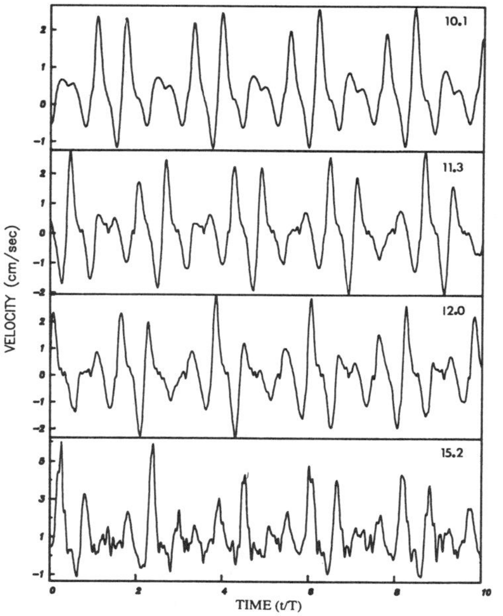

FIG. 2. The time dependence of the radial component of the velocity, measured at a point in the middle of the gap between the inner and outer cylinders, for two Reynolds numbers ($R/R_c=10.1$ and 11.3) corresponding to modulated wavy vortex flow, and two Reynolds numbers ($R/R_c=12.0$ and 15.2) corresponding to chaotic flow. The corresponding power spectra are shown in Fig. 3. The time T is the period of rotation of the inner cylinder; in seconds, $T=18.76/(R/R_c)$.

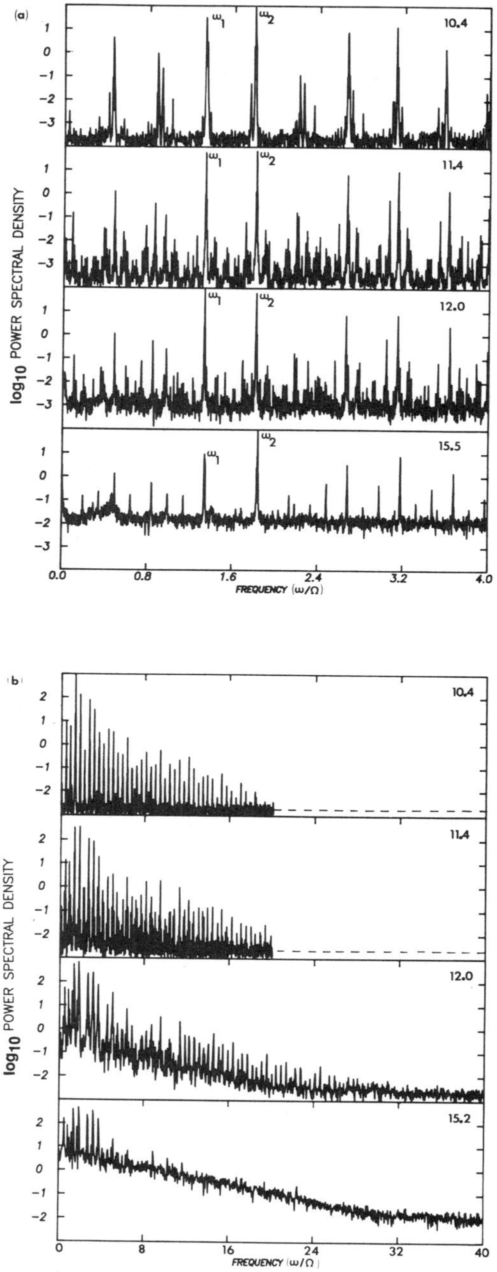

FIG. 3. Velocity power spectra showing (a) low-frequency and (b) high-frequency behavior for quasiperiodic modulated wavy vortex flow at $R/R_c=10.4$ and 11.4 and chaotic flow at $R/R_c=12.0$ and 15.5. The frequency scales are in units of the cylinder frequency Ω.

cannot be extracted from the reconstructed phase portraits.

Recently Fraser and Swinney[28,29] have shown that the optimum choice of τ in most systems corresponds to the first local minimum in the mutual information function,

$$I=\int\int P(X,Y)\log_2[P(X,Y)/P(X)P(Y)]dX\,dY\ ,$$

where, in the present case, $X=V(t)$ and $Y=V(t+\tau)$, so $I=I(\tau)$. $P(X)$ and $P(X,Y)$ are, respectively, the probability density and the joint probability density. The mutual information measures the dependence of two variables in a more general way than the autocorrelation function, which measures the *linear* dependence. The mutual information gives the accuracy in the prediction of the value of $V(t+\tau)$, given the value of $V(t)$. Now if $V(t)$ and $V(t+\tau)$ are to be used as coordinates in constructing a phase portrait, these coordinates should be as independent as possible. Therefore, the optimum choice in τ corresponds to a minimum in the mutual information function. The first minimum is preferable over later minima because as time passes the points on the attractor spread out over the invariant measure.

Figure 5 shows the mutual information as a function of τ for the Couette-Taylor flow data obtained at

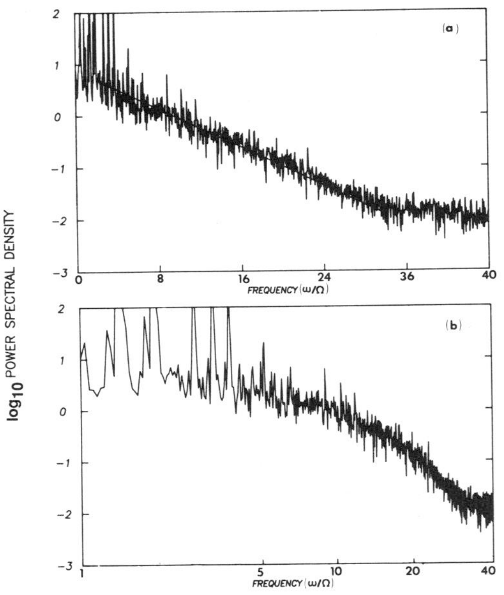

FIG. 4. (a) A semilogarithmic velocity power spectrum and (b) a log-log spectrum for data obtained at $R/R_c=15.5$. The exponential decay (straight-line behavior) down to the instrumental noise level (10^{-2}) is clear in (a).

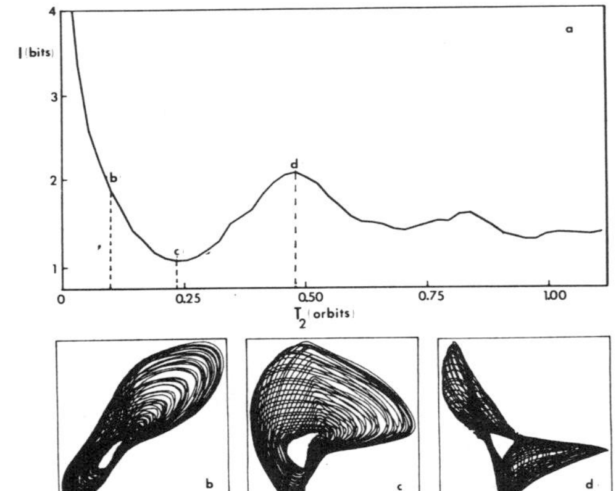

FIG. 5. (a) The mutual information is plotted as a function of time delay τ for data obtained at $R/R_c=10.1$. The time delay is given relative to the period T_2 of the second fundamental frequency ω_2. Two-dimensional projections of phase portraits, $V(t+\tau)$ versus $V(t)$, constructed for three different time delays, are shown below the graph of the mutual information: (b) $\tau/T_2=0.11$, which is too small a delay time; (c) $\tau/T_2=0.24$, corresponding to the first minimum in the mutual information, this is the optimum choice for τ; (d) $\tau/T_2=0.48$, which is too large a delay time.

$R/R_c=10.1$. For these data and for all other data presented in this paper the first minimum in the mutual information occurs for a delay time about equal to one-fourth of the period of the second fundamental frequency, ω_2. Delay times near this optimum value were used in our analysis.

Figure 6(a) shows phase portraits obtained for a sequence of Reynolds numbers, and Fig. 6(b) shows the corresponding Poincaré sections given by the intersection of orbits in three-dimensional phase portraits with planes. At $R/R_c=10.1$ and 11.3 the Poincaré sections clearly show that the orbits lie on the surface of a torus; the scatter presumably arises from instrumental noise. Even at $R/R_c=12.0$, where the photographs, time series, and power spectra indicate the presence of noise, the torus is still evident, although it is much noisier and it has developed prominent wrinkles. With our data files of about 300 orbits we were unable to quantify the folding at the wrinkles.

D. Lyapunov exponents

Lyapunov exponents characterize the exponentially fast divergence or convergence of nearby trajectories in phase space. The largest Lyapunov exponent is zero for a limit cycle, a 2-torus, or, more generally, an n-torus, but at the onset of chaos the largest Lyapunov exponent becomes positive. Wolf *et al.*[30] have developed a method for computing Lyapunov exponents from experimental data, and that method has been applied to our data for the Couette-Taylor system.[9] The conclusion was that the largest Lyapunov exponent becomes positive at $R/R_c=11.7\pm0.2$, which is in accord with the other evidence for the onset of chaos.

E. Circle maps and the onset of chaos

One-dimensional maps of circle, θ_{n+1} versus θ_n, can be constructed from the Poincaré sections of a 2-torus, as shown in Fig. 7: θ_n is the angle of the nth point measured with respect to a polar coordinate system whose origin is inside the closed loop formed by the Poincaré section. Figure 7 shows that this construction yields smooth curves for the function $\theta_{n+1}=f(\theta_n)$, even beyond the onset of chaos. This is further evidence that the nonperiodic flow is deterministic—given θ_n, the map *determines* θ_{n+1}.

Extensive recent studies of circle maps have been motivated by the interest in characterizing the transition from quasiperiodic flow on a 2-torus to chaos.[31–35] Circle maps with cubic inflection points have been found to exhibit *universal* behavior in the vicinity of the control parameter value at which the map becomes noninvertible. As the point of noninvertibility of the map is approached, there is an increasing probability that the winding number of the map, which corresponds to the ratio of the two frequencies of the quasiperiodic flow, will become locked at a rational number value. The theory predicts the scaling behavior of these frequency-locked bands. At the critical point, where the map becomes noninvertible, only frequency-locked states occur. Thus frequency locking plays a crucial role in the large body of theory that has

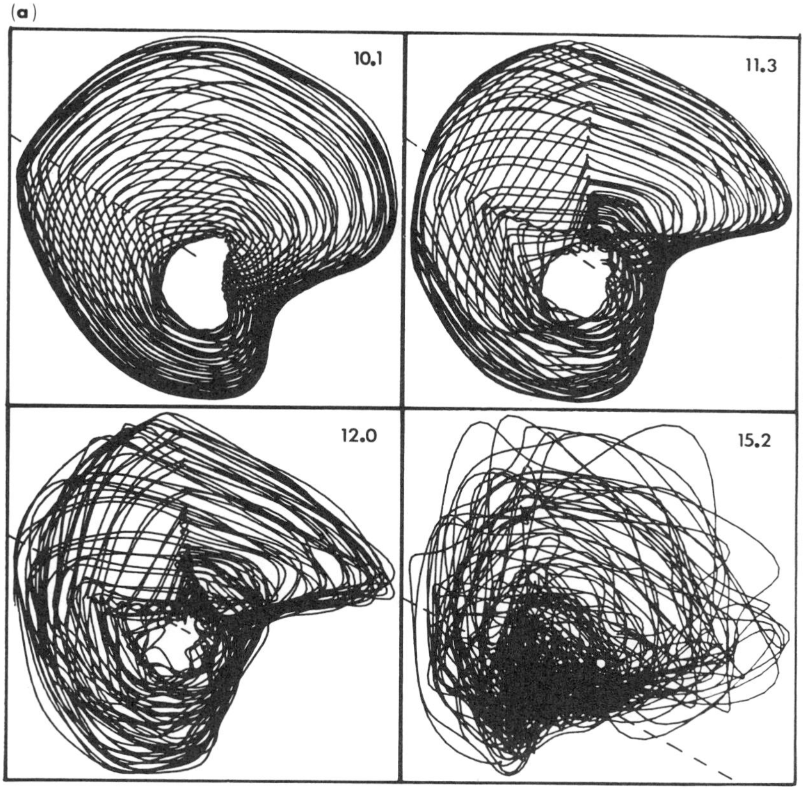

FIG. 6. (a) Two-dimensional phase portraits $V(t+\tau)$ versus $V(t)$ constructed for two Reynolds numbers ($R/R_c = 10.1$ and 11.3) corresponding to modulated wavy vortex flow and two Reynolds numbers ($R/R_c = 12.0$ and 15.2) corresponding to chaotic flow. (b) Poincaré sections given by the intersection of orbits of three-dimensional phase portraits [where the third axis is $V(t+2\tau)$] with planes normal to the paper through the dashed lines in (a). [The delay times τ at the four Reynolds numbers are 162, 144, 144, and 108 ms, respectively; about 40 of the 300 orbits observed at each Reynolds number are shown in (a).]

been developed for circle maps.

Unfortunately, the existent circle-map theory does not apply to the quasiperiodic to chaotic transition in the Couette-Taylor system, because frequency locking is not observed. In fact, Rand has shown that frequency locking *cannot* occur in systems with circular symmetry.[20] Thus the breakup of the 2-torus in systems such as the Couette-Taylor system, rotating concentric spheres, and the differentially heated rotating annulus must occur in a way that is different from the systems described by maps with a cubic inflection point.

Considering again the circle maps in Fig. 7, we should point out that although the loss of invertibility of the circle map appears to occur approximately at the Reynolds number corresponding to the onset of chaos, it is difficult to determine precisely the Reynolds number at which the map becomes noninvertible. This difficulty arises because the appearance of the map depends to some degree on the choice of the delay time and the location of the Poincaré section and on the origin of the polar coordinate system. With the delay time chosen to be far from the value corresponding to the first minimum in the mutual information or with the coordinate system origin purposely located far off center, we found that we could obtain noninvertible maps in the quasiperiodic flow regime, below the onset of chaos. Therefore, we can only say that the association of the loss of invertibility of the circle map with the onset of chaos is suggestive rather than conclusive in our experiments. It is certainly clear, however, that further theoretical and experimental study of circle maps in systems with rotational symmetry is warranted.

IV. ATTRACTOR DIMENSION

As we have shown, the photographs, power spectra, and phase portraits indicate that the nonperiodic Couette-Taylor flow is deterministic. We will now show that the strange attractor is low dimensional, and in doing this we will examine several effects that could influence the result of dimension calculations.[5]

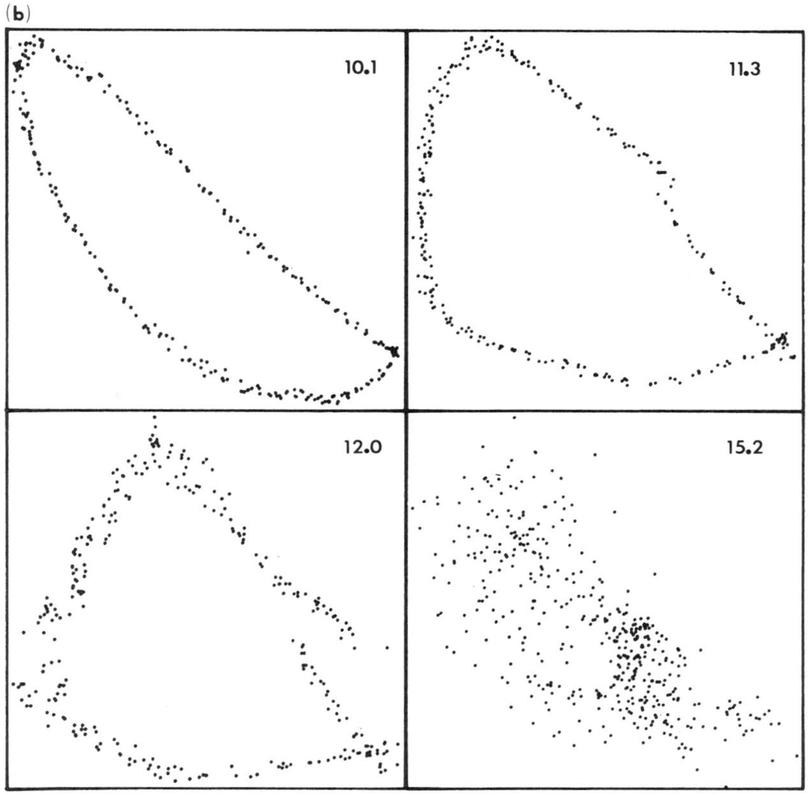

FIG. 6. (*Continued*).

We have computed the pointwise dimension d_μ (Farmer *et al.*[1]) and the correlation dimension ν (Grassberger and Procaccia[36]), both of which are based on the idea that the number of points $N(\epsilon)$ in a ball of radius ϵ scales with the dimension d, $N(\epsilon)\propto\epsilon^d$. The pointwise and correlation dimensions are given by

$$d_\mu=\lim_{\epsilon\to 0}\frac{\langle \ln N(\epsilon)\rangle_x}{\ln\epsilon}, \tag{1}$$

$$\nu=\lim_{\epsilon\to 0}\frac{\ln\langle N(\epsilon)\rangle_x}{\ln\epsilon}, \tag{2}$$

where the ball of radius ϵ is centered at the reference points x on the attractor, and the embedding dimension m satisfies $m\geq 2d_\mu+1$. The quantities d_μ and ν are related by the inequality $\nu\leq d_\mu$; they can be obtained from the same algorithm simply by interchanging the logarithm and the average over different reference points x.

We find that the dimension values determined from the Couette-Taylor data satisfy the inequality $\nu\leq d_\mu$, but the difference between ν and d_μ is within the uncertainty in our dimension values; therefore, we will refer to the calculated dimension values simply as d.

For an m-dimensional embedding we usually use 2^{m+2} reference points because the reference points then fill the phase space equally densely for each value of the embedding dimension. The number of data points used in calculating the dimension is discussed in Sec. IV D below.

A. Different regions of scaling

According to Eqs. (1) and (2), the dimension should be given (in the limit $\epsilon\to 0$, for sufficiently large m) by the slopes of graphs of $\log_{10}N(\epsilon)$ versus $\log_{10}\epsilon$; Fig. 8(a) shows such a graph. Graphs of the local slope, $d[\log_{10}N(\epsilon)]/d(\log_{10}\epsilon)$, such as shown in Fig. 8(b), are very helpful in understanding the ϵ dependence of $N(\epsilon)$ for laboratory data, since the limit $\epsilon\to 0$ obviously cannot be taken due to experimental noise and the finite amount of data.

Consider the four distinct regions shown in Fig. 8(b): In region A, the number of points in a ball and the slope both approach zero as ϵ approaches zero because of the finite number of data points. For slightly larger ϵ, region B, the instrumental or measurement noise is dominant—the balls are smaller than the smallest temporal and spatial scales that can be resolved in the experiment. Since random noise fills all dimensions of phase space, the slope in this region approaches the value of embedding dimension m as ϵ is decreased. With a very large number of points the slope would attain a value equal to embedding dimension before the data-limited region, region A, would be reached. However, in the example in Fig. 8(b), the data file size limitation is reached at a slope of about 3.7, well before a slope equal to the embedding dimension ($m=6$) is attained; with further decrease in ϵ, the slope then begins to decrease. This transition from region B to region A would occur at smaller ϵ if the number of data points were increased, but there would always be a transition from region B to region A for any data file with a finite number of points.

Region C is the region of primary interest: the constant non-integer-valued slope reflects the fractal structure of the strange attractor. The value of the slope is the dimension of the attractor if the embedding dimension is sufficiently large (see next subsection).

Finally, in region D the ball size ϵ approaches the size

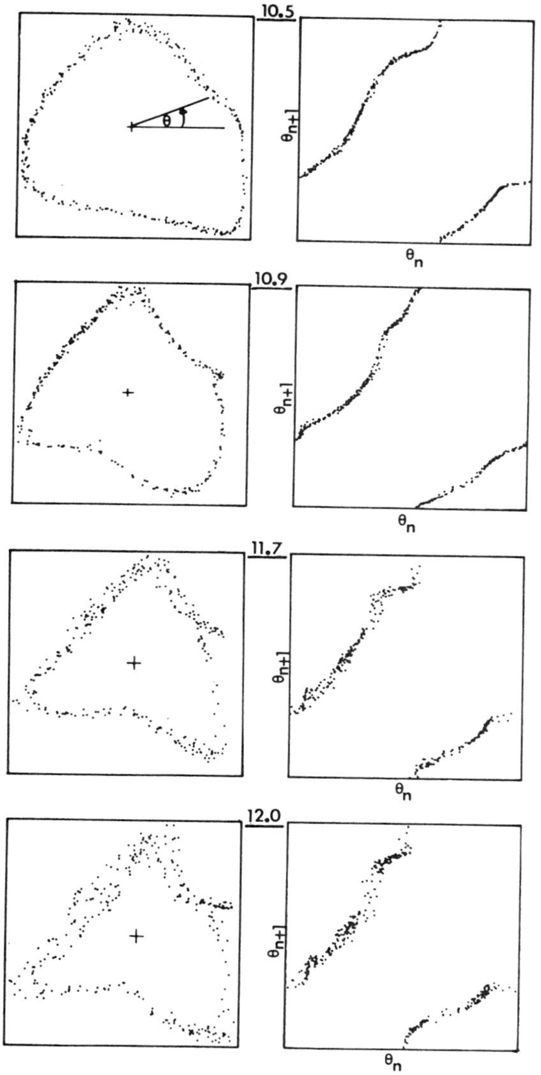

FIG. 7. The Poincaré sections shown on the left were used to construct maps of the circle shown on the right, θ_{n+1} versus θ_n, $\theta \epsilon [0, 2\pi]$. The crosses in the Poincaré sections show the origins of the polar coordinate systems. At $R/R_c = 10.5$ and 10.9 the maps are invertible; at $R/R_c = 11.7$ the map has become fuzzy and the slope appears to be about zero for two values of θ_n; and at $R/R_c = 12.0$ the map has developed a relative maximum and is no longer invertible.

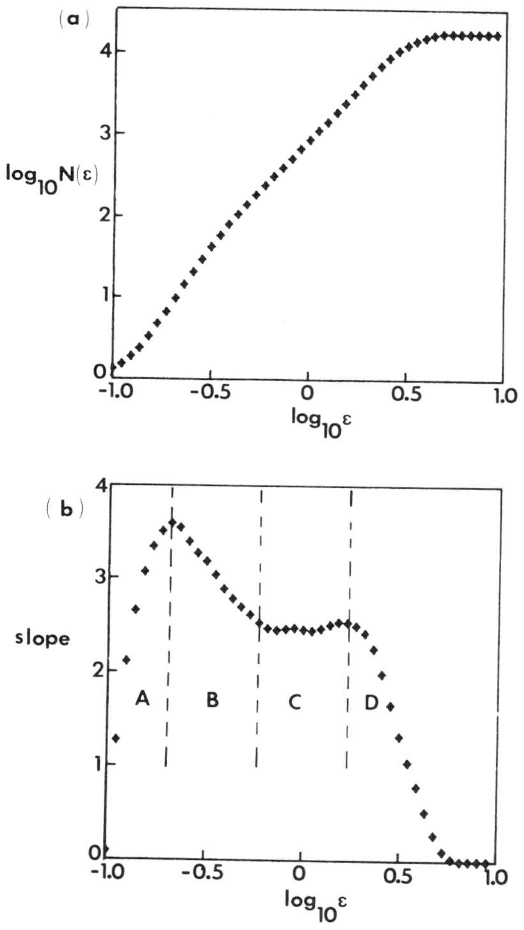

FIG. 8. (a) The number of points $N(\epsilon)$ inside a ball of radius ϵ, averaged over 256 reference points, for a strange attractor with 16 384 points. (The ball radius ϵ is the velocity in cm/s.) (b) The local slope of (a), $d[\log_{10} N(\epsilon)]/d(\log_{10}\epsilon)$ as a function of the ball radius ϵ. Regions A, B, C, and D are discussed in the text. The slope in the scaling region, region C, yields $d=2.4$. ($R/R_c = 12.4$; $m=6$; $\tau = 182$ ms)

of the attractor, and N saturates at a value corresponding to the total number of data points; hence the slope approaches zero. Around the transition from region C to D the slope increases for some systems. The reason for this is that the ball radius approaches the edge of the attractor where the curvature of the attractor contributes a larger amount of data.

A major problem in extracting dimension values from laboratory data is that the noise level can be so large that it is difficult to distinguish regions B and C. Reliable values of d can be obtained only if region C is fairly broad. Indeed, the idea of a self-similar fractal structure of an attractor is meaningful only if it occurs for a fairly wide range in ϵ. Wide scaling regions can be achieved only with a high signal-to-noise ratio and large numbers of data points (see Sec. IV D); otherwise, region B will extend to such large ϵ that no scaling region will be discernible. In our experiments region C extends over more than an order of magnitude range in ϵ at small Reynolds numbers, but for $R/R_c > 17$ the plateau corresponding to the self-similar structure is hardly discernible. Longer data files and more precise data are needed to obtain wider scaling regions.

Another source of error arises in averaging over several reference points: different regions on the attractor can show different scaling behavior at a given length scale.[5(c)] Then averaging $N(\epsilon)$ over all reference points with ϵ fixed yields erroneous values of d. This problem is most serious when the scaling region (region C) is very narrow. In this case the scaling about individual reference points should be examined before averaging. We have done this at $R/R_c = 19.0$, where the averaged data show no well-defined scaling region, and we have found that about half of the randomly selected reference points have a scaling region. Not surprisingly, the scaling regions occur at different ϵ for different reference points. However, it is not simple to define a reliable procedure for choosing the scaling regions for individual reference points in the case of experimental data; therefore, our dimension values for $R/R_c > 17$ (see Sec IV G) have a large uncertainty.

B. Embedding dimension

The effect of varying the embedding dimension m is illustrated in Fig. 9. According to the embedding theorem,[27] the reconstructed phase space must have a dimension m of at most $2d + 1$ to ensure an embedding of the attractor. For m too small, projection effects could reduce the apparent dimensionality of an attractor. For each set of data we computed d as a function of m, and we found that usually we had to increase m to nearly $2d + 1$ before the value deduced for d became independent of m; see Fig. 9.

C. Delay time

A poor choice of delay time τ can result in a very narrow scaling region, or there can be an apparent scaling region that does not actually correspond to the dimension of the attractor, or there may even be no scaling region at all. For example, if τ is too small, the attractor can be so flat that a ball with a radius equal to the instrumental noise scale will also span the thickness of the entire attractor. The effect of varying delay times is illustrated in Fig. 10, which shows graphs of the slope $d[\log_{10}N(\epsilon)]/d(\log_{10}\epsilon)$ computed for delay times corresponding to the attractors shown in Figs. 5(b), 5(c), and 5(d). For the optimum delay time, corresponding to the first minimum in the mutual information, the width of the scaling region [see Fig. 10(b)] is larger than for shorter [Fig. 10(a)] or longer [Fig. 10(c)] delay times.

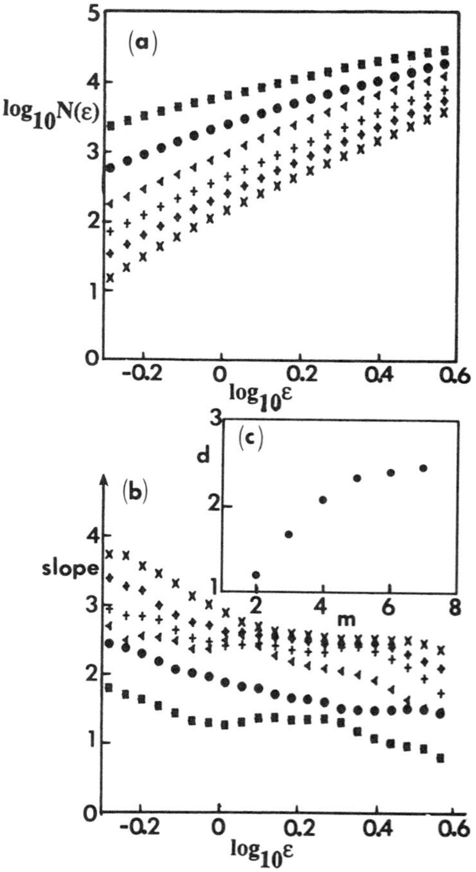

FIG. 9. The effect of varying the embedding dimension m is illustrated with data obtained at $R/R_c = 16.0$. (■, $m=2$; ●, $m=3$; ◀, $m=4$; +, $m=5$; ♦, $m=6$; and ×, $m=7$). (a) The dependence of the average number of points $N(\epsilon)$ in a ball on the radius ϵ of the ball. The data file has 32 768 points, and 2^{m+2} reference points were used in obtaining the average of $\log_{10}N(\epsilon)$. (b) The slope of the curves shown in (a) as a function of ϵ. (c) The slope of the curves in the scaling region as a function of m. The value of the dimension d is given by the asymptotic value of the slope at large m, which is 2.5.

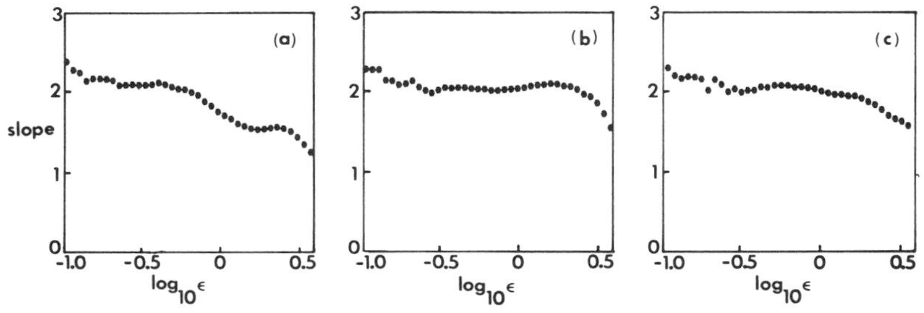

FIG. 10. The slope $d[\log_{10}N(\epsilon)]/d(\log_{10}\epsilon)$ for the data shown in Fig. 5: (a) $\tau/T_2=0.11$, which is smaller than the optimum value of τ; (b) $\tau/T_2=0.24$, which is the optimum τ; and (c) $\tau/T_2=0.48$, which is larger than the optimum τ. The scaling region where the slope is constant is wider for the optimum choice of τ.

D. Data requirements

The number of data points necessary to determine the dimension of an attractor depends on the structure of the attractor, the dimension of the attractor, and the sampling rate. To our knowledge there have been no systematic numerical or analytic studies of the data requirements for determining the dimension of strange attractors, but the number of points necessary to resolve an attractor down to a length scale ϵ increases exponentially with increasing dimension.[3] Therefore, the analysis of high-dimensional attractors by the techniques we have used will require extremely large data files. However, Sornerjai[5(e)] has recently proposed that "*calibrated*" algorithms could be used to obtain reasonable estimates of dimension from fairly small data sets. This intriguing possibility should be carefully examined in a future study.

We have found that the accuracy in the value of d deduced for a given number of points depends on the sampling rate. Broomhead and King[37] give a lower bound on the sampling time, but we know of no criterion for choosing the optimum sampling rate. We have tested the influence of sampling time for model systems such as the Rössler and Lorenz attractors,[38] and have found that the broadest scaling regions (region C in Fig. 8) are obtained for about 10—30 points per orbit; the size of the scaling region shrinks for larger or smaller sampling rates. In our experiment we have used rates of 10—100 points per orbit.

We have examined the number of data points n necessary to determine the dimension of one of our attractors for Couette-Taylor flow, and the results are shown in Fig. 11. The sampling rate was held fixed at about 35 points/orbit. It is evident that, as n is increased, the size of the scaling region extends to lower and lower ϵ. Only 400 data points provide enough resolution to resolve the attractor at large ϵ. However, n should also be large enough to resolve scales down to the noise scale, and this requires several thousand points for this 2.4-dimensional attractor, as Fig. 11 illustrates.

E. Analysis of random numbers

Guckenheimer has used the order statistics of independent random variables to examine problems attendant with obtaining accurate dimension estimates for randomly distributed points.[3(b)] Here we illustrate some of these problems by applying our dimension algorithm to randomly distributed data.

The number of points required to achieve an average separation ϵ in a d-dimensional space of linear extent L is $(L/\epsilon)^d$. Since the determination of dimension from Eq. (1) or (2) requires taking the limit $\epsilon\to 0$, the average separation should be small, say 3% or less of the linear

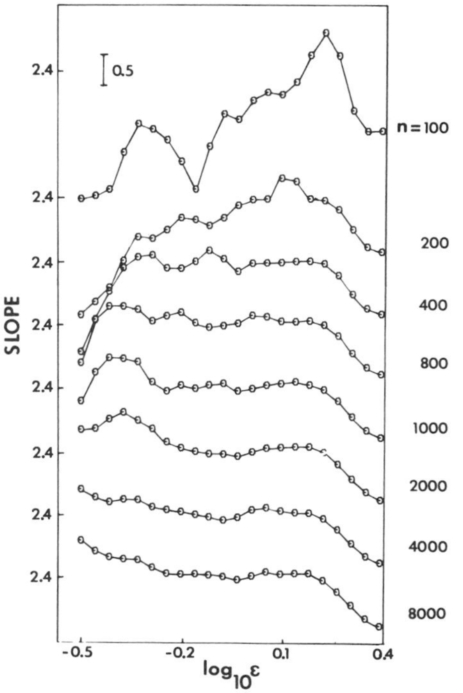

FIG. 11. The dependence of the slope $d[\log_{10}N(\epsilon)]/d(\log_{10}\epsilon)$ on the number of data points n for data obtained at $R/R_c=12.0$. A file with 32 768 points yields $d=2.4$. ($m=7$)

extent. This would require 4×10^4 points in a three-dimensional space and 5×10^{10} points in a seven-dimensional space.

We have determined numerically the approximate width of the scaling range, $\Delta\log_{10}\epsilon$, for randomly distributed points in hypercubes with dimension $m=2$, 3, and 4. Independently generated random numbers were used for each coordinate. The number of points n necessary to achieve scaling ranges $\Delta\log_{10}\epsilon$ of widths 0.4 and 0.8 was determined as a function of m, as shown in Fig. 12. Although the determination of the width of the scaling region is imprecise, the results suggest that the required number of data points increases slightly faster than 10^m. We should mention that even for the largest number of points (10^7), the value of d given by the slopes of the $\log_{10}N$ versus $\log_{10}\epsilon$ graphs was a few percent smaller than m.

In the next numerical experiment we investigated the number of points required to determine the dimension of randomly distributed points in a ten-dimensional space. The value of d obtained for 10^7 points was about 9, but d was still increasing with increasing n, as Fig. 13 illustrates. Both for the present case and for the two-, three-, and four-dimensional spaces discussed in the previous paragraph we thought that the difference between the dimension deduced from the data and the dimension of the space could perhaps arise from a deterministic property of the random-number generators. However, we tried two different 64-bit random-number generators (the Cray Fortran routine RANF and the IMSL routine GGUW) on a Cray X-MP computer, and no difference was found.

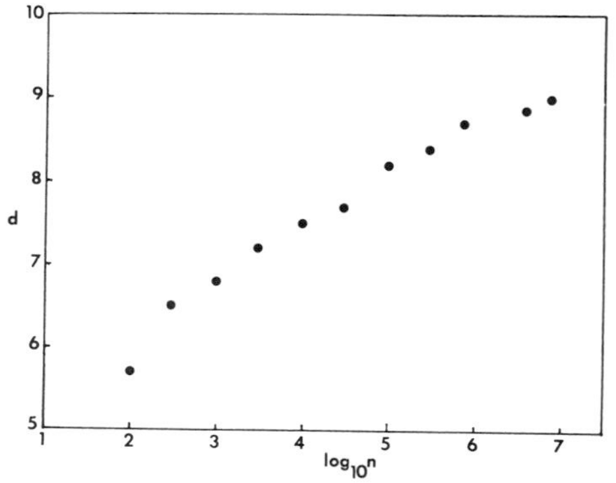

FIG. 13. The value of d determined from the dimension algorithm for a ten-dimensional hypercube containing n randomly distributed points. The analysis used 100 randomly chosen reference points.

The dimension estimates in Fig. 13 are low primarily because of edge effects [see Ref. 3(b), p. 361]. That is, the increase in the number of points within a ball of radius ϵ is slower for a ball whose center is near an edge of the space than for a ball located far from any edge.[3,10] Figure 14 shows that for 300 000 points randomly distributed in a ten-dimensional space, the value of$d$ increases from 8.4 when the location of the reference points is unrestricted, to $d=9.2$ when the reference points are required to be at least 12% (of the linear extent) from an edge.

These considerations of dimension calculations for randomly distributed points provide at best only a qualitative guide to the importance of different effects for strange at-

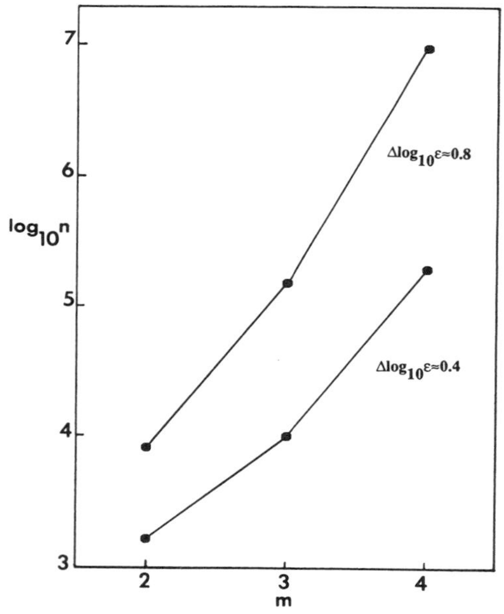

FIG. 12. The number of points n required to achieve scaling ranges (region C in Fig. 8) of width 0.4 and 0.8 for two-, three-, and four-dimensional hypercubes containing randomly distributed points. The analysis used 100 randomly chosen reference points; 1000 reference points were found to yield smoother curves in the scaling region, but the dimension estimates were not significantly improved.

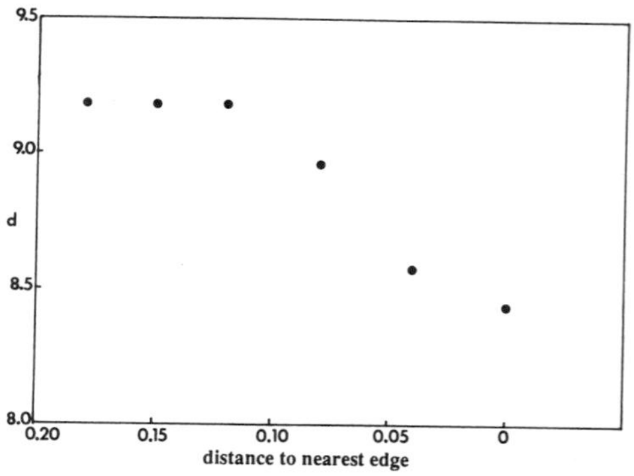

FIG. 14. The dependence of d on the location of the reference points for 300 000 randomly distributed points in a ten-dimensional hypercube. The abscissa gives the maximum distance from the reference points to the edge of the box, expressed as a fraction of the size of the box; the 100 reference points were randomly located inside this smaller box.

tractors, which of course have highly complicated structures. The effect of the location of the reference points will vary for different deterministic systems. For example, a 2-torus has no edge, while for other systems the influence of edges increases with the dimension of the attractor.

F. Filtering effect

Experimental data are often low-pass filtered, either in the process of acquiring the data or later by software. Low-pass filtering can not only influence the noise scale, but also, if the filtering is very severe, it can reduce the number of degrees of freedom that can be detected. The effect on Couette-Taylor data obtained at $R/R_c = 12.0$ is shown in Fig. 15, where reducing the filter cutoff from $20\omega/\Omega$ to $5\omega/\Omega$ reduces the value determined for d. In our experiment we did not apply any filter other than those given by the acquisition process.

G. Dimension of the Couette-Taylor attractor

Figure 16 shows the dimension of the Couette-Taylor attractor as a function of the Reynolds number. The uncertainty in d, which arises from the various effects we have discussed, increases from 0.1 for $d \approx 2$, to 0.4 for $d \approx 3$ ($R/R_c \approx 17$). The error increases rapidly with further increase in Reynolds number because, as discussed in Sec. IV A, it becomes difficult to discern a scaling region; the dimension values shown for $R/R_c > 17$ are included only to indicate the continuing increase in the dimension with increasing Reynolds number.

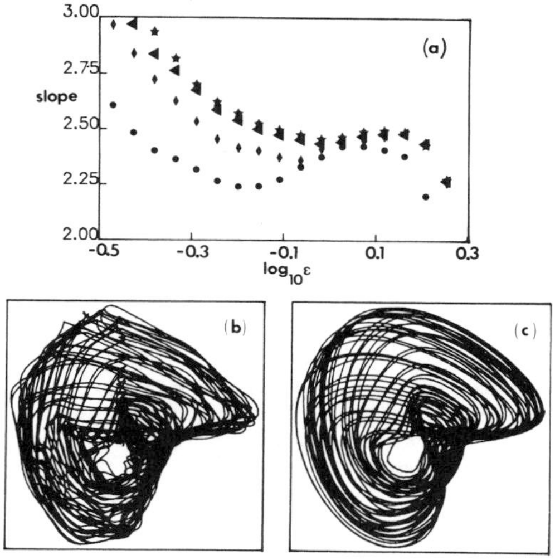

FIG. 15. The effect of low pass filtering of Couette-Taylor data obtained at $R/R_c = 12.0$. (a) The slopes of $\log_{10} N(\epsilon)$ vs $\log_{10}\epsilon$ graphs are shown for four cutoff frequencies: ●, $5/\Omega$; ♦, $10/\Omega$; ◀, $15/\Omega$; and ★, $20/\Omega$. (b) Phase portrait for a cut-off frequency of $20/\Omega$. (c) Phase portrait for a cut-off frequency of $5/\Omega$.

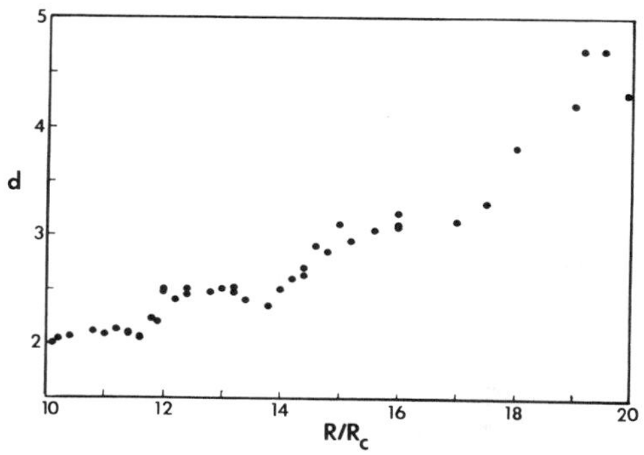

FIG. 16. The dimension of the Couette-Taylor attractor as a function of the Reynolds number.

The results for the dimension show that the onset of chaotic flow occurs at $R/R_c = 11.7 \pm 0.2$.[19] When the Reynolds number is increased beyond the onset of chaos, the value of the dimension increases monotonically except for small dips at $R/R_c = 11.7$ and 13.7. These dips are possibly a consequence of the fact that at these Reynolds numbers the ratio of the two fundamental frequencies happens to be near simple rational numbers; therefore, the torus is not covered well by the data—longer data files may be needed at these Reynolds numbers.

The reproducibility of the values determined for the dimension was quite good. Measurements were made over a period of six months, during which time the cylinder system was disassembled and reassembled several times, and the dimension values were found to reproduce to a much better precision than that indicated by the quoted uncertainties.

V. DISCUSSION

We have shown that the dimension of the attractors in the Couette-Taylor experiment is 2.0 for modulated wavy vortex flow. When the Reynolds number is increased beyond $R/R_c = 11.7 \pm 0.2$, the dimension becomes noninteger, increasing above the value 2.0. At this same Reynolds number broadband noise appears in the power spectrum and the largest Lyapunov exponent becomes positive, thus providing further evidence that $R/R_c = 11.7 \pm 0.2$ marks the onset of chaos.[19] The exponential decay of the power spectrum provides additional evidence that the observed nonperiodic behavior corresponds to low-dimensional deterministic chaos, not stochastic behavior.

The dimension values should be viewed as characterizing the flow in the *entire* annulus: measurements that were made at different spatial points for the same Reynolds number yielded (within the quoted uncertainties) the same value for the dimension. The dimension values seem remarkably small when one examines the photographs of the flow, particularly at the larger Reynolds numbers, say $R/R_c = 17$ (see Fig. 1). The dimension calculation suggests that, even for this noisy looking flow, only a small number of degrees of freedom may be relevant to the dynamics.

We have emphasized some of the pitfalls in the determination of dimension. It is not difficult to develop an algorithm that will yield numbers that can be called dimension, but it is far more difficult to be confident that those numbers truly represent the dynamics of the system under study. We have examined briefly some of the requirements for data to be used in determining dimension and have found that the number of data points needed increases dramatically with increasing d, at least for the techniques for computing dimension [Eqs. (1) and (2)] that we have used. There is clearly a great need for systematic numerical and analytic studies of the potential and the limitations of dimension calculations.

Another problem that warrants further study is the question of the route to chaos in systems with rotational symmetry. The well-studied routes to chaos—period doubling, intermittency (tangent bifurcations), and the breakup of a 2-torus as described by circle maps—apparently do not occur in this system.[39] As we have shown, circle maps can be obtained from the data for the Couette-Taylor system, but the behavior of these maps is apparently quite different from the maps that exhibit frequency locking. Thus there is a need for numerical and analytic studies of models with rotational symmetry and for further experiments on the transition to chaos for different conditions in rotationally symmetric systems.

ACKNOWLEDGMENTS

This work was supported in part by National Science Foundation (NSF) Grant No. MSM-82-06889 and in part by the U.S. Office of Naval Research and U.S. National Aeronautics and Space Administration (NASA) through the NASA-Ames Research Center, Moffet Field, California under Interchange No. NCA2-1R781-401.

---

[1]J. D. Farmer, E. Ott, and J. A. Yorke, Physica **7D**, 153 (1983).
[2]P. Grassberger and I. Procaccia, Phys. Rev. Lett. **50**, 346 (1983).
[3](a) J. Guckenheimer, Annu. Rev. Fluid Mech. **18**, 15 (1986); (b)Contemp. Math. **28**, 357 (1984).
[4]J. P. Eckmann and D. Ruelle, Rev. Mod. Phys. **57**, 617 (1986).
[5]Several papers in *Dimensions and Entropies in Chaotic Systems,* edited by G. Mayer-Kress (Springer, Berlin, 1986), treat problems in determining dimensions from time-series data: (a) R. Badii and A. Politi, *ibid.* p. 67; (b) F. Hunt and F. Sullivan, *ibid.* p. 74; (c) F. Holzfuss and G. Mayer-Kress, *ibid.* p. 114; (d) W. E. Caswell and J. A. Yorke, *ibid.* p. 123; (e) R. L. Somerjai, *ibid.* p. 137; (f) J. G. Caputo, B. Malraison, and P. Atten, *ibid.* p. 180.
[6]C. C. Foias, O. P. Manley, and R. Teman, J. Fluid Mech. **150**, 427 (1985).
[7]O. P. Manley and Y. M. Treve, Phys. Lett. **82A**, 88 (1981).
[8]J.-M. Ghidaglia (unpublished).
[9]A. Brandstater, J. Swift, H. L. Swinney, A. Wolf, J. D. Farmer, E. Jen, and J. P. Crutchfield, Phys. Rev. Lett. **51**, 1442 (1983); A. Brandstater, J. Swift, H. L. Swinney, and A. Wolf, in *Turbulence and Chaotic Phenomena in Fluids,* edited by T. Tatsumi (Elsevier North-Holland, Amsterdam, 1984), p. 179; A. Brandstater and H. L. Swinney, in *Fluctuations and Sensitivity in Nonequilibrium Systems,* edited by W. Horsthemke and D. Kondepudi (Springer, Berlin, 1984), p. 166; A. Brandstater, Ph.D. dissertation Universität Kiel, 1984.
[10]B. Malraison, P. Atten, P. Berge, and M. Dubois, J. Phys. (Paris) Lett. **44**, L987 (1983); P. Atten, J. G. Caputo, B. Malraison, and Y. Gagne, J. Mec. Theor. Appl., special issue (1984), p. 133; see also Ref. 5(f).
[11]J. Guckenheimer and G. Buzyna, Phys. Rev. Lett. **51**, 1438 (1983). The transition to chaos in a related system, a two-layer baroclinic fluid, has been studied by J. E. Hart, Physica **20D**, 350 (1986).
[12]R. C. DiPrima and H. L. Swinney, in *Hydrodynamic Instabilities and the Transition to Turbulence,* edited by H. L. Swinney and J. P. Gollub (Springer, Berlin, 1985), p. 139.
[13]P. Matisse and M. Gorman, Phys. Fluids **27**, 759 (1984).
[14]R. G. W. Brown, in *Photon Correlation Techniques,* edited by E. O. Schulz-DuBois (Springer, Berlin, 1982), p. 66.
[15]R. G. W. Brown and R. Jones, Opt. Lett. **8**, 449 (1983).
[16]P. R. Fenstermacher, H. L. Swinney, and J. P. Gollub, J. Fluid Mech. **94**, 103 (1979).
[17]R. S. Shaw, C. D. Andereck, L. A. Reith, and H. L. Swinney, Phys. Rev. Lett. **48**, 1172 (1981).
[18]M. Gorman and H. L. Swinney, J. Fluid Mech. **117**, 123 (1982).
[19]The uncertainty of ± 0.2 in the value of R/R_c at onset is estimated from an analysis of the photographs of the flow, velocity power spectra, phase portraits, and circle maps, and from the values deduced for the dimension and Lyapunov exponents of the attractors.
[20]D. Rand, Arch. Rat. Mech. Anal. **79**, 1 (1982); see also M. Gorman, D. Rand, and H. L. Swinney, Phys. Rev. Lett. **46**, 992 (1981).
[21]The component called ω_2 here was called ω_3 in Ref. 16.
[22]H. S. Greenside, G. Ahlers, P. C. Hohenberg, and R. W. Walden, Physica **5D**, 322 (1982).
[23]D. Sigeti and W. Horsthemke, Phys. Rev. A. **35**, xxxx (1987).
[24]G. Ahlers and R. P. Behringer, Phys. Rev. Lett. **40**, 712 (1978); G. Ahlers and R. W. Walden, *ibid.* **44**, 445 (1980).
[25]P. Atten, J. C. Lacroix, and B. Malraison, Phys. Lett. **79A**, 255 (1980).
[26]N. H. Packard, J. P. Crutchfleld, J. D. Farmer, and R. S. Shaw, Phys. Rev. Lett. **45**, 712 (1980); J. C. Roux, R. H. Simoyi, and H. L. Swinney, Physica **8D**, 257 (1983).
[27]F. Takens, in *Dynamical Systems and Turbulence,* Vol. 898 of *Lecture Notes in Mathematics,* edited by D. A. Rand and L. S. Young (Springer, Beriin, 1981), p. 366.
[28]A. M. Fraser and H. L. Swinney, Phys. Rev. A **33**, 1134 (1986).
[29]A. M. Fraser, in *Dimensions and Entropies in Chaotic Systems,* edited by G. Mayer-Kress (Springer, Berlin, 1986), p. 82; IEEE Trans. Inf. Theory (to be published).
[30]A. Wolf, J. B. Swift, H. L. Swinney, and J. A. Vastano, Physica **16D**, 285 (1985).
[31]P. Coullet, C. Tresser, and A. Arneodo, Phys. Lett. **77A**, 327 (1980).
[32]M. J. Feigenbaum, L. P. Kadanoff, and S. J. Shenker, Physica

5D, 370 (1983).

[33]S. Ostlund, D. Rand, J. Sethna, and E. Siggia, Physica **8D**, 303 (1983).

[34]M. H. Jensen, P. Bak, and T. Bohr, Phys. Rev. A **30**, 1960 (1984); T. Bohr, M. H. Jensen, and P. Bak, *ibid.* **30**, 1970 (1984).

[35]P. Cvitanovic, M. H. Jensen, L. P. Kadanoff, and I. Procaccia, Phys. Rev. Lett. **55**, 343 (1985).

[36]P. Grassberger and I. Procaccia, Physica **9D**, 189 (1983).

[37]D. S. Broomhead and G. P. King, Physica **20D**, 217 (1986).

[38]These systems are defined and discussed in Ref. 30.

[39]However, G. Pfister observed period doubling when the rotational symmetry of his Couette-Taylor system was broken by tilting one end very slightly, Proceedings of the Second International Symposium on Applications of Laser Anemometry in Fluid Mechanics, Lisbon, 1984, p. 3.2. Also, see G. Pfister, in *Flow of Real Fluids,* Vol. 235 of *Lecture Notes in Physics,* edited by E. E. A. Meier and F. Obermeier (Springer, Berlin, 1985), p. 199.

Dimension Measurements for Geostrophic Turbulence

John Guckenheimer
Division of Natural Sciences, University of California, Santa Cruz, California 95064

and

George Buzyna
Geophysical Fluid Dynamics Institute, Florida State University, Tallahassee, Florida 32306
(Received 20 July 1983)

The transition to geostrophic turbulence in a rotating annulus experiment has characteristics which differ from the scenarios for the transition to turbulence which have been previously described. Estimates of dimension from the experimental data suggest a mechanism in which a discrete symmetry of the preturbulent state plays an important role. The techniques used for the dimension estimates are new and substantially more efficient than those used previously.

PACS numbers: 47.25.Qv, 47.10.+g

A rotating, differentially heated annulus of fluid is a classical laboratory model for the large-scale midlatitude circulation of the Earth's atmosphere.[1] Extensive investigation has classified large portions of the stability diagram for these laboratory systems in a parameter plane representing Rossby and Taylor numbers.[2,3] Figure 1 illustrates the behavior from one set of experiments conducted at the Geophysical Fluid Dynamics Institute, and described in detail elsewhere.[4] This Letter reports the results of additional analysis of the data from these experiments, analysis directed towards characterizing the transition to geostrophic turbulence.

Experiments with Rayleigh-Bénard convection[5] have revealed substantial qualitative differences in the transition to aperiodic flow in small- and large-aspect-ratio containers. With small-aspect-ratio containers, there is a close parallel between the routes to chaos found in nonlinear dynamical systems with a low-dimensional state space and the transitions observed experimentally.[6] We were motivated to examine the transition to geostrophic turbulence in a rotating annulus because the experimental data did not appear consistent with the mechanisms by which a quasiperiodic motion becomes aperiodic in simulations of low-dimensional systems.[7] Our analysis confirms this conclusion, but we discuss our methods before presenting these results and speculating on their significance.

Previous efforts to calculate the dimension of attractors[8] have required significant amounts of computation. The methods employed here are much more efficient though they seem subject to larger statistical fluctuations when applied to a given amount of data. The theory underlying the method will be developed in more detail elsewhere, but we give a brief sketch here. Strange attractors are complicated geometric objects, typically with a fractal structure,[9] which appear in the state space of chaotic flows. The problem which confronts us is the specification of a reasonable definition of the dimension of an attractor

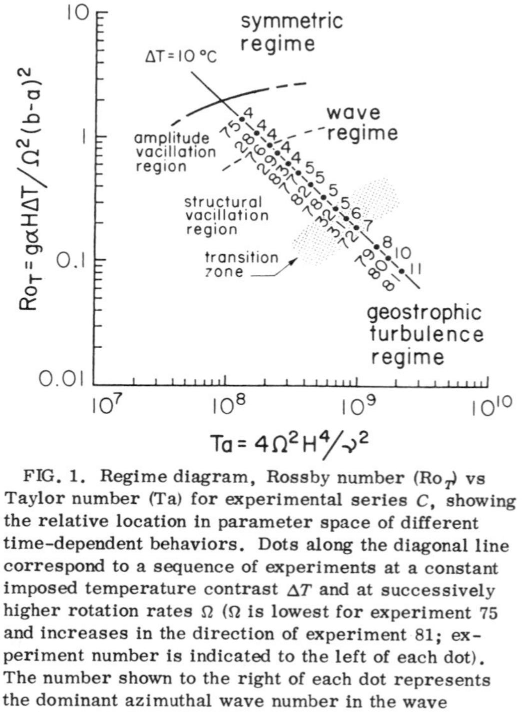

FIG. 1. Regime diagram, Rossby number (Ro_T) vs Taylor number (Ta) for experimental series *C*, showing the relative location in parameter space of different time-dependent behaviors. Dots along the diagonal line correspond to a sequence of experiments at a constant imposed temperature contrast ΔT and at successively higher rotation rates Ω (Ω is lowest for experiment 75 and increases in the direction of experiment 81; experiment number is indicated to the left of each dot). The number shown to the right of each dot represents the dominant azimuthal wave number in the wave regime, and the peak of the broad wave-number spectrum in the geostrophic turbulence regime.

which can be readily estimated. Intuitively, the dimension should roughly correspond to the number of independent measurements which must be made to distinguish distinct states of the fluid *on the attractor*. In these measurements of dimension, we weigh regions of the attractor by the frequency with which a typical trajectory visits them through the introduction of an invariant measure μ supported on the attractor.

Our strategy for computing dimension is adapted from the work of Billingsley[10] and Young.[11] Let μ be a measure on R^n and Λ be a set of positive μ measure. Denote the ball of radius r centered at x by $B_x(r)$ and its μ volume by $V_x(r) = \mu(B_x(r))$. Young[11] proves that if for $x \in \Lambda$,

$$\underline{\delta} \leq \liminf_{r\to 0} \frac{\ln V_x(r)}{\ln r} \leq \limsup_{r\to 0} \frac{\ln V_x(r)}{\ln r} \leq \overline{\delta},$$

then the Hausdorff dimension of Λ, $D_{\mathrm{H}}(\Lambda)$, satisfies

$$\underline{\delta} \leq D_{\mathrm{H}}(\Lambda) \leq \overline{\delta}.$$

We adapt this technique for dimension calculations by *assuming* that for almost all points x of the attractor (relative to μ), the limits in the above inequalities exist and are independent of x. An estimate for the dimension of the attractor is then given by calculating the function $V_x(r)$ for a typical point x and plotting $\ln V_x(r)$ vs $\ln r$. The calculation of $V_x(r)$ proceeds by *assuming* that an experimentally observed trajectory gives a good randon sample of points on the attractor relative to the measure μ. For attractors whose dimension is not very small, this assumption is nontrivial because very long trajectories are needed to sample all regions of the attractors. We return to this point in later discussion. Our dimension estimates accept the estimate that $V_x(r)$ is the proportion of points in the observed trajectory which lie within distance r of the point x.

The calculations themselves proceed efficiently, as follows. One begins with an experimentally observed trajectory of $N+1$ points which is assumed to be on an attractor and lack transient behavior. From this trajectory, one selects a reference point x and computes the distance d_t from x to the other N points y_t of the trajectory. The list of distances d_t are then sorted numerically. From the sorted list, one reads that the ith largest distance gives a value of r for which $V_x(r) = i/N$. We have plotted these values for $i = 2^j$ ($j = 1, \ldots, k$; $j = \log_2 N$) on a log-log plot of r vs $V_x(r)$. The dimension is estimated from such a plot as the inverse of the slope of this curve. We caution, however, that the function $V_x(r)$ need not be a smooth curve, so that an accurate estimation of the dimension may require a much wider range of values of r than one can hope to obtain experimentally or numerically.

To check the suitability of this strategy for calculating dimension, we performed several numerical tests. The results of one are shown in Fig. 2, where we display the log-log plots for seven separate collections of 5000 points on tori T^n of dimensions $n = 2$, 3, 5, 10, 25, 50, and 99. The straight lines in these plots have slopes which correspond to the different values of n. These experiments give us confidence that this method of calculating dimension gives good rough estimates with amounts of data that are readily obtainable from experiments and require only modest amounts of computation, on the order of $N(\ln N + n)$ operations for N observations of n-variate data.

We have also tested the method on several modes of chaotic attractors, including the Hénon map,[12] the Lorenz equations,[13] and a map with an attracting nowhere differentiable torus.[14] These tests give results consistent with estimates obtained with use of other methods.[8,15] However, as noted by Pfeffer,[16] the method must be used with great caution and does not always give reli-

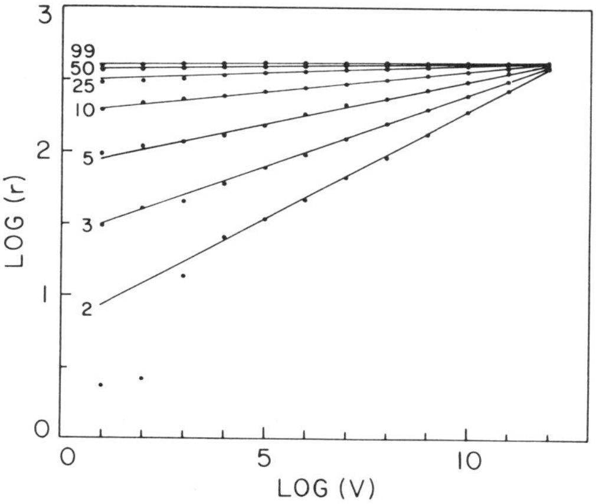

FIG. 2. Log-log plots of r vs $V(r)$ for several numerical examples on tori T^n of dimension $n = 2$, 3, 5, 10, 25, 50, and 99. [Numerical values of $\ln(r)$ represent a relative scale; numerical values shown on the $\ln V$ axis represent j of the $i = 2^j$ value from the sorted list of data; see text.] The inverse of the slope is proportional to the dimension.

able results for reasonable models of some of the experiments analyzed here (cf. the discussion of amplitude vacillation below.)

Let us return now to the experimental data. The results of our dimension computation are displayed in Fig. 3. The apparatus contained 2016 thermistors in an annulus with inner and outer radii 30.48 ± 0.01 and 60.96 ± 0.01 cm. The working fluid, 1.52-cS ($1\ S = 1\ cm^2/sec$) silicone fluid (Prandtl number $N_{Pr} = 21$), had a free top surface and a depth of 15.00 to 0.05 cm. To perform our computations, we selected recordings from 27 thermistors located near middepth and well away from the walls of the rotating annulus. The number 27 was arbitrarily chosen to be large enough to represent the dimensions of the attractors we expected to find. The "distance" d_t represents here a maximum absolute value of the 27 temperature differences between the reference point x and another point y_t of the trajectory. Dimension calculations based upon a larger number of thermistors are in progress.

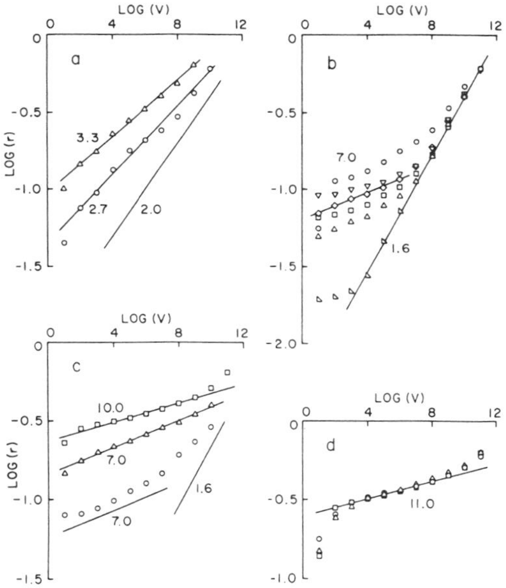

FIG. 3. Log-log plots of r vs $V(r)$ for the experimental series (coordinate axes scaled as in Fig. 2). (a) Amplitude vacillation region; symbols and experiment numbers in order of increasing rotation rate are, respectively, circles, 75; and triangles, 28. (b) Structural vacillation region: right triangles, 76; triangles, 29; squares, 83; inverse triangles, 77; diamonds, 82; and circles, 78. (c) Transition zone: circles, 32; triangles, 31; and squares, 72. (d) Geostrophic turbulence regime; circles, 79; triangles, 80; and squares, 81.

The stability diagram of Fig. 1 shows four distinct regimes, three of which represent time-dependent flows. In the region of amplitude vacillation, the fluid motion is quasiperiodic. If one models this quasiperiodic motion as a superposition of traveling waves, then Fourier analysis indicates that some 95% of the variance of the data is contained in four Fourier modes. Pfeffer[16] observed that our dimension estimates underestimate the dimension corresponding to a superposition of n-linear traveling waves, and we can show that the method yields a dimension estimate of $\frac{1}{2}(n+1)$ if the temporal frequencies of the waves are independent. For the experimental data, our dimension estimates are consistent with a model represented by the superposition of four traveling waves. The data are insufficient to determine whether there are two, three, or four independent frequencies.

The dynamics of the structural vacillation regime are still not fully characterized. The estimated dimension of 1.6 for the structural vacillation regime is surprisingly small and requires further investigation. The dimension measurements shown in Fig. 3 indicate that, as the rotation rate of the appartus increases, there is a jump in the dimension of the attractor which begins at small amplitude in phase space and grows continuously. As the amplitude of this high-dimensional motion approaches the amplitude of the dominant wave itself, the fluid undergoes the transition to geostrophic turbulence. Conservative estimates for the dimension of the attractor in these regimes are within the range of 7 to 12.

The jump in dimension from the amplitude vacillation regime through the structural vacillation regime is a departure from the scenarios for the transition to chaos which have been described for low-dimensional dynamical systems that deserves explanation. There is a dynamical mechanism which appears consistent with our measurements and other information about this flow. The observed flow in the structural vacillation regime has an (approximate) discrete spatial symmetry which corresponds to rotation by $2\pi/4$ or $2\pi/5$ in the experiments analyzed in the structural vacillation regime. If a new asymmetric oscillatory mode of instability appears in the fluid motion, then the symmetry of the fluid equations and the motion prior to the instability force the instability to be degenerate. The translates of the asymmetric oscillatory mode by the symmetry group would be distinct oscillatory modes. Thus, it is plausible that a quasiperiodic state with fivefold symmetry would evolve into an attractor with a jump in dimension from 2 to 7.

An alternative hypothesis to the scenario described above is that random effects or noise are responsible for the growing disorder of the fluid in the structural vacillation regime. Guckenheimer[17] presents methods for distinguishing the effects of noise and deterministic chaos upon experimental data by analyzing the short-term evolution of nearby initial states. These methods do not seem to be practical here because the dimension of the attractors is too large. Recalling our assumption in the dimension computations that the trajectory represented by the experimental observations was randomly distributed on the attractor, one can estimate the expected recurrence time for a trajectory. This recurrence time grows exponentially with dimension and is experimentally unreasonable for attractors of even moderate dimension. Detailed reconstructions of the dynamics of attractors from experimental data are feasible only for attractors of low dimension. Geostrophic turbulence does not appear to be a fluid state represented by a low-dimensional attractor, suggesting that further understanding of this fluid state will require a statistical methodology.

We wish to thank R. L. Pfeffer for his support and encouragement. One of us (J.G.) would also like to thank Uriel Frisch for conversations which prompted the development of the methods used here. This research was supported in part by the U. S. Office of Naval Research (Contract No. N00014-77-C-0265), the U. S. Air Force Office of Scientific Research (Contract No. 83-0143), and the National Science Foundation (Grant No. MCS92-002260).

---

[1]D. Fultz *et al.*, Meteorol. Monogr. No. 21, 96, 97 (1959).

[2]R. Hide and P. J. Mason, Adv. Phys. 24, 47–100 (1975).

[3]R. L. Pfeffer, G. Buzyna, and R. Kung, J. Atmos. Sci. 37, 2129–2149, 2577–2599 (1980).

[4]G. Buzyna, R. L. Pfeffer, and R. Kung, to be published.

[5]G. Ahlers and R. P. Behringer, Phys. Rev. Lett. 40, 712–716 (1978).

[6]J. P. Gollub and S. V. Benson, J. Fluid Mech. 100, 449–470 (1980).

[7]J. P. Eckmann, Rev. Mod. Phys. 53, 643–654 (1981).

[8]D. Russel, J. Hansen, and E. Ott, Phys. Rev. Lett. 45, 1175 (1980).

[9]B. Mandelbrot, *The Fractal Geometry of Nature* (Freeman, San Francisco, 1982).

[10]P. Billingsley, *Ergodic Theory and Information* (Wiley, New York, 1965).

[11]L.-S. Young, "Dimension, Entropy and Lyapunov Exponents" (to be published).

[12]M. Hénon, Commun. Math. Phys. 50, 69 (1976).

[13]E. Lorenz, J. Atmos. Sci. 20, 130 (1963).

[14]J. D. Farmer, E. Ott, and J. Yorke, to be published.

[15]P. Grassberger and I. Procaccia, Phys. Rev. Lett. 50, 346 (1983).

[16]R. Pfeffer, private communication.

[17]J. Guckenheimer, Nature (London) 298, 358–361 (1982).

CHAPTER 8

Calculation of Lyapunov Exponents

J.-P. Eckmann, S.O. Kamphorst, D. Ruelle, S. Ciliberto
Liapunov exponents from time series
Phys. Rev. A **34**, 4971 (1986).

P. Bryant, R. Brown, H. Abarbanel
Lyapunov exponents from observed time series
Phys. Rev. Lett. **65**, 1523 (1990).

U. Parlitz
Identification of true and spurious Lyapunov exponents from time series
Int. J. Bif. Chaos **2**, 155 (1992).

Editors' Notes

In Wolf et al. (1985), a computational method is described for approximating the largest Lyapunov exponents directly from the rate of separation of neighboring points. A computationally viable method for computing Lyapunov exponents from experimental data was described in Eckmann and Ruelle (1985) and Sano and Sawada (1985). The basic idea begins with embedding the data and reconstructing the local linear dynamics. Then the evolution of an m-frame of perpendicular vectors is followed, using Gram-Schmidt orthogonalization (as implemented by the QR-factorization) on each step to preserve the perpendicularity of the m-frame. A great deal of research on computing Lyapunov exponents from experimental time series has followed, largely improvements and variants of this basic idea. Results of the application of this algorithm to artificially generated data and experimental data are described in **Eckmann et al.** (1986).

Bryant et al. (1990) report on a variant in which the local linear dynamics are approximated by hypothesizing higher degree polynomial models to reconstruct the dynamics; the linear terms which result are in some cases more accurate. Full details are found in Brown et al. (1991). Stoop and Parisi (1991) advocate the use of the singular value decomposition to restrict the dynamics to the tangent plane of the attractor prior to approximating the local linear dynamics.

Parlitz (1992) suggests a simple way to detect and reject spurious Lyapunov exponents, that is, ones that arise as artifacts of the embedding process. Upon reversal of the direction of time, the Lyapunov exponents should, in theory, change sign. The exponents that respect this fact when the algorithm is repeated on the reversed time series are accepted as true Lyapunov exponents.

Further reading

Reviews of research on the calculation of correlation dimension and Lyapunov exponents are contained in the following useful surveys. Both cover a wide range of issues of time series analysis and exploitation for chaotic systems.

GRASSBERGER, P., SCHREIBER, T., SCHAFFRATH, C., Nonlinear time sequence analysis. Int. J. Bif. Chaos **1**, 512 (1991).

ABARBANEL, H., BROWN, R., SIDOROWICH, J., TSIMRING, L., The analysis of observed chaotic data in physical systems. Rev. Mod. Phys. (1993).

Liapunov exponents from time series

J. -P. Eckmann and S. Oliffson Kamphorst
Département de Physique Théorique, Université de Genève, CH-1211 Genève 4, Switzerland

D. Ruelle
Institut des Hautes Etudes Scientifiques (IHES), F-91440 Bures-sur-Yvette, France

S. Ciliberto
Istituto Nazionale di Ottica, I-50125 Arcetri (Firenze), Italy
(Received 24 April 1986)

We analyze in detail an algorithm for computing Liapunov exponents from an experimental time series. As an application, a hydrodynamic experiment is investigated.

I. INTRODUCTION

In Ref. 1 two of us proposed a method to compute Liapunov exponents from an experimental time series. Here we report on a detailed analysis of this algorithm for numerical and laboratory experiments. Note that a very similar proposal has been made independently by Sano and Sawada.[2] In the course of the discussion, we shall also point out some divergences between Refs. 1 and 2.

Before discussing our algorithm, we briefly state what we are trying to do. A time evolution is realized, in Nature, in the laboratory, or on the computer, and it is assumed that this time evolution can be described by a differentiable dynamical system in a phase space of possibly infinite dimensions. We want to obtain Liapunov exponents corresponding to the large-time behavior of the system. On a more mathematical level, the large-time behavior defines an ergodic measure in phase space for the time evolution, and we are interested in the corresponding Liapunov exponents. For a discussion of these concepts and precise definitions see, for instance, Ref. 1. What we know is a time series $(x_i)_{1\le i\le N}$ obtained by monitoring a scalar signal for a finite time T and with finite precision. Clearly, thus, there are limitations on how much we can say about the characteristic exponents—it is the aim of this paper to discuss some of these limitations. Certainly, we have to assume that the recording time T is long, that the noise level is low, and that the measurements are made with good precision (viz., 10^{-3} or 10^{-4} if we want to determine one or two positive characteristic exponents). From a sufficiently good time series, one can in principle obtain all non-negative characteristic exponents, and it may or may not be possible to obtain also some negative ones (cf. Ref. 1).

A complete list of other methods for computing Liapunov exponents is given in Ref. 1. To our knowledge, the proposals in Refs. 1 and 2 are the only ones which allow a systematic computation of several Liapunov exponents.

II. THE ALGORITHM

It is convenient to present the measured time series in the form of a sequence of integers $x_1, x_2, \ldots, x_N$, with $0 \le x_i \le 10\,000$. (The choice of integer values speeds up the computation without sacrificing experimental precision.) The upper bound 10 000 is in accordance with a precision of 10^{-4} and can easily be modified, if required. We assume that the time interval τ between measurements is fixed, so that $x_i = x(i\tau)$. Note that the recording time is $T = N\tau$. The present paper deals specifically with the case of a scalar signal, but the method can easily be extended to multidimensional signals.

Conceptually, the algorithm (a copy of the computer program implementing this algorithm can be obtained from the authors) to be discussed involves the following steps: (a) reconstructing the dynamics in a finite dimensional space, (b) obtaining the tangent maps to this reconstructed dynamics by a least-squares fit, (c) deducing the Liapunov exponents from the tangent maps. We now consider these different steps in detail.

(a) We choose an embedding dimension d_E and construct a d_E-dimensional orbit representing the time evolution of the system by the time-delay method.[3] This means that we define

$$\vec{\mathbf{x}}_i = (x_i, x_{i+1}, \ldots, x_{i+d_E-1}) \tag{1}$$

for $i = 1, 2, \ldots, N - d_E + 1$. In view of step (b), we have to determine the neighbors of $\vec{\mathbf{x}}_i$, i.e., the points $\vec{\mathbf{x}}_j$ of the orbit which are contained in a ball of suitable radius r centered at $\vec{\mathbf{x}}_i$,

$$\|\vec{\mathbf{x}}_j - \vec{\mathbf{x}}_i\| \le r \tag{2}$$

with

$$\|\vec{\mathbf{x}}_j - \vec{\mathbf{x}}_i\| = \max_{0\le\alpha\le d_E-1} \{ |x_{j+\alpha} - x_{i+\alpha}| \} . \tag{3}$$

The use of (3) rather than the Euclidean norm allows a fast search for the $\vec{\mathbf{x}}_j$ which satisfy (2). We first sort the x_i (using "Quicksort," see, e.g., Knuth[4]) so that

$$x_{\Pi(1)} \le x_{\Pi(2)} \le \cdots \le x_{\Pi(N)}$$

and store the permutation Π and its inverse Π^{-1}. Then, to find the neighbors of x_i in dimension 1, we look at $k = \Pi^{-1}(i)$ and scan the $x_{\Pi(s)}$ for $s = k+1, k+2, \ldots$ until $x_{\Pi(s)} - x_i > r$, and similarly for $s = k-1, k-2, \ldots$.

For an embedding dimension $d_E > 1$, we first select the values of s for which $|x_{\Pi(s)} - x_i| \le r$, as above, and then impose the further conditions

$$|x_{\Pi(s)+\alpha} - x_{i+\alpha}| \le r ,$$

for $\alpha = 1,2,\ldots,d_E - 1$.

(b) Having embedded our dynamical system in d_E dimensions (it would be more correct to say that we have projected our dynamical system to R^{d_E}), we want to determine the $d_E \times d_E$ matrix T_i which describes how the time evolution sends small vectors around $\vec{\mathbf{x}}_i$ to small vectors around $\vec{\mathbf{x}}_{i+1}$. The matrix T_i is obtained by looking for neighbors $\vec{\mathbf{x}}_j$ of $\vec{\mathbf{x}}_i$ and imposing

$$T_i(\vec{\mathbf{x}}_j - \vec{\mathbf{x}}_i) \approx \vec{\mathbf{x}}_{j+1} - \vec{\mathbf{x}}_{i+1} . \tag{4}$$

The vectors $\vec{\mathbf{x}}_j - \vec{\mathbf{x}}_i$ may not span R^{d_E} (think, for instance, of an embedding of the three-dimensional Lorentz system in four dimensions). Therefore, the matrix T_i may only be partially determined. This indeterminancy does not spoil the calculation of the positive Liapunov exponents, but is nevertheless a nuisance because it introduces parasitic exponents which confuse the analysis, in particular with respect to zero or negative exponents which otherwise might be recoverable from the data. The way out of this difficulty is to allow T_i to be a $d_M \times d_M$ matrix with a matrix dimension $d_M \le d_E$, corresponding to the time evolution from $\vec{\mathbf{x}}_i$ to $\vec{\mathbf{x}}_{i+m}$.

Specifically, we assume that there is an integer $m \ge 1$ such that

$$d_E = (d_M - 1)m + 1 , \tag{5}$$

and associate with $\vec{\mathbf{x}}_i$ a d_M-dimensional vector

$$\begin{aligned}\mathbf{x}_i &= (x_i, x_{i+m}, \ldots, x_{i+(d_M-1)m}) \\ &= (x_i, x_{i+m}, \ldots, x_{i+d_E-1}) ,\end{aligned} \tag{6}$$

in which some of the intermediate components of (1) have been dropped. When $m > 1$ we replace (4) by the condition

$$T_i(\mathbf{x}_j - \mathbf{x}_i) \approx \mathbf{x}_{j+m} - \mathbf{x}_{i+m} . \tag{7}$$

Taking $m > 1$ does not mean that we delete points from the data file, i.e., all points are acceptable as $\mathbf{x}_j$, and the distance measurements are still based on d_E, not on d_M. Note that, in view of (6) and (7), the matrix T_i has the form

$$T_i = \begin{bmatrix} 0 & 1 & 0 & \cdots & 0 \\ 0 & 0 & 1 & \cdots & 0 \\ \vdots & \vdots & \vdots & & \vdots \\ 0 & 0 & 0 & \cdots & 1 \\ a_1 & a_2 & a_3 & \cdots & a_{d_M} \end{bmatrix} .$$

If we define by $S_i^E(r)$ the set of indices j of neighbors $\vec{\mathbf{x}}_j$ of $\vec{\mathbf{x}}_i$ within distance r, as determined by (2), then we obtain the a_k by a least-squares fit

$$\sum_{j \in S_i^E(r)} \left[\sum_{k=0}^{d_M-1} a_{k+1}(x_{j+km} - x_{i+km}) - (x_{j+d_Mm} - x_{i+d_Mm}) \right]^2 = \text{minimum} .$$

The least-squares fit is the most time-consuming part of our algorithm when $S_i^E(r)$ is large. We limit ourselves therefore typically to the first 30–45 neighbors of the a point. We use the least-squares algorithm by Householder.[6] This algorithm may fail for several reasons, the most prominent being that card $S_i^E(r) < d_M$. We therefore choose r sufficiently large so that $S_i^E(r)$ contains at least d_M elements.

In fact, we make a new choice of $r = r_i$ for every i. This choice is a compromise between two conflicting requirements: take r sufficiently small so that the effect of nonlinearities can be neglected, take r sufficiently large so that there are at least d_M neighbors of $\vec{\mathbf{x}}_i$, and in fact somewhat more than d_M to improve statistical accuracy.

For the specific examples discussed in Sec. IV we have selected r as follows. Count the number of neighbors of x_i corresponding to increasing values of r from a preselected sequence of possible values, and stop when the number of neighbors exceeds for the first time $\min(2d_M, d_M + 4)$. If with this choice the matrix T_i is singular, or, more generally, does not have a previously fixed minimal rank, we again increase r_i. It should be noted that this last criterion only seems to come into operation for time series obtained for low-dimensional computer experiments (such as maps of the interval). We stress that the singularity of T_i in itself is not catastrophic for the algorithm and the first p positive Liapunov exponents are not affected provided the rank of the T_i is at least p (which may be a lot less than d_M). One should thus not stop the calculation, as suggested in Ref. 2 when the map is singular, since information about the expanding direction(s) will be lost.

(c) Step (b) gives a sequence of matrices $T_i, T_{i+m}, T_{i+2m}, \ldots$. One determines successively orthogonal matrices $Q_{(j)}$ and upper triangular matrices $R_{(j)}$ with positive diagonal elements such that $Q_{(0)}$ is the unit matrix and

$$\begin{aligned} T_1 Q_{(0)} &= Q_{(1)} R_{(1)} , \\ T_{1+m} Q_{(1)} &= Q_{(2)} R_{(2)} , \\ &\ldots , \\ T_{1+jm} Q_{(j)} &= Q_{(j+1)} R_{(j+1)} , \\ &\ldots . \end{aligned} \tag{8}$$

This decomposition is unique except in the case of zero diagonal elements. Then the Liapunov exponents λ_k are given by

$$\lambda_k m = \frac{1}{\tau K} \sum_{j=0}^{K-1} \ln R_{(j)kk} ,$$

where $K \le (N - d_M m - 1)/m$ is the available number of matrices, and τ is sampling time step. Obviously, fewer matrices can be taken to shorten the computing time. [See Ref. 1 for a justification of the algorithm of Eq. (8).]

III. REMARKS ON THE ALGORITHM

(a) Let us comment again on the usefulness of taking the matrix dimension d_M different from the embedding dimension d_E. As we have said, if d_M is not sufficiently low, there is some numerical indeterminacy in the coefficients on the T_i which, combined with noise, produces undesirable parasitic Liapunov exponents (examples of this phenomenon will be shown in Sec. IV). It is thus natural to take d_M relatively low (a little bigger than the expected number of positive Liapunov exponents). But if one takes d_E too small, the embedding (or rather projection) of the dynamics in R^{d_E} would not be well defined; orbits with different directions might go through the same point. The cure is to take $d_E > d_M$. This is, admittedly, a nonrigorous prescription, and leaves some "intuitive" freedom. We try to overcome this by examining the result for several d_M and d_E. Note that for disentangling the dynamics, the important thing is the embedding time $d_E\tau$ rather than d_E; this is a first indication that it is not wise to take τ very small.

(b) As already discussed, the choice of the radius r at $\vec{\mathbf{x}}_i$ is a compromise between limitations due to nonlinearities and limitations due to noise. In fact, we have chosen the smallest ball around $\vec{\mathbf{x}}_i$ which contains enough neighbors for an unambiguous determination of T_i (note that the algorithm becomes impractically slow when there are more than about 45 neighbors). In principle, i.e., with very good experimental data one can do a little better.

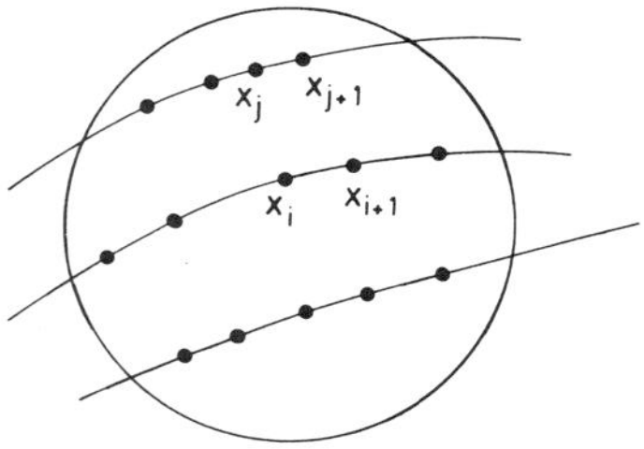

FIG. 1. Neighbors are picked up mostly on the orbit itself.

Since the effect of errors is the worst for the short vectors $\vec{\mathbf{x}}_j - \vec{\mathbf{x}}_i$ one could replace the ball

$$\{ \vec{\mathbf{x}}_j :\ \|\vec{\mathbf{x}}_j - \vec{\mathbf{x}}_i\| \le r \}$$

by a shell

$$\{ \vec{\mathbf{x}}_j :\ r_{\min} \le \|\vec{\mathbf{x}}_j - \vec{\mathbf{x}}_i\| \le r \} .$$

(c) To obtain good statistics it is, of course, desirable to have a time series with a large number N of measurements. However, the really important thing is the total recording time $T = N\tau$, and increasing N at fixed T by making τ very small would be useless. Actually, the experimental studies of Sec. IV show that for large embedding dimension d_E (hence large r), and small τ, many of the neighbors of $\vec{\mathbf{x}}_i$ are in factor of the form $\vec{\mathbf{x}}_{i\pm1}, \vec{\mathbf{x}}_{i\pm2}, \ldots$ (see Fig. 1). Attempts at numerical projection onto the supplement of this line have given bad results.

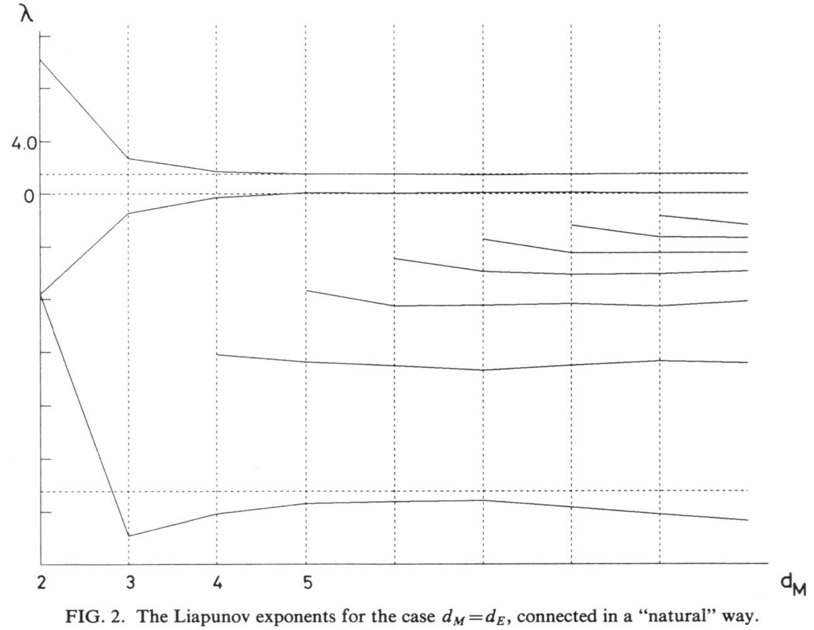

FIG. 2. The Liapunov exponents for the case $d_M = d_E$, connected in a "natural" way.

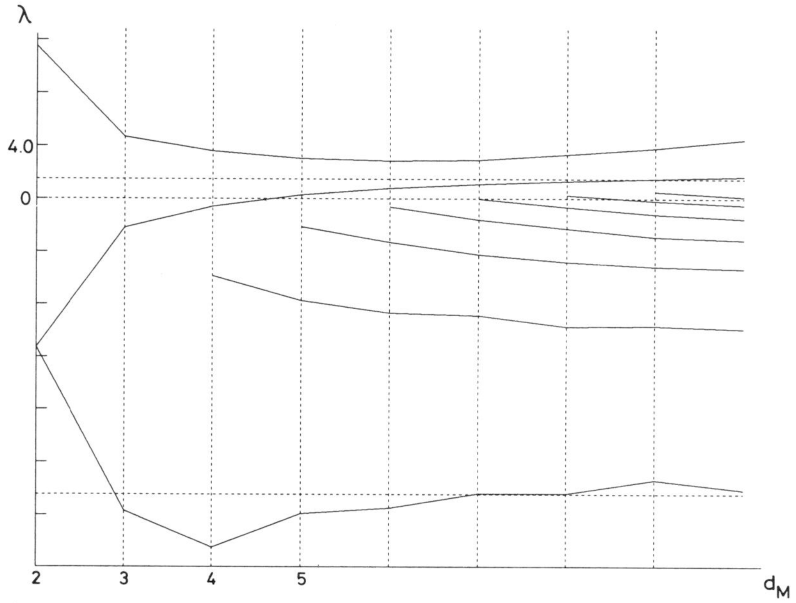

FIG. 3. Typical behavior when too few neighbors are chosen: card $S_E^i(r) \geq d_M$.

(d) Summary of advice.

(1) Use long recording time T, but not very small time step τ.

(2) Use large embedding dimension d_E.

(3) Use a matrix dimension d_M somewhat larger than the expected number of positive Liapunov exponents.

(4) Choose r such that the number of neighbors is greater than $\min(2d_M, d_M + 4)$.

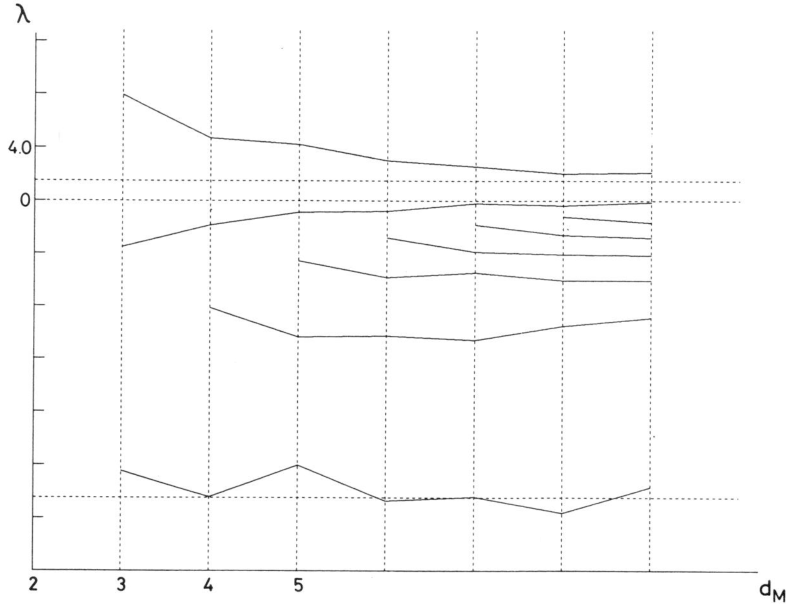

FIG. 4. A noise level of 0.4% has been added.

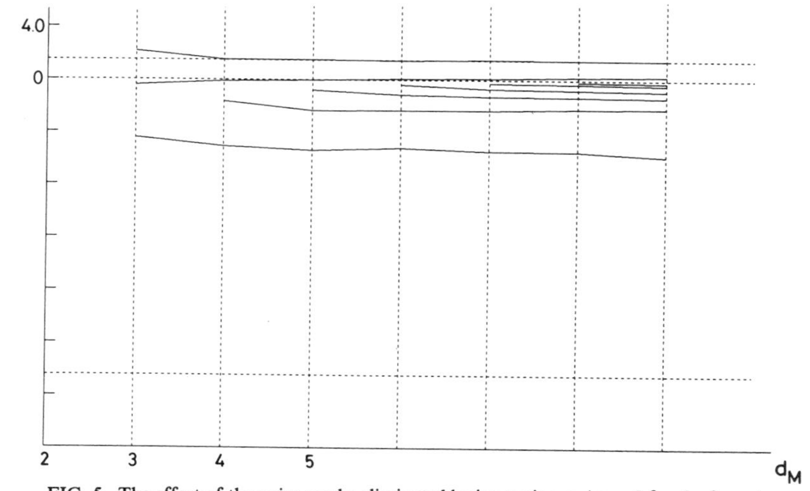

FIG. 5. The effect of the noise can be eliminated by increasing m ($m=5$ for the figure).

(5) Otherwise keep r as small as possible.

(6) Do not step the calculation if a singular matrix arises.

(7) Take a product of as many matrices as possible to determine the Liapunov exponents.

In particular this procedure eliminates the difficulties encountered by Vastano and Kostelich.[6]

IV. EXAMPLES

We begin with the Lorentz equations

$$\frac{d}{dt}\begin{bmatrix} x \\ y \\ z \end{bmatrix} = \begin{bmatrix} -\sigma x+\sigma y \\ -xz+rx-y \\ xy-bz \end{bmatrix},$$

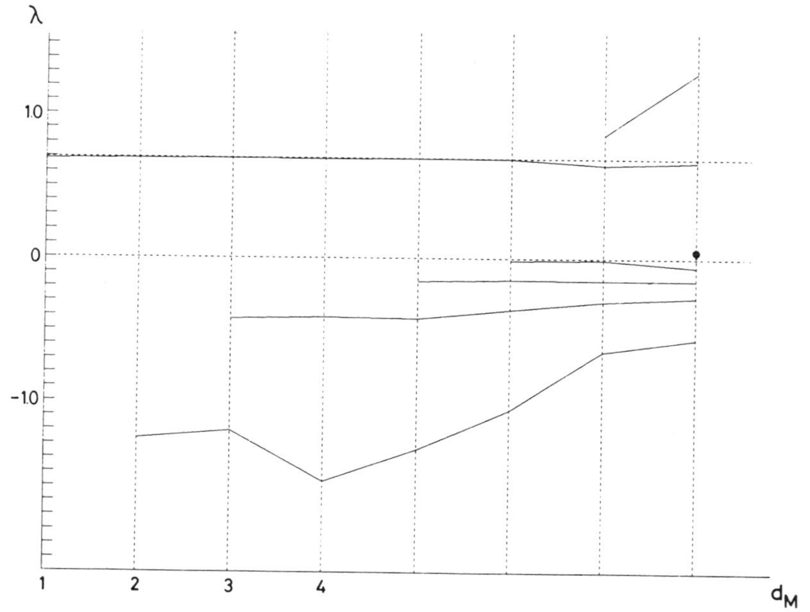

FIG. 6. Analysis of the map $x \rightarrow 1-2x^2$ with a resolution of 10^{-4}.

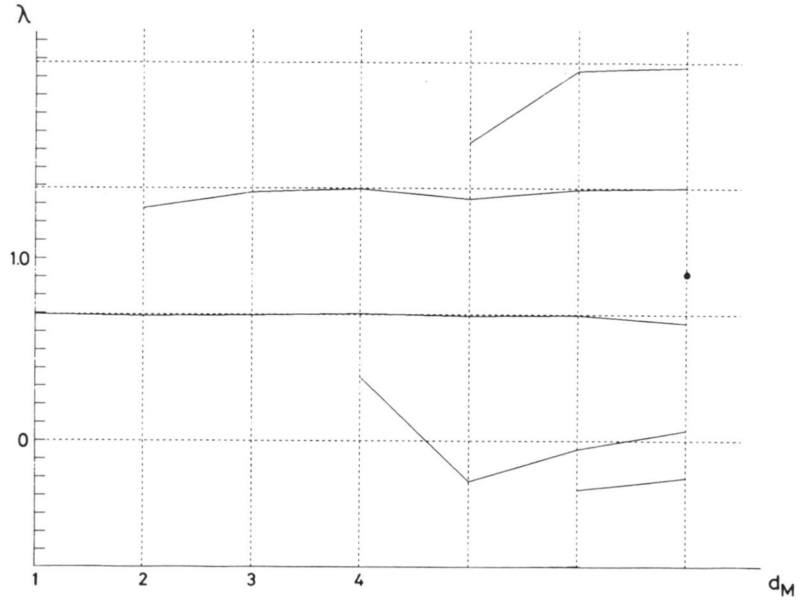

FIG. 7. Analysis of the map $x \to 1-2x^2$ with a resolution of 2^{-31}. A spurious positive Liapunov exponent appears at $\approx 2\ln 2$.

which we study for the parameter values $\sigma=16$, $b=4$, and $r=45.92$. (These parameter values give the usual picture.) We take $\tau=0.03$ and 64 000 data points. In Fig. 2, we take $m=1$, i.e., $d_M=d_E$ and we require card $S_E^i(r) \geq 2d_M$. In this case the agreement with the numerically known non-negative Liapunov exponents (dashed lines) is very good for $d_E \geq 5$. Note that there is a large deviation at $d_E=2$. This serves as an indication that the system lives in a space with more than two dimensions. Also, as observed in Sec. III, $d_E=3$ is not a sufficiently large embedding dimension for precise values of the Liapunov exponents. Finally, it should be noted that increasing the minimal number of neighbors does not change the above observations. In Fig. 3 we illustrate the effect of taking too few neighbors. With the same parameters as in Fig. 2, we have required only card $S_E^i(r) \geq d_M$. The increase of the curves is a typical signature of a lack of sufficiently many neighbors. In Fig. 4 we analyze the influence of (artificially added) noise. We have added 0.4% noise (in terms of the total data latitude) and we observe that the prediction of the Liapunov exponent is wiped out. A typical signature of noise is the decrease of the Liapunov exponent with d_M. Note also that the effect of the noise can be essentially eliminated if we increase m, that is by increasing d_E while keeping d_M fixed. This is shown in Fig. 5, where we have chosen $m=5$ and card $S_E^i(r) \geq d_M + 4$. The results are usually good for the positive Liapunov exponents, but the zero exponent tends to increase with d_M. The collection of Figs. 2–5 is a clear illustration of the summary of advice of Sec. III.

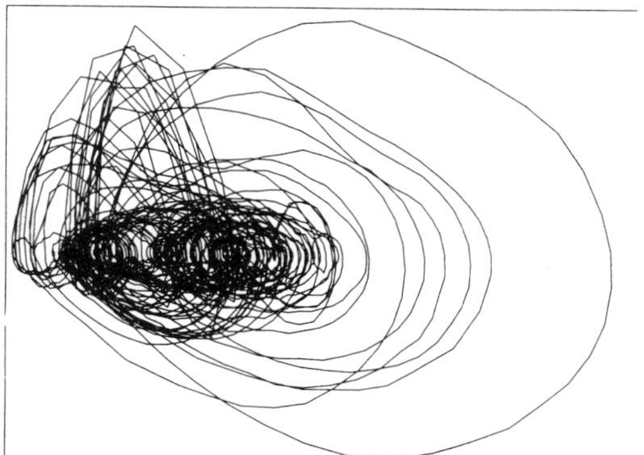

FIG. 8. An experimental orbit. Horizontal axis is x_i, vertical axis $x_i - x_{i-1}$.

We next illustrate in more detail the effect of having too few data points. For this we shall deal with the very simple system defined by the map $x \to 1-2x^2$ of the interval $(-1,1)$. It has a Liapunov exponent ln2. Figure 6 shows the results analogous to Fig. 2, with a resolution of the data of 10^{-4} (obtained by multiplying each x by 5000 and truncating). To make the statistical errors smaller, we have insisted on a minimum of 20 neighbors per point. In Fig. 7 we have applied the same algorithm, but with a precision of $2^{-31} \approx 0.5 \times 10^{-9}$. While this situation is unlikely to occur in laboratory experiments, it is typical for the appearance of spurious Liapunov exponents which are

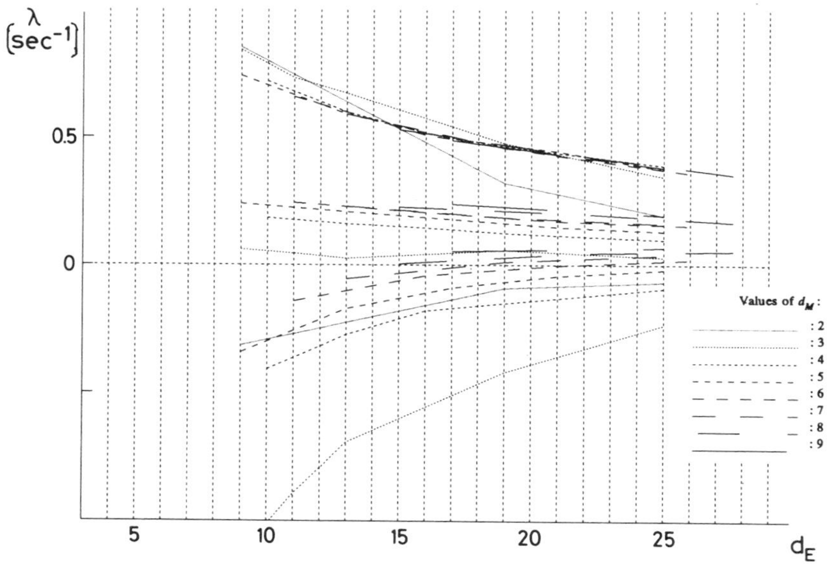

FIG. 9. Largest three Liapunov exponents as a function of d_E for the signal of Fig. 8, for different values of d_M.

about twice (in principle even thrice) the real ones. This phenomenon is generated by the finiteness of the data set. This means that we cannot achieve the limit $r \to 0$ to determine the matrices T_i. Therefore one expects the nonlinearities to be important. A simple calculation shows that if one carries along second-order effects in the equations leading to the T_i, for the map $x \to 1-2x^2$ in dimension $d_M = 2$, one obtains two Liapunov exponents, one at ln2 and one at 2 ln2. In the absence of noise, the nonlinear terms desingularize the equations for the T_i when d_M is larger than the true dimension of the system, and they tend to generate multiples of the "true" Liapunov ex-

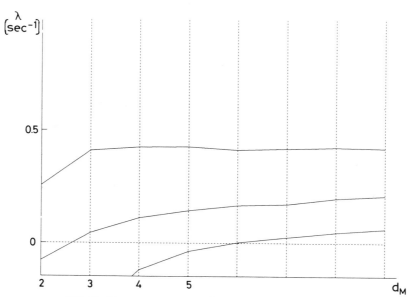

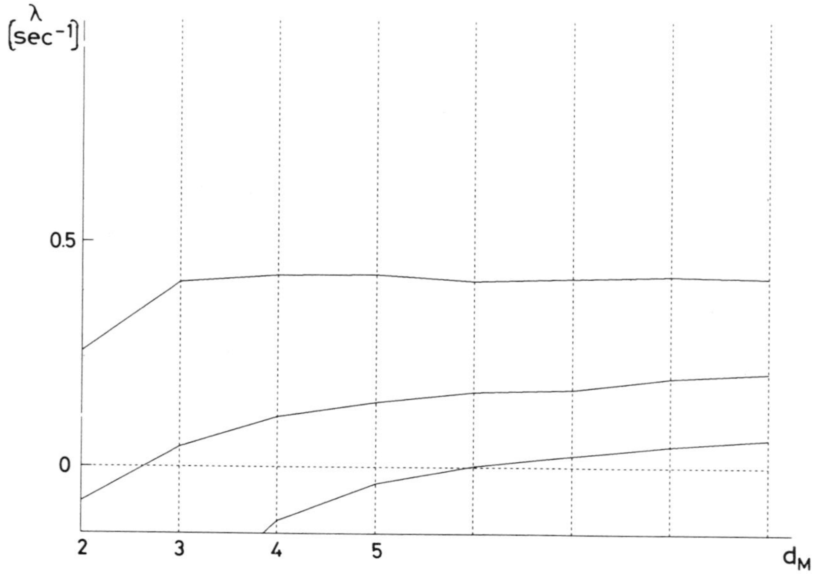

FIG. 10. Liapunov exponents as a function of d_M at fixed $d_E = 22$.

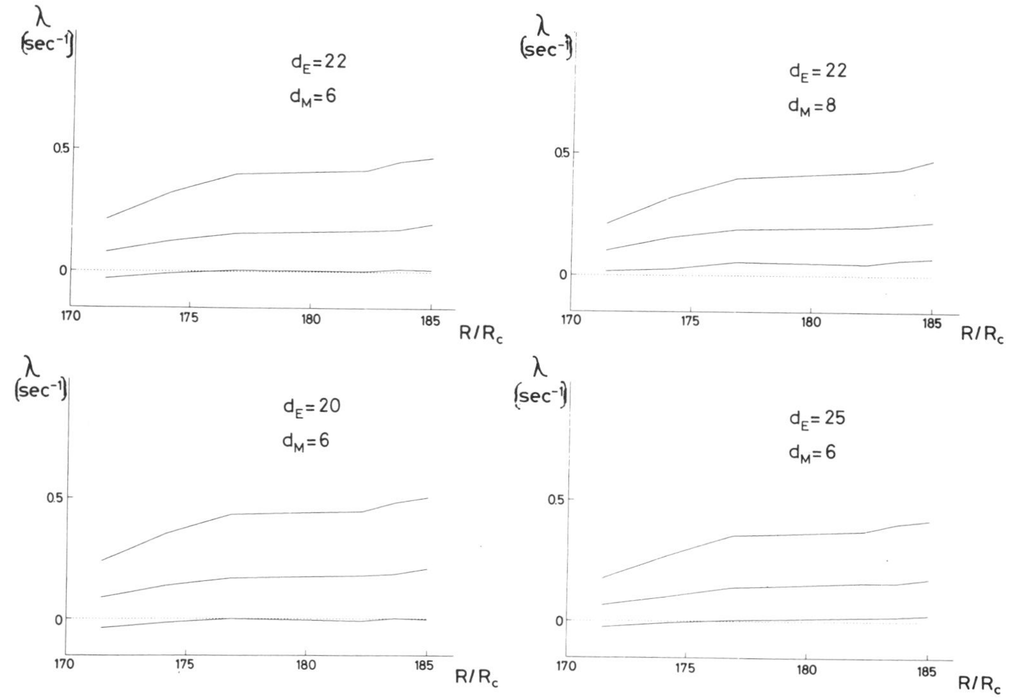

FIG. 11. Three largest Liapunov exponents as a function of Rayleigh number, for different d_M and d_E. From theoretical arguments we know that one Liapunov exponent is equal to zero.

ponents. Having gone to very high precision in the calculations leading to Fig. 7, we have produced explicitly the generation of a "double" Liapunov exponent. (We suspect that Table I in Ref. 5 has the same origin.) In third-order terms are above the numerical imprecision one will in principle see a "triple" Liapunov exponent, and so on. For data coming from laboratory experiments, the equations for the T_i are in general rather desingularized by the noise of the data. In that case a model calculation can be done, based on some independence assumptions of the noise (which may not be justified in general). For the map $x \to 1-2x^2$ and $d_M=2$, this calculation predicts that the two Liapunov exponents will be ln2 and ≈ -1.2.

Let us summarize. All of the above effects only concern the spurious Liapunov exponents, because they are caused by the desingularization of the equations for the T_i. If the noise is not too small relative to the precision and the density of the data, one will see the true Liapunov exponents and the spurious ones will all be negative. This seems to be the usual situation in laboratory experiments.

We now discuss the more difficult problem of analyzing an experimental time series. In particular, we have made measurements coming very close to the required desiderata of Sec. III. In this section we only analyze one of these runs, from the same point of view as the numerical experiments. In the Appendix, the experiments and the complete results are described, as function of a varying parameter. To give a certain feeling of what is involved, we draw a piece of the experimental orbit (Fig. 8).

In Fig. 9 we summarize the results of varying the lags m and matrix dimension d_M in such a way that d_E varies between 9 and 26. We limit d_M between 2 and 9. One observes a relatively flat section in the region above $d_E=20$ for $d_M>3$. This limit is more visible in Fig. 10, where we plot, by interpolation, a section of Fig. 8 at $d_E=22$. In view of the preceding discussions, we conclude that there are two positive Liapunov exponents. In the Appendix, we show the complete results for a series of experiments with varying parameters. (The main body of this paper was done by the first three authors, and the experiment, as well as the Appendix, have been provided by the fourth author. Our collaboration has made an optimization of the experiment and the analysis possible.)

Noted added in proof. An interesting recent paper by A. M. Fraser and H. L. Swinney [Phys. Rev. A 33, 1134 (1986)] indicates how to choose time delays optimally in the reconstruction of dynamics from a time series. For other general literature see Ref. 1, which contains in particular a reference to a paper by Wolf *et al.* We remain unconvinced that the method described in that paper allows the systematic computation of several Liapunov exponents from noisy experimental data.

ACKNOWLEDGMENTS

This work has been made possible by various institutions and organizations. We gratefully acknowledge the support of Institut des Hautes Etudes Scientifique (IHES), Bures-sur-Yvette (JPE), Coordenação de Aperfeiçoamento de Pessoal do Ensino Superior (CAPES), Brazil (SOK), Fonds National Suisse (DR,SC), European Economic Community Contract STI-082-J-C(CD), (SC).

APPENDIX

The transition from a regular to a chaotic behavior has been widely studied both theoretically and experimentally in many physical, chemical, and other natural phenomena. In fluid mechanics one of the most used systems to investigate the onset of turbulence is thermal convection in a horizontal fluid layer heated from below, that is, Rayleigh-Benard convection (RB).[7] When the fluid is confined in a cell whose horizontal dimensions are of the same order of the fluid height (small aspect ratio cells), it has been found that the transition to the chaotic behavior can be explained in terms of the nonlinear interaction of a small number of degrees of freedom. This has been verified either by checking if the observed route to chaos was equal to one of the standard routes to chaos for low-dimensional systems[7] or more quantitatively with the measurement of the fractal dimension of the attractor.[8,9]

In a RB experiment the determination of the positive Liapunov exponents, that are indeed an important sign of the existence of a strange attractors, has been done only in Ref. 2 with a method similar to that outlined in the previous pages. Nevertheless, as pointed out in this text, there are some differences between the two methods and so we have applied the algorithm here proposed to evaluate the Liapunov exponents from a series of data recorded in the chaotic regime of a RB cell.

The fluid layer has horizontal sizes $l_x = 4$ cm, $l_y = 1$ cm, and height $d = 1$ cm (aspect ratios $\Gamma_x = 4, \Gamma_y = 1$). The fluid is silicon oil with Prandtl number 30. The bottom and top plates of the cell are made of copper and the temperature stability is about 1 mK. The lateral walls are made of glass to allow optical inspection. The detection system allows a semilocal measurement to be made. In fact, it consists of a laser beam, with a diameter of 1 mm, that crosses the fluid layer parallel to the rolls axis and is deflected by the thermal gradients inside the fluid. By measuring the deflection of the laser beam outside the cell we can measure the thermal gradient averaged along the optical path. We record the horizontal component of the gradient because it usually has the largest time-dependent amplitude.

In order to have a signal-to-noise ratio within the requirements specified in this paper, particular attention has been paid to reduce the environmental noise produced, for example, by the air convection along the optical path of the laser beam, by the vibrations of the mirrors, and of the laser cavity. Furthermore, to eliminate high-frequency noise, the signal has also been filtered at a suitable cutoff frequency to avoid that the rising time of the filter influencing the evaluation of the Liapunov exponents. This way a signal-to-noise ratio of about 10^{-4} has been achieved.

Analyzing the fluid behavior as a function of R/R_c (R_c is the critical value of the Rayleigh number), we find, except for a small region at $81 < R/R_c < 90$ where the convective motion is time dependent, a stable four-rolls structure for $R/R_c < 141$. Above this threshold a periodic oscillation at a frequency of 75 mHz is observed. Increasing R, the fluid crosses many different periodic and biperiodic states and it goes into the chaotic region via intermittency at $R/R_c > 170$. We have characterized this chaotic behavior by measuring the Liapunov exponents as a function of the Rayleigh number.

To satisfy all other requirements of the algorithm, 40 000 points, with a sampling frequency of 5 Hz, have been recorded for each measurement. This way, the time evolution of the system is followed for about 600 periods of the main oscillation. Many tests have been done to verify how the Liapunov exponents depend on d_E and d_M. It has been found that the value of λ is sufficiently stable in the interval $20 < d_E < 25$ and $5 < d_M < 8$. The results are reported in Fig. 11, where the values of the positive Liapunov exponents are shown as a function of R for different d_E and d_M. We see that the qualitative behavior of the curves is similar and the difference between them is about 10%. The measurements were done for $R/R_c = 171.41$, 174.08, 176.75, 182.10, 183.44, and 184.79. Figures 8–10 show details for $R/R_c = 182.10$.

By moving the detection point inside the cell by about 1 cm and keeping R at the last value shown in Fig. 11 we find that the Liapunov exponents change by less than 5%. As a conclusion, the positive Liapunov exponents in the chaotic regime of a RB convection experiment have been determined using the method proposed in (Ref. 1). Even though the error of the measurement is not small (about 10%) it is still possible to follow how the number of the Liapunov exponents and their values change as a function either of the control parameter R or of the position where the measurement has been taken inside the fluid.

[1]J.-P. Eckmann and D. Ruelle, Rev. Mod. Phys. **57**, 617 (1985).
[2]M. Sano and Y. Sawada, Phys. Rev. Lett. **55**, 1082 (1985).
[3]N. H. Packard, J. P. Crutchfield, J. D. Farmer, and R. S. Shaw, Phys. Rev. Lett. **45**, 712 (1980).
[4]D. E. Knuth, *The Art of Computer Programming* (Addison-Wesley, New York, 1973), Vol. 3.
[5]J. A. Vastano and E. J. Kostelich, in *Entropies and Dimensions,* edited by G. Mayer-Kress (Springer-Verlag, Berlin, in press).
[6]J. H. Wilkinson, and C. Reinsch, *Linear Algebra* (Springer-Verlag, Berlin, 1971).
[7]For a recent review, see, e.g., R. P. Behringer, Rev. Mod. Phys. **57**, 657 (1985).
[8]B. Malraison, P. Atten, P. Bergé, and M. Dubois, J. Phys. Lett. **44**, L897, (1983).
[9]M. Giglio, S. Musazzi, and V. Perini, Phys. Rev. Lett. **53**, 240 (1984).

Lyapunov Exponents from Observed Time Series

Paul Bryant and Reggie Brown
Institute for Nonlinear Science, University of California at San Diego, La Jolla, California 92093-0402

Henry D. I. Abarbanel[(a)]
Department of Physics and Marine Physical Laboratory, Scripps Institution of Oceanography, University of California at San Diego, La Jolla, California 92093-0402

(Received 11 June 1990)

We examine the question of accurately determining Lyapunov exponents for a time series. We find that it is advantageous to use local mappings with higher-order Taylor series, rather than linear maps as done earlier. We demonstrate this procedure for the Ikeda map and the Lorenz system. We present methods for identifying spurious exponents by analyzing data-set singularities and by determining the Lyapunov direction vectors. The behavior of spurious exponents in the presence of noise is also investigated, and found to be different from that of the true exponents.

PACS numbers: 05.45.+b

Determining the Lyapunov exponents of a nonlinear system from measurements of a time series is an important challenge for any analysis of the dynamics. Positive exponents are generally regarded as equivalent to the presence of real dynamical chaos, and the Lyapunov exponents are classifiers of the dynamics since they are characteristic of the attractor and independent of any given orbit or initial condition.[1,2] If the governing equations are known, then there are reliable methods for determining all of the exponents. If one only has a time series, then the problem becomes much more difficult. There are several reported efforts to provide algorithms for the determination of the Lyapunov exponents from observations alone.[3-6] Our own experience[7] with these algorithms is that they are reliable only for the largest exponent and not for the others. The importance of Lyapunov exponents in the study of physical systems has led us to provide an improvement on these previous efforts and to address several other questions of importance in the determination of Lyapunov experiments from data. In this Letter we report on the basic outline of our methods. We leave details and numerous other examples to our longer paper.[8]

Earlier work and ours assume that a scalar time series $x(t_0+n\tau)=x(n)$, $n=1,2,\ldots,N_D$, has been observed. From this the phase space of the system has been reconstructed by the familiar time-delay method[8-12] to produce data vectors in d dimensions:

$$\mathbf{y}(n)=[x(n),x(n+T),\ldots,x(n+T(d-1))]\,.$$

In discretized time one takes the dynamics to be a map of R^d to itself which evolves the vectors $\mathbf{y}(n)$: $\mathbf{y}(n+T_2)=\mathbf{F}(\mathbf{y}(n))$, where the time delay T_2 is independent of T. The product of the Jacobians of this map $\mathbf{DF}(\mathbf{y})=\partial\mathbf{F}(\mathbf{y})/\partial\mathbf{y}$ evaluated along an orbit contains the information required for the Lyapunov exponents.[1,2]

The first step in the analysis is to find the neighboring points of a given point in the data set. Our choice of neighbors is limited by the finite size of the data set, by stochastic noise, and most importantly by the fractal nature of the attractor. These limitations are the main source of difficulties in the analysis. Finding legitimate neighbors of a given point is one of the most critical tasks in obtaining accurate results. For this reason it is often advisable to maintain two different dimensions for converting the scalar data set into time-delay vectors. The first is a "local dimension" d which is equal to the number of Lyapunov exponents that the calculation will produce and is the dimension of the Jacobian matrices. The second is a "global dimension" d_G which is used in the process of identifying neighbors. d_G can be made larger than d in order to insure that we do not have any false neighbors entering the calculation as might be the case if the attractor is folded in such a way that it crosses itself. One way of choosing values for d and d_G is to start by computing a rough estimate of the fractal dimension d_A of the attractor. If an object of dimension d_A is mapped in a very general way into a space of dimension d_G it can typically have self-intersections of dimension $2d_A - d_G$. Thus we would like to make d_G larger than $2d_A$. It is not advisable to make d much greater than d_A, since locally the attractor has very little extension into these additional directions. A good first choice is to try making d the next integer greater than or equal to the fractal dimension d_A. This choice will always give at least one negative exponent for a chaotic system. One is then free to further increase d and see what effect this has on the results.

Earlier work attempts to find the required Jacobians by making local *linear* maps of neighborhoods near the orbit $\mathbf{y}(n)$ to neighborhoods at a subsequent time. We depart from these earlier works by making local polynomial maps, allowing for a more accurate determination of $\mathbf{DF}(\mathbf{y})$. While there have been previous uses of higher-order mappings in studies of dynamical systems (see, e.g., Refs. 13 and 14) this is, we believe, the first application of this approach to the calculation of the full Lyapunov spectrum. The vector from the rth neighbor to an orbit point $\mathbf{y}(n)$ is denoted by $\mathbf{z}^r$. Using a least-squares fit to the data we obtain a polynomial map from this vector at time 0 to the same vector on time step T_2 later: $\mathbf{z}^r(n;0) \to \mathbf{z}^r(n;T_2)$, including all terms up to some specified order N_{Tay}. We have examined the effects of retaining terms up to fifth order in $\mathbf{z}^r(n;0)$. We use at least twice the number of neighbors as parameters to be determined in order to insure reliable results. We then proceed to calculate the Lyapunov exponents using the QR decomposition technique discussed by Eckmann *et al.*[3,15,16] As we show with our methods, one can determine the positive, zero, and often one or more negative exponents.

In this Letter we report on results from the Ikeda[17] map of the complex plane to itself,

$$z(n+1) = p + Bz(n)\exp\{i\kappa - i\alpha/[1+|z(n)|^2]\}, \quad (1)$$

where $p = 1.0$, $B = 0.9$, $\kappa = 0.4$, and $\alpha = 6.0$. For these parameters we calculate (using the map) that λ_1 and λ_2 are 0.503 and -0.719, respectively. We also study the Lorenz system of three ordinary differential equations:[18]

$$\begin{aligned} \frac{dx_1}{dt}(t) &= \sigma[x_2(t) - x_1(t)], \\ \frac{dx_2}{dt}(t) &= -x_1(t)x_3(t) + rx_1(t) - x_2(t), \\ \frac{dx_3}{dt}(t) &= x_1(t)x_2(t) - bx_3(t), \end{aligned} \quad (2)$$

where we take $\sigma = 16$, $b = 4$, and $r = 45.92$. For these parameters the accepted values for the Lyapunov exponents are 1.50, 0.00, and -22.5, respectively. The large negative exponent makes this system a particularly challenging test for our time-series method, and also requires the use of very accurate data, as will be shown.

The difficulty in determining the negative exponents from a time series comes primarily from the fact that the attractor is often very "thin" at many locations in the directions associated with certain negative exponents. Even when there is a reasonably large and accurate data set, this will often make curvature effects within a given neighborhood become significant. A linear analysis becomes totally inaccurate when the displacement due to local data-set curvature is comparable to the thickness of the data set. Going to a higher-order approximation of the mapping can correct this.

The Ikeda map is an excellent example of a system for which the use of separate local and global dimensions is important. Examination of the two-dimensional time-delay representation Fig. 1 shows clearly the self-intersection effect which was discussed previously. Having determined the fractal dimension d_A to be about 1.8 we would choose d_G to be at least 3 and preferably 4, and an appropriate value for d is 2. Using the incorrect values $d_G = d = 2$ in a third-order calculation we obtain $\lambda_1 = 0.565$ and $\lambda_2 = -0.426$, while if we use $d_G = 4$ and $d = 2$ we obtain $\lambda_1 = 0.512$ and $\lambda_2 = -0.736$ which are much closer to the correct values (0.503 and -0.719). If we keep d equal to d_G but increase both of their values to 3, the overlap problem is reduced but we now obtain 3 exponents (0.554, -0.262, and -0.821) and so we are faced with the problem of deciding which, if any, are valid exponents.

We move now to the Lorenz system. In the case of data from the Lorenz equations we have used two slightly different settings for the evolution time lag. We display the results of both our calculations in Table I.

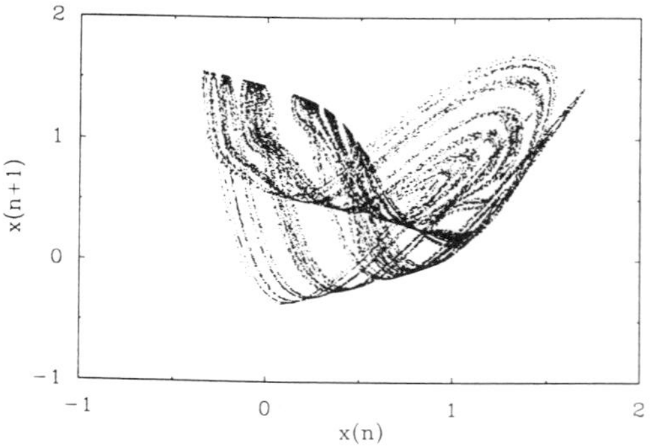

FIG. 1. Reconstructed phase portrait of the Ikeda map in $d_G = 2$. The lack of a one-to-one projection of the attractor onto the $[x(n), x(n+1)]$ plane is clear to the eye.

TABLE I. In the top part of the table we display the Lyapunov exponents for the Lorenz system computed from 50 000 data points evaluated with a sampling time $\tau = T_2 = 0.02$ and a time delay $T = 5\tau = 0.1$. The data have five digits of accuracy and are analyzed using $d = d_G = 3$ for varying orders N_{Tay} of the mapping. In the bottom part we display the Lyapunov exponents for the Lorenz system computed from 20 000 data points evaluated with a sampling time $\tau = 0.05$ and a time delay $T = T_2 = 2\tau = 0.1$. The data have 9+ digits of accuracy and are analyzed using $d = 3$ and $d_G = 7$ for mapping orders 1-5.

N_{Tay}	λ_1	λ_2	λ_3
1	1.4504	−0.005 712 3	−13.999
2	1.5027	−0.046 041	−19.448
3	1.5121	0.006 964 1	−22.925
4	1.5561	0.032 219	−23.465
1	1.549	−0.094 70	−14.31
2	1.519	−0.026 47	−20.26
3	1.505	−0.005 695	−22.59
4	1.502	−0.002 847	−22.63
5	1.502	−0.000 387	−22.40

As the reader will observe, the negative exponent is very difficult to obtain, and here we see it dramatically "snapping into place" as we increase the order of the calculation to 3 and above. Also note the improvement in accuracy of the zero exponent in the lower part of Table I. The greater accuracy of the lower part of Table I is due primarily to the higher accuracy of the data set.

In Table II, we analyze the Lorenz equations with a local dimension $d = 4$, which we know must generate at least one spurious exponent. When using a second-order fitting to our local map, the results are poor for all of the Lyapunov exponents. Increasing the polynomial fitting to third order we find that the last three exponents are very close to the true exponents, while the first is 10 times larger than the true value of the largest exponent.

In addition to obtaining the Lyapunov exponents, one can also obtain the direction vectors $\mathbf{L}_i$ associated with these exponents. The $\mathbf{L}_i$ are defined by the requirement that a small displacement along any one of these directions followed forward or backward in time will expand or contract on average at the rate given by the corresponding exponent. Although they are different at every location on the attractor, their calculation requires a knowledge of the orbit far into the past and future for a given point on the attractor. The details of their calculation are to be found in our longer paper.[8] They should be examined to see if two or more of them are nearly collinear. This can occur if a poor choice was made for the delay time (probably too small) or if nonlinear effects are generating a spurious exponent. The spurious nature of λ_1 in our Lorenz-system calculations can be rapidly identified by examining the local data thickness Th_1 in the $\mathbf{L}_1$ direction, which is over 5 orders of magnitude smaller than the thickness Th_2 for $\mathbf{L}_2$. A valid positive exponent should not exhibit any significant "thinness," and it is preferable for all exponents to have thickness levels above the intrinsic noise level of the data set. The thickness Th_i is essentially the rms displacement of the data points within a local neighborhood in the $\mathbf{L}_i$ direction, with corrections for data-set curvature; more details can be found elsewhere.[8]

Although we have shown that it is possible to include singular directions in the calculation and later identify the questionable exponents, the presence of relatively small amounts of noise makes this more difficult. This is illustrated in Fig. 2 for the Lorenz system. We have added Gaussian white noise to the data points with the indicated standard deviation. In Fig. 2(a) we have used $d = 3$, while in 2(b) we used $d = 4$, which gives one spurious exponent. In both cases we used a third-order expansion for the local mappings. The spurious exponent in 2(b) drops rapidly as the added noise is increased, going from +19 down to −6. This behavior is in fact another way of identifying a spurious exponent in extremely accurate data. However, the absence of such a drop does not guarantee that spurious exponents do not exist.

We conclude by noting that a simple extension to higher order of earlier methods for determining the Lyapunov spectrum for a dynamical system from observations alone works strikingly well when tested on familiar systems such as the Ikeda map and Lorenz attractor. We have also suggested and tested on these examples ways to determine which of the exponents are valid and which are spurious. Finally, we explored the effect of numerical accuracy and external noise on the determina-

TABLE II. Lyapunov exponents and thicknesses (Th) of the attractor along the corresponding Lyapunov direction vectors for the Lorenz system, through order $N_{Tay} = 4$. The calculation was done with 20 000 data points using $d = 4$ and $d_G = 7$ so that there is one spurious exponent. For $N_{Tay} = 2$ and above, the spurious exponent separates from the true ones and can be identified by its extremely small thickness value.

N_{Tay}	λ_1	Th_1	λ_2	Th_2	λ_3	Th_3	λ_4	Th_4
1	1.936	0.467	0.802	1.083	−1.137	1.111	−13.44	0.162
2	4.36	0.002 56	1.401	0.412	−0.656	0.454	−20.57	0.001 16
3	18.04	1.89×10^{-7}	1.502	0.129	−0.0005	0.091	−22.77	2.7×10^{-4}
4	26.96	2.5×10^{-10}	1.503	0.066	−0.0048	0.066	−22.55	5.6×10^{-5}

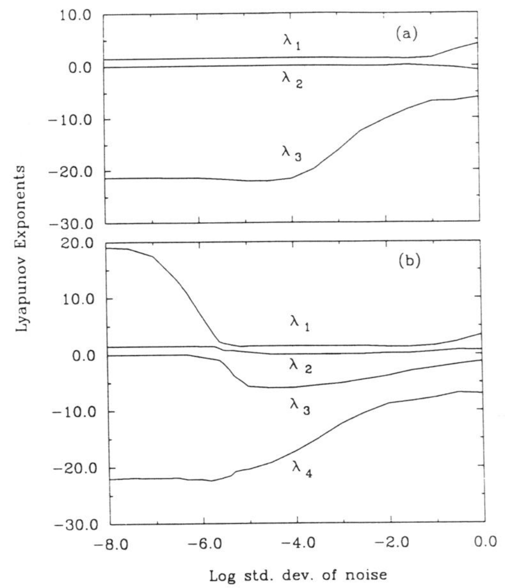

FIG. 2. The effect of external noise on the determination of Lyapunov exponents for the Lorenz system. (a) Here local dimension is $d=3$. (b) Here $d=4$. The spurious exponent wanders from about +19 to nearly −6 as the noise level is varied. Note that the exponents do not cross each other but prefer to switch roles as they become close. In the $d=3$ case the correct exponents are more robust against the addition of noise.

tion of these exponents. We found the calculation to be quite sensitive to noise, which reflects the fact that fluctuations in phase-space points make determination of the required local Jacobians quite sensitive. Further examples and extensive details on the methods outlined here will be found in our longer paper.[8] Also, the algorithms used here are available on request from the authors.

We are most appreciative for productive conversations with Andy Fraser, Ed Lorenz, and Jorma Rissanen about the material covered in this Letter. This work was supported in part under Defense Advanced Research Projects Agency-University Research Initiative, URI Contract No. N00014-86-K-0758. Some of the computation reported in this paper was carried out at the NASA Ames Research Center's Numerical Aerodynamic Simulation Program under the auspices of the Joint Program in Nonlinear Science between the University of California and NASA Ames.

[(a)]Also at Institute for Nonlinear Science.

[1]V. I. Oseledec, Tr. Mosk. Mat. Obšč. **19**, 17 (1968).

[2]J.-P. Eckmann and D. Ruelle, Rev. Mod. Phys. **57**, 617 (1985).

[3]J.-P. Eckmann, S. O. Kamphorst, D. Ruelle, and S. Ciliberto, Phys. Rev. A **34**, 4971 (1986).

[4]G. Benettin, L. Galgani, A. Giorgilli, and J.-M. Strelcyn, Meccanica **15**, 9 (1980).

[5]A. Wolf, J. B. Swift, H. L. Swinney, and J. A. Vastano, Physica (Amsterdam) **16D**, 285 (1985).

[6]M. Sano and Y. Sawada, Phys. Rev. Lett. **55**, 1082 (1985).

[7]H. D. I. Abarbanel, R. Brown, and J. B. Kadtke, Phys. Lett. A **138**, 401 (1989); Phys. Rev. A **41**, 1782 (1990).

[8]R. Brown, P. Bryant, and H. D. I. Abarbanel, "Computing the Lyapunov Spectrum of a Dynamical System from Observed Time Series," Institute for Nonlinear Science-University of California at San Diego report, 1990 (to be published).

[9]F. Takens, in *Dynamical Systems and Turbulence, Warwick, 1980*, edited by D. Rand and L. S. Young, Lecture Notes in Mathematics Vol. 898 (Springer-Verlag, Berlin, 1981), p. 366.

[10]R. Mañé, in *Dynamical Systems and Turbulence, Warwick, 1980* (Ref. 9), p. 230.

[11]A. M. Fraser and H. L. Swinney, Phys. Rev. A **33**, 1134 (1986); A. M. Fraser, IEEE Trans. Inf. Theory **35**, 245 (1989).

[12]H. D. I. Abarbanel and J. B. Kadtke, "Information Theoretic Methods for Choosing the Minimum Embedding Dimension for Strange Attractors," University of California at San Diego-Institute for Nonlinear Science report, 1989 (to be published).

[13]J. D. Farmer and J. J. Sidorowich, in *Evolution Learning and Cognition*, edited by Y. C. Lee (World Scientific, Singapore, 1988).

[14]P. Bryant and C. Jeffries, Physica (Amsterdam) **25D**, 196 (1987).

[15]W. H. Press, B. Flannery, S. Teulkolsky, and W. Vetterling, *Numerical Recipes* (Cambridge Univ. Press, New York, 1986).

[16]J. J. Dongarra, J. R. Bunch, C. B. Moler, and G. W. Stewart, *Linpack Users' Guide* (Society for Industrial and Applied Mathematics, Philadelphia, 1979).

[17]K. Ikeda, Opt. Commun. **30**, 257 (1979).

[18]E. N. Lorenz, J. Atmos. Sci. **20**, 130 (1963).

IDENTIFICATION OF TRUE AND SPURIOUS LYAPUNOV EXPONENTS FROM TIME SERIES

ULRICH PARLITZ
Institut für Angewandte Physik, TH Darmstadt, Schloßgartenstraße 7, D-6100 Darmstadt, Germany

Received July 29, 1991; Revised August 7, 1991

A new method for the identification of true and spurious Lyapunov exponents computed from time series is presented. It is based on the observation that the true Lyapunov exponents change their signs upon time reversal whereas the spurious exponents do not. Furthermore by comparison of the spectra of the original data and the reversed time series suitable values for the free parameters of the algorithm used for computing the Lyapunov exponents (e.g., the number of nearest neighbors) are determined. As an example for this general approach an algorithm using local nonlinear approximations of the flow map in embedding space by radial basis functions is presented. For noisy data a regularization method is applied in order to get smooth approximating functions. Numerical examples based on data from the Hénon map, a four-dimensional analog of the Hénon map, a quasiperiodic time series, the Lorenz model, and Duffing's equation are given.

1. Introduction

Lyapunov exponents belong to the most important quantities used in Nonlinear Dynamics to distinguish chaotic from nonchaotic behavior. When the dynamics is chaotic, positive exponents occur that quantify the rate of separation of neighboring (initial) states and give the period of time where predictions are possible. For experimental work, methods for computing the Lyapunov exponents of a given time series are of great importance. The different algorithms that have been proposed during the last years for this purpose are all based on the reconstruction of the corresponding attractor by means of a suitable embedding with time delay coordinates [Packard *et al.*, 1980; Takens, 1981]. When the Lyapunov exponents are computed directly from the rate of separation of the neighboring state points along a fiducial orbit [Wolf *et al.*, 1985] in general, only the largest exponents can be determined. Therefore most of the methods are based on approximations of the unknown flow (or map) governing the dynamics on the reconstructed attractor in embedding space. From the sequence of Jacobians of this map at the state points along the reconstructed orbit, the Lyapunov exponents can be computed by standard methods [Sano & Sawada, 1985; Eckmann & Ruelle, 1985; Geist *et al.*, 1990]. The crucial point of this approach is the accuracy of the approximation of the Jacobian of the unknown flow. The most often used algorithms to achieve this goal are based on linear approximations [Eckmann *et al.*, 1986; Sato *et al.*, 1987; Holzfuss & Lauterborn, 1989, Stoop & Meier, 1988; Buzug *et al.*, 1990; Stoop & Parisi, 1991]. Even for noiseless data, the results of these methods may become incorrect when the number of data points is small and/or the degree of nonlinearity of the flow to be approximated is high so that the neighboring state points exceed the range of the linear ansatz. The straightforward extensions of this approach using approximations with higher order polynomials [Bryant *et al.*, 1990; Briggs, 1990; Brown *et al.*, 1991] or radial basis functions [Holzfuss & Parlitz, 1991] have been proposed only recently and yield much better results than the linear methods.

A drawback common to all algorithms proposed so far is the fact that they do not provide any error estimates for their results. Furthermore spurious exponents occur when the reconstructed attractor is contained in a (smooth) submanifold of the embedding space possessing a dimension m that is smaller than the embedding dimension d. This situation occurs when the embedding dimension is chosen too large or when, due to the global topology of the attractor to be reconstructed, a high dimensional embedding is required. A two-dimensional torus for example can only be reconstructed in three or more dimensions. The local fits for approximating the flow in such situations are based on clouds of points along the m directions of the submanifold containing the attractor. For the remaining $d - m$ directions, no reconstructed state points are available and the shape of the graph of the approximating function depends only on the ansatz (basis functions) used. As a consequence, there appear in general $d - m$ spurious Lyapunov exponents that are unrelated to the dynamics on the attractor. For noisy data the reconstructed state points do not lie exactly on the m-dimensional submanifold but scatter also to some extent in the other $d - m$ directions. Therefore the spurious Lyapunov exponents are very sensitive to noise. When the number of spurious exponents is known one can distinguish between a *global dimension d* that is used for the embedding and the selection of neigboring state points and the *local dimension m* of the submanifold containing the reconstructed attractor. The computation of the Lyapunov exponents can in this case be restricted to m-dimensional subspaces resulting in the m true exponents [Eckmann *et al.*, 1986; Bryant *et al.*, 1990; Brown *et al.*, 1991]. Another method to avoid the spurious exponents was proposed by Stoop & Parisi [1991]. They use singular value decompositions to determine at each reference point along the trajectory a linear approximation of the m-dimensional subspace containing the state points. Then linear approximations of the flow restricted to these "tangent planes" of the attractor are computed that yield only the m true Lyapunov exponents. Both methods are based on a knowledge of the number of spurious Lyapunov exponents which is, in general, not known *a priori*. Only Bryant *et al.* [1990] and Brown *et al.* [1991] proposed a method for identifying spurious Lyapunov exponents by estimating the local thickness of the embedded attractor along the corresponding Lyapunov vectors. This criterion allows one to decide whether a computed Lyapunov exponent is true or spurious but it gives no error bounds for the (true) results.

Another difficulty occurring upon computation of Lyapunov exponents from time series consists in the right choice of the free parameters of the algorithms used (e.g., the number of neighboring state points used for the fits). In many cases the results have a rather sensitive dependence on these parameters and convergence does not, in general, imply correctness. In this paper, the following questions are addressed: how to choose the parameters of the algorithm and how to decide which of the Lyapunov exponents computed by a given method are true and correct and which are spurious. The key for our analysis is a check for self-consistency based on the results from the given time series and the Lyapunov exponents computed from the time reversed data. As an example for this general approach, an algorithm will be considered where the unknown flow is approximated by radial basis functions (Hardy's multiquadrics). For noisy data or data with low resolution (e.g., from an A/D-converter) this approximation scheme is combined with a regularization technique (Tikhonov-Phillips-regularization) to smooth the graphs of the approximating functions. The sensitivity to noise of the resulting highly nonlinear approximations depends on a stiffness parameter and a free regularization parameter that are determined by the comparison of the Lyapunov spectra of the original and the time reversed data. Differentiating the locally fitted radial basis functions yields the desired approximations of the Jacobians of the flow that are then used for the computation of the Lyapunov exponents. As examples, the Lyapunov exponents from the data of the Hénon map, a four-dimensional analog of the Hénon map, a quasiperiodic time series, the Lorenz model, and noisy time series from Duffing's equation are presented.

2. Approximation of the Flow in Embedding Space

Let $\{s^i\}$ be a given *time series* that is supposed to originate from a chaotic or quasiperiodic attractor of the dynamical system under investigation. The scalar *samples* $s^i = s(t_i)$ are taken at times $t_i = i \cdot t_s$ where t_s denotes the *sampling time*. This time series is embedded in an d-dimensional Euclidean space $\mathbb{R}^d$ by the method of delay coordinates [Packard *et al.*, 1980; Takens, 1981]. The distribution of *state points* $\mathbf{x}^i = (s^i, s^{i+n_d}, \ldots, s^{i+(d-1)n_d})$ in the *embedding space* $\mathbb{R}^d$ depends on the choice of the *delay time* $t_d = n_d \cdot t_s$. Criteria for determining the optimal value of d and t_d (or n_d) can for example, be found in Liebert *et al.* [1991] and the references cited therein. The state points $\mathbf{x}^i \in \mathbb{R}^d$ are mapped by some (unknown) *flow* (or *map)* $\Phi^{t_e} : \mathbf{x}^i \mapsto \mathbf{x}^{i+n_e}$ where $t_e = n_e \cdot t_s$ denotes the *evolution time*. For a given evolution time t_e this flow has to be approximated in order to determine the temporal evolution of neighboring state points and to compute the Lyapunov exponents. Some authors consider only the case $t_d = t_e$ (i.e., $n_d = n_e$) where the images of the state points $\mathbf{x}^i = (x_1^i, x_2^i, \ldots, x_d^i)$ are given as $\mathbf{x}^{i+n_e} = (x_2^i, x_3^i, \ldots, x_d^i, g(\mathbf{x}^i))$ and the approximation of Φ^{t_e} is reduced to the determination of a good representation of the function $g : \mathbb{R}^d \to \mathbb{R}$. We do not use this simplifying assumption here in order to keep the approximation scheme as flexible as possible.

The components f_l of the function $f : \mathbb{R}^d \to \mathbb{R}^d$ approximating the flow Φ^{t_e} around a given state point $\mathbf{x}^i \in \mathbb{R}^d$ can be chosen as superpositions $f_l(\mathbf{x}) = \sum_{k=1}^{M} c_{kl} X_k(\mathbf{x})$ of suitable *basis functions* X_k ([Press *et al.*, 1986]). The basis functions used for the computations presented in this paper are *radial basis functions* (*Hardy's multiquadrics*) $X_k(\mathbf{x}) = \sqrt{r^2 + \|\mathbf{x} - \mathbf{z}^k\|^2}$ centered at $\mathbf{z}^1 = \mathbf{x}^i$ and at the $M-1$ nearest neighbors $\{\mathbf{z}^k\}_{k=2,\ldots,M}$ of $\mathbf{x}^i$ where $\|\cdot\|$ denotes the Euclidean norm [Franke, 1982; Powell, 1985; Broomhead & Lowe, 1988; Casdagli, 1989; Poggio & Girosi 1990]. When the *stiffness parameter* r equals zero the function f is piecewise linear, for all $r > 0$ it becomes smooth. For the numerical computations discussed below, r was considered as a function of $\mathbf{x}^i$ and replaced by $r_i = r \cdot \max_{k=1,\ldots,M} \|\mathbf{x}^i - \mathbf{z}^k\|_\infty$. The neighboring points have been selected using the maximum norm $\|\cdot\|_\infty$ in the embedding space and the search algorithm described by Eckmann *et al.* [1986].

For the computation of the coefficients c_{kl} one has to distinguish between *interpolation* where the graph of the function is forced to pass through the images $\mathbf{y}^k = \Phi^{t_e}(\mathbf{z}^k)$ of the M state points $\mathbf{z}^k$ and *approximation* where this condition is relaxed in favor of smoothness to reduce the influence of noise.

In the case of interpolation [Holzfuss & Parlitz, 1991] the following system of linear equations has to be solved for the coefficients $\{c_{kl}\}$ of each component f_l of the interpolating function f where y_l^k is the lth component of $\mathbf{y}^k = \Phi^{t_e}(\mathbf{z}^k)$.

$$\mathbf{y}_l = \begin{pmatrix} y_l^1 \\ \vdots \\ y_l^M \end{pmatrix} = \begin{pmatrix} f_l(\mathbf{z}^1) \\ \vdots \\ f_l(\mathbf{z}^M) \end{pmatrix} = \begin{pmatrix} \sum_{k=1}^{M} c_{kl} X_k(\mathbf{z}^1) \\ \vdots \\ \sum_{k=1}^{M} c_{kl} X_k(\mathbf{z}^M) \end{pmatrix} \tag{1}$$

Using matrix notation these d systems of linear equations can be summarized as

$$Y = \begin{pmatrix} y_1^1 & \cdots & y_d^1 \\ \vdots & \ddots & \vdots \\ y_1^M & \cdots & y_d^M \end{pmatrix} = \begin{pmatrix} X_1(\mathbf{z}^1) & \cdots & X_M(\mathbf{z}^1) \\ \vdots & \ddots & \vdots \\ X_1(\mathbf{z}^M) & \cdots & X_M(\mathbf{z}^M) \end{pmatrix} \times \begin{pmatrix} c_{11} & \cdots & c_{1d} \\ \vdots & \ddots & \vdots \\ c_{M1} & \cdots & c_{Md} \end{pmatrix} = A \cdot C \tag{2}$$

The vectors $\mathbf{y}_l \in \mathbb{R}^M$ and $\mathbf{y}^k \in \mathbb{R}^d$ are the columns and rows of the matrix Y, respectively. Micchelli [1986] proved for a large class of radial basis functions (including Hardy's multiquadrics used here) that the *design matrix* A is invertible when all points $\mathbf{z}^k \in \mathbb{R}^d$ are distinct. (For polynomial basis functions A may become singular, e.g., when data from the Hénon map are used with $d > 2$ [Parlitz, 1992]). Although being analytically nonsingular the matrix A is often very close to singular and stable numerical algorithms based on singular-value or QR decompositions are recommended to solve Eq. (2) for the coefficient matrix C [Press *et al.*, 1986]. Furthermore it turned out to be advantageous to substract from all images y_l^i the image value y_l^1 of the (current) reference point before performing the fit. (The Jacobian matrix to be computed remains invariant with respect to this translation.) The transpose of the desired Jacobian of f at the point $\mathbf{x}^i$ is given by

$$Df^{tr}(\mathbf{x}^i) = \begin{pmatrix} \frac{\partial X_1}{\partial x_1}(\mathbf{x}^i) & \cdots & \frac{\partial X_M}{\partial x_1}(\mathbf{x}^i) \\ \vdots & \ddots & \vdots \\ \frac{\partial X_1}{\partial x_d}(\mathbf{x}^i) & \cdots & \frac{\partial X_M}{\partial x_d}(\mathbf{x}^i) \end{pmatrix} \cdot C = D \cdot C \tag{3}$$

The Lyapunov exponents of the examples given in this paper are computed by the treppen-iteration algorithm [Eckmann & Ruelle, 1985; Geist *et al.*, 1990] briefly sketched in Appendix A. The number N of Jacobians computed along the orbit in order to estimate the Lyapunov exponents will be called *number of iterations*.

In the case of data that are polluted by noise or are of limited resolution because of truncation effects (e.g., from an A/D-converter) it is not favorable to force the graph of the approximating function f to pass through the images of the given state points. Instead of the interpolation condition, small deviations of the graph of f from the data are allowed. To determine a smooth approximating function, usually least squares fits with more data points than basis functions are computed.

parameters of the approximation algorithm (i.e., d, n_d, n_e, M, r) in a self-consistent way is based on a comparison of the Lyapunov exponents computed from the given time series and those derived from the time reversed series. Geometrically, time reversal of the data series results in a reflection for even dimensions, and in a rotation of the corresponding attractor in embedding space for odd dimensions. Thus the algorithm for computing the Lyapunov exponents in both directions of time experiences essentially the same distribution of state points with the same outliers and thickness due to noise and the same (local) curvature. This means that any approximation scheme for the flow works equally well (or badly) in both directions. If it works well it turns out that the true exponents change their signs upon time reversal (as expected) but the spurious do not. The reason for this behavior is that the true exponents depend on the (reversed) dynamics on the attractor whereas the spurious exponents are a result of the attempt to fit an approximating function to directions in embedding space where (almost) no state points (informations) are available. Since the attractor is only reflected, its (local) geometry remains essentially the same and the spurious exponents are thus not affected by the time reversal.

Besides these geometrical properties of the attractor, the noisy part of the data is often invariant under time reversal too. This means that (at least for additive noise due to the measurement) the noise level is the same in both directions and can be taken into account by adjusting the parameters of the algorithm (e.g., M, r and μ) in a way that both spectra become consistent.

When the reversed time series is embedded according to the modified rule $\mathbf{x}^i_{\text{rev}} = (s^{i+(d-i)n_d}_{\text{rev}}, \ldots, s^i_{\text{rev}})$ every state point of the original time series equals a state point of the reversed data. When computing approximating functions at these reference points the same M neighbors are selected and the same matrix A is computed. The only differences are the future and past images of the neighboring points given by the matrices Y_{orig} and Y_{rev}, respectively. The coefficient matrices C_{orig} and C_{rev} corresponding to these fits can be computed simultaneously by the algorithm described in Appendix B and lead to the Jacobians Df_{orig} and Df_{rev} at the given state point. When following the original trajectory from the past to the future, the Jacobians of the reversed orbit are computed in the "wrong" (reversed) temporal order and a modified version of the treppen-iteration algorithm has to be used to compute the Lyapunov exponents of the reversed time series (see Appendix A). Therefore the most time consuming parts of the fit procedure (selection of neighbors and solution of the regularization problem) have to be carried out only once (per fit). These considerations show how closely related the approximations for both directions of time are — a fact that may help in understanding more deeply why the spurious exponents remain almost unchanged upon time reversal.

As already mentioned, the method of time reversal for identifying true and spurious Lyapunov exponents is completely independent from the algorithm used for the computation of the exponents. It will be demonstrated below in connection with the approximations by radial basis functions and the regularization technique described above. All Lyapunov exponents of the numerical examples are given in bits per time unit and are computed for $n_d = 1 = n_e$.

4. Numerical Examples

The first example is a time series of the *Hénon map* (1024 samples, a = 1.4 and b = 0.3):

$$x_{n+1} = 1 - ax_n^2 + y_n$$
$$y_{n+1} = bx_n \,. \tag{7}$$

Based on a five-dimensional embedding and approximations with stiffness parameter $r = 1$ and $M = 20$ state points, the original and the reversed time series yield the Lyapunov spectra $S_{\text{orig}} = \{0.61, -2.35, -4.88, -11.96, -17.07\}$ and $S_{\text{rev}} = \{2.38, -0.60, -5.70, -11.11, -15.67\}$, respectively. As can be easily seen, the results corresponding to the exact Lyapunov exponents $\lambda_1 = 0.60$ and $\lambda_2 = -2.34$ change their signs whereas the other (spurious) exponents are in both cases negative and of comparable magnitude. Thus the spectrum S_{orig} of the orginal time series and the negative $-S_{\text{rev}}$ of the spectrum of the reversed data have just the true exponents as common first elements. This overlap of the spectra is illustrated in Fig. 2a. For an arbitrary time series with three true Lyapunov exponents the first three elements of the two spectra should coincide up to their signs as shown in Fig. 2b. Figure 2c shows a situation with inconsistent results. In this case, the approximations made are probably rather poor and the Lyapunov exponents obtained are thus suspected to be inexact. Often, it is possible to achieve consistent spectra by varying the free parameters of the algorithm until the spectra overlap properly. For the general case of m true and $d - m$ spurious Lyapunov exponents the spectra $S_{\text{orig}} = (\lambda_1^{\text{orig}}, \ldots, \lambda_d^{\text{orig}})$

An alternative to this approach is the so called *Tikhonov–Phillips regularization* method [Louis, 1989] where a function G is minimized that consists of an Euclidean norm measuring the deviation of the data from the approximating function and a term controlling the smoothness of the approximation

$$G(\mathbf{c}_l) = \|\mathbf{y}_l - A\mathbf{c}_l\|^2 + \mu^2\|B\mathbf{c}_l\|^2 \ . \tag{4}$$

Each column $\mathbf{c}_l$ of the coefficient matrix C is (then) computed by minimizing $G(\mathbf{c}_l)$ where the stabilizer $\|B\mathbf{c}_l\|$ is given by some suitable matrix B. The influence of the stabilizer on the solution $\mathbf{c}_l$ depends on the value of the *regularization parameter* μ. To visualize this influence, Fig. 1 shows a one-dimensional example where noise has been added to a parabola (dotted line) and the resulting points (black squares) are used for an approximation with radial basis functions. The stabilizer B used was the identity matrix I. For small values of the regularization parameter μ the resulting graph (solid line) passes through the data points, for medium μ it approximates the parabola and (too) large values of μ yield a very flat curve. For the examples discussed below μ has been replaced by 10^α to be able to tune μ over several magnitudes conveniently.

The coefficients $\mathbf{c}_l$ minimizing $G(\mathbf{c}_l)$ can be computed by solving the corresponding normal equations

$$(A^{tr}A + \mu^2 B^{tr}B)\mathbf{c}_l = A^{tr}\mathbf{y}_l \ . \tag{5}$$

Since the direct solution of Eq. (5) may be numerically unstable an algorithm based on Householder transformations is used to compute the solutions $\mathbf{c}_l$ (see Appendix B). For some particular choices of the stabilizer B the solution $\mathbf{c}_l$ can be expressed in terms of the singular value decomposition $A = U \cdot W \cdot V^{tr}$ of the design matrix A [Louis, 1989]. As an example we shall consider here the stabilizer $B = I$ that has also been used in the following for computing the Lyapunov exponents from noisy data. For $B = I$ the solution $\mathbf{c}_l$ of (5) can be written as

$$\mathbf{c}_l = \sum_{i=1}^{M} \frac{w_i}{w_i^2 + \mu^2} < \mathbf{y}_l,\ \mathbf{u}_i > \mathbf{v}_i \tag{6}$$

where $\mathbf{u}_i$ and $\mathbf{v}_i$ are the column vectors of the orthogonal matrices U and V, respectively. For vanishing μ, perturbations of $\mathbf{y}_l$ (i.e., noise) are amplified at most by those terms of the sum with small singular values w_i. The fraction in Eq. (6) can thus be viewed as a filter that reduces the influence of the most noise susceptible terms contributing to the solution $\mathbf{c}_l$.

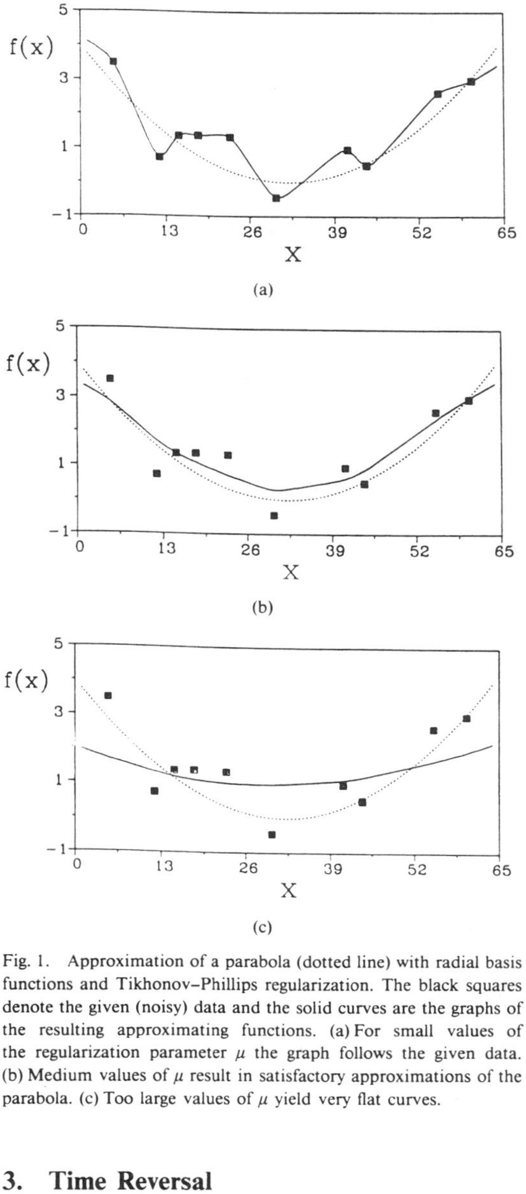

Fig. 1. Approximation of a parabola (dotted line) with radial basis functions and Tikhonov–Phillips regularization. The black squares denote the given (noisy) data and the solid curves are the graphs of the resulting approximating functions. (a) For small values of the regularization parameter μ the graph follows the given data. (b) Medium values of μ result in satisfactory approximations of the parabola. (c) Too large values of μ yield very flat curves.

3. Time Reversal

The right choice of the regularization parameter μ is, in the case of time series analysis, a difficult task because, in general, for a given set of state points, (almost) no information about the actual deviations from the true dynamics due to noise and the local curvature of the underlying flow is available. The method proposed in this paper in order to estimate μ and the other free

(a)	(b)	(c)
$-\lambda_d^{rev}$	$-\lambda_d^{rev}$	$-\lambda_d^{rev}$
$\vdots$	$\vdots$	$\vdots$
$-\lambda_3^{rev}$	$-\lambda_4^{rev}$	$-\lambda_3^{rev}$
$\lambda_1^{orig} = -\lambda_2^{rev}$	$\lambda_1^{orig} = -\lambda_3^{rev}$	$\lambda_1^{orig} = -\lambda_2^{rev}$
$\lambda_2^{orig} = -\lambda_1^{rev}$	$\lambda_2^{orig} = -\lambda_2^{rev}$	λ_2^{orig}
λ_3^{orig}	$\lambda_3^{orig} = -\lambda_1^{rev}$	$\lambda_3^{orig} = -\lambda_1^{rev}$
$\vdots$	λ_4^{orig}	λ_4^{orig}
λ_d^{orig}	$\vdots$	$\vdots$
	λ_d^{orig}	λ_d^{orig}

Fig. 2. Overlap of the Lyapunov spectra $S_{\text{orig}} = \{\lambda_1^{\text{orig}}, \ldots, \lambda_d^{\text{orig}}\}$ and $S_{\text{rev}} = \{\lambda_1^{\text{rev}}, \ldots, \lambda_d^{\text{rev}}\}$ of the original time series and the reversed data. (a) Consistent overlap for two true exponents. (b) Consistent overlap for three true exponents. (c) Inconsistent overlap.

and $S_{\text{rev}} = (\lambda_1^{\text{rev}}, \ldots, \lambda_d^{\text{rev}})$ have to fulfill condition (8) to be considered as a consistent result.

$$\begin{aligned} \forall i = 1, \ldots, m: \quad & \lambda_i^{\text{orig}} = -\lambda_{m+1-i}^{\text{rev}} \\ \forall i = m+1, \ldots, d: \quad & \lambda_i^{\text{orig}} \sim \lambda_i^{\text{rev}} (< 0) \, . \end{aligned} \tag{8}$$

For this condition we have assumed that both spectra are ordered in a way that the true exponents occur first in the spectrum and the spurious results have (large) negative values. It turned out that, for radial basis functions, such an ordering can always be achieved by varying the free parameters of the fit procedure as will be shown in the following examples. When using polynomial approximations the spurious exponents very often occur between the true exponents [Briggs, 1990; Brown *et al.*, 1991]. This "disorder" is caused by the fact that the corresponding least squares problem turns out to be ill-posed. Using suitable regularization methods the ordering needed for condition (8) can be achieved for polynomial approximations too [Parlitz, 1991].

The consistency condition (8) is used below to estimate suitable parameter values of the algorithm for computing Lyapunov exponents. As an example, Fig. 3 shows both spectra S_{orig} and $-S_{\text{rev}}$ versus the stiffness parameter r used upon the approximation. For values of r between 1 and 3, the spectra are stable with respect to changes in r and overlap in the way described above. The number of state points used was $M = 15$. This and the following figure are based on N = 1000 iterations.

Figure 4 shows the spectra S_{orig} and $-S_{\text{rev}}$ of the same time series being dependent on the embedding dimension d for $M = 15$ and $r = 1$. As expected the embedding dimension of $d = 1$, which is too small,

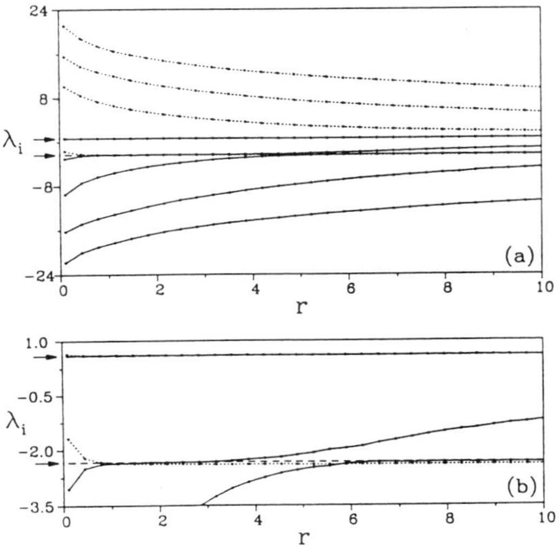

Fig. 3. Lyapunov exponents from 1k data of the Hénon map (7) for embedding dimension $d = 5$ and fits based on $M = 15$ state points and $N = 1000$ iterations. Plotted are the spectrum S_{orig} of the original time series (solid curves) and the negative $-S_{\text{rev}}$ of the spectrum of the reversed data (dotted curves) versus the stiffness parameter r used upon the approximation of the flow. The arrows and the dashed lines give the exact Lyapunov exponents. (a) Complete spectra (b) Enlargement of (a).

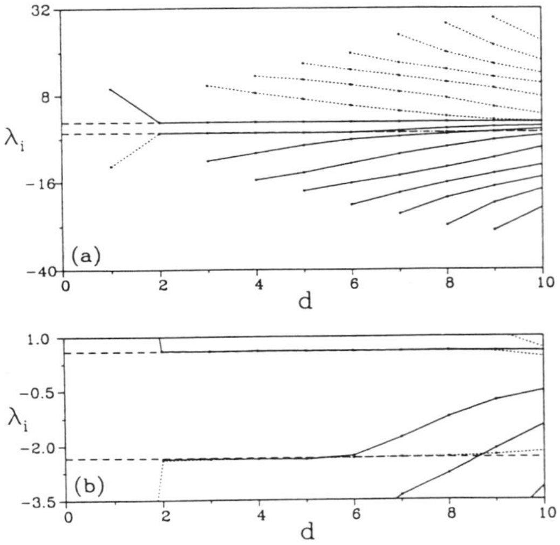

Fig. 4. Lyapunov exponents from $1k$ data of the Hénon map (7). Plotted are the spectrum S_{orig} of the original time series (solid curves) and the negative $-S_{\text{rev}}$ of the spectrum of the reversed data (dotted curves) versus the embedding dimension d for $r = 1$, $M = 20$, and $N = 1000$. The dashed lines give the exact Lyapunov exponents. (a) Complete spectra (b) Enlargement of (a).

yields wrong and inconsistent results. For $d = 2 - 6$, the spectra approximately fulfill condition (8), i.e., the true exponents of the time series are reliably identified. For $d > 6$, more neighbors have to be used for the fit in order to get a proper overlap of the spectra. Of course, longer time series yield even better results but we have restricted ourselves here to $1k$ data in order to demonstrate that the described method already works for small data sets.

As an example for a high dimensional chaotic attractor the data of a four-dimensional analog of the Hénon map [Baier & Klein, 1990]

$$\begin{aligned} (x_1)_{n+1} &= a - (x_{D-1})_n^2 - b(x_D)_n \\ (x_i)_{n+1} &= (x_{i-1})_n \qquad i = 2, \ldots, D = 4 \end{aligned} \tag{9}$$

have been investigated. The chaotic attractor that occurs for $a = 1.76$ and $b = 0.1$ possesses three positive Lyapunov exponents $\lambda_1 = 0.226$, $\lambda_2 = 0.207$, $\lambda_3 = 0.171$ and one negative $\lambda_4 = -3.92$. For this example, the number M of state points for the local fits is chosen as control parameter. Figure 5 shows the spectra S_{orig} and $-S_{\text{rev}}$ versus M for $16k$ data embedded in a five-dimensional state space. The stiffness parameter

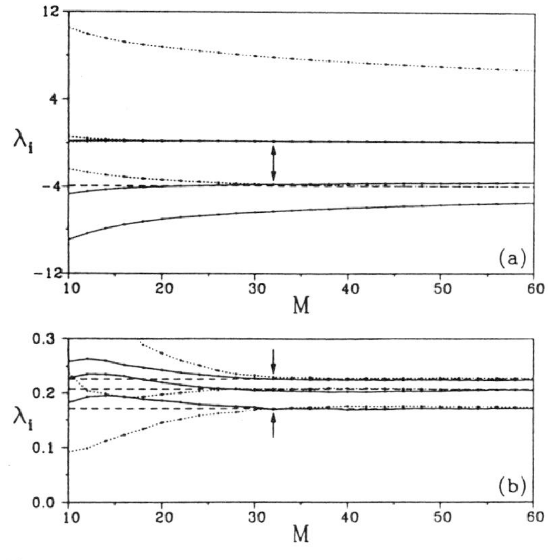

Fig. 5. Lyapunov exponents from $16k$ data of the four-dimensional analog of the Hénon map (9) for $d = 5$, $r = 2$, and $N = 10000$. Plotted are the spectrum S_{orig} of the original time series (solid curves) and the negative $-S_{\text{rev}}$ of the spectrum of the reversed data (dotted curves) versus the number of state points M used in the approximation of the flow. The dashed lines give the exact Lyapunov exponents. The arrows denote the overlap of the spectra fulfilling condition (8). (a) Complete spectra (b) Enlargement of (a).

for the local fits equalled $r = 2$ and the number of iterations was $N = 10000$. For $M = 31$ neighbors, the spectra overlapped in the way described above and the four Lyapunov exponents were detected with relative high precision.

The next example is $8k$ data from a quasiperiodic signal generated by Eq. (10) with $\omega_1 = 1$ and $\omega_2 = (\sqrt{5} - 1)/2$. The sampling time was $t_s = 2$.

$$s(t) = \sin(\omega_1 t) + 2 \sin(\omega_2 t) \tag{10}$$

Figure 6 shows the spectra S_{orig} and $-S_{\text{rev}}$ versus the embedding dimension d for $M = 20$, $r = 5$, and $N = 2000$. For $d = 1$, criterion (8) is not fulfilled. The two-dimensional embedding yields spectra that overlap near zero, i.e., one of the two vanishing Lyapunov exponents may be viewed as detected. For $d = 3$ the second exponent already comes rather close to zero and for four-dimensional embedding two vanishing Lyapunov exponents occur. When d is increased further the exponents are almost constant, i.e., the results are robust with respect to changes of the control parameter d. This example also shows that, for the difficult task of identifying quasiperiodic dynamics by Lyapunov exponents, the method proposed is useful in improving the reliability of the results.

As an example for a chaotic continuous dynamical system, $8k$ data from the Lorenz model (11) with

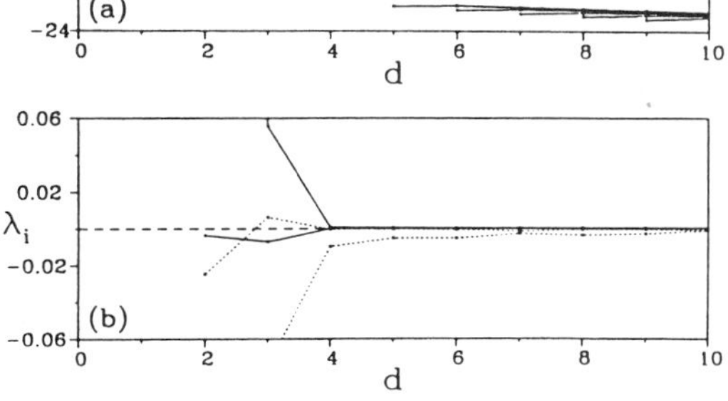

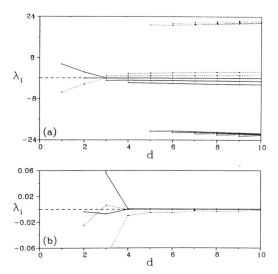

Fig. 6. Lyapunov exponents from $8k$ data of a quasiperiodic time series generated with Eq. (10) and sampled with $t_s = 2$. Plotted are the spectra S_{orig} (solid curves) and $-S_{\text{rev}}$ (dotted curves) versus the embedding dimension d for $M = 20$, $r = 5$ and $N = 1000$. The dashed line denotes zero. (a) Complete spectra (b) Enlargement of (a).

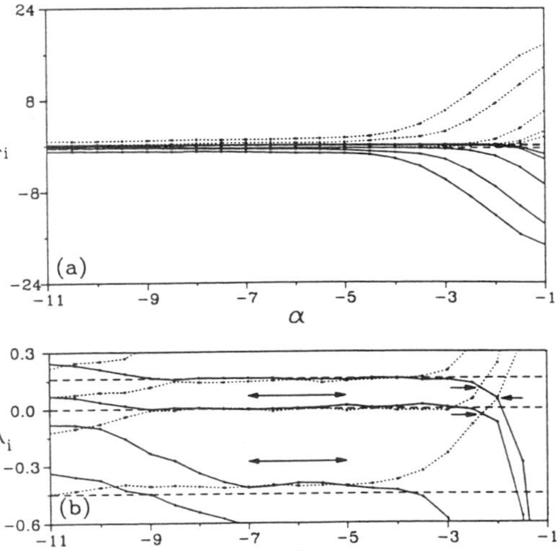

Fig. 9. Lyapunov exponents from $8k$ data of the Duffing Equation (12). The data were artificially polluted by Gaussian noise with variance σ equal to 0.1% of the amplitude of the signal and then truncated to 10 bits. The parameter values used were $t_s = 1$, $d = 5$, $M = 30$, $r = 10$ and $N = 2000$. Plotted are the spectrum S_{orig} of the original time series (solid curves) and the negative $-S_{\text{rev}}$ of the spectrum of the reversed data (dotted curves) versus the number of state points M used in the approximation of the flow. The dashed lines give the exact Lyapunov exponents and the arrows denote three different overlaps of the spectra fulfilling condition (8). (a) Complete spectra (b) Enlargement of (a).

Lyapunov exponents. The parameter values chosen for this example were $d = 5$, $M = 20$, $r = 5$, and $N = 2000$.

Besides truncation additive noise may pollute experimental data. To simulate this problem, Gaussian noise with a standard deviation σ of 0.1% of the amplitude of the signal has been added to the (exact) Duffing data. The resulting noisy time series has been truncated to10 bits. Due to the noise (and the truncation) both spectra S_{orig} and S_{rev} are shifted to positive values, i.e., S_{orig} and $-S_{\text{rev}}$ move into opposite directions. To compensate this drift (i.e., the noise) the functions approximating the flow are smoothed by the regularization method. Figure 9 shows the dependence of the two spectra S_{orig} and $-S_{\text{rev}}$ on the exponent α of the regularization parameter $\mu = 10^{\alpha}$ for $d = 5$, $M = 30$, $r = 10$, and $N = 2000$. As indicated by the arrows, condition (8) is fulfilled for three values of α where three, two and one true Lyapunov exponents are detected, respectively. But only for α between -7 and -5 this overlap is stable with respect to changes of α and approximates the exact results.

As a rule of thumb we recommend large values of M and r for the evaluation of noisy data. Furthermore condition (8) should be fulfilled for some interval of the control parameter(s) with plateaus of the corresponding Lyapunov exponents.

The number of significant bits necessary to leave the results unchanged has also been investigated for the other models and was found to lie between 10 and 12 bits.

5. Conclusion

The examples show that the proposed check for self-consistency based on a time reversal of the data is a useful tool for estimating free parameters of the algorithm used for computing the Lyapunov spectrum and for identifying true and spurious exponents. Whenever consistent spectra occurred for the cases investigated, their true Lyapunov exponents approximated the exact values. Therefore, it seems reasonable to tune the free parameters of the algorithm also for experimental data until an overlap fulfilling condition (8) occurs. If this overlap is stable with respect to small changes of the free parameters, there is some evidence that the computed Lyapunov exponents are true and probably lie near the exact values. These consistency considerations can in principle be applied to all methods for computing Lyapunov exponents proposed during the last years. Many of these algorithms are able to estimate the exact Lyapunov exponents of the time series with sufficient accuracy, but the true exponents are very often scattered between spurious results. Since true and spurious Lyapunov exponents both converge in general with the same rates, it is very difficult for an experimentalist to decide which results are really true and exact and which are only spurious. Time reversal provides a criterion in this case to select the only relevant true exponents from the computed spectra. The knowledge of the number of spurious exponents can then be used for a second evaluation of the data where global and local dimensions are distinguished. In this way, not only are spurious elements avoided but the accuracy of the computed true exponents will also be improved. In principle the determination of suitable values of the free parameters resulting in stable overlap regions can be done automatically. In this way, a black box algorithm for computing Lyapunov exponents from time series is conceivable, that can be used even by unexperienced users. The methods discussed above also apply to other problems where reliable approximations of the flow in embedding space are needed (e.g., prediction, noise reduction, controlling).

$\sigma = 16$, $b = 4.0$ and $R = 45.92$ have been sampled with $t_s = 0.1$, and then embedded in a five-dimensional state space

$$\begin{aligned} \dot{x} &= -\sigma x + \sigma y \\ \dot{y} &= -xz + Rx - y \\ \dot{z} &= xy - bz \ . \end{aligned} \tag{11}$$

The Lyapunov exponents of the corresponding strange attractor are $\lambda_1 = 2.164$, $\lambda_2 = 0$ and $\lambda_3 = -32.46$ bits per unit time. Figure 7 shows the original and the (negative) reversed spectrum being dependent on the stiffness parameter r. The number of iterations and the number of state points used in the approximation were $N = 2000$ and $M = 30$, respectively. The only value of r where the overlap condition (8) is fulfilled is $r = 3.2$, as denoted by the arrows. The positive and the vanishing Lyapunov exponent agree very well with the exact values and are stable for values of r between 3 and 9. The result for the large negative exponent at $r = 3.2$ does not coincide with the exact value given by the dashed line. Only for values of r larger than 9 do the smallest elements of the spectra S_{orig} and $-S_{\text{rev}}$ converge to $\lambda_3 = -32.46$. But for this range of r-values, condition (8) is no longer fulfilled. Up to this violation of Eq. (8), the time reversed data obviously yield better results for the negative Lyapunov exponents of the time series than the original data. Similar results have also been observed for the other examples investigated (see for example, Figs. 3, 4, 5, 8, and 9) and may serve as a hint for future improvements of the algorithm. Nevertheless this example also shows that, for noiseless data, negative Lyapunov exponents of very large magnitude may in principle be reliably estimated.

Experimental data are usually the output of an A/D-converter. To simulate this situation 8k data from a chaotic attractor of the *Duffing equation* (12) with $d = 0.2$, $a = 40$ and $\omega = 1$ have been sampled with $t_s = 1$ and truncated to a given number of bits

$$\ddot{x} + d\dot{x} + x + x^3 = a \cos(\omega t) \ . \tag{12}$$

The exact values of the Lyapunov exponents are $\lambda_1 = 0.16$, $\lambda_2 = 0$ and $\lambda_3 = -0.45$ bits per unit time. Figure 8 shows the spectra S_{orig} and $-S_{\text{rev}}$ versus the number of significant bits resulting from the truncation of the exact data. As can be seen about 10 bits is the minimum resolution needed to reproduce the exact

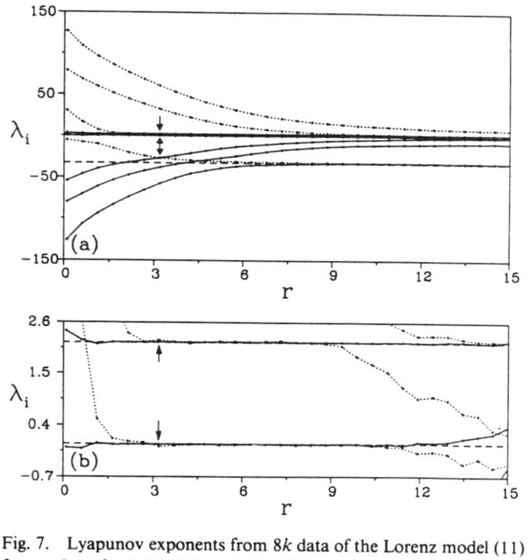

Fig. 7. Lyapunov exponents from $8k$ data of the Lorenz model (11) for $t_s = 0.1$, $d = 5$, $M = 30$, and $N = 2000$. Plotted are the spectrum S_{orig} of the original time series (solid curves) and the negative $-S_{\text{rev}}$ of the spectrum of the reversed data (dotted curves) versus the stiffness parameter r used in the approximation of the flow. The dashed lines give the exact Lyapunov exponents. The arrows denote an overlap of the spectra fulfilling condition (8). (a) Complete spectra (b) Enlargement of (a).

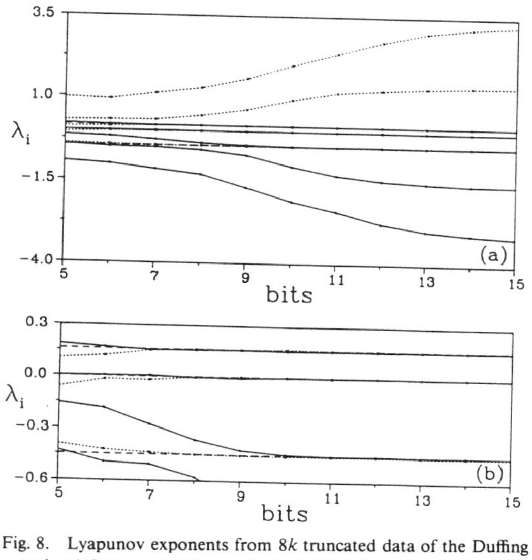

Fig. 8. Lyapunov exponents from $8k$ truncated data of the Duffing equation (12) computed for $t_s = 1$, $d = 5$, $M = 20$, $r = 5$, and $N = 2000$. Plotted are the spectrum S_{orig} of the original time series (solid curves) and $-S_{\text{rev}}$, the negative of the spectrum of the reversed data (dotted curves), versus the number of significant bits used for computing the Lyapunov exponents. The dashed lines give the exact Lyapunov exponents. (a) Complete spectra (b) Enlargement of (a).

Acknowledgement

The author would like to thank Prof. W. Lauterborn and the members of his nonlinear dynamics group, especially J. Holzfuss, T. Kurz and M. Wiesenfeldt, for valuable discussions. Stimulating conversations with U. Dreßler, S. Ergezinger, A. Cenys and H. D. I. Abarbanel are acknowledged, too.

This work was supported by the Deutsche Forschungsgemeinschaft (SFB 185).

References

Baier, G. & Klein, M. [1990] "Discrete steps up the dynamic hierarchy," *Phys. Lett.* **A151**(6), 281–284.

Briggs, K. [1990] "An improved method for estimating Liapunov exponents of chaotic time series," *Phys. Lett.* **A151**(1, 2), 27–32.

Broomhead, D. S. & Lowe, D. [1988], "Multivariable functional interpolation and adaptive networks," *Complex Systems* **2**, 321–355.

Brown, R., Bryant, P. & Abarbanel, H. D. I. [1991] "Computing the Lyapunov spectrum of a dynamical system from an observed time series," *Phys. Rev.* **A43**(6), 2787–2806.

Bryant, P., Brown, R. & Abarbanel, H. D. I. [1990] "Lyapunov exponents from observed time series," *Phys. Rev. Lett.* **65**(13), 1523–1526.

Buzug, Th., Reimers, T. & Pfister, G. [1990] "Optimal reconstruction of strange-attractors from purely geometrical arguments," *Europhys. Lett.* **13**(7), 605–610.

Casdagli, M. [1989] "Nonlinear prediction of chaotic time series," *Physica* **D35**, 335–356.

Eckmann, J.-P. & Ruelle, D. [1985] "Ergodic theory of chaos and strange attractors," *Rev. Mod. Phys.* **57**(3), 617–656.

Eckmann, J.-P., Kamphorst, S. O., Ruelle, D. & Ciliberto, S. [1986] "Lyapunov exponents from time series," *Phys. Rev.* **A34**(6), 4971–4979.

Franke R. [1982] "Scattered data interpolation: test of some methods," *Mathematics and Computing* **38**, 181–200.

Geist, K., Parlitz, U. & Lauterborn, W. [1990] "Comparison of different methods for computing Lyapunov exponents," *Prog. Theor. Phys.* **83**(5), 875–893.

Holzfuss, J. & Lauterborn, W. [1989], "Liapunov exponents from a time series of acoustic chaos," *Phys. Rev.* **A39**(4), 2146–2152.

Holzfuss, J. & Parlitz, U. [1991] "Lyapunov exponents from time series," Proceeding of the Conference *Lyapunov Exponents*, Oberwolfach 1990, eds. Arnold, L., Crauel, H., Eckmann, J.-P. in: *Lecture Notes in Mathematics* (Springer-Verlag).

Liebert, W., Pawelzik, K. & Schuster, H. G. [1991] "Optimal embeddings of chaotic attractors from topological considerations," *Europhys. Lett.* **14**(6), 521–526.

Louis, A. K. [1989] "Inverse und schlecht gestellte Probleme," *Teubner* 1989.

Miccheli, C. A. [1986] "Interpolation of scattered data: Distance matrices and conditionally positive definite functions," *Constr. Approx.* **2**, 11–22.

Packard, N. H., Crutchfield, J. P., Farmer, J. D. & Shaw, R. S. [1980] "Geometry from a time series," *Phys. Rev. Lett.* **45**, 712–716.

Parlitz, U. [1992] "Nonlinear approximations and the computation of Lyapunov exponents from time series," preprint.

Poggio, T. & Girosi, F. [1990] "Regularization algorithms for learning that are equivalent to multilayer networks," *Science* **247**, 978–982.

Powell, M. J. D. [1985] "Radial basis functions for multivariable interpolation: A review," in: *Algorithms for approximation* Based on the proceedings of the IMA conference on Algorithms for Approximation of Functions and Data held at the Royal Military College of Science, Shrivenham, July 1985, Ed. by J. C. Mason and M. G. Cox. Clarendon Press, Oxford, 1987.

Press, W. H., Flannery, B. P., Teukolsky, S. A. & Vetterlin, W. T. [1986] "Numerical recipes," Cambridge University Press.

Sano, M. & Sawada, Y. [1985] "Measurement of the Lyapunov spectrum from a chaotic time series," *Phys. Rev. Lett.* **55**, 1082–1085.

Sato, S., Sano, M. & Sawada, Y. [1987] "Practical methods of measuring the generalized dimension and largest Lyapunov exponent in high dimensional chaotic systems," *Prog. Theor. Phys.* **77**(1), 1–5.

Stoer, J. [1989] "Numerische Mathematik 1," 5. Auflage (Springer-Verlag).

Stoop, R. & Meier, P. F. [1988] "Evaluation of Lyapunov exponents and scaling functions from time series," *J. Opt. Soc. Am.* **B5**(5), 1037–1045.

Stoop, R. & Parisi, J. [1991] "Calculation of Lyapunov exponents avoiding spurious elements," *Physica* **D50**, 89–94.

Takens, F. [1981] "Detecting strange attractors in turbulence," in *Dynamical Systems and Turbulence*, eds. Rand, D. A. & Young, L.-S. (Springer-Verlag, Berlin) 366–381.

Wolf, A., Swift, J. B., Swinney, L., Vastano, J. A. [1985] "Determining Lyapunov exponents from a time series," *Physica* **D16**, 285–317.

Appendix A

The most effective way to compute the Lyapunov spectrum from the sequence of Jacobians $Df(\mathbf{x}^i)$ along an orbit is based on the factorization (A.1)

$$Df^N(\mathbf{x}^1) = \prod_{i=1}^{N} Df(\mathbf{x}^i) = Q^N \cdot R^N \cdot \ldots \cdot R^1 \qquad \text{(A.1)}$$

where Q^N is an orthogonal matrix and the R^i are upper triangular matrices with positive diagonal elements. The Lyapunov exponents are then given as

$$\lambda_k = \lim_{N\to\infty} \frac{1}{Nt_e} \ln \prod_{i=1}^{N} R^i_{kk} = \lim_{N\to\infty} \frac{1}{Nt_e} \sum_{i=1}^{N} \ln R^i_{kk} \qquad \text{(A.2)}$$

where t_e is the evolution time and N is the number of iterations (time steps) used. The QR-decomposition

(A.1) of $Df^N(\mathbf{x}^1)$ can be computed recursively by the so-called treppen-iteration algorithm (A.3) [Eckmann & Ruelle, 1985; Geist *et al.*, 1990].

$$Q^i \cdot R^i = Df(\mathbf{x}^i) \cdot Q^{i-1} \quad (i = 1, \ldots, N), \quad Q^0 = I \qquad (A.3)$$

When the Jacobians $Df_{\text{rev}}(\mathbf{x}^j)$ of the reversed time series are computed simultaneously with those of the original time series they occur in the "wrong" temporal order, i.e., the matrix $Df_{\text{rev}}(\mathbf{x}^{j+1})$ is computed *before* $Df_{\text{rev}}(\mathbf{x}^j)$ with j counting the time running backwards. To achieve an iterative decomposition like (A.3) one may consider the factorization (A.4) of the product of Jacobians.

$$DF^N_{\text{rev}}(\mathbf{x}^N) = \prod_{j=1}^{N} Df_{\text{rev}}(\mathbf{x}^j) = L^N \cdot \ldots \cdot L^1 \cdot Q^1 \qquad (A.4)$$

where Q^1 is again orthogonal but the L^j are lower triangular matrices with positive diagonal elements that yield the Lyapunov exponents (of the reversed time series) analogous to Eq. (A.2). The factorization (A.4) can be computed by iterated LQ-decompositions

$$L^j \cdot Q^j = Q^{j+1} \cdot Df_{\text{rev}}(\mathbf{x}^j) \quad (j = N, \ldots, 1), \quad Q^{N+1} = I \qquad (A.5)$$

that are equivalent to QR-decompositions of the transpose of the product $Q^{j+1} \cdot Df_{\text{rev}}(\mathbf{x}^j)$

$$(Q^j)^{\text{tr}} \cdot (L^j)^{\text{tr}} = Df^{\text{tr}}_{\text{rev}}(\mathbf{x}^j) \cdot (Q^{j+1})^{\text{tr}} . \qquad (A.6)$$

Appendix B

For computing a minimizing solution $\mathbf{c}_l$ of the function

$$\begin{aligned} G(\mathbf{c}_l) &= \|\mathbf{y}_l - A\mathbf{c}_l\|^2 + \mu^2 \|B\mathbf{c}_l\|^2 \\ &= \left\| \begin{pmatrix} \mathbf{y}_l \\ 0 \end{pmatrix} - \begin{pmatrix} A \\ \mu B \end{pmatrix} \mathbf{c}_l \right\|^2 \\ &= \|\tilde{\mathbf{y}}_l - \tilde{A}\mathbf{c}_l\|^2 \end{aligned} \qquad (B.1)$$

the vector $\tilde{\mathbf{y}}_l - \tilde{A}\mathbf{c}_l$ is subjected to a sequence of Householder transformations $P_1, \ldots, P_M$ resulting in an orthogonal transformation P that preserves the Euclidean norm $\|\cdot\|$ and leads to a set of equations that can be solved by backsubstitution. With the following abbreviations

$$\begin{aligned} &P = P_M \cdot \ldots \cdot P_1 \\ &P \cdot \tilde{A} = \begin{pmatrix} R \\ 0 \end{pmatrix} \quad \text{with:} \quad R = \begin{pmatrix} R_{11} & \ldots & R_{1M} \\ \vdots & \ddots & \vdots \\ 0 & \ldots & R_{MM} \end{pmatrix} \\ &P\tilde{\mathbf{y}}_l = \begin{pmatrix} \mathbf{h}_l \\ \mathbf{g}_l \end{pmatrix} \quad \text{with:} \quad \mathbf{h}_l \in \mathbb{R}^M \end{aligned} \qquad (B.2)$$

the function G can be written as

$$G(\mathbf{c}_l) = \|P(\tilde{\mathbf{y}}_l - \tilde{A}\mathbf{c}_l)\|^2 = \left\| \begin{pmatrix} \mathbf{h}_l - R\mathbf{c}_l \\ \mathbf{g}_l \end{pmatrix} \right\|^2 \qquad (B.3)$$

G takes its minimum when $\mathbf{h}_l - R\mathbf{c}_l = 0$, i.e., Eq. (B.4) has to be solved for the coefficient vector $\mathbf{c}_l$.

$$R\mathbf{c}_l = \mathbf{h}_l \; . \qquad (B.4)$$

If the upper triangular matrix R contains vanishing diagonal elements it is singular and the solution $\mathbf{c}_l$ of Eq. (B.4) is not unique. Equation (B.5) in those cases selects the solution $\mathbf{c}_l$ with the smallest norm.

$$\forall i = M, \ldots, 1 : c_{il} = \begin{cases} (h_{il} - \Sigma^M_{k=i+1} r_{ik} c_{kl}) / r_{ii} & \text{if } r_{ii} \neq 0 \\ 0 & \text{if } r_{ii} = 0 \end{cases} \qquad (B.5)$$

Practically the columns $\mathbf{c}_l$ of the coefficient matrix C can be computed simultaneously by applying the Householder transformations P_i to all columns $\mathbf{y}_l$ of the matrix Y and solving Eq. (B.4) for the resulting vectors $\mathbf{h}_l$ ($l = 1, \ldots, M$). Thus only a single QR decomposition of the design matrix A is necessary to solve the regularization problem. For further details and stability properties of the algorithm see chapter 4.8.2 in Stoer [1989].

CHAPTER 9

Periodic Orbits and Symbolic Dynamics

D.P. Lathrop, E. J. Kostelich
Characterization of an experimental strange attractor by periodic orbits
Phys. Rev. A **40**, 4028 (1989).

L. Flepp, R. Holzner, E. Brun, M. Finardi, R. Badii
Model identification by periodic-orbit analysis for NMR-laser chaos
Phys. Rev. Lett. **67**, 2244 (1991).

J.C. Sommerer, W.L. Ditto, C. Grebogi, E. Ott, M. Spano
Experimental confirmation of the theory for critical exponents of crises
Phys. Lett. A **153**, 105 (1991).

F. Papoff, A. Fioretti, E. Arimondo, G.B. Mindlin, H. Solari, R. Gilmore
Structure of chaos in the laser with saturable absorber
Phys. Rev. Lett. **68**, 1128 (1992).

Editors' Notes

Lathrop and Kostelich (1989), in an analysis of experimental data from the Belousov-Zhabotinskii chemical reaction, use embedding techniques to obtain periodic orbits, entropy and symbolic dynamics. Similar goals are achieved in the paper by **Flepp et al.** (1991) for a very different experimental system, namely a nuclear magnetic resonance laser.

Sommerer et al. (1991) consider the transition to "crisis-induced intermittency" in the chaotic gravitational bucking and unbuckling of a magnetoelastic ribbon in an oscillating magnetic field. Since the crisis is mediated by a special unstable periodic orbit in the attractor, a prime task of this paper is the determination of this orbit and its Jacobian matrix from measured data.

The paper of **Pappof et al.** (1992) carries out a topological analysis of experimental data from a laser with saturable absorber. Their analysis utilizes the idea of a "knot-holder" to constrain the orbit topology.

Characterization of an experimental strange attractor by periodic orbits

Daniel P. Lathrop
Center for Nonlinear Dynamics and Department of Physics, University of Texas, Austin, Texas 78712

Eric J. Kostelich
Institute for Physical Science and Technology and Department of Mathematics, University of Maryland, College Park, Maryland 20742
and Center for Nonlinear Dynamics and Department of Physics, University of Texas, Austin, Texas 78712
(Received 18 January 1989)

We describe a general procedure to locate periodic saddle orbits in a chaotic attractor reconstructed from experimental data. The method is applied to data from a Belousov-Zhabotinskii chemical reaction. The eigenvalues associated with the saddle orbits are used to estimate the Lyapunov exponents. An analysis of the next amplitude map determines the allowable periodic orbits and yields an estimate of the topological entropy.

Recent theoretical work suggests that periodic saddle orbits determine much of the dynamics on typical attractors.[1-3] In hyperbolic attractors, for example, the natural measure, fractal dimension, and Lyapunov exponents can be expressed as limits involving the periodic saddle orbits.[1,2] Moreover, the saddle orbits can provide a useful characterization of the structure and dynamics of the attractor as a parameter varies.[3] In this paper, we show how to extract the periodic saddle orbits from an attractor reconstructed from experimental data. These orbits are used to estimate the Lyapunov exponents, information dimension, and topological entropy of the attractor. The estimates compare well with those obtained by conventional methods.

The attractor is obtained from a time series from an oscillating Belousov-Zhabotinskii (BZ) chemical reaction.[4,5] The experiment consists of a continuously stirred tank reactor into which various chemical species are fed at a constant rate. The bromide ion concentration in the reactor is recorded at equally spaced intervals, giving a time series of 65 000 values.

A phase-space attractor is reconstructed from the data using the standard method of time delays.[6,7] The experimental data consist of a scalar time series, denoted $\{s_i\}_{i=1}^{n}$. The attractor is the set of points $\mathbf{x}_i = (s_i, s_{i+\tau}, \ldots, s_{i+(d-1)\tau})$ where d is the embedding dimension and τ is the time delay. In this discussion, we choose $d=3$ and $\tau=124$ on the basis of the mutual information criterion discussed in Ref. 8. Figure 1(a) shows a two-dimensional projection of the reconstructed attractor.

The saddle orbits that we consider are periodic, and they appear to have one repelling direction. In this three-dimensional reconstruction, they also have one attracting direction. Trajectories approach the saddle orbit along this direction and remain nearby for a time before they are pushed away. While a point remains near the saddle orbit, it moves with a frequency which is approximately the same as that of the saddle orbit. The saddle orbits therefore can be located as follows. Let $\epsilon > 0$, and let $\mathbf{x}_i$ be a point on the reconstructed attractor. We follow the observed images $\mathbf{x}_{i+1}, \mathbf{x}_{i+2}, \ldots$ of $\mathbf{x}_i$ until we find the smallest index $k > i$ such that $\|\mathbf{x}_k - \mathbf{x}_i\| < \epsilon$. If such a k exists, we define $m = k - i$ and say that $\mathbf{x}_i$ is an (m, ϵ) recurrent point.

In this analysis, we fix $\epsilon = 0.005$ and compute distances in terms of the maximum norm. (The time series is normalized to the unit interval.) The value of m can be calculated using the above procedure for 51 000 out of 65 000 attractor points. We find that over 95% of the recurrence times m are clustered in small intervals around $m = 125$, 250, 375, 500, 625, 750, 875, and 1000, as illustrated by the histogram in Fig. 2. The scatter in the m

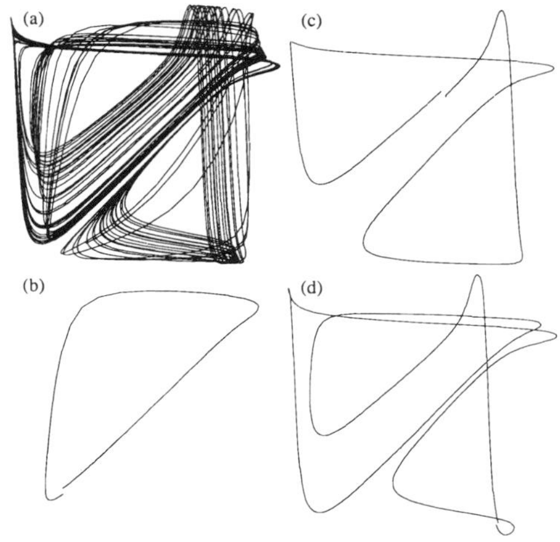

FIG. 1. (a) The BZ attractor. (b)–(d) Trajectories near the period-1, -2, and -3 saddles, respectively.

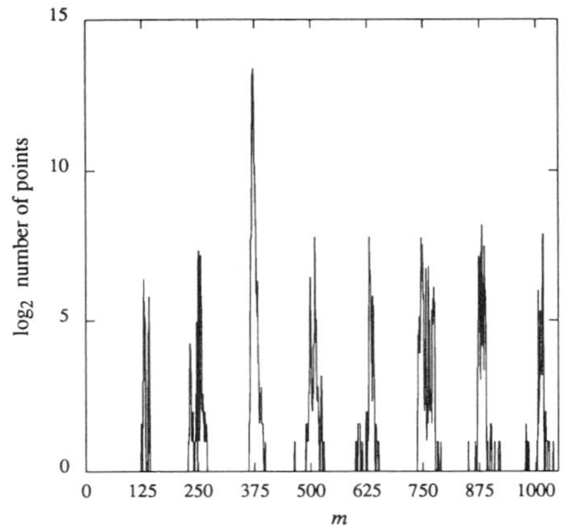

FIG. 2. Histogram showing the number of attractor points associated with each recurrence time m.

TABLE I. Eigenvalues associated with the periodic saddle orbits for the BZ attractor.

Period	Eigenvalue
1	3.66
2	5.34
3	3.18
4	4.17
5	7.10
6	3.40
7	9.56
8	5.56

values occurs for two reasons. First, the recurrence time of a trajectory depends on how closely it approaches a saddle orbit. Second, we record the time when the trajectory first returns to the ϵ neighborhood of the reference point $\mathbf{x}_i$. This time may differ slightly from the time the trajectory most closely approaches $\mathbf{x}_i$.

The trajectory through each point $\mathbf{x}_i$ in Fig. 1(b) comes back within $\epsilon=0.005$ of $\mathbf{x}_i$ approximately 125 time steps later. This orbit has the shortest period, so we call it a period-1 saddle. (In what follows we define a "period" as 125 time steps.) Figures 1(c) and 1(d) show portions of the trajectories which lie near the period-2 and -3 saddles, respectively. In this paper we consider orbits up to period 8. It is important to include periodic saddles of sufficiently high period in order to capture most of the attractor points.

The stability of each saddle orbit is estimated from a linear approximation of the dynamics at points on nearby trajectories. Let $\mathbf{x}_{\text{ref}}$ be an (m,ϵ) point, and let $\{\mathbf{x}_i\}_{i=1}^{n}$ be a collection of points in a 6ϵ neighborhood of $\mathbf{x}_{\text{ref}}$.[9] We assume that the dynamics in this neighborhood is nearly linear; that is, we write the map f which takes $\mathbf{x}_i$ to $\mathbf{x}_{i+m}$ as $f(\mathbf{x})=A\mathbf{x}+\mathbf{b}$ for some 3×3 matrix A and a three-vector $\mathbf{b}$. A least-squares procedure similar to that described in Refs. 7 and 10 is used to calculate A and b. Here A is an approximation of the Jacobian matrix $Df(\mathbf{x}_{\text{ref}})$. The absolute value of the largest eigenvalue of A provides an estimate of the stability (more precisely, the strength of the repulsion) of the saddle orbit near $\mathbf{x}_{\text{ref}}$.

Table I contains a list of the periodic saddle orbits and their associated eigenvalues, which are calculated as follows. Each (m,ϵ) point associated with the orbit of period p is used as a reference point $\mathbf{x}_{\text{ref}}$. The Jacobian matrix $Df^{p}(\mathbf{x}_{\text{ref}})$ and the magnitude L of its largest eigenvalue is computed using least squares whenever 50 or more points can be found in a 6ϵ neighborhood of $\mathbf{x}_{\text{ref}}$.[11] The median value of L over the total number of reference points is listed in the table.

One's ability to estimate the eigenvalues associated with a given saddle orbit depends on the number of trajectories that lie nearby. Saddle orbits in densely populated regions of the attractor (where the natural measure is large) are easier to characterize than those in regions which are rarely visited.

The Lyapunov exponents (also called characteristic exponents) measure the rate of separation of nearby initial conditions on the attractor (see Ref. 7 for a precise discussion and additional references). An attractor is *chaotic* if the largest Lyapunov exponent λ_1 is positive. Roughly speaking, this implies the distance between a typical pair of nearby points on the attractor grows as $2^{\lambda_1 t}$ for small t.

The basic idea behind existing algorithms[12,13] to estimate Lyapunov exponents from experimental data is to follow sets of trajectories for short intervals to measure the observed rates of separation and average them. (Another, more ambitious method is described in Ref. 14.) We now consider a different approach, where we evaluate the Lyapunov exponents in terms of the eigenvalues of the periodic orbits.[2]

Most of the points on the BZ attractor are within $\epsilon=0.005$ of the periodic saddles listed in Table I. Thus an estimate of the Lyapunov exponents can be obtained from a weighted average of the eigenvalues of the saddle orbits. In this case we weight the eigenvalues according to the number of associated (m,ϵ) points. This yields the estimate $\lambda_1=0.56$ bits/period, which agrees well with the estimate $\lambda_1=0.50$ bits/period using the algorithm of Wolf *et al.*[12]

In this experiment it appears that an embedding dimension of 3 is sufficient to reconstruct the attractor (no trajectory crosses another). Under this assumption, there are three Lyapunov exponents for the BZ attractor: a positive exponent (λ_1, estimated above), a zero exponent, and a negative exponent (λ_3).[7] The negative exponent measures the rate at which points near the attractor approach it. Negative Lyapunov exponents are difficult to estimate from experimental data often because one cannot observe how different parts of the attractor (or the associated return map) contract onto each other. If we suppose that the data in this experiment are accurate to 0.1%, then two points whose initial separation contracts by a factor of 1000 after one period will be indistinguishable because of the noise. In other words, we cannot measure the negative Lyapunov exponent if $\lambda_3 \lesssim -10$ bits/period.

This consideration leads to an approximate value of the information dimension D_I.[15] Assuming that the Kaplan-Yorke[16] conjecture holds for this attractor, then

$$D_I = 2 + (\lambda_1 + \lambda_2)/|\lambda_3| \ .$$

Using the estimates for the Lyapunov exponents obtained above, we have $D_I \sim 2 + 0.6/10 = 2.06$. Although this estimate of λ_3 is somewhat speculative, it is consistent with calculations of the information dimension using the method of nearest neighbors,[17] which gives $D_I = 2.12 \pm 0.04$. (The estimates depend on which nearest neighbors are used; this is reflected in the variance.)

Additional information about the periodic saddle orbits can be obtained from an analysis of the next amplitude map $f(x_n) = x_{n+1}$, shown in Fig. 3. Here we have plotted the $(n+1)$st relative minimum in the time series as a function of the nth relative minimum. For convenience, we normalize the time series to the unit interval so that f is defined on [0,1]. This return map has a single critical point: an absolute maximum at x_c. Let $A = [0, x_c)$ and $B = (x_c, 1]$. A careful examination of the experimental data reveals that points in A map only to points in B [i.e., if $x < x_c$ then $f(x) > x_c$], but points in B map to points in $A \cup B$. We represent this rule with the transition matrix[18]

$$M = \begin{array}{c} \\ A \\ B \end{array} \begin{array}{c} \begin{array}{cc} A & B \end{array} \\ \begin{bmatrix} 0 & 1 \\ 1 & 1 \end{bmatrix} \end{array}$$

containing a zero entry for the disallowed transition AA.

This observation yields an estimate of the topological entropy. Let N_p be the number of periodic orbits of period p in the attractor. The topological entropy h_t is given as[19]

$$h_t = \lim_{p \to \infty} \frac{1}{p} \log_2 N_p \ .$$

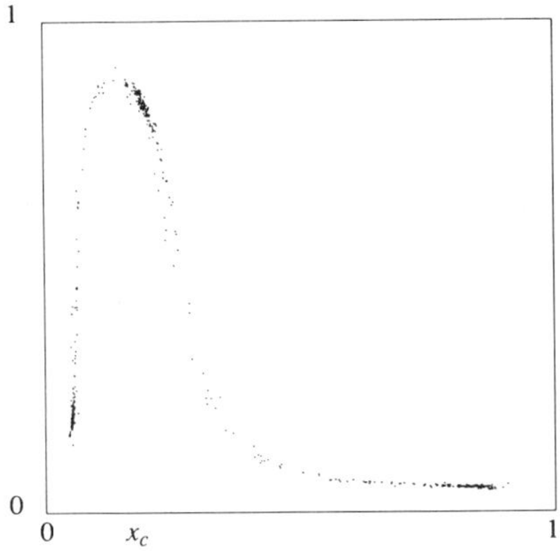

FIG. 3. Return map for the BZ attractor.

For example, the map $f(x) = 4x(1-x)$ has 2^p periodic orbits of period p.[18,20] (The initial conditions can be represented as binary strings and the dynamics as a left shift by one bit. The number of points of period p is the number of different binary strings of length p.) The topological entropy of this map is 1.

In this BZ attractor, not all binary sequences (strings of A's and B's) are possible; in particular, the sequence AA is disallowed. (This exclusion principle is called *pruning*[21].) The easiest way to determine the possible sequences is to use the transition matrix M given above. It can be shown[18] that the number of orbits of period p for a map of the interval is given by $\mathrm{tr} M^p$.[22] Assuming that the dynamics is described by M, then

$$h_t = \lim_{p \to \infty} \frac{1}{p} \log_2 \mathrm{tr} M^p \ .$$

In this case, $h_t \to \log_2 \lambda_{max}$ as $p \to \infty$, where $\lambda_{max} = (1+\sqrt{5})/2$ is the largest eigenvalue of M. Hence $h_t \approx 0.696$ bits/orbit.

This estimate of the topological entropy of the BZ attractor is an upper bound because we have used only the lowest-order transition matrix. A better upper bound can be obtained by examining longer sequences of strings. For example, if we examine the data further to determine which pairs of three-digit strings occur, we find the transition matrix[23]

$$M' = \begin{array}{c} \\ ABA \\ ABB \\ BAB \\ BBA \\ BBB \end{array} \begin{array}{c} \begin{array}{ccccc} ABA & ABB & BAB & BBA & BBB \end{array} \\ \begin{bmatrix} 0 & 0 & 1 & 1 & 0 \\ 1 & 1 & 1 & 0 & 0 \\ 1 & 1 & 1 & 1 & 0 \\ 0 & 0 & 1 & 1 & 1 \\ 1 & 1 & 1 & 0 & 1 \end{bmatrix} \end{array} .$$

Proceedings as before, we find $\lambda_{max} \approx 1.48$, so that $h_t \approx 0.563$ bits/orbit.

The topological entropy h_t is an upper bound for the metric entropy, which in this case should be equal to the positive Lyapunov exponent λ_1 since there is only one expanding direction on the attractor.[7] Our estimate of $\lambda_1 = 0.56$ bits/orbit is in excellent agreement with the value of h_t obtained from the transition matrix. (Another approach to the estimation of topological entropy is discussed in Refs. 24 and 25.)

The determination of M' depends somewhat on the choice of the critical point x_c in Fig. 3, because there are a few orbits which pass very close to x_c. (In addition, there are two orbits which lie away from the main curve of the return map. We attribute them to experimental artifacts.) Although the dynamics described by M' suggests that at least two different period-6 orbits are possible, only one is observed in the experiment. This implies that either additional pruning occurs (which is not apparent from the binary triples), or the other period-6 orbit is visited too infrequently to be detected in these data.

The determination of the periodic saddle orbits from experimental data is possible in principle as long as the underlying dynamics is relatively low dimensional. In addition, a low noise level and a long time record are important so that the recurrent points can be located easily. Finally, the stability of the saddle orbits determines their visibility; saddle orbits whose positive eigenvalues are small can be found more easily than those whose eigenvalues are large.

We thank K. Coffman and N. Kreisberg for access to their experimental data; B. Scott for assistance in computing Lyapunov exponents; and R. F. Williams, P. Cvitanović, and H. L. Swinney for helpful discussions. This research was supported in part by the DARPA Applied and Computational Mathematics Program, and the U.S. Department of Energy Office of Basic Energy Sciences.

---

[1]D. Auerbach, P. Cvitanović, J.-P. Eckmann, G. H. Gunaratne, and I. Procaccia, Phys. Rev. Lett. **58**, 2387 (1987).

[2]C. Grebogi, E. Ott, and J. A. Yorke, Phys. Rev. A **37**, 1711 (1988).

[3]P. Cvitanović, Phys. Rev. Lett. **61**, 2729 (1988).

[4]J. C. Roux, R. H. Simoyi, and H. L. Swinney, Physica D **8**, 257 (1983).

[5]K. Coffman, W. D. McCormick, Z. Noszticzius, R. H. Simoyi, and H. L. Swinney, J. Chem. Phys. **86**, 119 (1987).

[6]F. Takens, in *Dynamical Systems and Turbulence,* Vol. 898 of *Springer Lecture Notes in Mathematics,* edited by D. A. Rand and L.-S. Young (Springer-Verlag, New York, 1981), p. 366.

[7]J.-P. Eckmann and D. Ruelle, Rev. Mod. Phys. **57**, 617 (1985).

[8]A. M. Fraser and H. L. Swinney, Phys. Rev. A **33**, 1134 (1986).

[9]The distance 6ϵ is chosen so that the neighborhood usually contains at least 50 other points for the linear regression.

[10]E. J. Kostelich and J. A. Yorke, Phys. Rev. A **38**, 1649 (1988).

[11]The Jacobian is computed from the collection of nearest neighbors and their images $125p$ time steps later. The eigenvalue of the period-3 orbit is estimated from a random sample of the associated (m,ϵ) points.

[12]A. Wolf, J. B. Swift, H. L. Swinney, and J. A. Vastano, Physica D **16**, 285 (1985).

[13]J.-P. Eckmann, S. O. Kamphorst, D. Ruelle, and S. Ciliberto, Phys. Rev. A **34**, 4971 (1986); M. Sano and Y. Sawada, Phys. Rev. Lett. **55**, 1082 (1985).

[14]J. P. Crutchfield and B. McNamara, Complex Systems **1**, 417 (1987).

[15]J. D. Farmer, E. Ott, and J. A. Yorke, Physica D **7**, 153 (1983).

[16]J. L. Kaplan and J. A. Yorke, in *Functional Differential Equations and Approximations of Fixed Points,* Vol. 730 of *Springer Lecture Notes in Mathematics,* edited by H. O. Peitgen and H. O. Walther (Springer-Verlag, New York, 1979).

[17]R. Badii and A. Politi, J. Stat. Phys. **40**, 725 (1985); E. J. Kostelich and H. L. Swinney, in *Chaos and Related Nonlinear Phenomena,* edited by I. Procaccia and M. Shapiro (Plenum, New York, in press).

[18]R. L. Devaney, *An Introduction to Chaotic Dynamical Systems* (Benjamin, Menlo Park, CA, 1986).

[19]R. Bowen, Trans. Am. Math. Soc. **154**, 377 (1971); A. B. Katok, Publ. Math. IHES **51**, 137 (1980). This result is proved only for the case of Axiom *A* attractors, but it is conjectured to hold more generally.

[20]P. Collet and J.-P. Eckmann, *Iterated Maps on the Interval as Dynamical Systems* (Birkhäuser, Boston, 1981).

[21]P. Cvitanović, G. Gunaratne, and I. Procaccia, Phys. Rev. A **38**, 1503 (1988).

[22]This is the trace of M^p, defined as the sum of the diagonal entries.

[23]A 0 entry means that the corresponding pair is not allowed; a 1 means that it is allowed. Thus the first three entires in the first row mean that the sequences ABAABA and ABAABB are not allowed, but the sequence ABABAB is allowed.

[24]J. P. Crutchfield and N. H. Packard, Physica D **7**, 201 (1983).

[25]P. Collet, J. P. Crutchfield, and J.-P. Eckmann, Commun. Math. Phys. **88**, 257 (1983).

Model Identification by Periodic-Orbit Analysis for NMR-Laser Chaos

L. Flepp, R. Holzner, and E. Brun
Physik-Institut der Universität, CH-8001 Zürich, Switzerland

M. Finardi and R. Badii
Paul-Scherrer Institut, CH-5232 Villigen, Switzerland
(Received 2 May 1991)

It is shown that the conventional Bloch-Kirchhoff description of NMR-laser activity needs the inclusion of a nonlinear relaxation term for the transverse magnetization in order to account for the experimental observation. The validity of the modified equations is demonstrated by comparing in a two-parameter space the Poincaré sections and periodic-orbit structures of model and experiment. All unstable orbits up to order nine have been extracted and employed to obtain an approximation to a generating partition.

PACS numbers: 05.45.+b, 76.60.−k

In this Letter, we present a detailed analysis of high-quality (i.e., low-noise, drift-free, long) time series from an NMR-laser system with modulated parameters and propose an extension of the conventional Bloch-Kirchhoff description which yields very good quantitative agreement with the experimental observation. We investigated the bifurcation structure of the system in a two-parameter space, constructed Poincaré sections, and extracted all unstable periodic orbits up to order nine. This allowed us to demonstrate the validity of the extended Bloch-type equations and evaluate the topological entropy K_0 by means of explicit construction of the symbolic dynamics directly from the embedded time series.

The NMR-laser [1] activity is provided by the nuclear spins of the ^{27}Al in a ruby crystal, placed at a temperature of 4.2 K in a static magnetic field $\mathbf{B}_0$ of magnitude 1.1 T. The total nuclear magnetization $\mathbf{M}=(M_x,M_y,M_z)$ precesses with the NMR frequency $\nu_a=12.3$ MHz. The population inversion is obtained by means of a microwave pump (dynamic nuclear polarization) [1] and the resonance by enclosing the active medium in a cavity: in our case an LC circuit, tuned to ν_a for single-mode selection, which provides the feedback radiation field $\mathbf{B}$ (proportional to the current in the circuit) necessary for coherent spin-flip behavior. Furthermore, the cavity is forced to operate with a modulated quality factor $Q(t)=Q_0(1+A\cos\Omega t)$, where $\Omega\in(350,800)$ s^{-1} and $A\in(0,0.03)$. The "laser output" is proportional to the transverse nuclear-magnetization amplitude $M_t=(M_x^2+M_y^2)^{1/2}$.

The extended Bloch-type laser (EBL) model in the rotating-frame approximation [2] reads

$$\begin{aligned}\dot{x}&=\sigma[y-x/f(\tau)]\,,\\ \dot{y}&=-y(1+ay)+rx-xz\,,\\ \dot{z}&=-bz+xy\,,\end{aligned}\qquad(1)$$

where $x\propto B_t$, $y\propto M_t\geq 0$, $z\propto M_z-M_e$: B_t represents the rotating field amplitude and M_e the pump magnetization. The proportionality factors, as well as the parameters $\sigma=4.875$, $r=1.807\propto M_e$, and $b=2\times10^{-4}$, depend on various physical constants [2]. The function $f(\tau)=1+A\cos(\omega\tau)$ describes the modulation with frequency $\omega\in(0.014,0.034)$ and τ is a rescaled time. Equations (1) reduce to the Lorenz system [3] for $a=A=0$. Under these conditions, the asymptotic attractor is the fixed point $x=y=x^*=[b(r-1)]^{1/2}$, $z=z^*=r-1$.

The difference with the conventional model consists of the term $-ay^2$, where $a\approx0.2621$, which describes a nonlinear damping for the transverse magnetization y (recall that $y\geq0$). This term is essentially the first nonlinear contribution in a series expansion [4] and has been introduced to account for the observed decay of the laser output (relaxation oscillation) to a stable fixed point in the absence of the modulation ($A=0$), after Q switching. The corresponding time evolution of $y(t)$ vs t is displayed in Fig. 1 for both experiment [Fig. 1(a)] and model [Fig. 1(b)]: In the latter case, we also marked the position of the peaks (circles) and indicated (dashed line) the envelope of the damped oscillation obtained from the conventional model ($a=0$). The manifest discrepancy is not surprising, since the phenomenological Bloch equations often provide incorrect predictions in solids [5]. It must be stressed that the correction term is actually very small: In fact, by setting $(X,Y,Z)=(x/x^*,y/x^*,(z-z^*)/x^*)$ and rescaling time τ to $\tau'=\tau x^*$, we obtain the term $-Y(1+a'Y)/x^*$, where $a'=3.33\times10^{-3}$ and $Y\in(0,3)$ for all considered A values. This transformation also shows that X can be adiabatically eliminated, since $\sigma\rightarrow\sigma'\approx384\gg1$, so that $X\approx Yf(\tau')$ and Y never changes sign [4].

The study of the transient behavior after Q switching provides a first confirmation of the validity of our approach, although limited to a short time interval (0.2 s) and to a single-parameter choice. In order to extend our analysis to generic asymptotic motion, we considered the Q-modulated laser which can exhibit chaotic behavior [6]. The correspondence between model and experiment

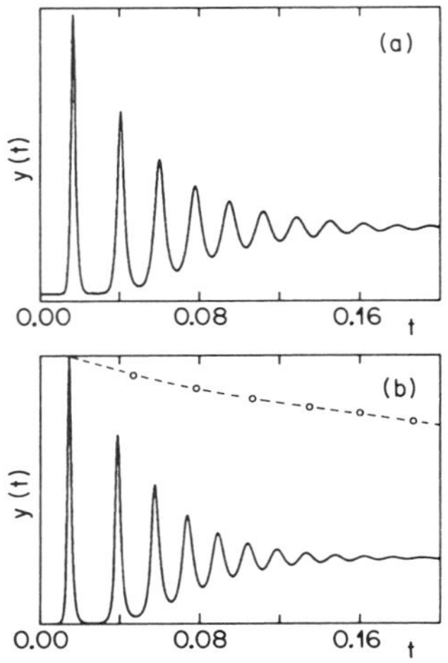

FIG. 1. Time evolution for the transverse magnetization $y(t)$ (in arbitrary units) vs time t (in s), after Q switching: (a) experiment; (b) EBL model. The circles represent the peaks of the corresponding curve (the dashed line is its envelope) obtained from the conventional model: In this case, both decay rate and relaxation frequency are incorrectly reproduced.

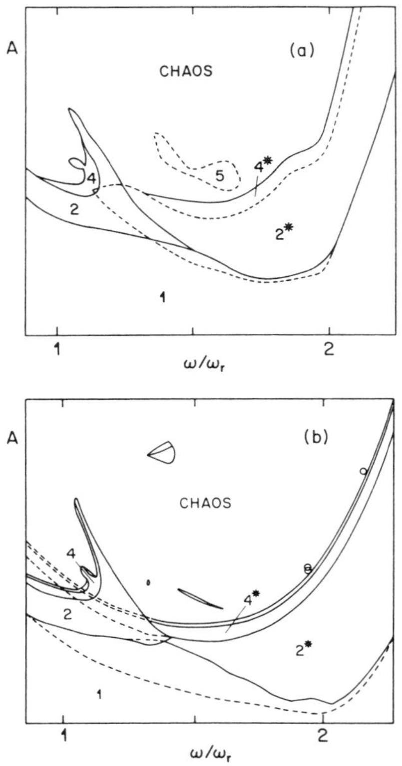

FIG. 2. Regions of existence of the main stable periodic orbits for the NMR-laser system in the parameter space ω-A: (a) experiment; (b) EBL model. The frequency is reported in units of the rescaled relaxation frequency $\omega_r = 0.0163$. Two different periods two and four (marked as $2,2^*$ and $4,4^*$) coexist. Multistability regions are delimited by dashed lines. The circles in (b) indicate the points (1.944,0.018), (1.944,0.0185), and (2.15,0.027) at which the periodic-orbit analysis has been carried out.

is shown by performing a sequence of increasingly more stringent tests: comparison between bifurcation diagrams, stability regions in the ω-A parameter space, Poincaré sections, and sets of unstable periodic orbits.

The phenomenology of the EBL system (1) can be discussed with reference to Fig. 2 [7], where the ω-A parameter space is displayed. Various domains of stable periodic behavior are shown for both experiment [Fig. 2(a)] and model [Fig. 2(b)]. The two pictures are in very good agreement to within the effect of experimental noise, which destroys thin stability regions in 2(a). The finiteness of the measurement resolution for the parameter values (2 or 3 digits) also contributes to the uncertainty of the comparison: The period-5 stable orbit visible in 2(a), for example, has been detected only as unstable in 2(b). The conventional Bloch-Kirchhoff equations, on the other hand, exhibit a behavior which is only qualitatively similar to that shown in Fig. 2 [6].

Several bifurcation diagrams for the variable y have been experimentally recorded as a function of A, at different ω values. The accuracy of model (1) has been confirmed in the whole parameter range shown in Fig. 2 by comparison with the numerical results. Five crisis lines [8] have been detected in the chaotic region where the strange attractors collide with different unstable periodic orbits.

A more severe test for the EBL equations has been performed by studying chaotic attractors for various parameter values, both in the full embedding space and on Poincaré sections [9]. Scalar time series $\{\xi_1, \xi_2, \ldots, \xi_N\}$, consisting of $N = 8 \times 10^5$ twelve-bit integers, have been recorded by sampling the transverse magnetization $M_t(t)$ with a frequency $\nu = 25/T$, where $T = 2\pi/\Omega$ is the period of the forcing term: i.e., $\xi_i = M_t(i/\nu)$. The data are then embedded in an E-dimensional space by constructing vectors of the form $\mathbf{v}_k = \{\xi_k, \xi_{k+5}, \ldots, \xi_{k+5(E-1)}\}$, where $5/\nu$ is the appropriate delay time [9]. Portions of trajectory which lie close to the unstable periodic orbits of the system have then been identified by requiring them to return in certain (spherical) regions of radius R within selected time intervals (chosen around multiplets of T, up to $9T$). The precision R with which the detected recurrent orbits shadow the actually periodic ones has been chosen to vary with the position in phase space, in order to minimize the relative error in the search. Further-

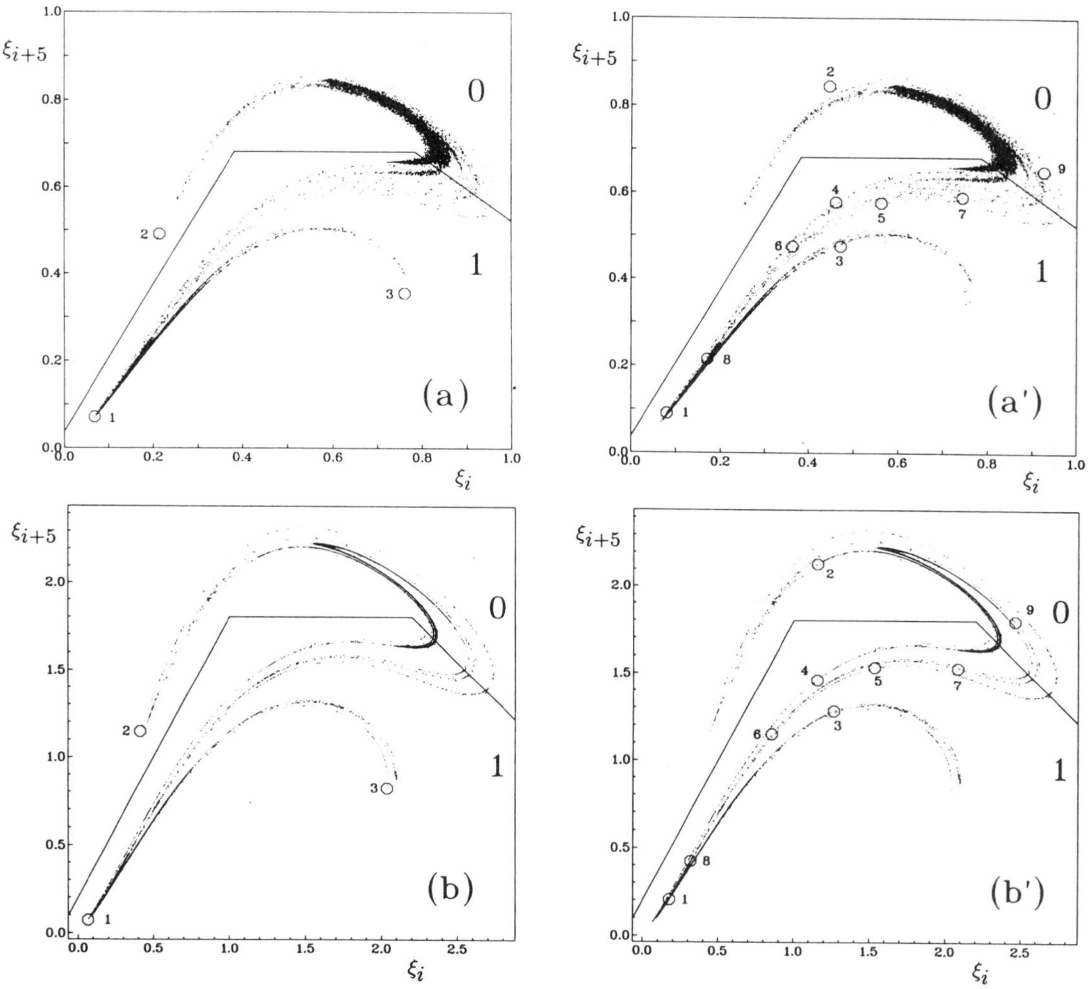

FIG. 3. Two-dimensional projections of the Poincaré section for $A=0.018$ and $\omega=0.031\,68$, and binary partition with elements labeled 0 and 1. The intersection points of the (a),(b) period-3 and of the (a'),(b') period-9 cycle are labeled in order of occurrence: (a) experiment; (b) model.

more, R is proportional to $\sqrt{E}$ and independent of the sampling time $1/\nu$, so that the results are consistent throughout the physically meaningful variation range [9] for E and ν. A Newton method has been used to locate the periodic orbits of the differential model (1). All cycles up to order nine have been compared with the experimental ones and found to agree very well in embedding spaces of dimension E between 6 and 16. In fact, the information dimension D_1 [9] for all considered strange attractors (the corresponding parameter values are indicated by circles in Fig. 2) is between 2.1 and 2.4 [10], so that $2D_1+1<6$.

In order to obtain a more complete characterization of the dynamics, we have constructed Poincaré sections Σ by considering embedded points $\mathbf{v}_k$ which lie 25 time units apart from each other (the system is, in fact, externally forced). The intersection points of the unstable cycles with Σ have then been assigned a different symbolic labeling through a binary partition [9] $\mathcal{D}=\{\Delta_0,\Delta_1\}$. Recall that a generating partition associates a unique phase-space point to each infinitely long symbolic sequence (among which are the periodic ones). Hence, the partition $\mathcal{D}$ constitutes a good approximation to a generating one, since it distinguishes all periodic points up to order nine [11,12(b),12(c)]. Of course, separating periodic points is a necessary (although not, in general, sufficient) condition for $\mathcal{D}$ to be generating. In Fig. 3, we report experimental [3(a) and 3(a′)] and numerical [3(b) and 3(b′)] strange attractors on a two-dimensional projection of the Poincaré section Σ for $A=0.018$ and $\omega=0.031\,68$, together with the curve defining $\mathcal{D}$. Notice the intersection points of a period-3 orbit [3(a) and 3(b)] and of a period-9 orbit [3(a′) and 3(b′)] with Σ. The former one lies "on the border" of the attractor and gives rise to a crisis for $A_c \approx 0.018\,02$.

The analysis of the symbolic dynamics, performed with the unfolding method of Ref. [12], leads to a description in terms of a full binary tree over the two primitive words $w_1=1$, $w_2=01$ for both experiment and model up to hierarchical level 4 (the longest orbit of which is 01010101). In fact, the only experimental forbidden sequence is 00, whereas numerical calculations show that also three strings of length 9 are not allowed. The topological entropy is therefore approximated by $K_0=0.481=\ln[(1+\sqrt{5})/2]$ [12(b)] at the experimental resolution. The prime cycles [13] have the following encodings: w_1, w_2, w_2w_1, $w_2w_1^2$, $w_2^2w_1$, $w_2w_1^3$, $w_2^2w_1^2$, $w_2w_1^4$, $w_2^2w_1^3$, $w_2w_1^5$, $w_2^3w_1^2$, $w_2^2w_1^4$, $w_2w_1^6$, $w_2^2w_1^5$, and $w_2w_1^7$, where w^n denotes the nth repetition of word w. The logic-tree representation of the symbolic dynamics constitutes an invariant topological characterization of the system. The full coincidence found between measured and numerical data (confirmed for the other two time series) clearly shows that the EBL model indeed describes the experimental observation up to the available resolution. A more complete characterization of the dynamics, which includes metric invariants as well (sequence probabilities, scaling functions, thermodynamic averages [13], and complexity [12]), will be presented elsewhere [4].

We acknowledge useful discussions with F. Waldner, P. F. Meier, and P. Talkner and financial support by the Swiss National Science Foundation.

[1] P. Boesiger, E. Brun, and D. Meier, Phys. Rev. Lett. **38**, 602 (1977).

[2] E. Brun, B. Derighetti, D. Meier, R. Holzner, and M. Ravani, J. Opt. Soc. Am. B **2**, 156 (1985).

[3] E. N. Lorenz, J. Atmos. Sci. **20**, 130 (1963).

[4] A more detailed discussion is presented in L. Flepp, R. Holzner, E. Brun, M. Finardi, and R. Badii (to be published).

[5] A. Abragam, *Principles of Nuclear Magnetism* (Clarendon, Oxford, 1989).

[6] M. Ravani, B. Derighetti, E. Brun, G. Broggi, and R. Badii, J. Opt. Soc. Am. B **5**, 1029 (1988).

[7] The rescaled equations for (X,Y,Z) have been actually integrated.

[8] C. Grebogi, E. Ott, and J. A. Yorke, Phys. Rev. Lett. **48**, 1507 (1982).

[9] J.-P. Eckmann and D. Ruelle, Rev. Mod. Phys. **57**, 617 (1985).

[10] They have been evaluated by means of the nearest-neighbor method; R. Badii and A. Politi, J. Stat. Phys. **40**, 725 (1984); G. Broggi, J. Opt. Soc. Am. B **5**, 1020 (1988).

[11] D. Auerbach, P. Cvitanović, J. P. Eckmann, G. Gunaratne, and I. Procaccia, Phys. Rev. Lett. **58**, 2387 (1987).

[12] (a) R. Badii, Europhys. Lett. **13**, 599 (1990); (b) R. Badii, M. Finardi, and G. Broggi, in *Chaos, Order and Patterns*, edited by P. Cvitanović *et al.* (Plenum, New York, 1991); (c) R. Badii, in *Measures of Complexity and Chaos*, edited by N. B. Abraham *et al.* (Plenum, New York, 1990).

[13] P. Cvitanović, Phys. Rev. Lett. **61**, 2729 (1988).

Experimental confirmation of the theory for critical exponents of crises

John C. Sommerer [a,b], William L. Ditto [c], Celso Grebogi [a,c], Edward Ott [a] and Mark L. Spano [c]

[a] *The University of Maryland, College Park, MD 20742, USA*
[b] *Applied Physics Laboratory, Laurel, MD 20707, USA*
[c] *Naval Surface Warfare Center, Silver Spring, MD 20903, USA*

Received 14 November 1990; accepted for publication 14 December 1990
Communicated by A.P. Fordy

We investigate the scaling of the average time τ between intermittent bursts for a chaotic system that undergoes a homoclinic tangency crisis, which causes a sudden expansion in the attractor. The system studied is a periodically driven (frequency f), nonlinear, magnetoelastic ribbon. The observed behavior of τ is well fit by a power-law scaling $\tau \sim |f-f_c|^{-\gamma}$, where $f=f_c$ at the crisis. We identify the unstable periodic orbit mediating the crisis, and determine its linearized eigenvalues from experimental data. The critical exponent γ found from the scaling of τ is shown to agree with that theoretically predicted for a two-dimensional map on the basis of the eigenvalues of the mediating periodic orbit.

In dissipative, nonlinear dynamical systems it is often found that small changes in a system parameter lead to sudden changes in chaotic attractors. This phenomenon, termed *crisis* [1], has been observed experimentally in a number of systems [2–5]. Theoretically, a crisis occurs when, as the system parameter f reaches its critical value f_c, the chaotic attractor collides with an unstable periodic orbit; if the unstable orbit lies inside the attractor's basin of attraction, the characteristic change in the attractor is an increase in size. After the crisis, orbits of the dynamical system typically move chaotically for a time as though on the smaller, pre-crisis attractor, then burst into chaotic motion over a larger region of phase space, then return to the region of the smaller attractor, and so on. Such *crisis-induced intermittency* [6] has an associated characteristic time τ, the average time between bursts. For a large class of low-dimensional systems, and for f just past its critical value, the characteristic time τ is predicted to have a power-law scaling [6,7], $\tau \sim |f-f_c|^{-\gamma}$. This scaling behavior has been observed experimentally [3–5].

For some low-dimensional systems, the critical exponent γ can be predicted theoretically: for one-dimensional maps with generic, quadratic maxima, γ is restricted to be $\frac{1}{2}$; for a large class of two-dimensional maps (and therefore three-dimensional flows), γ is determined by the eigenvalues of the unstable periodic orbit mediating the crisis [6,7]. Of the experimental studies of scaling in crisis-induced intermittency to date, one [3] showed the generic one-dimensional result $\gamma=\frac{1}{2}$. Later studies [4,5] reported crises where $\gamma \neq \frac{1}{2}$; in one of these [5], an unstable periodic orbit appeared to be involved in the crisis, but no connection between the critical exponent and the eigenvalues of the periodic orbit was made. We have observed crisis-induced intermittency in a nonlinear mechanical system, and have for the first time confirmed the theoretically-predicted key role played by an unstable orbit in mediating the crisis. We identify the unstable periodic orbit mediating the crisis and compare the critical exponent predicted on the basis of the orbit's eigenvalues with the scaling behavior of the time between bursts, τ. We find good agreement between the predicted and measured exponents.

The experimental system was a gravitationally buckled, amorphous magnetoelastic ribbon [8]

From *Phys. Lett. A* **153** (1991). Reprinted with permission from Elsevier Science Publishers.

(transversely annealed $Fe_{81}B_{13.5}Si_{3.5}C_2$; 3 mm×65 mm× 25 μm), clamped vertically at its base, and driven parametrically by a sinusoidally time-varying component to an imposed magnetic field (fig. 1). The ribbon is made from a new class of amorphous materials that exhibit very large reversible changes in their Young modulus E with the application of a small magnetic field (inset, fig. 1). The oscillating magnetic field changes the stiffness of the ribbon, which therefore buckles to a correspondingly greater or lesser degree. The degree to which the ribbon is buckled is measured by an MTI Fotonic sensor near the base of the ribbon. Additional details of the apparatus are discussed in refs. [5,9]. The sensor output, monotonically related to the ribbon curvature, was used as the single measured dynamical variable in the experiments. The rest of the system's phase portrait was reconstructed by delay-coordinate embedding [10]. The sensor output voltage time series was $V_i = V(t_i)$ ($t_i = i\Delta t$, $i = 1, 2, 3, \ldots$; Δt is the forcing period). This choice of Δt was made to obtain a stroboscopic Poincaré section in a $(d+1)$-dimensional continuous-time phase space. For sufficiently large d, this choice of Δt induces a discrete d-dimensional map $\boldsymbol{x}_{i+1} = \boldsymbol{F}(\boldsymbol{x}_i)$ for the system, where $\boldsymbol{x}_i = (V_i, V_{i+1}, V_{i+2}, \ldots, V_{i+d})$. It should be noted that a complete model of the ribbon's behavior requires

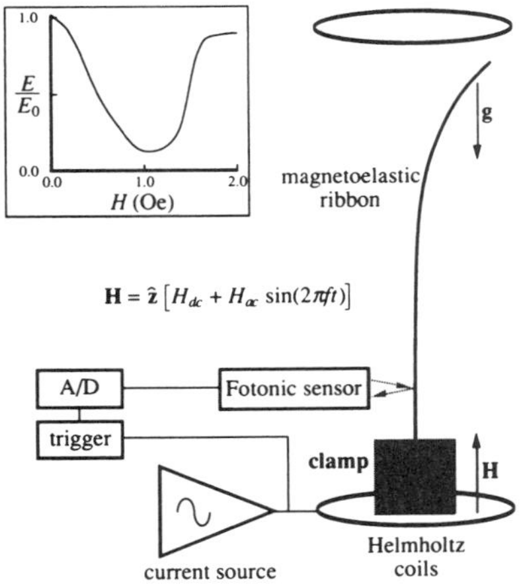

Fig. 1. Experimental setup. Inset: ratio of Young modulus E of the ribbon to zero-field modulus E_0 versus applied magnetic field.

a partial differential (i.e., infinite-dimensional) equation. It is therefore remarkable that essential features of the ribbon's dynamics can be captured in a low-dimensional mapping, as has been shown previously [5,9,11].

We observed that for an imposed magnetic field $H = H_{dc} + H_{ac}\sin(2\pi ft)$, (where $H_{dc} = 0.82$ Oe and $H_{ac} = 2.05$ Oe), as f decreases through about 0.97 Hz, the system undergoes a crisis. Before the crisis, the system state moves chaotically on a strange attractor, which as expected is somewhat smeared by system noise [12] (fig. 2a). After the crisis (fig. 2b), the attractor is much larger. The dynamics during the bursts is more complicated than that between bursts. In fact, the embedding dimension d must be increased to 3 in order to preserve the invertibility of the mapping $\boldsymbol{F}$ (i.e., prevent overlap of different regions of the attractor [#1]; see below). However, the part of the attractor corresponding to the pre-crisis attractor (the *core*) is adequately represented in a two-dimensional phase space, and so we expect the theory of refs. [6,7] to apply.

The scaling exponent γ was determined by measuring the characteristic time τ as a function of f. Time series of three hour duration were obtained for closely spaced values of f near the crisis. Defining the end of a burst was complicated by the higher dimension of the return dynamics mentioned above. Projecting the three-dimensional post-crisis attractor onto a two-dimensional phase space meant that an orbit returning from a burst could appear to fall on the core attractor, only to immediately initiate another burst. Therefore, τ was estimated using only times between bursts longer than five forcing periods. These interburst times were found to be exponentially distributed as predicted by theory [1,13]. The scaling behavior for such a conditional estimate $\hat{\tau}$ (the mean of all samples greater than a cutoff) is almost identical to that of τ, as long as the $\hat{\tau}$ is over twice as long as the conditioning cutoff. Thus γ can be estimated by fitting the observed values of $\hat{\tau}$ to the predicted scaling for τ.

The theoretical scaling law for τ assumes that the system is deterministic; for such a system, the characteristic time goes to infinity at the crisis. Real sys-

[#1] It may be that there is a different two-embedding for which such overlap is absent.

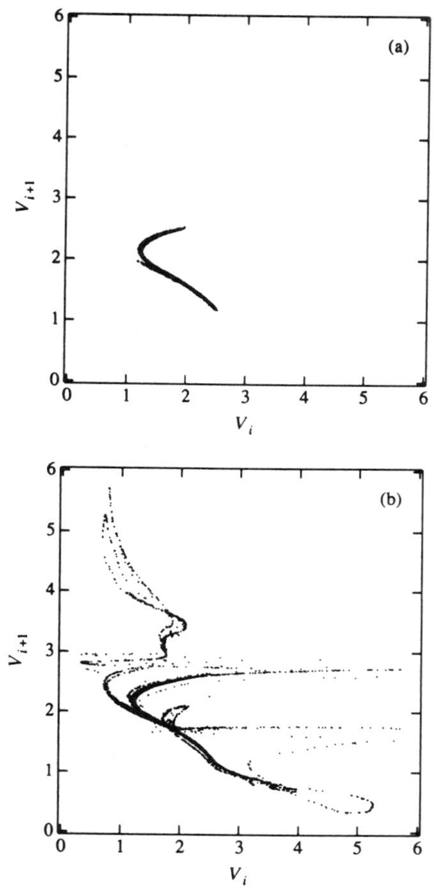

Fig. 2. Delay coordinate embedding of time series taken before (a) and after (b) the crisis. The time delay Δt is $1/f$, the stroboscopic sampling period. The core attractor of (a) is enlarged by burst dynamics in (b). Isolated data in (a) are part of the initial transient.

tems, which are always accompanied by noise, show a finite [14] characteristic time at $f=f_c$. This makes determining the value of f_c more complicated than in an ideal system, and somewhat limits the accuracy with which it can be determined. The value of f_c for this crisis was determined by minimizing the degree to which a plot of $\log(\hat{\tau})$ versus $\log(f_c-f)$ deviated from a straight line. Given f_c, γ was determined from the best nonlinear least-squares fit of the functional form $\hat{\tau}=k(f_c-f)^{-\gamma}$ to the data. The data and corresponding fit are shown in fig. 3. The statistical estimate of the critical exponent is $\gamma=1.12\pm0.02$.

Careful observation of the system state just before a burst indicated that system trajectories closely approached a period-three orbit before leaving the core attractor. Fig. 4 shows V_{i+3} versus V_i for the phase-space region of the core attractor after crisis; several

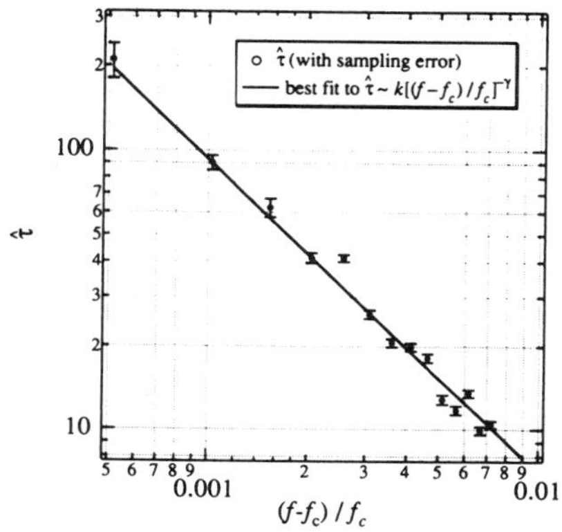

Fig. 3. Estimated mean time between bursts $\hat{\tau}$ versus $|f_c-f|/f_c$. Nonlinear least-squares fit to $\hat{\tau}=k(f_c-f)/f_c]^{-\gamma}$ gives $\gamma=1.12\pm0.02$.

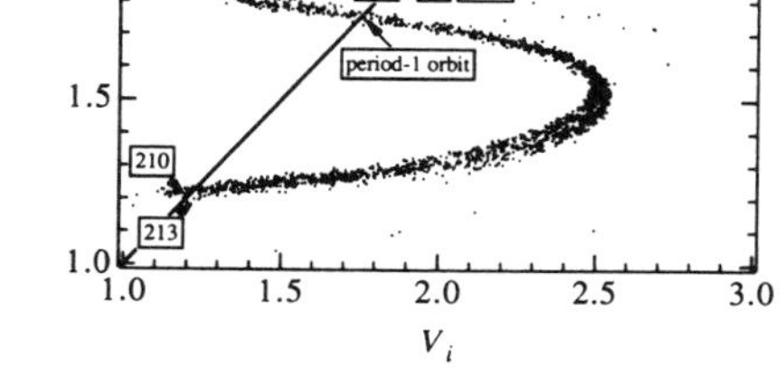

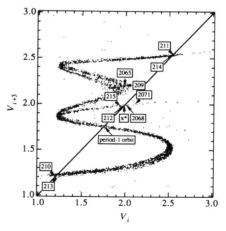

Fig. 4. Delay-coordinate embedding of time series taken after crisis. The boxed numbers indicate the iterate number i.

iterates passing near the saddle are indicated. Any period-three orbit appears as fixed points of the three-times-iterated mapping $F^{(3)}$ and, because we use a delay-coordinate embedding, these must lie along the 45° line (i.e., $V_{i+3}=V_i$). The attractor intersects the 45° line in four places. One of these intersections is a period-one point; the other three are elements of the same period-three saddle. For $f>f_c$ (before the crisis), the attractor only intersects the 45° line at the period-one point. This indicates the absence of any period-three orbits on the attractor before crisis. Crises in two-dimensional maps are limited to two varieties, depending on whether the mediating periodic orbit is on the attractor before (heteroclinic), or only just at and after (homoclinic) the crisis [6,7]. The absence of a period-three orbit on the attractor before the crisis means that the crisis is of the homoclinic type. The theoretically predicted critical exponent for a homoclinic crisis is given [6,7] by $\gamma=\log|\beta_2|/2\log|\beta_1\beta_2|$, where β_1 and β_2 are the expanding ($|\beta_1|>1$) and contracting ($|\beta_2|<1$) eigenvalues, respectively, of the periodic orbit.

Estimating these eigenvalues from the experimental data is complicated by the facts that: (i) the map F is clearly very dissipative, (ii) the system is contaminated by noise, and (iii) the linear neighborhood of the period-three point is relatively small. Therefore, following an orbit as it passes near the period-three point, as suggested in ref. [6], is not a good way to estimate the eigenvalues. Instead, we use pairs of iterates to estimate the Jacobian matrix of the map around the period-three point, and from this estimate obtain the linearized eigenvalues of the periodic point.

Our general eigenvalue estimation procedure is outlined as follows. To obtain the eigenvalues of a period-n orbit, we approximate the map $F^{(n)}$ (the n-times iterated map F) by a Taylor series

$$\boldsymbol{x}_{i+n}+\delta\boldsymbol{x}_{i+n}=\boldsymbol{F}^{(n)}(\boldsymbol{x}_i)+\mathrm{D}\boldsymbol{F}^{(n)}|_{\boldsymbol{x}_i}\cdot\delta\boldsymbol{x}_i +\mathrm{O}(|\delta\boldsymbol{x}_i|^2)\,,$$

and let $\boldsymbol{x}_i$ be the period-n point, which we denote $\boldsymbol{x}^*$. Because $\boldsymbol{F}^{(n)}(\boldsymbol{x}^*)=\boldsymbol{x}^*$, we see that the Jacobian $\mathrm{D}\boldsymbol{F}^{(n)}|_{\boldsymbol{x}^*}$ takes tangent vectors into tangent vectors near the saddle. We can use the difference between a pair of iterates on the same orbit to approximate a tangent vector, and watch its propagation near the period-n point:

$$\boldsymbol{x}_{i+n}-\boldsymbol{x}_{j+n}\approx\mathrm{D}\boldsymbol{F}^{(n)}|_{\boldsymbol{x}^*}\cdot(\boldsymbol{x}_i-\boldsymbol{x}_j),\quad i\neq j\,.$$

The key advantage of this method over that of ref. [6] is that here we can take advantage of widely (temporally) separated recurrent iterates (i.e., i very different from j). Furthermore we do not require many sequential iterates which are all within the linear neighborhood of the saddle and whose differences are also significantly above the system noise level.

Because we use a delay-coordinate embedding, the surface of section map for $\boldsymbol{x}=(y,z)$ is of the form

$$y_{i+n}=g(y_i,z_i),\quad z_{i+n}=y_i\,,$$

and we therefore know that the form of the Jacobian is particularly simple:

$$\mathrm{D}\boldsymbol{F}^{(n)}|_{\boldsymbol{x}^*}=\begin{pmatrix} r & s\\ 1 & 0\end{pmatrix},$$

where $r=\partial g/\partial y|_{(y^*,z^*)}$ and $s=\partial g/\partial z|_{(y^*,z^*)}$.

Calculating the eigenvalues of $\mathrm{D}\boldsymbol{F}^{(n)}|_{\boldsymbol{x}^*}$ explicitly and using the fact that $\beta_1\beta_2=\det(\mathrm{D}\boldsymbol{F}^{(n)}|_{\boldsymbol{x}^*})$, we have

$$\gamma=\frac{\log[\frac{1}{2}r-\frac{1}{2}(r^2+4s)^{1/2}}{2\log|s|}\,.$$

Thus, estimation of the eigenvalues of the period-n orbit is reduced to the statistical estimation of the quantities r and s, from the linear multiple regression model

$$y_{i+n}-y_{j+n}=r(y_i-y_j)+s(z_i-z_j)\,,$$

where a set of ($i\neq j$) pairs is chosen from the vicinity of the saddle. Constraints can be placed on the regression, because the sign of s will be known from the type of period-n saddle involved (an ordinary saddle implies $s<0$; a flip saddle implies $s>0$). Furthermore, the ($i\neq j$) pairs can be chosen in such a way as to maximize the statistical confidence of the estimates, i.e., chosen such that the contributions of the r and s terms in the regression are about equal. (Use of the restricted form of the Jacobian and a careful choice of data for the regression analysis is essential in a noisy system. Least-squares estimation of the general Jacobian with a comprehensive data set was done successfully with a lower-noise system in ref. [15], but for our system, this results in a very poor estimate of the stable eigenvalue, and consequently, a poor estimate of γ.) Finally, propagation of errors using the $\gamma(r, s)$ formula allows the statistical uncertainties in r and s to be translated into uncertainty in γ. A nearly identical analysis pertains to the heteroclinic case.

We take the point $\boldsymbol{x}^*$ to be as shown in fig. 4. Observation of iterates passing near $\boldsymbol{x}^*$ shows that the period-three saddle is an ordinary one ($s<0$), that the stable direction is nearly vertical, and that the unstable direction has a slope of about 0.1. The second row of $\mathrm{D}\boldsymbol{F}$ indicates that the eigenvectors must be of the form $(\beta, 1)$. We therefore know that $\beta_2 \ll 1$, while $\beta_1 \approx 10$, indicating that $|s| \ll |r|$, i.e., $\mathrm{D}\boldsymbol{F}$ is ill-conditioned. This information allows the determination that the best regression estimates of r and s will be obtained for pairs of iterates whose vector difference is nearly vertical. Using 150 such pairs from a single orbit, we estimated r, s and their statistical uncertainties, and conclude that $\gamma = 1.07 \pm 0.15$. This estimate of the critical exponent, based on the eigenvalues of the period-three saddle, clearly is statistically consistent with the exponent ($\gamma = 1.12 \pm 0.02$) in the observed scaling of $\hat{\tau}$. Such consistency strongly supports the key role played by unstable periodic points in mediating crisis phenomena in nonlinear dynamical systems.

The first author was supported by the U.S. Air Force Office of Scientific Research. Additional support was provided by the U.S. Office of Naval Research (Physics Division), and the U.S. Department of Energy (Office of Scientific Computing).

References

[1] C. Grebogi, E. Ott and J.A. Yorke, Phys. Rev. Lett. 48 (1982) 1507; Physica D 7 (1983) 181.

[2] C. Jeffries and J. Perez, Phys. Rev. A 27 (1983) 601;
S.K. Brorson, D. Dewey and P.S. Linsay, Phys. Rev. A 28 (1983) 1201;
H. Ikezi, J.S. deGrasse and T.H. Jensen, Phys. Rev. A 28 (1983) 1207;
E.G. Gwinn and R.M. Westervelt, Phys. Rev. Lett. 54 (1985) 1613;
D. Dangoisse, P. Glorieux and D. Hannequin, Phys. Rev. Lett. 55 (1985) 746.

[3] R.W. Rollins and E.R. Hunt, Phys. Rev. A 29 (1984) 3327.

[4] T.L. Carroll, L.M. Pecora and F.J. Rachford, Phys. Rev. Lett. 59 (1987) 2891.

[5] W.L. Ditto et al., Phys. Rev. Lett. 63 (1989) 923.

[6] C. Grebogi, E. Ott, F. Romeiras and J.A. Yorke, Phys. Rev. A 36 (1987) 5365.

[7] C. Grebogi, E. Ott and J.A. Yorke, Phys. Rev. Lett. 57 (1986) 1284.

[8] H.T. Savage and C. Adler, J. Magn. Magn. Mater. 58 (1986) 320;
H.T. Savage and M.L. Spano, J. Appl. Phys. 53 (1982) 8002.

[9] H.T. Savage et al., J. Appl. Phys. 67 (1990) 5619.

[10] F. Takens, in: Dynamical systems and turbulence, eds. D.A. Rand and L.-S. Young (Springer, Berlin, 1980) pp. 366ff;
N.H. Packard et al., Phys. Rev. Lett. 45 (1980) 712.

[11] W.L. Ditto et al., Phys. Rev. Lett. 65 (1990) 533.

[12] E. Ott, E.D. Yorke and J.A. Yorke, Physica D 16 (1985) 62;
E. Ott and J.D. Hanson, Phys. Lett. A 85 (1981) 20.

[13] J.A. Yorke and E.D. Yorke, J. Stat. Phys. 21 (1979) 263.

[14] J.C. Sommerer, E. Ott and C. Grebogi, to be published.

[15] D.P. Lathrop and E.J. Kostelich, Phys. Rev. A 40 (1989) 4028.

Structure of Chaos in the Laser with Saturable Absorber

F. Papoff,[1] A. Fioretti,[2] E. Arimondo,[2] G. B. Mindlin,[3],[a] H. Solari,[3],[b] and R. Gilmore[3]

[1]*Scuola Normale Superiore, Piazza dei Cavalieri 2, 56100 Pisa, Italy*

[2]*Dipartimento di Fisica, Piazza Torricelli 6 56100 Pisa, Italy*

[3]*Department of Physics and Atmospheric Science, Drexel University, Philadelphia, Pennsylvania 19104*

(Received 7 August 1991)

We carry out a topological analysis on an experimental data set from the laser with saturable absorber. This analysis is based on the topological organization of low period orbits extracted from chaotic time series data. This allows us to determine for the first time that previously proposed models are compatible with the data.

PACS numbers: 42.50.Lc, 05.45.+b, 42.55.−f

It has recently become possible to provide a topological classification of strange attractors. In this classification scheme the topological structure of a strange attractor is given by a set of integers. These integers describe the structure of the "knot-holder" [1] or "template" which supports the strange attractor [2]. The template describes the stretching and compressing mechanisms responsible for creating the strange attractor. These stretching and compressing mechanisms are also responsible for organizing the unstable periodic orbits which are embedded in the strange attractor in a unique way.

This topological classification is in contrast to the classification of strange attractors according to their metric properties (e.g., Lyapunov exponents, various dimensions). Metric properties are invariant under a coordinate transformation but not under control parameter variation. Topological properties remain invariant under both coordinate transformations and control parameter variations, or change in experimental conditions. This means, in particular, that chaotic data sets taken for a physical system under different experimental conditions will exhibit the same topological classification.

Furthermore, it is possible to determine the topological classification of a strange attractor by carrying out an analysis ("topological analysis") on scalar time series data [3]. This provides, for the first time, a test to determine whether a model which is proposed to describe a chaotic process is in fact compatible with that process. Topological analyses are carried out on the experimental time series data and data generated by the model. If the topological analyses identify different templates (sets of integers), the model can be rejected as not compatible with the data. Otherwise, the model is compatible with the data.

In this Letter we apply this topological analysis, for the first time, to determine whether models proposed to described the laser with saturable absorber (LSA) [4–7] are compatible with a number of experimental data sets which have been taken from the LSA [6–9].

The experimental setup consists of an infrared cavity containing a discharge CO_2 amplifier and an absorber cell. We have used CH_3I:He and OsO_4:He in the ratio 1:20 as absorbers. The laser output intensity I is digitized and discretely sampled at a rate of about 80 samples per period for fixed values of the control parameters. These include the discharge current, the absorber pressure, and the laser frequency detuning. A number of long time series, up to 32×10^3 8-bit data, were stored in a microcomputer by use of a digital oscilloscope.

In the region of the LSA where instabilities and chaos are found, the laser intensity pulse starts below threshold, developing a large peak L followed by a variable number of small peaks S. Each minimum is followed by a maximum; the deeper the minimum, the more intense the following maximum. A segment of a typical data set is shown in Fig. 1.

Three [4,6], four [5], and five [7] variable models have been proposed to describe the LSA. The mechanism responsible for the existence of chaotic behavior in these models is the following. An unstable limit cycle (saddle cycle) or its degenerate limit, an unstable focus, has stable and unstable invariant manifolds which approach tangency [10] [cf. Fig. 2(a)]. In the three-, four-, and five-dimensional models the unstable manifold of the saddle cycle is two dimensional, while the stable manifold is 2D, 3D, and 4D, respectively. As the tangency is approached, a number of stable periodic orbits are created by saddle-node bifurcations, which then undergo period-doubling cascades. This generates a complicated dynamics even before the tangency occurs [10,11]. When the manifolds behave as shown in Fig. 2(b) the flow is hyperbolic and exhibits a Smale horseshoe [10] with zero global torsion. A point in the strange attractor near the unstable invariant manifold W^u will evolve during one period along W^u (stretching) while at the same time being compressed along the direction of the stable manifold W^s towards the invariant set W^u (folding). The stretching and folding mechanisms, responsible for the creation of chaos, are represented schematically by the template for this flow, shown in Fig. 2(c).

To determine whether these models are compatible with the experimental data, we must identify the template underlying the experimental strange attractor, and compare it with the template shown in Fig. 2(c). The topological analysis of the chaotic data was carried out in a number of simple steps.

First, the unstable periodic orbits embedded in the strange attractor were determined by the method of close

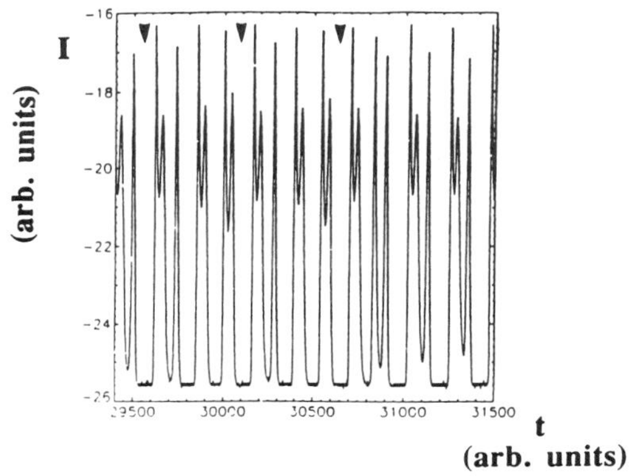

FIG. 1. A segment of the time series data. The intensity is plotted vs time. In this case a close return after 7 periods occurs (indicated by the arrows), as the strange attractor is closely following a period-7 orbit, staying close to it during approximately 14 periods.

returns [3]. If the phase-space trajectory of the system enters the neighborhood of an unstable periodic orbit with a relatively small Lyapunov exponent, it may remain in the neighborhood of that orbit sufficiently long so that it returns near its starting point and evolves near an earlier point of its trajectory (cf. Fig. 1). Such close return segments are easily identified by plotting $|I(i)-I(i+p)|$, where i indexes the discretely sampled intensity ($1 \le i \le 32\times 10^3$). Segments $i_{\min} < i < i_{\max}$ where the difference above is relatively small indicate segments of chaotic time series data that can then be used as representations of unstable periodic orbits of period p, measured in units of the sample rate. In this way, four or more periodic orbits were located in each data file taken under different experimental conditions for the LSA.

Second, an embedding of these segments as well as the entire strange attractor in R^3 must be constructed so that topological organization of orbit pairs can be determined. We have adopted a differential phase-space embedding

$$I(i) \to [y_1(i), y_2(i), y_3(i)] ,$$

where $y_1(i) = \sum_{j=1}^{i} I(j) e^{-(i-j)/N}$, $y_2(i) = I(i)$, $y_3(i) = I(i) - I(i-1)$. Here N is taken as about two cycle times ($N \sim 2\times 80$) [3]. This embedding of the period-1 and -3 orbits is shown in Fig. 3.

The embedded strange attractor had a hole in the center, making possible the construction of a Poincaré section. In each section we were able to construct a return map. For each map we found an orientation-preserving and an orientation-reversing branch. This allowed the development of a symbolic dynamics for all orbits in each data set: x (y) for passage through the orientation-preserving (-reversing) branch. A segment of data which closely follows the unstable period-1 orbit (y) is shown in Fig. 4. This clearly identifies the orbit y as a

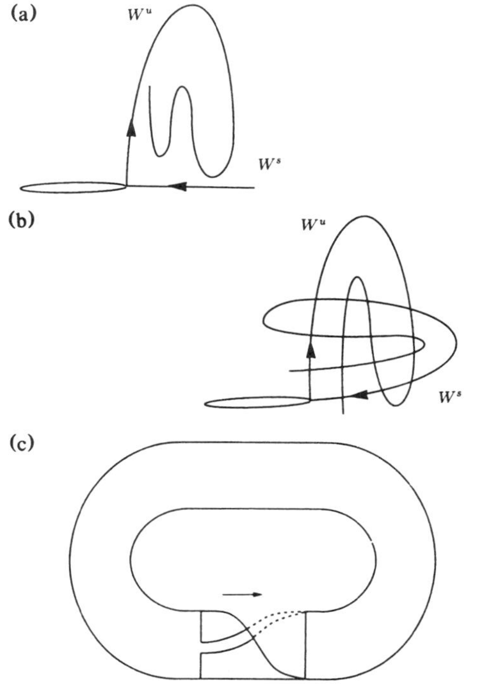

FIG. 2. (a) A saddle cycle whose unstable and stable manifolds are approaching a tangency. (b) The stable and unstable manifolds of a saddle cycle after the last tangency. There is a hyperbolic invariant set coexisting with the saddle cycle. (c) The horseshoe template. For each periodic orbit held by this manifold, there is an orbit in the invariant set of (b) with identical topological properties.

flip saddle, as the orbit segment yy forms the boundary of a nonorientable strip. This allows the identification x with the small peaks, $x \sim S$, and y with the large peaks, $y \sim L$. The unstable limit cycle y describes unstable periodic behavior of the LSA consisting of a series of large peaks.

Third, the linking numbers of all pairs of periodic orbits identified in each data set were determined [3,12]. The linking number of two orbits is defined by a Gauss integral [3] and is roughly the number of times the two orbits wind around each other. Computation of the linking number in the differential phase-space embedding is particularly easy and was done by counting crossings of the orbits. This is shown in Fig. 3, where the over and under crossings of the period-1 and period-3 orbits are clearly indicated.

Fourth, a template was identified on the basis of the linking numbers of a small set of orbits. Each template has a unique signature in terms of the linking numbers of pairs of periodic orbits. In fact, the linking numbers of

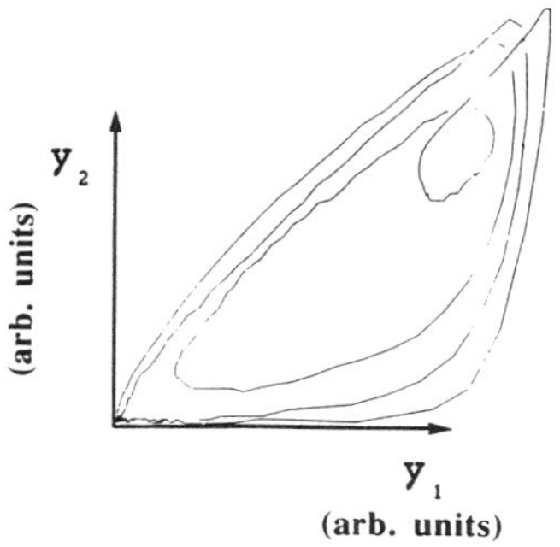

FIG. 3. A two-dimensional projection of the three-dimensional embedding for the segment close to the period-1 orbit and for a segment close to a period-3 orbit. The over and under crossings keep track of the third dimension. The segments of orbits in the outer part of the figure have higher values of the third coordinate than the segments of orbits in the inner part. Two orbit segments which cross each other are interchanging their order on the third coordinate and, for this reason, are on the orientation reversing branch. The period-1 and period-3 orbits cross each other 4 times, so their linking number is $\frac{1}{2}(4)=2$.

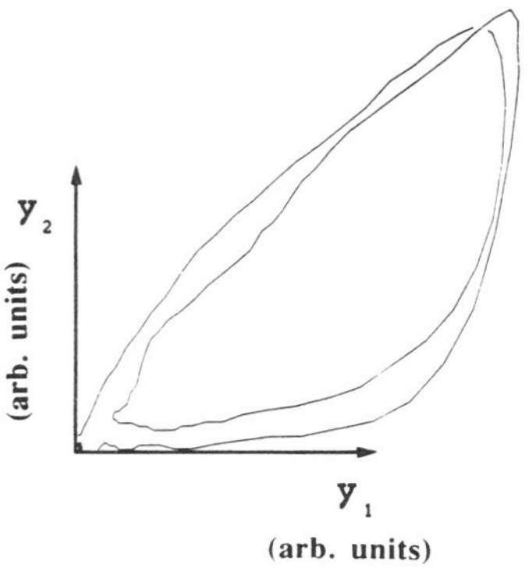

FIG. 4. Two cycles around the period-1 flip saddle orbit. As in Fig. 3, the segments of orbits in the outer part of the figure have higher values of the third coordinate than the segments in the inner part. The strip whose boundary is approximated by these orbits is nonorientable as it is a rectangle with two sides joined with a half twist.

only the period-1 and period-2 orbits suffice to distinguish different templates. When not all of these orbits can be extracted from the data, alternative subsets of orbits can be used to identify the template. Additional orbits then provide confirmation of the template identification [3].

For each of the experimental LSA data files studied, a minimum of four periodic orbits was reconstructed. These orbits had symbolic names, y, yx^n, $yx^{n-1}y$, with n ranging from 1 to 15. Orbits of higher period, which are compositions of different pulses with $n=1,2$ have also been extracted from some files. For each of the files studied, the template identified was the horseshoe template with zero global torsion. In every case, the template was overdetermined by the periodic orbits extracted.

We point out here that we cannot entirely reconstruct the phase space, since the intensity of the LSA goes below our detection limit between most successive pulses. Thus, in principle we cannot exclude the possibility that part of the large peaks which are not completely reconstructed lie on a third branch of a template. However, all the orbits extracted from our data set are compatible with a template having only two branches, which is the simplest template compatible with the data.

This analysis reveals that as the control parameters in the experiments are changed, the template supporting the strange attractor remains unchanged. What does change is the set of unstable periodic orbits which are present in the strange attractor and/or their degree of instability. These changes have consequences which can be easily visualized. This can be done by plotting the cumulative number of times the phase-space trajectory passes through the orientation-preserving (or -reversing) branch of the template (p_0 or p_1) as a function of the cumulative number of passes through the template (p_0+p_1). For each of the files studied, this plot was nearly a straight line, with slope P/Q. For each of the files, the higher period orbits extracted had a "rotation number" $p_1/(p_0+p_1)$, well approximated by the ratio P/Q. In Fig. 5 we show such a plot for three LSA data files taken at different control parameter values.

One of the benefits of a topological analysis is that the recovery of a "badly ordered" or non-well-ordered orbit from the data set is a sufficient condition to show that the topological entropy of the flow is positive, and that the temporal behavior is chaotic [13]. This requires much less work than an analysis based on the computation of Lyapunov exponents, which can be problematic for relatively small data sets. We have found different badly ordered orbits in LSA data files. One of these orbits of period 7 ($yxyyxyx$) is shown in Fig. 1.

In conclusion, for the first time a topological analysis of chaotic time series data was used to determine if models proposed to describe the dynamics of a physical system are in fact compatible with the dynamics responsible for generating the data. The large number of experimental data sets which were recorded and analyzed for the LSA indicate that the dynamics is governed by a flow organized by a Smale horseshoe with zero global torsion. Variation of the control parameters restricts the flow to different parts of the underlying template, which remains unchanged as the control parameters are varied. Previously proposed models of dimensions three, four, and five

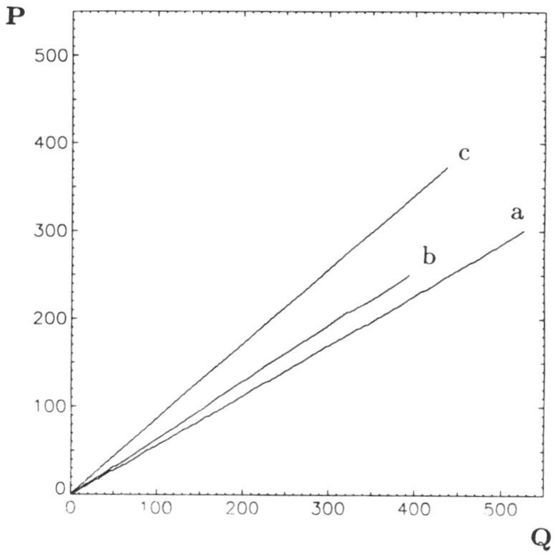

FIG. 5. Cumulative number of passes P through the orientation reversing branch vs the cumulative number of passes Q through the template for three different control parameter values. (Curve a) Absorber CH_3I:He with amplifier current discharge of 10.9 mA and pressure 0.1 mbar; (curves b and c) absorber OsO_4:He with amplifier current discharge of 9.3 mA and pressure 0.73 and 0.47 mbar, respectively.

exhibit the same underlying template, and are therefore not inconsistent with the experimental data sets.

This work was partially performed during the visit of one author (F.P.) to Drexel University. We wish to thank Professor J. R. Tredicce for helpful discussions and Dr. M. A. Natiello for help in the preparation of the computer codes. One of us (F.P.) thanks Foundazione A. Della Riccia for financial support. This work was partially supported by NSF Grant No. PHY88-43235.

[a]Present address: Departamento de Física y Matematica Aplicada, Facultad de Ciencias, Universidad de Navarra, 31080 Pamplona, Navarra, Spain.

[b]Present address: Departamento de Física, FCEN-Universidad de Buenos Aires, Pabellón I, Ciudad Universitaria, 1428 Buenos Aires, Argentina.

[1] P. Holmes, in *New Directions in Dynamical Systems*, edited by T. Bedford and J. Swift (Cambridge Univ. Press, Cambridge, 1988), p. 150.

[2] G. B. Mindlin, X. J. Hou, H. G. Solari, R. Gilmore, and N. B. Tufillaro, Phys. Rev. Lett. **64**, 2350 (1990).

[3] G. B. Mindlin, H. G. Solari, M. A. Natiello, R. Gilmore, and X. J. Hou, J. Nonlinear Sci. **1**, 147 (1991).

[4] B. Zambon, Phys. Rev. A **44**, R688 (1991).

[5] F. L. Hong, M. Tachikawa, T. Oda, and T. Shimizu, J. Opt. Soc. Am. B **6**, 1378 (1989).

[6] M. Lefranc, D. Hennequin, and D. Dangoisse, J. Opt. Soc. Am. B (to be published).

[7] D. Hennequin, F. de Tomasi, B. Zambon, and E. Arimondo, Phys. Rev. A **37**, 2243 (1988).

[8] D. Dangoisse, A. Bekkali, F. Papoff, and P. Glorieux, Europhys. Lett. **6**, 335 (1988).

[9] M. Tachikawa, F. L. Hong, K. Tanii, and T. Shimizu, Phys. Rev. Lett. **60**, 2266 (1988); **61**, 1042 (1988).

[10] J. Guckenheimer and P. Holmes, *Nonlinear Oscillations, Dynamical Systems, and Bifurcations of Vector Fields* (Springer-Verlag, New York, 1983).

[11] F. Papoff, A. Fioretti, and E. Arimondo, Phys. Rev. A **44**, 4639 (1991).

[12] L. H. Kaufman, *On Knots* (Princeton Univ. Press, Princeton, NJ, 1987).

[13] J. M. Gambaudo, S. Van Strien, and C. Tresser, Ann. Inst. Henri Poincaré **49**, 335 (1989).

PART III

Prediction, Filtering, Control and Communication in Chaotic Systems

All stable processes we shall predict. All unstable processes we shall control.

— paraphrase of a statement by J. von Neumann [see *Infinite in All Directions,* by Freeman Dyson]

CHAPTER 10

Prediction

J.D. Farmer, J. Sidorowich
Predicting chaotic time series
Phys. Rev. Lett. **59**, 845 (1987).

M. Casdagli
Nonlinear prediction of chaotic time series
Physica D **35**, 335 (1989).

G. Sugihara, R. M. May
Nonlinear forecasting as a way of distinguishing chaos from measurement error in time series
Nature **344**, 734 (1990).

T. Sauer
Time series prediction using delay coordinate embedding
In: *Time Series Prediction: Forecasting the Future and Understanding the Past*, Santa Fe Institute Studies in the Science of Complexity XV, A.S. Weigend and N.A. Gershenfeld, Eds., 175. Addison-Wesley, Reading, MA (1993).

Editors' Notes

Beginning around the time of Takens' embedding theorem in 1980, reconstructing chaotic attractors from time series became a common technique. The geometric structure of the attractor could be studied in this way, and in particular approximations of correlation dimension were attempted using the method of Grassberger and Procaccia (1983). The next major step was taken in 1985, when it was proposed to reconstruct not only the geometric attractor, but the dynamics on that attractor. In the survey of Eckmann and Ruelle (1985), the technique of reconstructing the local linear dynamics of a deterministic attractor from an experimental time series was introduced. This survey, and the new ideas it contained, became very influential in extending the concepts of chaos to experimental science. At about the same time, Sano and Sawada (1985) were using similar ideas for approximating Lyapunov exponents.

One of the many applications that follows from reconstructing the dynamics locally in phase space is time series prediction. **Farmer and Sidorowich** (1987) report on a direct implementation of this idea, and discuss the scaling of prediction error for several artificial and experimental time series.

Casdagli (1989) compares local linear approximations to the dynamics with global approximations, in the form of radial basis functions.

Sugihara and May (1990) use a simple local linear method to predict short time series, including infected population series for measles and chickenpox. As well as being short, these series are especially prone to measurement error, making conclusions difficult and controversial.

Sauer (1993) follows the Eckmann-Ruelle prescription, using the singular value decomposition to restrict the dynamics to the tangent plane of the attractor, and using Fourier interpolation to counteract undersampling difficulties. Predictions are made for the Lorenz attractor and for time series generated by a laser experiment.

Further reading

Activities at the Santa Fe Institute have resulted in two volumes containing extensive research reports on prediction:

Nonlinear Modeling and Forecasting, Santa Fe Institute Studies in the Science of Complexity XII, M. Casdagli and S. Eubank, Eds. Addison-Wesley, Reading MA (1992).

Time Series Prediction: Forecasting the Future and Understanding the Past, Santa Fe Institute Studies in the Science of Complexity XV, A.S. Weigend and N.A. Gershenfeld, Eds. Addison-Wesley, Reading, MA (1993).

Predicting Chaotic Time Series

J. Doyne Farmer and John J. Sidorowich[a]
Theoretical Division and Center for Nonlinear Studies, Los Alamos National Laboratory, Los Alamos, New Mexico 87545
(Received 22 April 1987)

We present a forecasting technique for chaotic data. After embedding a time series in a state space using delay coordinates, we "learn" the induced nonlinear mapping using a local approximation. This allows us to make short-term predictions of the future behavior of a time series, using information based only on past values. We present an error estimate for this technique, and demonstrate its effectiveness by applying it to several examples, including data from the Mackey-Glass delay differential equation, Rayleigh-Bénard convection, and Taylor-Couette flow.

PACS numbers: 05.45.+b, 02.60.+y, 03.40.Gc

One of the central problems of science is forecasting: Given the past, how can we predict the future? The classic approach is to build an explanatory model from first principles and measure initial data. Unfortunately, this is not always possible. In fields such as economics, we still lack the "first principles" necessary to make good models. In other cases, such as fluid flow, the models are good, but initial data are difficult to obtain. We can derive partial differential equations that allow us to predict the evolution of a fluid (at least in principle), but specification of an initial state requires the measurement of functions over a three-dimensional domain. Acquisition of such large amounts of data is usually impossible. Typical experiments employ only a few probes, each of which produces a single time series. Partial differential equations simply cannot operate on such data. In either case, when we lack proper initial data or when we lack a good model, we must resort to alternative approaches.

Such an alternative is exemplified by the work of Yule,[1] who in 1927 attempted to predict the sunspot cycle by building an *ad hoc* linear model directly from the data. The modern theory of forecasting[2] as it has evolved since then views a time series $x(t_i)$ as a realization of a random process. This is appropriate when effective randomness arises from complicated motion involving many independent, irreducible degrees of freedom.

An alternative cause of randomness is chaos,[3] which can occur even in very simple deterministic systems. While chaos places a fundamental limit on long-term prediction,[3] it suggests possibilities for *short-term* prediction: Random-looking data may contain simple deterministic relationships, involving only a few irreducible degrees of freedom. In chaotic fluid flows, for instance, experimental[4] and theoretical results[5] indicate that in some cases the state space collapses onto an attractor of only a few dimensions.

In this paper we present a method to make predictions about chaotic time series. These ideas were originally inspired by efforts to beat the game of roulette, in collaboration with Packard.[6]

If the data are a single time series, the first step is to embed it in a state space. Following the approach introduced by Packard *et al.*,[7] and put on a firm mathematical basis by Takens,[7] we create a state vector $\mathbf{x}(t)$ by assigning coordinates $x_1(t)=x(t)$, $x_2(t)=x(t-\tau)$, ..., $x_d(t)=x(t-(d-1)\tau)$, where τ is a delay time. If the attractor is of dimension D, a minimal requirement is that $d \ge D$.

The next step is to assume a functional relationship between the current state $\mathbf{x}(t)$ and the future state $\mathbf{x}(t+T)$,

$$\mathbf{x}(t+T)=f_T(\mathbf{x}(t)). \tag{1}$$

We want to find a predictor F_T which approximates f_T.

If the data are chaotic, then f_T is necessarily nonlinear. There are several possible approaches: One can assume a standard functional form, such as an mth-order polynomial in d dimensions, and fit the coefficients to the data set using least squares.[8] Forecasts for longer times $2T, 3T, \ldots$, can then be made by composing F_T with itself. This approach has the disadvantage that errors in approximation grow exponentially with composition. An alternative is to fit a new function F_T for each time T. This has the advantage that global approximation techniques only work well for smooth functions—and higher iterates of chaotic mappings are *not* smooth. Yet another approach is to recast Eq. (1) as a differential equation and write $\mathbf{x}(t+T)$ as its integral. All of these approaches suffer from the problem that the number of free parameters for a general polynomial is $(m+d)!/(m!d!) \approx d^m$, which is intractable for large d.

Our preliminary results suggest that a more effective approach is the *local approximation*, using only nearby states to make predictions. To predict $x(t+T)$ we first impose a metric on the state space, denoted by $\| \; \|$, and find the k nearest neighbors of $\mathbf{x}(t)$, i.e., the k states $\mathbf{x}(t')$ with $t' < t$ that minimize $\|\mathbf{x}(t) - \mathbf{x}(t')\|$. We then construct a local predictor, regarding each neighbor $\mathbf{x}(t')$ as a point in the domain and $x(t'+T)$ as the corresponding point in the range. The simplest approach to construct a local predictor is approximation by nearest neighbor, or *zeroth-order approximation*, i.e., $k=1$ and $x_{\text{pred}}(t,T) = x(t'+T)$. A superior approach is the first-order, or linear, approximation, with our taking k greater than d, and fitting a linear polynomial to the pairs $(\mathbf{x}(t'), x(t'+T))$. For convenience we usually treat the range as a scalar, mapping d-dimensional states into one-dimensional values, although for some purposes it is desirable to let the range be d dimensional. The fit can be made in any of several ways; for the work reported here we did least squares by singular-value decomposition. When $k = d+1$ this is equivalent to linear interpolation, but to ensure stability of the solution it is frequently advantageous to take $k > d+1$. We have also experimented with approximation using higher-order polynomials, but in higher dimensions our results are not significantly better than those obtained with first order.

If done in the most straightforward manner, finding a neighboring value in a data base of N points requires the order of N computational steps. This can be reduced to $\log N$ by the partitioning of the data in a decision tree.[9] Furthermore, once the neighbors are found predictors for many times T can be computed in parallel. With these speedups the computations reported here can be done on small computers.

To facilitate the comparison of results, in this paper we simply build the data base from the first part of the time series, and hold it fixed as we make predictions on the remainder. Alternatively, it is possible for one to optimize the performance with respect to either memory or data limitations by dynamically updating the data base.

To evaluate the accuracy of our predictions, we compute the root-mean-square error, $\sigma_\Delta(T) = \langle [x_{\text{pred}}(t,T) - x(t+T)]^2 \rangle^{1/2}$. For convenience we normalize this by the rms deviation of the data $\sigma_x = \langle (x - \langle x \rangle)^2 \rangle^{1/2}$, forming the *normalized error* $E = \sigma_\Delta(T)/\sigma_x$. If $E=0$, the predictions are perfect; $E=1$ indicates that the performance is no better than a constant predictor $x_{\text{pred}}(t,T) = \langle x \rangle$. To estimate E we make as many predictions as we need for reasonable convergence, typically on the order of 1000.

We have applied our method to several artificial and experimental systems, including the logistic map,[3] the Hénon map,[3] the Mackey-Glass delay-differential equation,[10] Taylor-Couette flow,[11] and Rayleigh-Bénard convection in an ^3He-^4He mixture.[12] Our results are summarized in Table I.

An illustration of the performance of the local linear approximation is given in Fig. 1, with use of convection data obtained by Haucke and Ecke.[12] The dimension of this time series is $D \approx 3.1$ (see Ref. 12). To compare with a "standard forecasting technique," we also show

TABLE I. A summary of forecasts using local linear approximation for several different data sets. D is an estimate of the attractor dimension, d is the embedding dimension, N is the number of data points used, $T_{\max}$ is the rough prediction time (Ref. 13) at which the normalized error approached 1, and t_{char} is the "characteristic time" for the time series estimated as the inverse of the mean frequency in the power spectrum. In comparison, a standard forecasting technique (global linear autoregression) gave $T_{\max}$ values that were typically about one characteristic time.

	Differential delay[a] $t_d =$				Rayleigh-Bénard[b] $R/R_c =$			Couette[c] $R/R_c =$		
	17	23	30	100	10.55	12.19	12.24	10.2	12.9	13.7
D	2.1	2.7	3.5	10	2.0	2.6	3.1	2.0	2.7	3.1
d	4	4	6	18	5	6	6	10	6	6
N	10^4	2×10^4	2×10^4	10^5	10^4	10^4	3×10^4	10^4	3×10^4	3×10^4
$T_{\max}$	600	300	300	150	∞	1000 s	100 s	∞	3 s	1 s
t_{char}	50	55	60	65	3 s	2 s	1.5 s	0.5 s	0.4 s	0.1 s

[a]Reference 10.
[b]Reference 12.
[c]Reference 11.

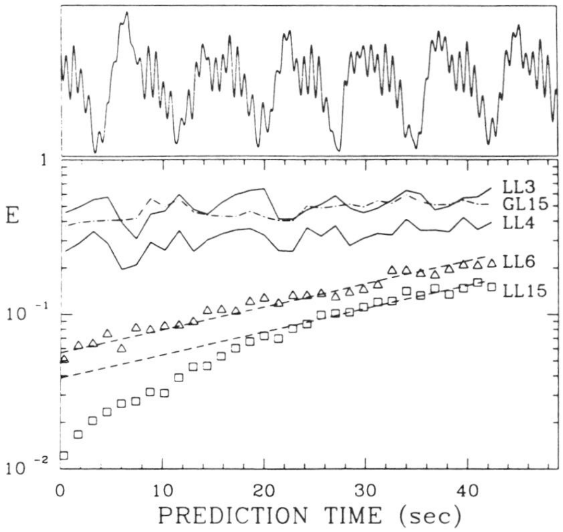

FIG. 1. Top: A time series obtained from Rayleigh-Bénard convection in an ^3He-^4He mixture (Ref. 12), with Rayleigh number $R/R_c = 12.24$, and dimension $D \approx 3.1$. Bottom: The normalized error $E(T) = \sigma_\Delta(T)/\sigma_x$. The top and bottom time scales are the same. We show results for the local linear (LL) and global linear (GL) methods; numbers following the initials indicate the embedding dimension. The dashed lines are from Eq. (2), with k equal to the computed metric entropy from Ref. 12, and C determined by a least-squares fit.

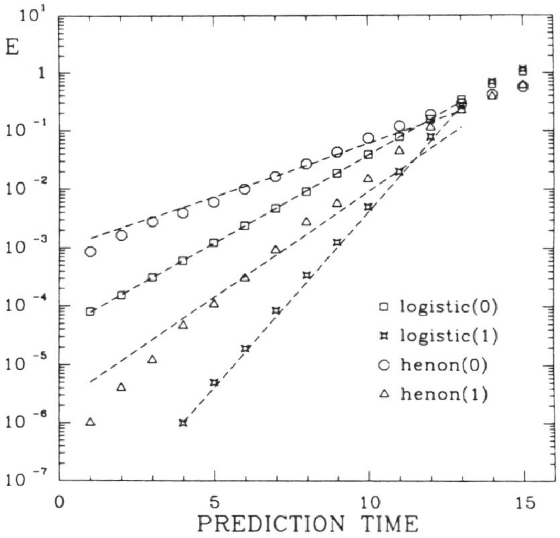

FIG. 2. The normalized error as a function of the prediction time T, for the logistic and Hénon maps (Ref. 3). Results are shown using zeroth-order (0) and first-order (1) local approximation. The dashed lines are from Eq. (2), with a least-squares fit for C and the positive Lyapunov exponents, $k = \log 2$ for the logistic map, and $k \approx 0.42$ for the Hénon map (Ref. 14).

results obtained using a global linear approximation (linear autoregression[2]). When $d < D$, the quality of prediction for the local approximation is roughly the same as that obtained with the global linear approach, but for d sufficiently large the predictions are significantly better.

How well does this local approximation work? This depends on the parameters of the problem, including the number of data points N, the attractor dimension D, the metric entropy h, the signal-to-noise ratio S, and the prediction time T. There are two distinct regimes: If the typical spacing between data points, $\epsilon \approx N^{-1/D} < S^{-1}$, then the forecast is limited by noise. Following Shaw,[3] the average information in a prediction is $\langle I(T)\rangle \approx \ln S - hT$. For a narrowly peaked distribution with $E \ll 1$, to first order $\langle I(T)\rangle$ is proportional to $-\ln E$.

The second regime occurs when $\epsilon > S^{-1}$ and the accuracy of forecasts is limited by the number of data points. In this case, providing d is sufficiently greater than D, in the limit that $E \ll 1$ we propose the following error estimate:

$$E \approx Ce^{(m+1)kT} N^{-(m+1)/D}, \tag{2}$$

where m is the order of approximation and C is a constant. k equals the largest Lyapunov exponent when $m = 0$, and equals the metric entropy otherwise. The arguments leading to this formula are too involved to report here, but they are based on the following facts: The error of interpolation in one dimension is proportional to $f^{(m+1)}\epsilon^{m+1}$; to leading order the mth derivative grows under iteration as the mth power of the first derivative, and the average derivatives along the unstable manifold grow according to the positive Lyapunov exponents. Detailed arguments leading to this result will be presented elsewhere.[14]

The scalings predicted by Eq. (2) are illustrated in Figs. 2 and 3. The exponential increase of $E(T)$ is demonstrated in Fig. 2, for numerical experiments on the logistic and Hénon maps.[3] For the logistic map the slopes are very close to those predicted. For the Hénon map, a least-squares fit gives slopes about 10% greater than those expected with the positive Lyapunov exponent, indicating a possible correction[14] to Eq. (2). For the convection data on Fig. 1, agreement with computed values of the metric entropy[12] is very good as indicated in the figure.

Note that setting $E(T_{max}) = 1$ in Eq. (2) yields $T_{max} = (\ln N)/kD$, independent of the order m. Thus zeroth-order interpolation is less effective than first order, except when E is the order of 1. Equation (2) suggests that higher-order polynomial interpolation might be more effective, but this is difficult in more than two dimensions.

The power-law variation of E with N is illustrated in

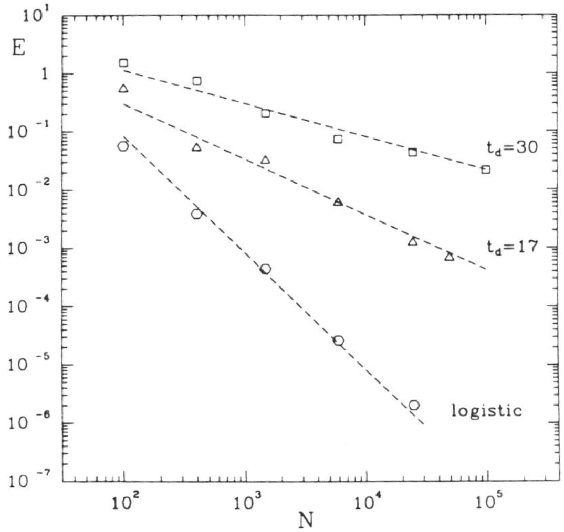

FIG. 3. The normalized error with use of local linear approximation as a function of the number of data points N, at fixed prediction time T. For the logistic map (Ref. 3) $r=4$ and $T=3$; for the Mackey-Glass delay-differential equation (Ref. 10) $T=40$, with two values of the delay parameter t_d. The dashed lines are from Eq. (2), with $D=1$ for the logistic map and $D=D_L$, the Lyapunov dimension, from Ref. 10 for Mackey-Glass equation.

Fig. 3, where we show the behavior for the logistic map and the Mackey-Glass equation. The agreement of the slopes with the expected values of $(m+1)/D$ based on computations of the Lyapunov dimension[10] is quite good. This scaling law breaks down for large N, on account of an approach to the noise floor (when $\epsilon < S^{-1}$).

In addition to the obvious practical applications of forecasting, the construction of approximate models can be a useful diagnostic tool to investigate chaos. Equation (2) demonstrates how forecasting can be used to estimate dimension and entropy; Lyapunov exponents are also easily obtained. Forecasting provides a way to determine whether the resulting numerical values of dimension and entropy are reliable. Ultimately, the ability to forecast successfully with deterministic methods may be the strongest test of whether or not low-dimensionality chaos is present.

At this point, this work is still in a preliminary stage and many possibilities remain to be investigated. In a future paper we plan to compare the local approximation in more detail to some other approaches, in particular recursive-ordinary-differential-equation models such as neural nets.

In this paper we have shown that a forecasting approach based on deterministic chaos can be quite effective in predicting low- to moderate-dimensionality time series. Furthermore, this can be done with reasonable amounts of data and computer time. The most important point is not the specific technique, but rather the demonstration that approaches based on deterministic chaos can be effective; we expect that new and better techniques will emerge rapidly as more attention is focused on this problem. Such methods should be effective for problems in fluid dynamics, control theory, artificial intelligence, and possibly even economics.

We would like to thank Scott Konishi for assistance on numerical procedures, Glenn Carter for assistance in reading nonstandard computer tape formats, and Ian Percival, Josh Deutsch, and particularly Norman Packard for valuable conversations. We would also like to thank Steve Omohundro for making us aware of the literature on decision trees. In a different context, he has recently independently proposed a local linear approximation as an efficient means of performing artificial-intelligence tasks.[15]

---

[a]Permanent address: Department of Physics, University of California at Santa Cruz, Santa Cruz, CA 95064.

[1]G. U. Yule, Philos. Trans. Roy. Soc. London A **226**, 267 (1927).

[2]For example, S. M. Pandit and S.-M. Yu, *Time Series and System Analysis with Applications* (Wiley, New York, 1983).

[3]For reviews see, for example, R. S. Shaw, Z. Naturforsch. **36a**, 80 (1981); J. P. Crutchfield, J. D. Farmer, N. H. Packard, and R. S. Shaw, Sci. Am. **254** (No. 12), 46 (1986); J. Ford, Phys. Today **36** (No. 4), 40 (1983).

[4]For example, G. Mayer-Kress, *Dimensions and Entropies in Chaotic Systems* (Springer-Verlag, Berlin, 1986).

[5]N. Aubry, P. Holmes, J. L. Lumley, and E. Stone, Cornell University Report No. FDA-86-15, 1986 (to be published).

[6]T. A. Bass, *The Eudaemonic Pie* (Houghton-Mifflin, New York, 1985).

[7]N. H. Packard, J. P. Crutchfield, J. D. Farmer, and R. S. Shaw, Phys. Rev. Lett. **45**, 712 (1980); F. Takens, in *Dynamical Systems and Turbulence*, edited by D. A. Rand and L.-S. Young (Springer-Verlag, Berlin, 1981).

[8]D. Gabor, W. P. Wilby, and R. Woodcock, Proc. IEEE **108B**, 422 (1960).

[9]J. L. Bentley and J. H. Friedman, A. C. M. Comput. Surv. **11**, 297 (1979).

[10]M. C. Mackey and L. Glass, Science **197**, 287 (1977); J. D. Farmer, Physica (Amsterdam) **4D**, 366 (1982).

[11]A. Brandtstater, J. Swift, H. L. Swinney, A. Wolf, J. D. Farmer, E. Jen, and J. P. Crutchfield, Phys. Rev. Lett. **51**, 1442 (1983); A. Brandtstater and H. L. Swinney, Phys. Rev. A **35**, 2207 (1987).

[12]H. Haucke and R. Ecke, Physica (Amsterdam) **25D**, 307 (1987).

[13]T_{max} is a crude estimate of our ability to forecast. In some cases $E(T)$ reaches a plateau at a level less than 1. In these cases we estimate T_{max} by extrapolating the initial rate of increase. (In some cases such as GL15 in Fig. 1, it is necessary to expand the T axis to see the initial increase.)

[14]Detailed arguments suggest that there may be a correction to the T scaling in Eq. (2), but we have not yet resolved this.

[15]S. Omohundro, "Efficient algorithms with neural network behavior" (to be published).

NONLINEAR PREDICTION OF CHAOTIC TIME SERIES

Martin CASDAGLI*
School of Mathematical Sciences, Queen Mary College, University of London, Mile End Road, London E1 4NS, UK

Received 2 January 1988
Revised manuscript received 4 June 1988
Accepted 6 June 1988
Communicated by R. S. MacKay

Numerical techniques are presented for constructing nonlinear predictive models directly from time series data. The accuracy of the short-term predictions is tested using computer-generated time series, and comparisons are made of the effectiveness of the various techniques. Scaling laws are developed which describe the data requirements for reliable predictions. It is also shown how to use the models to convincingly distinguish low-dimensional chaos from randomness, and to make statistical long-term predictions.

1. Introduction

There are several well-known algorithms available for the analysis of chaotic time series, see [12, 22, 23, 33, 40] and references therein. The purpose of these algorithms is to calculate geometric and dynamical invariants of an underlying strange attractor, such as information dimension and Lyapunov exponents. If this can be done consistently, then the chaos has in some sense been quantified. This approach leaves much to be desired from an experimentalist's point of view. Firstly, the data requirements for the algorithms can be prohibitive. It can be hard to tell in practice if they have been met, and this may lead to some time series being erroneously diagnosed as chaotic [21]. Secondly, even if the algorithms are successful, the calculated invariants are of limited practical use. It can be deduced that the time series is deterministic, so that it should be possible in principle to build a predictive model for it. The largest Lyapunov exponent gives an indication of how far into the future reliable predictions can be made, and the information dimension gives an indication of how complex such a model must be. However no idea is given as to how to construct the predictive model itself.

*Present address: Center for Nonlinear Studies, Los Alamos National Laboratory, Los Alamos, NM 87545, USA.

It is the purpose of this paper to address the above problems using a radically different approach. Instead of calculating invariants, we attempt to construct a predictive model directly from time series data. A related problem is, given a time series generated from a finite-dimensional strange attractor of a specified partial differential equation, construct an ordinary differential equation describing the dynamics restricted to the strange attractor. Recent ideas on this problem motivated our research [6, 34]. However, we do not require an explicit partial differential equation as an ingredient for our approach. As described in detail in section 2, we treat prediction as an "inverse problem" in dynamical systems. The standard problem in dynamical systems is, given a nonlinear map, describe the asymptotic behaviour of iterates. The inverse problem is, given a sequence of iterates, construct a nonlinear map that gives rise to them. This map would then be a candidate for a predictive model. There are a vari-

From *Physica D* **35** (1989). Reprinted with permission from Elsevier Science Publishers.

ety of numerical techniques for solving the inverse problem which we describe in section 3. These techniques essentially involve interpolating or approximating unknown functions from scattered data points. The performance of the resulting predictive models is evaluated by calculating their "predictor errors"; thus some iterates are used to construct the predictive model, other iterates are used to test the predictions. We have compared the predictor errors of the various techniques using chaotic time series generated by model dynamical systems, such as the Ikeda map [24], the Lorenz equations [26] and the Mackey–Glass delay-differential equation [28]. The results are encouraging; for low-dimensional systems good predictions can be obtained using relatively few data points. On a more theoretical level, we have attempted to understand the results by developing scaling laws describing how many data points are required for a reliable prediction.

Other topics we address are as follows. In section 4 we show that the inverse approach can be used to convincingly distinguish low-dimensional chaos from randomness. We compare this approach to that of the algorithms referred to above, using data from the Mackey–Glass equation as an example. In section 5 we describe two alternative methods for making short-term predictions, and identify the best method by deriving scaling laws, which are tested numerically. In section 6 we show that it is sometimes possible to accurately reconstruct fractal invariant measures given remarkably few data points. Thus long-term prediction may be possible in a statistical sense. We also consider the problem of predicting which parameters in a family of dynamical systems give rise to stable periodic behaviour. Finally, in section 7 we summarize the advantages and limitations of the inverse approach, and mention directions for future research.

When this research was underway, the only literature that we were aware of on an inverse approach to dynamical systems was the work of Barnsley [1], which addresses the problem of data compression rather than prediction. Soon after obtaining encouraging results (in particular those of section 6), we received very recent preprints from three separate groups [9, 15, 25], which use similar approaches to ours. The work of Bayly et al. [2] also contains some related ideas. These preprints contain references to earlier work on reconstruction of one-dimensional Poincaré return maps, and on more general work by statisticians about the nonlinear forecasting of time series (see for example Gabor et al. [20] and Tong [38]). The detailed problems addressed and techniques used by these groups are all different. The research presented in this paper was influenced by these preprints as follows. In section 3 we have performed comparison tests of our originally proposed technique (radial basis functions) with the techniques proposed by other groups. The radial basis function technique was found to have several desirable features. In sections 3 and 5 we were influenced by the work of Farmer and Sidorowich [15] to derive scaling laws. However some of our scaling laws are different, and indicate an improved technique for short-term prediction. In fact Farmer and Sidorowich have now reached similar conclusions about short-term prediction in work done concurrently with ours [16]. Further references will be made to these preprints where appropriate.

In this paper we have concentrated on demonstrating the feasibility of the inverse approach to prediction, and on understanding the numerical results obtained for model dynamical systems. We have not remarked upon potential applications to science and engineering. Some of the references mentioned above contain speculations on this exciting topic, and illustrate their techniques by analysing time series generated by laboratory experiments. There is also convincing statistical evidence that the sunspot cycle and some population dynamics data are better modelled by limit cycles of nonlinear systems (see [38] and references therein) than by the well-known linear ARMA approach [3]. It is a challenge to both develop the techniques presented here further, and identify real world problems to which they apply.

2. The inverse problem

In section 2.1 a theoretical framework is developed for the inverse approach to dynamical systems referred to in the introduction. In section 2.2 a general description is given of how to apply this approach to time series, leaving details to section 3.

2.1. *Theoretical framework*

Let f: $\mathbb{R}^m \to \mathbb{R}^m$ be a smooth map with a strange attractor α, and let $x_n = f^n(x_0)$, $1 \le n < \infty$ be a typical sequence of iterates under f lying on α (i.e. assume that α has an ergodic natural invariant measure ν, and that x_0 is a generic point of ν). An *inverse problem* is to construct a smooth map $\tilde{f}_\infty$: $\mathbb{R}^m \to \mathbb{R}^m$ in terms of the x_n such that $x_{n+1} = \tilde{f}_\infty(x_n)$, $1 \le n < \infty$. This inverse problem has an essentially unique solution: since the x_n are dense in α, and $\tilde{f}_\infty$ is continuous, it follows that $\tilde{f}_\infty|_\alpha = f|_\alpha$. Note that there are no restrictions on the behaviour of $\tilde{f}_\infty$ outside α, because the functions are not required to be analytic. However $\tilde{f}_\infty$ is a perfect predictive model for f on the strange attractor α, and the theory of this inverse problem is trivial.

A much more interesting and realistic inverse problem is to construct a smooth map $\tilde{f}_N$: $\mathbb{R}^m \to \mathbb{R}^m$ using only a finite number of iterates x_n, $1 \le n \le N$, for which $x_{n+1} = \tilde{f}_N(x_n)$, $1 \le n \le N-1$. This inverse problem has several solutions. To see this, let $m = 1$, and f be a map with a strange attractor $\alpha = [a, b]$. To find a solution to the inverse problem, any smooth function $\tilde{f}_N$ can be chosen with a graph which interpolates through the data points (x_n, x_{n+1}), $1 \le n \le N-1$. To pick out an appropriate solution $\tilde{f}_N$, an interpolant is required which is a good approximation to f on the interval $[a, b]$. For $m = 1$, this is a standard problem in approximation theory, and suitable interpolation techniques are well known, for example Lagrange interpolation or splines [30]. For $m > 1$, the inverse problem amounts geometrically to fitting m smooth functions or "hyper-surfaces" $\pi_i \tilde{f}_N$: $\mathbb{R}^m \to \mathbb{R}$ through the data points $(x_n, \pi_i x_{n+1})$, $1 \le n \le N-1$ for $i = 1, \ldots, m$ (π_i denotes the projection onto the ith coordinate). Numerical techniques for doing this will be reviewed in section 3. For a given interpolation technique, $\tilde{f}_N$ is well defined, and will be referred to as a *predictor*.

In order to quantify how good $\tilde{f}_N$ is as a predictor for f, we define the *predictor error* $\sigma(\tilde{f}_N)$ of $\tilde{f}_N$ by (2.1),

$$\sigma^2(\tilde{f}_N) = \lim_{m \to \infty} \frac{1}{M} \times \sum_{n=N}^{N+M-1} \left\| x_{n+1} - \tilde{f}_N(x_n) \right\|^2 / \operatorname{Var}, \tag{2.1}$$

where $\operatorname{Var} \equiv \lim_{M \to \infty} M^{-1} \Sigma_{m=1}^{M} \| x_m - (\lim_{M \to \infty} M^{-1} \Sigma_{m=1}^{M} x_m) \|^2$ is a normalizing factor. The above limit is guaranteed to exist since x_0 is a generic point of the ergodic measure ν. In fact $\sigma^2(\tilde{f}_N) = \int \| \tilde{f}_N - f \|^2 \, d\nu / \operatorname{Var}$, so that the predictor error is a natural distance to define between the predictor $\tilde{f}_N$ and f. The fundamental theoretical problem of the inverse approach, as we see it, is to establish order of convergence properties for $\sigma(\tilde{f}_N)$, which we call *scaling laws*. We refer to a predictor $\tilde{f}_N$ being *of order* γ if $\sigma(\tilde{f}_N) = \mathcal{O}(N^{-\gamma/D})$ as $N \to \infty$, where D is the information dimension of α. It is straightforward to make reasonable conjectures as to what the scaling laws might be, if there is an order of convergence result for the interpolation technique used. Thus if f is twice differentiable and piecewise linear interpolation, an order 2 technique, is used to construct $\tilde{f}_N$, (ignoring the fact that $\tilde{f}_N$ will not then be smooth), then $\sigma(\tilde{f}_N) = \mathcal{O}(h^2)$ as $N \to \infty$, where h is the typical distance between neighbouring points x_n, $1 \le n \le N-1$. But $h = \mathcal{O}(N^{-1/D})$ [13], so that we are lead to conjecture the scaling law (2.2), first proposed by Farmer and Sidorowich [15]:

$$\sigma(\tilde{f}_N) = \mathcal{O}(N^{-2/D}). \tag{2.2}$$

Thus piecewise linear predictors are expected to be of order 2. This may not be true exactly, one difficulty being that there are likely to be large fluctuations in the individual error terms $x_{n+1} - \tilde{f}_N(x_n)$ due to irregular spacing of the points x_n. In fact ongoing work indicates that the arithmetic mean in (2.1) must be replaced by a geometric mean in order for (2.2) to hold. However, if (2.2) is approximately true, it can be used to gain insight into data requirements for reliable predictions. In section 3.2 we will numerically test such scaling laws, and explore to what extent they are relevant in practice.

2.2. *Practical considerations*

We now indicate how the analysis of time series fits into the above framework. Suppose that $s(t)$ is a scalar variable sampled at discrete intervals of time $t = n\tau$, $1 \le n < \infty$, and that the underlying dynamics is that of a strange attractor lying on an I-dimensional invariant manifold of a dynamical system. In such a case we refer to the sequence $s(n\tau)$, $1 \le n < \infty$ as a *chaotic time series*. Then the Takens embedding theorem indicates that for generic τ (in practice this seems to mean almost all τ), and some $m \le 2I + 1$, there is a smooth map f: $\mathbb{R}^m \to \mathbb{R}$ satisfying (2.3) for $1 \le n < \infty$.

$$f(s((n+m-1)\tau), \dots, s(n\tau)) = s((n+m)\tau). \tag{2.3}$$

For a precise formulation of this theorem, which also requires genericity assumptions on the underlying dynamical system and measurement function s, see [36]. A trial value of m above is referred to as an *embedding dimension*. We will refer to the minimal embedding dimension for which (2.3) holds as m^*. Note that m^* is not an invariant of the underlying dynamical system, and is specific to the Takens embedding process. The inverse problem is to compute a smooth function $\tilde{f}_N$ satisfying (2.3) for $1 \le n \le N - m$, given the finite time series $s(n\tau)$, $1 \le n \le N$. This evidently falls into the class of problems described above.

There are, however, considerable practical difficulties to be overcome. We have adopted the following, by no means optimal, strategy as a first approach to these problems. Firstly, to test if the inverse approach has been successful, an estimate is required for $\sigma(\tilde{f}_N)$. This is obtained by using a time series of length $N + M$, and estimating $\sigma(\tilde{f}_N)$ by (2.4):

$$\sigma^2(\tilde{f}_N) = \frac{1}{M} \sum_{n=N}^{N+M-1} \left(s((n+1)\tau) - \tilde{f}_N(s(n\tau), \dots, s((n-m+1)\tau))\right)^2 / \operatorname{Var}, \tag{2.4}$$

where Var is the variance of the time series. Secondly, m^* is not known a priori. This difficulty is overcome by starting the above procedure with $m = 1$, and repeating it for increasing values of m until an acceptably small value is found for $\sigma(\tilde{f}_N)$. This will only be possible if the value of m^* is not too high (see section 4). Thirdly, there is the problem of which interpolation technique to use. This is addressed in detail in section 3. Essentially, we suggest using a technique which gives the smallest value for $\sigma(\tilde{f}_N)$, without using too much computational resources. Finally, there is the important problem of noise pollution of the time series. This is briefly considered in section 4. For reasonably small noise levels the techniques presented here would seem to be adequate. For a different approach, see Tong [38] and Crutchfield and McNamara [9].

We end this section with a simple description of how the above techniques for making predictions can be interpreted directly on the graph of the time series data. Consider the time series in fig. 1, which is generated by the Mackey–Glass equation (3.8) with $\Delta = 30$ and $\tau = 6$. Suppose a prediction is required at time $t = N\tau$, where $N = 500$, and that $m = 6$. If a local linear approximation technique is used to calculate $\tilde{f}_N$ (see section 3), then the prediction is based on finding 14 neighbouring points to $s((N-5)\tau), \dots, s(N\tau)$ of the form

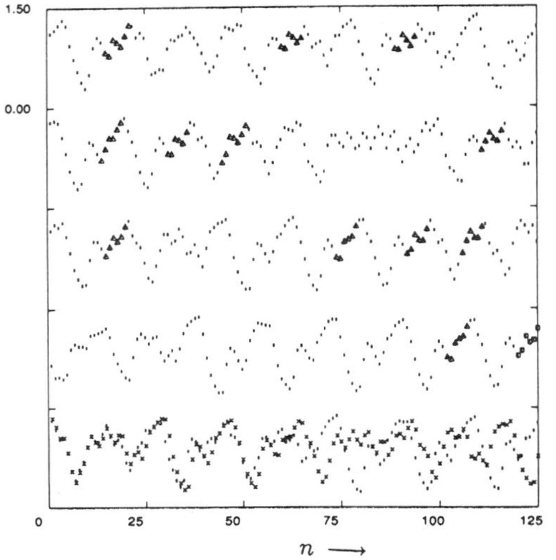

Fig. 1. A time series of length 625 generated by the Mackey–Glass equation with $\Delta = 30$. The crosses represent 125 predictions based on only the first 500 data points using a piecewise linear predictor. The triangles represent neighbouring points to the squares in an embedding space of dimension six, and are used to make the first prediction. The subsequent predictions are made by iterating this procedure using the predicted values as new starting points.

$s((n-5)\tau), \ldots, s(n\tau)$ in $\mathbb{R}^6$. These neighbouring points are illustrated in fig. 1. The prediction for $s((N+1)\tau)$ is simply a weighted average of the $s((n+1)\tau)$. This procedure has been repeated 125 times in fig. 1, using the predicted value as a new starting point. The first 20 predictions are quite good, but as expected the predictions get successively worse. We will return to this example in subsequent sections, in a attempt to quantify how good the predictions are expected to be, and for how long they remain good.

Geometrically, the predictions obtained using the above local technique are based on finding local portions of the curve $s(t)$ in the past, which closely resemble the present, and basing future predictions on what occurred immediately after these past events. It should be emphasised that no attempt was made to fit a function to all of the curve $s(t)$ at once. There is nothing mysterious or deep about the inverse approach to prediction; it has a firm basis in common sense. What we find exciting is that the approach can be automated to give striking results, assuming that there is a sufficient amount of historical data of reasonable accuracy, and that the attractor dimension is not too high. In fact it has been pointed out to us that a similar approach to weather forecasting was suggested several years ago by Lorenz [27], but found to be unsuccessful, presumably due to lack of data and the high dimensionality of the weather.

3. Comparison of approximation techniques

In section 3.1 several techniques are reviewed for approximating functions from scattered data points, listing some of their known advantages and limitations. In section 3.2 the predictor errors of these techniques are compared using a variety of dynamical systems as case studies.

3.1. *Review of approximation techniques*

Although we stated above that a predictor $\tilde{f}_N$ must be an interpolant ($\tilde{f}_N(x_n) = x_{n+1}$, $1 \leq n \leq N-1$), in this section we also allow $\tilde{f}_N$ to be an *approximant* ($\tilde{f}_N(x_n) \approx x_{n+1}$, $1 \leq n \leq N-1$). This widens the range of numerical techniques available, and indeed approximants may be more suitable for noise polluted time series. We also relax the requirement that $\tilde{f}_N$ be a smooth function. There are then several techniques at our disposal, which we divide into three categories: global techniques, local techniques and radial basis functions. Of course there are many techniques which we have not considered.

Global techniques

In this approach the coordinate functions $\pi_i \tilde{f}_N$: $\mathbb{R}^m \to \mathbb{R}$, $i = 1, \ldots, m$ are chosen from a standard function basis. For example for a *polynomial predictor* $\pi_i \tilde{f}_N$ is chosen to be a polynomial in m

variables of degree at most d, d being given. The $\binom{m+d}{m} \equiv (m+d)!/m!d!$ free parameters of the polynomial are chosen to minimize (3.1):

$$\sum_{n=1}^{N-1} \left(\pi_i x_{n+1} - \pi_i \tilde{f}_N(x_n)\right)^2. \tag{3.1}$$

This is a linear least-squares problem, which can be solved efficiently by the QR algorithm [39]. In this way an optimal polynomial predictor of a given degree d is obtained. This technique has been considered in [9, 15, 20].

Advantages. The resulting predictor $\tilde{f}_N$ is in a "standard form", and is conceptually easy to think about. Also, $\tilde{f}_N$ is guaranteed to converge to f as N and d increase, by the Weierstrass approximation theorem.

Limitations. There are a very large number of free parameters to be chosen for moderately large m and d. The QR algorithm will then require extensive computational resources, and becomes unmanageable on standard workstations for $\binom{m+d}{m} \approx 500$. Also if N is large (it can typically range from 10^4 to 10^5 in laboratory experiments), then the $\binom{m+d}{m} \times N$ matrix input into the QR algorithm may exceed storage capacity. Finally, there are no known order of convergence results for $m > 1$, and polynomials of high degree have an undesirable tendency to oscillate wildly.

Following Bayly et al. [2], we have also considered *rational predictors*, where $\pi_i \tilde{f}_N$ is chosen to be a ratio p/q of polynomials, and the coefficients of p and q are chosen to minimize (3.2):

$$\sum_{n=1}^{N-1} \left(\pi_i x_{n+1} q(x_n) - p(x_n)\right)^2. \tag{3.2}$$

We fix the constant term of q to be equal to 1, to remove a degeneracy in this minimization problem. It is again a linear least-squares problem to determine the coefficients of p and q. There are known to be distinct advantages of rational approximation over polynomial approximation if $m = 1$, but in higher dimensions little is known. We have not tested any other global techniques.

Local techniques

To construct a *local predictor*, instead of "globally" fitting the graph of $\pi_i \tilde{f}_N$ to all the data points at once, the graph of $\pi_i \tilde{f}_N$ is constructed by piecing together "local" graphs. Following [15], we do this as follows. If the value of $\pi_i \tilde{f}_N$ is required at a point x, the k nearest neighbours of $x_1, \ldots, x_{N-1}$ to x are found, and a polynomial of degree at most d (d small) is fitted through the corresponding data points, by least squares using the QR algorithm as above. k must be at least as big as $\binom{m+d}{m}$ for the least-squares problem to have a unique solution. In practice, choosing k larger than this number may improve the predictor error $\sigma(\tilde{f}_N)$; in the numerical experiments presented below we somewhat arbitrarily chose $k = 2\binom{m+d}{m}$. If $d = 1$, a piecewise linear approximation is obtained; if in addition $k = m + 1$, a piecewise linear interpolant is obtained (but this is not true linear interpolation, a triangulation of the data points is required to achieve continuity). Variations of this technique have been considered in [9, 15, 38], and is implicit in the work of [12, 33].

Advantages. The algorithm which implements the above procedure consumes much less computational resources than that for global approximation. This feature can be improved even more if data trees are used to organize nearest neighbour searches, however the implementation is rather involved (see [15] and references therein). It is also straightforward to conjecture scaling laws for the predictor error. Thus using Taylor's theorem, the scaling law (3.3) is expected for a degree d local predictor if f is $d+1$ times differentiable. This scaling law was first proposed by Farmer and Sidorowich [15]:

$$\sigma(\tilde{f}_N) = \mathcal{O}(N^{-(d+1)/D}). \tag{3.3}$$

Limitations. The resulting predictor $\tilde{f}_N$ is in general a discontinuous function, and is not in any standard form that we are used to thinking about. The discontinuities give rise to some undesirable behaviour when long-term iterates are computed for $\tilde{f}_N$ (see section 6).

Radial basis functions

This technique is a global interpolation technique with good localization properties. It provides a smooth interpolation of scattered data in arbitrary dimensions, and has proven useful in practice [17]. There are essentially no spline techniques available in this situation. Another technique which seems to have some features in common with this is approximation by *neural networks* [25], but the theory behind these is much less well understood, and the algorithms to solve the resulting nonlinear least-squares problems take much longer to run. We now briefly review what is known abut radial basis functions; for more information see the review by Powell [31]. In this approach $\pi_i \tilde{f}_N$ is chosen to be of the form (3.4):

$$\pi_i \tilde{f}_N(x) = \sum_{n=1}^{N-1} \lambda_n \phi(\|x - x_n\|) + \sum_{n=1}^{\hat{d}} \mu_n p_n(x). \tag{3.4}$$

Here ϕ is a function from $\mathbb{R}^+$ to $\mathbb{R}$, $\|\cdot\|$ denotes the Euclidean norm on $\mathbb{R}^m$ and $\{p_n; n = 1, \ldots, \hat{d}\}$ is a basis of the space of polynomials of degree at most d from $\mathbb{R}^m$ to $\mathbb{R}$, d being given. The functions $\phi(\|x - x_n\|)$ are called *radial basis functions*. Frequently the polynomial term is not included. In this case the coefficients λ_n are chosen to satisfy the interpolation conditions (3.5):

$$\pi_i \tilde{f}_N(x_n) = \pi_i(x_{n+1}), \quad n = 1, \ldots, N-1. \tag{3.5}$$

If the x_n are distinct (which they are for the examples here), then it follows from a result of Michelli [29] that the linear system (3.5) has a unique solution if $\phi(r)$ has the form $\phi(r) = (r^2 + c^2)^{-\beta}$, where $\beta > -1$ and $\beta \neq 0$. This interpolant evidently has localization properties if $\beta > 0$, but surprisingly this can also be true for increasing functions $\phi(r)$ if the polynomial term is added. In this case it is also necessary to add the constraints (3.6):

$$\sum_{n=1}^{N-1} \lambda_n p_j(x_n) = 0, \quad j = 1, \ldots, \hat{d}. \tag{3.6}$$

Under similar hypotheses to the above, the resulting system of linear equations for the coefficients λ_n and μ_n can be proven to have a unique solution [29]. Moreover, if $m = 1$, $\phi(r) = r^3$ and $d = 1$, then it can be shown that the resulting interpolant $\pi_i \tilde{f}_N$ is identical to that obtained using cubic splines [31]. Also, for any $m > 1$, if $\phi(r) = r^2 \log r$ and $d = 1$, then the interpolant $\tilde{f}_N$ minimizes the "bending energy" of a thin plate constrained to go through the data points [10]. Thus some increasing radial basis functions can be considered to be generalizations of splines to higher dimensions, and hence have good localization properties.

Advantages. The technique is very easy to implement numerically, as the algorithm is essentially independent of the dimension m (if $d \leq 1$). We used an IMSL routine for solving the indefinite symmetric systems of linear equations (3.5) and (3.6). The resulting predictor $\tilde{f}_N$ is a smooth interpolant, and thus elegantly solves the inverse problem as originally stated. For data points regularly spaced on a grid there are some order of convergence results [32].

Limitations. If standard algorithms are used the solution of the linear systems (3.5) and (3.6) is very costly for large N, and becomes unfeasible for $N \approx 500$ on standard workstations. This can be avoided by "localizing" the technique, thus sacrificing smoothness. In section 3.2 this was done using the 50 nearest neighbours to a point to compute localized radial basis predictors. A more elegant method is to exploit the structure of the linear systems (3.5) and (3.6) to produce a faster and more stable iterative algorithm. Such ideas are also relevant for using radial basis functions to smooth noise polluted data, and this has proven successful in low-dimensional examples [11], but the algorithms are somewhat involved, and we have yet to implement them. However, ongoing research indicates that more straightforward least-squares approaches can be used to construct radial basis functions which smooth noise polluted data.

3.2. *Comparison tests*

In table I we have summarized computations of predictor errors for some model dynamical systems using the above techniques. We chose $M = 10^3$ to estimate the sums (2.1) and (2.4). The computations were performed on a Sun3 workstation. The results are to some extent dependent on the precise distribution of the N data points used to construct the predictors, but other distributions give predictor errors of the same order of magnitude. The dynamical systems were chosen in order of increasing complexity as follows.

Firstly, we chose the Ikeda map [24] (3.7):

$$f(x, y) = (1 + \mu(x\cos t - y\sin t),\ \mu(x\sin t + y\cos t)), \tag{3.7}$$

where $t = 0.4 - 6.0/(1 + x^2 + y^2)$, with the parameter $\mu = 0.7$ (this is $Ikeda^1$ in table I). This map was chosen because it has a complicated functional form which is not of the type used in any of the above approximation techniques, but is also well behaved and slowly varying. The parameter $\mu = 0.7$ was chosen to give a strange attractor with relatively simple geometry (see section 6). We also considered the fourth iterate of the Ikeda map as an example of a dynamical system with the same attractor but much more complicated dynamics ($Ikeda^4$ in table I). Secondly, as an example of a differential equation, we chose the Lorenz equations [26], with the usual parameters, and sampling rates $\tau = 0.05$ and 0.20 ($Lorenz^{0.05}$ and $Lorenz^{0.20}$ in table I). The resulting time-τ map is quite close to a quadratic map for $\tau = 0.05$, but much more complicated for $\tau = 0.20$. Finally, we chose the Mackey–Glass delay-differential equation [28] (3.8) with parameters $\Delta = 17$, 30 and 100, and embedding dimensions $m = 4$, 6 and 17 respectively (MG_Δ in table I).

$$\frac{dx(t)}{dt} = -0.1x(t) + \frac{0.2x(t-\Delta)}{1 + x(t-\Delta)^{10}}. \tag{3.8}$$

The initial condition for (3.8) was taken to be $x(t) = 0.9$ for $0 \le t \le \Delta$. The sampling rate $\tau = 6$ was used in this and subsequent sections. This example was chosen to facilitate comparisons with other work [15, 25], and can exhibit high-dimen-

Table I
Estimated values of $\log_{10}\sigma(\tilde{f}_N)$ are tabulated for predictors $\tilde{f}_N$ calculated using a variety of techniques, from time series generated by various dynamical systems. The degrees used for the best polynomial and rational predictors are shown in small figures next to the values of the predictor errors. Also tabulated is the information dimension D, estimated by the Kaplan–Yorke formula [13].

	D	N	Poly.	Rational	Loc.$^{d=1}$	Loc.$^{d=2}$	Radial	N.Net
$Ikeda^1$	1.32	100	-5.57^{12}	-8.01^{8}	-1.71	-2.34	-2.95	
$Ikeda^4$	1.32	500	-1.05^{14}	-1.39^{14}	-1.26	-1.60	-2.10	
$Lorenz^{0.05}$	2.0	500	-4.62^{6}	-4.30^{3}	-2.00	-3.48	-3.54	
$Lorenz^{0.20}$	2.0	500	-0.61^{5}	-2.16^{6}	-0.88	-0.87	-1.35	
MG_{17}	2.1	500	-1.95^{7}	-1.14^{2}	-1.48	-1.89	-1.97	-2.00
MG_{30}	3.6	500	-1.40^{4}	-1.33^{2}	-1.24	-1.42	-1.60	-1.50
MG_{30}	3.6	9×10^4			-1.89	-1.96	-2.04[a]	
MG_{100}	10	9×10^4			-0.99		-1.12[a]	

[a] Localized.

sional chaos [14]. Note that all of the above dynamical systems have been derived to model physical processes, so that our results should be relevant to the analysis of at least some time series generated by natural processes.

We used the following techniques to construct predictors. Firstly, we used both polynomial and rational approximation to construct global predictors. The degrees of the polynomials were chosen to give the smallest predictor error. This was done by systematically increasing the degrees of the polynomials until results deteriorated. Secondly, we used local approximation with $d=1$ and $d=2$ to construct piecewise linear and piecewise quadratic predictors. Finally, we used a radial basis function technique with $\phi(r)=r^3$ and $d=1$; other choices for $\phi(r)$ and d gave comparable results. Also included in table I are results from a neural network technique taken from Lapedes and Farber [25].

From a scrutiny of table I, some tentative observations as to the relative merits of the various techniques can be made. Firstly, all the techniques give acceptable results, even in the case of MG_{100}, which has a high-dimensional attractor. Secondly, the performance of the global predictors seems highly system-dependent. Strikingly good results are obtained for the simple systems $Ikeda^1$ and $Lorenz^{0.05}$, but these deteriorate badly for $Ikeda^4$ and $Lorenz^{0.20}$. Thirdly, piecewise quadratic predictors seem superior to piecewise linear predictors. However, surprisingly, these advantages are not so striking for complicated systems with large numbers of data points (this was also observed in the case of local predictors by Farmer and Sidorowich [15]). Finally, radial basis predictors seem superior to piecewise quadratic predictors, and perform roughly like neural network predictors for the limited data we have in that case.

In order to refine and understand the above observations, we have investigated how the predictor error $\sigma(\tilde{f}_N)$ scales with the number of data points N. Firstly, we considered an artificial situation in which the data points x_n were chosen to be evenly spaced on a two-dimensional grid $[0.0, 1.5] \times [-1.0, 0.5]$. The predictor error was evaluated for the Ikeda map with parameter $\mu=0.7$, using 900 points on a smaller grid $[0.13, 1.37] \times [-0.87, 0.37]$ to compute the sum (2.1). The results are shown in fig. 2(a). In fig. 2(b) corresponding results are shown for the second iterate of the Ikeda map. The scaling laws for the piecewise linear and quadratic predictors agree with eq. (3.3). The scaling law for the radial basis predictor appears to be of order 6. In one dimension it would have been of order 4, so the situation has improved in higher dimensions, as expected from the results of Powell [32]. The localized radial basis predictor ultimately obeys the same scaling law as the piecewise linear predictor, though is superior due to initially obeying the same scaling law as the global radical basis predictor. The predictor error for the polynomial technique appears to depend exponentially on N, though the exponent is highly system-dependent.

Secondly, we consider the natural situation in which the data points x_n lie on a strange attractor. Results for the Ikeda map with parameter $\mu=0.7$ are shown in fig. 2(c). There are evidently some problems with the convergence of the predictor error for large N. This does not appear to be due to numerical errors in the least-squares fitting and matrix inversion algorithms, although for the localized radial basis predictors the resulting matrices were too ill conditioned to invert for large N. Rather, the lack of convergence seems to be due to very infrequent bad predictions, and the problem disappears if 10% of the worst predictions are ignored, as is shown in fig. 2(d). Ongoing research indicates that it is possible to anticipate in advance where most of these bad predictions will occur, so throwing away 10% of the worst predictions is not unrealistic. Alternatively, as mentioned in section 2.1, there are theoretical reasons for replacing the arithmetic mean (2.1) by a geometric mean. If this is done then since the infrequent bad predictions have a much less severe effect on the geometric mean, throwing away 10% of the worst predictions is unnecessary; however we have not implemented this approach here. The

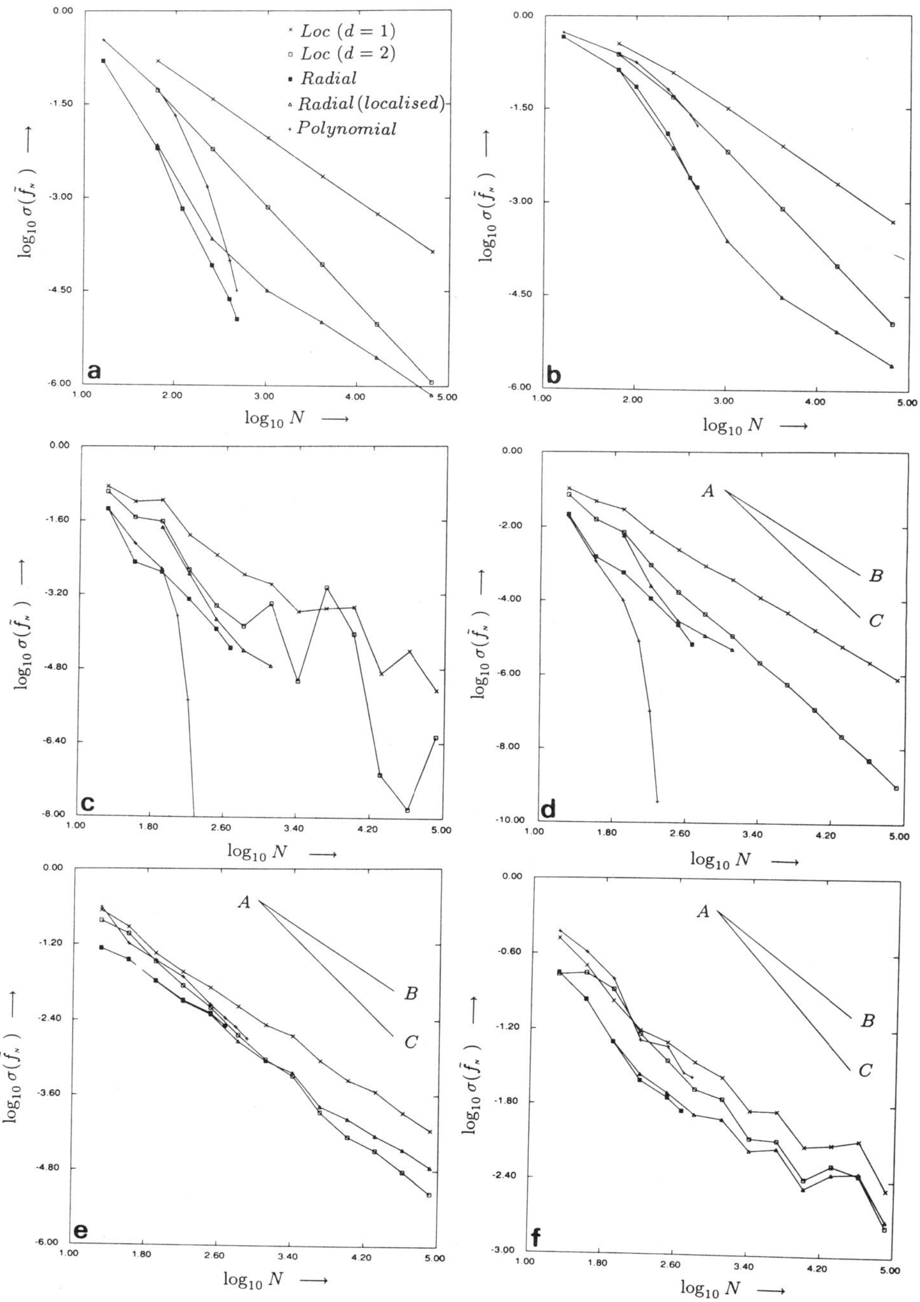

Fig. 2. Predictor errors calculated for a variety of different techniques. (a) Ikeda map with $\mu = 0.7$ on a grid. (b) Second iterate of Ikeda map with $\mu = 0.7$ on a grid. (c) Ikeda map with $\mu = 0.7$. (d)–(f) 10% of the worst predictions are ignored. (d) Ikeda map with $\mu = 0.7$. (e) Mackey–Glass equation with $\Delta = 17$. (f) Mackey–Glass equation with $\Delta = 30$. The slopes AB and AC represent expected scaling laws for predictors of order 2 and 3 respectively.

scaling laws are as expected, except for the global radial basis predictor. This now appears to be of order about 3, which indicates that global approximation on a fractal may be more difficult than global approximation on a grid. On the other hand, the performance of the polynomial predictor does not seem to be impaired.

Finally, results for the Mackey–Glass equation with parameters $\Delta = 17$ and 30 are shown in figs. 2(e) and 2(f) respectively. We have also ignored the worst 10% of the predictions in computing the predictor errors. The scaling laws for the local predictors are as expected, except for the piecewise quadratic predictor in fig. 2(f). Ongoing research indicates that much better agreement to the anticipated scaling law is obtained using a geometric mean to compute the predictor error, as mentioned above. The scaling law for the radial basis predictor seems to only be of order about 2. The polynomial predictor seems to behave roughly like the piecewise quadratic predictor, for the limited range of values of N tested. Its superiority is presumably lost because the degree of the polynomial cannot be chosen to be large at these values of the embedding dimension.

We summarize the above results as follows. For small values of N, radial basis predictors are superior to local predictors. For large values of N, there is little to choose between the local predictors. However, if 10% of the worst prediction are ignored, then piecewise quadratic predictors are superior to localized radial basis predictors, which are superior to piecewise linear predictors. Polynomial predictors can give strikingly superior results for very low-dimensional systems, but do not perform so well otherwise. The results for the local predictors can be understood in terms of scaling laws, except that for high-dimensional systems, the scaling law for piecewise linear predictors is more robust than that for piecewise quadratic predictors. We do not have a good understanding of the scaling laws for radial basis predictors, and it remains a challenge to investigate why results deteriorated when points were sampled from a fractal attractor instead of a regular lattice. The conclusions presented here are consistent with concurrent work of Farmer and Sidorowich on piecewise linear and quadratic predictors [16].

4. Detecting chaos in a time series

In this section we describe how the inverse approach can be used to detect chaos in a time series, and compare this to other methods which have proven useful in practice. For related ideas see [2, 8, 9, 15, 16].

Having settled on a suitable technique for constructing predictors from section 3, the method is straightforward. Given a time series $s(n\tau)$, $1 \leq n \leq N + M + 1$, as in section 2.2, predictors $\tilde{f}_{N,m}$ are constructed satisfying (2.3), for a range of values of trial embedding dimensions m. If the time series is chaotic, it is expected that the predictor errors $\sigma(\tilde{f}_{N,m})$ will suddenly decrease to a value close to zero as m is increased to the correct minimal embedding dimension m^*, and remain close to zero for values of m above m^*. If the time series is random, no such decrease should be observed.

To illustrate this method, we chose chaotic time series generated by the Mackey–Glass equation with parameters $\Delta = 30$ and 60 for which $D = 3.6$ and 6.4 respectively. These values of D were estimated by Farmer [14] using the Kaplan–Yorke formula [18]. The 6.4-dimensional time series "looks" much more chaotic than that illustrated in fig. 1. For convenience, we calculated the predictors $\tilde{f}_{N,m}$ using a localized radial basis technique with $\phi(r) = (r^2 + c^2)^{-1}$, c equal to the standard deviation of the data, no polynomial term in (3.4) and 50 nearest neighbours to solve eqs. (3.6). The predictors $\tilde{f}_{N,m}$ were evaluated for several different values of N, so that data requirements for the method could be investigated. The resulting predictor errors $\sigma(\tilde{f}_{N,m})$ were estimated taking $M = 100$ in (2.4), and are displayed in figs. 3(a) and 3(c). It is apparent that minimal embedding dimensions $m^* = 6$ and 10 have been successfully identified. Moreover the slopes of the

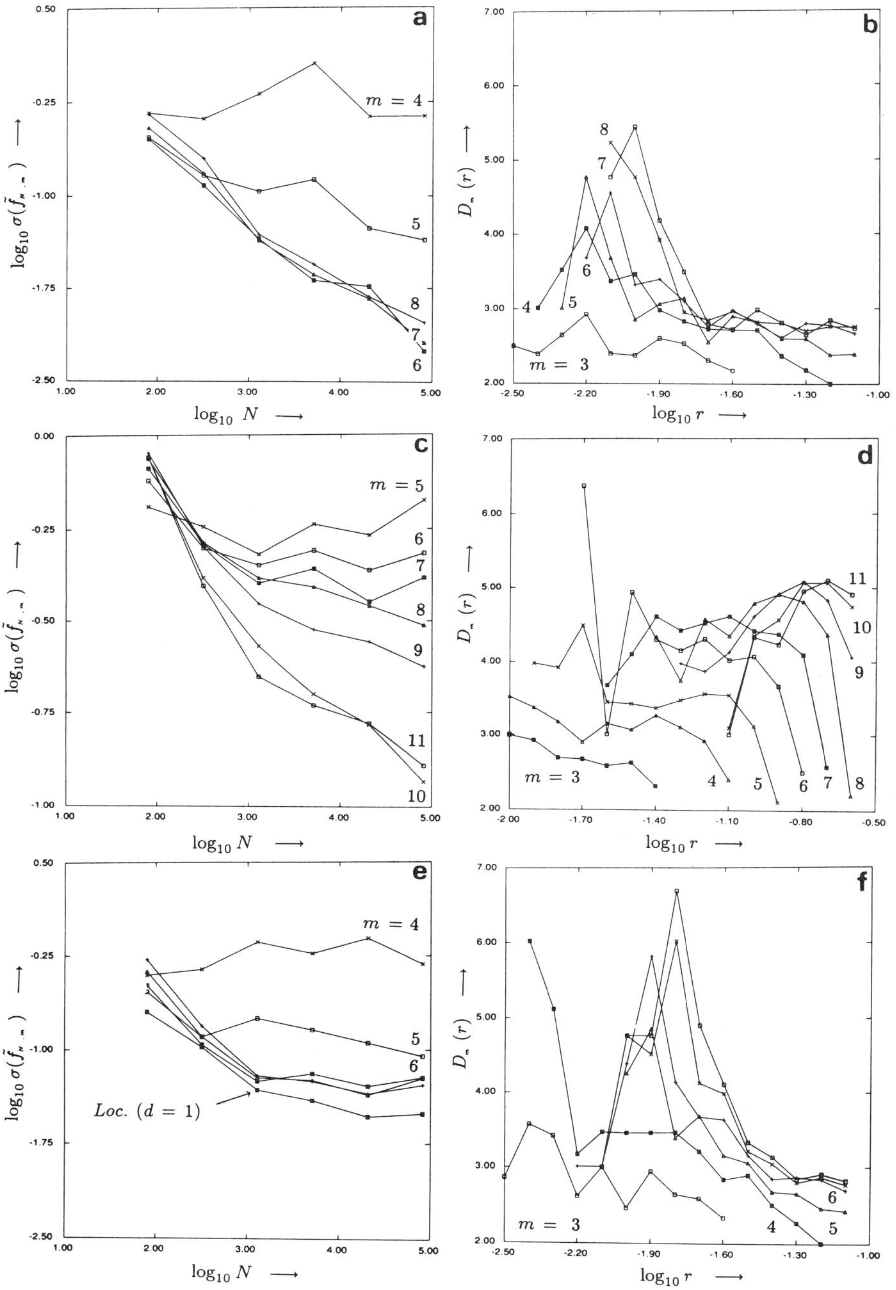

Fig. 3. Predictor error and dimension estimates at various embedding dimensions m. (a), (b) for the Mackey–Glass equation with $\Delta = 30$. (c), (d) for the Mackey–Glass equation with $\Delta = 60$. (e), (f) for the Mackey–Glass equation with $\Delta = 30$ and noise added.

lines at these values of m are consistent with the scaling law $\sigma(\tilde{f}_{N,m}) = \mathcal{O}(N^{-2/D})$, which was expected from the results of section 3.2. However, note that as in dimension calculations (see below), the slope varies somewhat over the level to which the attractor is sampled. It is not clear to us whether better estimates for the dimension of attractors can be obtained by exploiting the scaling law (2.1). We repeated the above procedure for a randomly generated time series, and observed no values of m giving rise to decreasing predictor errors, as expected.

To compare the above method to well-known methods of dimension calculations, we have performed correlation dimension calculations on the same time series using an implementation of the Grassberger–Procaccia algorithm [22] which involves considerably less computations, suggested to us by Caputo [7], as follows. At a given trial embedding dimension m, define the functions $C_m(r, i)$ by (4.1):

$$C_m(r, i) = (\text{number of } x(j) \text{ such that } \|x(i) - x(j)\| \leq r), \quad (4.1)$$

where $x(n) \equiv s((n+m-1)\tau), \ldots, s(n\tau) \in \mathbb{R}^m$. Define the function $C_m(r)$ by

$$C_m(r) = M^{-1} \sum_{i=1}^{M} C_m(r, n_i),$$

where the n_i are M numbers chosen at random between 1 and N. Then the correlation dimension is estimated by plotting $\log C_m(r)$ against $\log r$. If the time series is chaotic, and the value of m is larger than D, the slope of the graph converges to D as N tends to infinity and r tends to zero. The values $D_m(r)$ of these slopes were plotted in figs. 3(b) and 3(d) for the time series considered above, taking $N = 8 \times 10^4$ and $M = 100$. The correlation dimensions were crudely estimated by eye at about 3 and 5 respectively. Note that these values are significantly less than the information dimension as estimated using the Kaplan–Yorke formula. This has also been observed in [23], and may be partly due to the Kaplan–Yorke formula over estimating the information dimension.

We now summarize the advantages of the inverse approach over the dimension calculations. Firstly, we can be confident that the estimated values of the predictor errors presented in fig. 3 are more reliable than those of the estimated values of the dimension, because the form of the scaling law (2.2) allows for data requirements to be easily investigated. This should be compared to the following scaling law of [19] for dimension calculations. Suppose the exact correlation dimension of an attractor is D, and the dimension estimate using N points as above is D_N. Then it is expected that $|D - D_N| = \mathcal{O}(N^{-1/D})$. One problem with this scaling law is that it is difficult to use, because the value of D on the left-hand side is an unknown. Thus a simple log–log plot as for the predictor error will not reveal if the law is satisfied. Indeed it is only very recently that such data requirement scaling laws have been investigated numerically [35] (these results indicate that there can be other problems with formulating such scaling laws). Also, the above scaling law indicates more stringent data requirements for dimension calculations than for an order 2 predictor technique, such as piecewise linear approximation. We are thus lead to expect that the latter should in some sense be able to reliably detect chaos of twice the dimension as the well-known dimension calculations, although we have not succeeded in making this idea precise. Secondly, the minimal embedding dimension m^* can be obtained using the predictor method, but is not accessible to dimension calculations, for which any value of m larger than D should in theory lead to an acceptable estimate for D [13]. The value of m^* is also in general different to the topological dimension of the attractor as obtained using the algorithms of [4] or [19]. Finally, from a pragmatic point of view it should be more convincing and useful to have a predictive model for a time series which gives predictions to within a known tolerance (the predictor error), than to have a good estimate of the information dimension of an underlying attractor.

Other well-known methods for detecting chaos in a time series are Lyapunov exponent calculations. There are two alternative algorithms to perform such calculations, due to Wolf et al. [40], and Eckmann et al. [12] and Sano and Sawada [33]. These methods are more closely related to the predictor method than the dimension calculations, in the sense that dynamical information rather than geometric information is extracted from the time series. However, we expect that the Lyapunov exponent methods will suffer from most of the disadvantages of the dimension calculations mentioned above. For example, it is also difficult to test for a data requirement scaling law in practice (we are aware of none that have been published), and it is not clear how much practical use a Lyapunov exponent calculation is without the ability to make predictions. On a more theoretical level, we expect the data requirements for reliable Lyapunov exponent calculations to be more severe than for predictor error calculations, though we have not tested this numerically. In the case of the algorithm due to Wolf et al., it can be argued that the error in the Lyapunov exponent estimates scale as $N^{-1/D}$ [40]. In the case of the algorithm due to Eckmann et al. and Sano and Sawada, a local linear approximation is used to estimate the derivative of an underlying map (rather than the map itself), so that we expect the corresponding error to scale as $N^{-1/D}$ (rather than $N^{-2/D}$), because one order of approximation is lost in approximating a derivative. Thus the scaling laws are very similar to those for dimension calculations, and inferior to (2.2).

There are, however, some limitations to the predictor method. An obvious limitation is that the sampling rate τ must be kept reasonably small, so that the function being approximated does not have wildly varying slopes (note that this may not be possible in practice for intermittent phenomena exhibiting fluctuations on two time scales [37]). This remark is also applicable to Lyapunov exponent calculations, but should be less relevant in dimension calculations. Also, we do not claim to be able to distinguish high-dimensional chaos from randomness with the predictor method. However we have not fully investigated over what range of dimensions the method is limited to. Ideally, we propose that all of the above methods should be used to analyze experimental time series. At the very least, as suggested by [15], the predictor method should provide a valuable consistency check on the more well-known methods, which have sometimes been applied beyond their known domains of applicability.

To end this section, we briefly consider the problem of detecting chaos in noise-polluted time series. This problem has been addressed in detail by [9] using a different technique. To illustrate our more simple-minded approach, we took the time series used to produce figs. 3(a) and 3(b), and added a sequence of independent normally distributed random numbers, to simulate measurement noise. The standard deviation of the noise was roughly 0.0042, and the standard deviation of the resulting signal was roughly 0.28, giving a noise to signal ratio of roughly $10^{-1.8}$. In figs. 3(e) and 3(f), the analogues of figs. 3(a) and 3(b) are shown for the noise-polluted data, except that in fig. 3(e) a local linear technique is also shown for the minimal embedding dimension $m^* = 6$. In fig. 3(e) a noise floor is reached, below which the predictor error does not fall. We expect this noise floor to be proportional to the noise to signal ratio. The noise floor is roughly $10^{-1.5}$ for the local linear predictor, and roughly $10^{-1.3}$ for the radial basis predictor. This is presumably because the least-squares fitting used to construct the local linear predictor averaged out some of the noise. In fig. 3(b) the noise causes the estimated dimension to change significantly below about $r = 0.025$, which is roughly 6 times the standard deviation of the noise. We conclude that, for this example, the effects of noise on the predictor error calculations are somewhat less severe than for the dimension calculations. Note that we have not yet attempted to optimize the performance of noise smoothing predictors. For the local linear predictors in embedding dimension m, better results may be obtained by choosing more than $2(m+1)$ neighbours to compute the local linear maps in the presence of large noise levels. Also as mentioned in section 3.1 there are techniques which can be used to construct radial basis functions for smoothing noise-polluted data. There is evidently much further work to be done to decide on the best smoothing technique when confronted with experimental or real world data.

5. Short-term prediction

In this section two alternative methods for making short-term predictions are described, and the best method is identified by deriving and testing scaling laws.

The objective is to construct a short-term predictor $\tilde{f}_N^{(i)}$ from data x_n, $1 \le n \le N$, for which $x_{n+i} = \tilde{f}_N^{(i)}(x_n)$, $1 \le n \le N-i$. One method of achieving this which we refer to as the *iterative method*, is to construct a "one step" predictor $\tilde{f}_N$ as usual, and take $\tilde{f}_N^{(i)}$ to be the ith iterate of $\tilde{f}_N$, $\tilde{f}_N^{(i)} \equiv (\tilde{f}_N)^i$. The predictor error $\sigma(\tilde{f}_N^{(i)})$ of $\tilde{f}_N^{(i)}$ is defined in the usual way. Since errors grow under iteration at a rate equal to the largest Lyapunov exponent of f, $\lambda_{\max}$, we expect the scaling law (5.1) to hold:

$$\sigma\left(\left(\tilde{f}_N\right)^i\right) = e^{i\lambda_{\max}} \sigma\left(\tilde{f}_N\right). \tag{5.1}$$

There is a second method of constructing short-term predictors, due to Farmer and Sidorowich [15], which we refer to as the *direct method*. Their method is to construct a predictor $\tilde{f}_N^{(i)}$ which satisfies $x_{n+i} = \tilde{f}_N^{(i)}(x_n)$, $1 \le n \le N-i$ directly. Thus the ith iterate of f is approximated directly, $\tilde{f}_N^{(i)} \equiv (\tilde{f^i})_N$. If $\tilde{f}_N$ is a degree d local predictor, the scaling law (5.2) is expected to hold:

$$\sigma\left(\left(\tilde{f^i}\right)_N\right) = e^{i(d+1)h} \sigma\left(\tilde{f}_N\right), \tag{5.2}$$

where h is the metric entropy of f [15]. Since the metric entropy is equal to the sum of the positive Lyapunov exponents, these scaling laws strongly indicate that the first method will give better predictions. In fact the arguments behind (5.2) are somewhat more involved than for (5.1), and we have found that in some circumstances (5.2) overestimates the errors for direct prediction [16]. However, we do not expect that the metric entropy can in general be replaced by a quantity smaller than the largest Lyapunov exponent in (5.2), so that the iterative method is still expected to give better predictions.

We have performed the following numerical tests to see to what extent the above arguments hold in practice. In fig. 4(a) results for various short-term predictors of the Ikeda map with parameter $\mu = 0.7$ are shown. It is clear that the iterative method is superior. In fig. 4(b), in which 10% of the worst predictions are ignored, the scaling laws are also reasonably well obeyed. For the iterative method, the slopes are about 10% greater than expected. We believe that this is due to fluctuations in the local expansion rates on the attractor. The largest expansion rates are overemphasized, both by taking a root mean square average to compute the predictor error, and because there tend to be fewer data points available in regions of large expansion to make reliable predictions with. For the direct method the slopes are about 25% greater than expected. This effect was also observed by Farmer and Sidorowich [15].

In figs. 4(c) and 4(d) results for the Mackey–Glass equation with parameter $\Delta = 30$ are shown. The iterative method is again superior, except in the case of the polynomial predictor, which eventually blows up under iteration. We believe this is because polynomial predictors give bad approximations to the true dynamics except very close to the attractor. Thus only one bad prediction needs to be made for the predictions to quickly blow up. The blow up of iterated polynomial predictors was also observed in [15] and [25]. The superiority of the iterative method for neural networks was observed independently by Lapedes and Farber [25]. The superiority of the iterative method for piecewise linear predictors has been observed in work concurrent with ours by Farmer and Sidorowich [16]. In fig. 4(d), in which 10% of the worst predictions are ignored, the scaling laws are obeyed to within about 10%. The results for the iterated

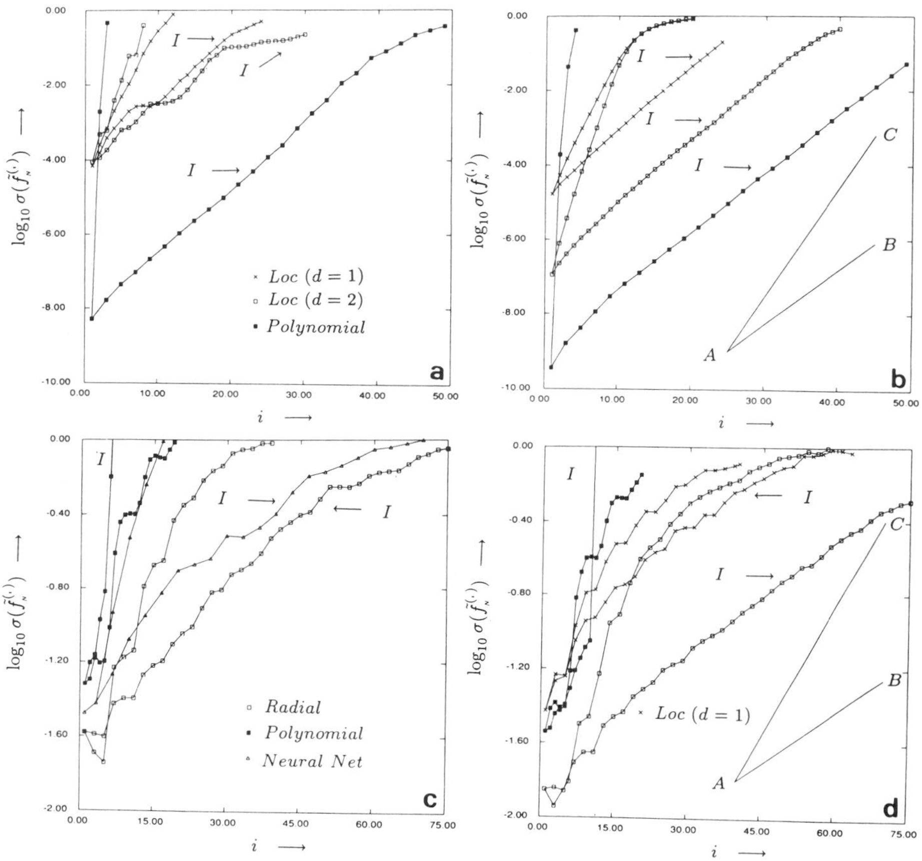

Fig. 4. Short-term predictor errors calculated for a variety of different techniques. (a) Ikeda map with $\mu = 0.7$. (c) Mackey–Glass equation with $\Delta = 30$. (b), (d) same as (a), (c) but 10% of the worst predictions are ignored. The graphs labelled I are obtained using the iterative method. The slopes AB and AC represent scaling laws expected from eqs. (5.1) and (5.2) respectively.

piecewise linear predictor show that the maximum expected time for which predictions are significantly better than a random guess (for which $\log_{10}\sigma(\tilde{f}_N^{(i)}) = 0$), is about 45 iterations. This should be compared to fig. 1. The corresponding time for the iterated radial basis predictor is much longer. The above scaling laws, together with those of section 3, indicate that maximum prediction times of at least twice those obtained by the direct method should be achievable using the iterative method. These results also confirm that the sampling rate τ chosen in a physical experiment should not be too large to obtain the best results with the inverse approach in general. Thus with reference to figs. 4(c) and 4(d), results will certainly deteriorate if τ is chosen greater than about 30. On the other hand, τ must not be chosen too small, or measurement noise may cause results to deteriorate, and the attractor may not be sufficiently explored.

6. Long-term prediction

The results presented in this section are perhaps the most visually striking we have. However, we have only carried out the numerical experiments for one- and two-dimensional maps, and have not tested for scaling laws, so the results may not be as general as those of previous sections. In section 6.1 we illustrate that predictors can be used to accurately reconstruct fractal invariant measures from data that appears to contain no fractal structure. In section 6.2 we illustrate that predictors can be used to predict the location of period-doubling cascades and periodic windows in bifurcation diagrams of parametrized systems. This application of the inverse approach was also suggested in [2] and [9], but not specifically implemented.

6.1. *Predicting invariant measures*

Let $\tilde{f}_N$ be a predictor for a dynamical system f with a natural invariant measure ν (we shall refer to natural invariant measures as *invariant measures* in this section). Then if the predictor error $\sigma(\tilde{f}_N)$ is small, it might be expected that $\tilde{f}_N$ has an invariant measure $\tilde{\nu}_N$ which is close to ν. This intuition comes mostly from numerical experiments, where of necessity invariant measures (i.e. "strange attractors") are calculated by iterating an approximation to a dynamical system many times. Although the precise order in which the iterates visit various parts of the phase space will not be at all similar to that of the true dynamical system (or for an independently carried out experiment), due to positive Lyapunov exponents, the overall density of the points does appear to be reproducible. The reasons behind this are well understood for Axiom A systems, but a general theory is lacking.

In fig. 5(a), we took $N = 25$ data points from the Ikeda map with $\mu = 0.7$ to construct a predictor $\tilde{f}_N$, using a radial basis technique with $\phi(r) = (r^2 + c^2)^{-1}$, c equal to the standard deviation of the data, and no polynomial term in (3.4). The predictor error was roughly 10^{-2}, and the map $\tilde{f}_N$ was iterated 10^4 times with initial condition one of the data points, producing the fractal object $\tilde{\nu}_N$ illustrated. Remarkably, the correct invariant measure ν, illustrated in fig. 5(b), is almost identical to $\tilde{\nu}_N$, despite virtually none of this structure being apparent in the initial data. Note that the density of points is also almost identical, as illustrated by the proportions of points falling in the small box. The reason that this is possible is that the 25 data points give enough information to construct a good approximation to f, and hence indirectly ν. For a more complicated example, we repeated the above procedure for the Ikeda map with $\mu = 0.83$ in figs. 5(c) and 5(d). 50 data points were chosen so that the predictor error was also roughly 10^{-2}.

In fig. 5(e), we chose a local quadratic predictor for the Ikeda map with $\mu = 0.7$, and 75 data points, ensuring that the predictor error was also roughly 10^{-2}. It is evident that the discontinuities in $\tilde{f}_N$ have destroyed some of the fractal structure, and that the density of points in the small box is not as accurately reproduced. Thus local predictors may not be suitable for predicting invariant measures. To illustrate what can go wrong if the predictor error is not small enough, we reduce the number of data points in the $\mu = 0.7$ calculation to 10, for which the predictor error is about 10^{-1} and obtained fig. 5(f). On the other hand, a small predictor error does not guarantee that $\tilde{\nu}_N$ is close to ν, as was illustrated by the results for a polynomial predictor of the Mackey–Glass equation in section 5. We have also observed that piecewise linear predictors can blow up under iteration in some cases.

The above ideas may be useful in predicting time averages of a function $\phi: \mathbb{R}^m \to \mathbb{R}$ as follows. By the ergodic theorem, the time average $\langle \phi \rangle$ of ϕ is given by $\int \phi \, d\nu$. If the data points x_n, $1 \le n \le N$, are used to estimate $\langle \phi \rangle$ by $\langle \phi \rangle \approx N^{-1} \sum_{n=1}^{N} \phi(x_n)$, then by the law of large numbers, the error is expected to be of order $N^{-1/2}$, which may be

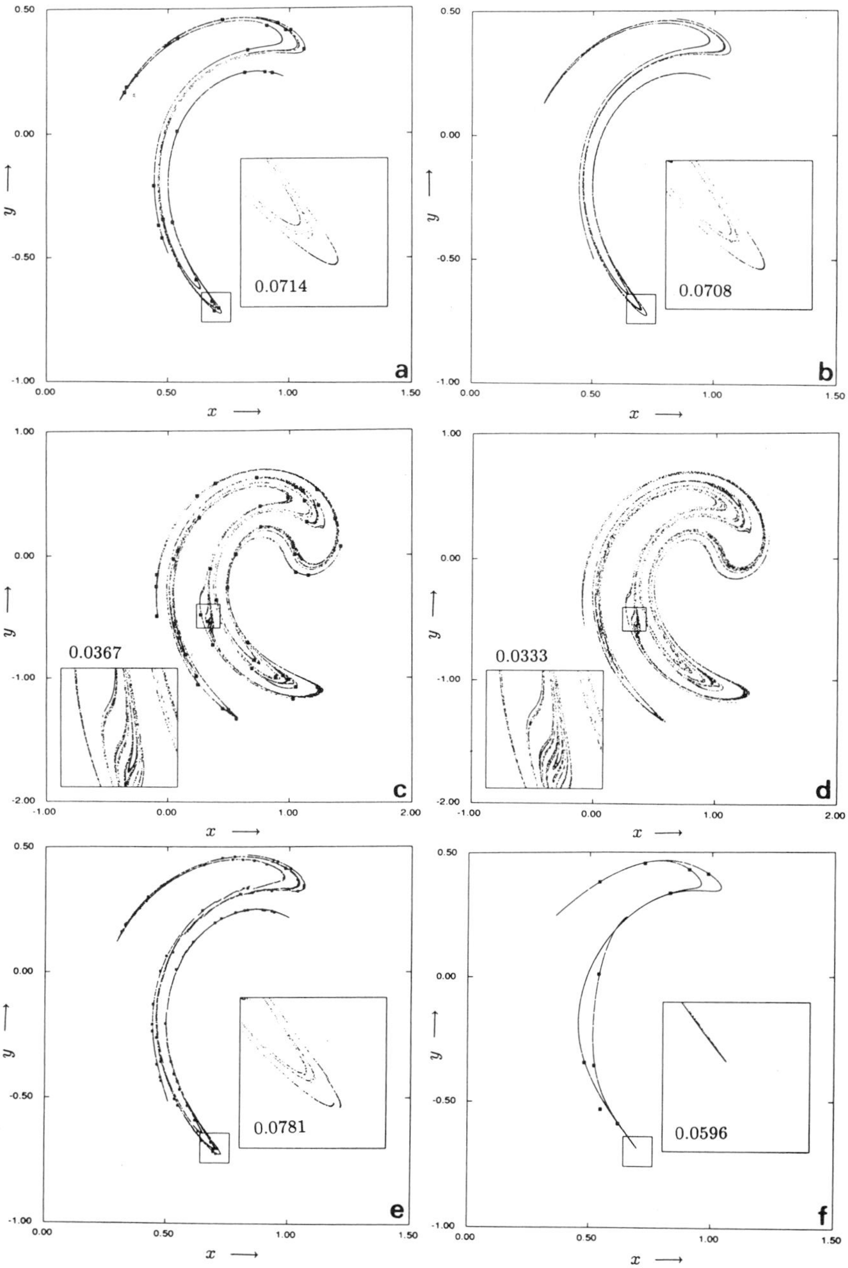

Fig. 5. Invariant measures for the Ikeda map with $\mu = 0.7$ and 0.83, and some of its predictors. The data points used to construct the predictors are represented by solid squares. The numbers in the boxes represent the proportion of points falling in the boxes. (a) $\tilde{\nu}_N$ with $N = 25$ for $\mu = 0.7$. (b) ν for $\mu = 0.7$. (c) $\tilde{\nu}_N$ with $N = 50$ for $\mu = 0.83$. (d) ν for $\mu = 0.83$. (e) $\tilde{\nu}_N$ with $N = 75$ and a piecewise quadratic predictor for $\mu = 0.7$. (f) $\tilde{\nu}_N$ with $N = 10$ for $\mu = 0.7$.

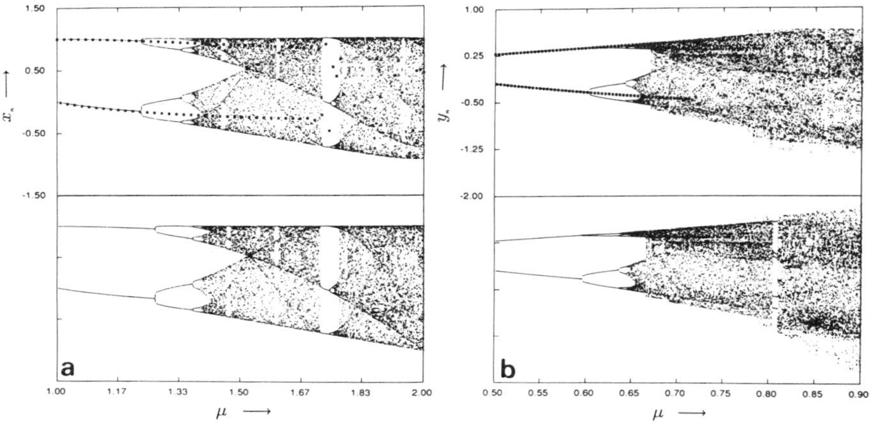

Fig. 6. Bifurcation diagrams for (a) eq. (6.1) and (b) the Ikeda map. The correct bifurcation diagrams are shown in the lower half of the figure. The predicted bifurcation diagrams are shown in the upper half of the figure, together with the data points used to construct them.

quite large. Another method of estimating $\langle\phi\rangle$ is to use the points x_n to construct a predictor $\tilde{f}_N$ and its invariant measure $\tilde{\nu}_N$. The time average $\langle\phi\rangle$ would then be estimated by $\langle\phi\rangle \approx \int\phi \, d\tilde{\nu}_N \approx (N+M)^{-1}\Sigma_{n=1}^{N+M}\phi(\tilde{f}_N^n(x_1))$ for M large. This method is evidently successful for predicting the time averages of the characteristic functions on the small boxes in figs. 5(a) and 5(c). However, in general, if the predictor $\tilde{f}_N$ is of order γ, we expect the error in the estimated time average to be of order $N^{-(\gamma-1)/D}$. This is because for $\tilde{\nu}_N$ to be close to ν, we expect that the derivative of $\tilde{f}_N$ must be close to the derivative of f, so that one order of approximation is lost. Thus, since most predictors are in practice of order 2, the above method for predicting time averages may only be advantageous if $D < 2$.

6.2. *Predicting bifurcations*

For simplicity, we will assume that the bifurcations of a one-parameter family of dynamical systems are the object of interest. Let $F: \mathbb{R}^m \times \mathbb{R} \to \mathbb{R}^m$ be a smooth map denoting such a family. The bifurcations of the family can be neatly encapsulated in a *bifurcation diagram*. For example, let $m = 1$, and F be given by (6.1):

$$F(x,\mu) = 1 - \mu\left(1 - \cos\left(\frac{\pi}{2}x\right)\right). \qquad (6.1)$$

The bifurcation diagram is shown in the lower half of fig. 6(a). At each of 500 values of μ between 1 and 2, an orbit of the dynamical system is plotted in the vertical direction, after letting transients die out. In fig. 6(b), an analogous bifurcation diagram is shown for the Ikeda map (3.7), except that in this case, the y coordinate of the orbit is plotted in the vertical direction. The bifurcations of interest revealed in these bifurcation diagrams are period-doubling cascades, and periodic windows (ranges of parameter values above the transition to chaos where stable periodic orbits exist).

The above bifurcations are unlikely to be observed at the correct parameter values in a physical experiment, because it is inconvenient to repeat the experiment at each parameter value separately. A more convenient approach is to sweep through the parameters, producing a single data set x_n, $1 \le n \le N$, corresponding to the process $x_{n+1} = F(x_n, \mu_n)$, $\mu_{n+1} = \mu_n + \epsilon$, where ϵ is small. Transients are not allowed to die out as the parameter is varied. This results in period-doubling cascades being delayed, and periodic windows may be bypassed altogether, as illustrated in the upper halves of fig. 6 by the solid squares.

The inverse approach can be applied to the data to recover much of the structure of the correct bifurcation diagram as follows. One of the techniques of section 3.1 is used to construct a map $\tilde{F}_N: \mathbb{R}^m \times \mathbb{R} \to \mathbb{R}^m$ such that $x_{n+1} = \tilde{F}_N(x_n, \mu_n)$ for $1 \leq n \leq N-1$, where $\mu_{n+1} = \mu_n + \epsilon$. The bifurcation diagram of the family $\tilde{F}_N$ is then computed in the usual way. This is shown in the upper halves of fig. 6. A radial basis function technique with $\phi(r) = r^3$ and $d = 1$ was used to construct $\tilde{F}_N$. The number of data points was taken to be 100 in fig. 6(a) and 200 in fig 6(b). Evidently much of the structure of the correct bifurcation diagram has been recovered, particularly the period-doubling cascades. We have not attempted to derive scaling laws quantifying how close the bifurcations of the family $\tilde{F}_N$ are to those of F as $N \to \infty$. However, as in section 6.1, it is important that the derivatives of $\tilde{F}_N$ are close to those of F, so that the stability of the corresponding periodic orbits are well approximated.

7. Conclusions

In this paper, we have demonstrated numerically that an inverse approach to dynamical systems can be successfully used to identify chaos in a time series, make short-term predictions, and also some restricted forms of long-term prediction. Comparison tests have been made with other techniques, and scaling laws have been derived which interpret most of the results. For small sets of data points, radial basis predictors have been shown to give results which compare favourably to those obtained using neural network predictors [25], though using much less computational resources. For large sets of data points, we have not been able to improve on results obtained by piecewise linear predictors [15]. However, if a few predictions are ignored, piecewise quadratic predictors have been found to give significantly better results. The usefulness of global polynomial and rational predictors seem limited to relatively simple examples. For the model dynamical systems considered here, the inverse approach has been shown to be superior to dimension calculations for identifying chaos, and to be less susceptible to the effects of noise. The method of iterating predictors to make short-term predictions has been shown to be superior to the more direct method of Farmer and Sidorowich [15]. Finally, we have shown that in certain cases near the transition to chaos, it is possible to predict invariant measures and bifurcations.

Some limitations of the inverse approach, at least using the numerical techniques presented here, are as follows. Firstly, we do not expect good results for time series of dimension greater than about 10, for data sets of length typically generated in laboratory experiments. Thus we do not expect much success with the techniques presented here in situations involving fully developed turbulence. Secondly, it is important that sampling rates can be chosen which are not too large relative to the characteristic time scales of variation of the system being sampled. Thirdly, for systems exhibiting intermittent bursts [37], we do not expect to predict a burst in the time series due to the complicated dynamics and rare occurrence of these events. Finally, if extra information is available about a system in addition to a time series, such as explicit underlying partial differential equations or symmetries, then the inverse approach as presented here does not exploit such information. Consequently, for such systems it may be possible to improve significantly upon the inverse approach using other techniques.

Some problems which we are concerned with are as follows. The comparison of the dimension and Lyapunov exponent methods for detecting chaos with the inverse approach should be done in more detail, and data requirement scaling laws for the former should be tested. Comparison tests should be made not just between white noise and a suspected chaotic time series, but also for correlated noise generated by, for example, ARMA models [3] of the time series under consideration.

There is also a need for a much wider range of time series, especially experimentally generated, to be used in evaluating which of the different types of predictors presented here give the smallest predictor errors. We do not understand the scaling laws observed for the radial basis predictors. It may be possible to investigate these in more detail using the numerical techniques of Dyn et al. [11]. Nor have we tested scaling laws for long-term prediction. It would be desirable to provide a more theoretical foundation for the scaling laws presented here, by proving mathematical results, at least for Axiom A systems. On a more practical level, we have not tried to improve the predictors by attempting to find optimal coordinate systems in which to embed the time series. This may be especially relevant for vector valued time series of high dimension [6, 34], or for reducing the effects of noise [5]. finally, it would be of great interest if some of the above ideas could be put to use in science and engineering problems. Although we expect that superior techniques for constructing predictors may be found in the future, the best of the techniques presented here already seem to us to be capable of yielding results of practical value.

Acknowledgements

It is a pleasure to thank Jean-Guy Caputo for several stimulating discussions. I would also like to thank Ariah Iserles for making me aware of the literature on radial basis functions, and the University of Arizona, where this research was initiated, for its warm hospitality. Finally, I have benefitted from the comments of the referees in clarifying some of the ideas presented here. This work was supported by the US ONR under grant NOOO14-85-K-0412, and by a UK Science and Engineering Research Council award.

References

[1] M. F. Barnsley and S. Demko, Iterated function systems and the global construction of fractals, Proc. R. Soc. Lond. A 399 (1985) 243.

[2] B. J. Bayly, I. Goldhirsch and S. A. Orszag, Independent degrees of freedom of dynamical systems, preprint, Princeton University (1986).

[3] G. E. P. Box and G. M. Jenkins, Time Series Analysis, Forecasting and Control (Holden-Day, San Francisco, 1970).

[4] D. S. Broomhead, R. Jones and G. P. King, Topological dimension and local coordinates from time series data, J. Phys. A 20 (1987) L563.

[5] D. S. Broomhead, and G. P. King, Extracting qualitative dynamics from experimental data, Physica D 20 (1987) 217.

[6] D. S. Broomhead, A. C. Newell, D. Rand and J. C. Lerman, A finite dimensional description of turbulent flows, University of Arizona (1987), in preparation.

[7] J. G. Caputo, PhD thesis, Grenoble.

[8] J. Cremers and A. Hubler, Construction of differential equations from experimental data, Z. Naturforsch. A 42 (1987) 797.

[9] J. P. Crutchfield and B. S. McNamara, Equations of motion from a data series, Complex Systems 1 (1987) 417.

[10] J. Duchon, Spline minimising rotation-invariant semi-norms in Sobolev spaces, in: Constructive Theory of Several Variables, Lecture Notes in Mathematics 571, W. Shempp and K. Zeller, eds. (Springer, Berlin, 1977).

[11] N. Dyn, D. Levin and S. Rippa, Numerical procedures for surface spline fitting of scattered data by radial basis functions, SIAM J. Sci. Stat. Comput. 7 (1986) 639.

[12] J. P. Eckmann, S. O. Kamporst, D. Ruelle and S. Ciliberto, Lyapunov exponents from a time series, Phys. Rev. A 34 (1986) 4971.

[13] J. P. Eckmann and D. Ruelle, Ergodic theory of chaos and strange attractors, Rev. Mod. Phys. 57 (1985) 617.

[14] J. D. Farmer, Chaotic attractors of an infinite dimensional system, Physica D 4 (1982) 366.

[15] J. D. Farmer and J. J. Sidorowich, Predicting chaotic time series, Phys. Rev. Lett. 59 (1987) 845.

[16] J. D. Farmer and J. J. Sidorowich, Exploiting chaos to predict the future and reduce noise, preprint, Los Alamos (1988).

[17] R. Franke, Scattered data interpolation: tests of some methods, Math. Comp. 38 (1982) 181.

[18] P. Frederickson, J. L. Kaplan, E. D. Yorke and J. A. Yorke, The Lyapunov dimension of strange attractors, J. Diff. Equ. 49 (1983) 185.

[19] H. Froeling, J. P. Crutchfield, J. D. Farmer, N. H. Packard and R. S. Shaw, On determining the dimension of chaotic flows, Physica D 3 (1981) 605.

[20] D. Gabor, W. P. Wilby and R. Woodcock, A universal nonlinear filter, predictor and simulator which optimizes itself by a learning process, Proc. IEEE B 108 (1960) 422.

[21] P. Grassberger, Do climate attractors exist? Nature 323 (1986) 609.

[22] P. Grassberger and I. Procaccia, Characterisation of strange attractors, Phys. Rev. Lett. 50 (1983) 346.

[23] P. Grassberger and I. Procaccia, Dimensions and entropies of strange attractors from a fluctuating dynamics approach, Physica D 13 (1984) 34.

[24] K. Ikeda, Multiple-valued stationary state and its instabil-

ity of the transmitted light by a ring cavity system, Opt. Commun. 30 (1979) 257.
[25] A. Lapedes and R. Farber, Nonlinear signal processing using neural networks: Prediction and signal modelling, preprint, Los Alamos (1987).
[26] E. N. Lorenz, Deterministic non-periodic flow, J. Atmos. Sci. 20 (1963) 130.
[27] E. N. Lorenz, Atmospheric predictability as revealed by naturally occurring analogues, J. Atmos. Sci. 26 (1969) 636.
[28] M. C. Mackey and L. Glass, Oscillation and chaos in physiological control systems, Science 197 (1977) 287.
[29] C. A. Michelli, Interpolation of scattered data: distance matrices and conditionally positive definite functions, Constr. Approx. 2 (1986) 11.
[30] M. J. D. Powell, Approximation Theory and Methods (Cambridge University Press, Cambridge, 1981).
[31] M. J. D. Powell, Radial basis functions for multivariate interpolation: a review, preprint, Univ. of Cambridge (1985).
[32] M. J. D. Powell, Radial basis function approximations to polynomials, preprint, Univ. of Cambridge (1987).
[33] M. Sano and Y. Sawada, Measurement of the Lyapunov spectrum from chaotic time series, Phys. Rev. Lett. 55 (1985) 1082.
[34] L. Sirovich and J. D. Rodriguez, Coherent structures and chaos: a model problem, Phys. Lett. A 120 (1987) 211.
[35] L. Smith, private communication.
[36] F. Takens, Detecting strange attractors in turbulence, in: Dynamical Systems and Turbulence, Warwick 1980, Lecture Notes in Math. 898 (Springer, Berlin, 1981), pp. 366–381.
[37] R. K. Tavakol and A. S. Tworkowski, Fluid intermittency in low dimensional systems, Phys. Lett. A 126 (1988) 318.
[38] H. Tong, Threshold Models in Non-linear Time Series Analysis, in: Lecture Notes in Statistics, vol. 21 (Springer, New York, 1983).
[39] J. H. Wilkinson and C. Reinsch, Linear Algebra (Springer, Berlin, 1971).
[40] A. Wolf, J. B. Swift, H. L. Swinney and J. A. Vastano, Determining Lyapunov exponents from a time series, Physica D 16 (1985) 285.

Nonlinear forecasting as a way of distinguishing chaos from measurement error in time series

George Sugihara* & Robert M. May†

* Scripps Institution of Oceanography, University of California, San Diego, La Jolla, California 92093, USA
† Department of Zoology, Oxford University, Oxford, OX1 3PS, UK

An approach is presented for making short-term predictions about the trajectories of chaotic dynamical systems. The method is applied to data on measles, chickenpox, and marine phytoplankton populations, to show how apparent noise associated with deterministic chaos can be distinguished from sampling error and other sources of externally induced environmental noise.

TWO sources of uncertainty in forecasting the motion of natural dynamical systems, such as the annual densities of plant or animal populations, are the errors and fluctuations associated with making measurements (for example, sampling errors in estimating sizes, or fluctuations associated with unpredictable environmental changes from year to year), and the complexity of the dynamics themselves (where deterministic dynamics can easily lead to chaotic trajectories).

Here we combine some new ideas with previously developed techniques[1-7,16,24-26], to make short-term predictions that are

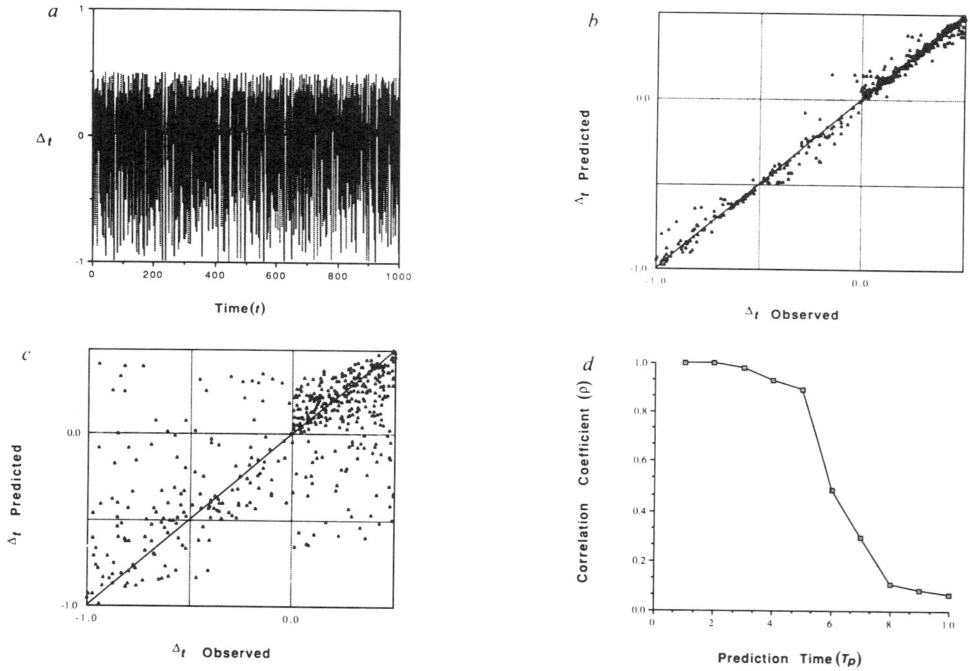

FIG. 1 *a*, Time series of 1,000 points (which in many ways is indistinguishable from white noise) generated by taking first-differences, $\Delta_t = x_{t+1} - x_t$, of the tent map: $x_{t+1} = 2x_t$ for $0.5 > x_t > 0$; $x_{t+1} = 2 - 2x_t$ for $1 > x_t > 0.5$. *b*, Predicted values two steps into the future ($T_p = 2$) versus observed values for the tent delta time series depicted in *a*. Specifically, the first 500 points in the series were used to generate a library of patterns, which were then used as a basis for making predictions for each of the second 500 points. As described in the text, the predictions were made using a simplex projection method, and in this figure the embedding dimension and lag time are $E = 3$ and $\tau = 1$, respectively. Here the coefficient of correlation between predicted and actual values is $\rho = 0.997$ ($N = 500$). For comparison, we note that the corresponding correlation coefficient obtained using the first half of the series to predict the second half with an autoregressive linear model (where the predictions are based on the weighted average of three linear maps, one for each of the three different τ-values that give the best results in such a linear scheme) is $\rho = 0.04$. *c*, Exactly as for Fig. *b*, except here the predictions are five time steps into the future ($T_p = 5$). The correlation coefficient between predicted and actual values is now $\rho = 0.89$ ($N = 500$). *d*, Summary of the trend between *b* and *c*, by showing ρ between predicted and observed values in the second half (second 500 points) of the time series of *a*, as a function of T_p. As in *b* and *c*, the simplex projection method here uses $E = 3$ and $\tau = 1$. That prediction accuracy (as measured by the coefficient of correlation between predicted and observed values) falls as predictions extend further into the future is a characteristic signature of a chaotic attractor.

based on a library of past patterns in a time series[1]. By comparing the predicted and actual trajectories, we can make tentative distinctions between dynamical chaos and measurement error: for a chaotic time series the accuracy of the nonlinear forecast falls off with increasing prediction-time interval (at a rate which gives an estimate of the Lyapunov exponent[3]), whereas for uncorrelated noise, the forecasting accuracy is roughly independent of prediction interval. For a relatively short time series, distinguishing between autocorrelated noise and chaos is more difficult; we suggest a way of distinguishing such 'coloured' noise from chaos in our scheme, but questions remain, at least for time series of finite length.

The method also provides an estimate of the number of dimensions, or 'active variables', of the attractor underlying a time series that is identified as chaotic. Unlike many current approaches to this problem (for example, that of Grassberger and Procaccia[8]), our method does not require a large number of data points, but seems to be useful when the observed time series has relatively few points (as is the case in essentially all ecological and epidemiological data sets).

Forecasting for a chaotic time series

Below, we outline the method and show how it works by applying it to a chaotic time series generated artificially from the deterministic 'tent map'. We then apply it to actual data on measles and chickenpox in human populations (which have been previously analysed using different techniques[9-13]) and on diatom populations. We conclude that the method may be capable of distinguishing chaos from measurement error even in such relatively short runs of real data.

As an example of the difficulties in short-range forecasting, we consider the chaotic time series shown in Fig. 1*a*. This time series was generated from the first-difference transformation ($x_{t+1} - x_t$) on the deterministic tent map or triangular 'return map' (described in detail in the legend to Fig. 1*a*). Here and elsewhere is this report, we first-difference the data partly to give greater density in phase space to such chaotic attractors as may exist, and partly to clarify nonlinearities by reducing the effects of any short-term linear autocorrelations. It should be noted, however, that both in our artificial examples and in our later analysis of real data, we obtain essentially the same results if we work with the raw time series (without first-differencing). With the exception of a slight negative correlation between immediately adjacent values, the sequence in Fig. 1*a* is uncorrelated, and is in many ways indistinguishable from white noise: the null hypothesis of a flat Fourier spectrum cannot be rejected using Bartlett's Kolmogorov-Smirnov test, with $P = 0.85$. Because nonadjacent values in the time series are completely uncorrelated, standard statistical methods (that is, linear autoregression) cannot be used to generate predictions two or more steps into the future that are significantly better than the mean value (that is, zero) for the series.

Figure 1*b* and *c* show the results of local forecasting with the above data. The basic idea here, as outlined below, is that if deterministic laws govern the system, then, even if the dynamical behaviour is chaotic, the future may to some extent be predicted from the behaviour of past values that are similar to those of the present.

Specifically, we first choose an 'embedding dimension', E, and then use lagged coordinates to represent each lagged sequence of data points $\{x_t, x_{t-\tau}, x_{t-2\tau}, \ldots, x_{t-(E-1)\tau}\}$ as a point in this E-dimensional space; for this example we choose $\tau = 1$, but the results do not seem to be very sensitive to the value of τ, provided that it is not large[14,15]. For our original time series, shown in Fig. 1*a*, each sequence for which we wish to make a prediction—each 'predictee'—is now to be regarded as an E-dimensional point, comprising the present value and the $E - 1$ previous values each separated by one lag time τ. We now locate all nearby E-dimensional points in the state space, and choose a minimal neighbourhood defined to be such that the predictee is contained within the smallest simplex (the simplex with minimum diameter) formed from its $E + 1$ closest neighbours; a simplex containing $E + 1$ vertices (neighbours) is the smallest simplex that can contain an E-dimensional point as an interior point (for points on the boundary, we use a lower-dimensional simplex of nearest neighbours). The prediction is now obtained by projecting the domain of the simplex into its range, that is by keeping track of where the points in the simplex end up after p time steps. To obtain the predicted value, we compute where the original predictee has moved within the range of this simplex,

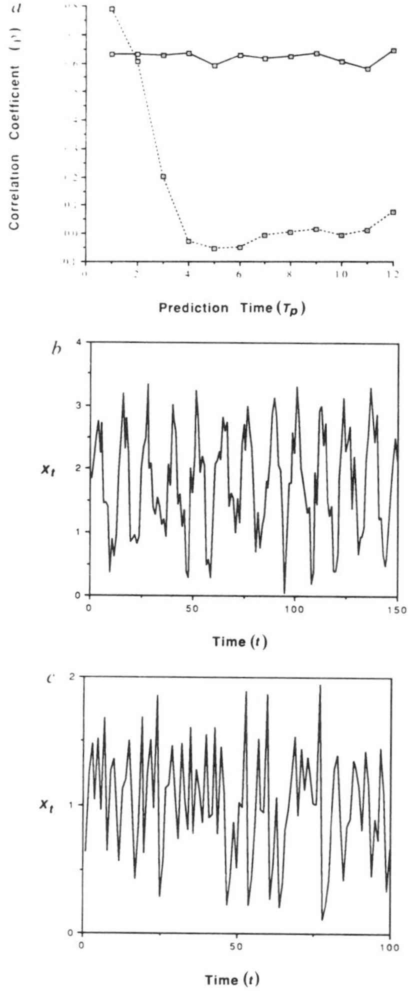

FIG. 2 *a*, Solid line shows ρ between predicted and observed values for the second half of the time series defined in *b* (which is, in fact, a sine wave with additive noise) as a function of T_p. As discussed in the text, the accuracy of the prediction, as measured by ρ, shows no systematic dependence on T_p. By contrast, the time series shown in *c* (which is the sum of two separate tent map series) does show the decrease in ρ with increasing T_p, as illustrated by the dashed line, that is characteristic of a chaotic sequence. Both curves are based on the simplex methods described in the text, with $E = 3$ and $\tau = 1$. *b*, First 150 points in the time series generated by taking discrete points on a sine wave with unit amplitude ($x_t = \sin(0.5t)$), and adding a random variable chosen (independently at each step) uniformly from the interval [−0.5, 0.5]. That is, the series is generated as a 'sine wave + 50% noise'. *c*, Time series illustrated here is generated by adding together two independent tent map sequences.

giving exponential weight to its original distances from the relevant neighbours. This is a nonparametric method, which uses no prior information about the model used to generate the time series, only the information in the output itself. It should apply to any stationary or quasi-ergodic dynamic process, including chaos. This method is a simpler variant of several more complicated techniques explored recently by Farmer and Sidorowich[3] and by Casdagli[7].

Figure 1*b* compares predicted with actual results, two time steps into the future. Figure 1*c* makes the same comparison, but at five time steps into the future. There is obviously more scatter in Fig. 1*c* than in Fig. 1*b*. Figure 1*d* quantifies how error increases as we predict further into the future in this example, by plotting the conventional statistical coefficient of correlation, ρ, between predicted and observed values as a function of the prediction-time interval, T_p (or the number of time steps into the future, p). Such decrease in the correlation coefficient with increasing prediction time is a characteristic feature of chaos (or equivalently, of the presence of a positive Lyapunov exponent, with the magnitude of the exponent related to the rate of decrease of ρ with T_p). This property is noteworthy, because it indicates a simple way to differentiate additive noise from deterministic chaos: predictions with additive noise that is uncorrelated (in the first-differences) will seem to have a fixed amount of error, regardless of how far, or close, into the future one tries to project, whereas predictions with deterministic chaos will tend to deteriorate as one tries to forecast further into the future. Farmer and Sidorowich[3,16] have derived asymptotic results (for very long time series, $N \gg 1$) that describe how this error typically propagates, over time, in simple chaotic systems. The standard correlation coefficient is one of several alternative measures of the agreement between predicted and observed values; results essentially identical to those recorded in Figs 1-6 can be obtained with other measures (such as the mean squared difference between predicted and observed values as a ratio to the mean squared error).

Forecasting with uncorrelated noise

Figure 2*a* (solid line) shows that, indeed, this signature of ρ decreasing with T_p does not arise when the erratic time series is in fact a noisy limit cycle. Here we have uncorrelated additive noise superimposed on a sine wave (Fig. 2*b*). Such uncorrelated noise is reckoned to be characteristic of sampling variation. Here the error remains constant as the simplex is projected further into the future; past sequences of roulette-wheel numbers that are similar to present ones tell as much or little about the next spin as the next hundredth spin. By contrast, the dashed line in Fig. 2*a* represents ρ as a function of T_p, for a chaotic sequence generated as the sum of two independent runs of tent map; that is, for the time series illustrated in Fig. 2*c*. Although the two time series in Fig. 2*b* and Fig. 2*c* can both look like the sample functions of some random process, the characteristic signatures in Fig. 2*a* distinguish the deterministic chaos in Fig. 2*c* from the additive noise in Fig. 2*b*.

The embedding dimension

The predictions in Figs 1 and 2 are based on an embedding dimension of $E = 3$. The results are, however, sensitive to the choice of E. Figure 3*a* compares predicted and actual results for the tent map two time steps ahead ($T_p = 2$), as in Fig. 1*b*, except that now $E = 10$ (versus $E = 3$ in Fig. 1*b*). Clearly the predictions are less accurate with this higher embedding dimension. More generally, Fig. 3*b* shows ρ between predicted and actual results one time step into the future ($T_p = 1$) as a function of E, for two different choices of the lag time ($\tau = 1$ and $\tau = 2$). It may seem surprising that having potentially more information—more data summarized in each E-dimensional point, and a higher-dimensional simplex of neighbours of the predictee—reduces the accuracy of the predictions; in this respect, these results differ from results reported by Farmer and Sidorowich for parametric forecasting involving linear interpolation to construct local polynomial maps[3]. We think this effect is caused by contamination of nearby points in the higher-dimensional embeddings with points whose earlier coordinates are close, but whose recent (and more relevant) coordinates are distant. If this is so, our method may have additional applications as a trial-and-error method of computing an upper-bound on the embedding dimension, and thence on the dimensionality of the attractor (see also refs 2, 6, 7).

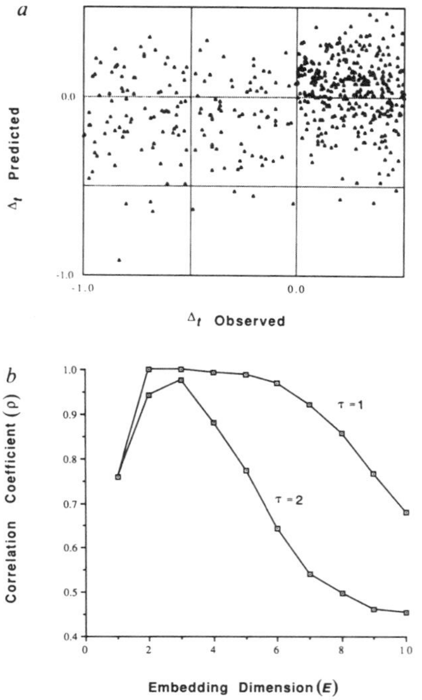

FIG. 3 *a*, Similar to Fig. 1*b*, this figure shows predictions one time step into the future ($T_p = 1$) versus observed values, for the second 500 points in the tent map time series of Fig 1*a*, with the difference that here we used an embedding dimension $E = 10$ (in contrast to $E = 3$ in Fig. 1*b*; the lag time remains unchanged at $\tau = 1$). As discussed in the text, the accuracy of the prediction deteriorates as E gets too large $\rho = 0.25$, $N = 500$). *b*, Correlation coefficient between predicted and observed results, ρ, is shown as a function of E for predictions one time step into the future ($T_p = 1$). The relationship is shown for $\tau = 1$ and $\tau = 2$. The figure indicates how such empirical studies of the relation between ρ and E may be used to assess the optimal E.

Problems and other approaches

We have applied these ideas to a variety of other 'toy models', including the quadratic map along with other first-order difference equations and time series obtained by taking points at discrete time intervals from continuous chaotic systems such as those of the Lorenz and Rossler models (in which the chaotic orbits are generated by three coupled, nonlinear differential equations). The results for ρ as a function of T_p are in all cases very similar to those shown in Fig. 1*d*. Even in more complicated cases, such as those involving the superposition of different chaotic maps, we observe a decline in ρ versus T_p; here, however, the signature can show a step pattern, with each step corresponding to the dominant Lyapunov exponent for each map.

So far, we have compared relationships between ρ and T_p for chaotic time series with the corresponding relations for white noise. More problematic, however, is the comparison with ρ-T_p relationships generated by coloured noise spectra, in which there are significant short-term autocorrelations, although not long-term ones. Such autocorrelated noise can clearly lead to correlations, ρ, between predicted and observed values that decrease as T_p lengthens. Indeed, it seems likely that a specific pattern of autocorrelations could be hand-tailored, to mimic any given relationship between ρ and T_p (such as that shown in Fig. 1*d*) obtained from a finite time series. We conjecture, however, that such an artifically designed pattern of autocorrelation would in general give a flatter ρ-versus-E relationship than those of simple chaotic time series corresponding to low-dimensional attractors (for example, see Fig. 3*b*). Working from the scaling relations for error versus T_p in chaotic systems[3,16], Farmer (personal communication) has indeed suggested that asymptotically (for very large N), the ρ-T_p relationships generated by autocorrelated noise may characteristically scale differently from those generated by deterministic chaos. Although we have no solution to this central problem—which ultimately may not have any general solution, at least for time series of the sizes found in population biology—we suggest that an observed time series may tentatively be regarded as deterministically chaotic if, in addition to a decaying ρ-T_p signature, the correlation, ρ, between predicted and observed values obtained by our methods is significantly better than the corresponding correlation coefficient obtained by the best-fitting autoregressive linear predictor (see also, ref. 16). For the tent map, as detailed in the legend to Fig. 1, *b* and *c*, our nonlinear method gives ρ values significantly better than those from autoregressive linear models (composed of the weighted average of the three best linear maps).

Most previous work applying nonlinear theory to experimental data begins with some estimate of the dimension of the underlying attractor[2-7]. The usual procedure (for exceptions, see refs 2, 6, 7, 25) is to construct a state-space embedding for the time series, and then to calculate the dimension of the putative attractor using some variant of the Grassberger-Procaccia algorithm[8]. A correlation integral is calculated that is essentially the number of points in E space separated by a distance less than l, and the power-law behaviour of this correlation integral (l^{ν}) is then used to estimate the dimension, D, of the attractor ($D \geqslant \nu$). This dimension is presumed to give a measure of the effective number of degrees of freedom or 'active modes' of the system. An upper bound on a minimal embedding dimension (which can be exceeded when the axes of the embed-

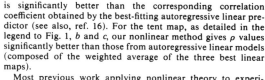

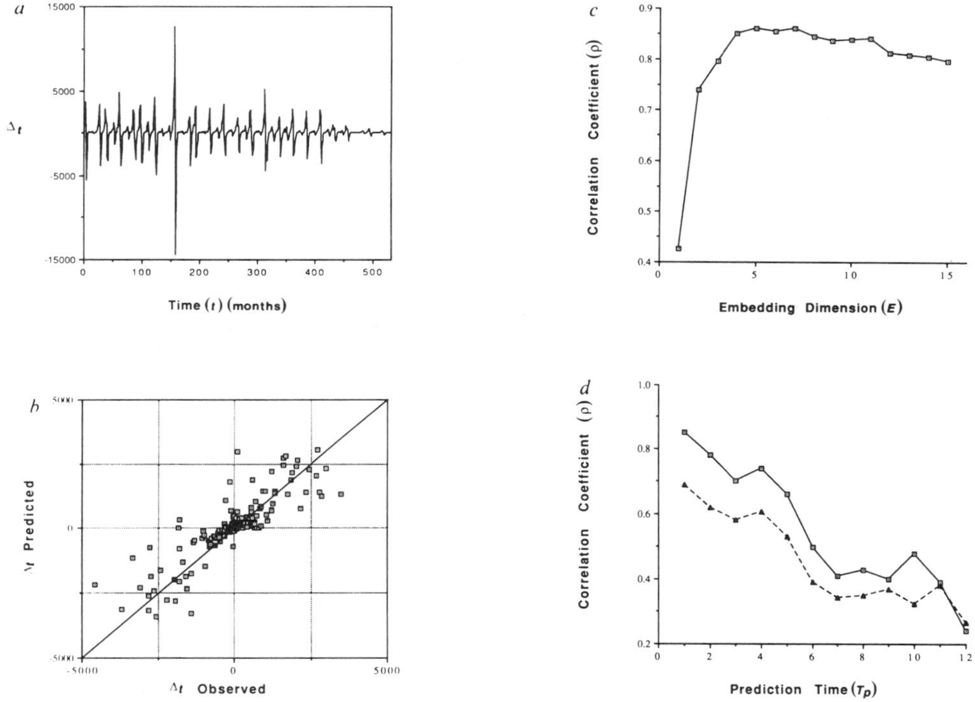

FIG. 4 *a*, Time series generated by taking first differences, $x_{t+1} - x_t$, of the monthly number of cases of measles reported in New York City between 1928 and 1972 (the first 532 points in the sequence shown here). After 1963, the introduction of immunization against measles had a qualitative effect on the dynamics of infection; this can be seen in the later part of the sequence illustrated here. *b*, Using the methods described earlier, the first part of the series in Fig. 4*a* (216 points, 1928 to 1946)) was used to construct a library of past patterns, which were then used as a basis for predicting forward from each point in the second part of the series, from 1946 to 1963. Predicted and observed values are shown here for predictions one time step into the future ($T_p = 1$), using $E = 6$ and $\tau = 1$. The correlation coefficient between predicted and observed values is $\rho = 0.85$ ($P < 10^{-5}$ for $N = 216$). For comparison, the corresponding prediction based on an autoregressive linear model (composed of five optimal linear maps, compare Fig. 1*b*) gives $\rho = 0.72$ (which is significantly different from $\rho = 0.85$ at the $P < 0.0005$ level). *c*, As in Fig. 3*b*, ρ between predicted and observed results, is shown as a function of E (for $T_p = 1$ and $\tau = 1$). This figure indicates an optimal embedding dimension of $E \sim 5$–7, corresponding to a chaotic attractor with dimension 2–3. *d*, Here ρ, between predicted and observed results for measles, is shown as a function of T_p (for $E = 6$ and $\tau = 1$). For the points connected by the solid lines, the predictions are for the second half of the time series (based on a library of patterns compiled from the first half). For the points connected by the dashed lines, the forecasts and the library of patterns span the same time period (the first half of the data). The similarity between solid and dashed curves indicates that secular trends in underlying parameters do not introduce significant complications here. The overall decline in prediction accuracy with increasing time into the future is, as discussed in the text, a signature of chaotic dynamics as distinct from uncorrelated additive noise.

ding are not truly orthogonal) is $E_{min} < 2D+1$, where D is the attractor dimension[8,15]. The scaling regions used to estimate power laws by these methods are typically small and, as a consequence, such calculations of dimension involve only a small fraction of the points in the series (that is, they involve only a small subset of pairs of points in the state space). In other words, the standard methods discard much of the information in a time series, which, because many natural time series are of limited size, can be a serious problem. Furthermore, the Grassberger-Procaccia and related methods are somewhat more qualitative, requiring subjective judgement about whether there is an attractor of given dimensions. Prediction methods, by contrast, have the advantage that standard statistical criteria can be used to evaluate the significance of the correlation between predicted and observed values. As Farmer and Sidorowich[3,16], and Casdagli[7], have also emphasized, prediction methods should provide a more stringent test of underlying determinism in situations of given complexity. Prediction is, after all, the *sine qua non* of determinism.

Time series from the natural world

Measles. For reported cases of measles in New York City, there is a monthly time series extending from 1928 (ref. 17). After 1963, immunization began to alter the intrinsic dynamics of this system, and so we use only the data from 1928 to 1963 ($N = 432$). These particular data have received a lot of attention recently, and they are the focus of a controversy about whether the dynamics reflect a noisy limit cycle[9] or low-dimensional chaos superimposed on a seasonal cycle[10-13]. In particular, the data have been carefully studied by Schaffer and others[13-16,27], who have tested for low-dimensional chaos using a variety of methods, including the Grassberger-Procaccia algorithm[13], estimation of Lyapunov exponents[13], reconstruction of Poincare return maps[10,11], and model simulations[12,13]. Although it is not claimed that any of these tests are individually conclusive, together they support the hypothesis that the measles data are described by a two- to three-dimensional chaotic attractor.

Figure 4*a* shows the time series obtained by taking first differences, $X_{t+1} - X_t$, of these data. As discussed above, the first difference was taken to 'whiten' the series (that is, reduce autocorrelation) and to diminish any signals associated with simple cycles (a possibility raised by proponents of the additive noise hypothesis[9]). We then generated our predictions by using the first half of the series (216 points) to construct an ensemble of points in an E-dimensional state space, that is, to construct a library of past patterns. The resulting information was then used to predict the remaining 216 values in the series, along the lines described above, for each chosen value of E. Figure 4*b*, for example, compares predicted and observed results, one time step into the future ($T_p = 1$ month), with $E = 6$. Figure 4*c* shows ρ between predicted and observed results as a function of E for $T_p = 1$. Taking the optimal embedding dimension to be that yielding the highest correlation coefficient (or least error) between prediction and observation in one time step, it is seen from Fig. 4*c* that $E \simeq 5$–7. This accords with previous estimates[10-13] made using various other methods, and is consistent with the finding of an attractor with dimension $D = 2$–3.

The points joined by the solid lines in Fig. 4*d* show ρ as a function of T_p (for $E = 6$). Prediction error seems to propagate in a manner consistent with chaotic dynamics. This result, in

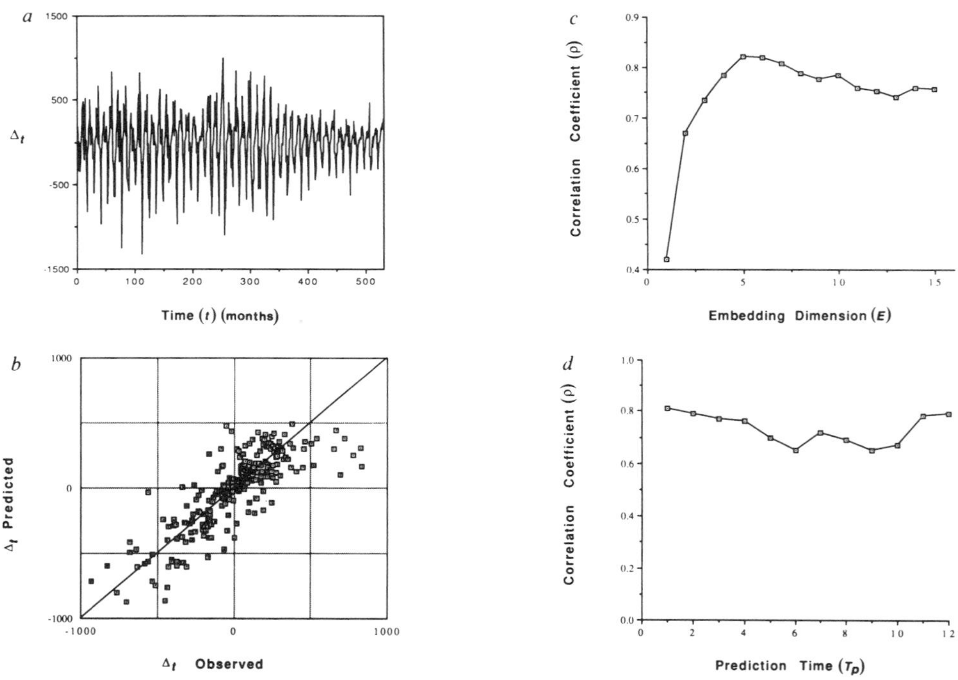

FIG. 5 *a*, As for Fig. 4*a*, except the time series comes from taking first-differences of the monthly numbers of reported cases of chickenpox in New York City from 1928 to 1972. *b*, As in Fig. 4*b*, predicted and observed numbers of cases of chickenpox are compared, the predictions being one time step into the future, $T_p = 1$ (here, $E = 5$ and $\tau = 1$). The correlation coefficient between predicted and observed values is $\rho = 0.82$; an autoregressive linear model alternatively gives predictions which have $\rho = 0.84$. In contrast to Fig. 4*b* for measles, here there is no significant difference between our prediction technique and standard linear autoregressive methods. *c*, Correlation coefficient between predicted and observed results for chickenpox, ρ, shown as a function of E for predictions one time step into the future ($T_p = 1$ and $\tau = 1$). *d*, Compare with Fig. 4*d*; ρ, between predicted and observed values, as a function of T_p (with $E = 5$ and $\tau = 1$) is shown. Here the lack of dependence of ρ on T_p, which is in marked contrast with the pattern for measles in Fig. 4*d*, indicates pure additive noise (superimposed on a basic seasonal cycle).

combination with the significantly better performance ($P <$ 0.0005) of our nonlinear predictor as compared with an optimal linear autoregressive model (see legend to Fig. 4*b*) agrees with the conclusion that the noisy dynamics shown in Fig. 4*a* are, in fact, deterministic chaos[10-13].

For data from the natural world, as distinct from artificial models, physical or biological parameters, or both, can undergo systematic changes over time. In this event, libraries of past patterns can be of dubious relevance to an altered present and even-more-different future. In a different context, there is the example of how secular trends in environmental variables can complicate an analysis of patterns of fluctuation in the abundance of bird species[18,19]. We can gauge the extent to which secular trends might confound our forecasting methods in the following way. Rather than using the first half of the time series to compile the library of patterns, and the second half to compute correlations between predictions and observations, we instead investigate the case in which the library and forecasts span the same time period. Therefore we focus our predictions in the first half of the series, from which the library was drawn. To avoid redundancy, however, between our forecasts and the model, we sequentially exclude points from the library that are in the neighbourhood of each predictee (specifically, the $E\tau$ points preceding and following each forecast). The points connected by the dashed lines in Fig. 4*d* show the ρ versus T_p relationship that results from treating the measles data in this way (again with $E = 6$). The fairly close agreement between these results (for which the library of patterns and the forecasts span the same time period) and those of the simpler previous analysis (the solid line in Fig. 4*d*) indicates that within these time frames, secular trends in underlying parameters are not qualitatively important.

Chickenpox. Figure 5*a–d* repeat the process just described for measles, but now for monthly records of cases of chickenpox in New York City from 1949 to 1972 (ref. 20). Figure 5*a* shows the time series of differences, $X_{t+1} - X_t$. The 532 points in Fig. 5*a* are divided into two halves, with the first half used to construct the library, on which predictions are made for the second 266 points. These predictions are compared with the actual data points, as shown for predictions one time step ahead ($T_p = 1$ month) in Fig. 5*b*. In Fig. 5*b*, $E = 5$; Fig. 5*c* shows that an optimum value of E, in the sense just defined, is about 5 to 6. By contrast with Fig. 4*d* for measles, ρ between predicted and observed results for chickenpox shows no dependence on T_p: one does as well at predicting the incidence next year as next month. Moreover, the optimal linear autoregressive model performs as well as our nonlinear predictor. We take this to indicate that chickenpox has a strong annual cycle (as does

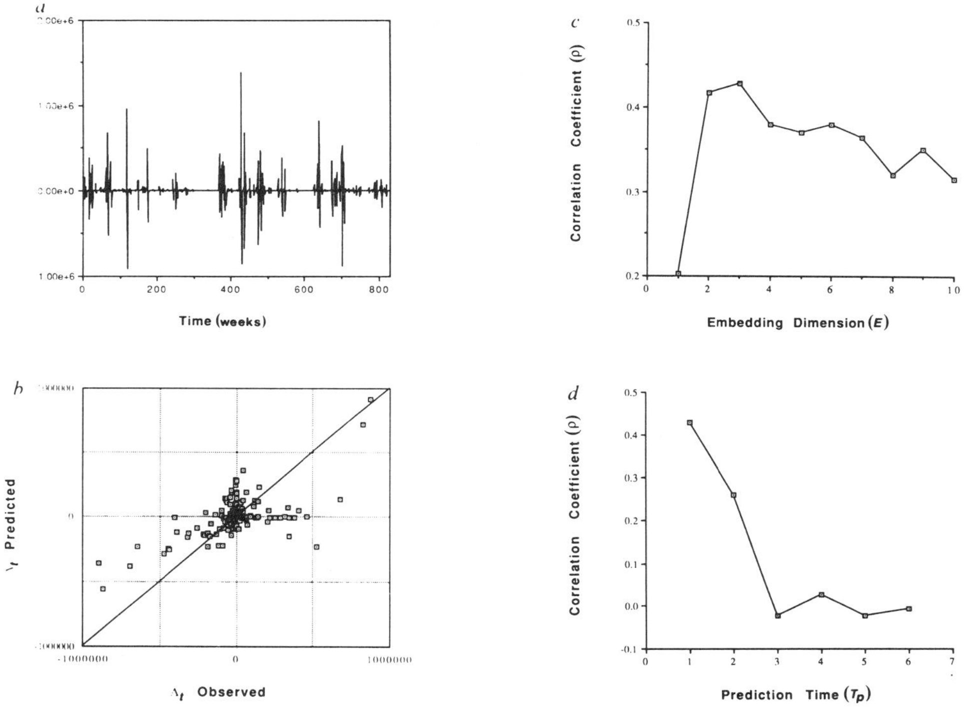

FIG. 6 *a*, Time series of first differences, $x_{t+1} - x_t$, of the weekly numbers of diatoms in seawater samples taken at Scripps Pier, San Diego, from 1929 to 1939 ($N = 830$). *b*, Using the first half of the time series in *a* to construct a library of patterns, we use the simplex projection methods described in the text to predict one week ($T_p = 1$) into the future from each point in the second half of the series ($N = 415$); here $E = 3$, and $\tau = 1$. The correlation coefficient between predicted and observed values is $\rho = 0.42$ ($P < 10^{-4}$ for $N = 415$); the best autoregressive linear predictions (composed of three optimal linear maps) give $\rho = 0.13$, which is significantly less than the nonlinear result ($P < 0.0005$). *c*, As in Figs 4*c* and 5*c*, ρ between predicted and observed values is shown as a function of the choice of E for predictions two time steps into the future ($T_p = 2$ and $\tau = 1$). This figure indicates an optimal E of about 3, consistent with an attractor of dimension about 2. *d*, As in Figs 4*d* and 5*d*, ρ is shown as a function of the T_p (for $E = 3$ and $\tau = 1$). Here the correlation coefficient decreases with increasing prediction interval, in the manner characteristic of chaotic dynamics generated by a some low-dimensional attractor. That ρ is about 50% at best, however, indicates that roughly half the variance in the time series comes from additive noise. The dynamics of this system therefore seem to be intermediate between those of measles (for which Fig. 4*d* indicates deterministically chaotic dynamics) and chickenpox (for which Fig. 5*d* indicates purely additive noise superimposed on a seasonal cycle).

measles), with the fluctuations being additive noise (in contrast to measles, for which the fluctuations derive mainly from the dynamics).

The contrast between measles and chickenpox can be explained on biological grounds[21]. Measles has a fairly high 'basic reproductive rate' ($R_0 = 10$–20), and, after a brief interval of infectiousness, recovered individuals are immune and uninfectious for life; these conditions tend to produce long-lasting 'interepidemic' oscillations, with a period of about 2 years, even in the simplest models[22]. This, in combination with seasonal patterns, makes it plausible that measles has complex dynamics. Chickenpox is less 'highly reproductive' (with R_0 values of about 8 to 10), and may recrudesce as shingles in later life; this makes for an infection less prone to show periodicities other than basic seasonal ones associated with schools opening and closing, and therefore indicates seasonal cycles with additive noise. Furthermore, reporting was compulsory for measles but not for chickenpox over the time period in question, which itself would be likely to make sampling error greater for chickenpox. Whatever the underlying biological explanation, the patterns in Figs 4*d* and 5*d* differ in much the same way as those illustrated in Fig. 2*a* for the artificially generated time series of Fig. 2, *a* and *b*.

Marine plankton. A time series is provided by Allen's weekly record of marine planktonic diatoms gathered at Scripps Pier, San Diego, between 1920 and 1939 ($N = 830$). With the exception of the work of Tont[23], this collection of information has been little analysed, and not at all in the light of contemporary notions about nonlinear dynamics. The data comprise weekly totals of the numbers of individuals of all diatom species, tallied in daily seawater samples collected over ~20 years. As for our analysis of the measles and chickenpox data above, we do not 'smooth' the diatom data in any of the usual ways, although we take first-differences for reasons stated earlier. The resulting time series is shown in Fig. 6*a*.

The results of using the first half of the diatom series to predict the second half are shown in the usual way in Fig. 6*b*. Figure 6*c* shows ρ between predicted and observed results looking one time step ahead ($T_p = 1$), as a function of E. The optimum embedding dimension seems to be about 3. This value for E is consistent with our independent analysis of the data using the Grassberger–Procaccia algorithm, which indicates that $D \simeq 2$. Figure 6*d* shows ρ as a function of T_p (for $E = 3$). The consistent decay in predictive power as one extrapolates further into the future is consistent with the dynamics of the diatom population being partly governed by a chaotic attractor. This view is supported by the significantly better fit of the nonlinear predictor as compared with the optimal linear autoregressive model ($P < 0.0005$). We note, however, that deterministic chaos at best accounts for about 50% of the variance, with the rest presumably deriving from additive noise; the relatively low dimension of the attractor for diatoms compared with measles makes it plausible that the noisier fit of predicted weekly fluctuations in diatoms, versus the predicted monthly fluctuations in measles, reflects a much higher sampling variance for diatoms than for reported measles cases.

Conclusion

The forecasting technique discussed here is phenomenological in that it attempts to assess the qualitative character of a system's dynamics—and to make short-range predictions based on that understanding—without attempting to provide an understanding of the biological or physical mechanisms that ultimately govern the behaviour of the system. This often contrasts strongly with the laboratory and field-experiment approaches that are used to elucidate detailed mechanisms by, for example, many population biologists. The approach outlined here splits the time series into two parts, and makes inferences about the dynamical nature of the system by examining the way in which ρ (the correlation coefficient between predicted and observed results for the second part of the series) varies with prediction interval, T_p, and embedding dimension, E; given the low densities of most time series in population biology, we share Ruelle's[28] lack of confidence in a direct assessment of the dimension of any putative attractor by Grassberger–Procaccia or other algorithms. Our approach works with artificially generated time series (for which we know the actual dynamics, and the underlying mechanisms, by definition), and it seems to give sensible answers with the observed time series for measles, chickenpox and diatoms (deterministic chaos in one case, seasonal cycles with additive noise in another, and a mixture of chaos and additive noise in the third). We hope to see the approach applied to other examples of noisy time series in population biology, and in other disciplines in which time series are typically sparse. □

Received 12 July 1989; accepted 19 February 1990.

1. Lorenz, E. N. *J. atmos. Sci.* **26,** 636–646 (1969).
2. Tong, H. & Lim, K. S. *Jl. R. statist. Soc.* **42,** 245–292 (1980).
3. Farmer, J. D. & Sidorowich, J. J. *Phys. Rev. Lett.* **62,** 845–848 (1987).
4. Priestly, M. B. *J. Time Series Analysis* **1,** 47–71 (1980).
5. Eckman, J. P. & Ruell, D. *Rev. mod. Phys.* **57,** 617–619 (1985).
6. Crutchfield, J. P. & MacNamara, B. S. *Complex Systems* **1,** 417–452 (1987).
7. Casdagli, M. *Physica D,* **35,** 335–356 (1989).
8. Grassberger, P. & Procaccia, I. *Phys. Rev. Lett.* **50,** 346–369 (1983).
9. Schwartz, I. *J. math. Biol.* **21,** 347–361 (1985).
10. Schaffer, W. M. & Kot, M. *J. theor. Biol.* **112,** 403–407 (1985).
11. Schaffer, W. M. & Kot, M. in *Chaos: An Introduction* (ed. Holden, A. V.) (Princeton Univ. Press, 1986).
12. Schaffer, W. M., Ellner, S. & Kot, M. *J. math. Biol.* **24,** 479–523 (1986).
13. Schaffer, W. M., Olsen, L. F., Truty, G. L., Fulmer, S. L. & Graser, D. J. in *From Chemical to Biological Organization* (eds Markus, M., Muller, S. C. & Nicolis, G.) (Springer-Verlag, New York, 1988).
14. Yule, G. U. *Phil. Trans. R. Soc.* **A226,** 267–278 (1927).
15. Takens, F. in *Dynamical Systems and Turbulence* (Springer-Verlag, Berlin, 1981).
16. Farmer, J. D. & Sidorowich, J. J. in *Evolution, Learning and Cognition* (ed. Lee, Y. C.) (World Scientific, New York, 1989).
17. London, W. P. & Yorke, J. A. *Am. J. Epidem.* **98,** 453 (1973).
18. Pimm, S. L. & Redfearn, A. *Nature* **334,** 613–614 (1988).
19. Lawton, J. H. *Nature* **334,** 563 (1988).
20. Helsenstein, U. *Statist. Med.* **5,** 37–47 (1986).
21. Anderson, R. M. & May, R. M. *Nature* **318,** 323–329 (1985).
22. Anderson, R. M., Grenfell, B. T. & May, R. M. *J. Hyg.* **93,** 587–608 (1984).
23. Tont, S. A. *J. mar. Res.* **39,** 191–201 (1981).
24. Varosi, F., Grebogi, C. & Yorke, J. A. *Phys. Lett.* **A124,** 59–64 (1987).
25. Abarbanel, H. D., Kadtke, J. B. & Brown, R. *Phys. Rev.* **B41,** 1782–1807 (1990).
26. Mees, A. I. Research Report No. 8 (Dept Mathematics, University of Western Australia, 1989).
27. Drepper, F. R. in *Erodynamics* (eds Wolff, W., Soeder, C. J. & Drepper, F. R.) (Springer-Verlag, New York, 1988).
28. Ruelle, D. *Proc. R. Soc. A* (in the press).

ACKNOWLEDGEMENTS. We thank Henry Abarbanel, Martin Casdagli, Sir David Cox, Doyne Farmer, Arnold Mandell, John Sidorowich and Bill Schaffer for helpful comments, John McGowan and Sargun Tont for directing us to Allen's phytoplankton records, Alan Trombla for computing assistance, and the United States NSF and the Royal Society for supporting this research.

Tim Sauer
Department of Mathematical Sciences, George Mason University, Fairfax, VA 22030

Time Series Prediction by Using Delay Coordinate Embedding

We present a numerical algorithm for short-term prediction of a time series based on delay coordinate embedding. Filtered delay coordinates are used to reconstruct the time series in a space large enough to unfold the dynamical attractor. Within this space, local linear models are built of dimension at or less than the dimension of the attractor, by projecting the embedded data down to the top few singular value decomposition modes. These low-dimensional linear models are used for prediction ahead in time.

The algorithm is used to make short-term predictions for Data Set A of the Santa Fe Time Series Prediction and Analysis Competition, as well as for artificial data generated by the Lorenz attractor.

1. INTRODUCTION

The analysis of time series produced by linear processes is based on the fact that a finite-dimensional linear system produces a signal characterized by a finite number of frequencies. There exist highly successful methods of time series prediction based on the exploitation of this fact in the frequency domain or the time domain (such as ARMA models). The key considerations in this pursuit are the effect of noise on the model development and use of the model for prediction and other analysis. Brockwell and Davis (1987) is an example of this direction in time series analysis.

These conventional methods are generally ineffective for solving problems such as prediction and filtering for time series produced by nonlinear processes. In particular, a time series measured from a chaotic process typically has a continuous Fourier spectrum instead of a discrete set of frequencies. As a result of this fact, frequency domain methods lack applicability. Time domain methods such as ARMA models become inappropriate because a single global model no longer applies to the entire state space underlying the signal; see Gershenfeld and Weigend (this volume). Some beginning work on state-dependent models for nonlinear signals can be found in Priestley (1988) and Tong (1990).

Along with the discovery and recognition of chaotic systems during the past quarter-century, researchers have looked for an effective means of analyzing the time series produced by these systems. The major goals have been to use the measured time series for system identification and reconstruction, and for prediction and control of future behavior of the system.

One line of investigation that has become very successful is the reconstruction of the state space of a dynamical system using delay coordinates. Packard et al. (1980), in a letter on the reconstruction of chaotic attractors using independent coordinates from a time series, attribute the suggestion of using delay coordinates to D. Ruelle. Mathematical results on delay coordinates for nonlinear systems were first published by Takens (1981).

Eckmann and Ruelle (1985) took the idea one step further and suggested examining not only the delay coordinates of a point, but also the relation between the delay coordinates of a point and the point which occurs a number of time units later. In principle, one can then approximate not only the attractor, but also its dynamics. Their discussion centers on the computation of Lyapunov exponents from experimental data, but there are many other applications, including time series prediction. Implementations of prediction algorithms which followed include those of Farmer and Sidorowich (1985) and Casdagli (1989).

In this paper we describe an implementation which builds on the Eckmann-Ruelle suggestion by making use of conventional filters, interpolation, and the singular value decomposition to optimize prediction accuracy. We demonstrate the use of this algorithm on a time series produced by the Lorenz attractor, and on Data Set A of the Santa Fe Time Series Prediction and Analysis Competition, which consists of laser intensity data collected from a laboratory experiment as described by Hübner et al. (this volume). This prediction algorithm follows closely along the lines of the noise reduction algorithm in Sauer (1992).

Our implementation uses the information from a length w window of the time series to compute an m-dimensional vector with which to represent the current state of the system. This process is called a *low-pass embedding.* The dynamics in the m-dimensional reconstruction space is then modeled locally by a l-dimensional linear map, where the best l-dimensional subspace is chosen for the local map domain. The numbers w, m, and l, respectively, should be chosen according to the sampling rate of the time series, the approximate dimension needed to embed (unfold) the dynamical attractor in the reconstructed state space, and the approximate dimension of the attractor, respectively.

In most cases, we expect $w \geq m \geq l$. For best results, we recommend l significantly smaller than m (by at least a factor of two; see the theory in Section 2). We also recommend m smaller than w except in the case of a poorly sampled flow. In particular, the equality $w = m = l$ is not required. This "decoupling" of the three parameters w, m, and l allows more accurate predictions than the case $w = m = l$ implicit in previous implementations (see, for example, Farmer & Sidorowich, 1988).

Section 2 of the paper provides background on delay coordinate embedding of attractors, including the low-pass embedding used in the algorithm. In Section 3, the details of the prediction algorithm are described, and the choices that were made to arrive at our implementation are discussed informally. Section 4 contains the computational results of the algorithm.

2. EMBEDOLOGY

The *state* of a deterministic dynamical system is the information necessary to determine the entire future evolution of the system. A useful paradigm to keep in mind is a system of n ordinary differential equations in n variables. Under minimal requirements on the smoothness of the equations, the existence and uniqueness properties hold. That means that through any point $\mathbf{a} = (x_1, \ldots, x_n)$ in R^n, there exists a unique trajectory with $\mathbf{a}$ as the initial condition. Therefore, the future of the system that unfolds from state $\mathbf{a}$ is uniquely determined by $\mathbf{a}$. In this case the set of all possible states, or *state space*, is the Euclidean space R^n.

A time series is a list of numbers which are assumed to be measurements of an observable quantity over time. The system on which the observable is being measured is evolving with time: that is, it is a dynamical system. An observable quantity, such as the one generating the time series, is a function only of the state of the underlying system. If the system returns to the same state later, the observable will yield the same value for the time series.

This is the foundation of our approach to prediction. We identify the present state of the system which is producing the time series, and search the past history for similar states. By studying the evolution of the observable following the similar states, information about the future can be inferred.

To be more precise about our approach to prediction, we introduce some notation. Denote the present state of the system by $\mathbf{a}$, and call the measured quantity at the present time $h(\mathbf{a})$. By assumption, the present state $\mathbf{a}$ contains all information needed to produce the state t time units into the future, which we will call

$F_t(\mathbf{a})$. In these terms, the prediction problem is to calculate the observed quantity t time units in the future knowing only the present state. That is, given $\mathbf{a}$, find $P_t(\mathbf{a}) = h(F_t(\mathbf{a}))$. We might call P_t the *prediction function*.

Viewed in this way, the prediction problem breaks down into two subproblems: the *representation problem* and the *function approximation problem*. The representation problem consists of transforming the present state $\mathbf{a}$, which is an abstraction, into something concrete enough to be manipulated in a computer program. The function approximation problem consists of doing the same for the prediction function P_t.

In the representation problem, one attempts to reconstruct the present state from the time series or, in other words, to use the available data to find a vector $\mathbf{b}$ with which to replace the theoretical state $\mathbf{a}$. A particularly elegant solution to this problem is *delay coordinate embedding* (see Takens, 1981, and Sauer, Yorke, & Casdagli, 1991). Each state $\mathbf{a}$ is identified with a unique vector $\mathbf{b}$ in a Euclidean space R^m. A prediction of the time series t units into the future can be made by computing $P_t(\mathbf{b})$. We describe this approach to the representation problem in the remainder of Section 2.

Solving the second problem, the function approximation problem, means finding an efficient approximation of the function $P_t : R^m \to R$ using available data. If a good approximation can be found, then prediction involves locating the present state vector $\mathbf{b}$, and evaluating P_t there. We treat the function approximation problem in Section 3.

2.1 DELAY COORDINATES

The technique of delay coordinate reconstruction is to reproduce the set of dynamical states of a system using vectors derived from a time series measured from the system. Let A denote a compact finite-dimensional set of states of a system. For example, this set may consist of an equilibrium, periodic orbit, or chaotic attractor. Let $g : A \to R$ be an observation function which is a measurement of some quantity of the system, and let τ be a real number greater than zero. For each state $\mathbf{a} \in A$, one can define the m-dimensional vector

$$\mathbf{b} = [g(\mathbf{a}), g(F_{-\tau}(\mathbf{a})), \ldots, g(F_{-(m-1)\tau}(\mathbf{a}))]. \tag{1}$$

This vector is called a *delay coordinate vector* because its components consist of time-delayed versions of the observable of the system. It is an interesting fact that under quite reasonable conditions on the dynamics F_t of the system, this correspondence $D(\mathbf{a}) = \mathbf{b}$ is a one-to-one correspondence, as long as m is greater than twice the box-counting dimension of A, and the observation function g is chosen generically. This fact is called the Fractal Delay Coordinate Embedding Prevalence Theorem in Sauer, Yorke, and Casdagli (1991).

The vector **b** above is a segment of a time series with equally spaced data produced by the measurement function g. That is,

$$\mathbf{b} = [x_t, x_{t-\tau}, \ldots, x_{t-(m-1)\tau}] \tag{2}$$

where $x_t = g(\mathbf{a})$ is the value of the time series at time t and **a** is the state. Thus **b** is readily available.

Any tangent manifold structure that exists on A is also carried over by the correspondence D. In the original formulation of the theorem by Takens (1981), A is assumed to be a smooth manifold, and the conclusions are that the image $D(A)$ is also a smooth manifold and that D is a diffeomorphism. It is shown in Sauer, Yorke, and Casdagli (1991) that more generally, if A is a fractal attractor, then not only does the one-to-one correspondence hold, but also any manifold structure that exists, such as an unstable manifold, is faithfully represented on $D(A)$. This motivates the belief, for example, that the positive Lyapunov exponents of A can be measured on the reconstructed set $D(A)$, even when A is fractal and not a smooth manifold.

The one-to-one correspondence is useful because the state **a** of a deterministic dynamical system, and thus its future evolution, is completely specified by the corresponding vector **b**. Suppose, at a given time, one observes the vector **b** in reconstruction space R^m, and that this is followed one second later by a particular event. If the correspondence D of vectors with states is one-to-one, then each appearance of the measurements represented by **b** will be followed one second later by the same event. There is predictive power in finding a one-to-one map.

In practice, the precise measurements represented by **b** may not be repeated exactly. Moreover, noise in the measuring apparatus may prevent the exact reconstruction vector **b** that corresponds to **a** from being known. Yet if the correspondence is reasonably smooth (the correspondence is as smooth as the dynamics), similar measurements will predict similar events. This solution to the representation problem, giving concrete proxies for the invisible states, provides a foundation for attempting prediction.

2.2 FILTERED DELAY COORDINATES

Correspondences can be devised that are more sophisticated than Eq. (1). These are more adept at handling noise in practical situations, yet can be mathematically proved to yield a one-to-one correspondence. A long-term goal is to merge, as much as possible, the useful filtering techniques used in conventional signal processing with embedding ideas. In this work, we demonstrate a first step, using very simple filters based on the Fourier transform.

Defining a *filtered delay coordinate embedding* consists of replacing Eq. (1) with the more general

$$\mathbf{b} = M[g(\mathbf{a}), g(F_{-\tau}(\mathbf{a})), \ldots, g(F_{-(w-1)\tau}(\mathbf{a}))]^T. \tag{3}$$

In this formulation, M is an $m \times w$ matrix of rank m (so $w \geq m$). As before, $\mathbf{b}$ is an m-dimensional vector. Translating back to the time series, which is defined by $x_t = g(\mathbf{a})$, we are setting

$$\mathbf{b} = M[x_t, x_{t-\tau}, \ldots, x_{t-(w-1)\tau}]^T. \tag{4}$$

The vector $\mathbf{b}$ that corresponds to the state $\mathbf{a}$ has entries which are linear combinations of the entries of Eq. (2).

It can be shown that this filtered delay coordinate map, like the original delay coordinate map, is virtually assured to give a one-to-one correspondence of state $\mathbf{a}$ with vector $\mathbf{b}$. Specifically, if m is greater than twice the box-counting dimension of the attractor A, then almost every observation function g will lead to a one-to-one correspondence, except for possible collapsing of periodic points by M (see details in Sauer, Yorke, & Casdagli, 1991).

The theoretical results state that given a series of data that is noise-free, the hidden attractor A and the reconstruction $D(A)$ will have identical topological and dynamical properties. This fact motivates using the same approach in practice, where the data are collected with noise.

In the present case, we recommend defining $M = M_3 M_2 M_1$ to be the composition of the following three linear operations:

1. M_1 = FFT (discrete Fourier transform) of order w;
2. M_2 sets to zero all but the lowest $m/2$ frequency contributions; and
3. M_3 = inverse FFT of order m, using the remaining $m/2$ frequencies.

Multiplication by the matrix M has the effect of a low-pass filter on the length w window of the time series. This type of filtered delay coordinate map could be called a *low-pass embedding.* One starts with a length w section of the signal, and represents it in the reconstruction by a real vector of length m. For example, if $w = 32$ and $m = 16$, the information that is contained in Eq. (2) but missing from Eq. (4) is essentially the upper half of the Fourier spectrum. If the sampling rate is not too low, it is likely that the lower half of the spectrum will be sufficient to give a good approximation to the length w section of the clean signal underlying the noise.

The basic philosophy of the low-pass embedding is to use all available data for the purpose of attractor representation. Typically, a window containing a number of oscillations of the time series is used to develop a representation of the current state of the system. The window length w, in terms of samples of the time series, is dependent on the sampling rate and should be treated independently of the dimension m of the representation space.

The low-pass embedding is a way to intelligently "downsample" the data by a factor of w/m. In the absence of noise, there is no advantage of this approach over simply decimating the series by a factor of w/m (choosing every w/m point). In the presence of high-frequency noise (through digitization error, for example), however, increased representation accuracy is achieved by this type of downsampling.

2.3 INTERPOLATION

In the next section, we describe the use of local linear models of the dynamics in the reconstructed state space for prediction. The local model of the dynamics near the present state vector is built using the nearest neighbors of the present state. However, if the available data set is sparse, there may not be enough neighbors within the "radius of validity" of the linear model with which to fit the parameters of the model, and large errors will result. A further difficulty is that near neighbors that do exist will be translated in time an amount that is less than one sampling step.

A solution to this problem, illustrated by Figure 1, is upsampling: to artificially increase the sampling rate by interpolating the time series. This augments the existing reconstructed states by filling in the sampling interval. There are many types of interpolation that could be used. We chose to fit a section of the time series with a Fourier polynomial, and then sample the polynomial s equally spaced times each sampling period. A choice of a power of 2 for s makes the interpolation particularly convenient using the Fast Fourier Transform; we used $s = 16$ in all results reported in this paper.

Although in principle this interpolation is done prior to the filtering step of the previous section, in practice it may be inconvenient in terms of storage to explicitly interpolate and store the entire expanded series. As a matter of implementation, the interpolation can be done on an as-needed basis. Moreover, since Fourier interpolation uses the same frequency component information as the filtering step of the previous subsection, some efficiency can be gained by combining the two steps.

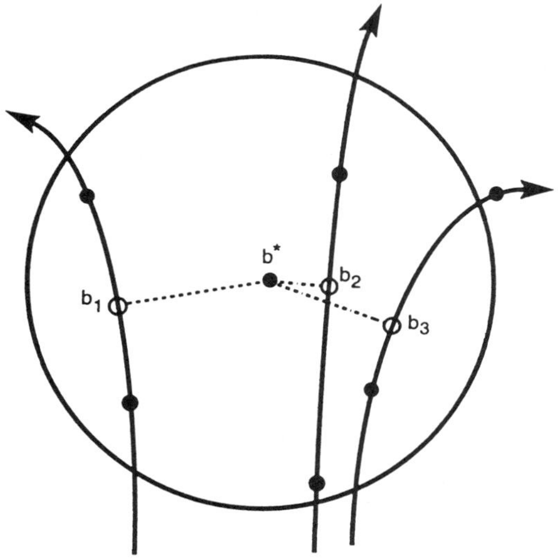

FIGURE 1 The reference point ($\mathbf{b}^*$) is shown together with its nearest neighbors (black dots) in reconstructed state space. By interpolating the time series, the more appropriate neighbors $\mathbf{b}_1, \mathbf{b}_2, \mathbf{b}_3$ (open dots) can be found. Our algorithm uses the nearest interpolated neighbor from each pass of the trajectory.

3. PREDICTION

In the last section we presented a solution to the first half of the prediction problem, that of the representation of the present state by a computable entity. Next, we proceed to the remaining problem of estimating the prediction function, which has as input the present state and as output a future value of the time series.

We refer to the known data series as the *training set*. Through filtered delay coordinate embedding, we can reconstruct the underlying dynamics. In fact, for each reconstructed state $\mathbf{b}$ in the training set, we can look up the value of the series t time units later, and call it $X = P_t(\mathbf{b})$.

To continue a time series, we find the reconstruction vector $\mathbf{b}^*$ corresponding to the end of the series, and use our knowledge of the training set to estimate $P_t(\mathbf{b}^*)$. There are many possible ways in which to do this estimation. We discuss one approach in the next section.

3.1 EVALUATION OF THE PREDICTION FUNCTION

Assume that $\mathbf{b}^*$ is a vector in R^m representing a state of the system. For example, $\mathbf{b}^*$ could be a filtered delay coordinate vector. We want to evaluate $P_t(\mathbf{b}^*)$, where P_t is the function that gives the value of the time series t time units into the future. To do this, search the training set for the nearest k neighbors $\mathbf{b}_1, \ldots, \mathbf{b}_k$ of $\mathbf{b}^*$ in R^m, subject to the condition that only one neighbor, the one nearest to $\mathbf{b}^*$, is chosen from each nearby trajectory segment. The interpolation described in the previous section should be fine enough that the direction from $\mathbf{b}^*$ to the nearest point on the trajectory should be virtually perpendicular to the trajectory, as in Figure 1.

Since each $\mathbf{b}_i$ was found from the training set, we also know the value $X_i = P_t(\mathbf{b}_i)$, which is the value of the observable t time units after the state of the system was $\mathbf{b}_i$. Note that X_i is a value (or an interpolated value) from the original time series, and not a filtered value. Even if filtered delay coordinates are being used for $\mathbf{b}_i$, the raw time series value (or its interpolate) should be used for X_i. Information will be lost if the entire time series is low-pass filtered in advance.

The remaining task is to use the X_i to calculate a corresponding value X^* for the observable t units after the state was $\mathbf{b}^*$. The way in which the k nearest neighbors of $\mathbf{b}^*$ are used to estimate the prediction $P_t(\mathbf{b}^*)$ is summarized in four steps:

1. Find the center of mass $\mathbf{c}$ of the neighbors $\mathbf{b}_1, \ldots, \mathbf{b}_k$.
2. For a fixed dimension $l \leq m$, find the best low-dimensional linear space R^l passing through the point $\mathbf{c}$. By "best," we mean the subplane R^l of R^m which minimizes the squared distances to the neighbors. This can be easily found using the singular value decomposition (SVD). For details of the SVD, see Golub and Van Loan (1989). To be precise, define the matrix A whose rows consist of the vectors $\mathbf{b}_1 - \mathbf{c}, \ldots, \mathbf{b}_k - \mathbf{c}$. The SVD of A will have form $A = U\Sigma V^T$, where U and

V are orthogonal matrices and Σ is diagonal with nonincreasing nonnegative entries. The l dominant right singular vectors (the first l columns of V) span the desired space R^l.

3. Project the data points $\mathbf{b}_1 - \mathbf{c}, \ldots, \mathbf{b}_k - \mathbf{c}$ down to this R^l, and form the affine (constant + linear) model $L : R^l \to R$ which best fits the data points

$$(\Pi(\mathbf{b}_1 - \mathbf{c}), X_1), \ldots, (\Pi(\mathbf{b}_k - \mathbf{c}), X_k)$$

where $\Pi : R^m \to R^l$ is the projection onto the best R^l. The model L has form

$$L(\mathbf{x}) = \mathbf{a} \cdot \mathbf{x} + d$$

where $\mathbf{a}$ is an l-vector and d is a constant scalar.

4. Project $\mathbf{b}^* - \mathbf{c}$ onto R^l and evaluate the model there. The result will be $L(\Pi(\mathbf{b}^* - \mathbf{c})) = X^*$, an estimate for $P_t(\mathbf{b}^*)$.

3.2 DISCUSSION

There are many unresolved issues relating to the way in which P_t is estimated. To the extent that the training set is large and noise is low, these issues become of lesser importance. But, in realistic situations, these optimal conditions are often absent, and predictions will be sensitive to the estimation method for P_t.

THE BIAS/VARIANCE DILEMMA AND THE CHOICE OF LOCAL MODEL. The challenge of continuing Data Set A of the Competition brings these issues into the foreground. The given time series has *length* 1,000 during which the amplitude undergoes periods of growing oscillations followed by collapses and reinsertion into a growth period. There are only four such collapses in the available training set. In this sense, there are only four examples with which to predict the next collapse and reinsertion level. By this measure, the training set is rather short.

Noise is also a major issue with Data Set A, which consists of 8-bit data. The known series consists of 1,000 integers between 3 and 255. At best, we can assume there is digitization noise of 0.5 units, which corresponds to 10–20% noise during the low-intensity period just after the collapse. Assuming that this part of the signal is important for determination of the reinsertion level at which the next growth period begins, this noise level is significant.

These twin problems, lack of data and presence of noise, have a large impact on the method used for approximating the prediction function P_t. Once it has been decided to use local models, there is a choice between linear and nonlinear local models. Even if linear models are chosen, the question remains "what should be the dimension l of the model?"

Severe constraints are put on this choice if very few independent neighbors (four, for example, in the case of Data Set A) of the reference point $\mathbf{b}^*$ representing the present state are available in the training set. A linear model $L : R^l \to R$ is of form

$$L(\mathbf{x}) = \mathbf{a} \cdot \mathbf{x} + d$$

and as such has $l + 1$ free parameters. If a nonlinear model were used in place of L, presumably far more parameters would be needed. Since each independent neighbor in the training set can contribute one constraint on the parameters, there are strict upper bounds on the sophistication of the model that can be used.

Of course, one can widen the search for neighbors, and accept training points that are farther away from the reference point for the purposes of determining the local model $L : R^l \to R$. In fact, this is where our method meets the bias/variance dilemma that is inherent in the study of nonparametric functional inference. (See Casdagli & Weigend, this volume, and Geman, Bienenstock, & Doursat, 1992, for a discussion.) Using a small number of neighbors increases the variance of the model's estimation, because of the presence of noise. If more far-flung neighbors are accepted in order to decrease the variance, the assumption that the model is locally valid is compromised (for example, the approximation of a curved manifold by the linear tangent plane) and the bias is increased. To summarize, noise causes high variance, but because of the restricted amount of data, any attempt to decrease the variance will tend to increase the bias.

These considerations motivated our choice of embedding dimension m and model dimension l. It is important for m to be high enough so that the delay coordinate vector $\mathbf{b}$ locates the true state $\mathbf{a}$ as closely as possible. It is also important, in the presence of noise and small data sets, to have the freedom to choose l relatively low.

The use of the singular value decomposition helps to concentrate the usefulness of the available data. The idea of sorting out principal directions on the reconstructed attractor using the SVD appears in the papers of Broomhead and King (1986) and Broomhead, Jones, and King (1987). By finding the dominant directions of the data and projecting onto them, the regression step is restrained from trying to fit parameters in the less relevant directions. This decoupling of the reconstruction dimension, represented by m, and the model dimension l, allows prediction to proceed in the presence of small data sets without the ill-conditioning problems associated with requiring $m = l$.

In this report we emphasize the importance of allowing independent choice of the signal window length w, the dimension m of the representation space of the reconstructed attractor, and the model dimension l. Roughly speaking, the choice of w depends on the sampling rate of the time series, the choice of m depends on the degrees of freedom explored by the dynamical attractor, and the choice of l is constrained, because of the bias/variance dilemma, by the amount of available data. This is one of the major differences between the present method and previous implementations of nonlinear forecasting using delay coordinates, for which the choice $w = m = l$ is commonly made.

WEIGHTED REGRESSION. Once a linear map $L : R^l \to R$ is chosen for the local model and the number of neighbors that will determine L are fixed, there are further choices of how to use the data to fit the model. Our experiments showed that weighted regression improved prediction performance. We weighted the contribution of neighbor $\mathbf{b}_i$ to the regression by the factor $(1 - (1/2)d_i^2)^3$, where

$$d_i = \frac{\text{dist }(\mathbf{b}_i, \mathbf{b}^*)}{\text{dist }(\mathbf{b}_k, \mathbf{b}^*)},$$

and where $\mathbf{b}_k$ is the furthest of the k neighbors from $\mathbf{b}^*$. This weighting strategy was ad hoc and is not vital to the success of the prediction method. We did not do extensive testing of different weighting strategies, but we can say that this was slightly superior to flat weights $d_i = 1$. No claim is made that this weighting scheme is optimal.

There are two approaches to the weighting issue that we have to suggest, and in fact, these approaches are also relevant to the other choices referred to in the preceding discussion. One approach is to use a combination of theory, heuristic, and exhaustive search to find the optimal weighting strategy. A second is to take an adaptation approach at this stage of the process, and to use the training set to optimize a simple model whose few weights vary locally through reconstruction space. Neither of these approaches have been systematically explored by this author. The same general procedure could also be applied to the model selection discussion above.

DIRECT VERSUS ITERATED PREDICTION. Assume we are given a time series $x_1, \ldots, x_N$ and asked to provide a continuation. We apply our method to predict one time unit ahead and get an estimate $\hat{x}_{N+1}$. To get an estimate for x_{N+2}, there are two obvious choices. The *direct prediction* method means that the original method is applied to $x_0, \ldots, x_N$ to predict two time units ahead. In contrast, *iterated prediction* means applying the method to $x_0, \ldots, x_N, \hat{x}_{N+1}$ to predict one unit ahead.

Much discussion has ensued over which choice is superior. The reliability of direct prediction is suspect because it is forced to predict farther ahead. On the other hand, iterated prediction uses $\hat{x}_{N+1}$, which is possibly corrupted data. Farmer and Sidorowich (1988), for example, argue that iterated prediction is superior, although under ideal conditions that may not be realized in practice.

We would suggest that the discussion should be refocused. The question should not be which is better, but how both approaches can be used to optimize prediction. One suggestion would be to average the two approximations for x_{N+2}; that is essentially what is done in our algorithm. This is a clear opportunity to exploit the availability of two approximations to minimize the variance of the estimate.

3.3 ALGORITHM

In this section we put together the pieces from the previous sections and describe the prediction algorithm. There are five parameters to be chosen ahead of time. They are:

1. s = interpolation steps per sample period;
2. w = window length (length of input window for filtered embedding);
3. m = low-pass embedding dimension;
4. l = model dimension (number of dominant SVD modes projected onto); and
5. k = number of neighbors used to fit a one-dimensional linear model.

We begin with a scalar time series $\{x_1, \ldots, x_N\}$. By the interpolation step discussed above, we can instead consider the time series to have length sN, where s denotes the number of interpolation steps per sample period. Then each window of length w, in the original time units, can be made into a filtered delay coordinate vector, of dimension m, as discussed in Section 2.2. There will be approximately sN such vectors.

We will call the time interval over which the prediction is made the *prediction horizon.* For each prediction horizon t and each filtered delay vector $\mathbf{b}$, one can look up the value $P_t(\mathbf{b})$ of the interpolated, unfiltered time series corresponding to the time t units after the state vector was $\mathbf{b}$. These are used as in Section 3.1 to evaluate the prediction function P_t when predictions are needed.

Continuing the time series $\{x_1, \ldots, x_N\}$ means producing $x_{N+1}, x_{N+2}, \ldots$ which are extrapolations of the given series. The method for producing a value x_i, for $i > N$, of the series continuation takes advantage of the fact that there are many possible estimators for x_i. In fact, for some previous time $j < i$, let $\mathbf{b}_j$ be the filtered delay coordinate vector that represents the state at time j. Then $P_{i-j}(\mathbf{b}_j)$ is an estimator for x_i. Our suggestion is to average many of these available estimators to continue the series.

We calculate each series continuation value x_i for $i > N$ as

$$x_i = \frac{1}{w} \sum_{j=i-w}^{i-1} P_{i-j}(\mathbf{b}_j)$$

where $\mathbf{b}_j$ is the filtered delay coordinate vector at time j. In other words, the prediction for time i is the average of the predictions for time i from the previous w time steps. This is a mixture of direct and iterated prediction.

4. EXAMPLES

In this section we exhibit the result of applying the algorithm outlined above to two different time series. The first is computer-generated data from the Lorenz attractor. The second is laboratory data measured from a pumped far-infrared laser in a chaotic state, provided as Data Set A of the Competition. (The relation between the Lorenz equations and Data Set A is explored by Hübner et al., this volume.)

We report on several runs of the algorithm in the following. For comparison purposes, the parameters of the algorithm were set identically on all runs:

- s = interpolation steps per sample period = 16;
- w = window length = 32;
- m = low-pass embedding dimension = 16;
- l = model dimension = 1; and
- k = number of neighbors used to fit a one-dimensional linear model = 4.

4.1 LORENZ EQUATIONS

The Lorenz attractor time series was generated by solving the Lorenz equations (Lorenz, 1963):

$$\begin{aligned} \dot{x} &= \sigma(y - x) \\ \dot{y} &= \rho x - y - xz \\ \dot{z} &= -\beta z + xy \end{aligned} \tag{6}$$

where the parameters are set at the standard values $\sigma = 10$, $\rho = 28$, $\beta = 8/3$. Solutions to this system of three differential equations exhibit the sensitive dependence on initial conditions which is characteristic of chaotic dynamics. In realistic situations, knowledge of the true state of a system can be done only in finite precision. In such cases, sensitivity to initial conditions rules out long-term prediction. On the other hand, short-term prediction is possible to the extent that the current position can be estimated and that the dynamics can be approximated.

A long trajectory of the Lorenz attractor was generated using a differential equation solver, and the x-coordinate of the trajectory was sampled (with a sampling period of $\Delta t = 0.05$) to create a univariate time series. At this sampling rate, the x-coordinate completes an oscillation every 15 to 20 samples.

For the purposes of this exercise, no dynamical information about the Lorenz equations was used by the algorithm. The time series up to a certain point was input to the algorithm, and the continuation was produced.

In Figure 2, we show an example of using 10,000 training points to predict the continuation of the clean Lorenz time series. The dashed curve is the true continuation of the Lorenz data for the next 200 time steps (.05 per time step), and the solid curve is the series predicted by the algorithm of this paper.

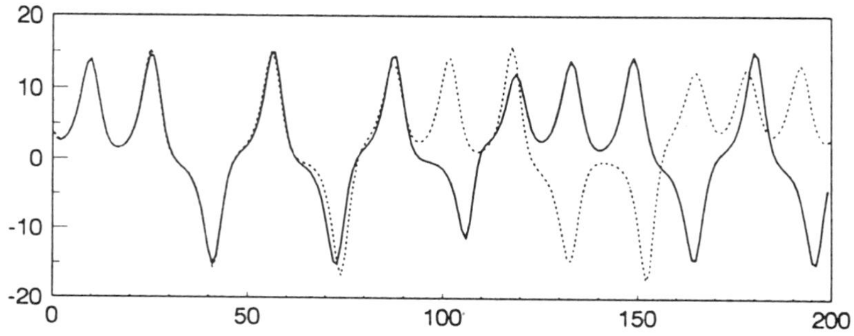

FIGURE 2 Time series prediction for signal from Lorenz attractor. Solid curve is predicted time series, and dashed curve is true time series. Predictions start at time 0 and stay near correct signal for around 90 time units.

Knowledge of the dynamics of the Lorenz system shows us that each time the signal passes through zero, the corresponding state on the attractor passes through a region of high sensitivity, and the predicted signal becomes susceptible to error. Most of the time, we see the divergence between predicted and true at these points. A positive aspect of the predictions in Figure 2, which is also true for the predicted continuations of the experimental data in the next section, is that even after the predictions diverge from the true trajectory determined by the given time series, the predictions seem to follow *some* true trajectory of the system. Thus, this algorithm may be useful not only for short-term prediction, but for long-term simulation.

Fifty examples of continuations as in Figure 2 were run, each using 10,000 training points, which corresponds to approximately 600 oscillations of the Lorenz signal. The results are summarized in Figure 3, where the root mean square error of the prediction is graphed against the prediction horizon. For the purposes of this graph, the average errors over ten neighboring predictions have been gathered together. Each data point plotted at t on the prediction horizon is the RMS of the errors between $t - 4$ and $t + 5$ over all 50 runs.

The graph shows a monotonic increase in error up to around 8, the (root mean square) size of the Lorenz attractor, after which the prediction is uncorrelated with the true time series. The error bars represent the sample standard deviations, and are quite large, considering that they represent an aggregate of 50 repetitions. Their size points to the large amount of variance in the accuracy of the predicted continuations.

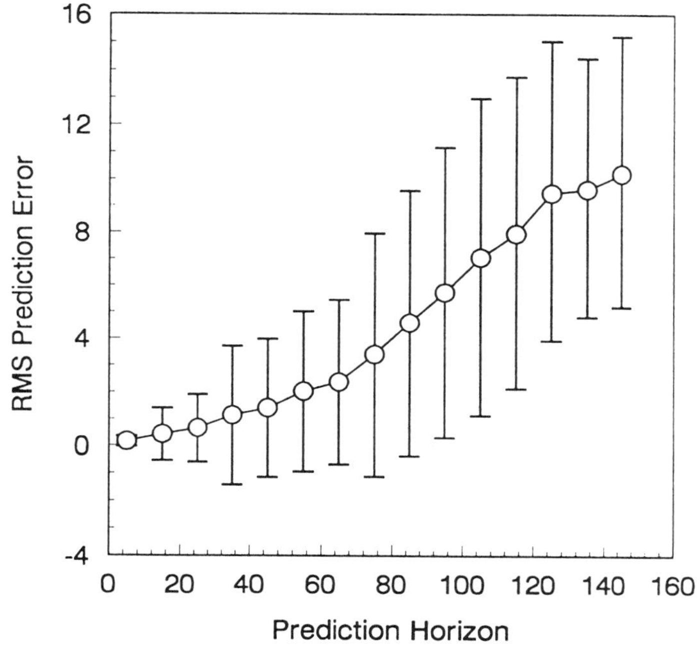

FIGURE 3 Root mean square prediction error is plotted as a function of the time interval of the prediction. The errors are an average over 50 runs. A training set of 10,000 points, sampled at $\Delta t = 0.05$, was used.

4.2 EXPERIMENTAL DATA

Our algorithm was also applied to the laser data analyzed in Hübner et al. (1989). Data Set A provided by the organizers of the Competition consisted of a time series of this data of length 1,000. The goal of the competition was to produce a continuation of the provided data of length 100.

One of the motivations of the laser experiment described in Hübner et al. (1989) was to develop a physical system that was closely modeled by the Lorenz equations. The parameter values of the Lorenz equations needed for the correspondence are different than those used in the previous subsection, so no direct comparison can be made with those results. (See Hübner et al., this volume.)

We show the continuation of the length 1,000 series produced by our algorithm in Figure 4. As before, the dashed curve is the true continuation, and the solid curve is given by the algorithm. A continuation of 400 points is given. The oscillatory signal grows slowly in intensity until it reaches a transitional state, followed

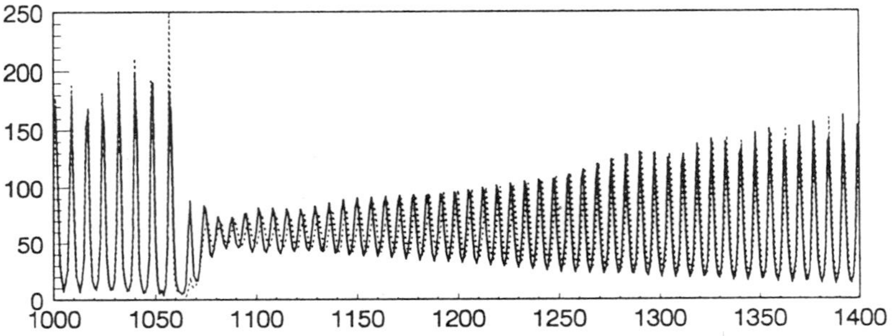

FIGURE 4 A continuation of length 400 of Data Set A of the Competition. The solid curve is the predicted continuation, and the dashed curve is the correct continuation.

by a reinsertion into the slowly varying region of phase space. The transition time can be well estimated, but the phase and magnitude beyond the transition can be problematic. More details are given by Gershenfeld and Weigend (this volume), comparing our prediction with predictions obtained with other algorithms such as neural networks.

After the close of the competition, an expanded version of the laser data was released. The longer data set contained 10,000 points, including the original set of 1,000 points and its continuation. Using this extra data, more continuations could be attempted, to test algorithm performance in a more statistically significant way. Figure 5 shows the results of four further attempts at continuing the time series, using the same training set, but with continuations beginning at other points. We emphasize that the training set for the four runs in Figure 5 is the same as for the first (Figure 4), namely, the original 1,000 points provided for the competition. Predicted continuations of 200 points are given by the solid curve, compared with the true continuation in the dashed curve.

A measure of prediction accuracy is given by the *normalized mean-squared error* (Gershenfeld & Weigend, this volume)

$$\text{NMSE} = \frac{1}{\sigma^2 N} \sum_{i=1}^{N} (x_i - \hat{x}_i)^2, \tag{7}$$

where x_i is the true value of the ith point of the series of length N, $\hat{x}_i$ is the predicted value, and σ is the standard deviation of the true time series during the prediction interval. In other words, NMSE is the ratio of mean squared errors of the prediction method in question and the method which predicts the mean at every step.

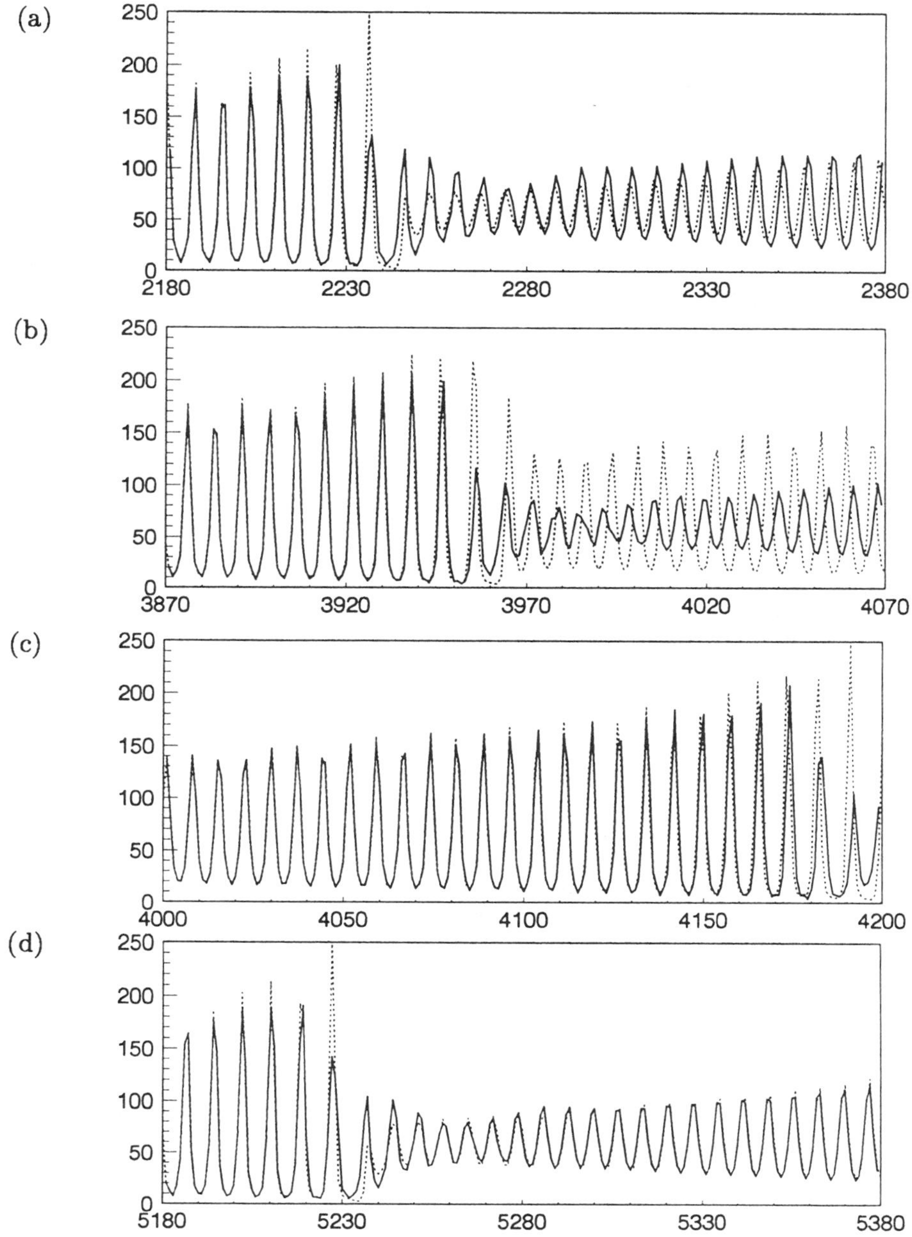

FIGURE 5 Four continuations of length 200 of Data Set A, each with a different starting point. The training set is the same 1,000 points provided as part of the competition. The solid curve is the predicted continuation, and the dashed curve is the true continuation.

TABLE 1 Performance Statistics for Predictions of Set A, Using Original 1,000 Points as the Training Set[1]

Starting point	Duration	NMSE	*Wan, this volume*[2]
1000	100	0.077	(0.027)
	200	0.199	
2180	100	0.174	(0.065)
	200	0.366	
3870	100	0.183	(0.487)
	200	0.617	
4000	100	0.006	(0.023)
	200	0.254	
5180	100	0.111	(0.160)
	200	0.093	

[1] Normalized mean-squared error (NMSE) is given for ten different prediction runs with five different starting points. The first two runs correspond to the first half of Figure 4. The remaining runs are displayed in Figure 5.

[2] The corresponding values of the neural network by Wan (this volume) are provided for comparison by the editors.

Table 1 summarizes the prediction error measured by NMSE of the algorithm on the laser data. Prediction errors are given for continuations of length 100 and of length 200. The first two runs of Table 1, starting at 1,000, refer to the first quarter and first half of Figure 4, respectively. The remainder of Table 1 refers to the continuations of Figure 5, which are of length 200. Therefore, the NMSE of the continuations from point 2180 of the laser data set are given in Table 1, for the 100-point and 200-point continuation, which correspond to the first half and full length of Figure 5(a), respectively. The other three starting points for continuation are 3870, 4000, and 5180, and similarly, the errors in Table 1 refer to the respective continuations pictured in Figure 5.

As in the case of prediction of the Lorenz data, there is a large variance in the results, depending on the success with which the continuation handles the reinsertion (or, more accurately, mode switching, in the physical system). An example with particularly large error for prediction horizon over 100 is the predicted continuation beginning at 3870 in Figure 5(b). The probable cause of the large error in these predictions presumably is due to the fact that the behavior of the time series in the range 3960–3980 appears atypical, compared to the behavior seen in the training set composed of the range 1–1000.

5. SUMMARY

Eckmann and Ruelle (1985) explained how the dynamics of a process can be reconstructed from an experimental time series. It follows that, in principle, short-term prediction of a deterministic time series is possible, even in the case that the system producing the time series is chaotic. The implementation described in this paper follows their suggestion of using delay coordinate embedding to reconstruct the attractor, and local models to fit the dynamics of the system.

In contrast to earlier implementations of their idea, we allow the parameters w = window length, m = embedding dimension, and l = model dimension to be chosen independently. This is essential for successful prediction because the three parameters govern quite separate aspects of state space representation and prediction.

The prediction algorithm is applied to artificial data generated from the Lorenz system of differential equations and to laboratory data from a laser experiment. The summary data collected in Figure 3 and Table 1 document the prediction accuracy attained.

6. REFERENCES

P. Brockwell, R. Davis, *Time Series: Theory and Methods*. Springer-Verlag, New York (1987).

D.S. Broomhead, G.P. King, Extracting qualitative dynamics from experimental data. Physica **20D** 217-236 (1986).

D.S. Broomhead, R. Jones, and G.P. King, Topological dimension and local coordinates from time series data. J. Phys. A **20** L563-L569 (1987).

M. Casdagli, Nonlinear prediction of chaotic time series. Physica **35D** 335-356 (1989).

J.-P. Eckmann and D. Ruelle, Ergodic theory of chaos and strange attractors. Rev. Mod. Phys. **57** 617-656 (1985).

J. D. Farmer, J. Sidorowich, Exploiting chaos to predict the future and reduce noise, in: Evolution, Learning and Cognition, ed. Y.C. Lee (World Scientific, Singapore, 1988).

S. Geman, E. Bienenstock, R. Doursat, Neural networks and the bias/variance dilemma. Neural Comp. **4** 1-58 (1992).

G. Golub and C. Van Loan, *Matrix Computations*. Second edition, The Johns Hopkins University Press (1989).

U. Huebner, N.B. Abraham, and C.O. Weiss, Dimensions and entropies of chaotic intensity pulsations in a single-mode far-infrared NH_3 laser. Phys. Rev. A **40** 6354 (1989).

E. Lorenz, Deterministic non-periodic flow. J. Atmos. Sci. **20** 130 - 141 (1963).

M. Priestley, *Nonlinear and Nonstationary Time Series Analysis*. Academic Press (1988).

N. Packard, J. Crutchfield, D. Farmer, R. Shaw, Geometry from a time series. Physical Review Letters **45**, 712 (1980).

T. Sauer, A noise reduction method for signals from nonlinear systems. Physica D **58**, 193 - 201 (1992).

T. Sauer, J.A. Yorke, M. Casdagli, Embedology. J. Stat. Phys. **65** 579-616 (1991).

F. Takens, Detecting strange attractors in turbulence. Lecture Notes in Math. **898**, Springer-Verlag (1981).

H. Tong, *Nonlinear Time Series*. Clarendon Press, Oxford (1990).

CHAPTER 11

Noise Reduction

E.J. Kostelich, J.A. Yorke
Noise reduction in dynamical systems
Phys. Rev. A **38**, 1649 (1988).

P. Grassberger, R. Hegger, H. Kantz, C. Schaffrath, T. Schreiber
On noise reduction methods for chaotic data
Chaos **3**, 127 (1993).

S. Hammel
A noise reduction method for chaotic systems
Phys. Lett. A **148**, 421 (1990).

Editors' Notes

Conventional approaches to noise filtering are largely inapplicable to signals measured from chaotic systems, because of the fact that chaotic signals typically exhibit a continuous frequency power spectrum. **Kostelich and Yorke** (1989) described an approach to noise reduction for chaotic signals corrupted with measurement noise, which explicitly uses the local linear dynamics reconstructed from the signal. This method allowed significant noise reduction to be achieved when applied to experimental data. Grassberger and Schreiber (1991) describe a simpler algorithm with comparable power. They show that correlation dimension calculations can be made with much improved accuracy when a noise reduction algorithm is used.

Cawley and Hsu (1992) and Sauer (1992) describe methods which enhance the accuracy of the local linear dynamics by employing standard filters, including the singular value decomposition, and in the latter case, bandpass filtering as a prefilter. These methods allowed significantly larger amounts of noise to be handled. The paper of **Grassberger et al.** (1993) surveys these two methods and suggests a few modifications that result in a powerful noise reduction algorithm.

The paper of **Hammel** (1990) solves the problem of reducing noise from a signal produced by a system whose local linear dynamics are known. The structure of the local stable and unstable manifolds is used to iteratively refine the orbit producing the signal. This direction was also taken up in Farmer and Sidorowich (1991).

Further reading

A survey of all of these methods, which identifies common features and analyzes the goals and limitations of the particular methods, is the following:

KOSTELICH, E.J., SCHREIBER, T., Noise reduction in chaotic time-series data: A survey of common methods. Phys. Rev. E **48**, 1752 (1993).

Noise reduction in dynamical systems

Eric J. Kostelich
Center for Nonlinear Dynamics and Department of Physics, University of Texas, Austin, Texas 78712

James A. Yorke
Institute for Physical Science and Technology and Department of Mathematics, University of Maryland, College Park, Maryland 20742
(Received 29 October 1987)

A method is described for reducing noise levels in certain experimental time series. An attractor is reconstructed from the data using the time-delay embedding method. The method produces a new, slightly altered time series which is more consistent with the dynamics on the corresponding phase-space attractor. Numerical experiments with the two-dimensional Ikeda laser map and power spectra from weakly turbulent Couette-Taylor flow suggest that the method can reduce noise levels up to a factor of 10.

The ability to extract information from time-varying signals is limited by the presence of noise. Methods of noise reduction are a subject of widespread interest in communication,[1] physical systems,[2] and experimental measurements.[3] Recent experiments to study the transition to turbulence in systems far from equilibrium, like those by Fenstermacher *et al.*,[4] Behringer and Ahlers,[5] and Libchaber *et al.*,[6] succeeded largely because of instrumentation that enabled them to quantify and reduce the noise.

In recent years, traditional methods of time series analysis like power spectra have been augmented by new methods. In many cases, the time series can be viewed as a dynamical system with a low-dimensional attractor that can be reconstructed from the time series using time delays.[7] Because the dynamics of the phase-space attractor are not localized in a time or frequency domain, traditional noise-reduction methods like Wiener[8] and Kalman[9] filters are not applicable. In this paper we describe a noise-reduction procedure that works by taking many nearby points in phase space (corresponding to widely varying times in the original signal) to find a local approximation of the dynamics. These approximations can be used collectively to produce a new time series whose dynamics are more consistent with those on the phase-space attractor. We demonstrate the efficacy of the method using chaotic attractors obtained from the Ikeda laser map[10] and a Couette-Taylor fluid flow experiment.[11]

The discrete sampling of the original signal means that the points on the reconstructed attractor can be treated as iterates of a nonlinear map f whose exact form is unknown. However, we assume that f is nearly linear in a small neighborhood about each attractor point x_i and write

$$x_{i+1}=f(x_i)\approx A_i x_i+b_i\equiv L(x_i)$$

for some matrix A_i and vector b_i. (The matrix A_i is the Jacobian of f at x_i.) This can be done with least-squares procedures similar to those described in Ref. 12. Let $\{x_j\}_{j=1}^{n}$ be a collection of points in a small ball around the ith reference point, and let $y_i=f(x_j)$ denote the observed image of x_j. The kth row $a_i^{(k)}$ of A_i and kth component $b_i^{(k)}$ of b_i are given by the least-squares solution of the equation

$$y_j^{(k)}=b_i^{(k)}+(a_i^{(k)}\,|\,x_j)\ , \tag{1}$$

where $y_j^{(k)}$ is the kth component of y_j and (|) denotes the dot product. (Farmer and Sidorowich[13] have generated similar approximations for the different purpose of forecasting chaotic time series.)

We remark that Eq. (1) can be ill-conditioned, for example, when the unstable manifold at x_i is nearly one dimensional and A_i is 2×2. We detect this situation by computing the singular values and right singular vectors[14] of the matrix X whose jth row is x_j to find the condition number of X, which is defined as the ratio of the largest to the smallest singular value. When the condition number is sufficiently large, we solve Eq. (1) using the components of x_j contained in the subspace spanned by the singular vectors corresponding to the largest singular values. (For instance, we find a one-dimensional linear approximation of f wherever the points x_j fall nearly along a single line.) Moreover, because error exists both in the points x_j and their observed images y_j, a modified least-squares procedure as described in Ref. 15 often gives better estimates of A_i and b_i.

In the second stage of the method, we use the linear (more precisely, linear + constant) maps L to correct errors in the observed trajectories as follows. Given a "window" of consecutive points $\{x_i, x_{i+1}, \ldots, x_{i+p}\}$ on the observed trajectory, we find the collection of points $\{\hat{x}_i, \hat{x}_{i+1}, \ldots, \hat{x}_{i+p}\}$ closest to the observed ones which also best satisfy the corresponding linear maps. More precisely, the new trajectory $\{\hat{x}_{i+k}\}_{k=0}^{p}$ minimizes the sum of squares

$$\sum_{k=0}^{p} \|\hat{x}_{i+k} - L(\hat{x}_{i+k-1})\|^2 + \|\hat{x}_{i+k} - x_{i+k}\|^2 + \|\hat{x}_{i+k+1} - L(\hat{x}_{i+k})\|^2 \quad (2)$$

(terms with subscripts outside $[i,i+p]$ are omitted). This procedure can be iterated by replacing the original trajectory $\{x_i\}$ with the most recent least-squares trajectory $\{\hat{x}_i\}$, then finding a new solution to Eq. (2).[16] Moreover, the windows can overlap; for instance, the second window can begin in the middle of the first.

We have conducted numerical experiments using the attractor produced by the Ikeda map $f(z)=\rho + c_2 z \exp\{i[c_1 - c_3/(1+|z|^2)]\}$, which models the dynamics of a bistable laser cavity.[10] We consider the attractor for the mapping $z_{j+1}=f(z_j)$, where $\rho=1$, $c_1=0.4$, $c_2=0.9$, $c_3=6$. (The complex number z_j is identified with the 2-vector x_j.) Numerical evidence suggests that initial conditions in $[0.5,1.8]\times[-2,1]$ are in the basin of a chaotic attractor whose numerically calculated Lyapunov exponents are $0.7296, -1.034$ (logarithms base 2) and whose Lyapunov dimension is 1.71.

We measure the noise level in terms of the *pointwise error* $e_j = \|x_{j+1} - L(x_j)\|$, i.e., the distance between the observed image and the predicted one [using the linear maps from Eq. (1)]. The *mean error* is $E = (\sum_j e_j^2/N)^{1/2}$, the root-mean-square value of the pointwise error over all N points on the attractor. We define the *noise reduction* $R = 1 - E_{\text{fitted}}/E_{\text{noisy}}$, where the mean errors are computed for the adjusted and original noisy attractor, respectively. (R is a measure of the self-consistency of the time series, assuming that the linear maps are accurate approximations of the true dynamics.)

The numerical experiments on the Ikeda attractor use 65 536 iterates, to which 0.1% uniformly distributed random noise is added. The noise is independent of the dynamics, i.e., the input to the computer program is the series $\{z_j + \eta_j : \eta_j \text{ random}\}$, for which $E_{\text{noisy}} = 7.588 \times 10^{-4}$. The linear maps L are computed using at least 50 points about each attractor point. Points are collected until the condition number of Eq. (1) is less than ten.[17] Trajectory adjustment is done in windows of 24 points, and the windows overlap by two points. After noise reduction, $E_{\text{fitted}} = 1.178 \times 10^{-4}$, so that the total noise reduction R is 84%. When 1% noise is added, we find $R = 83\%$.

We have performed similar numerical trials with the Hénon attractor,[18] for which the $(j+1)$st time series value is given by

$$x_{j+1} = f(x_j, x_{j-1}) = 1 - 1.4x_j^2 + 0.3x_{j-1} .$$

In this case the pointwise error can be measured exactly by replacing L with f (the mean error E then becomes a "correctness index"). When 1% noise is added to the in-

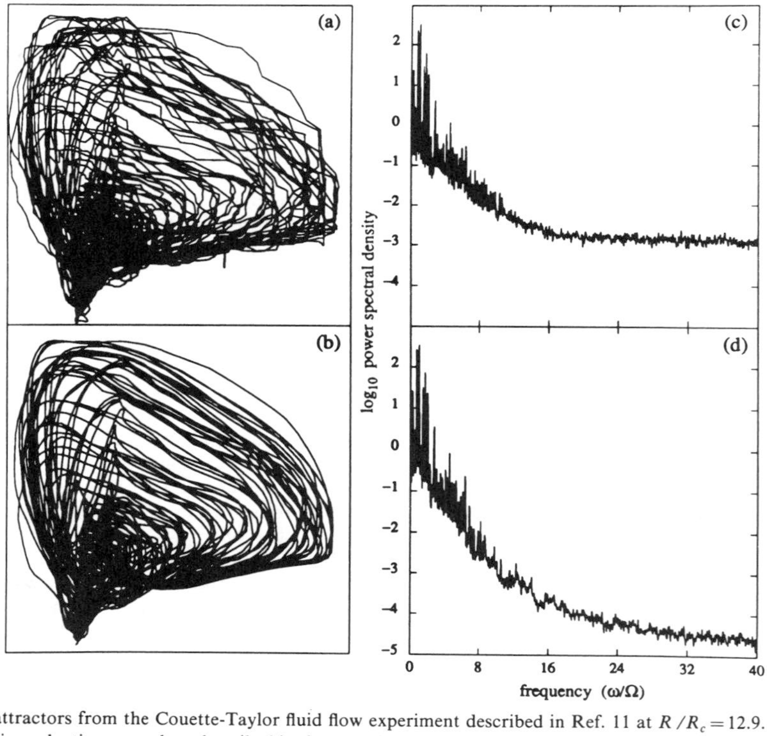

FIG. 1. Chaotic attractors from the Couette-Taylor fluid flow experiment described in Ref. 11 at $R/R_c = 12.9$. (a) Raw data. (b) Attrator after the noise reduction procedure described in the text. (c) and (d) Power spectra corresponding to (a) and (b), respectively. The units of frequency and power spectral density are as described in Ref. 11.

put as described above, the noise reduction (measured with the actual map) is 79%.[19] In addition, noise levels can be reduced almost as much in cases where the noise is added to the dynamics, i.e., where the input is of the form

$$\{x_{j+1}:x_{j+1}=f(x_j+\eta_j,x_{j-1}+\eta_{j-1}),\eta_j,\eta_{j-1}\text{ random}\}\ .$$

Next we consider the application of the method to data from the Couette-Taylor fluid flow experiment described in Ref. 11. Figure 1(a) shows a two-dimensional phase portrait of the raw time series at a Reynolds number $R/R_c=12.9$, which corresponds to weakly chaotic flow.[11] The corresponding phase portrait from the filtered time series[20] is shown in Fig. 1(b). The noise reduction, using the above criterion with the linear maps, is 63%.

Figure 1(c) and 1(d) show the power spectra for the corresponding time series. We emphasize that the dynamical information used to adjust the trajectories (viz., the motion of ensembles of points which are close together in phase space) corresponds to portions of the original signal that are widely and irregularly spaced in time. One question therefore is whether reducing the high-frequency noise corresponds to discovering the true dynamics which have been masked by noise. We believe that the answer is yes, based on those cases where there is an underlying low-dimensional dynamical system. However, in chaotic process some high-frequency components remain, because they are appropriate to the dynamics.

The method is particularly useful in calculations of dynamical quantities such as metric entropy and attractor dimension from experimental data. As an example, we consider the correlation dimension.[21] Let $C(x_i,\epsilon)$ denote the fraction of points on the attractor that fall within a distance ϵ of a randomly chosen (with respect to the natural measure) reference point x_i. Let $C(\epsilon)$ be the average values of $C(x_i,\epsilon)$ over the reference points x_i. Then $C(\epsilon)\sim\epsilon^d$ for small ϵ, where d is the correlation dimension.[21]

The dimension calculation illustrated in Fig. 2 is for the Ikeda attractor described above. The value of $C(\epsilon)$ is estimated from 1000 reference points using 48 values of ϵ, equally spaced on a logarithmic scale from 2^{-10} to 2^{-2} (only the range $2^{-6}\le\epsilon\le 2^{-3}$ is shown). Distances are normalized so that the total attractor extent is 1. The dimension, which is estimated as the derivative of $\log C(\epsilon)$ with respect to $\log\epsilon$, is taken as the slope of the regression line through six consecutive $(\log\epsilon,\log C(\epsilon))$ pairs. Although the noise level in the input is only 1%, noise inflates the dimension estimate even at ball sizes which are 3% of the attractor extent (top curve). However, the

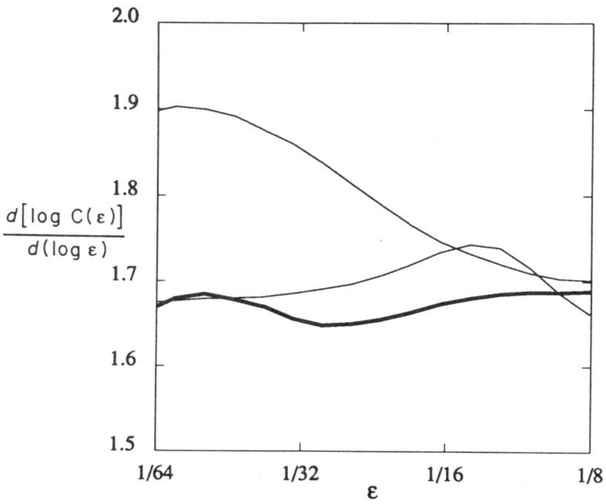

FIG. 2. Grassberger-Procaccia correlation dimension d for the Ikeda attractor, using a data set with 1% uniformly distributed random noise (top curve), the same data set after the noise-reduction procedure (middle curve), and the original noiseless attractor (bold curve). The Lyapunov dimension of the attractor is 1.71.

dimension estimate for the fitted attractor (middle thin curve) compares favorably to that obtained from the noiseless attractor (bold line).

Since accurate linear approximations are essential for the success of the method, there must be an ample number of points in a small neighborhood about each point on the attractor. Thus, the data requirements depend on the dimension of the attractor. This method is best suited to situations where large amounts of data can be collected but the measurement precision is limited. This method promises to be of considerable value in the analysis of experimental data when the time series can be viewed as arising from a dynamical system with a low-dimensional attractor.

We thank Anke Brandstäter for providing data from her Couette-Taylor experiment. E. K. thanks Bill Jefferys, Andrew Fraser, Randy Tagg, and Harry Swinney for helpful discussions. This research is supported by the U.S. Department of Energy Office of Basic Energy Sciences, U.S. Air Force Office of Scientific Research, U.S. Defense Advanced Research Projects Agency–Applied and Computational Mathematics Program.

[1]For instance, see R. G. Gallager, *Information Theory and Reliable Communication* (Wiley, New York 1968).

[2]For instance, see *Sixth International Conference on Noise in Physical Systems,* edited by P.H.E. Meijer, R. D. Mountain, and R. J. Soulen (U.S. GPO, Washington, D.C., 1981), and similar volumes.

[3]For instance, see D. R. Brillinger, *Time Series: Data Analysis and Theory,* (Holden-Day, San Francisco, 1980); M. Priestly, *Spectral Analysis and Time Series* (Academic, London, 1981), two volumes.

[4]P. R. Fenstermacher, H. L. Swinney, and J. P. Gollub, J. Fluid Mech. **94**, 103 (1979).

[5]R. P. Behringer and G. Ahlers, J. Fluid Mech. **125**, 219 (1982); G. Ahlers and R. P. Behringer, Phys. Rev. Lett. **40**, 712 (1978).

[6]A. Libchaber, S. Fauve, and C. Laroche, Physica D **7**, 73 (1983).

[7]Examples of the use of this method for the analysis of experimental data are given in *Dimensions and Entropies in Chaotic Systems*, edited by G. Mayer-Kress (Springer-Verlag, Berlin, 1986).

[8]For example, see L. R. Rabiner and B. Gold, *Theory and Application of Digital Signal Processing* (Prentice-Hall, Englewood Cliffs, N.J., 1975).

[9]For example, see A. E. Bryson and Y. C. Ho, *Applied Optimal Control* (Ginn, Waltham, Mass., 1969).

[10]S. M. Hammel, C. K. R. T. Jones, and J. V. Moloney, J. Opt. Soc. Am. B **2**, 552 (1985).

[11]A. Brandstäter and H. L. Swinney, Phys. Rev. A **35**, 2207 (1987).

[12]J.-P. Eckmann and D. Ruelle, Rev. Mod. Phys. **54**, 617 (1985); J.-P. Eckmann, S. O. Kamphorst, D. Ruelle, and S. Ciliberto, Phys. Rev. A **34**, 4971 (1986); M. Sano and Y. Sawada, Phys. Rev. Lett. **55**, 1082 (1985).

[13]J. D. Farmer and J. J. Sidorowich, Phys. Rev. Lett. **59**, 845 (1987).

[14]J. J. Dongarra, C. B. Moler, J. R. Bunch, and G. W. Stewart, *LINPACK User's Guide* (Society of Industrial and Applied Mathematics, Philadelphia, 1979).

[15]W. H. Jefferys, Astron. J. **85**, 177 (1980); *ibid.* **86**, 149 (1981); M. G. Kendall and A. Stuart, *The Advanced Theory of Statistics* (Griffin, London, 1961), Vol. 2, p. 375.

[16]Because the input is a scalar time series, we constrain the trajectory adjustment to yield a scalar time as output. For instance, if $x_1=(\xi_i,\xi_2)$ consists of the first two values in the time series, then the output begins with values $\hat{\xi}_1,\hat{\xi}_2,\hat{\xi}_3$ such that $\hat{x}_1=(\hat{\xi}_1,\hat{\xi}_2),\hat{x}_2=(\hat{\xi}_2,\hat{\xi}_3)$.

[17]To save CPU time, a maximum of 200 points is used to compute the linear maps. Despite these constraints, it is never necessary to compute a one-dimensional map approximation for this attractor as described in the text.

[18]M. Hénon, Commun. Math. Phys. **50**, 69 (1976).

[19]When the program is run on noiseless input, the mean error in the output is 0.025% of the attractor extent, which suggests that errors arising from small nonlinearities are negligible when the input contains enough points.

[20]In the noise reduction procedure, the attractor is reconstructed in four dimensions from a time series containing 32 768 values. Linear maps are computed using at least 50 points in each ball. Trajectories are fitted using windows of 24 attractor points which overlap by 6 points.

[21]P. Grassberger and I. Procaccia, Physica D **9**, 189 (1983).

On noise reduction methods for chaotic data

Peter Grassberger, Rainer Hegger, Holger Kantz, Carsten Schaffrath, and Thomas Schreiber[a)]
Physics Department, University of Wuppertal, Gauss-Strasse 20, D-5600 Wuppertal 1, Germany

(Received 11 September 1992; accepted for publication 18 January 1993)

Recently proposed noise reduction methods for nonlinear chaotic time sequences with additive noise are analyzed and generalized. All these methods have in common that they work iteratively, and that in each step of the iteration the noise is suppressed by requiring locally linear relations among the delay coordinates, i.e., by moving the delay vectors towards some smooth manifold. The different methods can be compared unambiguously in the case of strictly hyperbolic systems corrupted by measurement noise of infinitesimally low level. It was found that all proposed methods converge in this ideal case, but not equally fast. Different problems arise if the system is not hyperbolic, and at higher noise levels. A new scheme which seems to avoid most of these problems is proposed and tested, and seems to give the best noise reduction so far. Moreover, large improvements are possible within the new scheme and the previous schemes if their parameters are not kept fixed during the iteration, and if corrections are included which take into account the curvature of the attracting manifold. Finally, the fact that comparison with simple low-pass filters tends to overestimate the relative achievements of these nonlinear noise reduction schemes is stressed, and it is suggested that they should be compared to Wiener-type filters.

I. INTRODUCTION

Recently, there has been an increased interest in nonlinear time series analysis (see, e.g., Ref. 1 for a review). In particular, algorithms have been given for constructing dynamical models generating measured time sequences,[2,3] and for forecasting such sequences.[4–9] Closely related to these is the problem of noise reduction.

Any noise reduction method must assume that the time series to be cleaned can be unambigously seperated into "noise" and "signal" on the basis of some objective criterion. Conventional methods like linear filters use for this the spectrum: Low- or high-pass filters assume that the signal has much lower resp. much higher typical frequencies. Alternatively, if such a distinction on the basis of characteristic time scales is not possible, broad band components of the spectra are assumed to be related to noise, while sharp lines are due to the signal. This is fine for signals from regular sources which give periodic or quasi-periodic signals, but if the signal is not (quasi-) periodic, linear filters cannot remove the noise without also distorting the signal. This is due to the fact that already pure signals from chaotic systems show broad band spectra, and no Fourier-based method is able to distinguish this from random noise. For the latter, one has somehow to use the fact that deterministically chaotic (and dissipative) motion takes place on attractors which are smooth submanifolds of the total available phase space. This implies that state vectors constructed from delay variables are constrained to fall onto geometrical objects which are locally linear.

When adding noise to an otherwise deterministic system, we have to distinguish between "dynamic" and "measurement" noise. Assume that the noise-free dynamics would be

$$\mathbf{y}_{n+1}=\mathbf{f}(\mathbf{y}_n). \tag{1.1}$$

We speak of measurement noise if there exists a trajectory satisfying this exact dynamics, but the measured trajectory is corrupted by additive (or possibly multiplicative) noise,

$$\mathbf{x}_n=\mathbf{y}_n+\mathbf{r}_n. \tag{1.2}$$

Dynamic noise, in contrast, is added already during the evolution,

$$\mathbf{x}_{n+1}=\mathbf{f}(\mathbf{x}_n)+\mathbf{r}_n \tag{1.3}$$

so that no near-by trajectory satisfying the exact dynamics needs to exist *a priori*. The "shadowing problem" deals just with the question whether such a trajectory does exist nevertheless, and how to find it.

Obviously the shadowing problem is harder than the problem of removing measurement noise. For a noise reduction scheme to be practically useful, one should demand that it works at least in the latter case, and that it at least does not produce nonsense in the case of dynamic noise. In particular, this is claimed for the noise reduction schemes recently proposed by several authors.[10–16] All these methods are supposed to work also when the data are sampled with too low frequency for conventional low-pass or Wiener[17] filters.

Let us assume that we have observed only a univariate time series corrupted by measurement noise, and construct state vectors by means of delay coordinates. Then the observed signal x_n is obtained from a "true" trajectory y_n by

$$x_n=y_n+r_n. \tag{1.4}$$

[a)]Current address: Niels Bohr Institute, Blegdamsvej 17, DK-2100 Copenhagen ø, Denmark.

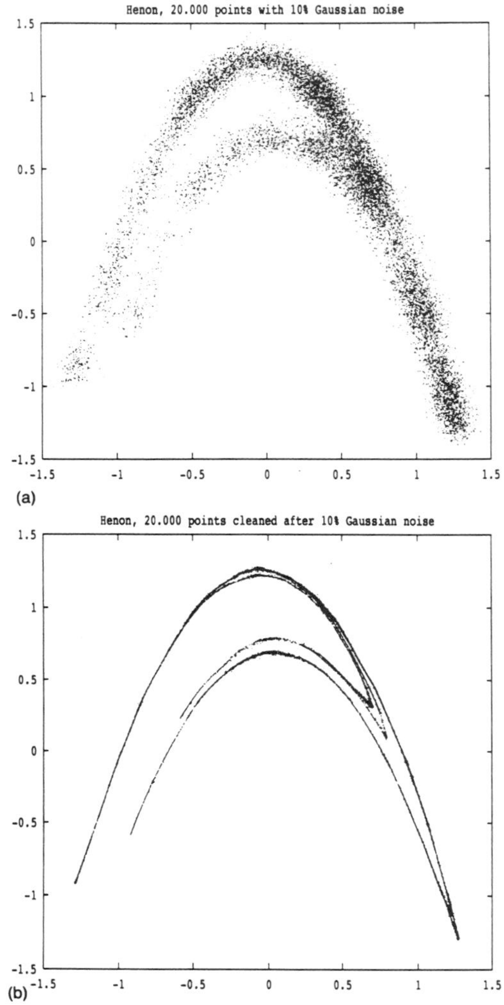

FIG. 1. Hénon attractor with 10% Gaussian white noise before (panel a) and after (panel b) noise reduction. Here, the exact form of the Hénon map was not assumed to be known, but was fitted from the noisy time series of 20 000 points shown in panel (a). The noise reduction algorithm used was that described in Secs. IV and V.

Let m be a positive integer. If it is sufficiently much larger than the attractor dimension,[18] then an m-dimensional embedding will give in general a faithful representation of it, and Eq. (1.1) can be written either as

$$y_{n+m}=f(y_n,y_{n+1},\ldots,y_{n+m-1}) \tag{1.5}$$

or implicitly as

$$F(y_n,y_{n+1},\ldots,y_{n+m})=0. \tag{1.6}$$

The embedding dimension m will however not be known *a priori*, and for practical reasons it will often be convenient to use an embedding dimension larger than strictly needed. In this case, there will not be a single relation of this type, but several constraints

$$F_q(y_n,y_{n+1},\ldots,y_{n+m})=0, \quad q=1,2,\ldots,Q\leqslant m, \tag{1.7}$$

where $m-Q+1$ is the dimension of the manifold.

Most of the above authors assume that the true dynamics is sufficiently smooth so that the functions f and F can be approximated locally by linear functions. Thus, for points $\mathbf{y}_k=(y_k,\ldots,y_{k+m})$ in the neighborhood of the point $\mathbf{x}_n=(x_n,x_{n+1},\ldots,x_{n+m})$ [Notice that we change here the notation with respect to Eqs. (1.1)–(1.3). While there $\mathbf{x}_n$ was an m dimensional vector, now and in the following it will be $m+1$ dimensional.] they assume for the noise-free system either a single constraint

$$F^{(n)}(\mathbf{y}_k)=\sum_{i=0}^{m} a_i^{(n)}y_{k+i}+b^{(n)}=0 \tag{1.8}$$

or, if the embedding dimension m is larger than the minimally necessary one, a set $\{F_q^{(n)}\,|\,q=1,2,\ldots,Q\}$ of such constraints. In the former case, the point is confined to a hyperplane, in the latter it is in a lower-dimensional linear subspace.

The most straightforward attempt to use Eq. (1.8) would be to write it in an explicit "forward time" form, and to define a "corrected" time sequence as

$$x_{n+m}^{\text{corr}}=\frac{1}{a_m^{(n)}}\left[\sum_{i=0}^{m-1} a_i^{(n)}x_{n+i}+b^{(n)}\right]. \tag{1.9}$$

After this has been done for all n, we would replace x_n by the new x_n^{corr}, and repeat the procedure. Thus we would obtain an iterative scheme which hopefully would converge towards the correct trajectory y_n.

Unfortunately, this would not work for a chaotic system, since the "correction" would be ill defined along the unstable manifold. Any error in x_0 would induce a bigger error in x_1, etc. In order to stabilize the corrections in the direction of the unstable manifold, we must also use information from the future.

In Refs. 10–12 this is done in two separate steps. In the first step, for each point the local map (1.8) or a higher order polynomial is constructed. In the second step, a new trajectory is found which on the one hand satisfies this map (at least approximately) and on the other hand is close to $\{x_n\}$. In all three papers this is implemented differently, but we do not want to discuss the details here. We just point out that the second step is the crucial one and numerically nontrivial in all three approaches.

In contrast to this, the authors of Refs. 13–16 virtually eliminate the second step by essentially projecting (in different ways) the vectors $\mathbf{x}_n$ onto the subspaces defined by the constraints $F_q^{(n)}=0$. Actually the latter is not precisely what they do. Since the delay vectors $\mathbf{x}_n$ for different n are not independent (they involve partly the same elements of the time series), projecting all of them exactly onto their hyperspaces would in general be mutually inconsistent. Therefore one has to find a compromise between the dif-

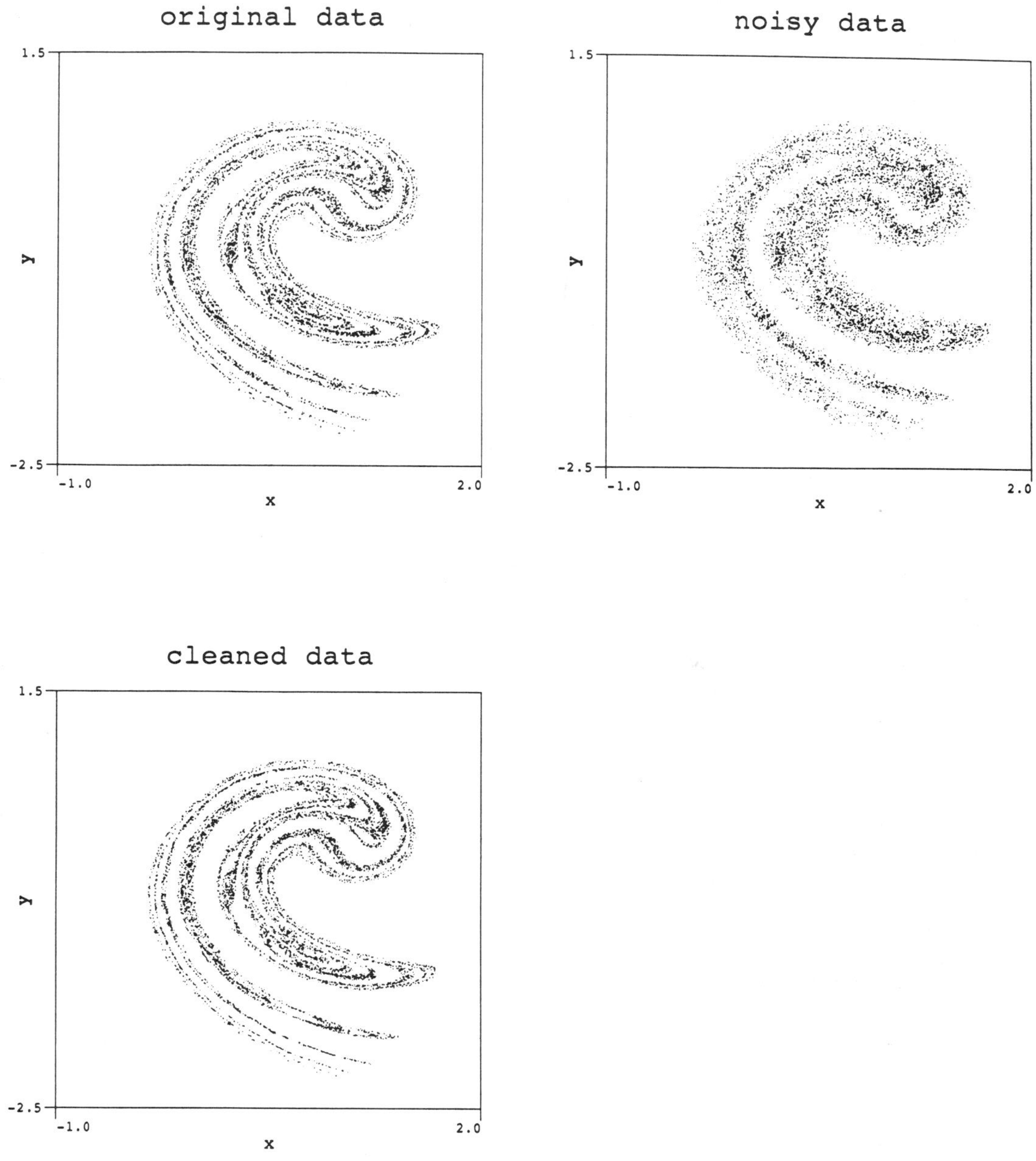

FIG. 2. Ikeda attractor $z' = 1 + 0.9ze^{0.4i - 6i/(1+|z|^2)}$: clean (panel a), with 5% uniform white noise (panel b), and after noise reduction (panel c) using again only information from the noisy signal. The algorithm used here was indeed somewhat simpler than in Fig. 1, and used the two-variate time series of points $z_n = x_n + iy_n$ as described in Ref. 13.

ferent projections. Since the results of Refs. 13–16 are quite promising, in the following we shall concentrate on this class of noise reduction schemes.

In Sec. II, we shall analyze in more detail the method proposed independently by Cawley and Hsu[14,15] and by Sauer.[16] In Sec. III, we shall discuss that by Schreiber and Grassberger,[13] and in Sec. IV we shall discuss a method which can be understood as a compromise between both, and which we consider as optimal. The methods discussed in Secs. II and III can be obtained as special cases of the latter. Details of the implementation and numerical results are given in Sec. V.

Two typical examples of the performance of our algorithm are shown in Figs. 1 and 2. They used time sequences of 20 000 points of the well known Hénon and Ikeda maps. More details on these figures will also be given in later sections.

Before leaving this introduction, we should however stress that comparisons with linear filtering methods should be done in a fair way, by comparing with *optimal* linear filters. These are Wiener filters[17] and closely related "Wiener-type" filters using singular value decomposition (SVD).[19,20] In general, they can give much better results than the simple low-pass filters used for comparison in previous analyses.[10,13,14] Since they are useful also as preliminary first steps in nonlinear noise reduction methods, we discuss them in Appendix A. Another useful "zeroth order" method with which our results should be compared is obtained by correcting each point towards the center of gravity of its neighbors.[24] Results of this and of SVD-Wiener filters will be discussed also in Sec. V.

II. CAWLEY–HSU–SAUER METHOD

Apart from first applying a low-pass linear filter and some other tricks which we do not want to discuss here in detail (some will be discussed in Sec. V), the noise reduction method used in Refs. 14–16 consists essentially in a compromise between *orthogonal* projections onto the subspaces defined by the linear constraints $F_q^{(n)}=0$.

Let us be more precise and consider the correction of the nth value x_n in the case of embedding dimension m and a single constraint $F=0$. The discussion will be simpler if we assume for the moment that this constraint is not linear and thus only locally valid, but is globally valid and and hence nonlinear. The total time series consists of the T values $(x_1,\ldots,x_T)$. Due to the noise, the measured time sequence does not satisfy Eq. (6), but rather

$$F(\mathbf{x}_j)=F(x_j,x_{j+1},\ldots,x_{j+m})=\epsilon_j\neq 0, \quad 1\leqslant j\leqslant T-m. \tag{2.1}$$

If ϵ_j is small [or $F(\mathbf{x})$ is linear], then an orthogonal projection of $\mathbf{x}_j$ onto the surface $F(\mathbf{x}_j)=0$ would be given by

$$\mathbf{z}_j=\mathbf{x}_j+\boldsymbol{\theta}_j \tag{2.2}$$

with

$$\boldsymbol{\theta}_j=\text{const}\ \nabla_j F, \tag{2.3}$$

where $\nabla_j F=\nabla F(\mathbf{x}_j)$. Inserting this ansatz into Eq. (2.1) and Taylor expanding, we obtain

$$\epsilon_j=F(\mathbf{z}_j-\boldsymbol{\theta}_j)=0-\boldsymbol{\theta}_j\cdot\nabla_j F \tag{2.4}$$

which gives us finally

$$\boldsymbol{\theta}_j=-\epsilon_j\frac{\nabla_j F}{\|\nabla_j F\|^2}. \tag{2.5}$$

If the constraint were satisfied everywhere except for one time j so that all ϵ_j were zero except for one single ϵ_{j_0}, then Eq. (2.2) would give corrections only to x_{j_0}, $x_{j_0+1},\ldots,x_{j_0+m}$. But this is not realistic, and we must expect that the corrections for x_n, say, obtained from projecting onto $F(\mathbf{x}_{n-m})=0$, $F(\mathbf{x}_{n-m+1})=0,\ldots,F(\mathbf{x}_n)=0$ are all different. The natural assumption made in Refs. 16, 14, and 15 is that the finally accepted correction is proportional to the sum of all these corrections,

$$x_n'=x_n+\alpha\sum_{j=n-m}^{n}\theta_{j,n-j}. \tag{2.6}$$

Inserting this into $F(\mathbf{x}_k)$ will give us the new errors $\epsilon_k'=F(\mathbf{x}_k')$ which are then used to get the second correction x_k'', etc. Again Taylor expanding $F(\mathbf{x}_k')$, we obtain after some algebra our first result

$$\epsilon_k'=(1-\alpha)\epsilon_k-\alpha\sum_{j\neq k}\epsilon_j\frac{\nabla_k F\cdot\mathbf{B}^{j-k}\nabla_j F}{\|\nabla_j F\|^2}. \tag{2.7}$$

Here $\mathbf{B}$ is the $(m+1)\times(m+1)$ backshift matrix defined as $B_{ij}=\delta_{i,j+1}$. Its nth power is simply $(\mathbf{B}^n)_{ij}=\delta_{i,j+n}$.

Using the normalized vectors

$$\mathbf{e}_k=\frac{\nabla_k F}{\|\nabla_k F\|} \tag{2.8}$$

and normalized errors

$$\eta_k=\frac{\epsilon_k}{\|\nabla_k F\|}, \tag{2.9}$$

Eq. (2.7) can be written as

$$\eta_k'=(1-\alpha)\eta_k-\alpha\sum_{j\neq k}(\mathbf{e}_k\cdot\mathbf{B}^{j-k}\mathbf{e}_j)\eta_j. \tag{2.10}$$

If there are several constraints, then Eq. (2.3) has to be replaced by the ansatz $\boldsymbol{\theta}_j=\sum_{q=1}^{Q}\alpha_q\nabla_j F_q$. The generalization of Eq. (2.7) is straightforward but somewhat cumbersome, and will not be given here. For realistic numerical applications where the global constraints F_q are replaced by linear fits which hold only locally, we will not use it anyhow but will use instead the formalism of Sec. IV.

Let us now study in some examples the convergence of the iterative scheme based on Eq. (2.10).

(a) 1-dimensional maps: First we consider 1-d maps $F(y_n,y_{n+1})\equiv y_{n+1}-f(y_n)=0$. Here,

$$\nabla_j F=(-df(x_j)/dx_j,1) \tag{2.11}$$

and Eq. (2.10) reduces to

$$\eta_k'=(1-\alpha)\eta_k+\alpha[\beta_k\eta_{k-1}+\beta_{k+1}\eta_{k+1}] \tag{2.12}$$

with

$$\beta_k=-\mathbf{e}_{k-1}\cdot\mathbf{B}\mathbf{e}_k=\frac{f_k'}{\sqrt{(1+[f_{k-1}']^2)(1+[f_k']^2)}},$$

$$f_k'=df(x_k)/dx_k. \tag{2.13}$$

During the iteration of Eq. (2.12), the errors spread essentially like they do in a reactive diffusion process with frozen randomness. For a sawtooth map $f(x)=ax+c \bmod 1$

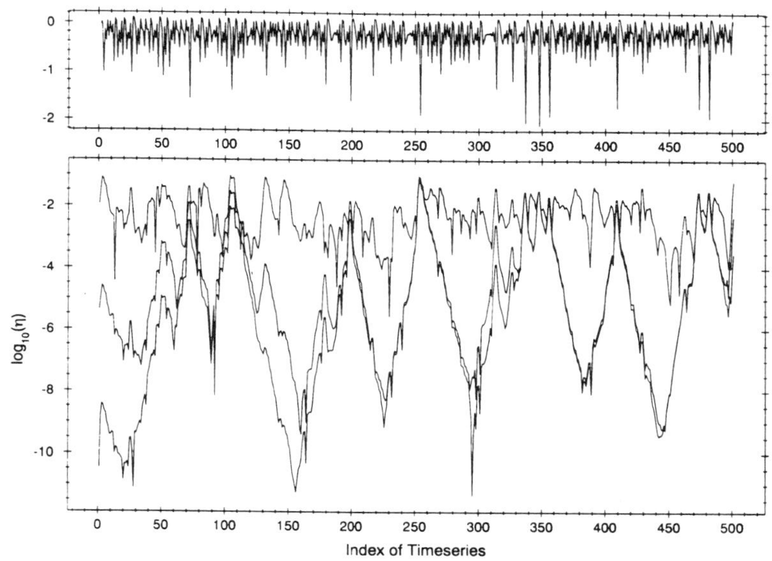

FIG. 3. Lower part: logarithm of η_k plotted against k after 50, 1000, and 2000 iterations of Eq. (2.14) for the Hénon map. The iteration had started with $\eta_k = 1 \forall k$. On the upper part, the distance of (x_k, x_{k+1}) from the nearest primary tangency point as determined in Ref. 22 is plotted on a logarithmic scale. We see that the convergence of the iteration becomes arbitrarily slow near these tangency points.

with $a > 1$, we can easily see that this diffusion process is damped for sufficiently small α, since the largest eigenvalue of Eq. (2.12) is $\mu = 1 - \alpha(a-1)^2/(1+a^2)$. Thus in this case the iteration converges exponentially, though less fast than with the trivial scheme $\eta_k' = (1-\alpha)\eta_k + \alpha\eta_{k+1}/a$ which can be considered as an application of the method discussed in the next section.

For maps with $f'(x) \neq \text{const}$ we cannot in general show convergence analytically, even if they are hyperbolic. But numerical simulations very clearly show that Eq. (2.12) always converged for chaotic maps to $\eta_k = 0$, for sufficiently small α. But for maps with a quadratic critical point (like the logistic map), the convergence is neither uniform nor exponential. More precisely, the convergence seems to be arbitrarily slow if x comes close to a critical point of $f(x)$. The average error seems to decrease only algebraically with the number of iterations.

Thus we conclude that the above represents indeed a viable noise reduction scheme for 1-d maps, though not necessarily a very efficient one.

(b) 2-d maps: This is very similar for 2-d maps $y_{n+2} = f(y_n, y_{n+1})$. There, the rhs of the analogon to Eq. (2.12) contains 4 terms,

$$\eta_k' = (1-\alpha)\eta_k + \alpha[\sigma_k\eta_{k-1} + \sigma_{k+1}\eta_{k+1} + \tau_{k-1}\eta_{k-2} + \tau_{k+1}\eta_{k+2}]. \tag{2.14}$$

Again we can show for maps with constant derivatives,

$$y_{n+2} = a_0 y_n + a_1 y_{n+1} + c \mod c', \tag{2.15}$$

that the iteration converges exponentially with an eigenvalue

$$\mu = 1 - \alpha \frac{(1-a_0-a_1)^2}{1+a_0^2+a_1^2}. \tag{2.16}$$

which is again larger than the analogous eigenvalue of the method discussed in the next section.

But for nonlinear maps like the Hénon[21] map $y_{n+2} = a + by_n - y_{n+1}^2$, exponential convergence seems again to be prevented by homoclinic tangencies. For the Hénon map, we have

$$\sigma_k = -\mathbf{e}_{k-1} \cdot \mathbf{B}\mathbf{e}_k = \frac{2(bx_k - x_{k+1})}{\sqrt{(1+b^2+4x_k^2)(1+b^2+4x_{k+1}^2)}}, \tag{2.17}$$

$$\tau_k = -\mathbf{e}_{k-1} \cdot \mathbf{B}^2\mathbf{e}_{k+1} = \frac{b}{\sqrt{(1+b^2+4x_k^2)(1+b^2+4x_{k+2}^2)}}. \tag{2.18}$$

In simulations for randomly chosen trajectories with the standard parameters $a = 1.4$, $b = 0.3$, Eq. (2.14) never diverged for $\alpha < 0.9$, but convergence was arbitrarily slow when the point (x_n, x_{n+1}) was near the line of "primary" tangencies[22] (see Fig. 3). This agrees with the fact that "realistic" application of the Cawley–Hsu–Sauer method (which involves also local linearizations, use of fitted maps instead of the exact map, and eventually linear prefiltering) always gave somewhat poorer results for the Hénon map[14] than the algorithm discussed in the next section. It seems to result from the fact that also the first and last components of delay vectors are corrected. As we had pointed out in the Introduction, we should expect that these components are hard to correct due to instabilities along the stable resp. unstable manifolds.

(c) Higher dimensions: For this same reason, we should expect also slow convergence when applying the method to higher dimensional systems. But applications of Sauer's algorithm to the Lorenz and Rössler models[16] gave very good results, much better, e.g., than linear filtering and the methods of Refs. 10 and 13.

This might be due to the fact that the time series tested in Ref. 16 were embedded in very high dimensions and measured with rather short delays. This could imply that the final slowness of convergence would become apparent only after very many iterations, much more than needed for realistically acceptable noise reduction. Another reason might be that in autonomous flows (as opposed to maps) there exists always one vanishing Lyapunov exponent. It corresponds to uncertainties in the flow direction, and implies that errors in this direction cannot be corrected very well anyhow. This affects strongly the methods of Refs. 10 and 13 which rely on hyperbolicity, but it might affect much less the Cawley–Hsu–Sauer method.

Thus we conclude that this method is very efficient for oversampled time series, in spite of its theoretical drawbacks. The latter can make it somewhat slower than the method discussed in Sec. III, but they make it also more robust. Indeed, we found that in many cases its performance can be improved by modifications which will be discussed in Sec. V together with other details of the implementation.

III. SCHREIBER–GRASSBERGER METHOD

Instead of projecting *orthogonally* onto the surfaces $F(\mathbf{x}_j)=0$, in Ref. 13 we projected by correcting only a single delay coordinate. In the examples shown in Ref. 13, we used even m, $m=2r$, and corrected the central coordinate x_{j+r}. More generally we write $m=r+s$ and correct x_{j+r}. The coordinates x_j, $x_{j-1},\ldots,x_{j+r-1}$ are called "past coordinates," the coordinates $x_{j+r+1},\ldots,x_{j+m}$ are called "future coordinates."

In this way, we avoided on the one hand the problem of conflicting projections discussed above. On the other hand, since this correction is well controlled both in the stable and unstable directions, we should have faster convergence when dealing with hyperbolic systems. The latter and the fact that the resulting correction is very simple and intuitive,

$$x_n'=x_n-\alpha\,\frac{F(\mathbf{x}_{n-r})}{\partial F(\mathbf{x}_{n-r})/\partial x_n} \tag{3.1}$$

are the main advantages.

Due to the simplicity of the method, it can also be applied directly to multivariate time sequences.[23] In the multivariate case, x_n in the above is a vector, F is a scalar function, and $\partial F/\partial x_n$ is its gradient. In practice, F will be taken as linear and its fit is straightforward along the lines of Ref. 13. Figure 4 was obtained in this way, with $m=2$ and $k=l=1$. More details about the multivariate method will be given in a forthcoming paper.[23]

The main disadvantages of Eq. (3.1) are the following:

(i) For each index n, we can take into account only one constraint, i.e., the attractor is always constrained to be on a codimension 1 manifold, even if the embedding dimension m is chosen large. *A priori*, one might have thought this to be a serious problem which should ruin the applicability of the method. To our surprise this was not the case. We obtained, e.g., very good results when embedding the Hénon map in 6 dimensions. In all cases where the exact dimension of the attractor was known, considerable "overembedding" improved the performance.[13] Nevertheless we have the suspicion that the mediocre performance for the Lorenz and Rössler systems might be due to this.

(ii) As seen from Eq. (3.1), we run into problems if $\partial F/\partial x_n$ is very small, i.e., if the manifold is roughly parallel to the x_n axis. In this case, the proposed correction is very large [unless the error $F(\mathbf{x}_{n-r})$ happens to be small too], and we should not accept it. Indeed, we simply discarded such corrections completely in Ref. 13, by not correcting x_n at all if the proposed correction was larger than some given threshold. In the case of linear hyperbolic mappings like Eq. (2.15), the iteration of Eq. (3.1) converges indeed exponentially with leading eigenvalues $\mu=1-\alpha \pm\alpha(1-a_0)/a_1$.[13] For nonlinear maps the situation is again less simple, even if they are hyperbolic. Let us assume for simplicity a 2-d map of the Hénon type. Thus, for any pair (r,s) of delays the exact dynamics can be represented by a constraint

$$F(y_{n-r},y_n,y_{n+s})=0 \tag{3.2}$$

which we can also write explicitly as

$$y_n=f(y_{n-r},y_{n+s}). \tag{3.3}$$

For the Hénon map with $r=s=1$, this reads for instance

$$y_n=\pm\sqrt{a+by_{n-1}-y_{n+1}}. \tag{3.4}$$

Equation (3.1) reduces then to

$$x_n'=(1-\alpha)x_n+\alpha f(x_{n-r},x_{n+s}). \tag{3.5}$$

Assume now that the values of x_{n-r} and x_{n+s} had errors. [For some maps such as, e.g., the Hénon map, it may then happen that Eq. (3.3) has no real-valued solutions if x_{n-r} and x_{n+s} have errors. In such cases we do not correct at all.] These errors are then multiplied by $\partial f/\partial x_{n-r}$ resp. $\partial f/\partial x_{n+s}$ in the correction of x_n. For r, $s\to\infty$, one easily sees that these partial derivatives behave for nearly all x's as

$$\partial f/\partial x_{n-r}\sim e^{-r|\lambda_2|},\quad \partial f/\partial x_{n+s}\sim e^{-s\lambda_1}, \tag{3.6}$$

where $\lambda_1>0>\lambda_2$ are the Lyapunov exponents. Indeed, using, e.g., x_{n-r} and x_{n-r+1} as independent variables we can write

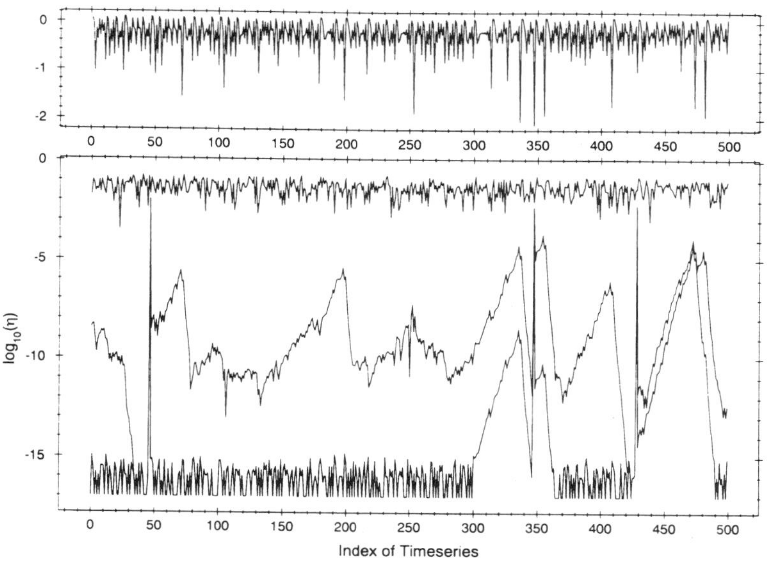

FIG. 4. Similar to Fig. 3, but using the noise reduction scheme given by Eq. (3.5) with $r=s=1$. In the lower curves, the logarithm of $x_{n+1}-a-bx_{n-1}+x_n^2$ is plotted after 0, 100, and 500 iterations. The iteration started with measurement errors distributed uniformly in $[-0.05,0.05]$. As in Fig. 3, the exact form of the Hénon map was assumed to be known, and the sign in $x_n' = \frac{1}{2}(x_n \pm \sqrt{a+bx_{n-1}-x_{n+1}})$ was chosen as that of x_n. This cannot be applied in points where the square root would become imaginary, and in such points nothing was done at all. The iteration converges to machine precision except at these points. In the upper curve, we also plotted the logarithms of the distances from the nearest primary tangency point, as in Fig. 3.

$$\frac{\partial f}{\partial x_{n+s}}=\left.\frac{\partial x_n}{\partial x_{n+s}}\right|_{x_{n-r}}=\frac{\partial x_n/\partial x_{n-r+1}|_{x_{n-r}}}{\partial x_{n+s}/\partial x_{n-r+1}|_{x_{n-r}}}\sim\frac{e^{r\lambda_1}}{e^{(r+s)\lambda_1}}, \tag{3.7}$$

and similarly for $\partial f/\partial x_{n-r}$.

For sufficiently large delays k and l and for sufficiently low noise level, our method should thus converge. But for finite k and l the above partial derivatives can be larger than unity even for hyperbolic systems, and we must expect problems. This was indeed seen in the Hénon case with r and s up to 2. There neither Eq. (3.5) nor its linearized version

$$\theta_n'=(1-\alpha)\theta_n+\alpha\left[\theta_{n-r}\frac{\partial f}{\partial y_{n-r}}+\theta_{n+s}\frac{\partial f}{\partial y_{n+s}}\right] \tag{3.8}$$

converged exponentially. But unlike Sauer's method discussed in Sec. II, it seemed that tangencies between the stable and unstable manifold are not the main reason when using the exact nonlinear equation (3.5) (see Fig. 4). Instead, there the lack of convergence happens only at points where we do not correct at all since the square root in Eq. (3.5) would be negative [we have not found a better method to correct such points; for the linearized version, Eq. (3.8), the iteration even diverged at these points].

We should finally point out that the choice $r=s$ used in Ref. 13 is in general not optimal. Since in general $\lambda_{\text{unstable}} < |\lambda_{\text{stable}}|$ (since the system is dissipative), Eqs. (3.6) and (3.8) suggest that in general $r \leqslant s$ should be optimal.

IV. OPTIMIZED METHOD

After the preceding discussions, the points to be improved should be obvious: in order to avoid the drawbacks of both Secs. II and III, we should neither make large corrections for the outer components of the delay vectors, nor should we project by changing only a single coordinate.

Let us first discuss the construction of the "proposal vectors" $\boldsymbol{\theta}_k$, and deal later with the problem of properly weighting them.

One possibility consists in making projections orthogonal to the attractor *in the subspace spanned by some of the central coordinates*. Thus we would choose a pair of positive integers (r,s) with $p:=m+1-r-s>1$, and project in the subspace spanned by the p coordinates $x_{n+r},\ldots,x_{n+m-s}$ of the $(m+1)$-dimensional delay vector $(x_n,\ldots,x_{n+m})$. Technically, this would be obtained from the formalism of Sec. II by replacing Eq. (2.3) by

$$\boldsymbol{\theta}_j=\text{const }\mathbf{P}\nabla F(\mathbf{x}_j), \tag{4.1}$$

where $\mathbf{P}$ is the projection operator onto the subspace spanned by the coordinates $x_{j+r},\ldots,x_{j+m-l}$,

$$\mathbf{P}_{ii'}=\begin{cases}\delta_{ii'}, & \text{for } r\leqslant i\leqslant m-s\\ 0, & \text{else.}\end{cases} \tag{4.2}$$

Actually we shall stay more general and simply assume Eq. (4.1) with some non-negative symmetric matrix $\mathbf{P}$ which will in general not be a projection matrix. We only require that for nearly all $\mathbf{x}_j$ the corrections $\boldsymbol{\theta}_j$ will be transverse to the manifold $F(\mathbf{x}_j)=0$. If we have Q such constraints, i.e., a codimension Q manifold, then $\mathbf{P}$ should be of rank $\geqslant Q$. In practice, we shall use diagonal matrices, $P_{ik}=\delta_{ik}P_i$, with P_i maximal in the middle of the delay window and decreasing towards its ends.

From the ansatz Eq. (4.1) we obtain

$$\theta_j=-\epsilon_j\frac{\mathbf{P}\nabla_jF}{(\nabla_jF\cdot\mathbf{P}\nabla_jF)}. \tag{4.3}$$

Since the different correction proposals $\theta_{n,i}$ for the same coordinate x_i will have different errors we can also generalize the ansatz Eq. (2.6) so that not all proposed corrections $\theta_{j,k}$ are weighted equally:

$$x_n'=x_n+\sum_{j=n-m}^{n}(\mathbf{W}\boldsymbol{\theta}_j)_{n-j}, \tag{4.4}$$

where $\mathbf{W}$ is another diagonal non-negative $(m+1)\times(m+1)$ matrix. With this ansatz, we obtain

$$\epsilon_k'=\epsilon_k-\alpha\sum_j\epsilon_j\frac{(\nabla_kF\cdot\mathbf{B}^{j-k}\mathbf{WP}\nabla_jF)}{(\nabla_jF\cdot\mathbf{P}\nabla_jF)}. \tag{4.5}$$

The algorithm thus depends on two non-negative diagonal matrices $\mathbf{P}$ and $\mathbf{WP}$ which should be chosen such that convergence of Eq. (4.5) is fastest. In the following we shall argue for a certain choice of $\mathbf{P}$ which is close to optimal for most purposes. Since we found no similiar general results for the matrix $\mathbf{W}$ we fix it to the unit matrix.

A. *Locally linear constraints.* In realistic cases we use locally linear constraints of the form of Eq. (1.8) which have to be fitted from the very time series to be cleaned. Instead of a single constraint at each point, we also should expect $Q>1$ such constraints F_q, so that the attractor is constrained to be on a manifold of codimension Q. One could modify Eqs. (4.3) and (4.5) such that they could still be used in this case. But the resulting formalism would be somewhat inconvenient, and we use a different approach.

The basic idea is to formulate a minimization problem, from which both the linear constraints and the corrections are obtained in one step. We first define for each point $\mathbf{x}_n$ a neighborhood $\mathscr{U}_n$ on which the linearized constraints are supposed to be valid.

Each constraint can be written as $(\mathbf{a}_q^{(n)}\cdot\mathbf{y})+b_q^{(n)}=0$, where $\mathbf{y}$ is again the noise-free signal corresponding to $\mathbf{x}$, and $\mathbf{a}_q^{(n)}=\nabla_nF_q$. Assume there are $|\mathscr{U}_n|$ points in the neighborhood $\mathscr{U}_n$. For each of them, the constraints can be written as

$$(\mathbf{a}_q^{(n)}\cdot(\mathbf{x}_k+\boldsymbol{\theta}_k))+b_q^{(n)}=0,\quad q=1,\ldots,Q,\quad \mathbf{x}_k\in\mathscr{U}_n, \tag{4.6}$$

where the Q vectors $\mathbf{a}_q^{(n)}$ should be linearly independent and properly normalized. The latter can be achieved by requiring

$$(\mathbf{a}_q^{(n)}\cdot\mathbf{P}\mathbf{a}_{q'}^{(n)})=\delta_{qq'}. \tag{4.7}$$

With these two requirements we guarantee the consistency of the corrections with the dynamics. Finally, as the new trajectory should be as close as possible to the old one, we have to impose

$$\epsilon=\sum_{k:\mathbf{x}_k\in\mathscr{U}_n}(\boldsymbol{\theta}_k\cdot\mathbf{P}^{-1}\boldsymbol{\theta}_k)\overset{!}{=}\min. \tag{4.8}$$

[Instead of Eq. (4.8), we could also use a weighted sum where each contribution is multiplied by a weight factor w_k with $w_k>0$ and $\Sigma_k w_k=1$. We then would have to replace Eqs. (4.9) and (4.10) below by $\xi_i^{(n)}=\Sigma_{\mathbf{x}_k\in\mathscr{U}_n}w_kx_{k+i}$ resp. $C_{i,j}^{(n)}=\Sigma_{\mathbf{x}_k\in\mathscr{U}_n}w_kx_{k+i}x_{k+j}-\xi_i^{(n)}\xi_j^{(n)}$. In Ref. 14 it is claimed that distance-dependent weights $w_k\propto|\mathbf{x}_n-\mathbf{x}_k|^{-2}$ are indeed better than the uniform weight we use in this paper.] Here we have of course assumed that $\mathbf{P}^{-1}$ exists, i.e., that all P_i are different from zero. If not, then the corresponding component of $\boldsymbol{\theta}_k$ has to vanish, which would be formally obtained by using $1/P_i=\infty$ in Eq. (4.8). Notice that Eq. (4.8) indeed forces the ith component of $\boldsymbol{\theta}_k$ to be proportional to P_i, as we want it to be.

The minimization problem defined by Eqs. (4.8), (4.6), and (4.7) is solved in Appendix B, with the following result.

From the points $\mathbf{x}_k\in\mathscr{U}_n$ we first construct the average

$$\xi_i^{(n)}=\frac{1}{|\mathscr{U}_n|}\sum_{k:\mathbf{x}_k\in\mathscr{U}_n}x_{k+i},\quad i=0,1,\ldots,m \tag{4.9}$$

and the $(m+1)\times(m+1)$ covariance matrix

$$C_{i,j}^{(n)}=\frac{1}{|\mathscr{U}_n|}\sum_{\mathbf{x}_k\in\mathscr{U}_n}x_{k+i}x_{k+j}-\xi_i^{(n)}\xi_j^{(n)}. \tag{4.10}$$

We then define

$$R_i=1/\sqrt{P_i} \tag{4.11}$$

and

$$\Gamma_{ij}^{(n)}=R_iC_{i,j}^{(n)}R_j. \tag{4.12}$$

The Q orthonormal eigenvectors of the matrix $\boldsymbol{\Gamma}^{(n)}$ with smallest eigenvalues are called $\mathbf{e}_q^{(n)}$, $q=1,\ldots,Q$. The projector onto the subspace spanned by these vectors is denoted as $\mathscr{Q}^{(n)}$, i.e.,

$$\mathscr{Q}_{ij}^{(n)}=\sum_{q=1}^{Q}e_{q,i}^{(n)}e_{q,j}^{(n)}. \tag{4.13}$$

Then the ith component of $\boldsymbol{\theta}_n$ is given by

$$\theta_{n,i}=\frac{1}{R_i}\sum_{j=0}^{m}\mathscr{Q}_{ij}^{(n)}R_j(\xi_j^{(n)}-x_{n+j}). \tag{4.14}$$

Intuitively, the $\mathbf{a}_q^{(n)}$ are orthogonal to a linear subspace, in which the neighborhood of the considered point is embedded. Formally this subspace is given by the leading eigenvectors of the covariance matrix. The correction simply consists in projecting the point onto this hyperplane of codimension Q, but in a metric given by $\mathbf{P}$ rather than orthogonally.

We notice that the cases studied in Secs. II and III are just special cases of the present algorithm, and are most easily implemented using Eq. (4.14). In the case of Sec. II, we have simply $R_i=1$ for all i. The case of Sec. III is obtained formally from the above (with $Q=1$) by using $R_i=1$ for $i=r$ and $R_i=R\gg 1$ for all $i\neq r$. In the limit $R\to\infty$, a single eigenvalue of $\Gamma^{(n)}$ will stay finite, and Eq. (4.14) gives exactly the correction proposed in Ref. 13.

Before leaving this section, we should point out that a simplified version of the methods of this and the previous sections consists in replacing $\theta_{n,i}$ by $P_i(\xi_i^{(n)}-x_{n+i})$. In this way we would correct each point such that the coordinates for which $P_i\neq 0$ move towards the center of mass of its neighbors. Just like the original algorithms correspond to locally linear fits, this corresponds to assuming locally constant fits. It will in general give worse results, but it is much simpler to implement and might thus be worth while. A detailed description (including a complete FORTRAN code) and some numerical results are given in Ref. 24.

In Ref. 12 a second order polynomial was fitted to the local dynamics. Especially if the density of points is low the radii of the neighborhoods may be so large that a higher than first order approximation could be reasonable. In our algorithm one could include nonlinear terms in Eq. (4.6). But unfortunately then the corrections would appear in a nonlinear way in the minimization problem. The criterion for the constraints to be independent would be much more complicated than Eq. (4.7), and finally we would have to determine much more parameters of the dynamics. Instead, we restrict ourselves to linear constraints and in real applications subtract a local trend of the suggested corrections, which should correct curvature effects (see Sec. V).

V. IMPLEMENTATION AND RESULTS

In this last section, we shall present details of the implementation and some more numerical results. We compare in particular the algorithms based on Secs. II–IV. We also give results obtained from two simpler algorithms: SVD-Wiener filtering as described in Appendix A, and a fit with locally constant functions instead of locally linear functions.

In all test cases considered below, we only use white measurement noise with Gaussian or uniform distribution, i.e., the noisy signal is written as $x_n=y_n+\epsilon_n$, where ϵ_n is Gaussian or uniform white noise. The noise level is defined as $\sqrt{\langle\epsilon_n^2\rangle/\langle y_n^2\rangle}$. Furthermore, in all cases the exact deterministic model is known for performance estimation, though it is not used during noise reduction (for an application to signals with unknown underlying dynamics, see Ref. 13). The final trajectory after noise reduction is written $x_n^{\text{corr}} = y_n + \epsilon_n^{\text{red}}$. In this case, we have (at least) two measures for performance:

- The increase of the signal-to-noise ratio,

$$r_0=\sqrt{\frac{\Sigma_n (\epsilon_n)^2}{\Sigma_n (\epsilon_n^{\text{red}})^2}}. \tag{5.1}$$

The logarithmic improvement of the signal-to-noise ratio, measured in decibels (dB), is defined as

$$\delta SNR=20 \log_{10} r_0. \tag{5.2}$$

- The decrease of the violation of determinism. Assume the signal obeys the equation $y_n=f(y_{n-1},\ldots)$. Then we define

$$r_{\text{dyn}}=\sqrt{\frac{\Sigma_n (x_n-f(x_{(n-1)},\ldots))^2}{\Sigma_n (x_n^{\text{corr}}-f(x_{(n-1)}^{\text{corr}},\ldots))^2}}. \tag{5.3}$$

Though this could, in principle, be estimated for all our examples, its evaluation is easiest for a series whose dynamics is already formulated in terms of delay coordinates, as in the Hénon map—or in the case when all coordinates are dealt with simultaneously. Thus we evaluated it only in the latter cases.

A. SVD-Wiener filtering

Let us first discuss the results of SVD-Wiener filters as defined in Appendix A, case (c). In applying these filters, we need an estimate for the noise spectrum. The most natural assumption is that it is white (which is indeed true in our examples). We pretend however that we do not know the correct noise level. Thus we assume in applying Wiener filters that $S^{\text{noise}}=\min_k S_k$, i.e., that the minimum of S^{signal} is zero.

For low noise level and low sampling frequency it does not give any improvements since the distortion it involves is more important than the noise reduction. Thus for the Hénon map, we do not get any improvement in r_0 for a noise level less than ca. 45%, for any embedding dimension (we would get noise reduction for arbitrarily small noise level if we would use knowledge of this level). For higher noise levels, the best results are obtained with embedding dimension $m\approx 17$, but the results depend only weakly on m. For 100% noise level, we obtain $r_0=1.49$.

The situation is very different for frequently sampled flows. In Fig. 5 we show results for the x and z components of the Lorenz model[25] (with parameters $\sigma=10$, $\rho=28$, $\beta=8/3$), and for the x component of the Rössler model[26] (with parameters $a=0.36$, $b=0.4$, $c=4.5$). In panel (a), the noise level is 10%, while in panel (b) it is 100%. We see that the results improve greatly with the sampling frequency. We also see that the method works much better for the Rössler model, reflecting the rather sharply peaked spectrum in this model. On the other hand, there is little difference between the results for the x and z coordinates of the Lorenz attractor, in spite of the fact that the spectrum of the z component is much larger at large frequencies, and a simple low-pass filter would do much worse for the z component than for the x component.

The results shown in Fig. 5 were obtained with embedding dimensions $m=40$, but again they depended very little on the precise value of m. Practically the same results were obtained for all m between 30 and 50.

When comparing Fig. 5 with the results of Ref. 14, we see that there is very little difference at high sampling frequencies (> 100 points per cycle). Indeed, for such high sampling rates we found an improvement of the SNR even for very low-level noise with SNR < 1%.

We should finally point out that Wiener and SVD-Wiener filters based on finite window lengths m should in principle not increase the dimension of chaotic attractors,

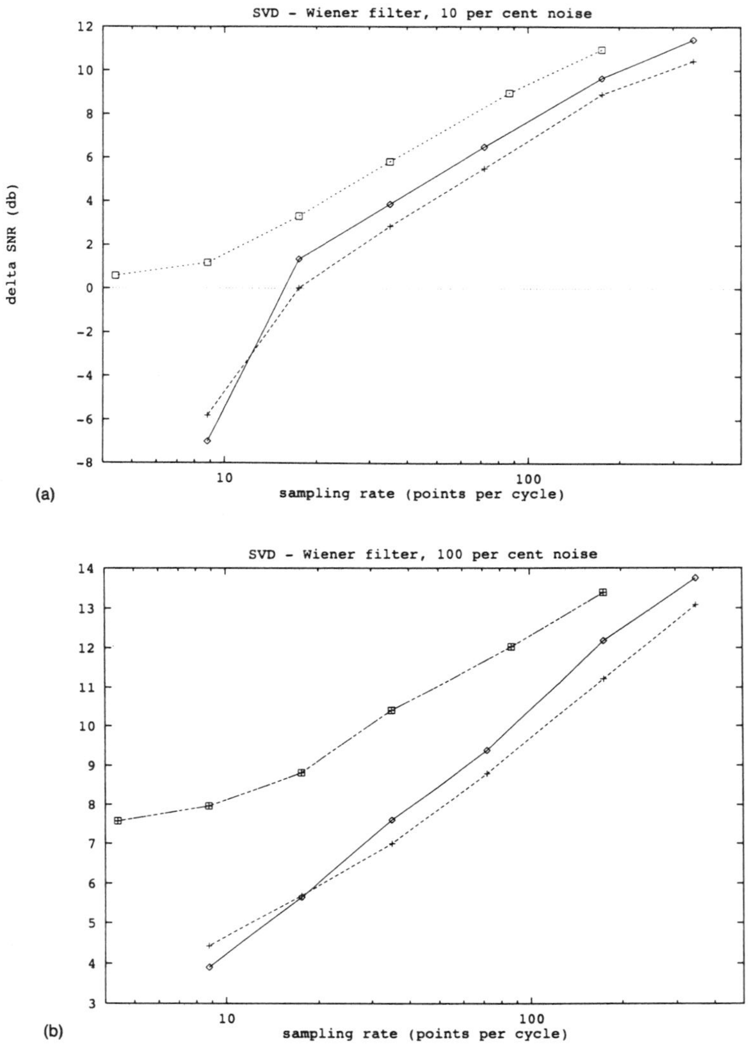

FIG. 5. Improvements of the signal-to-noise ratio, measured in decibels, and achieved by the SVD-Wiener method described in Appendix A. Each data point is based on a time series of 20 000 points. Embedding dimension was $m=40$. In panel (a), the noise level is 10%, while it is 100% in panel (b). Squares are for the x coordinate of the Rössler system, diamonds for the x coordinate of the Lorenz model, and crosses for the z coordinate of the latter.

in contrast to autoregressive filters.[27] In practice, they can however make dimension estimates harder when applied to too coarsely sampled time series.

B. Implementation of locally linear noise reduction

From the derivation in Sec. IV A it should be clear how to proceed:

• Embed the time series in an $m+1$-dimensional phase space using delay coordinates.

• For each embedding vector $\mathbf{x}_n$ find a neighborhood containing at least K points. For efficiency it is advisable to use a fast neighbor search algorithm, e.g., as described in Ref. 1.

• Compute the center of mass $\xi^{(n)}$, Eq. (4.9), the covariance matrix $\mathbf{C}^{(n)}$, Eq. (4.10), and its weighted form $\mathbf{\Gamma}^{(n)}$, Eq. (4.12).

• Determine all eigenvectors of $\mathbf{\Gamma}^{(n)}$. The correction is then obtained by projecting $\xi^{(n)}-\mathbf{x}_n$ into the subspace spanned by the Q eigenvectors with the smallest eigenvalues according to Eq. (4.14). Note that each element of the time series appears as a component in $m+1$ delay vectors. Therefore its correction is built up by several contributions.

• When all corrections are added the time series is replaced by the corrected one and the procedure is repeated.

When applying this scheme to a given time series one has to choose a suitable set of parameters, which are discussed in the following.

• For maps, the embedding dimension should be roughly twice the dimension for which the dynamics in delay coordinates gets deterministic. For finely sampled flows it could even be much higher because of redundancy of the information. Alternatively, data could be compressed as suggested in Ref. 16.

• The matrix **P** is conveniently chosen diagonal giving very small weights to the first and last components of the delay vectors and weight one to the center components (see discussion in Sec. IV).

• The number of constraints can be either kept fixed or increased during the iterations as the separation between signal and noise gets clearer. The remaining subspace has to be of a dimension large enough to cover the attractor.

• For optimal results the proper choice of the neighborhoods is essential. It has to be a compromise between the need for good statistics for the linear fits (large K), and the approximate linearity in the neighborhood (small radius). Assume that the dimension of the attractor is D, its overall size is normalized to 1, and that the entire time series contains T points. Then a box with K points will have a size $\approx (K/T)^{1/D}$. Assuming typical curvatures to be also of order 1, the nonlinearity will thus give errors of order $(K/T)^{2/D}$. Statistical fluctuations in the fitted vectors $\mathbf{a}^{(n)}$ on the other hand will be of order $\langle\epsilon\rangle/\sqrt{K-m}$, where $\langle\epsilon\rangle$ is the noise level. For the optimal value of K, both errors should be comparable. For large noise level, we get $K \gg m$ and thus

$$K \approx \{\langle\epsilon\rangle T^{2/D}\}^{2D/(4+D)}. \tag{5.4}$$

For vanishing noise level, K should decrease to $m+2-Q$, in which case the linear subspace is defined by a simplex. Since the noise level decreases during the iterations, K should be decreased as well. In regions of the phase space with higher than average density of points the estimate (5.4) leads to a neighborhood with a radius smaller than the noise level. In this case we take all neighbors closer than the assumed noise level.

• After a number of iterations the noise reduction does not improve further and the nonlinearities give cumulative systematic errors which distort the attractor. This can be partly avoided by shrinking the neighborhoods during iterations. However in spite of this, systematic shifts of the orbit due to nonlinearity of the attractor remain. In the noise-free case, points typically would be shifted perpendicularly to the manifold by an amount proportional to its curvature. This can be corrected by a trick suggested in Ref. 16. It is based on the assumed smoothness of the attracting manifold, due to which all neighboring points will feel essentially the same shift. This shift can thus be greatly reduced by subtracting from the proposed corrections $\boldsymbol{\theta}_n$ the average of the corrections for the neighboring points.

• For a proper choice of the matrix **P** and the number Q of constraints, large spurious corrections occur less frequently than in the algorithm described in Sec. III. It turned out that the distribution of the magnitude of the corrections was Poissonian in all cases. We interpret the tail as spurious and discard the corresponding corrections.

The following prefiltering may be useful, especially in cases with high noise level and high sampling frequencies (small delays): When searching for the neighbors defining the linearization neighborhood, filtered data are used (SVD filtering or the method described in Ref. 24). The corrections are added to the raw data and not to the prefiltered data. Thus the local dynamics may be defined in a better way, but on the other hand no information about the original time series is lost.

C. Numerical results

For a detailed comparison of the different algorithms we have chosen several typical test cases. The time series of the Hénon, the Ikeda, and the Lorenz system were contaminated with different noise levels. The performance is reported in Table I. The parameters we have used for the SVD method are quoted in Sec. V A. The other methods are special cases of the general formalism of Sec. IV. Therefore for a better comparison apart from the matrices **P** all the parameters were the same. For the embedding dimension we have chosen 5 for the Hénon, 7 for the Ikeda map, and 7, respectively, 13 for low and high sampled Lorenz data. In all cases we required between 20 and 50 neighbors taking into account the estimate in Eq. (5.4). The curvature effects were corrected. Furthermore, too big correction vectors θ were rescaled to the average magnitude of corrections, using as a criterion the distribution of their size. This cutoff is essential in the case of small noise, as "wrong corrections" have a big destructive effect in this case. For the Cawley–Hsu–Sauer algorithm the matrix **P** is unity, whereas for the new proposal we have chosen $P_i = 0.001$ for the two outer entries and $P_i = 1$ for all others. In the Schreiber–Grassberger case **P** contains only a single 1 in the center, all other entries on the diagonal are 0.001. In all cases we increased Q during the iteration, namely, from 2 to 3 for the Hénon, from 3 to 5 for the Ikeda, from 2 to 4 for the Lorenz with low sampling rate and from 6 to 10 with high sampling rate. In contrast to Ref. 13 and Sec. III, where only one constraint is possible, in the Schreiber–Grassberger limit of the generalized algorithm we used the above values of Q, which improved the performance. (See Fig. 6.)

The above parameters were in general not those which gave optimal performance for the specific time sequence. Instead, we preferred to quote results for a "compromise" set of parameters. Systematic studies of the dependence on single parameters were made in Refs. 14 and 15, but we found such studies not very useful due to the strong correlations between optimal parameters. Also, sets of param-

TABLE I. Numerical results of the different algorithms introduced in the paper. For explanation of the parameters see Sec. V.

			SVD	Sec. II		Sec. III		Sec. IV		0th order		Ref. 24	
System	T	ϵ	r_0	r_0	r_{dyn}	r_0	r_{dyn}	r_0	r_{dyn}	r_0	r_{dyn}	r_0	r_{dyn}
Hénon	5 000	100%	1.5	1.7	3.6	1.7	4.1	1.9	8.3	1.7	4.0		
		10%	···	2.6	4.5	3.4	7.5	4.2	10.7	2.8	5.6		
		1%	···	2.9	5.7	3.4	5.3	3.4	6.6	1.7	2.5		
	20 000	1%	···	3.2	6.5	4.1	6.8	4.2	8.4	2.4	3.8		
	100 000	0.3%	···	2.9	4.0	4.7	6.8	5.0	8.3	2.4	4.4		
Ikeda	20 000	10%	···	3.4	···	2.2	···	3.7	···	2.5	4.4[c]	3.6	6.6
(x coord.)		1%	···	1.5	···	1.7	···	2.1	···	1.2	1.7[c]	1.8	2.6
Lorenz[a]	20 000	10%	1.2	3.5	···	2.9	···	3.3	···	2.0	7.2[c]	3.1	8.1
	20 000	1%	···	2.3	···	1.8	···	2.3	···	1.1	1.3[c]	1.9	2.3
Lorenz[b]	5 000	10%	···	1.8	···	1.5	···	1.8	···	1.5	4.3[c]	2.4	4.8

[a]Delay $\tau=0.05$, x coordinate.
[b]Delay $\tau=0.2$, -"-.
[c]Obtained from the vector-valued time series using the 0th order multivariate version (Ref. 24).

eters optimal for one time series often gave mediocre results for other time series. Such differences occurred even for time series belonging to the same model but using different noise realizations. The above set of parameters gave consistently good results. Therefore the numbers reported in Table I must not be taken literally, but their relative magnitude is characteristic for the different algorithms. The only quantitative comparison with a noise reduction algorithm different from this class can be obtained from Ref. 10: For a Hénon trajectory of length 32 000 and 1% additive noise the authors of Ref. 10 find $r_{dyn}=4.8$.

By "0th order" we denote the algorithm explained in Ref. 24, i.e., substituting each point of the noisy trajectory by the average value of its neighbors. The optimal parameters are reported in Ref. 24.

For applications to data sets with unknown dynamics (experimental data) one first has to have at least a vague idea about the correct embedding dimension. Then the parameters should be chosen following the above examples. Our experience with "toy models" shows that suboptimal choices of the parameters simply lead to suboptimal results but do not destroy the data structures. One possible diagnostics is a statistical analysis of the corrections made. Especially they should not be strongly correlated to the signal.

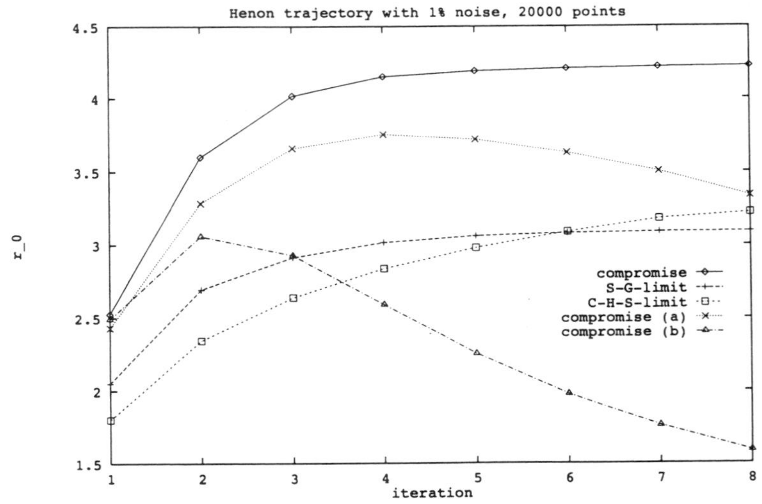

FIG. 6. Improvements of r_0 vs the number of iterations for three different choices of the matrix **P**: Cawley–Hsu–Sauer limit, Schreiber–Grassberger limit, and compromise with $P_i=0.001$ for $i=1$ and $i=5$ and $P_i=1$ else. The original series consisted of 20 000 points on the Hénon attractor, corrupted with 1% uniformly distributed measurement noise, and embedded with $m=4$. The two nonmonotonous curves are the results of the compromise without cutoff of too large corrections (a) and, in addition, without subtraction of the curvature effects (b). Both are crucial in this case [and (b) even more so for shorter trajectories], as the noise level is so small. For large noise levels and long trajectories both effects become negligible.

VI. CONCLUSIONS

We have seen that the two seemingly rather different classes of locally linear noise reduction methods developed by Cawley, Hsu, and Sauer[14–16] on the one hand, and by Schreiber and Grassberger[13] on the other, are indeed closely related. The formalism of Sec. IV is a generalization of both methods. It supplies the possibility to find a compromise between these two algorithms avoiding their disadvantages.

A basic problem with any noise reduction scheme—and thus also with the present ones—is that we have to *assume* that a separation into noise plus signal is meaningful. In real-life applications this need not always be the case. Also, no clear-cut criterion for the amount of noise to be subtracted can be given in such cases, and no performance measure is available. In view of this, robustness of the algorithm is of utmost importance.

The main problem with respect to applications is the choice of the different parameters. However, a quite robust scheme is obtained with the parameters suggested in Sec. V. The tabular presentation of the results shows the overall superiority of the new scheme for maps. For high sampled flows the performance of our compromise is as good as the Cawley–Hsu–Sauer limit. Nevertheless, the optimal algorithm and the best parameters depend on the noise level, the sampling rate, and the length of the time series.

For high noise levels and for large sampling rates, linear prefiltering as used by Sauer gives good results. We found that filtering based on SVD is for that purpose about as efficient as Fourier-based Wiener filtering, and usually better than simple low-pass filtering. For maps, sparsely sampled flows, or low noise levels, the distortions involved in linear prefiltering overwhelm its advantages.

Another trick (due to Sauer) which is not fundamental for the method to work but which can improve results considerably, is a correction which takes into account curvature effects.

As implemented in Sec. V, our algorithm is rather robust and not too time consuming. Typically, time sequences of ca. 10^4 points can be handled easily on work stations. It is much slower than conventional (linear) filters but it gives much better results, in particular for low noise levels.

ACKNOWLEDGMENTS

This work was supported by the Deutsche Forschungsgemeinschft, SFB 237. T.S. receives a European Communities grant within the framework of the SCIENCE programme, Contract No. B/SCI*-900557.

APPENDIX A: WIENER AND SVD–WIENER FILTERING

In this appendix we shortly discuss Wiener filtering[17] and its generalization[19,20] obtained by replacing the Fourier basis by the SVD basis.

In the following we assume that the time series is a randomly chosen realization of a stochastic stationary process $X = X_1 X_2 \cdots$ with zero mean.

(a) Let us first write the discrete Fourier transform of the noisy time series as

$$\tilde{x}_k = \sum_{j=1}^{T} x_j e^{2\pi i k j/T}. \tag{A1}$$

The spectrum S_k of the random process X is defined as

$$S_k = \langle |\tilde{x}_k|^2 \rangle. \tag{A2}$$

We assume it to be a sum of two contributions,

$$S_k = S_k^{(y)} + S_k^{\text{noise}} \tag{A3}$$

corresponding to the clean signal and the noise, resp. (we leave out the exact normalization of the spectrum for which several conventions exist in the literature). Notice that there is no cross term from the additive ansatz Eq. (2) for the *amplitudes* since we assume that signal and noise are statistically independent. We also assume for the moment that we know (or can guess) both S_k and the noise spectrum S_k^{noise}.

We look for a linear filter, i.e., for a linear relationship between the random variables X_n (whose realization are the x_n) and a filtered random variables X_n^{corr}, such that the spectrum of X_n^{corr} is $S_k - S_k^{\text{noise}}$ and such that

$$\langle |X_n^{\text{corr}} - X_n|^2 \rangle \tag{A4}$$

is minimal. This filter is uniquely given by[17]

$$\tilde{x}^{k,\text{corr}} = \left(1 - \frac{S_k^{\text{noise}}}{S_k}\right) \tilde{x}_k. \tag{A5}$$

If S_k^{noise} is white and S_k is monotonously decreasing, then this is essentially a low-pass filter. On the other hand, if S_k is not monotonous then it mainly filters away the regions where S_k is small (since it should be noise dominated there), and leaves the peak regions essentially unchanged. With such a filter, one can, e.g., clean completely periodic and quasiperiodic signals, provided the time series is sufficiently long.

(b) Assume we have only a single time series available, without any prior estimate of the spectrum, and we want to apply Fourier–Wiener filtering. Thus we have first to estimate the Fourier spectrum S_k from $\tilde{x}_k$. As is well known, we cannot simply use $S_k \approx |\tilde{x}_k|^2$ as an estimator since the statistical fluctuations of $|\tilde{x}_k|^2$ do not decrease in the limit $T \to \infty$. One way to circumvent this problem consists in modifying the above procedure as follows. We Fourier transform only short pieces $\mathbf{x}_n = (x_n, \ldots, x_{n+m})$ of length $m+1$, calling the results $\tilde{x}_k^{(n)}$, $k = 0, \ldots, m$, and estimate a coarse grained spectrum as

$$S_k^{(\text{coarse})} \approx \frac{1}{T-m} \sum_{n=1}^{T-m} |\tilde{x}_k^{(n)}|^2. \tag{A6}$$

In the filtering procedure it is then natural to apply Eq. (A5) to the same pieces $\mathbf{x}_n$, which gives $m+1$ "clean" candidates $z_j^{(n)}$ for each x_j,

$$z_j^{(n)} = \frac{1}{m+1} \sum_{k=0}^{m} \left(1 - \frac{S_k^{\text{noise,coarse}}}{S_k^{\text{coarse}}}\right) \tilde{x}_k^{(n)} e^{2\pi i k j/(m+1)}. \tag{A7}$$

In general, they will not be independent of n, i.e., Eq. (A7) would give $m+1$ mutually inconsistent "cleaned" values for each j. We propose that a natural compromise between them is just their arithmetic average,

$$x_j^{\text{corr}} = \frac{1}{m+1} \sum_{n=j-m}^{j} z_j^{(n)}. \tag{A8}$$

(c) In the SVD case, we proceed exactly as in case (b). We first form the $(m+1)\times(m+1)$ covariance matrix

$$C_{jk} = \frac{1}{T-m} \sum_{n=1}^{T-m} x_{n+j} x_{n+k}, \tag{A9}$$

its eigenvalues μ_i, and its (orthonormal) eigenvectors $\mathbf{e}_i$ ($i=0,\ldots,m$). We then assume that the eigenvalues (which correspond to the spectrum in the Fourier case) are additively composed of signal and noise contributions,

$$\mu_i = \mu_i^{(y)} + \mu_i^{\text{noise}}, \tag{A10}$$

and we form the matrix

$$P_{jk} = \sum_{i=0}^{m} \left(1 - \frac{\mu_i^{\text{noise}}}{\mu_i}\right) e_{i,j} e_{i,k}. \tag{A11}$$

If some of the "signal" eigenvalues $\mu_i^{(y)}$ are zero, then $\mathbf{P}$ projects into the orthogonal subspace. If all μ_i^{noise} were zero, then $\mathbf{P}$ would be just the unit matrix. The method proposed in Ref. 20 is obtained by assuming that the smallest μ_i's are pure noise and the others are pure signal, in which case $\mathbf{P}$ is a projection matrix. Notice however that in general $\mathbf{P}^2 \neq \mathbf{P}$, i.e., $\mathbf{P}$ is in general not a projection matrix. From each delay vector $\mathbf{x}_n = (x_n,\ldots,x_{n+m})$ we can form candidates $z_j^{(n)}$ for cleaned values of x_j with $j=n,\ldots,n+m$ as

$$z_j^{(n)} = \sum_{k=n}^{n+m} P_{j-n,k-n} x_k. \tag{A12}$$

Again, they will not be independent of n in general. We again propose as a compromise their arithmetic average,[20]

$$x_j^{\text{corr}} = \frac{1}{m+1} \sum_{n=j-m}^{j} z_j^{(n)} = \sum_{k=0}^{m} x_k \frac{1}{m+1} \sum_n P_{j-n,k-n}. \tag{A13}$$

Just like Wiener filtering, this method gives perfect cleaning for periodic and quasiperiodic signals, provided the time series and the window length m are long enough. Since the eigenvectors $\mathbf{e}_i$ become sines resp. cosines in the limit $m \to \infty$,[19] in this limit both methods essentially coincide.

APPENDIX B: A MINIMIZATION PROBLEM

In this appendix we solve the minimization problem defined by Eqs. (4.8), (4.6), and (4.7). We also show that its solution is a superposition of the form

$$\boldsymbol{\theta}_n = \sum_q \mu_{nq} \mathbf{P} \nabla F_q^{(n)}, \tag{B1}$$

i.e., it is exactly of the form of Eq. (4.1), generalized to several constraints.

We use a Lagrange multiplier formalism, with multipliers $\mu_{k,q}$ for the constraints (4.6) and with multipliers $\lambda_{qq'}$ for the constraints (4.7). The Lagrange function thus reads

$$L = \frac{1}{2} \sum_{k:\mathbf{x}_k \in \mathcal{U}_n} (\boldsymbol{\theta}_k \cdot \mathbf{P}^{-1} \boldsymbol{\theta}_k) - \sum_{\substack{k:\mathbf{x}_k \in \mathcal{U}_m \\ q<Q}} \mu_{kq} [\mathbf{a}_q^{(n)} \cdot (\mathbf{x}_k + \boldsymbol{\theta}_k) + b_q^{(n)}] + \frac{1}{2} \sum_{q,q' \leq Q} \lambda_{qq'} [\delta_{qq'} - (\mathbf{a}_q^{(n)} \cdot \mathbf{P} \mathbf{a}_{q'}^{(n)})]. \tag{B2}$$

Varying L with respect to $\boldsymbol{\theta}_k$, we obtain first

$$\boldsymbol{\theta}_k = \sum_q \mu_{kq} \mathbf{P} \mathbf{a}_q^{(n)}. \tag{B3}$$

Observing that $\mathbf{a}_q^{(n)}$ is just the gradient of the qth constraint function, this proves already our first claim. Multiplying Eq. (B3) with $\mathbf{a}_{q'}^{(n)}$ gives next

$$\mu_{kq} = \mathbf{a}_q^{(n)} \cdot \boldsymbol{\theta}_k = -b_q^{(n)} - (\mathbf{a}_q^{(n)} \cdot \mathbf{x}_k). \tag{B4}$$

Minimizing L with respect to $b_q^{(n)}$ gives $\Sigma_k \mu_{kq} = 0$ or

$$b_q^{(n)} = -(\mathbf{a}_q^{(n)} \cdot \boldsymbol{\xi}^{(n)}). \tag{B5}$$

Finally, varying L with respect to $\mathbf{a}_q^{(n)}$ and multiplying the derivative by $\mathbf{a}_{q''}^{(n)}$, we find that $\lambda_{qq'}=0$, as might have been anticipated. Dropping thus the term $\propto \lambda_{qq'}$ in this derivative and using Eqs. (B3), (B4), and (B5), we obtain

$$\mathbf{C}^{(n)} \mathbf{a}_q^{(n)} = \mathbf{P} \sum_{q'} M_{qq'} \mathbf{a}_{q'}^{(n)}. \tag{B6}$$

Here we have defined $M_{qq'} = \Sigma_k \mu_{kq} \mu_{kq'}$, and the matrix $\mathbf{C}^{(n)}$ is defined in Eq. (4.10). The matrix $\mathbf{M}$ is symmetric and can thus be diagonalized as $\mathbf{V}^{-1} \widetilde{\mathbf{M}} \mathbf{V}$ with orthogonal $\mathbf{V}$. Defining $\widetilde{\mathbf{a}}_q^{(n)} = \Sigma_{q'} V_{qq'} \mathbf{a}_{q'}^{(n)}$, and inserting this into Eq. (B6), it simplifies to

$$\mathbf{C}^{(n)} \widetilde{\mathbf{a}}_q^{(n)} = \mathbf{P} \widetilde{M}_{qq} \widetilde{\mathbf{a}}_q^{(n)}. \tag{B7}$$

Since $\mathbf{P}$ is symmetric and positive, $R_i = P_i^{-1/2}$ exists. We can define $\boldsymbol{\Gamma}^{(n)}$ as in Eq. (4.12), find that the vectors $\mathbf{e}_q^{(n)} = \sqrt{\mathbf{P}} \widetilde{\mathbf{a}}_q^{(n)}$ are its orthonormal eigenvectors, and see that $\boldsymbol{\theta}_k$ is given by Eq. (4.14). Since the covariance matric C has rank $m+1$, each set of Q of the $m+1$ eigenvectors is a solution of this problem, but the absolute minimum is reached when choosing the vectors with the smallest eigenvalues.

[1] P. Grassberger, T. Schreiber, and C. Schaffrath, Int. J. Bifurcation Chaos **1**, 521 (1991).

[2] J. P. Crutchfield and B. S. McNamara, Complex Systems **1**, 417 (1987).

[3] J. Cremers and A. Hübler, Z. Naturforsch. **42a**, 797 (1987).

[4] P. Grassberger, talk at Fritz Haber Conference on "Chaos and Related Non-Linear Phenomena," Kiryat Anavim, 1986; Physica Scripta **40**, 346 (1989).

[5] J. D. Farmer and J. Sidorowich, Phys. Rev. Lett. **59**, 845 (1987).

[6] J. D. Farmer and J. Sidorowich, "Exploiting Chaos to Predict the Future and Reduce Noise," in *Evolution, Learning and Cognition*, edited by Y. C. Lee (World Scientific, Singapore, 1988).

[7] M. Casdagli, Physica D **35**, 335 (1989).
[8] K. Stockbro, D. K. Umberger, and J. Hertz, J. Complex Systems (1991).
[9] K. Stockbro, "Predicting chaos with weighted maps," NORDITA preprint (1990).
[10] E. J. Kostelich and J. A. Yorke, Phys. Rev. A **38**, 1649 (1988).
[11] S. M. Hammel, Phys. Lett. A **148**, 421 (1990).
[12] J. D. Farmer and J. Sidorowich, Physica D **47**, 373 (1991).
[13] T. Schreiber and P. Grassberger, Phys. Lett. A **160**, 411 (1991).
[14] R. Cawley and G.-H. Hsu, Phys. Rev. A **46**, 3057 (1992).
[15] R. Cawley and G.-H. Hsu, Phys. Lett. A **166**, 188 (1992).
[16] T. Sauer, Physica D **58**, 193 (1992).
[17] W. H. Press, B. P. Flannery, S. A. Teukolski, and W. T. Vetterling, *Numerical Recipes* (Cambridge University, Cambridge, 1988).
[18] T. Sauer, J. A. Yorke, and M. Casdagli, J. Stat. Phys. **65**, 579 (1991).
[19] R. Vautard and M. Ghil, Physica D **35**, 395 (1989).
[20] R. Vautard, P. Yiou, and M. Ghil, Physica D **58**, 95 (1992).
[21] M. Hénon, Commun. Math. Phys. **50**, 69 (1976).
[22] P. Grassberger and H. Kantz, Phys. Lett. A **113**, 235 (1985).
[23] R. Hegger and T. Schreiber, Phys. Lett. A **170**, 305 (1992).
[24] T. Schreiber, Phys. Rev. E (in press) (1993).
[25] E. N. Lorenz, J. Atmos. Sci. **20**, 130 (1963).
[26] O. E. Rössler, Phys. Lett. **57**, 397 (1976).
[27] R. Badii *et al.*, Phys. Rev. Lett. **60**, 979 (1988).

A noise reduction method for chaotic systems

Stephen M. Hammel [1]
Naval Surface Warfare Center, 10901 New Hampshire Avenue, Silver Spring, MA 20903-5000, USA

Received 4 January 1990; revised manuscript received 29 June 1990; accepted for publication 29 June 1990
Communicated by D.D. Holm

A method is presented which can reduce the noise of a chaotic orbit on an attractor by more than ten orders of magnitude. This method is simple and fast: its performance is analyzed for several two-dimensional systems at moderate noise levels, including the Ikeda map. For this analysis, the underlying maps are assumed to be known.

Techniques of chaotic nonlinear dynamics have been applied to a variety of physical processes. An experimental measurement of a physical system typically produces a discrete time history of a scalar quantity. The system itself often resides in an infinite-dimensional phase space; however, the long term behaviour may settle onto a finite-dimensional attractor. For such systems, an n-dimensional delay coordinate construction creates a phase space which can reveal the multiple dimensionality of the dynamics [1]. At this stage of the data analysis, the given data set has been replaced by a discrete sequence of points in an n-dimensional space, which will be called a noisy orbit.

In this setting, the full data analysis problem is a transformation of the noisy orbit into a clean, reduced-noise orbit, together with the underlying maps. Thus, noise reduction is naturally related to the problem of forecasting or prediction, whereby the underlying map is learned, as the other primary component of the data analysis problem. In this work I will not discuss the prediction problem. I will assume that the underlying map is known, and I will examine the behavior of a noise reduction algorithm in this context. In particular, it is much easier to discern the performance and the convergence behavior of the noise reduction algorithm in isolation from the map-learning problem.

[1] Office of Naval Technology postdoctoral fellow in the NSWC Navy Dynamics Institute Program.

The processes of interest are those for which the underlying dynamics has a chaotic component. I will assume that the noisy orbit has some components due to deterministic chaos, and some components due to noise. The goal is to distinguish these two different types of components, and to remove the noise-induced components, thus making the measurement of other dynamical parameters more accurate.

The problem of noise reduction for chaotic systems has been a focus of attention recently. Kostelich and Yorke [2] compute linear maps at each point of the reconstructed attractor, and then implement a least-squares ("trajectory adjustment") procedure to create a new trajectory. Farmer and Sidorowich [3] discuss the performance of a variety of prediction procedures. For noise reduction, they describe an idea that exploits local expanding and contracting directions, and they give a demonstration of the technique when the map is known. Other relevant work in this direction is that of Abarbanel et al. [4] and Casdagli [5].

There are several ways by which a dynamical process, represented by f, can be obscured by noise. A sequence $\{\boldsymbol{x}_k\}_{k=0}^N$ satisfying $\boldsymbol{x}_{k+1}=f(\boldsymbol{x}_k)$, $0\leqslant k\leqslant N-1$ will be called a *true* orbit. If a *noisy orbit* $\{\boldsymbol{p}_k\}_{k=0}^N$ is generated by $\boldsymbol{p}_k=\boldsymbol{x}_k+\eta_k$ for small $\|\eta_k\|<\delta$, this will be termed *additive noise*. This method of noise generation can be imagined to occur when a true (physical) process is obscured by errors in measurement. The obvious goal is to reduce $\|\eta_k\|$.

A second way by which the dynamics can be ob-

From *Phys. Lett. A* **160** (1990). Reprinted with permission from Elsevier Science Publishers.

scured is associated with the shadowing problem [6,7]. As the noisy orbit $\{p_k\}_{k=0}^N$ is generated by the process f, errors are made upon each iteration of the process. The resultant points of the orbit then satisfy

$$\|p_{k+1}-f(p_k)\|<\delta\,,\quad 0\leqslant k\leqslant N-1\,, \tag{1}$$

for an appropriately small δ. This type of noise will be called *dynamic noise*, and the noise reduction problem is quite different for this case.

The noise reduction problem is to find a less noisy orbit near $\{p_k\}_{k=0}^N$. To make a new orbit which is less noisy, and to make a new orbit which is nearby, are often conflicting constraints which must be balanced in some reasonable way. At the limit of nearness lies the old original noisy $\{p_k\}_{k=0}^N$; at the other extreme, one can seek a noise-free true orbit $\{x_k\}_{k=0}^N$ which lies somewhat near $\{p_k\}_{k=0}^N$. This latter alternative is associated with the shadowing property: the orbit of x_0 is said to *ϵ-shadow* $\{p_k\}_{k=0}^N$ if $\|f^k(x_0)-p_k\|<\epsilon$ for $0\leqslant k\leqslant N$.

Note that in both the dynamic noise and additive noise problems, the existence of a prescribed map f is assumed. The remainder of this paper will be devoted to a description of a noise reduction method applied to two-dimensional plane diffeomorphisms f: $x_{k+1}=f(x_k)$, $x_k\in\mathbb{R}^2$. Henceforth the map is assumed to be known.

The noise reduction process is an algorithm applied to an entire noisy orbit to yield a new, less noisy orbit. This process, which perturbs slightly each point of the orbit, is called a refinement of the orbit. The refinement process is then applied to this new, cleaner orbit to produce a third orbit, expected to be less noisy than its predecessor. The refinement process is then iterated between 4 and 30 times, depending on the noise level and the orbit length N. I call the entire procedure *noise reduction by shadowing* since the proof of the shadow lemma is used to construct the scheme.

Noise can be introduced into the dynamics in different ways, and may be distributed in one of several ways. Thus, I replace the bound δ on absolute noise level by a more general measure. It is also desirable to use a relative noise measure, so I measure the initial noise level Δ, a ratio of noise level to signal strength:

$$\Delta=\frac{(\Sigma_{k=0}^{N-1}\|f(p_k)-p_{k+1}\|^2)^{1/2}}{(\Sigma_{k=0}^{N-1}\|f(p_k)-p_k\|^2)^{1/2}}\,. \tag{2}$$

Furthermore, it is necessary to monitor the performance of the noise reduction upon each application of the refinement process. The quantity Δ is not a discriminating indicator of the performance of the refinement process, since a single large value of $\|f(p_k)-p_{k+1}\|^2$ for one particular point p_k will dominate the sum, thus obscuring an average improvement for most points on an orbit. I use the following measure,

$$E=\frac{1}{N}\sum_{k=1}^{N}\log\epsilon_k\,, \tag{3}$$

where $\epsilon_k=\max\{\epsilon_\mu, \|f(x_{k-1})-x_k\|\}$. The quantity ϵ_μ is related to machine epsilon, indicating the relative machine precision; I use $\epsilon_\mu=10^{-16}$. E can be seen as a measure of how deterministic the orbit $\{x_k\}_{k=0}^N$ is, since no account is taken in this measure of the proximity of the original noisy orbit. Let ν_k denote the values of ϵ_k for the original noisy orbit; then a measure of the improvement given by the refinement process is

$$\Gamma=\frac{1}{N}\sum_{k=1}^{N}\log\nu_k-\frac{1}{N}\sum_{k=1}^{N}\log\epsilon_k\,. \tag{4}$$

If $\log_{10}\epsilon_k$ is used, then Γ gives the number of decades of improvement averaged over the orbit. I call Γ the improvement.

In order to illustrate the performance of noise reduction by shadowing, I will consider the dynamics of the Ikeda map. This is a dissipative map of the plane which is a model for an optically bistable ring cavity [8]. The complex signal $x_k\in\mathbb{C}$ represents the amplitude and phase of the electromagnetic field within a passive ring cavity. Within the cavity is a medium, the refractive index of which is a nonlinear function of the amplitude of the field. The field is assumed to propagate perpendicularly to the cavity mirrors (plane wave limit), and for each cavity round trip, x_{k+1} measures the field as a function of the value x_k at the previous pass:

$$f(x_k)=\gamma+0.9x_k\exp\{i[0.4-\alpha(1+\|x_k\|^2)^{-1}]\}\,. \tag{5}$$

For appropriate choice of the steady input field γ and nonlinear medium thickness α, the field in the

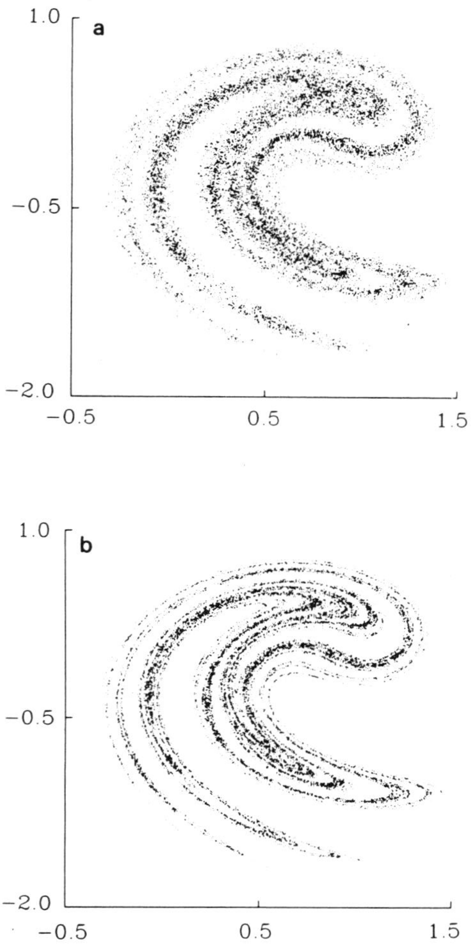

Fig. 1. (a) A noisy orbit of $N=10000$ points for the Ikeda map. Normally distributed noise with standard deviation ≈ 0.04 was added to a computed orbit, corresponding to an initial noise level $\Delta \approx 0.051$. (b) The cleaned orbit after 24 iterations of the refinement process. The fractal structure of the attractor is apparent in the folded leaves of the unstable manifold.

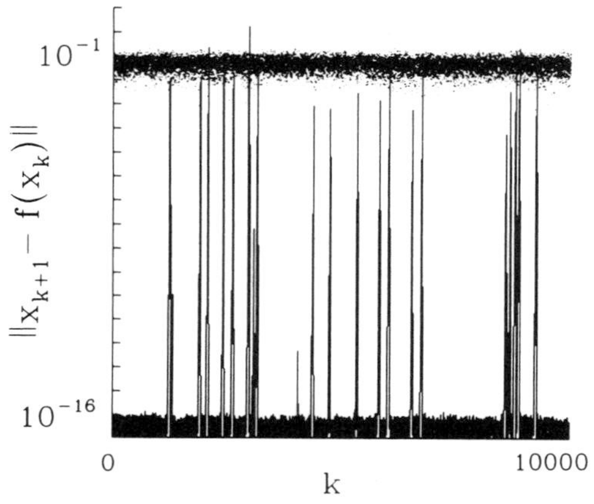

Fig. 2. The performance of the noise reduction process: $\|f(x_{k-1}) - x_k\|$ is measured with a logarithmic scale on the vertical axis for each of the 10000 points on the Ikeda orbit. The upper horizontal smear of points shows the original noisy orbit. After 24 refinements, the pointwise noise for the cleaned orbit is shown by the lower straight line graph. 92.1% of the points lie below 10^{-12}; fewer than 1.4% lie above 10^{-3}.

cavity is bistable. Thus, the intensity of the field within the cavity can behave like a binary switch: there are two competing attractors, one at a high amplitude, and one at a low amplitude.

The case of additive noise is now considered for an example of the process. In the numerical simulation, an orbit on the attractor associated with the low amplitude state is created for $\alpha=5.4$, and $\gamma=0.92$, values for which the associated orbit appears to be chaotic (there is a positive Lyapunov exponent for the numerical approximation.) An orbit $\{z_k\}_{k=0}^{N}$ of moderate length ($N=10000$) is generated to machine precision. The random normally distributed "noise" $\{\eta_k\}_{k=0}^{N}$ (simulating errors in measurement) is generated numerically, and the noisy orbit $\{p_k\}_{k=0}^{N}$ is created by $p_k = z_k + \eta_k$.

In fig. 1a, a noisy orbit $\{p_k\}_{k=0}^{N}$ of length $N=10000$ has been created for the Ikeda map, with an initial noise level $\Delta \approx 0.051$. In fig. 1b, the cleaned orbit is shown after 24 iterations of the refinement process. At the level of detail of a printed figure, this orbit is essentially indistinguishable from the "unsoiled" orbit $\{z_k\}_{k=0}^{N}$ except for 6 points which are visibly different.

I use the information portrayed in fig. 2 as the primary measure of the performance of the refinement process. The initial value of $\nu_k = \|f(p_{k-1}) - p_k\|$ is measured on the vertical axis for each of the 10000 points on the orbit, and is shown as a dot in the cloud

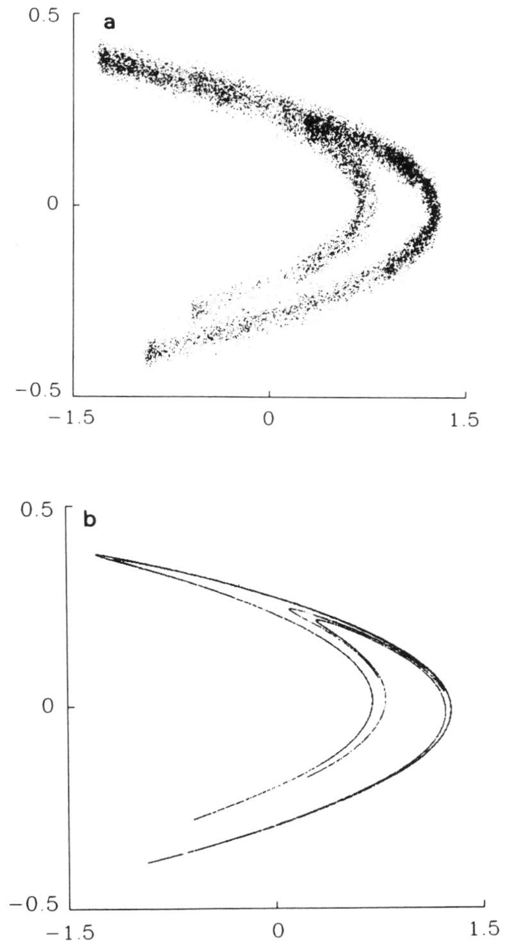

Fig. 3. (a) A noisy orbit for the Hénon map consisting of 10000 points, and $\Delta \approx 0.0455$. (b) The cleaned orbit after 20 refinements. The manifold structure of the Hénon attractor is apparent. A single point corresponding to the peak of the error spike in fig. 4 can be seen above the upper right portion of the attractor.

at the top of the figure. After 24 refinements, the measure $\epsilon_k = \|f(\boldsymbol{x}_{k-1}) - \boldsymbol{x}_k\|$ for each point of the resultant cleaned orbit is shown connected by straight lines. For almost all points, ϵ_k has been reduced to machine precision. The improvement is substantial: $\Gamma = 13.76$, corresponding to 13 orders of magnitude

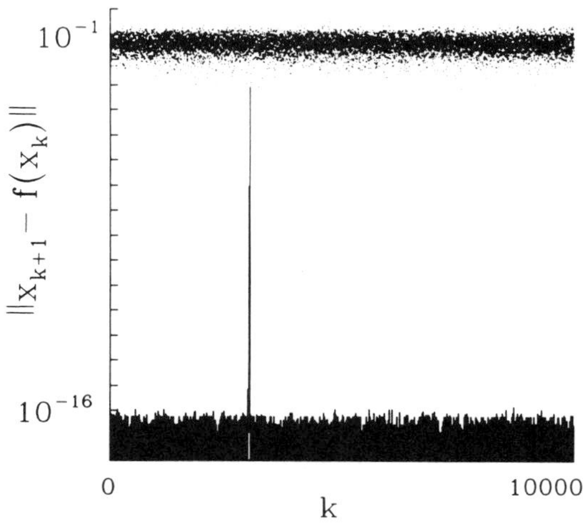

Fig. 4. The performance of the noise reduction process for a noisy orbit of length $N = 10000$ for the Hénon map. $\|f(\boldsymbol{x}_{k-1}) - \boldsymbol{x}_k\|$ is measured with a logarithmic scale on the vertical axis for each of the 10000 points on the orbit. The upper horizontal smear of points shows the original noisy orbit. After 24 refinements, the pointwise noise for the cleaned orbit is shown by the lower straight line graph.

for most points. The spikes in the graph of the cleaned orbit correspond to points where the angle θ_k between the expanding direction and the contracting direction for the linear map at that point is small.

For smaller values of Δ, the number of spikes in a graph of E similar to fig. 2 is reduced. The noise reduction algorithm quickly converges to an orbit with long segments which are machine-accurate representations of the dynamics of the map. Furthermore, this process also reveals the fractal structure of the unstable manifold associated with the attractor. The characteristics can be seen in the application of the refinement process to a noisy orbit of the Hénon map. The underlying map of the plane is defined by $(u_{k+1}, v_{k+1}) = (1 - 1.4u_k^2 + v_k, 0.3u_k)$. In fig. 3a the original noisy orbit $\{\boldsymbol{p}_k\}_{k=0}^{N}$ is shown. Fig. 3b shows the subsequent cleaned orbit $\{\boldsymbol{x}_k\}_{k=0}^{N}$ after 20 refinements, and the familar manifold structure of the Hénon attractor is evident.

The corresponding graph of E (compare to fig. 2) for the cleaned orbit $\{\boldsymbol{x}_k\}_{k=0}^{N}$ of length $N = 10000$ is shown in fig. 4.

Two portions of the orbit of length 2990 points and

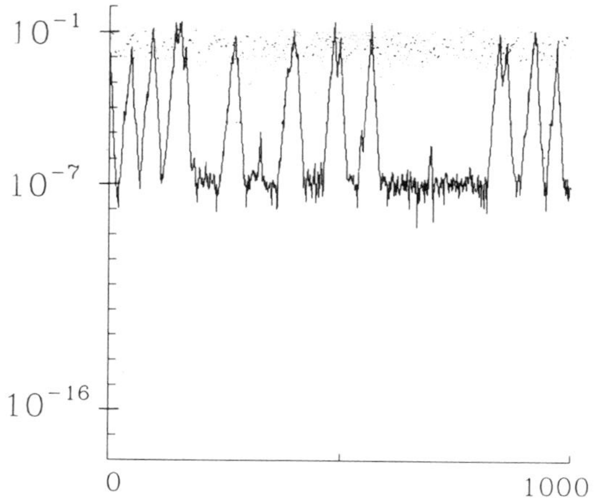

Fig. 5. A graph of $\{\|z_k - x_k\|\}_{k=0}^{1000}$, where $\{z^k\}_{k=0}^N$ is the original unsoiled single precision orbit to which noise is added. Only the first 1000 points are shown for clarity: the remaining 9000 points are qualitatively similar. The spikes show where the new cleaned orbit $\{x_k\}_{k=0}^N$ fails to converge to $\{z_k\}_{k=0}^N$. This does not contradict fig. 4 which shows $\{x_k\}_{k=0}^N$ to be quite low noise in the sense that it is strongly deterministic. The cleaned orbit $\{x_k\}_{k=0}^N$ can thus be essentially noise-free even over those short segments where it is not close to $\{\zeta_k\}_{k=0}^N$. Comparison with fig. 4 makes clear the sense in which the cleaned orbit is better than the original unsoiled orbit $\{z_k\}_{k=0}^N$ for each of the two long subsegments.

6975 points, which are machine accurate ($\|f(x_{k-1}) - x_k\| < 10^{-15}$), are interrupted by a single spike of 35 points for which $10^{-15} < \|f(x_{k-1}) - x_k\| < 10^{-1}$. It is also important to point out that the resultant cleaned orbit $\{x_k\}_{k=0}^N$ is substantially less noisy than the original unsoiled orbit $\{z_k\}_{k=0}^N$ which was used to generate the simulated noise via $p_k = z_k + \eta_k$. This is possible because $\{z_k\}_{k=0}^N$ is calculated with approximately 7 digit accuracy, whereas the cleaned orbit $\{x_k\}_{k=0}^N$ is accurate to essentially 16 digits. In fig. 5 I show the difference between the original unsoiled orbit $\{z_k\}_{k=0}^{1000}$ and the cleaned orbit $\{x_k\}_{k=0}^{1000}$ for the first 1000 points. Note that $\min\{\|z_k - x_k\|\} \approx 10^{-7}$, since $\{z_k\}_{k=0}^N$ itself is only calculated to single precision accuracy. This shorter segment is representative of the entirety of $\{\|z_k - x_k\|\}_{k=0}^N$, and the resultant change in horizontal scale reveals the characteristic slopes "up" the left side of a spike corresponding to the expanding Lyapunov exponent, and "down" the right side of a spike corresponding to the contracting Lyapunov exponent.

I shall describe the noise reduction by the shadowing method in more detail. The refinement process was presented previously in a proof of shadowing for non-hyperbolic systems [9], and it is derived from the construction used in the proof of the shadowing lemma [7]. A given two-dimensional noisy orbit is represented by $\{p_k\}_{k=0}^N$. A less noisy orbit is sought, and in the ensuing derivation, $\{x_k\}_{k=0}^N$ will represent this less noisy orbit. A noise-free orbit $\{x_k\}_{k=0}^N$ would satisfy $x_k = f(x_{k-1})$. Let $\Phi_k = x_k - p_k$, and $\Pi_k = f(p_{k-1}) - p_k$, and then $\Phi_k = f(x_{k-1}) - f(p_{k-1}) + \Pi_k$. Assume x_k is close to p_k, and with the linearized map at p_k denoted by L_k, make the approximation $L_k\Phi_k \approx f(x_k) - f(p_k)$, giving

$$\Phi_k \approx L_{k-1}\Phi_{k-1} + \Pi_k \tag{6}$$

as an iteration scheme for the unknown set $\{\Phi_k\}_{k=0}^N$.

Next, each Φ_k is represented by its components: $\{\hat{e}_k\}_{k=0}^N$ and $\{\hat{c}_k\}_{k=0}^N$ are unit vectors in the expanding and contracting directions, respectively,

$$\Phi_k = \alpha_k \hat{e}_k + \beta_k \hat{c}_k\,, \quad \Pi_k = \zeta_k \hat{e}_k + \xi_k \hat{c}_k\,. \tag{7}$$

The set $\{\Pi_k\}_{k=0}^N$ can be computed directly, so $\{\zeta_k\}_{k=0}^N$ and $\{\xi_k\}_{k=0}^N$ are known. It is now possible to define a bi-directional iterative scheme by treating (6) as an equality:

$$\Phi_k = L_{k-1}\Phi_{k-1} + \Pi_k\,. \tag{8}$$

Using (6) and (7),

$$\Phi_{k+1} = L_k(\alpha_k \hat{e}_k + \beta_k \hat{c}_k) + (\zeta_{k+1}\hat{e}_{k+1} + \xi_{k+1}\hat{c}_{k+1})\,. \tag{9}$$

This can be split into two equations – one on the expanding subspace, and one on the contracting subspace:

$$\alpha_{k+1} = \|L_k \hat{e}_k\| \alpha_k + \zeta_{k+1}\,,$$
$$\beta_{k+1} = \|L_k \hat{c}_k\| \beta_k + \xi_{k+1}\,. \tag{10}$$

Finally, these two equations can be cast into a stable iterative scheme:

$$\alpha_k = \frac{\alpha_{k+1} - \zeta_{k+1}}{\|L_k \hat{e}_k\|}\,, \quad \alpha_N = 0\,,$$
$$\beta_{k+1} = \beta_k \|L_k \hat{c}_k\| + \xi_{k+1}\,, \quad \beta_0 = 0\,. \tag{11}$$

Note that for the first equation, iteration begins at one end ($k=0$), while for the other iteration starts at the other end ($k=N$) of $\{\boldsymbol{p}_k\}_{k=0}^N$. It should be noted that this basic scheme must be modified if it is to work reliably on orbits for non-hyperbolic systems with high noise levels. Difficulties occur when the angle θ_k between stable and unstable directions becomes small. The small angles result in correction coefficients which are unacceptably large, and thus outside the domain within which the linear map is a good approximation.

The noise reduction by shadowing procedure makes large reductions in noise level with each of the first several refinements. A graph of the dependence of Γ on the number of refinements is a sigmoid curve with a rapid rise, so that for $n \leqslant 4$, say, and a given noise level Δ, one can observe that at the nth refinement, the improvement $\Gamma \approx 2n \log(1/\Delta)$. Roughly speaking, this means that the number of accurate digits doubles for each of the first few iterations. Of course, this depends on the size of Δ, and the practically achievable limit is determined by ϵ_μ, machine epsilon. This sigmoid curve describes the behavior averaged over all the points in the orbit; certain points do not behave as well. If at a point $\boldsymbol{p}_j$ the angle θ_j between expanding direction $\boldsymbol{e}_j$ and contracting direction $\hat{\boldsymbol{c}}_j$ is small, then the decomposition of the error vector $\boldsymbol{\Pi}_j$ results in large components ζ_j and ξ_j in both the expanding and contracting directions. This becomes a difficulty because of the bidirectional nature of the iteration schemes (eq. (11)): a large error vector component ζ_j makes α_{j-1} large, whereas a large error vector component ξ_j makes β_{j+1} large, while β_{j-1} is unaffected. Thus, one large deviation is carried upstream for one equation, and downstream for the other, and therefore the two components do not necessarily balance each other. It is this phenomenon that is responsible for the spikes in the graph of the refined orbit $\{\boldsymbol{x}_k\}_{k=0}^N$ in figs. 2 and 4. A further manifestation of this phenomenon is in the shape of the spikes: they are tent-shaped, and have sides which are essentially straight lines, with upslope roughly equal to the expanding Lyapunov exponent, and downslope nearly equal to the contracting Lyapunov exponent. The refinement process is adapted to prevent large deviations at points $\boldsymbol{p}_j$ where angle θ_j is small, and to perturb these points transversally to the line of tangency between stable and unstable directions.

The same refinement process is also applicable to the case of dynamic noise, and the same measures of noise level Δ and improvement Γ are used. However, for levels of Δ similar to the additive noise case, the improvement Γ is not as large. The reason for this is associated with the fundamental difference in nature of a *dynamically* noisy orbit. The cumulative nature of such noise deviations means that exponential separation of orbits in regions where θ_k is small can lead to values of $\|f(\boldsymbol{p}_k) - \boldsymbol{p}_{k+1}\|$ which are larger than they were for the original orbit. It is still possible to produce new orbits which are piecewise clean. As an example, the refinement process was applied to a dynamically noisy orbit of length $N=10000$ for the Ikeda map with a comparatively smaller noise level $\Delta \approx 0.001$. 18 refinements of the noise reduction process resulted in a new cleaned orbit for which the measure $\|f(\boldsymbol{x}_{k-1}) - \boldsymbol{x}_k\| < 10^{-12}$ for $\approx 85\%$ of the points of $\{\boldsymbol{x}_k\}_{k=0}^N$.

The procedure for noise reduction by shadowing that has been described has certain advantages when compared with reported results of Kostelich and Yorke, and Farmer and Sidorowich. Farmer and Sidorowich have developed a noise reduction method based on a maximum likelihood technique [3]. They assume independently distributed Gaussian noise, and compute joint probability functions at a given point by pulling back the distribution at several points in the future, and pushing forward the distribution at several points in the past. These distributions are distorted into ellipsoids with long axis along the appropriate stable or unstable direction. A resultant joint probability is then computed to improve the location of the given point.

There are several relevant comments concerning a comparison between the Farmer–Sidorowich method and the noise reduction by the shadowing method presented here. The sole experiment reported in ref. [3] shows the Farmer–Sidorowich noise reduction method applied to a time series of 200 points for the Hénon map at a noise level that was approximately 0.1%. This is a noise level some 45 times lower than the case reported in this Letter, and for an orbit which is 1/50 the length of the orbit used here. Their method is also more computationally intensive: the known Hénon map was used, and the method was iterated 10000 times for all 200 points. Their indicator of noise reduction is to measure the pointwise difference between the original double precision unsoiled Hénon orbit and their cleaned orbit. This is not a measure that can be applied when the exact true orbit is not known. I have applied the noise reduction by shadowing method to an orbit of length 10000 for the Hénon map for uniformly distributed noise at $\Delta \approx 0.001$ (0.1%). In fig. 6 is a graph of $\{\|z_k - x_k\|\}_{k=0}^{N}$ for this orbit which required 22 iterations of the refinement process.

Kostelich and Yorke [2] describe a noise reduction method which is coupled with a local linear approximation to the dynamics on the attractor. They use a sliding window of 24 successive points, and "adjust" the trajectory so as to minimize least square error between a linear image of the previous point and the chosen point. It is clear that their method achieves a smoothing of the original jitter in the data. They report a noise reduction of roughly one to two orders of magnitude. It is difficult to make a direct comparison with their results since any approximation of the dynamics for a given data set necessarily limits the effectiveness of any noise reduction technique, so that a reconstruction of a noise-free orbit can only be accurate to the approximated dynamics. However, the most important distinction between my method of noise reduction by shadowing and the least-squares trajectory adjustment technique of Kostelich and Yorke is that their method does not yield an exponential reduction in errors. I applied the noise reduction by shadowing method to a 10000 point orbit of the Hénon map at the same noise level that they investigated (1%, or $\Delta \approx 0.01$) [2]. (They apparently used a 65536 point orbit.) In this case when the map is known, they measure the mean error by

$$E = \left(\frac{1}{N} \sum_{k=1}^{N} \|f(x_{k-1}) - x_k\|^2 \right)^{1/2} . \tag{12}$$

They use this to define their noise reduction as $1 - E_{\text{fitted}}/E_{\text{noisy}} \approx 79\%$. They do not give the values of E_{fitted} or E_{noisy} for this experiment, but I found in the experiment I performed $E_{\text{noisy}} \approx 1.27 \times 10^{-2}$. Thus their noise reduction results in an E_{fitted} less than one order of magnitude smaller than their E_{noisy}. In contrast, my experiment resulted in $1 - E_{\text{fitted}}/E_{\text{noisy}} \approx 100\%$ (more precisely, $1 - 1.55 \times 10^{-15}$). This means an E_{fitted} 14 orders of magnitude smaller than E_{noisy}. This corresponds to an orbit for which $\|f(x_{k-1}) - x_k\| < 10^{-15}$ uniformly. However, a single spike in the pointwise measurement $\|f(x_{k-1}) - x_k\|$ will dominate their proposed noise reduction measure (e.g. reducing 10^{-15} to 10^{-5}), obscuring the exponential reduction to machine precision on the remainder of the orbit. This is why I do not use the initial noise level Δ to monitor the performance of the noise reduction by the shadowing algorithm. Thus their measure of noise reduction is not well suited to monitoring methods with exponential reduction in error.

In this paper I have shown that an algorithm based on techniques from the proof of the shadowing lemma can be used to reduce noise. This algorithm has been applied to several model two-dimensional chaotic systems, and for moderate levels of additive noise, the refinement process is capable of reducing noise by more than ten orders of magnitude for most points on a given noisy orbit. For lower levels of additive noise, the refinement process produces a deterministic less noisy orbit, accurate to machine ep-

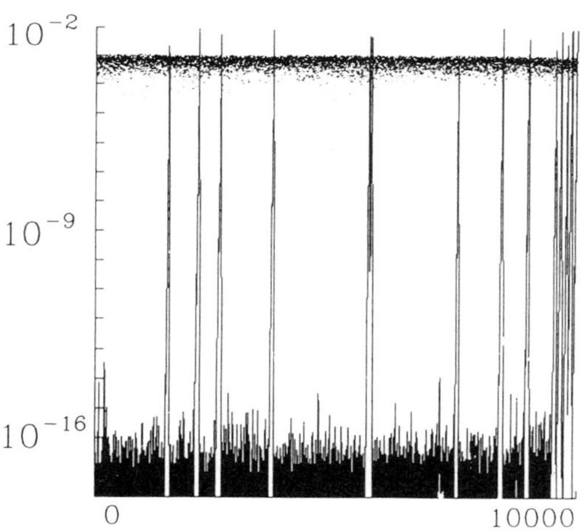

Fig. 6. A graph of $\{\|z_k - x_k\|\}_{k=0}^{10000}$ for a noisy orbit of the Hénon map for which $\Delta \approx 0.001$. For an experimental comparison with the method of Sidorowich and Farmer, in this case the original unsoiled orbit $\{z_k\}_{k=0}^{N}$ was calculated with double precision. Note that there are several long subsegments of more than 1000 successive iterates for which the cleaned orbit reproduces the original unsoiled orbit to machine precision.

silon. For higher noise, the algorithm will produce a new orbit which is piecewise clean: low noise along segments, interrupted by occasional deviations of the same order as the original noise. This means that more accurate measurement of dynamical parameters such as Lyapunov exponents can then be done on the cleaned orbit. The refinement process converges rapidly for most points, and compared to some other noise reduction schemes, it requires relatively few iterations. For the noise levels I have examined, between 4 and 30 applications of the refinement process have sufficed. It can easily be coupled to a prediction scheme which simply approximates by local linear maps.

I would like to thank Robert Cawley for numerous helpful conversations. This work was supported in part by the NSWC Navy Dynamics Institute Program with funding from the IR program at NSWC; by the Applied and Computational Mathematics Program of DARPA (orders 5830 and 6539); and the Office of Naval Research.

References

[1] J.-P. Eckmann and D. Ruelle, Rev. Mod. Phys. 57 (1985) 617.
[2] E. Kostelich and J.A. Yorke, Phys. Rev. A 38 (1988) 1649.
[3] J.D. Farmer and J.J. Sidorowich, in: Evolution, learning and cognition, ed. Y.C. Lee (World Scientific, Singapore, 1988).
[4] H.D.I. Abarbanel, R. Brown and J.B. Kadtke, preprint (1989).
[5] M. Casdagli, Physica D 35 (1989) 335.
[6] D.V. Anosov, Tr. Mat. Inst. Steklov 90 (1967) 1.
[7] R. Bowen, J. Diff. Eq. 18 (1975) 333.
[8] S.M. Hammel, C.K.R.T. Jones and J.V. Moloney, J. Opt. Soc. Am. B 2 (1985) 552.
[9] S.M. Hammel, J.A. Yorke and C. Grebogi, Bull. Am. Math. Soc. (NS) 19 (1988) 465.

CHAPTER 12

Control: Theory of Stabilization of Unstable Orbits

E. Ott, C. Grebogi, J.A. Yorke
Controlling chaos
Phys. Rev. Lett. **64**, 1196 (1990).

U. Dressler, G. Nitsche
Controlling chaos using time delay coordinates
Phys. Rev. Lett. **68**, 1 (1992).

F. Romeiras, C. Grebogi, E. Ott, W.P. Dayawansa
Controlling chaotic dynamical systems
Physica D **58**, 165 (1992).

H. Wang, E.H. Abed
Bifurcation control of chaotic dynamical systems
Proceedings of IFAC Nonlinear Control Systems Design Symposium, Bordeaux (1992).

Editors' Notes

The paper of **Ott et al.** (1990) puts forth a strategy for controlling systems whose uncontrolled orbits are chaotic, and points out that this can be done using small controls. The basic idea is to determine some of the low period unstable periodic orbits or steady states embedded in the attractor, and then use feedback to stabilize one of these, chosen as to yield improved performance. **Dressler and Nitsche** (1992) discuss how this strategy can be implemented using delay coordinate embedding. A more complete treatment of the theory, including delay coordinates, is given in the paper by **Romeiras et al.** (1992), which is based on the "pole placement technique," well-known in control theory.

Wang and Abed (1992) note that chaos often comes about through bifurcations of periodic orbits of steady states, and they seek to delay the onset of chaos by feedback control of these orbits.

Further reading

Two review articles on controlling chaotic systems have appeared recently:

CHEN, G., DONG, X., From chaos to order – perspectives and methodologies in controlling chaotic nonlinear dynamical systems. Int. J. Bif. Chaos **3**, 1363 (1993).

SHINBROT, T., GREBOGI, C., OTT, E., YORKE, J.A., Using small perturbations to control chaos. Nature **363**, 411 (1993).

Controlling Chaos

Edward Ott,[a],[b] Celso Grebogi,[a] and James A. Yorke[c]
University of Maryland, College Park, Maryland 20742
(Received 22 December 1989)

It is shown that one can convert a chaotic attractor to any one of a large number of possible attracting time-periodic motions by making only *small* time-dependent perturbations of an available system parameter. The method utilizes delay coordinate embedding, and so is applicable to experimental situations in which *a priori* analytical knowledge of the system dynamics is not available. Important issues include the length of the chaotic transient preceding the periodic motion, and the effect of noise. These are illustrated with a numerical example.

PACS numbers: 05.45.+b

The presence of chaos in physical systems has been extensively demonstrated and is very common. In practice, however, it is often desired that chaos be avoided and/or that the system performance be improved or changed in some way. Given a chaotic attractor, one approach might be to make some large and possibly costly alteration in the system which completely changes its dynamics in such a way as to achieve the desired behavior. Here we assume that this avenue is not available. Thus, we address the following question: Given a chaotic attractor, how can one obtain improved performance and a desired attracting time-periodic motion by making only *small* time-dependent perturbations in an *accessible* system parameter?

The key observation is that a chaotic attractor typically has embedded within it an infinite number of unstable periodic orbits.[1] Since we wish to make only small perturbations to the system, we do not envision creating new orbits with very different properties from the existing ones. Thus, we seek to exploit the already existing unstable periodic orbits. Our approach is as follows: We first determine some of the unstable low-period periodic orbits that are embedded in the chaotic attractor. We then examine these orbits and choose one which yields improved system performance. Finally, we tailor our small time-dependent parameter perturbations so as to stabilize this already existing orbit. In this Letter we describe how this can be done, and we illustrate the method with a numerical example. The method is very general and should be capable of yielding greatly improved performance in a wide variety of situations.

It is interesting to note that if the situation is such that the suggested method is practical, then the presence of chaos can be a great advantage. The point is that any one of a number of different orbits can be stabilized, and the choice can be made to achieve the best system performance among those orbits. If, on the other hand, the attractor is not chaotic but is, say, periodic, then small parameter perturbations can only change the orbit slightly. Basically we are then stuck with whatever system performance the stable periodic orbit gives, and we have no option for substantial improvement, short of making large alterations in the system.

Furthermore, one may want a system to be used for different purposes or under different conditions at different times. Thus, depending on the use, different requirements are made of the system. If the system is chaotic, this type of multiple-use situation might be accommodated without alteration of the gross system configuration. In particular, depending on the use desired, the system behavior could be changed by switching the temporal programming of the small parameter perturbations to stabilize different orbits. In contrast, in the absence of chaos, completely separate systems might be required for each use. Thus, when designing a system intended for multiple uses, purposely building chaotic dynamics into the system may allow for the desired flexibility. Such multipurpose flexibility is essential to higher life forms, and we, therefore, speculate that chaos may be a necessary ingredient in their regulation by the brain.

To simplify the analysis we consider continuous-time dynamical systems which are *three dimensional* and depend on one system parameter which we denote p [for example, $d\mathbf{x}/dt = \mathbf{F}(\mathbf{x},p)$, where $\mathbf{x}$ is three dimensional]. We assume that the parameter p is available for external adjustment, and we wish to temporally program our adjustments of p so as to achieve improved performance. We emphasize that our restriction to a three-dimensional system is mainly for ease of presentation, and that the case of higher-dimensional (including infinite-dimensional) systems can be treated by similar methods.[2]

We imagine that the dynamical equations describing the system are not known, but that experimental time series of some scalar-dependent variable $z(t)$ can be measured. Using delay coordinates[3,4] with delay T one can form a delay-coordinate vector,

$$\mathbf{X}(t) = [z(t), z(t-T), z(t-2T), \ldots, z(t-MT)] .$$

We are interested in periodic orbits and their stability properties, and we shall use $\mathbf{X}$ to obtain a surface of section for this purpose. In the surface of section, a continuous-time-periodic orbit appears as a discrete-time orbit cycling through a finite set of points. We require the dynamical behavior of the surface of section map in *neighborhoods* of these points in order to study the stability of the periodic orbits. To embed a small neighborhood of a point from $\mathbf{x}$ into $\mathbf{X}$, we typically only require as many dimensions as there are coordinates of the point. Thus, for our purposes, $M=D-1$ is generally sufficient. (This is in contrast with[3] $M+1=2D+1$, typically required for global embedding of the original phase space in the delay-coordinate space.) Hence, for the case considered ($D=3$), our surface of section is two dimensional.

We suppose that the parameter p can be varied in a small range about some nominal value p_0. Henceforth, without loss of generality, we set $p_0\equiv 0$. Let the range in which we are allowed to vary p be $p_* > p > -p_*$.

Using an experimental surface of section for the embedding vector $\mathbf{X}$, we imagine that we obtain many experimental points in the surface of section for $p=0$. We denote these points $\xi_1,\xi_2,\xi_3,\ldots,\xi_k$, where ξ_n denotes the coordinates in the surface of section at the nth piercing of the surface of section by the orbit $\mathbf{X}(t)$. For example, a common choice of the surface of section would be $z(t-MT)$ equals a constant, and $\xi_n=[z(t_n),\ldots,z(t_n-(M-1)T)]$, where $t=t_n$ denotes the time at the nth piercing. From such experimentally determined sequences it has been demonstrated that a large number of distinct unstable periodic orbits on a chaotic attractor can be determined.[5,6] We then examine these unstable periodic orbits and select the one which gives the best performance. Again using an experimentally determined sequence, we obtain the stability properties of the chosen periodic orbit (cf. Refs. 5 and 6 for discussion of how this can be done and for descriptions of its implementation in concrete experimental cases). For the purposes of simplicity, let us assume in what follows that this orbit is a fixed point of the surface of section map (i.e., period one; the case of higher period is a straightforward extension). Let λ_s and λ_u be the experimentally determined stable and unstable eigenvalues of the surface of section map at the chosen fixed point of the map ($|\lambda_u|>1>|\lambda_s|$). Let $\mathbf{e}_s$ and $\mathbf{e}_u$ be the experimentally determined unit vectors in the stable and unstable directions. Let $\xi=\xi_F\equiv 0$ be the desired fixed point. We then change p slightly from $p=0$ to some other value $p=\bar{p}$. The fixed-point coordinates in the experimental surface of section will shift from 0 to some nearby point $\xi_F(\bar{p})$ and we determine this new position. For small $\bar{p}$ we approximate $\mathbf{g}\equiv\partial\xi_F(p)/\partial p|_{p=0}\cong\bar{p}^{-1}\xi_F(\bar{p})$, which allows an experimental determination of the vector $\mathbf{g}$.

Thus, in the surface of section, near $\xi=0$, we can use a linear approximation for the map, $\xi_{n+1}-\xi_F(p)\cong\mathbf{M}\cdot[\xi_n-\xi_F(p)]$, where $\mathbf{M}$ is a 2×2 matrix. Using $\xi_F(p)\cong p\mathbf{g}$ we have

$$\xi_{n+1}\cong p_n\mathbf{g}+[\lambda_u\mathbf{e}_u\mathbf{f}_u+\lambda_s\mathbf{e}_s\mathbf{f}_s]\cdot[\xi_n-p_n\mathbf{g}]\,. \qquad (1)$$

[In the linearization (1), we have considered p_n to be small and of the same order as ξ_n.] We emphasize that $\mathbf{g}$, $\mathbf{e}_u$, $\mathbf{e}_s$, λ_u, and λ_s are all experimentally accessible by the embedding technique just discussed. In (1) $\mathbf{f}_u$ and $\mathbf{f}_s$ are contravariant basis vectors defined by $\mathbf{f}_s\cdot\mathbf{e}_s=\mathbf{f}_u\cdot\mathbf{e}_u=1$, $\mathbf{f}_s\cdot\mathbf{e}_u=\mathbf{f}_u\cdot\mathbf{e}_s=0$. Note that we have written the location of the fixed point as $p_n\mathbf{g}$ because we imagine that we adjust p to a new value p_n after each piercing of the surface of section. That is, we observe ξ_n and then adjust p to the value p_n. Thus p_n depends on ξ_n. Further, we only envision making this adjustment when the orbit falls near the desired fixed point for $p=0$.

Assume that ξ_n falls near the desired fixed point at $\xi=0$ so that (1) applies. We then attempt to pick p_n so that ξ_{n+1} falls on the stable manifold of $\xi=0$. That is, we choose p_n so that $\mathbf{f}_u\cdot\xi_{n+1}=0$. If ξ_{n+1} falls on the stable manifold of $\xi=0$, we can then set the parameter perturbations to zero, and the orbit for subsequent time will approach the fixed point at the geometrical rate λ_s. Thus, for sufficiently small ξ_n, we can dot (1) with $\mathbf{f}_u$ to obtain

$$p_n=\lambda_u(\lambda_u-1)^{-1}(\xi_n\cdot\mathbf{f}_u)/(\mathbf{g}\cdot\mathbf{f}_u)\,, \qquad (2)$$

which we use when the magnitude of the right-hand side of (2) is less than p_*. When it is greater than p_*, we set $p_n=0$. We assume in (2) that the generic condition $\mathbf{g}\cdot\mathbf{f}_u\neq 0$ is satisfied. Thus, the parameter perturbations are activated (i.e., $p_n\neq 0$) only if ξ_n falls in a narrow strip $|\xi_n^u|<\xi_*$, where $\xi_n^u=\mathbf{f}_u\cdot\xi_n$, and from (2) $\xi_*=p_*|(1-\lambda_u^{-1})\mathbf{g}\cdot\mathbf{f}_u|$. Thus, for small p_*, a typical initial condition will execute a chaotic orbit, unchanged from the uncontrolled case, until ξ_n falls in the strip. Even then, because of nonlinearity not included in (1), the control may not be able to bring the orbit to the fixed point. In this case the orbit will leave the strip and continue to wander chaotically as if there was no control. Since the orbit on the uncontrolled chaotic attractor is ergodic, at some time it will eventually satisfy $|\xi_n^u|<\xi_*$ and also be sufficiently close to the desired fixed point that attraction to $\xi=0$ is achieved. [In rare cases applying Eq. (2) when the trajectory enters the strip, but is still far from 0, may result in stabilizing the wrong periodic orbit which visits the strip.]

Thus, we create a stable orbit, but, for a typical initial condition, it is preceded in time by a chaotic transient in which the orbit is similar to orbits on the uncontrolled chaotic attractor. The length τ of such a chaotic transient depends sensitively on the initial condition, and, for randomly chosen initial conditions, has an exponential probability distribution[7] $P(\tau)\sim\exp[-(\tau/\langle\tau\rangle)]$ for large τ. The average length of the chaotic transient $\langle\tau\rangle$ increases with decreasing p_* and follows a power-law relation[7] for small p_*, $\langle\tau\rangle\sim p_*^{-\gamma}$.

We will now derive a formula for the exponent γ. Dotting the linearized map for ξ_{n+1}, Eq. (1), with $\mathbf{f}_u$, we obtain $\xi_{n+1}^u\cong 0$. In obtaining this result from (1) we have substituted p_n appropriate for $|\xi_n^u|<\xi_*$. We note that the result $\xi_{n+1}^u\cong 0$ is a linearization, and typically has a lowest-order nonlinear correction that is quadratic. In particular, $\xi_n^s = \mathbf{f}_s \cdot \xi_n$ is not restricted by $|\xi_n^u| < \xi_*$, and thus may not be small when the condition $|\xi_n^u| < \xi_*$ is satisfied. Hence the correction quadratic in ξ_n^s is most significant. Including such a correction we have $\xi_{n+1}^u \cong \kappa(\xi_n^s)^2$, where κ is a constant. Thus, if $|\kappa|(\xi_n^s)^2 > \xi_*$, then $|\xi_{n+1}^u| > \xi_*$, and attraction to $\xi = 0$ is not achieved, even though $|\xi_n^u| < \xi_*$. Attraction to $\xi = 0$ is achieved when the orbit falls in the small parallelogram P_c given by $|\xi_n^u| < \xi_*$, $|\xi_n^s| < (\xi_*/|\kappa|)^{1/2}$. For very small ξ_*, an initial condition will bounce around on the set comprising the uncontrolled chaotic attractor for a long time before it falls in the parallelogram P_c. At any given iterate the probability of falling in P_c is $\mu(P_c)$, the measure of the uncontrolled attractor contained in P_c. Thus, $\langle\tau\rangle^{-1} = \mu(P_c)$. The scaling of $\mu(P_c)$ with ξ_* is

$$\mu(P_c) \sim (\xi_*)^{d_u}[(\xi_*/|\kappa|)^{1/2}]^{d_s} \sim \xi_*^{d_u+(1/2)d_s},$$

where d_u and d_s are the partial pointwise dimensions for the uncontrolled chaotic attractor at $\xi = 0$ in the unstable direction and the stable direction, respectively. Thus, $\mu(P_c) = \xi_*^\gamma$, where $\gamma = d_u + d_s/2$. Since we assume the attractor to be effectively smooth in the unstable direction, $d_u = 1$. The partial pointwise dimension in the stable direction is given in terms of the eigenvalues[7] at $\xi = 0$, $d_s = \ln|\lambda_u|/\ln|\lambda_s|^{-1}$. Thus,

$$\gamma = 1 + \tfrac{1}{2}\ln|\lambda_u|/\ln|\lambda_s|^{-1}. \tag{3}$$

To study the effect of noise we add a term $\epsilon\boldsymbol{\delta}_n$ to the right-hand side of the linearized equations for $\boldsymbol{\xi}_{n+1}$, Eq. (1), where $\boldsymbol{\delta}_n$ is a random variable and ϵ is a small parameter specifying the intensity of the noise. The quantities $\boldsymbol{\delta}_n$ are taken to have zero mean ($\langle\boldsymbol{\delta}_n\rangle = 0$), be independent ($\langle\boldsymbol{\delta}_n\boldsymbol{\delta}_m\rangle = 0$ for $m \neq n$), and have a probability density independent of n. Dotting (1) with noise included with $\mathbf{f}_u$ we obtain $\xi_{n+1}^u = \epsilon\delta_n^u$, where $\delta_n^u \equiv \mathbf{f}_u \cdot \boldsymbol{\delta}_n$. Thus, if the noise is bounded, $|\delta_n^u| < \delta_{\max}$, then the stability of $\xi = 0$ will not be affected by the noise if the bound is small enough, $\epsilon\delta_{\max} < \xi_*$. If this condition is not satisfied, then the noise can kick an orbit which is initially in the parallelogram P_c into the region outside P_c. We are particularly interested in the case where such kickouts are caused by low-probability tails on the probability density and are thus rare. (If they are frequent, then our procedure is ineffective.) In such a case the average time to be kicked out $\langle\tau'\rangle$ will be long. Thus, an orbit will typically alternate between epochs of chaotic motion of average duration $\langle\tau\rangle$ in which it is far from $\xi = 0$, and epochs of average length $\langle\tau'\rangle$ in which the orbit lies in the parallelogram P_c. For small enough noise the orbit spends most of its time in P_c, $\langle\tau'\rangle \gg \langle\tau\rangle$, and one might then regard the procedure as being effective.

We now consider a specific numerical example. Our purpose is to illustrate and test our analyses of the average time to achieve control and the effect of noise. To do this we shall utilize the Henon map, $x_{n+1} = A - x_n^2 + By_n$, $y_{n+1} = x_n$, where we take $B = 0.3$. We assume that the quantity A can be varied by a small amount about some value A_0. Accordingly, we write A as $A = A_0 + p$, where p is the control parameter. For the values of A_0 which we investigate, the attractor for the map is chaotic and contains an unstable period-one (fixed-point) orbit. The coordinates (x_F, y_F) of the fixed point which is in the attractor for $p = 0$ along with the associated parameters and vectors appearing in Eq. (1) may be explicitly calculated. The quantity ξ_n appearing in (1) is $\xi_n = (x_n - x_F)\mathbf{x}_0 + (y_n - y_F)\mathbf{y}_0$. To test our prediction for the dependence of $\langle\tau\rangle$, the average time to approach $\xi = 0$, on the maximum allowed size of the parameter perturbation p_*, we proceed as follows. We iterate the map with $p = 0$ using a large number of randomly chosen initial conditions until all these initial conditions are distributed over the attractor (500 iterates were typically used). We then turn on the parameter perturbations and determine for each orbit how many further iterates τ are necessary before the orbit falls within a circle of radius $\tfrac{1}{2}\xi_*$ centered at the fixed point. We then calculate the average of these times. We do this for many different values of p_* and plot the results as a function of p_*. This is shown on the log-log plot in Fig. 1 along with the theoretical straight line of slope given by the exponent (3). We see that the agreement is

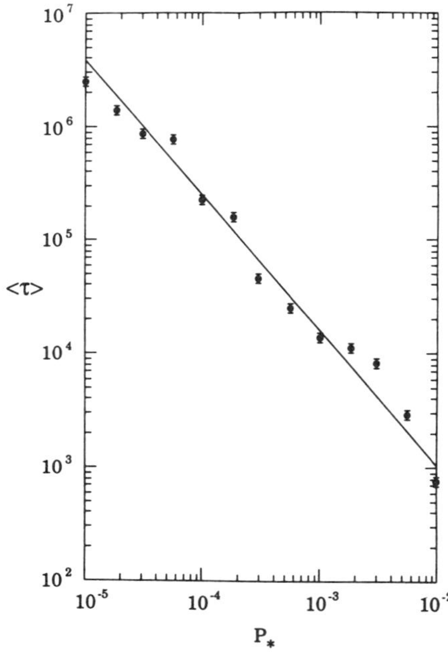

FIG. 1. $\langle\tau\rangle$ vs p_*. Points were computed using 128 randomly selected initial conditions. $A_0 = 1.4$.

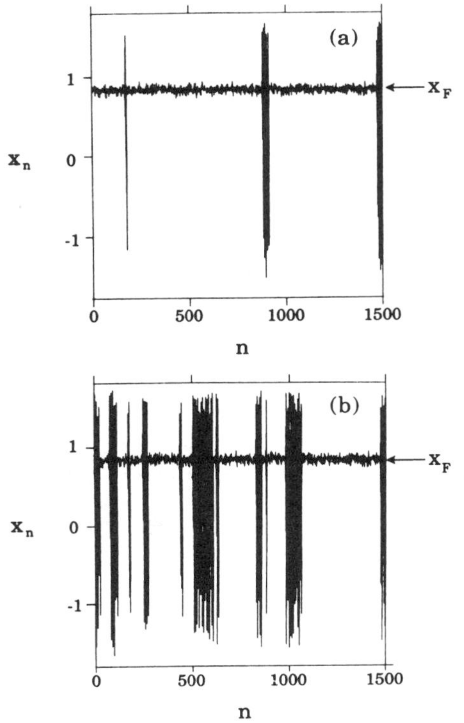

FIG. 2. x_n vs n for two cases with the same realization of the random vector δ. $p_* = 0.2$ and $A_0 = 1.29$ for both cases. (a) $\epsilon = 3.5 \times 10^{-2}$; (b) $\epsilon = 3.8 \times 10^{-2}$.

good although there are significant variations about the general power-law trend. These are to be expected due to the fractal nature of the attractor and have also been seen in numerical calculations of the pointwise dimension for points on chaotic attractors (cf. Grebogi, Ott, and Yorke[1]).

Next, we consider the issue of noise. We add terms $\epsilon\delta_{xn}$ and $\epsilon\delta_{yn}$ to the right-hand sides of the Henon map equations. The random quantities δ_{xn} and δ_{yn} are independent of each other, have mean value 0 and mean-squared value 1 ($\langle\delta_x^2\rangle = \langle\delta_y^2\rangle = 1$), and have a Gaussian probability density. Figure 2 shows orbit plots, x_n vs n for 1500 iterates of the noisy map with parameter perturbations given by (2), for two different noise levels and p_* held fixed at $p_* = 0.2$. As predicted the orbit stays near the fixed point with occasional bursts into the region far from $\xi = 0$, and these bursts are less frequent for small noise levels.

In conclusion, we have shown that there is great inherent flexibility in situations in which the dynamical motion is on a chaotic attractor. In particular, by using only small (carefully chosen) parameter perturbations it is possible to create a large variety of attracting periodic motions and to choose amongst these periodic motions the one most desirable.[8]

This research was supported by the U.S. Department of Energy (Scientific Computing Staff Office of Energy Research). The computation was done at the National Energy Research Supercomputer Center.

---

[a]Laboratory for Plasma Research.
[b]Departments of Electrical Engineering and of Physics.
[c]Institute for Physical Science and Technology and Department of Mathematics.

[1]The periodic orbits are dense in the attractor [i.e., periodic orbits pass through any neighborhood (however small) of any point on the attractor]. For discussions of the relation of ergodic properties of an attractor to its dense set of unstable periodic orbits, see, for example, C. Grebogi, E. Ott, and J. A. Yorke, Phys. Rev. A **37**, 1711 (1988); **36**, 3522 (1987); D. Auerbach *et al.*, Phys. Rev. Lett. **58**, 2387 (1987); H. Hata *et al.*, Prog. Theor. Phys. **78**, 511 (1987); A. Katok, Publ. Math. IHES **51**, 137 (1980); R. Bowen, Trans. Am. Math. Soc. **154**, 377 (1971).

[2]E. Ott, C. Grebogi, and J. A. Yorke, in *Chaos: Proceedings of a Soviet-American Conference* (American Institute of Physics, New York, 1990).

[3]F. Takens, in *Dynamical Systems and Turbulence*, edited by D. Rand and L. S. Young (Springer-Verlag, Berlin, 1981), p. 230.

[4]N. H. Packard *et al.*, Phys. Rev. Lett. **45**, 712 (1980).

[5]G. H. Gunaratne, P. S. Linsay, and M. J. Vinson, Phys. Rev. Lett. **63**, 1 (1989).

[6]D. P. Lathrop and E. J. Kostelich, "The Characterization of an Experimental Strange Attractor by Periodic Orbits" (to be published).

[7]C. Grebogi, E. Ott, and J. A. Yorke, Phys. Rev. Lett. **57**, 1284 (1986); P. Romeiras, C. Grebogi, E. Ott, and J. A. Yorke, Phys. Rev. A **36**, 5365 (1987).

[8]The general problem of controlling chaotic systems, while clearly very important, has, so far, received almost no attention. Two exceptions (which are quite different from our approach) are the papers of Hubler (who typically requires large controlling signals) and Fowler [A. Hubler, Helv. Phys. Acta **62**, 343 (1989); T. B. Fowler, IEEE Trans. Autom. Control **34**, 201 (1989)].

Controlling Chaos Using Time Delay Coordinates

Ute Dressler and Gregor Nitsche
Daimler-Benz Research Institute, Goldsteinstrasse 235, 6000 Frankfurt/Main 71, Germany
(Received 29 August 1991)

The Ott-Grebogi-Yorke control method is analyzed in the case that the attractor is reconstructed from a time series using time delay coordinates. It turns out that the control formula of Ott, Grebogi, and Yorke should be modified in order to apply to experimental systems if time delay coordinates are used. We reveal that the experimental surface of section map depends not only on the actual parameter but also on the preceding one. In order to meet this dependence two modifications are introduced which lead to a better performance of the control. To compare their control abilities they are applied to simulations of a Duffing oscillator.

PACS numbers: 05.45.+b, 03.20.+i

In 1990 Ott, Grebogi, and Yorke (OGY) [1] proposed a new method of controlling a chaotic dynamical system by stabilizing one of the many unstable periodic orbits embedded in a chaotic attractor, through only small time-dependent perturbations in some accessible system parameter. This makes OGY's approach quite different from other previously published methods on controlling chaos [2].

OGY's method has attracted the attention of many physicists interested in applications of nonlinear dynamics. One reason for this is that OGY stress that all values needed to achieve control can be obtained from an experimental signal starting with the well-known embedding technique [3,4]. Therefore the control method can in principle be applied to experimental systems where the dynamical equations are not known. Indeed, Ditto, Rauseo, and Spano demonstrated recently [5] a first control of a physical system using the method of Ott, Grebogi, and Yorke.

With regard to possible applications we investigate the OGY control method in the case that the attractor is reconstructed from a time series using time delay coordinates. It turns out that the control formula of OGY should be modified in order to apply to experimental systems if time delay coordinates are used. The main argument will be that during the control process one switches the control parameter p from p_{i-1} to p_i at times t_i (t_i is the time of the ith piercing of the surface of section by the trajectory). But, if one uses delay coordinates, the experimental surface of section map P does not only depend on the new actual parameter p_i (as OGY implicitly assume) but also on the old one p_{i-1}. In order to meet this dependence two modifications of the control algorithm are proposed. Their control abilities are compared with the original OGY formula by applying them to a time series obtained from simulations of a Duffing oscillator.

Let us briefly recall the OGY control idea. For simplicity we restrict ourselves to a two-dimensional discrete dynamical system (e.g., the surface of section map P of a three-dimensional continuous system). There also exist extensions of the method to higher-dimensional dynamical systems [6,7]. Let the system depend on some accessible parameter $p \in (p_0 - \delta p_{\max}, p_0 + \delta p_{\max})$ with maximal possible perturbation $\delta p_{\max}$, $\xi_{i+1} = P(\xi_i, p)$. Let $\xi_F = P(\xi_F, p_0)$ denote the unstable fixed point on the attractor which one wants to stabilize. The control idea is to monitor the system until it comes close to the desired fixed point and then change p by a small amount such that the next state ξ_{i+1} will fall into the stable direction of the fixed point. To do this one uses the first-order approximation of P near ξ_F and p_0,

$$\delta\xi_{i+1} \cong A\delta\xi_i + \mathbf{w}\delta p_i ,$$

with $\delta\xi_i = \xi_i - \xi_F$, $\delta p_i = p_i - p_0$, $A = D_\xi P(\xi_F, p_0)$, and $\mathbf{w} = \partial P/\partial p(\xi_F, p_0)$. Writing the linearization A as $A = \lambda_u \mathbf{e}_u \mathbf{f}_u + \lambda_s \mathbf{e}_s \mathbf{f}_s$, with $\mathbf{e}_u$ ($\mathbf{e}_s$) the unstable (stable) eigendirections of A with eigenvalues λ_u (λ_s) and $\mathbf{f}_u$ ($\mathbf{f}_s$) their contravariant basis vectors, i.e., $\mathbf{f}_s \cdot \mathbf{e}_u = \mathbf{f}_u \cdot \mathbf{e}_s = 0$ and $\mathbf{f}_s \cdot \mathbf{e}_s = \mathbf{f}_u \cdot \mathbf{e}_u = 1$, the condition that ξ_{i+1} falls on the local stable manifold of the fixed point can be formulated as $\mathbf{f}_u \cdot \delta\xi_{i+1} = 0$, which yields the control formula [8] for the new value of the control parameter $p_i = p_0 + \delta p_i$,

$$\delta p_i = -(\lambda_u / \mathbf{f}_u \cdot \mathbf{w}) \mathbf{f}_u \cdot \delta\xi_i \,. \tag{1}$$

The control is only activated if the resulting change in the parameter δp_i is less than the maximal allowed disturbance δp_{max}; otherwise δp_i is set to zero.

Let us now consider the case that the only information about the system is obtained by some measurement process which is mathematically realized by some scalar function Z on the state space M. If $Y(t) \in M$ is the state of the system at time t, the experimental time series $z(t) = Z(Y(t))$ is obtained. Using time delay coordinates with delay τ and embedding dimension d, a d-dimensional delay coordinate vector is formed, $\mathbf{X}(t) = (z(t), z(t-\tau), \ldots, z(t-(d-1)\tau)) \in \mathbf{R}^d$. The experimental surface of section is obtained by the common choice that one component of $\mathbf{X}(t)$ equals a constant, e.g., $[\mathbf{X}(t_i)]_1 \equiv z(t_i) = c$. This procedure gives the successive points $\xi_i \in \mathbf{R}^{d-1}$ in the surface of section and the surface of section map $\xi_{i+1} = P(\xi_i)$.

In what follows we focus our interest on the so obtained experimental surface of section map P. For the sake of simplicity let us assume that one wants to stabilize an unstable fixed point ξ_F of P which has been localized by the well-known technique of recurrent points [9-11]. Applying the OGY control algorithm implies that one (instantaneously) changes at the times t_i the parameter p from p_{i-1} to an appropriately chosen parameter p_i using (1). Let us now assume that the time between successive piercings of the surface of section is bigger than the lag window, i.e., $t_{i+1} - t_i > (d-1)\tau$. The reason that one hopes to be able to control the original system $Y(t)$ by observing $\mathbf{X}(t)$ is that for appropriately chosen embedding parameters d and τ [4] there exists a bijective relation $\mathbf{\Phi}$ between the states $\mathbf{X}(t)$ and $Y(t)$, i.e., $\mathbf{X}(t) = \mathbf{\Phi}(Y(t))$. The mapping $\mathbf{\Phi}$ is, however, closely related to the dynamical equations of the system and thus, in general, dependent on the actual value of the control parameter p_i. This will be taken into account by writing $\mathbf{\Phi}_{p_i}$ instead of $\mathbf{\Phi}$.

Our argumentation is now as follows. The point ξ_i at time t_i in the surface of section is related to the original state by $Y(t_i) = \mathbf{\Phi}_{p_{i-1}}^{-1}(c, z(t_i - \tau), \ldots, z(t_i - (d-1)\tau))$. Here we make use of our assumption that $(d-1)\tau < t_i - t_{i-1}$ which assures that p_{i-1} is the actual value of p during the whole time interval (t_{i-1}, t_i). The time development of the original system from time t_i to the time t_{i+1} is, in case of activated control, given by $\varphi_{p_i}^{t_{i+1}-t_i}$ with φ_p^t the flow map of the dynamical system depending on p. Thus the state of the system at time t_{i+1} is obtained by $Y(t_{i+1}) = \varphi_{p_i}^{t_{i+1}-t_i}(Y(t_i))$ and the corresponding state in the embedding space by $\mathbf{X}(t_{i+1}) = \mathbf{\Phi}_{p_i}(Y(t_{i+1}))$. Therefore we obtain $\mathbf{X}(t_{i+1}) = (\mathbf{\Phi}_{p_i} \circ \varphi_{p_i}^{t_{i+1}-t_i} \circ \mathbf{\Phi}_{p_{i-1}}^{-1})(\mathbf{X}(t_i))$. This gives our main conclusion. In the case of activated control (i.e., switching the parameter from p_{i-1} to p_i at time t_i) the experimental surface of section map P depends not only on the new actual value p_i but also on the preceding value p_{i-1}, i.e.,

$$\xi_{i+1} = P(\xi_i, p_{i-1}, p_i) \,.$$

Taking this as the starting point the algorithm of OGY is straightforwardly extended. The linearization which one has to consider now is given by

$$\delta\xi_{i+1} \cong A\delta\xi_i + \mathbf{v}\delta p_{i-1} + \mathbf{u}\delta p_i \,,$$

with $A = D_\xi P(\xi_F, p_0, p_0)$, $\mathbf{v} = \partial P / \partial p_{i-1}(\xi_F, p_0, p_0)$, and $\mathbf{u} = \partial P / \partial p_i(\xi_F, p_0, p_0)$. Demanding $\mathbf{f}_u \cdot \delta\xi_{i+1} = 0$ one obtains as a new control law

$$\delta p_i = -\frac{\lambda_u}{\mathbf{f}_u \cdot \mathbf{u}} \mathbf{f}_u \cdot \delta\xi_i - \frac{\mathbf{f}_u \cdot \mathbf{v}}{\mathbf{f}_u \cdot \mathbf{u}} \delta p_{i-1} \,. \tag{2}$$

When P is not influenced by the preceding perturbation δp_{i-1}, i.e., $\mathbf{v} = 0$, the original OGY control formula (1) is reobtained. To see this we note that the vector $\mathbf{w}$ in the control formula (1) is related to $\mathbf{u}$ and $\mathbf{v}$ by $\mathbf{w} = \mathbf{u} + \mathbf{v}$.

The new control formula (2) contains one possible instability. In the case that $|(\mathbf{f}_u \cdot \mathbf{v})/(\mathbf{f}_u \cdot \mathbf{u})| \geq 1$ holds the required perturbations δp_i will, in general, grow until they exceed the maximum allowed value δp_{max}, and the range of control will be left. To avoid this instability (i.e., the growing of δp_i) we propose an alternative approach. We try to find a control law for δp_i such that δp_{i+1} automatically will become zero. This is done by demanding that the system stabilizes only the next but one step, $i+2$, and that δp_{i+1} equals zero, i.e., by the requirements $\mathbf{f}_u \cdot \delta\xi_{i+2} = 0$ and $\delta p_{i+1} = 0$.

Using the linearization twice, these requirements yield the second modification of the control formula,

$$\delta p_i = -\frac{\lambda_u^2}{\lambda_u \mathbf{f}_u \cdot \mathbf{u} + \mathbf{f}_u \cdot \mathbf{v}} \mathbf{f}_u \cdot \delta\xi_i - \frac{\lambda_u \mathbf{f}_u \cdot \mathbf{v}}{\lambda_u \mathbf{f}_u \cdot \mathbf{u} + \mathbf{f}_u \cdot \mathbf{v}} \delta p_{i-1} \,. \tag{3}$$

The control formulas introduced above have been applied to simulations of a Duffing oscillator [12] given by $\ddot{x} + d\dot{x} + x + x^3 = f\cos\omega t$. This system has been numerically integrated. As a measurement function the displacement of the oscillator $z(t) \equiv x(t)$ is chosen. We use a three-dimensional embedding with delay time $\tau = T/4$ with $T = 2\pi/\omega$. The experimental surface of section was obtained by taking $[\mathbf{X}(t_i)]_1 = z(t_i) = \text{const}$. For the localization of fixed points, the standard method described in Refs. [9,10] is used (for details see also [13]). To obtain the vectors $\mathbf{u}$ and $\mathbf{v}$ the perturbations δp_i are alternately switched on and off at every piercing of the surface of section such that $\delta p_i = 0$ for i odd and $\delta p_i = \bar{p}$ for i even, $\bar{p}$ small, respectively. Regarding all pairs (ξ_i, ξ_{i+1}) with even i as one group and the pairs with odd i as another, it is now possible to fit affine mappings in the neighborhood of ξ_F to $P(\cdot, p_0 + \bar{p}, p_0)$ using only pairs

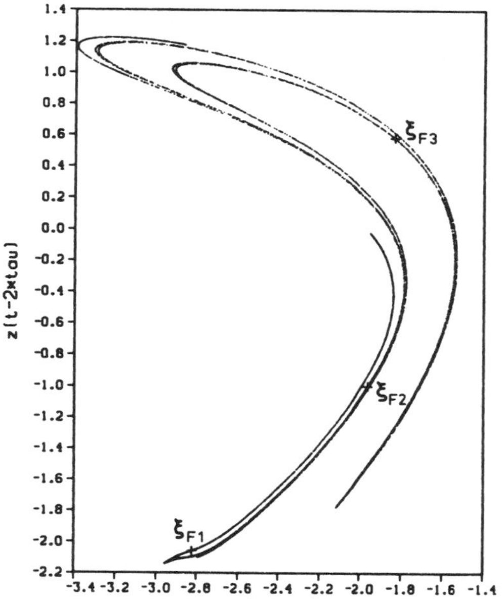

FIG. 1. A chaotic attractor of the Duffing oscillator ($d=0.2$, $f=p=36$, $\omega=0.661$) in the surface of section. The surface of section was obtained by the conditions $z(t_i)=1$, $\dot{z}(t_i)>0$, and $z(t_i-\tau)<0$. The three unstable fixed points observed are indicated by the crosses. For further reference they are called ξ_{F1}, ξ_{F2}, and ξ_{F3}.

(ξ_{r_i},ξ_{r_i+1}), r_i odd, and to $P(\cdot,p_0,p_0+\bar{p})$ using only pairs (ξ_{r_i},ξ_{r_i+1}), r_i even, respectively. These fits then determine $\mathbf{u}$ and $\mathbf{v}$ by the relations $P(\xi_F,p_0+\bar{p},p_0)\cong\xi_F+\mathbf{v}\bar{p}$ and $P(\xi_F,p_0,p_0+\bar{p})\cong\xi_F+\mathbf{u}\bar{p}$.

To compare the performances of the three different control formulas we tried to stabilize the three fixed points ξ_{F1}, ξ_{F2}, and ξ_{F3} which were determined embedded in a chaotic attractor of the Duffing oscillator (see Fig. 1). To stabilize these orbits we choose as accessible parameter p the amplitude of the driving f and a maximal allowed perturbation $\delta p_{\max}=0.5$. In Fig. 2, the three different control formulas are successively applied to stabilize ξ_{F1}. Only the second modification (3) was able to stabilize ξ_{F1}. The coefficients of the control formulas (see Table I) explain why our first modification (2) of the OGY algorithm did not work. The criterion for a stable control algorithm $|b_2|<1$ was hurt. The large absolute value of $b_2=(\mathbf{f}_u\cdot\mathbf{v})/(\mathbf{f}_u\cdot\mathbf{u})$, indicates further that the influence of the change of the preceding parameter p_{i-1} is relatively larger than that of the actual one p_i. But this is exactly what is neglected if one applies the original approach of OGY without considering the meaning of the time delay coordinates.

The stabilization of the second fixed point shows different features. Here the generic condition ($\mathbf{f}_u\cdot\mathbf{w}\neq 0$) of the OGY formula is almost violated. Because of the resulting large value of the coefficient a there were only rare cases where the control requirement $\delta p_i<\delta p_{\max}$ was met. But even then the control range was soon left without succeeding in control. The coefficient b_2 just violates the stability criterion. Indeed, the used perturbations δp_i increased at the beginning. But finally, probably due to nonlinear effects, the control procedure stabilized and the algorithm was able to achieve control. The second modification (3) was again able to achieve control but with perturbations drastically smaller than the one used for (2).

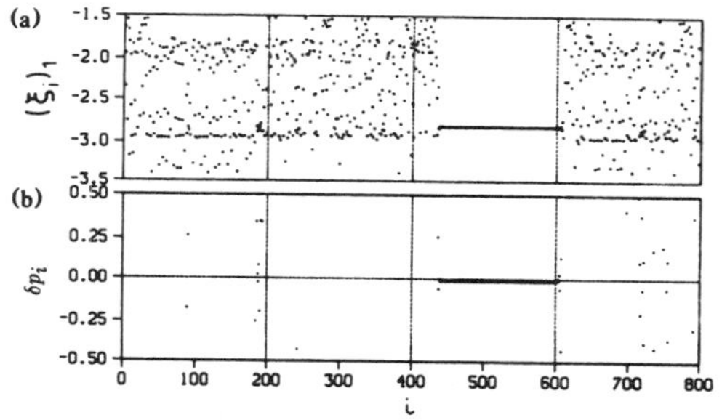

FIG. 2. (a) The first component $(\xi_i)_1$ of the points in the surface of section vs i. In order to stabilize the fixed points ξ_{F1} the OGY control formula (1) was switched on from $i=1$ to 200, the first modification (2) from $i=201$ to 400, the second (3) from $i=401$ to 600, and again OGY's control formula from 601 to 800. As can be seen only procedure (3) was able to stabilize ξ_{F1}. (b) The parameter perturbations δp_i vs i used for control. The maximal allowed disturbance was $\delta p_{\max}=0.5$ and $p_0=36$.

The third fixed point ξ_{F3} could be stabilized by any of the three versions of the control formula. For ξ_{F3} the coefficients of the control formulas are very similar (see Table I). The coefficient b_2 is relatively small which indicates the small influence of δp_{i-1} compared to δp_i. So one can expect that all three algorithms will work.

The algorithms were also tested using further surfaces of section. Among others we also investigated the stroboscopic surface of section map which was used by Ditto, Rauseo, and Spano in [5], i.e., as time series we took

TABLE I. The numerically obtained values of the coefficients in the control formulas for the three fixed points considered. The coefficients are introduced implicitly by writing the OGY control formula as $\delta p_i=a\hat{\mathbf{f}}_u\cdot\delta\xi_i$, the first modification (2) as $\delta p_i=b_1\hat{\mathbf{f}}_u\cdot\delta\xi_i+b_2\delta p_{i-1}$, and the second modification (3) as $\delta p_i=c_1\hat{\mathbf{f}}_u\cdot\delta\xi_i+c_2\delta p_{i-1}$ with $\hat{\mathbf{f}}_u=\mathbf{f}_u/\|\mathbf{f}_u\|$.

	a	b_1	b_2	c_1	c_2	λ_u
ξ_{F1}	−16.43	−228.4	−12.9	38.9	2.2	−1.87
ξ_{F2}	164.6	−7.35	1.04	−9.4	1.33	4.82
ξ_{F3}	−2.46	−1.97	0.20	−1.79	0.18	−1.85

a stroboscopic measurement $x(t_i)$, $t_i - t_{i-1} = T$, and obtained a surface of section with points $\xi_i = (x(t_i), x(t_{i-1}))$. In this surface of section the periodic motion corresponding to ξ_{F1} could be stabilized by all three algorithms. They were almost equivalent because b_2 and c_2 were nearly zero (of the order of 10^{-4}), so the other coefficients were practically the same ($a \approx b_1 \approx c_1 \approx 2.7$). The periodic motion corresponding to ξ_{F2} could not be stabilized because the embedding in the neighborhood of the fixed point was bad (not injective). The third fixed point finally could only be stabilized using the second modification (3).

Altogether the numerical investigations show that the possibility of stabilizing a fixed point is not an intrinsic property of a fixed point, as the eigenvalues λ_u and λ_s are, for example. The coefficients of the control formulas differ for different surfaces of section and so do their performances. We always observed that the performance of the first modification (2) is superior to the one of the original OGY formula and the second modification outperforms the latter two. However, their performances are similar whenever the influence of the preceding parameter perturbation δp_{i-1} is small which results in a small value of $\mathbf{f}_u \cdot \mathbf{v}$. But we did observe that the OGY formula failed and the applications of one of the modifications could stabilize the desired fixed point. As a rule this happened when the influence of the changes of the preceding parameter was noticeable, which resulted in a non-negligible value of $\mathbf{f}_u \cdot \mathbf{v}$.

In conclusion, we introduced two modifications of the control formula of OGY which can lead to a better performance of the control in the case that the dynamical system is reconstructed using time delay coordinates. Therefore these modifications extend the range of applicability of the OGY control method. With these modifications all remarkable advantages of the OGY control method are preserved; e.g., the dynamics equation is not required, the perturbations of the accessible parameter can be very small, different periodic points can be stabilized in the same parameter range for the same system, and after having determined the control coefficients the computational effort at every iteration is negligible which opens the possibility of real time applications. We expect that the OGY control method will yield important applications in the future for technical systems also.

We acknowledge fruitful discussions with C. Mohrdieck and U. Parlitz. One of us (G.N.) was supported by the Studienstiftung des Deutschen Volkes.

[1] E. Ott, C. Grebogi, and J. A. Yorke, Phys. Rev. Lett. **64**, 1196–1199 (1990).

[2] Here we want to mention the resonant control method (using no feedback) introduced by Hübler and Lücher [A. Hübler and E. Lücher, Naturwissenschaften **76**, 67 (1989); A. Hübler, Helv. Phys. Acta **62**, 291 (1989)]. Hübler suggests a modification of the underlying dynamics such that a goal dynamics becomes a stable solution. To do this one must have or construct a model equation for the dynamics and one must be able to modify experimentally the driving force of these equations with a possibly quite large modification. There are approaches to deal with these difficulties [J. Breeden and A. Hübler, Phys. Rev. A **42**, 5817 (1990)].

[3] N. H. Packard, J. P. Crutchfield, J. D. Farmer, and R. S. Shaw, Phys. Rev. Lett. **45**, 712–716 (1980).

[4] F. Takens, *Dynamical Systems and Turbulence* (Springer-Verlag, Berlin, 1981), p. 230.

[5] W. L. Ditto, S. N. Rauseo, and M. L. Spano, Phys. Rev. Lett. **65**, 3211–3214 (1990).

[6] F. J. Romeiras, E. Ott, C. Grebogi, and W. P. Dayawansa, in Proceedings of the American Control Conference (IEEE, New York, to be published).

[7] E. Ott, C. Grebogi, and J. A. Yorke, in *Chaos/XAOC Soviet-American Perspectives on Nonlinear Science*, edited by D. K. Campbell (American Institute of Physics, New York, 1990), pp. 153–172.

[8] We note that the OGY control formula looks different from the one in Ref. [1]. The reason for this is that we use the linearization around ξ_F and p_0 (as is used in [6]) and do not estimate the new position of the fixed point $\xi_F(p)$ as one changes the parameter p, $\xi_F(p) = \xi_F(p_0) + \mathbf{g}\delta p$, as OGY do. But it is easy to show that these approaches are equivalent and $\mathbf{g}$ and $\mathbf{w}$ are related through $\mathbf{g} = [1 - D_\xi P(\xi F, p_0)]^{-1}\mathbf{w}$.

[9] D. P. Lathrop and E. J. Kostelich, Phys. Rev. A **40**, 4028–4031 (1989).

[10] K. Pawelzik and H. G. Schuster, Phys. Rev. A **43**, 1808–1812 (1991).

[11] D. Auerbach, P. Cvitanovic, G. Gunaratne, J.-P. Eckmann, and I. Procaccia, Phys. Rev. Lett. **58**, 2387–2389 (1987).

[12] U. Parlitz and W. Lauterborn, Phys. Lett. **107A**, 351–355 (1985).

[13] G. Nitsche and U. Dressler (to be published).

Controlling chaotic dynamical systems

Filipe J. Romeiras[a,b], Celso Grebogi[a,c,d], Edward Ott[a,e,f] and W.P. Dayawansa[f,g]

[a]*Laboratory for Plasma Research, University of Maryland, College Park, MD 20742, USA*

[b]*Centro de Electrodinâmica (INIC) and Departamento de Matemática, Instituto Superior Técnico, 1096 Lisbon Codex, Portugal*

[c]*Department of Mathematics, University of Maryland, College Park, MD 20742, USA*

[d]*Institute for Physical Science and Technology, University of Maryland, College Park, MD 20742, USA*

[e]*Department of Physics, University of Maryland, College Park, MD 20742, USA*

[f]*Department of Electrical Engineering, University of Maryland, College Park, MD 20742, USA*

[g]*Systems Research Center, University of Maryland, College Park, MD 20742, USA*

Received 18 November 1991
Revised manuscript received 16 January 1992
Accepted 16 January 1992

We describe a method that converts the motion on a chaotic attractor to a desired attracting time periodic motion by making only small time dependent perturbations of a control parameter. The time periodic motion results from the stabilization of one of the infinite number of previously unstable periodic orbits embedded in the attractor. The present paper extends that of Ott, Grebogi and Yorke [Phys. Rev. Lett. 64 (1990) 1196], allowing for a more general choice of the feedback matrix and implementation to higher-dimensional systems. The method is illustrated by an application to the control of a periodically impulsively kicked dissipative mechanical system with two degrees of freedom resulting in a four-dimensional map (the "double rotor map"). A key issue addressed is that of the dependence of the average time to achieve control on the size of the perturbations and on the choice of the feedback matrix.

1. Introduction

It is common for systems to evolve with time in a chaotic way. In practice, however, it is often desired that chaos be avoided and/or that the system be optimized with respect to some performance criterion. Given a system which behaves chaotically, one approach might be to make some large (and possibly costly) alteration in the system which completely changes its dynamics in such a way as to achieve the desired objectives. Here we assume that this avenue is not available. Thus we address the following question: Given a chaotic system, how can we obtain improved performance and achieve a desired attracting time-periodic motion by making only *small* controlling temporal perturbations in an accessible system parameter.

The key observation is that a chaotic attractor typically has embedded densely within it an infinite number of unstable periodic orbits [1–5]. In addition, chaotic attractors can also sometimes contain unstable steady states (e.g., the Lorenz attractor has such an embedded steady state). Since we wish to make only small controlling perturbations to the system, we do not envision creating new orbits with very different properties from the already existing orbits. Thus we seek to exploit the already existing unstable periodic orbits and unstable steady states. Our approach is as follows: We first determine some of the unstable low-period periodic orbits and unstable steady states that are embedded in the chaotic attractor. We then examine these orbits and choose one which yields improved system performance. Finally, we apply small controls so as

From *Physica D* **58** (1992). Reprinted with permission from Elsevier Science Publishers.

to stabilize this already existing orbit.

Some comments concerning this method are the following:

(1) Before settling into the desired controlled orbit the trajectory experiences a chaotic transient whose expected duration diverges as the maximum allowed size of the control approaches zero.

(2) Small noise can result in occasional bursts in which the orbit wanders far from the controlled orbit.

(3) Controlled chaotic systems offer an advantage in flexibility in that any one of a number of different orbits can be stabilized by the small control, and the choice can be switched from one to another depending on the current desired system performance.

Although we describe the details only in the case of discrete time systems, this method is applicable in the continuous time case as well by considering the discrete time system obtained from the induced dynamics on a Poincaré section.

In order to illustrate the method we apply it to a periodically forced mechanical system (the kicked double rotor), which results in a four-dimensional map. Amongst the examples considered, we study cases where the unstable orbit of the uncontrolled system has two unstable eigenvalues and two stable eigenvalues, and the stabilization is achieved by variation of one control parameter characterizing the strength of the periodic forcing. The present paper generalizes our previous work [6] to the case of higher-dimensional systems [7] and also includes new material illustrating the effect of the choice of stabilization on the length of the chaotic transient experienced by the orbit before control is achieved. Other relevant references on the feedback stabilization of periodic or steady orbits embedded in chaotic attractors are the experiments of Ditto et al. [8], Singer et al. [9], and the paper of Fowler [10]. (Other works in the general field are listed in ref. [11].)

The plan of the paper is as follows. In section 2, we give an implementation of the method, initially developed in ref. [6], by using the "pole placement technique" [7, 12]. In particular, we address the problem of stabilization of periodic orbits with more than one unstable eigenvalue. We also discuss experimental implementation in the absence of an a priori mathematical system model and generalization of the method to deal with cases where delay coordinates embedding is used. In section 3 we present some results for the control of the Hénon map [13], a two-dimensional system that is used as a paradigm in the study of dynamical systems; these results extend those given in ref. [6] in directions relevant to our present study. In section 4 we present results for the control of the double rotor map [14], a four-dimensional system that describes a particular impulsively periodically forced mechanical system. Finally, in section 5 we present the main conclusions of the work.

2. Description of the method

2.1. Formulation

For the sake of simplicity we consider a discrete time dynamical system,

$$\boldsymbol{Z}_{i+1} = \boldsymbol{F}(\boldsymbol{Z}_i, p) , \tag{2.1}$$

where $\boldsymbol{Z}_i \in \mathbb{R}^n$, $p \in \mathbb{R}$ and $\boldsymbol{F}$ is sufficiently smooth in both variables. Here, p is considered a real parameter which is available for external adjustment but is restricted to lie in some small interval,

$$|p - \bar{p}| < \delta , \tag{2.2}$$

around a nominal value $\bar{p}$. We assume that the nominal system (i.e., for $p = \bar{p}$) contains a chaotic attractor. Our objective is to vary the parameter p with time i in such a way that for almost all initial conditions in the basin of the chaotic attractor, the dynamics of the system converge onto a desired time periodic orbit contained in the attractor. The control strategy is the following. We will find a stabilizing local feedback control law which is defined on a neighborhood of the desired periodic orbit. This is done by considering the first order approximation of the system at the chosen unstable periodic orbit. Here we assume that this approximation is stabilizable. Since stabilizability is a generic property of linear systems, this assumption is quite reasonable. The ergodic nature of the chaotic dynamics ensures that the state trajectory eventually enters into the neighborhood. Once inside, we apply the stabilizing feedback control law in order to steer the trajectory towards the desired orbit.

For simplicity we shall describe the method as applied to the stabilization of fixed points (i.e., period one orbits) of the map $\boldsymbol{F}$. The consideration of periodic orbits of period larger than one is straightforward and is discussed in section 2.5. Let $\boldsymbol{Z}_*(p)$ denote an unstable fixed point on the attractor. For values of p close to $\bar{p}$ and in the neighborhood of the fixed point $\boldsymbol{Z}_*(\bar{p})$ the map (2.1) can be approximated by the linear map

$$\boldsymbol{Z}_{i+1} - \boldsymbol{Z}_*(\bar{p}) = \mathbf{A}[\boldsymbol{Z}_i - \boldsymbol{Z}_*(\bar{p})] + \boldsymbol{B}(p - \bar{p}) , \tag{2.3}$$

where $\mathbf{A}$ is an $n \times n$ Jacobian matrix and $\boldsymbol{B}$ is an n-dimensional column vector,

$$\mathbf{A} = \mathrm{D}_Z \boldsymbol{F}(\boldsymbol{Z}, p) , \tag{2.4}$$

$$\boldsymbol{B} = \mathrm{D}_p \boldsymbol{F}(\boldsymbol{Z}, p) , \tag{2.5}$$

and these partial derivatives are evaluated at $\boldsymbol{Z} = \boldsymbol{Z}_*(\bar{p})$ and $p = \bar{p}$. We now introduce the time-dependence of the parameter p by assuming that it is a linear function of the variable $\boldsymbol{Z}_i$ of the form

$$p - \bar{p} = -\boldsymbol{K}^{\mathrm{T}}[\boldsymbol{Z}_i - \boldsymbol{Z}_*(\bar{p})] . \tag{2.6}$$

The $1 \times n$ matrix $\boldsymbol{K}^{\mathrm{T}}$ is to be determined so that the fixed point $\boldsymbol{Z}_*(\bar{p})$ becomes stable. Substituting (2.6) into (2.3) we obtain

$$\boldsymbol{Z}_{i+1} - \boldsymbol{Z}_*(\bar{p}) = (\mathbf{A} - \boldsymbol{B}\boldsymbol{K}^{\mathrm{T}})[\boldsymbol{Z}_i - \boldsymbol{Z}_*(\bar{p})] , \tag{2.7}$$

which shows that the fixed point will be stable provided the matrix $\mathbf{A} - \boldsymbol{B}\boldsymbol{K}^{\mathrm{T}}$ is asymptotically stable; that is, all its eigenvalues have modulus smaller than unity.

The solution to the problem of the determination of $\boldsymbol{K}^{\mathrm{T}}$, such that the eigenvalues of the matrix $\mathbf{A} - \boldsymbol{B}\boldsymbol{K}^{\mathrm{T}}$ have specified values, is well known from control systems theory and is called "pole placement technique" (see, for example, Ogata [12]). We summarize the relevant results.

2.2. *Review of the pole placement technique*

The eigenvalues of the matrix $\mathbf{A} - \boldsymbol{B}\boldsymbol{K}^{\mathrm{T}}$ are called the "regulator poles", and the problem of placing these poles at the desired locations by choosing $\boldsymbol{K}^{\mathrm{T}}$ with $\mathbf{A}$ and $\boldsymbol{B}$ given is the "pole placement problem".

Pole placement problem. Determine the matrix $\boldsymbol{K}^{\mathrm{T}}$ in such a way that the eigenvalues of the matrix $\mathbf{A} - \boldsymbol{B}\boldsymbol{K}^{\mathrm{T}}$ have specified (complex) values $\{\mu_1, \ldots, \mu_n\}$.

The following results [12] give a necessary and sufficient condition for a unique solution of the pole placement problem to exist, and also a method for obtaining it (Ackermann's method).

(1) The pole placement problem has a unique solution if and only if the $n \times n$ matrix

$$\mathbf{C} = (\boldsymbol{B} \,\vdots\, \mathbf{A}\boldsymbol{B} \,\vdots\, \mathbf{A}^2\boldsymbol{B} \,\vdots\, \ldots \,\vdots\, \mathbf{A}^{n-1}\boldsymbol{B}) ,$$

is of rank n. ($\mathbf{C}$ is called the controllability matrix).

(2) The solution of the pole placement problem is given by

$$\boldsymbol{K}^{\mathrm{T}} = (\alpha_n - a_n \ldots \alpha_1 - a_1)\mathbf{T}^{-1} ,$$

where $\mathbf{T} = \mathbf{CW}$, and

$$\mathbf{W} = \begin{pmatrix} a_{n-1} & a_{n-2} & \dots & a_1 & 1 \\ a_{n-2} & a_{n-3} & \dots & 1 & 0 \\ \vdots & \vdots & & \vdots & \vdots \\ a_1 & 1 & \dots & 0 & 0 \\ 1 & 0 & \dots & 0 & 0 \end{pmatrix}.$$

Here $\{a_1, \dots, a_n\}$ are the coefficients of the characteristic polynomial of $\mathbf{A}$,

$$|s\mathbf{I} - \mathbf{A}| = s^n + a_1 s^{n-1} + \cdots + a_n ,$$

and $\{\alpha_1, \dots, \alpha_n\}$ are the coefficients of the desired characteristic polynomial of $\mathbf{A} - \boldsymbol{BK}^{\mathrm{T}}$,

$$\Pi_{j=1}^{n}(s - \mu_j) = s^n + \alpha_1 s^{n-1} + \cdots + \alpha_n .$$

2.3. Control parameter

Our considerations so far are based on the linear eq. (2.7) and therefore only apply in the local region near $\boldsymbol{Z}_*(\bar{p})$. On the other hand, the limitation in the size of the parameter perturbations given by (2.2), when combined with (2.6), yields

$$|\boldsymbol{K}^{\mathrm{T}}[\boldsymbol{Z}_i - \boldsymbol{Z}_*(\bar{p})]| < \delta . \tag{2.8}$$

This defines a slab of width $2\delta/|\boldsymbol{K}^{\mathrm{T}}|$. We choose to activate the control according to (2.6) only for values of $\boldsymbol{Z}_i$ inside this slab, and we choose to leave the control parameter at its nominal value (i.e., $p = \bar{p}$) when $\boldsymbol{Z}_i$ is outside this slab. Other choices are possible.

In summary, the control is determined by

$$p - \bar{p} = -\boldsymbol{K}^{\mathrm{T}}[\boldsymbol{Z}_i - \boldsymbol{Z}_*(\bar{p})] \times u(\delta - |\boldsymbol{K}^{\mathrm{T}}[\boldsymbol{Z}_i - \boldsymbol{Z}_*(\bar{p})]|) , \tag{2.9}$$

for arbitrary $\boldsymbol{Z}_i$ [not necessarily close to $\boldsymbol{Z}_*(\bar{p})$], where u is the unit step function defined by

$$u(\alpha) = \begin{cases} 0, & \alpha < 0 , \\ 1, & \alpha > 0 . \end{cases}$$

At this stage it should be pointed out that the matrix $\boldsymbol{K}^{\mathrm{T}}$ can be chosen in many different ways. In principle, any choice of regulator poles inside the unit circle serves our purpose. In ref. [6], the authors made a very special, though quite natural, choice of the gain matrix $\boldsymbol{K}^{\mathrm{T}}$: the resulting value of $p - \bar{p}$ forces the orbit onto the (linear) stable manifold of the fixed point at each iteration. In terms of regulator poles this choice corresponds to setting n_s of these poles equal to the n_s stable eigenvalues of matrix $\mathbf{A}$ and the remaining $n - n_s$ to 0. In terms of the slab (2.8) this choice corresponds not only to orientating it parallel to the stable manifold but also taking an appropriate width.

The choice of the matrix $\boldsymbol{K}^{\mathrm{T}}$ will be discussed at some length in our applications of the method in sections 3 and 4.

2.4. Time to achieve control

The control is activated (i.e., $p \neq \bar{p}$) only if $\boldsymbol{Z}_i$ falls in the narrow slab (2.8). Thus, for small δ, a typical initial condition will execute a chaotic orbit, unchanged from the uncontrolled case, until $\boldsymbol{Z}_i$ falls in this slab. Even then, because of nonlinearity not included in the linearized eq. (2.7), the control may not be able to bring the orbit to the fixed point. In this case the orbit will leave the slab and continue to wander chaotically as if there was no control. Since the orbit on the uncontrolled chaotic attractor is ergodic, at some time it will eventually satisfy (2.8) and also be sufficiently close to the desired fixed point so that control is achieved.

Thus, we create a stable orbit, which, for a typical initial condition, is preceded by a chaotic transient [15–18] in which the orbit is similar to orbits on the uncontrolled chaotic attractor. The length τ of such chaotic transient depends sensitively on the initial condition of the particular orbit. For initial conditions randomly chosen in the basin of attraction the distribution of chaotic transient lengths is exponential [15, 16],

$$\phi(\tau) \simeq \frac{1}{\langle\tau\rangle} \exp\left(-\frac{\tau}{\langle\tau\rangle}\right) , \tag{2.10}$$

for large τ. The quantity $\langle \tau \rangle$ is the characteristic length of the chaotic transient, called in the present case the *average time to achieve control*. Estimates of the scaling of $\langle \tau \rangle$ with δ for small δ are given in Appendix A for the case of two-dimensional maps.

2.5. Control of periodic orbits of period greater than one

The analysis of periodic orbits given in sections 2.1–2.3 can be extended to nontrivial periodic orbits (i.e., orbits with period greater than one). The most direct way is to take the Tth iterate of the map, where T denotes the period of the orbit to be stabilized. For the T times iterated map, any point on the periodic orbit is a fixed point, and we can then apply the discussion sections 2.1–2.3. This method is, however, overly sensitive to noise, especially when long period periodic orbits are involved. Next we outline another method which we believe should, in general, be better. In terms of the treatment of section 2.2, the prescription we give below corresponds to placing the unstable eigenvalues of the uncontrolled problem at zero, while leaving the stable eigenvalues unchanged. (This is only one of many possibilities that could be given.)

We denote the periodic orbit by $\boldsymbol{Z}_{i*}(p)$, where $\boldsymbol{Z}_{(i+T)*}(p) = \boldsymbol{Z}_{i*}(p)$. In addition, we introduce the set of T matrices $\mathbf{A}_i$ which are $n \times n$ and the set of T column vectors $\boldsymbol{B}_i$ which are of dimension n, where

$$\mathbf{A}_i = \mathbf{A}_{i+T} = \mathbf{D}_Z \boldsymbol{F}(\boldsymbol{Z}, p) ,$$
$$\boldsymbol{B}_i = \boldsymbol{B}_{i+T} = \mathbf{D}_p \boldsymbol{F}(\boldsymbol{Z}, p) ,$$

and the partial derivatives are evaluated at $\boldsymbol{Z} = \boldsymbol{Z}_{i*}(\bar{p})$ and $p = \bar{p}$.

Linearizing as in eq. (2.3), we have

$$\boldsymbol{Z}_{i+1} - \boldsymbol{Z}_{(i+1)*}(\bar{p}) = \mathbf{A}_i[\boldsymbol{Z}_i - \boldsymbol{Z}_{i*}(\bar{p})] + \boldsymbol{B}_i(p_i + \bar{p}) . \tag{2.11}$$

Say that the periodic orbit has u unstable eigenvalues (i.e., u eigenvalues with magnitude greater than one) and s stable eigenvalues, where $u + s = n$. At each point $\boldsymbol{Z}_{i*}(\bar{p})$ on the $p = \bar{p}$ periodic orbit, determine vectors $\{\boldsymbol{v}_{i,1}, \boldsymbol{v}_{i,2}, \ldots, \boldsymbol{v}_{i,s}\}$ which span the linearized stable subspace. Now let

$$\boldsymbol{\phi}_{i,j} = \mathbf{A}_{i+u-1}\mathbf{A}_{i+u-2} \cdots \mathbf{A}_{i+j+1}\mathbf{A}_{i+j} ,$$

for $j = 1, 2, \ldots, (u-1)$ and

$$\mathbf{C}_i = (\boldsymbol{\phi}_{i,1}\boldsymbol{B}_i \vdots \boldsymbol{\phi}_{i,2}\boldsymbol{B}_{i+1} \vdots \cdots \vdots \boldsymbol{\phi}_{i,u-1}\boldsymbol{B}_{i+u-2} \vdots \boldsymbol{B}_{i+u-1} \vdots \boldsymbol{v}_{i+u,1} \vdots \boldsymbol{v}_{i+u,2} \vdots \cdots \vdots \boldsymbol{v}_{i+u,s})$$

(One choice of the vectors $\{\boldsymbol{v}_{i,1}, \boldsymbol{v}_{1,2}, \ldots, \boldsymbol{v}_{i,s}\}$ is the stable eigenvectors of $\mathbf{A}_i\mathbf{A}_{i-1} \ldots \mathbf{A}_{i-T+1}$.) The controllability condition (analogous to that in section 2.2) is that $\mathbf{C}_i$ be nonsingular. The desired result for the control is then specified by

$$p_i - \bar{p} = -\boldsymbol{K}_i^{\mathrm{T}}[\boldsymbol{Z}_i - \boldsymbol{Z}_{i*}(\bar{p})] , \tag{2.12a}$$

where

$$\boldsymbol{K}_i^{\mathrm{T}} = \boldsymbol{\kappa}\mathbf{C}_i^{-1}\boldsymbol{\phi}_{i,0} , \tag{2.12b}$$

and $\boldsymbol{\kappa}$ denotes an n-dimensional row vector whose first entry is one and all of whose remaining entries are zeros.

To derive eqs. (2.12) we iterate (2.11) u times,

$$\begin{aligned}\boldsymbol{Z}_{i+u} - \boldsymbol{Z}_{(i+u)*}(\bar{p}) &= \boldsymbol{\phi}_{i,0}[\boldsymbol{Z}_i - \boldsymbol{Z}_{i*}(\bar{p})] \\ &+ \boldsymbol{\phi}_{i,1}\boldsymbol{B}_i(p_i - \bar{p}) + \boldsymbol{\phi}_{i,2}\boldsymbol{B}_{i+1}(p_{i+1} - \bar{p}) \\ &+ \cdots + \boldsymbol{B}_{i+u-1}(p_{i+u-1} - \bar{p}) .\end{aligned} \tag{2.13a}$$

We then demand that $\boldsymbol{Z}_{i+u}$ land on the linearized stable manifold of the periodic orbit through the point $\boldsymbol{Z}_{(i+u)*}(\bar{p})$. That is, we choose the p's such that there exists s coefficients $\alpha_1, \alpha_2, \ldots, \alpha_s$ such that

$$\begin{aligned}\boldsymbol{Z}_{i+u} - \boldsymbol{Z}_{(i+u)*}(\bar{p}) &= \alpha_1\boldsymbol{v}_{i+u,1} + \alpha_2\boldsymbol{v}_{i+u,2} \\ &+ \cdots + \alpha_s\boldsymbol{v}_{i+u,s} .\end{aligned} \tag{2.13b}$$

Regarding (2.13a) and (2.13b) as $n = u + s$ equations in the n unknowns, $p_i, p_{i+1}, \ldots, p_{i+u-1}, \alpha_1, \alpha_2, \ldots, \alpha_s$, we then solve for p_i to obtain (2.12).

[Note from the above that at time i we could, once and for all, calculate all the control parameter values to be applied in the next u iterates, $p_i, p_{i+1}, \ldots, p_{i+u-1}$. In the presence of noise, however, this is not a good idea (assuming $u > 1$), since it does not take advantage of the opportunity to correct for the noise on each iterate. Therefore, we believe that, in the presence of noise, it is best to perform the calculation of p_i via eq. (2.12) on each iterate.]

2.6. Use of delay coordinates

In experimental studies of chaotic dynamical systems, delay coordinates are often used to represent the system state. This is sometimes useful because it only requires measurement of the time series of a *single* scalar state variable which, we denote $\xi(t)$. A delay coordinate vector can be formed as follows:

$$\boldsymbol{Z}(t) = (\xi(t), \xi(t - T_{\mathrm{D}}), \xi(t - 2T_{\mathrm{D}}), \ldots, \times \xi(t - MT_{\mathrm{D}})) ,$$

where T_{D} is some conveniently chosen delay time, and the time variable t is assumed continuous. Embedding theorems guarantee that for $M \geq 2n$, where n is the system dimensionality, the vector $\boldsymbol{Z}$ is generically a global one-to-one representation of the system state. (Actually, for our purposes, we do not require a global embedding; we only require $\boldsymbol{Z}$ to be one-to-one in the small region near the periodic orbit, and this can typically be achieved with $M = n - 1$.) To obtain a map, one can take a Poincaré surface of section. For the often encountered case of a system which is periodically forced at a period T_{F}, one can define a "stroboscopic surface of section" by sampling the state at discrete times $t_i = iT_{\mathrm{F}} + t_0$. In this case we have the discrete state variable

$$\boldsymbol{Z}_i = \boldsymbol{Z}(t_i) .$$

As pointed out by Dressler and Nitsche [19], in the presence of parameter variation, delay coordinates lead to a map of a different form than

$$\boldsymbol{Z}_{i+1} = \boldsymbol{F}(\boldsymbol{Z}_i, p_i) ,$$

which is the form assumed in sections 2.1–2.5. For example, in the periodically forced case, since the components of $\boldsymbol{Z}_i$ are $\xi(t_i - mT_{\mathrm{D}})$ for $m = 0, 1, \ldots, M$, the vector $\boldsymbol{Z}_{i+1}$ must depend not only on p_i, but also on all previous values of the parameter that were in effect during the time interval $t_i \leq t \leq t_i - MT_{\mathrm{D}}$. In particular, let r be the smallest integer such that $MT_{\mathrm{D}} < rT_{\mathrm{F}}$. Then the relevant map is in general of the form

$$\boldsymbol{Z}_{i+1} = \boldsymbol{G}(\boldsymbol{Z}_i, p_i, p_{i-1}, \ldots, p_{i-r}) . \tag{2.14a}$$

For $r = 1$ we have

$$\boldsymbol{Z}_{i+1} = \boldsymbol{G}(\boldsymbol{Z}_i, p_i, p_{i-1}) . \tag{2.14b}$$

We now discuss how the technique of section 2.2 can be applied in the case of delay coordinates, and, for simplicity, we limit the discussion to $r = 1$, eq. (2.14b). Linearizing as in eq. (2.3) and again restricting our attention to the case of a fixed point orbit, we have

$$\boldsymbol{Z}_{i+1} - \boldsymbol{Z}_*(\bar{p}) = \mathbf{A}[\boldsymbol{Z}_i - \boldsymbol{Z}_*(\bar{p})] + \boldsymbol{B}_{\mathrm{a}}(p_i - \bar{p}) + \boldsymbol{B}_{\mathrm{b}}(p_{i-1} - \bar{p}) , \tag{2.15}$$

where $\mathbf{A} = \mathbf{D}_Z \boldsymbol{G}(\boldsymbol{Z}, p, p')$, $\boldsymbol{B}_a = \mathbf{D}_p \boldsymbol{G}(\boldsymbol{Z}, p, p')$, and $\boldsymbol{B}_{\mathrm{b}} = \mathbf{D}_{p'} \boldsymbol{G}(\boldsymbol{Z}, p, p')$, and all partial derivatives are evaluated at $\boldsymbol{Z} = \boldsymbol{Z}_*(\bar{p})$ and $p = \bar{p} = p'$.

Now define a new state variable with one extra component by

$$\tilde{\boldsymbol{Z}}_{i+1} = \begin{pmatrix} \boldsymbol{Z}_{i+1} \\ p_i \end{pmatrix} , \tag{2.16}$$

and introduce the linear control law,

$$p_i - \bar{p} = -\boldsymbol{K}^{\mathrm{T}}[\boldsymbol{Z}_i - \boldsymbol{Z}_*(\bar{p})] - k(p_{i-1} - \bar{p}) . \tag{2.17}$$

Combining these equations, we obtain

$$\tilde{\boldsymbol{Z}}_{i+1} - \tilde{\boldsymbol{Z}}_*(\bar{p}) = (\tilde{\mathbf{A}} - \tilde{\boldsymbol{B}}\tilde{\boldsymbol{K}}^{\mathrm{T}})[\tilde{\boldsymbol{Z}}_i - \tilde{\boldsymbol{Z}}_*(\bar{p})] , \tag{2.18}$$

where

$$\tilde{\boldsymbol{Z}}_*(\bar{p}) = \begin{pmatrix} \boldsymbol{Z}_*(\bar{p}) \\ \bar{p} \end{pmatrix}, \quad \tilde{\mathbf{A}} = \begin{pmatrix} \mathbf{A} & \boldsymbol{B}_{\mathrm{b}} \\ \mathbf{0} & 0 \end{pmatrix},$$

$$\tilde{\boldsymbol{B}} = \begin{pmatrix} \boldsymbol{B}_{\mathrm{a}} \\ 1 \end{pmatrix}, \quad \tilde{\boldsymbol{K}} = \begin{pmatrix} \boldsymbol{K} \\ k \end{pmatrix}.$$

Since (2.18) is now of the same form as (2.3), the method of section 2.2 can be applied. (A similar result for any $r > 1$ also clearly holds.)

Another method of control for delay coordinates is to reduce (2.14b) directly to the form $\boldsymbol{Z}_{i+1} = \boldsymbol{F}(\boldsymbol{Z}_i, p_i)$ and then proceed as in sections 2.1 and 2.2. This reduction can be done by setting $p_i \equiv \bar{p}$ for every other time step. For example, say $p_i = 0$ for i odd, and $j = \frac{1}{2}i$ for even i. Then making the replacements $\boldsymbol{Z}_i \to \hat{\boldsymbol{Z}}_j$, $p_i \to \hat{p}_j$ for even i, and iterating (2.14b) twice we have

$$\begin{aligned} \hat{\boldsymbol{Z}}_{j+1} &= \boldsymbol{G}[\boldsymbol{G}(\hat{\boldsymbol{Z}}_j, \bar{p}, \hat{p}_j), \hat{p}_j, \bar{p}] \\ &\equiv \hat{\boldsymbol{F}}(\hat{\boldsymbol{Z}}_j, \hat{p}_j) , \end{aligned} \tag{2.19}$$

which is of the required form. We believe, however, that the first method we have given [i.e., that based on eq. (2.18)] should usually be capable of yielding superior results to the method based on (2.19) with respect to noise sensitivity and time to achieve control. This is because our second method does not take advantage of the opportunity to control on each time iterate while our first method does.

3. Controlling the Hénon map

In ref. [6], the authors used the Hénon map to illustrate the control method and, in particular, to test their theoretical predictions concerning the average time to achieve control. As already pointed out, their work is based on a particular choice of the gain matrix $\boldsymbol{K}^{\mathrm{T}}$. In this section we consider how different choices of $\boldsymbol{K}^{\mathrm{T}}$ affect the average time to achieve control for the Hénon map.

The Hénon map [13] is the two-dimensional map

$$\boldsymbol{Z} \mapsto \boldsymbol{Z}' = \boldsymbol{F}(\boldsymbol{Z}) ,$$

defined by

$$\begin{pmatrix} x \\ y \end{pmatrix} \mapsto \begin{pmatrix} x' \\ y' \end{pmatrix} = \begin{pmatrix} a - x^2 + by \\ x \end{pmatrix},$$

where $(x, y) \in \mathbb{R} \times \mathbb{R}$. We keep the parameter b fixed throughout ($b = 0.3$) and allow the control parameter a to vary around a nominal value $\bar{a}$ ($\bar{a} = 1.4$) for which the map has a chaotic attractor.

For $a = \bar{a} = 1.4$ there is an unstable saddle fixed point contained in the chaotic attractor. This fixed point is located at

$$\boldsymbol{Z}_*(a) = x_*(a)\begin{pmatrix} 1 \\ 1 \end{pmatrix},$$

$$x_*(a) = -c + (c^2 + a)^{1/2} , \quad c = \tfrac{1}{2}(1 - b) ,$$

for $a \geq -c^2$. Noting that the Jacobian matrix of partial derivatives of the map is

$$\mathbf{D}_Z\boldsymbol{F}(\boldsymbol{Z}) = \begin{pmatrix} -2x & b \\ 1 & 0 \end{pmatrix},$$

and that the stability of the fixed point is determined by the roots of the characteristic equation

$$|\mathbf{D}_Z\boldsymbol{F}[\boldsymbol{Z}_*(a)] - s\mathbf{I}| = 0 ,$$

one can easily check that the fixed point is stable for $-c^2 < a < 3c^2$ and unstable for $a > 3c^2$. (Hence the fixed point is unstable for $b = 0.3$, $a = \bar{a} = 1.4$ since $c = 0.35$.)

The quantities that appear in section 2.2 are as follows:

$$\mathbf{A} = \begin{pmatrix} -2\bar{x}_* & b \\ 1 & 0 \end{pmatrix}, \quad \boldsymbol{B} = \begin{pmatrix} 1 \\ 0 \end{pmatrix},$$

$$\mathbf{C} = (\boldsymbol{B} : \mathbf{A}\boldsymbol{B}) = \begin{pmatrix} 1 & -2\bar{x}_* \\ 0 & 1 \end{pmatrix},$$

$$\mathbf{W} = \begin{pmatrix} 2\bar{x}_* & 1 \\ 1 & 0 \end{pmatrix}, \quad \mathbf{T} = \begin{pmatrix} 0 & 1 \\ 1 & 0 \end{pmatrix},$$

$$\boldsymbol{K}^{\mathrm{T}} = (\alpha_1 - a_1 \quad \alpha_2 - a_2)\,,$$

where

$$a_1 = 2\bar{x}_* = -(\lambda_{\mathrm{u}} + \lambda_{\mathrm{s}})\,, \quad a_2 = -b = \lambda_{\mathrm{u}}\lambda_{\mathrm{s}}\,,$$

and

$$\alpha_1 = -(\mu_1 + \mu_2)\,, \quad \alpha_2 = \mu_1\mu_2\,.$$

Here $\bar{x}_* = x_*(\bar{a})$, and λ_{u} and λ_{s} are the eigenvalues of matrix $\mathbf{A}$,

$$\left.\begin{matrix}\lambda_{\mathrm{s}} \\ \lambda_{\mathrm{u}}\end{matrix}\right\} = -\bar{x}_* \pm (\bar{x}_*^2 + b)^{1/2}\,.$$

The quantities μ_1 and μ_2 are the regulator poles [i.e., the eigenvalues of $(\mathbf{A} - \boldsymbol{B}\boldsymbol{K}^{\mathrm{T}})$].

In order to better illustrate the different choices of regulator poles or, equivalently, of the matrix $\boldsymbol{K}^{\mathrm{T}}$, we have used the plane (α_1, α_2) [cf. fig. 1]. In this plane we have plotted the lines of marginal stability $\mu_1 = \pm 1$ $(1 \pm \alpha_1 + \alpha_2 = 0)$ and $\mu_1\mu_2 = 1$ $(\alpha_2 = 1)$; the bounded triangular region delimited by these lines (shown shaded in the figure) is the region where the regulator poles are stable. In addition, we have plotted as dashed lines the axes $(k_1, k_2) = \boldsymbol{K}^{\mathrm{T}}$ which are related to (α_1, α_2) by the translations

$$k_1 = \alpha_1 - a_1\,, \quad k_2 = \alpha_2 - a_2\,.$$

The straight solid line in the figure going through the origin of the (k_1, k_2) plane has slope $-\lambda_{\mathrm{s}}$ and intersects the line $\alpha_2 = 0$ at the point Q with coordinates $(\alpha_1, \alpha_2) = (-\lambda_{\mathrm{s}}, 0)$. To this point

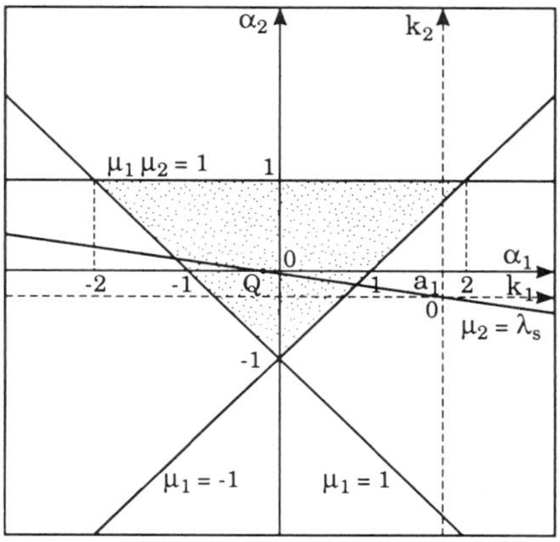

Fig. 1. Hénon map: choice of regulator poles.

corresponds the regulator poles

$$\mu_1 = 0\,, \quad \mu_2 = \lambda_{\mathrm{s}}\,,$$

and the matrix

$$\boldsymbol{K}^{\mathrm{T}} = \lambda_{\mathrm{u}}(1 \quad -\lambda_{\mathrm{s}}) \equiv \boldsymbol{K}_Q^{\mathrm{T}}\,.$$

$\boldsymbol{K}_Q^{\mathrm{T}}$ is the special choice of matrix $\boldsymbol{K}^{\mathrm{T}}$ made in ref. [6].

Before proceeding with the discussion, it is convenient to express the vector $\boldsymbol{K}^{\mathrm{T}}$ in polar coordinates

$$\boldsymbol{K}^{\mathrm{T}} = |\boldsymbol{K}^{\mathrm{T}}|(\cos\theta, \sin\theta)\,.$$

We consider the following two ways of varying the vector $\boldsymbol{K}^{\mathrm{T}}$ (inside the triangular region of stability):

(I) θ fixed, $|\boldsymbol{K}^{\mathrm{T}}|$ variable.

(II) $|\boldsymbol{K}^{\mathrm{T}}|$ fixed, θ variable.

In terms of the control slab defined by eq. (2.8) we have that in situation (I) the slab is kept orientated in a fixed direction while its width $w = 2\delta/|\boldsymbol{K}^{\mathrm{T}}|$ varies, whereas in situation (II) the direction of the slab is rotated while its width is kept fixed at $w = 2\delta/|\boldsymbol{K}_Q^{\mathrm{T}}|$. The choice of the $\boldsymbol{K}^{\mathrm{T}}$ in ref. [6] has $\theta = \theta_Q \equiv \tan^{-1}(-\lambda_{\mathrm{s}})$ and, as we shall see, this choice is optimal from the point of view of the time to achieve control. (To see that the choice of ref. [6] corresponds to $\theta = \theta_Q$, we note that with this choice one obtains a convergence rate to the periodic orbit of μ_s as in ref. [6].)

In the numerical experiments we calculated the average time to achieve control by the method described in Appendix B. We also allowed for different values of the maximum amplitude of the parameter perturbations, δ.

First we consider the case where θ is fixed (case I) at the value

$$\theta = \theta_Q .$$

This case has a simple interpretation in terms of regulator poles: $\mu_2 = \lambda_s$ is kept fixed while μ_1 is allowed to vary between -1 and $+1$. μ_1 and $|\boldsymbol{K}^T|$ are related by

$$|\boldsymbol{K}^T| = |\mu_1 - \lambda_u|(1 + \lambda_s^2)^{1/2} .$$

Fig. 2 shows results for $\langle\tau\rangle$ for this case. We see that the average time to achieve control increases with μ_1, although only moderately.

Fig. 3 shows results for $\langle\tau\rangle$ versus θ for $|\boldsymbol{K}|^T$ held fixed (case II) at

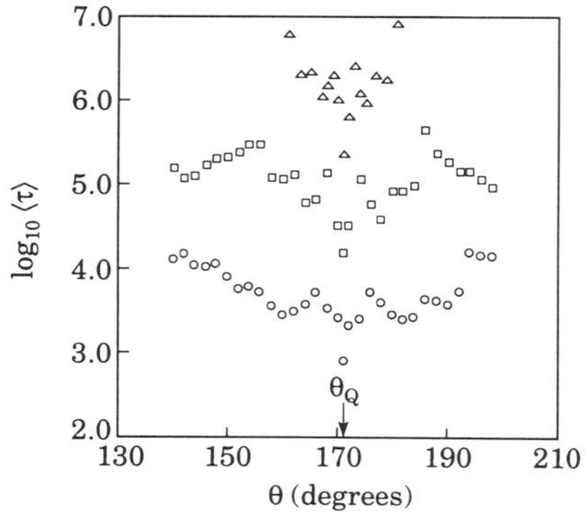

Fig. 3. Hénon map: $\log_{10}\langle\tau\rangle$ versus θ, with $|\mathbf{K}^T| = |\mathbf{K}_Q^T|$, for (○) $\delta = 10^{-2}$, (□) $\delta = 10^{-3}$, (△) $\delta = 10^{-4}$ ($\bar{a} = 1.4$, $b = 0.3$).

$$|\boldsymbol{K}^T| = |\boldsymbol{K}_Q^T| = |\lambda_u|(1 + \lambda_s^2)^{1/2} .$$

We see that the average time to achieve control has a strong minimum at $\theta = \theta_Q$.

Fig. 4 shows $\langle\tau\rangle$ versus $|\boldsymbol{K}^T|$ for three values of θ, $\theta = \theta_0$, $\theta = \theta_Q$, and $\theta = \theta_1$, where $\theta_0 < \theta_Q < \theta_1$ and θ_0 and θ_1 are close to θ_Q ($\theta_0 = 170.4°$, $\theta_Q = 171.1°$, $\theta_1 = 172.0°$). We observe that the $\theta = \theta_Q$ result is always below the results for $\theta = \theta_0$ and $\theta = \theta_1$ indicating that the average time

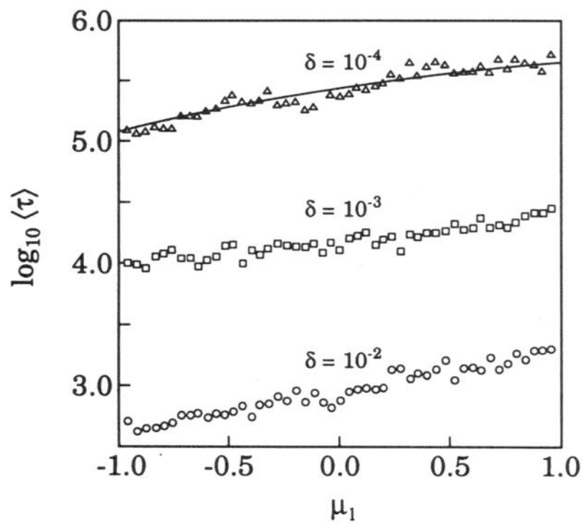

Fig. 2. Hénon map: $\log_{10}\langle\tau\rangle$ versus μ_1, with $\mu_2 = \lambda_s$, for (○) $\delta = 10^{-2}$, (□) $\delta = 10^{-3}$, (△) $\delta = 10^{-4}$. The theoretical curve was calculated using eq. (A.9) of appendix A. ($\bar{a} = 1.4$, $b = 0.3$).

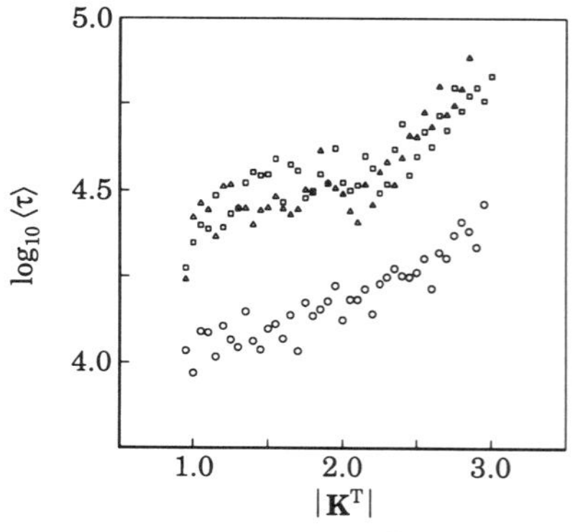

Fig. 4. Hénon map: $\log_{10}\langle\tau\rangle$ versus $|\boldsymbol{K}^T|$ for (○) $\theta = \theta_Q$, (□) $\theta = \theta_0$, (△) $\theta = \theta_1$ ($\theta_0 < \theta_Q < \theta_1$; $\theta_0 = \tan^{-1}[-\lambda_u\lambda_s/(\lambda_u + \lambda_s)] = 170.4°$, $\theta_Q = \tan^{-1}(-\lambda_s) = 171.1°$, $\theta_1 = 172.0°$), and $\delta = 10^{-3}$.

to achieve control has a strong minimum at $\theta = \theta_Q$ not only for $|K^T| = |K_Q^T|$ but for all values of $|K^T|$. Thus the condition $\theta = \theta_Q$ is optimal.

In Appendix A we show how the average time to achieve control can be obtained theoretically in the case of two-dimensional maps and verify that there is excellent agreement between the theoretical and experimental results in the case of the Hénon map.

4. Controlling the double rotor

In this section we apply the control method described in section 2 to a dynamical system known as the double rotor map. We start by deriving the map (section 4.1 and Appendix B), then study its fixed points (section 4.2) and its attractors (section 4.3), including chaotic ones, and finally proceed to control some of the fixed points embedded in one of the chaotic attractors (sections 4.4 and 4.5).

4.1. The double rotor map

The double rotor map is a four-dimensional map which describes the time evolution of a mechanical system known as the kicked double rotor [14]. This system is a four-dimensional extension of the kicked (single) rotor, a two-dimensional system that is described by the well-known dissipative standard map [20].

The double rotor is composed of two thin, massless rods connected as shown in fig. 5. The first rod, of length l_1, pivots about P_1 (which is fixed), and the second rod, of length $2l_2$, pivots about P_2 (which moves). The angles $\theta_1(t)$, $\theta_2(t)$ specify the orientations at time t of the first and second rods, respectively. A mass m_1 is attached at P_2, and masses $\frac{1}{2}m_2$ are attached to each end of the second rod (P_3 and P_4). Friction at P_1 (with coefficient ν_1) slows the first rod at a rate proportional to its angular velocity $\dot{\theta}_1(t) \equiv d\theta_1(t)/dt$; friction at P_2 (with coefficient ν_2) slows the second rod (and simultaneously accelerates the

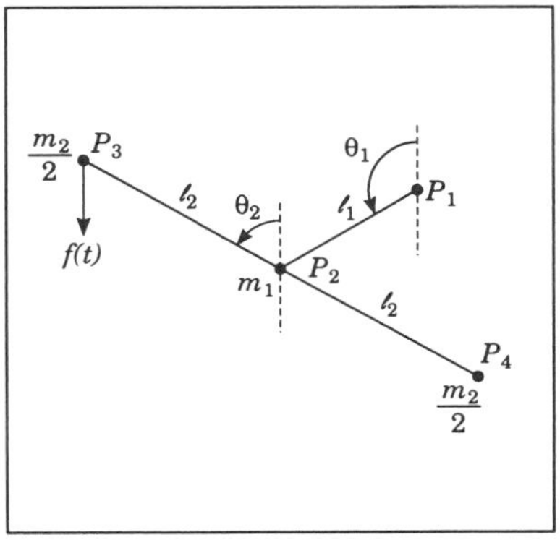

Fig. 5. The double rotor.

first rod) at a rate proportional to $\dot{\theta}_2(t) - \dot{\theta}_1(t)$. The end of the second rod marked P_3 receives periodic impulse kicks at times $t = T, 2T, \ldots$, always from the same direction and with constant strength f_0. There is no gravity.

In Appendix C we write the differential equations that describe the kicked double rotor and proceed to derive from them the *double rotor map* relating the state of the system just after consecutive kicks. We obtain the four-dimensional map

$$\boldsymbol{Z} \mapsto \boldsymbol{Z}' = \boldsymbol{F}(\boldsymbol{Z}) ,$$

defined by

$$\begin{pmatrix} \boldsymbol{X} \\ \boldsymbol{Y} \end{pmatrix} \mapsto \begin{pmatrix} \boldsymbol{X}' \\ \boldsymbol{Y}' \end{pmatrix} = \begin{pmatrix} \mathbf{M}\boldsymbol{Y} + \boldsymbol{X} \\ \mathbf{L}\boldsymbol{Y} + \boldsymbol{G}(\boldsymbol{X}') \end{pmatrix} , \tag{4.1}$$

where

$$\boldsymbol{X} = \begin{pmatrix} x_1 \\ x_2 \end{pmatrix} \in S^1 \times S^1 , \quad \boldsymbol{Y} = \begin{pmatrix} y_1 \\ y_2 \end{pmatrix} \in \mathbb{R} \times \mathbb{R} ,$$

and

$$\boldsymbol{G}(\boldsymbol{X}') = \begin{pmatrix} c_1 \sin x_1' \\ c_2 \sin x_2' \end{pmatrix} . \tag{4.2}$$

x_1, x_2 are the angular positions of the rods at the instant of the kth kick, $x_j = \theta_j(kT)$, while y_1, y_2

are the angular velocities of the rods immediately after the kth kick, $y_j = \dot{\theta}_j(kT^+)$. S^1 is the circle $\mathbb{R}(\mathrm{mod}\, 2\pi)$. $\mathbf{L}$ and $\mathbf{M}$ are constant 2×2 matrices defined by

$$\mathbf{L} = \sum_{j=1}^{2} \mathbf{W}_j \, \mathrm{e}^{\lambda_j T}, \quad \mathbf{M} = \sum_{j=1}^{2} \mathbf{W}_j \frac{\mathrm{e}^{\lambda_j T} - 1}{\lambda_j},$$

$$\mathbf{W}_1 = \begin{pmatrix} a & b \\ b & d \end{pmatrix}, \quad \mathbf{W}_2 = \begin{pmatrix} d & -b \\ -b & a \end{pmatrix},$$

$$a = \frac{1}{2}\left(1 + \frac{\nu_1}{\Delta}\right), \quad d = \frac{1}{2}\left(1 - \frac{\nu_1}{\Delta}\right), \quad b = -\frac{\nu_2}{\Delta},$$

$$\left.\begin{matrix}\lambda_1 \\ \lambda_2\end{matrix}\right\} = -\tfrac{1}{2}(\nu_1 + 2\nu_2 \pm \Delta),$$

$$\Delta = (\nu_1^2 + 4\nu_2^2)^{1/2}.$$

Finally, c_1 and c_2 are given by

$$c_j = \frac{f_0}{I} l_j, \quad j = 1, 2,$$

where

$$I = (m_1 + m_2)l_1^2 = m_2 l_2^2.$$

The following relation between matrices $\mathbf{L}$ and $\mathbf{M}$ will be useful below:

$$\mathbf{L} = \mathbf{I} + \mathbf{A}_\nu \mathbf{M}, \tag{4.3}$$

where

$$\mathbf{A}_\nu = \begin{pmatrix} -(\nu_1 + \nu_2) & \nu_2 \\ \nu_2 & -\nu_2 \end{pmatrix}.$$

(λ_1, λ_2 are precisely the eigenvalues of $\mathbf{A}_\nu$.) Note also that

$$|\mathbf{L}| = \mathrm{e}^{(\lambda_1 + \lambda_2)T},$$

$$|\mathbf{M}| = \frac{\mathrm{e}^{\lambda_1 T} - 1}{\lambda_1} \, \frac{\mathrm{e}^{\lambda_2 T} - 1}{\lambda_2},$$

$$|\mathbf{A}_\nu| = \nu_1 \nu_2.$$

From now on we assume that $\nu_1 = \nu_2 \equiv \nu$. This leads to

$$\left.\begin{matrix}\lambda_1 \\ \lambda_2\end{matrix}\right\} = -\tfrac{1}{2}\nu(3 \pm \sqrt{5}),$$

$$\left.\begin{matrix}a \\ d\end{matrix}\right\} = \tfrac{1}{2}(1 \pm \tfrac{1}{5}\sqrt{5}),$$

$$b = -\tfrac{1}{5}\sqrt{5}.$$

In all the numerical work described in the rest of this section the parameters ν, T, I, m_1, m_2, l_1, and l_2 were kept fixed at the values

$$\nu = T = I = m_1 = m_2 = l_2 = 1,$$
$$l_1 = 1/\sqrt{2}. \tag{4.4}$$

The only parameter which we shall vary is the forcing f_0 used as the control parameter.

4.2. Fixed points of the double rotor map

The fixed points $Z_* = (X_*, Y_*)$ of the map (4.1) are solutions of the system

$$\begin{aligned} X_* &= \mathbf{M}Y_* + X_* - 2\pi N, \\ Y_* &= \mathbf{L}Y_* + G(X_*), \end{aligned} \tag{4.5}$$

where the components of the vector $N = (n_1, n_2)$ are integer and are the rotation numbers in the x_1, x_2 variables. The rotation numbers n_1, n_2 are defined as the multiples of 2π by which x_{1*}, x_{2*} are increased in one iteration of the map before being brought to the interval $[0, 2\pi]$. From eqs. (4.5) we obtain, using (4.3),

$$\begin{aligned} Y_* &= 2\pi \mathbf{M}^{-1} N, \\ G(X_*) &= -2\pi \mathbf{A}_\nu N. \end{aligned} \tag{4.6}$$

Using the definitions of the matrices G and $\mathbf{A}_\nu$ we rewrite the second of the eqs. (4.6) in the form

$$\begin{pmatrix} \sin x_{1*} \\ \sin x_{2*} \end{pmatrix} = -\frac{2\pi\nu I}{f_0} \begin{pmatrix} (1/l_1)(-2n_1 + n_2) \\ (1/l_2)(n_1 - n_2) \end{pmatrix}$$
$$\equiv \frac{1}{f_0} \begin{pmatrix} f_{01} \\ f_{02} \end{pmatrix}, \tag{4.7}$$

where the identity on the right defines the two new quantities f_{01} and f_{02}. These equations show that for each pair of rotation numbers (n_1, n_2) a set of four possible solutions for (x_{1*}, x_{2*}) exists if $|f_0| \geq |f_{0c}|$, where $|f_{0c}| = \max(|f_{01}|, |f_{02}|)$. The four fixed points correspond to the four combinations of values of (x_{1*}, x_{2*}) that have the same pair of values of $(\sin x_{1*}, \sin x_{2*})$. When necessary we will use the notation

$$Z_*^{[N;q]} = (X_*^{[N;q]}, Y_*^{[N]}) ,$$

or more simply $[N; q]$, to identify the fixed points, where the index q labels the four possible solutions of (4.7) ($q = 1, 2, 3, 4$) and, as shown in fig. 6, corresponds to the ordering

$$x_{1*}^{[N;1]} = x_{1*}^{[N;2]} < x_{1*}^{[N;3]} = x_{1*}^{[N;4]} ,$$
$$x_{2*}^{[N;1]} = x_{2*}^{[N;3]} < x_{2*}^{[N;2]} = x_{2*}^{[N;4]}$$

Note that $Y_*^{[N]} = (y_{1*}^{[N]}, y_{2*}^{[N]})$ is the same for the four fixed points (i.e., it does not depend on q). Eqs. (4.7) also show that for $|f_0| \geq |f_{0c}|$, $(n_1, n_2) \neq (0, 0)$, another set of four fixed points exists with rotation numbers $(-n_1, -n_2)$. It is easy to see that to each point $(x_{1*}, x_{2*}, y_{1*}, y_{2*})$ of the first set corresponds a point of the second set given by $(2\pi - x_{1*}, 2\pi - x_{2*}, -y_{1*}, -y_{2*})$. This is a reflection of the fact that the double rotor map (4.1) itself is invariant under the change of variables $(x_1, x_2, y_1, y_2) \mapsto (2\pi - x_1, 2\pi - x_2, -y_1, -y_2)$.

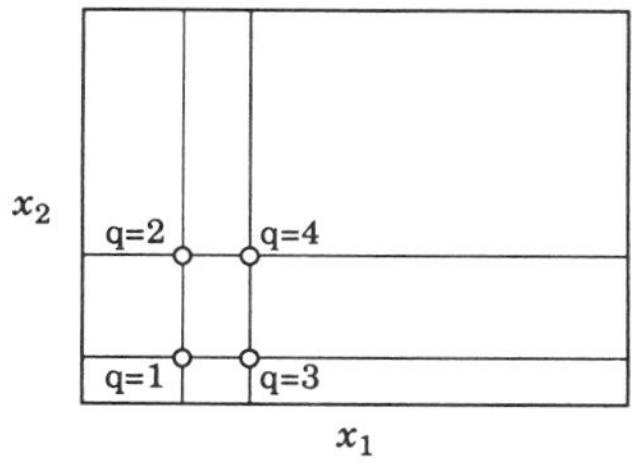

Fig. 6. Double rotor map: labeling of fixed points.

In table 1 we summarize the properties of the five sets of fixed points (36 fixed points) with smaller values of f_{0c} (when the other parameters of the map take the values specified by eqs. (4.4)), with rotation numbers $N = (0, 0)$, $\pm(1, 2)$, $\pm(0, 1)$, $\pm(1, 1)$, $\pm(2, 3)$. Note that the last three sets have the same value of f_{0c}. In fig. 7 we have plotted these fixed points in the plane (x_1, x_2). Their (y_1, y_2) coordinates are given by the first of eqs. (4.6).

Let us now turn our attention to the stability of the fixed points. The basic element of the analysis is the Jacobian (4×4) matrix of partial derivatives of the map (4.1),

$$\mathbf{D}_Z F(Z) = \begin{pmatrix} \mathbf{I}_2 & \mathbf{M} \\ \mathbf{H}(X') & \mathbf{L} + \mathbf{H}(X')\mathbf{M} \end{pmatrix} ,$$

where

$$\mathbf{H}(X') = \mathbf{D}_{x'} G(X') = \begin{pmatrix} c_1 \cos x_1' & 0 \\ 0 & c_2 \cos x_2' \end{pmatrix} ,$$

and $\mathbf{I}_n$ denotes the $n \times n$ identity matrix. The characteristic polynomial of $\mathbf{D}_Z F(Z_*)$ is

$$P(s) = |\mathbf{D}_Z F(Z_*) - s\mathbf{I}_4|$$
$$= |s^2 \mathbf{I}_2 - s(\mathbf{I}_2 + \mathbf{L} + \mathbf{HM}) + \mathbf{L}| , \tag{4.8a}$$

Table 1
Double rotor map: fixed points. The only stable fixed points are: $[(0, 0); 4]$ in the interval $0 < f_0 < 4.27\ldots$; $[(1, 2); 4]$ and $[(-1, -2); 1]$ in the interval $2\pi < f < 7.01\ldots$ [the other parameters are given by eq. (4.4)].

(n_1, n_2)	f_{01}	f_{02}	f_{0c}	h_{11c}	h_{22c}
$(0, 0)$	0	0	0	0	0
$\pm(1, 2)$	0	$2\pi\nu I/l_2$	$2\pi\nu I/l_2$	$\pm 2\pi\nu l_1/l_2$	0
$\pm(0, 1)$	$2\pi\nu I/l_1$	$2\pi\nu I/l_2$	$2\pi\nu I/l_1$	0	$\pm 2\pi\nu(l_2^2/l_1^2 - 1)^{1/2}$
$\pm(1, 1)$	$2\pi\nu I/l_1$	0	$2\pi\nu I/l_1$	0	$\pm 2\pi\nu l_2/l_1$
$\pm(2, 3)$	$2\pi\nu I/l_1$	$2\pi\nu I/l_2$	$2\pi\nu I/l_1$	0	$\pm 2\pi\nu(l_2^2/l_1^2 - 1)^{1/2}$

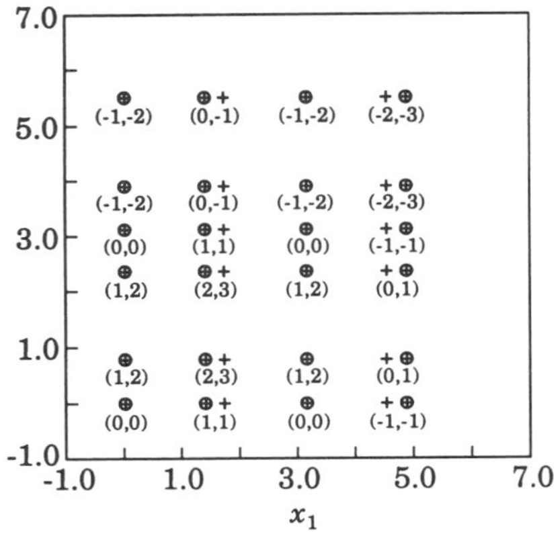

Fig. 7. Double rotor map: fixed points with rotation numbers (n_1, n_2). The symbol $(+)$ denotes fixed points with one unstable eigendirection, while the symbol $\oplus$ denotes fixed points with two unstable eigendirections. $[f_0 = 9.0$, other parameters given by eq. (4.4)].

where, for simplicity, we have set $\mathbf{H} \equiv \mathbf{H}(X_*)$. The characteristic equation

$$P(s) = 0 \,, \tag{4.8b}$$

determines the stability of the fixed points: if all the four roots have modulus smaller than one, the fixed point is stable. The stability of the fixed points as determined from eqs. (4.8) is discussed in appendix D.

For $f_0 = 9.0$, the nominal value in the control experiments of sections 4.4 and 4.5, all the fixed points are unstable. We have indicated in fig. 7 the number of unstable eigendirections at each fixed point.

We observe, from eq. (4.7), that as the forcing f_0 increases, the number of fixed points increases without bound. Not all these fixed points are necessarily embedded in the chaotic attractor, but those that are embedded in it are necessarily unstable. Furthermore, we find that the fixed points are roughly spread throughout the attractor, suggesting that there can be substantial flexibility to select among a variety of asymptotic behaviors by selecting different fixed points for control. (Even more flexibility can be achieved if we also consider periodic orbits of period greater than one.)

4.3. Bifurcation diagram

A bifurcation diagram shows how the attractors of a dynamical system change with a system parameter.

In figs. 8a, 8b we present a bifurcation diagram for the double rotor map, which was obtained in the following way. For each value of the parameter we took a large number of initial

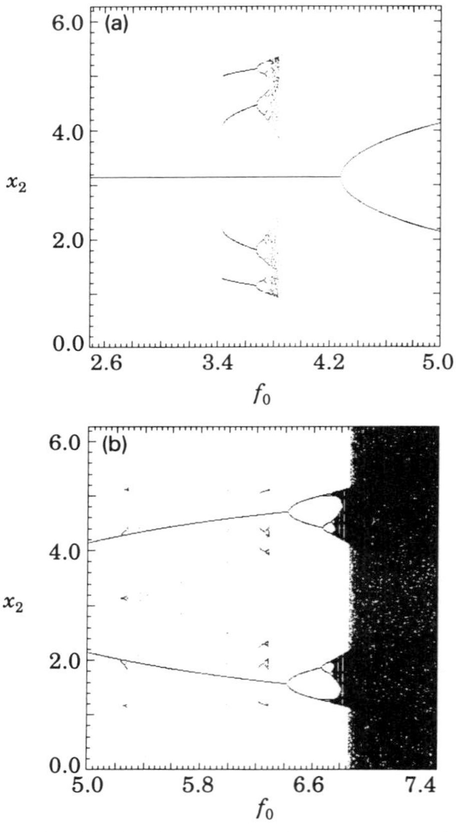

Fig. 8. (a), (b) Double rotor map: bifurcation diagram [parameters given by eq. (4.4); number of values of f_0 in each figure: 251; number of initial conditions for each f_0: 625; snapshot taken after 6000 iterations].

angles (x_1, x_2) with both x_1 and x_2 distributed uniformly in $[0, 2\pi]$ and iterated them starting with zero angular velocity [i.e., $(y_1, y_2) = (0,0)$]. After iterating a sufficient number of times so that the orbits are essentially on the attractor, we plotted the x_2 component of all the orbits.

The diagram clearly exhibits a main branch that develops continuously for all values of the parameter. This main branch illustrates a period-doubling bifurcation sequence to chaos: a period-1 periodic orbit bifurcates (at $f_0 \simeq 4.27$) to a period-2 periodic orbit which then bifurcates (at $f_0 \simeq 6.42$) to a period-4 periodic orbit which then bifurcates (at $f_0 \simeq 6.67$) to a period-8 periodic orbit, and so on, with an accumulation point of period doublings at $f_0 \simeq 6.75$ beyond which chaos appears. The period-2 periodic orbit in the sequence results from the bifurcation of the stable orbit $Z_*^{[(0,0);4]} = (\pi, \pi, 0, 0)$ discussed in section 4.2 which exists for $f_0 \geq 0$; at the value $f_{0u}^{[(0,0);4]}$ at which this orbit becomes unstable the stable period-2 periodic orbit is born.

Although it cannot be seen in the diagram, this period doubling sequence is peculiar in the following sense: what appears to be a period-2^m periodic orbit, $m \geq 2$, is in fact 2 period-2^{m-1} periodic orbits. This is a consequence of the symmetry of the double rotor map that forces the period-1 orbit to become unstable (at $f_0 \cong 4.2$) through an eigenvalue 1 instead through -1 as occurs in the normal period doubling bifurcation (an example of which is the bifurcation of the period-1 periodic orbit).

Besides this main branch, there are other period doubling sequences, one of which starts with a period-4 periodic orbit (at $f_0 \simeq 3.42$) and ends with a crisis (at $f_0 \simeq 3.84$). (A *crisis* is the sudden disappearance of a chaotic attractor by collision with an unstable periodic orbit [15, 16].)

It is convenient to have some quantitative characterization of the chaotic attractors revealed by the bifurcation diagram. For this purpose we introduce the spectrum of *Lyapunov exponents*, defined as follows [21, 22].

Consider an n-dimensional map $Z \mapsto F(Z)$ and its Jacobian matrix of partial derivatives $\mathbf{J}(Z) = \mathbf{D}_Z F(Z)$. Consider also the sequence $\{Z_0, Z_1, \ldots, Z_{k-1}\}$ generated by successive iteration of the initial condition Z_0. For this sequence introduce the matrix

$$\mathbf{J}_k = \mathbf{J}(Z_{k-1})\mathbf{J}(Z_{k-2}) \ldots \mathbf{J}(Z_1)\mathbf{J}(Z_0) .$$

Now let

$$\zeta_1(k) \geq \zeta_2(k) \geq \cdots \geq \zeta_n(k) ,$$

denote the n eigenvalues of $(\mathbf{J}_k^T \mathbf{J}_k)^{1/2}$, where $\mathbf{J}_k^T$ is the transpose of $\mathbf{J}_k$. The Lyapunov numbers of the map are then defined by

$$\eta_j = \lim_{k \to \infty} [\zeta_j(k)]^{1/k} , \quad j = 1, \ldots, n ,$$

where the positive real kth root is taken. They satisfy the same ordering as the $\zeta_j(k)$, $j = 1, \ldots, n$. The Lyapunov exponents are the logarithms of the Lyapunov numbers,

$$L_j = \log_e \eta_j , \quad j = 1, \ldots, n ,$$

satisfying the same ordering

$$L_1 \geq L_2 \geq \cdots \geq L_n .$$

Hence, for chaotic attractors of an n-dimensional map there are n Lyapunov exponents, L_j, $j = 1, \ldots, n$. A chaotic attractor is defined to be one which possesses a positive Lyapunov exponent, $L_1 > 0$.

For typical dynamical systems the Lyapunov exponents are the same for almost all initial conditions on the basin of attraction of the attractor. (This is true in particular for the chaotic attractors of the double rotor map for which we calculated Lyapunov exponents; these results are reported below.) Thus the spectrum of Lyapunov exponents may be indeed considered to be a property of the attractor. For maps such that the determinant of the Jacobian matrix is independent of the variable $\mathbf{Z}$ the Lyapunov exponents satisfy the identity

$$\sum_{j=1}^{n} L_j = \log_e |\mathbf{J}| .$$

This is true in the case of the double rotor map for which we have

$$\sum_{j=1}^{4} L_j = \log_e |\mathbf{L}| = (\lambda_1 + \lambda_2)T = -3\nu ,$$

the last equality applying when $\nu_1 = \nu_2 \equiv \nu$ and $T = 1$.

From the spectrum of Lyapunov exponents define the *Lyapunov dimension*.

$$d_L = k_L + \frac{\sum_{j=1}^{k_L} L_j}{|L_{k_L+1}|} ,$$

where $1 \le k_L \le n-1$ is the largest integer for which $\sum_{j=1}^{k_L} L_j \ge 0$. If $L_1 < 0$, we define $d_L = 0$; if $\sum_{j=1}^{n} L_j \ge 0$, we define $d_L = n$. (Note that $d_L = n$ is not possible in the case of dissipative systems for which $\log_e |\mathbf{J}| < 0$.) Kaplan and Yorke [23, 24] conjecture that d_L, as given above in terms of the Lyapunov exponents, is typically equal to the fractal dimension of the support of the measure of the attractor (the information dimension).

We have numerically calculated the Lyapunov exponents and the Lyapunov dimension of the chaotic attractor in the main branch of the bifurcation diagram as a function of the forcing f_0. We used the method described in refs. [21, 22] to calculate the exponents of a large number of orbits in the basin of attraction and then took the average of these values. The results of the calculation at evenly spaced values along the f_0 axis are shown in fig. 9. The Lyapunov dimension first becomes positive at the onset of chaos ($f_0 \simeq 6.75$). The attractor dimension goes through the integer values $d_L = 2$ and 3 at $f_0 \simeq 6.88$ and 12.7, respectively.

In the numerical experiments on control that we describe in sections 4.4 and 4.5 we took $\bar{f}_0 = 9.0$ as the nominal value of the control parameter. In Table 2 we list the corresponding values of the four Lyapunov exponents and the Lyapunov dimension. In order to illustrate the point made above regarding the fact that the Lyapunov exponents are the same for almost all initial conditions on the basin of attraction of the

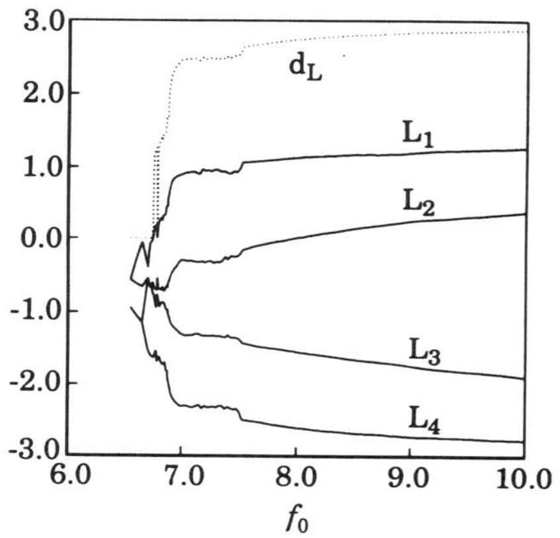

Fig. 9. Double rotor map: spectrum of Lyapunov exponents and Lyapunov dimension of chaotic attractors versus f_0 [eq. (4.4)].

Table 2
Double rotor map: calculation of Lyapunov exponents and Lyapunov dimension of chaotic attractor [$f_0 = 9.0$, other parameters given by eq. (4.4); number of initial conditions $= N_0 = 256$; number of iterations = 10 000]. $d_L = 2 + (L_1 + L_2)/|L_3| = 2.838$.

	j			
	1	2	3	4
$L_j = (1/N_0)\sum_{i=1}^{N_0} L_j^{(i)}$	1.205	0.256	−1.744	−2.717
$\min_{i=1,N_0} L_j^{(i)}$	1.182	0.228	−1.771	−2.734
$\max_{i=1,N_0} L_j^{(i)}$	1.229	0.284	−1.719	−2.693
$[(1/N_0)\sum_{i=1}^{N_0} (L_j^{(i)} - L_j)^2]^{1/2}$	0.00816	0.0102	0.00910	0.00724

attractor, we also give some details on the numerical calculation of these exponents.

We have now described in sufficient detail the two ingredients necessary to the application of the control method to the double rotor map: chaotic attractors and fixed points. It remains to be checked if the fixed points determined in section 4.2 are embedded in the chaotic attractor. By this we mean that any neighborhood of the fixed point contains an infinite number of points of the chaotic attractor. In order to check this, we consider the intersection of the attractor in its four-dimensional phase space with a three-dimensional hyperplane containing the fixed points $\mathbf{Z}_*$ that we wish to check. Numerically we approximate the hyperplane by a very narrow slab through each fixed point of the form

$$|\hat{\boldsymbol{K}}^{\mathrm{T}}(\boldsymbol{Z} - \boldsymbol{Z}_*)| < w . \tag{4.9}$$

Actually we took the slabs parallel to the plane (x_1, x_2) which implies that each slab contains the four fixed points with the same rotation number. We then examine a very long orbit and plot only those points satisfying (4.9). The intersection of our 2.8-dimensional attractor with a three dimensional hyperplane is a 1.8-dimensional cross-section. The small scale structure of this 1.8-dimensional intersection is somewhat fuzzed out due to the finite slab thickness. The results, for $f_0 = 9.0$, are given in figs. 10a–10e, which refer to the rotation numbers $\boldsymbol{N} = (0,0)$, $(1,2)$, $(0,1)$, $(1,1)$ and $(2,3)$, respectively. In these figures the relevant fixed points are denoted by a + symbol. The results indicate, with different degrees of certitude, that the first four sets of fixed points are indeed embedded in the attractor while the fifth is not. Note that fig. 10a nicely reveals the symmetry of the map with respect to the point $(\pi, \pi, 0, 0)$. Note also the fractal-like structure in this figure.

We conclude this discussion by mentioning what seems to be an interesting issue: the loss of hyperbolicity due to the existence of fixed points embedded in the attractor that have a number of unstable directions (that is, eigenvalues with magnitude bigger than one) different from the number of unstable directions of the attractor (that is, positive Lyapunov exponents). In fact, from the observation of fig. 7 and table 2, we see that while the chaotic attractor for $f_0 = 9.0$ has two positive Lyapunov exponents some of the unstable fixed points embedded in the attractor have only one unstable eigenvalue.

4.4. Control

We now proceed to apply the method developed in section 2 to control the fixed points of the double rotor map with control parameter f_0. Let us denote by $\bar{\boldsymbol{Z}}_*$ the fixed point to be controlled at the nominal value $\bar{f}_0$ of the parameter. The quantities that were introduced in section 2 now take the following particular form:

$$\mathbf{A} = \begin{pmatrix} \mathbf{I}_2 & \mathbf{M} \\ \mathbf{H}(\bar{X}_*) & \mathbf{L} + \mathbf{H}(\bar{X}_*)\mathbf{M} \end{pmatrix},$$

$$\mathbf{H}(\bar{X}_*) = \frac{\bar{f}_0}{I} \begin{pmatrix} l_1 \cos \bar{x}_{1*} & 0 \\ 0 & l_2 \cos \bar{x}_{2*} \end{pmatrix},$$

$$\boldsymbol{B}^{\mathrm{T}} = \begin{pmatrix} 0 & 0 & \frac{l_1}{I} \sin \bar{x}_{1*} & \frac{l_2}{I} \sin \bar{x}_{2*} \end{pmatrix},$$

$$\mathbf{C} = (\boldsymbol{B} \vdots \mathbf{A}\boldsymbol{B} \vdots \mathbf{A}^2\boldsymbol{B} \vdots \mathbf{A}^3\boldsymbol{B}),$$

$$\mathbf{T} = \mathbf{C}\mathbf{W},$$

$$\mathbf{W} = \begin{pmatrix} a_3 & a_2 & a_1 & 1 \\ a_2 & a_1 & 1 & 0 \\ a_1 & 1 & 0 & 0 \\ 1 & 0 & 0 & 0 \end{pmatrix},$$

$$\boldsymbol{K}^{\mathrm{T}} = (\alpha_4 - a_4 \quad \alpha_3 - a_3 \quad \alpha_2 - a_2 \quad \alpha_1 - a_1)\mathbf{T}^{-1} .$$

One immediate conclusion that can be drawn from these results is that the controllability matrix $\mathbf{C}$ is identically zero in the case of the fixed points with rotation numbers $\boldsymbol{N} = (0,0)$ for which $\sin \bar{x}_{1*} = \sin \bar{x}_{2*} = 0$. Hence these points are uncontrollable, at least when the control parameter is $\bar{f}_0$. We will show in the next subsection that this set of fixed points can be controlled if we modify the double rotor map to allow for

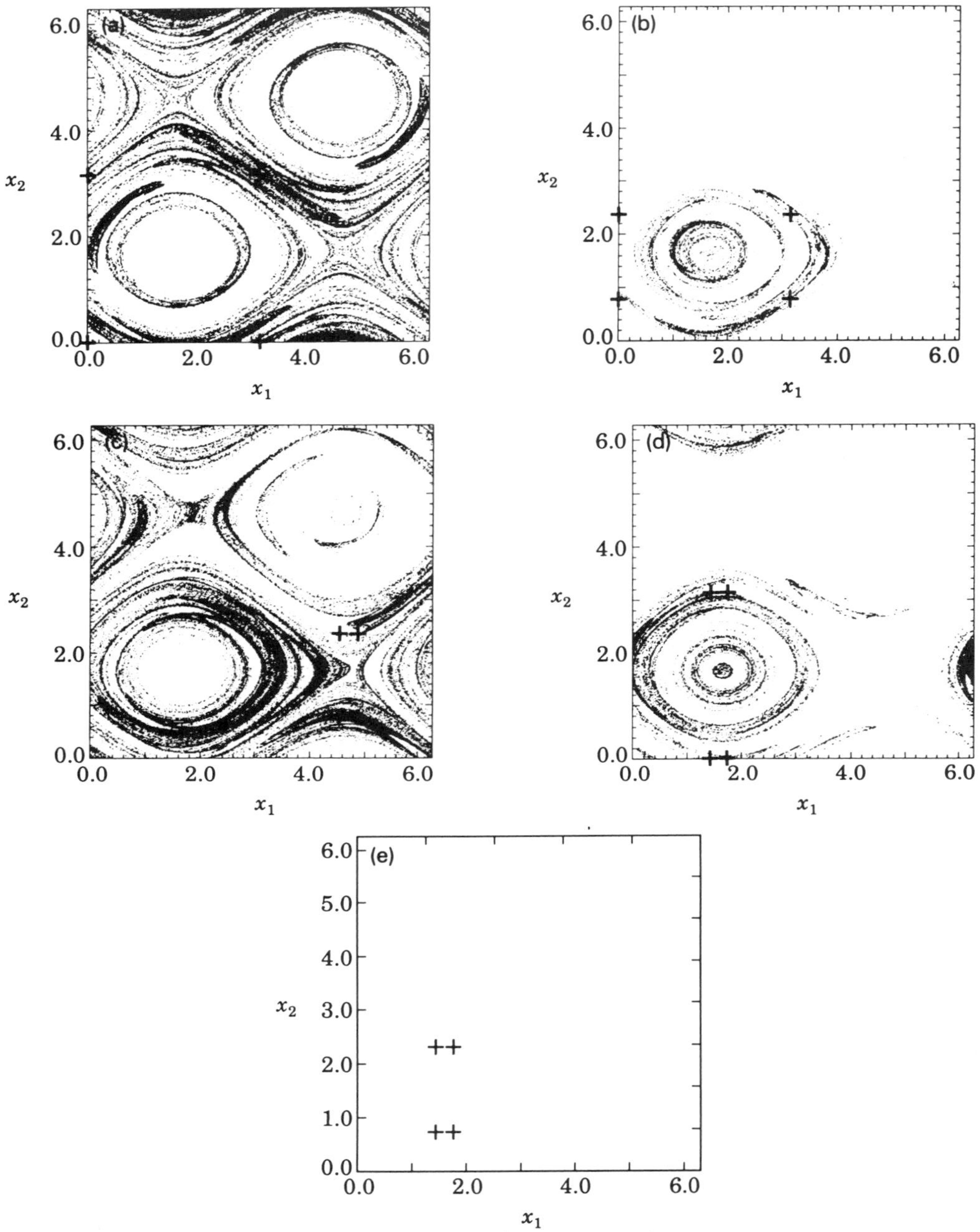

Fig. 10. Double rotor map: sections of chaotic attractor by slab $|\hat{\boldsymbol{K}}^{\mathrm{T}}(\boldsymbol{Z} - \boldsymbol{Z}_*)| < w$, $\hat{\boldsymbol{K}}^{\mathrm{T}} = (0, 0, 1, 1)$, $w = 10^{-2}$, through the fixed points (+) with rotation numbers (a) $\boldsymbol{N} = (0, 0)$, (b) $\boldsymbol{N} = (1, 2)$, (c) $\boldsymbol{N} = (0, 1)$, (d) $\boldsymbol{N} = (1, 1)$, (e) $\boldsymbol{N} = (2, 3)$. The map was iterated 10^8 times [$f_0 = 9.0$, eq. (4.4)].

kicks with variable direction and then take as control parameter the angle the kicks make with the vertical direction in fig. 5.

The method is illustrated in figs. 11a, 11b. The control of the first fixed point was turned on at $i=0$ with switches to control other fixed points occurring at later times. We plot the x_1 and x_2 coordinates of an orbit as a function of (discrete) time. The parameter perturbations were programmed to control successively four different fixed points of the set with rotation numbers $N = \pm(0, 1)$. The times at which we switched the control from stabilizing one fixed point to stabilizing another are labeled by the arrows in the figure. The figure clearly illustrates the flexibility offered by the method in controlling different periodic motions embedded in the attractor. The figure also shows that the time to achieve control is different from case to case.

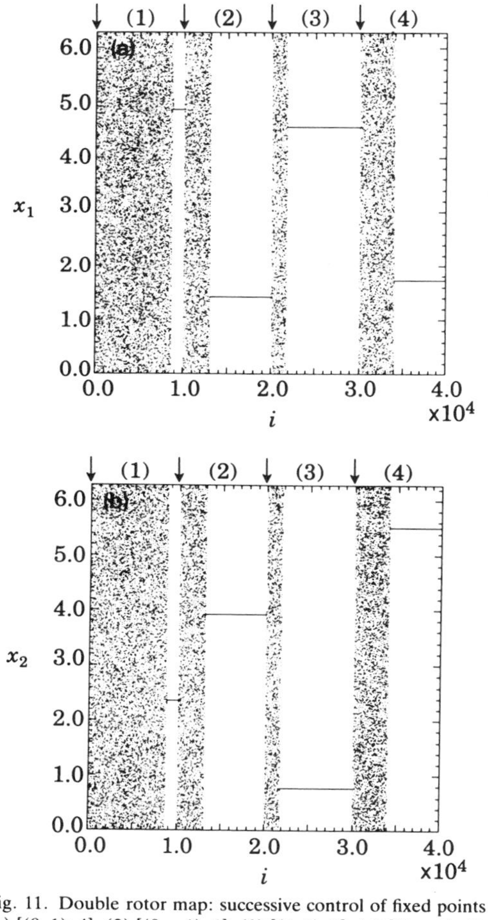

Fig. 11. Double rotor map: successive control of fixed points (1) [(0, 1); 4], (2) [(0, −1); 1], (3) [(0, 1); 1], (4) [(0, −1); 4]. The arrows indicate the times of switching. The regulator poles correspond to projection onto the stable manifold [$\delta = 1.0$, $\bar{f}_0 = 9.0$, eq. (4.4)].

We now report the results of several numerical experiments that were carried out with the purpose of understanding the behavior of the time to achieve control.

The first experiment was intended to confirm that the time to achieve control indeed follows an exponential probability distribution as indicated in section 2.4. We proceeded to control the fixed point [(0, 1); 4] by starting at a large number of different points on the attractor and measuring the time each orbit took to reach the fixed point. We then obtained the distribution function of the time to achieve control $\boldsymbol{\phi}(\tau)$ by plotting a histogram of τ using bins of constant size. The results are presented as a semilog plot in fig. 12 and show excellent agreement with the predicted fit to a straight line.

In our next experiment we looked at the dependence of the average time to achieve control on the size of the parameter perturbations, δ. The results are shown in fig. 13, where we have used logarithmic scales in both axes. The two fixed points [(0, 1); 4] and [(0, 1); 1] were controlled. (The first of these points has two unstable eigenvalues while the second has only one unstable eigenvalue.) We see that for the smaller values of δ the results closely follow straight lines indicating a power law dependence,

$$\langle \tau \rangle \sim \delta^{-\gamma},$$

in accord with the theoretical predictions of ref. [6] for two-dimensional maps.

In the experiments described until this point the choice of the regulator poles (eigenvalues of

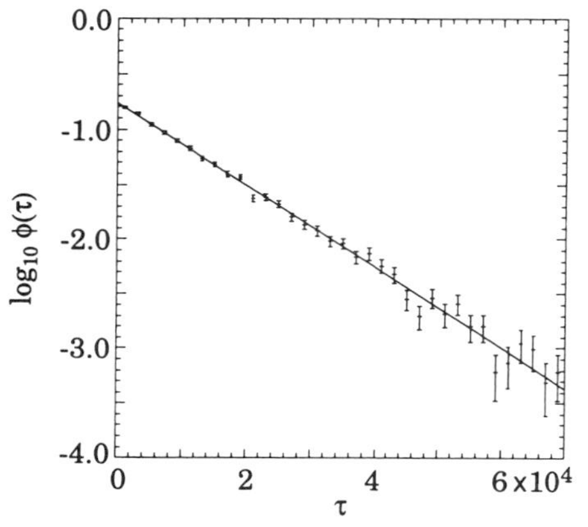

Fig. 12. Double rotor map: histogram of the time to achieve control τ of a sample of 8192 orbits. The fixed point controlled was [(0, 1); 4]. The regulator poles correspond to projection onto the stable manifold. [$\delta = 1.0$, $\bar{f}_0 = 9.0$, eq. (4.4)].

$\mathbf{A} - \boldsymbol{BK}^{\mathrm{T}}$) corresponded to projection onto the stable manifold of the fixed points. That is, the stable eigenvalues of matrix $\mathbf{A}$ were left unchanged, and the unstable eigenvalues were shifted to zero.

In our next set of experiments we looked at how different choices of regulator poles affect the average time to achieve control. We considered the fixed point [(0, 1); 4] with two unstable eigendirections and kept two of the regulator poles equal to the two stable eigenvalues of the fixed point. As regards the other two regulator poles, μ_1 and μ_2, three cases were considered:

$$\text{(I)} \quad \mu_2 = 0\,,$$
$$\text{(II)} \quad \mu_2 = \mu_1\,,$$
$$\text{(III)} \quad \mu_2 = -\mu_1\,.$$

μ_1 was then allowed to vary in the interval $(-1, 1)$. The results of the experiments are shown in fig. 14. In cases (I) and (II) the average time to achieve control essentially increases with μ_1, indicating behavior similar to that found for the Hénon map in fig. 2. In case (III) the average time to achieve control passes through a broad minimum. (Note that the point $\mu_1 = \mu_2 = 0$, which is common to the three cases, corresponds to projection onto the stable manifold.)

4.5. f_0-uncontrollable fixed points

We saw that the set of four fixed points with rotation numbers $N = (0, 0)$ could not be con-

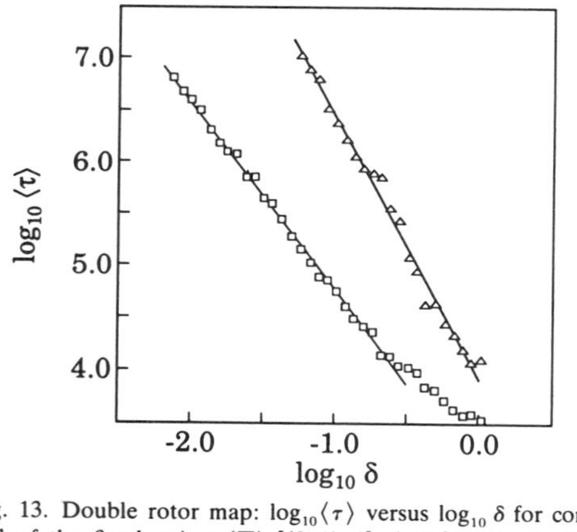

Fig. 13. Double rotor map: $\log_{10}\langle\tau\rangle$ versus $\log_{10}\delta$ for control of the fixed points (□) [(0, 1); 1], (△) [(0, 1); 4]. The regulator poles correspond to projection onto the stable manifold. The straight lines are least square fits to the data [excluding the last nine data points in the case of (□)], [$\bar{f}_0 = 9.0$, eq. (4.4)].

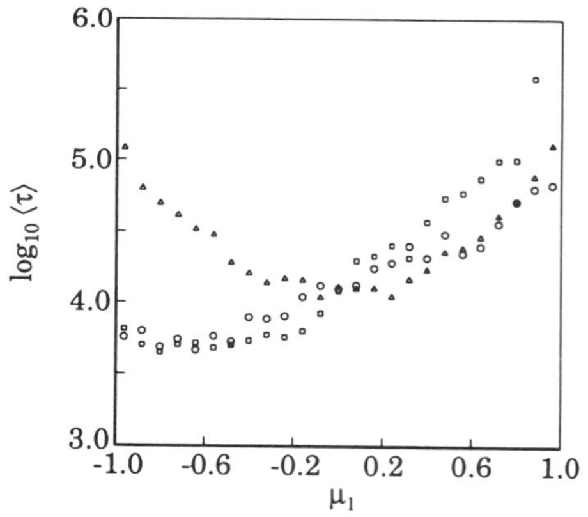

Fig. 14. Double rotor map: $\log_{10}\langle\tau\rangle$ versus μ_1 for (○) $\mu_2 = 0$, (□) $\mu_2 = \mu_1$, (△) $\mu_2 = -\mu_1$. The other two regulator poles were kept equal to the stable eigenvalues of the uncontrolled map. The fixed point controlled was [(0, 1); 4] [$\delta = 1.0$, $\bar{f}_0 = 9.0$, eq. (4.4)].

trolled by changes in the parameter f_0 because the controllability matrix at these points is identically zero.

We show now that these fixed points can be controlled by modifying the double rotor map to allow for kicks with variable direction and then taking the direction of the kicks to be the control parameter, with the nominal value corresponding to the previously fixed direction.

Let us assume that the direction of the kicks makes an angle ψ with the vertical downward direction. Going back to the derivation of the double rotor map in Appendix B, it is easy to verify that the introduction of kicks with variable direction can be taken into account by simply replacing the function $\boldsymbol{G}$ used in the definition of the map and given by eq. (4.2) by the new function

$$\tilde{G}(X) = \begin{pmatrix} c_1 \sin(x_1 - \psi) \\ c_2 \sin(x_2 - \psi) \end{pmatrix} .$$

Taking ψ to be the control parameter with variations around the nominal value $\bar{\psi} = 0$, the application of the method now involves the following quantities:

$$\mathbf{A} = \begin{pmatrix} \mathbf{I}_2 & \mathbf{M} \\ \tilde{\mathbf{H}}(\bar{X}_*) & \mathbf{L} + \tilde{\mathbf{H}}(\bar{X}_*)\mathbf{M} \end{pmatrix},$$

$$\tilde{\mathbf{H}}(\bar{X}_*) = \frac{f_0}{I} \begin{pmatrix} l_1 \cos x_{1*} & 0 \\ 0 & l_2 \cos \bar{x}_{2*} \end{pmatrix},$$

$$\boldsymbol{B}^{\mathrm{T}} = \begin{pmatrix} 0 & 0 & -\frac{f_0 l_1}{I} \cos \bar{x}_{1*} & -\frac{f_0 l_2}{I} \cos \bar{x}_{2*} \end{pmatrix}.$$

The fixed points are now all controllable by small perturbations of the parameter ψ around the nominal value $\bar{\psi} = 0$.

Figs. 15a, 15b illustrate the control of the fixed points [(0, 0); 3] and [(0, 0); 4] by kicks of variable direction. The parameter perturbations were programmed to control the first of these points from $i = 0$ to $i = 10^4$ and the second from $i = 10^4$ to $i = 2 \times 10^4$.

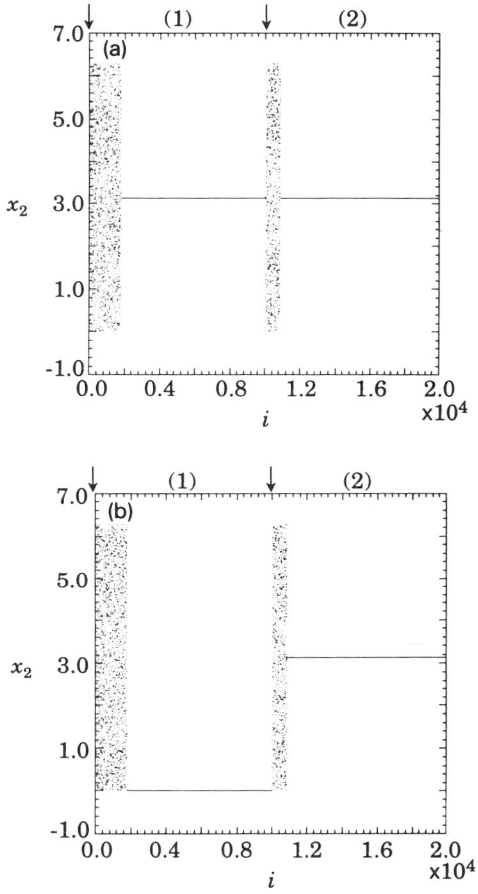

Fig. 15. Modified double rotor map: successive control of fixed points (1) [(0, 0); 3], (2) [(0, 0); 4] by kicks of variable direction. The arrows indicate the times of switching. The regulator poles correspond to projection onto the stable manifold [$\delta = 0.05$, $f_0 = 9.0$, eq. (4.4)].

5. Discussion

The transient phase where the orbit wanders chaotically before locking in to a controlled orbit can be greatly shortened by applying the technique discussed by Shinbrot et al. [25]. In the latter paper it was pointed out that orbits can be rapidly brought to a target region on the attractor (in the present case the neighborhood of the periodic orbit which we wish to stabilize) by using small control perturbations when the orbit is far from the neighborhood of the periodic orbit to be stabilized. The idea was that, since chaotic systems are exponentially sensitive to perturbations, careful choice of even small control perturbations can, after some time, have a large effect on the orbit location and can be used to guide it. Thus the time to achieve control can, in principle, be greatly shortened by properly applying small controls when the orbit is far from the neighborhood of the desired periodic orbit.

One issue which we have not addressed is the effect of noise. If the noise remains small, it may not be sufficient to kick the orbit out of the neighborhood of the chosen periodic orbit where the control is activated. In this case, the orbit remains near the desired periodic orbit indefinitely. However, it may be that the random noise is such that it may occasionally kick the orbit far enough away from the periodic orbit that the orbit falls outside the small controlled phase space region. In this case, after the orbit is kicked out of the controlled phase space region, it wanders chaotically over the attractor until it falls in the controlled region again. Thus there are epochs where the orbit is kept near the desired orbit interspersed with epochs wherein the orbit wanders chaotically far from the desired orbit. If the latter are, on average, relatively much shorter than the former, then one might still regard the control as being effective. See ref. [6] for numerical experiments on this effect using the Hénon map. We also remark that the procedure discussed in the previous paragraph [25] can be used to greatly reduce the duration of the noise induced epochs where the orbit bursts out of the controlled phase space region.

In this paper we have considered the case where there is only a single control parameter available for adjustment. While generically a single parameter is sufficient for stabilization of a desired periodic orbit, there may be some advantage to utilizing several control variables. Therefore, the single control parameter p becomes a vector (e.g., ref. [26] discusses the case where the number of control parameters is equal to the number of unstable eigenvalues). In particular, the added freedom in having several control parameters might allow better means of choosing the control so as to minimize the time to achieve control, as well as the effects of noise.

Finally we wish to point out that full knowledge of the system dynamics is not necessary in order to apply our technique (see also ref. [6]). In particular, we only require the location of the desired periodic orbit, the linearized dynamics about the periodic orbit, and the dependence of the location of the periodic orbit on small variation of the control parameter. Recently, delay coordinate embedding [19, 27] has been utilized in several experimental studies (refs. [8, 28–31]) to extract such information purely from observations of experimental chaotic orbits on the attractor without any a priori knowledge of the system of equations governing the dynamics, and such information has been utilized to control periodic orbits [9]. Hence, application of our method is not limited to cases where a complete knowledge of the system is available.

In conclusion, we have demonstrated that chaotic dynamics can often be converted, by using only a small feedback control, to motion on a desired periodic orbit. Furthermore, by switching the small control, one can switch the time asymptotic behavior from one periodic orbit to another. In some situations, where the flexibility offered by the ability to do such switching is desirable, it may be advantageous to design the system so that it is chaotic. In other situations, where one is presented with a chaotic system, the method may allow one to eliminate the chaos and achieve greatly improved behavior at relatively low cost.

Acknowledgements

This work was supported by the US Department of Energy (Scientific Computing Staff, Office of Energy Research), the Portuguese Junta Nacional de Investigação Científica e Tecnológica, and the National Science Foundation (Engineering Research Center Program). The computation was done at the National Energy Research Supercomputer Center.

Appendix A. Time to achieve control in the case of two-dimensional maps

We assume that control is achieved if the orbit remains in the slab (2.8) for two consecutive iterations of the map. The two conditions

$$|\boldsymbol{K}^{\mathrm{T}}(\boldsymbol{Z}-\boldsymbol{Z}_*(\bar{p}))|<\delta\,,\quad |\boldsymbol{K}^{\mathrm{T}}(\boldsymbol{Z}'-\boldsymbol{Z}_*(\bar{p}))|<\delta\,, \tag{A.1}$$

define a control "parallelepiped" P_{c}, where $\boldsymbol{Z}'=\boldsymbol{F}(\boldsymbol{Z},p)$. For small δ, an initial condition will bounce around on the set comprising the uncontrolled chaotic attractor for a long time before it falls in the control parallelepiped P_{c}. At any given iterate the probability of falling in P_{c} is approximately the natural measure (see, for example, [17, 18, 23]) of the uncontrolled chaotic attractor contained in P_{c}. If we follow many orbits this probability $\mu(P_{\mathrm{c}})$ also gives the rate at which these orbits fall into P_{c}. Thus $\mu(P_{\mathrm{c}})$ is the inverse of the average time for a typical orbit to first fall in P_{c},

$$\langle\tau\rangle^{-1}=\mu(P_{\mathrm{c}})\,. \tag{A.2}$$

An estimate for $\mu(P_{\mathrm{c}})$ can be given in the two-dimensional case [23]:

$$\mu(P_{\mathrm{c}})\sim\int_{P_{\mathrm{c}}}\rho|v_{\mathrm{s}}|^{d_{\mathrm{s}}-1}|v_{\mathrm{u}}|^{d_{\mathrm{u}}-1}\,\mathrm{d}v_{\mathrm{s}}\,\mathrm{d}v_{\mathrm{u}}\,, \tag{A.3}$$

where v_{s} and v_{u} denote linear coordinates in the stable and unstable directions. In here d_{u} and d_{s} are the pointwise dimensions [1] for the uncontrolled chaotic attractor at the fixed point in the unstable and the stable directions, respectively; ρ is a normalizing constant. Assuming that the attractor is smooth in the unstable direction we have $d_{\mathrm{u}}=1$, while d_{s} is given in terms of the eigenvalues at the fixed point [1, 17, 18] by

$$d_{\mathrm{s}}=\frac{\log_{\mathrm{e}}|\lambda_{\mathrm{u}}|}{\log_{\mathrm{e}}(1/|\lambda_{\mathrm{s}}|)}\,.$$

In order to determine the control parallelepiped, we need to obtain $\boldsymbol{Z}'$ in the neighborhood of the fixed point $\boldsymbol{Z}_*(\bar{p})$ with a better approximation than that provided by the linear map (2.3). We therefore take

$$\boldsymbol{V}'=\mathbf{A}\boldsymbol{V}+\boldsymbol{B}(p-\bar{p})+\tfrac{1}{2}\boldsymbol{Q}(\boldsymbol{V},\boldsymbol{V})+\tfrac{1}{2}\boldsymbol{D}(p-\bar{p})^2\,, \tag{A.4}$$

where

$$\boldsymbol{V}=\boldsymbol{Z}-\boldsymbol{Z}_*(\bar{p})\,,\quad \boldsymbol{V}'=\boldsymbol{Z}'-\boldsymbol{Z}_*(\bar{p})\,,$$

$\mathbf{A}$, $\boldsymbol{B}$ were defined by (2.4), (2.5) and $\boldsymbol{Q}$, $\boldsymbol{D}$ are two vectors with components q_k, d_k $(k=1,2)$ defined by

$$q_k=\sum_{i,j=1}^{2}\left.\frac{\partial^2 f_k}{\partial x_i\,\partial x_j}\right|_{(\boldsymbol{Z}_*(\bar{p}),\,\bar{p})}v_iv_j\,,$$

$$d_k=\left.\frac{\partial^2 f_k}{\partial p^2}\right|_{(\boldsymbol{Z}_*(\bar{p}),\bar{p})}\,;$$

in here x_k, f_k, and v_k $(k=1,2)$ denote the components of the vectors $\boldsymbol{X}$, $\boldsymbol{F}$ and $\boldsymbol{V}$. Using (2.6) to eliminate $p-\bar{p}$ from eq. (A.4), we obtain

$$\boldsymbol{V}'=\mathbf{A}\boldsymbol{V}+\tfrac{1}{2}\boldsymbol{Q}(\boldsymbol{V},\boldsymbol{V})-\boldsymbol{B}(\boldsymbol{K}^{\mathrm{T}}\boldsymbol{V})+\tfrac{1}{2}\boldsymbol{D}(\boldsymbol{K}^{\mathrm{T}}\boldsymbol{V})^2\,. \tag{A.5}$$

The control "parallelogram" P_{c} will therefore be defined by the two equations

$$|\boldsymbol{K}^{\mathrm{T}}\boldsymbol{V}|<\delta\,,\quad |\boldsymbol{K}^{\mathrm{T}}\boldsymbol{V}'|<\delta\,. \tag{A.6}$$

In order to compare with the numerical experimental results described in section 3 we have carried out the calculation of (A.3) in the case of the Hénon map. Writing this map in the form

$$\begin{pmatrix} x_1 \\ x_2 \end{pmatrix} \mapsto \begin{pmatrix} f_1(x_1, x_2) \\ f_2(x_1, x_2) \end{pmatrix} = \begin{pmatrix} a - x_1^2 + bx_2 \\ x_1 \end{pmatrix},$$

and taking a to be the control parameter while b is kept fixed, we obtain

$$\mathbf{A} = \begin{pmatrix} -2\bar{x}_* & b \\ 1 & 0 \end{pmatrix}, \quad \boldsymbol{B} = \begin{pmatrix} 1 \\ 0 \end{pmatrix},$$

$$\boldsymbol{D} = \begin{pmatrix} 0 \\ 0 \end{pmatrix}, \quad \boldsymbol{Q}(\boldsymbol{V}, \boldsymbol{V}) = \begin{pmatrix} -2v_1^2 \\ 0 \end{pmatrix},$$

$$\boldsymbol{K}^{\mathrm{T}}\boldsymbol{V} = k_1 v_1 + k_2 v_2 ,$$

$$\boldsymbol{K}^{\mathrm{T}}\boldsymbol{V}' = -k_1 v_1^2 + (k_2 - k_1^2 - 2\bar{x}_* k_1)v_1 + k_1(b - k_2)v_2 .$$

Also we note that for the Hénon map the variables (v_1, v_2) and (v_s, v_u) are related by

$$\begin{pmatrix} v_1 \\ v_2 \end{pmatrix} = \begin{pmatrix} \lambda_s\gamma_s & \lambda_u\gamma_u \\ \gamma_s & \gamma_u \end{pmatrix} \begin{pmatrix} v_s \\ v_u \end{pmatrix}, \qquad (A.7)$$

where

$$\gamma_s = (1 + \lambda_s^2)^{-1/2}, \quad \gamma_u = (1 + \lambda_u^2)^{-1/2} .$$

Letting

$$\xi = v_1 , \quad \eta = v_1 + tv_2 , \quad t = k_2/k_1 ,$$

and using (A.7) to change the variables of integration, eq. (A.3) can be written in the form

$$\mu(P_c) \sim \rho_0 \frac{|r|}{|1 - r|^{d_s}} \int_{P_c} |\xi - r\eta|^{d_s - 1}\, d\xi\, d\eta , \qquad (A.8)$$

where

$$\frac{1}{r} = 1 + \frac{t}{\lambda_u}$$

and

$$\frac{1}{\rho_0} = \frac{1}{\rho} \gamma_u |\lambda_u| [\gamma_s(\lambda_s - \lambda_u)]^{d_s} .$$

The integration in the variable in the direction of the straight lines $\boldsymbol{K}^{\mathrm{T}}\boldsymbol{V} = \pm\delta$ can be done exactly. On the contrary, except in the case $k_1 = 0$, the integration in the other variable does not seem to be possible in closed form. We have therefore resorted to numerical integration to obtain the results presented in fig. 16 (see below).

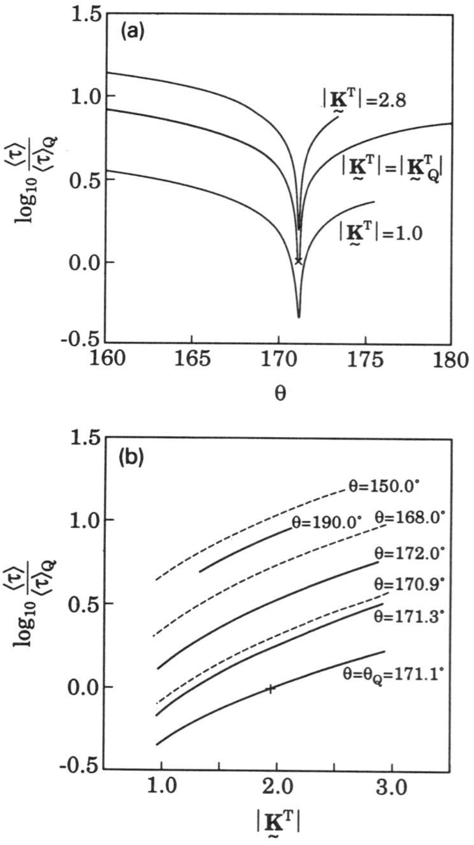

Fig. 16. Hénon map: (theoretical) curves of $\log_{10}(\langle\tau\rangle/\langle\tau\rangle_Q)$ versus $\theta = \arg(\boldsymbol{K}^{\mathrm{T}})$ with $|\boldsymbol{K}^{\mathrm{T}}|$ fixed (a) and of $\log_{10}(\langle\tau\rangle/\langle\tau\rangle_Q)$ versus $|\boldsymbol{K}^{\mathrm{T}}|$ with θ fixed (b). The $\times$ (+) denotes the reference value.

An accurate analytical approximation can be obtained in the case $t = -\lambda_s$ (slab parallel to the stable manifold) in the limit $\delta \to 0$:

$$\mu(P_c) \sim \rho_1 g(\mu_1)\, \delta^{\gamma}[1 + 0(\delta)]\,, \tag{A.9}$$

where g is the function defined by

$$g(\mu_1) = \frac{2}{\gamma}\, \frac{(1+\mu_1)^{\gamma} - (1-\mu_1)^{\gamma}}{\mu_1(|\lambda_u| + \mu_1)^{\gamma}}\,,$$

$$\mu_{1m} < \mu_1 < 0 \quad 0 < \mu_1 < \mu_{1M}$$

$$g(0) = 4|\lambda_u|^{-\gamma}\,.$$

and

$$\gamma = \tfrac{1}{2} d_s + 1\,,$$

$$\rho_1 = \frac{\rho_0}{d_s}\, \frac{|r|}{|1-r|^{d_s}}\,, \quad \frac{1}{r} = 1 - \frac{\lambda_s}{\lambda_u}\,,$$

$$\mu_{1m} = -1 + \frac{r^2}{|\lambda_u| - 1}\, \delta + \mathcal{O}(\delta^1)\,,$$

$$\mu_{1M} = 1 - \frac{r^2}{|\lambda_u| + 1}\, \delta + \mathcal{O}(\delta^2)\,.$$

The dependence of $\mu(P_c)$ on δ given by (A.9) is precisely that predicted in ref. [6]. The dependence of $\mu(P_c)$ on μ_1 shows very good agreement with the experimental results – see fig. 2. Note that in plotting the theoretical curve we used as normalizing constant ρ_1 that obtained by least square fitting the theoretical curve to the experimental points.

We have used eq. (A.8) to study the dependence of $\langle \tau \rangle = 1/\mu(P_c)$ on the gain vector $\boldsymbol{K}^{\mathrm{T}}$. In fig. 16a we have plotted curves of $\langle \tau \rangle$ versus $\theta = \arg(\boldsymbol{K}^{\mathrm{T}})$ with $|\boldsymbol{K}^{\mathrm{T}}|$ kept fixed and in fig. 16b curves of $\langle \tau \rangle$ versus $|\boldsymbol{K}^{\mathrm{T}}|$ with θ kept fixed. $\langle \tau \rangle$ was normalized to its value at the point Q (see section 3 and fig. 1). The results show that $\langle \tau \rangle$ exhibits a strong minimum at $\theta = \theta_Q$ for all values of $|\boldsymbol{K}^{\mathrm{T}}|$ and increases slowly with $|\boldsymbol{K}^{\mathrm{T}}|$ for all values of θ, in agreement with the experimental results of figs. 3 and 4.

Appendix B. Numerical method for calculating $\langle \tau \rangle$

In this appendix we describe the procedure used in sections 3 and 4 to numerically obtain the average time to achieve control, $\langle \tau \rangle$.

From (2.10) we obtain the fraction of chaotic transients with length smaller than some value $\tau_{\max}$,

$$\rho_{\tau_{\max}} = \int_0^{\tau_{\max}} \phi(\tau)\, \mathrm{d}\tau = 1 - \exp\left(-\frac{\tau_{\max}}{\langle \tau \rangle}\right),$$

and the average length of the chaotic transients with length smaller than $\tau_{\max}$,

$$\langle \tau \rangle_{\tau_{\max}} = \int_0^{\tau_{\max}} \tau \phi(\tau)\, \mathrm{d}\tau = \langle \tau \rangle \left[1 - \left(1 + \frac{\tau_{\max}}{\langle \tau \rangle}\right) \exp\left(-\frac{\tau_{\max}}{\langle \tau \rangle}\right)\right].$$

Combining these two equations we obtain

$$\langle \tau \rangle = \frac{\langle \tau \rangle_{\tau_{\max}}}{1 - (1 - \rho_{\tau_{\max}})[1 - \log_e(1 - \rho_{\tau_{\max}})]}\,, \tag{B.1}$$

which is the required formula. Note that $\rho_\infty = 1$, $\langle \tau \rangle_\infty = \langle \tau \rangle$.

The numerical procedure to calculate the average time to achieve control is as follows. Take a large number N_0 of randomly chosen initial conditions and iterate each of them with the uncontrolled map [i.e., with $\boldsymbol{Z} \mapsto \boldsymbol{F}(\boldsymbol{Z}, \bar{p})$] a sufficient number of times until they are all distributed over the attractor according to its natural measure. Then switch on the control as specified by (2.9) and determine how many further iterates are necessary for $N_f \leq N_0$ orbits to fall within a circle of small radius centered at the fixed point. Letting $\tau_{\max}$ be this number of iterates and $\{\tau_j\}$ with $j = 1, \ldots, N_f$ be the times required for the N_f orbits to fall within the small circle, we have

$$\rho_{\tau_{\max}} = \frac{N_f}{N_0}\,, \quad \langle \tau \rangle_{\tau_{\max}} = \frac{1}{N_f} \sum_{j=1}^{N_f} \tau_j\,.$$

Finally we use eq. (B.1) to obtain $\langle \tau \rangle$. In our numerical experiments described in sections 3 and 4, we took $N_0 = 192$, $N_f = 121$, values that led to a good compromise between accuracy and computation time.

Appendix C. Derivation of the double rotor map

The equations of motion of the kicked double rotor are

$$\frac{d}{dt}\left(\frac{\partial L}{\partial \dot\theta_j}\right) - \frac{\partial L}{\partial \theta_j} = -\frac{\partial F}{\partial \dot\theta_j}\,, \quad j = 1, 2\,, \tag{C.1}$$

where the Lagrangian function L is the difference between the kinetic energy,

$$K(\dot\theta_1, \dot\theta_2) = \tfrac{1}{2} I_1 \dot\theta_1^2 + \tfrac{1}{2} I_2 \dot\theta_2^2\,,$$

and the potential energy,

$$V(\theta_1, \theta_2, t) = (l_1 \cos\theta_1 + l_2 \cos\theta_2) f(t)\,,$$

i.e., $L = T - V$, and Rayleigh's dissipation function F is

$$F(\dot\theta_1, \dot\theta_2) = \tfrac{1}{2}\nu_1 I_1 \dot\theta_1^2 + \tfrac{1}{2}\nu_2 I_2 (\dot\theta_2 - \dot\theta_1)^2\,.$$

The sequence of forcing kicks is given by the semi-infinite comb of delta functions of period T and strength f_0,

$$f(t) = f_0 \sum_{k=1}^{\infty} \delta(t - kT)\,. \tag{C.2}$$

In here I_1 and I_2 are the moments of inertia,

$$I_1 = (m_1 + m_2) l_1^2\,, \quad I_2 = m_2 l_2^2\,,$$

and ν_1, ν_2 are the coefficients of friction.

Elimination of L and F from (C.1) yields

$$\frac{d}{dt}\begin{pmatrix}\dot\theta_1\\ \dot\theta_2\end{pmatrix} = \begin{pmatrix} -(\nu_1 + \nu_2 I_2/I_1) & \nu_2 I_2/I_1 \\ \nu_2 & -\nu_2 \end{pmatrix}\begin{pmatrix}\dot\theta_1\\ \dot\theta_2\end{pmatrix} + f(t)\begin{pmatrix}(l_1/I_1)\sin\theta_1\\ (l_2/I_2)\sin\theta_2\end{pmatrix}. \tag{C.3}$$

We now proceed to integrate eq. (C.3). For simplicity we take $I_1 = I_2 \equiv I$.

Since the effect of the kicks is instantaneous (i.e., $f(t) = 0$, for $t \neq kT$, $k = 1, 2, \ldots$) eqs. (C.3) are linear between successive kicks. In particular, for $0 < t < T$, eqs. (C.3) reduce to

$$\frac{d}{dt}\begin{pmatrix}\dot\theta_1\\ \dot\theta_2\end{pmatrix} = \mathbf{A}_\nu \begin{pmatrix}\dot\theta_1\\ \dot\theta_2\end{pmatrix},$$

$$\mathbf{A}_\nu = \begin{pmatrix} -(\nu_1 + \nu_2) & \nu_2 \\ \nu_2 & -\nu_2 \end{pmatrix}. \tag{C.4}$$

This system can be easily solved by the usual methods for linear differential equations with constant coefficients. Denoting by $\dot\theta_1(0)$, $\dot\theta_2(0)$ the initial angular velocities this solution is

$$\begin{pmatrix}\dot\theta_1(t)\\ \dot\theta_2(t)\end{pmatrix} = \mathbf{L}(t)\begin{pmatrix}\dot\theta_1(0)\\ \dot\theta_2(0)\end{pmatrix}, \tag{C.5}$$

where

$$\mathbf{L}(t) = \sum_{j=1}^{2} \mathbf{W}_j\, e^{\lambda_j t}\,.$$

λ_1, λ_2 are the eigenvalues of matrix $\mathbf{A}_\nu$,

$$\left.\begin{matrix}\lambda_1\\ \lambda_2\end{matrix}\right\} = -\tfrac{1}{2}(\nu_1 + 2\nu_2 \pm \Delta)\,,$$

$$\Delta = (\nu_1^2 + 4\nu_2^2)^{1/2}\,,$$

and $\mathbf{W}_1$, $\mathbf{W}_2$ are the constant matrices

$$\mathbf{W}_1 = \begin{pmatrix} a & b \\ b & d \end{pmatrix}, \quad \mathbf{W}_2 = \begin{pmatrix} d & -b \\ -b & a \end{pmatrix},$$

where

$$\left.\begin{matrix}a\\ d\end{matrix}\right\} = \tfrac{1}{2}\left(1 \pm \frac{\nu_1}{\Delta}\right), \quad b = -\frac{\nu_2}{\Delta}\,.$$

The position of the rods is obtained by integration of eq. (C.5). Denoting by $\theta_1(0)$, $\theta_2(0)$ the initial positions one obtains

$$\begin{pmatrix}\theta_1(t)\\ \theta_2(t)\end{pmatrix} = \mathbf{M}(t)\begin{pmatrix}\dot\theta_1(0)\\ \dot\theta_2(0)\end{pmatrix} + \begin{pmatrix}\theta_1(0)\\ \theta_2(0)\end{pmatrix}, \tag{C.6}$$

where

$$\mathbf{M}(t) = \int_0^t \mathbf{L}(\xi)\,\mathrm{d}\xi = \sum_{j=1}^{2} \mathbf{W}_j \frac{\mathrm{e}^{\lambda_j t} - 1}{\lambda_j}.$$

Eqs. (C.5), (C.6) completely describe the motion of the rotor for $0 < t < T$ (before the first kick).

At $t = T$ the kick instantaneously changes the angular velocity of each rod but not its position; that is, the angular velocity of each rod is discontinuous at $t = T$, while the position is continuous. Denoting by $\theta_j(T^{\pm})$, $\dot\theta_j(T^{\pm})$, $j = 1, 2$ the values of $\theta_j(t)$, $\dot\theta_j(t)$ just before and just after the kick at $t = T$, we therefore have

$$\theta_j(T^-) = \theta_j(T^+) = \theta_j(T), \tag{C.7}$$

$$\dot\theta_j(T^+) - \dot\theta_j(T^-) = \frac{f_0}{I} l_j \sin\theta_j(T), \tag{C.8}$$

for $j = 1, 2$.

The solution of eqs. (C.3) for $T < t < 2T$ is identical to the solution of the linear system eq. (C.4) for $0 < t < T$ except that the initial conditions $\theta_j(0)$, $\dot\theta_j(0)$, $j = 1, 2$ are replaced by $\theta_j(T)$, $\dot\theta_j(T^+)$, $j = 1, 2$.

The solution of eqs. (C.3) is a composition of the solution of eqs. (C.4) with the effect of the kicks at $t = T, 2T, \ldots$. To study the dynamics of the rotor it is natural to consider only the state of the system immediately after each kick. Thus we obtain from (C.5)–(C.8) the *double rotor map*,

$$\begin{pmatrix}\theta_1^{(k+1)}\\ \theta_2^{(k+1)}\end{pmatrix} = \mathbf{M}(T)\begin{pmatrix}\dot\theta_1^{(k)}\\ \dot\theta_2^{(k)}\end{pmatrix} + \begin{pmatrix}\theta_1^{(k)}\\ \theta_2^{(k)}\end{pmatrix}, \tag{C.9a}$$

$$\begin{pmatrix}\dot\theta_1^{(k+1)}\\ \dot\theta_2^{(k+1)}\end{pmatrix} = \mathbf{L}(T)\begin{pmatrix}\dot\theta_1^{(k)}\\ \dot\theta_2^{(k)}\end{pmatrix} + \frac{f_0}{I}\begin{pmatrix}l_1 \sin\theta_1^{(k+1)}\\ l_2 \sin\theta_2^{(k+1)}\end{pmatrix}, \tag{C.9b}$$

where

$$\theta_j^{(k)} = \theta_j(kT), \quad j = 1, 2$$

are the positions of the rods at the instant of the kth kick, and

$$\dot\theta_j^{(k)} = \dot\theta_j(kT^+), \quad j = 1, 2$$

are the angular velocities of the rods immediately after the kth kick.

Appendix D. Stability of fixed points for the double rotor map

The coefficients of the characteristic equation (4.8) depend on the fixed point and on the forcing f_0 only through the two non-zero elements of the matrix $\mathbf{H}$, which we are going to denote by h_{11} and h_{22}. The discussion of the stability of the fixed points is conveniently carried out in the plane (h_{11}, h_{22}) by considering the intersections between the lines of marginal stability of the characteristic equation (where one of the roots has modulus unity) and the "orbits" described by the paths followed by the fixed points as the forcing f_0 is varied.

The "orbits" of the fixed points can be obtained by first eliminating x_{j*} ($j = 1, 2$) between the two equations

$$f_0 \sin x_{j*} = f_{0j}, \quad h_{jj} = \frac{f_0}{I} l_j \cos x_{j*},$$

with the result

$$h_{jj} = \pm \frac{l_j}{I} (f_0^2 - f_{0j}^2)^{1/2}, \quad j = 1, 2,$$

and then eliminating f_0 between these two equations, with the result

$$\left(\frac{h_{22}}{l_2}\right)^2 - \left(\frac{h_{11}}{l_1}\right)^2 = \frac{1}{I^2}(f_{01}^2 - f_{02}^2),$$

which is the equation of the hyperbola described

by each fixed point in the plane (h_{11}, h_{22}) when f_0 is varied. It should be pointed out that symmetric fixed points with respect to $(x_{1*}, x_{2*}, y_{1*}, y_{2*}) \mapsto (2\pi - x_{1*}, 2\pi - x_{2*}, -y_{1*}, -y_{2*})$ describe the same "orbit." The lines of marginal stability are defined by the equation (see eq. (4.8b))

$$P(e^{i\alpha}) = 0\,, \tag{D.1}$$

where α can take values in the interval $[0, 2\pi)$. When $\alpha = 0$ or $\alpha = \pi$ this equation simplifies considerably. We obtain:

(i) $P(1) = 0 = |\mathbf{HM}| = |\mathbf{H}||\mathbf{M}|$; as $|\mathbf{M}| \neq 0$ this implies

$$h_{11}h_{22} = 0\,.$$

(ii) $P(-1) = 0 = |2(\mathbf{I} + \mathbf{L}) + \mathbf{HM}| = |\mathbf{H} + \mathbf{R}||\mathbf{M}|$, where

$$\mathbf{R} = 2(\mathbf{A}_\nu + 2\mathbf{M}^{-1})\,;$$

writing $\mathbf{R} = \{r_{ij}\}_{i,j=1,2}$ this leads to

$$h_{11}h_{22} + r_{22}h_{11} + r_{11}h_{22} + (r_{11}r_{22} - r_{12}r_{21}) = 0\,.$$

When $\alpha \neq 0, \pi$ it can be shown that the eq. (D.1) has no solutions in the (real) plane (h_{11}, h_{22}); that is, there are no lines of marginal stability with $\alpha \neq 0, \pi$.

In fig. 17 we have plotted in the plane (h_{11}, h_{22}) the lines of marginal stability $P(1) = P(-1) = 0$ (for the parameter values given by eq. (4.4)). The bounded region between these lines, which is the shaded region in the figure, is the only region of the plane where all the roots of the characteristic equation have modulus smaller than unity. We have also plotted the "orbits" of the first five sets of fixed points, the arrows indicating the direction the forcing increases; the critical values of (h_{11}, h_{22}) at $f_0 = f_{0c}$, which occur on the lines $P(1) = 0$, are given in table 1. We see that of all these orbits only two cross the shaded region: one corresponds to the fixed point $[(0,0); 4]$; the other to the fixed points $[(1,2); 4]$ and $[(-1,-2); 1]$. These fixed points are therefore stable while their "orbits" remain in the shaded region and become unstable when the "orbits" cross the line $P(-1) = 0$; that is, they are stable in a finite interval of values of f_0, $f_{0c} < f_0 < f_{0u}$. The other fixed points are unstable for all values of f_0. The values of f_{0u} are also given in table 2.

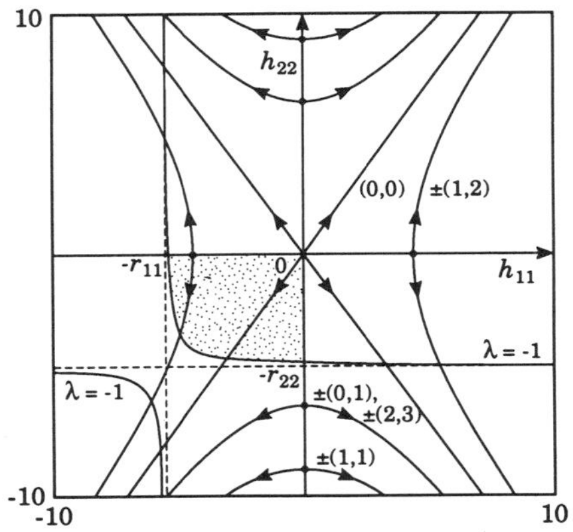

Fig. 17. Double rotor map: stability diagram of the fixed points with rotation numbers (n_1, n_2) $[f_0 = 9.0$, eq. (4.4)].

Fig. 17 only applies for the particular values of the parameters given by eqs. (4.4). For other values the relative positions of the lines of marginal stability and "orbits" of fixed points are different, and fixed points with other rotation numbers may be stable. In general, we can make the following statements regarding the stability of the fixed points: those with rotation numbers $(0,0)$ are stable over an interval $0 < f_0 < f_{0u}^{(0,0)}$; all the others are either stable over an interval $f_{0c}^{(n_1,n_2)} < f_0 < f_{0u}^{(n_1,n_2)}$ or are always unstable.

References

[1] C. Grebogi, E. Ott and J.A. Yorke, Phys. Rev. A 37 (1988) 1711.

[2] D. Auerbach, P. Cvitanovich, J.P. Eckmann, G. Gunaratne and I. Procaccia, Phys. Rev. Lett. 58 (1987) 2387.

[3] T. Morita, H. Hata, H. Mori, T. Horita and K. Tomita, Prog. Theor. Phys. 78 (1987) 511.
[4] A. Katok, Publ. Math. IHES 51 (1980) 137.
[5] R. Bowen, Trans. Am. Math. Soc. 154 (1971) 377.
[6] E. Ott, C. Grebogi and J.A. Yorke, Phys. Rev. Lett. 64 (1990) 1196.
[7] F.J. Romeiras, E. Ott, C. Grebogi and W.P. Dayawansa, Proc. 1991 Am. Control Conf. (American Automatic Control Council, IEEE Service Center, Piscataway, NJ, 1991), pp. 1112–1119.
[8] J. Singer, Y.-Z. Wang and H.H. Bau, Phys. Rev. Lett. 66 (1991) 1123.
[9] W.L. Ditto, S.N. Rauseo and M.L. Spano, Phys. Rev. Lett. 65 (1990) 3211.
[10] T.B. Fowler, IEEE Trans. on Automatic Control 34 (1989) 201.
[11] A.M. Block and J.E. Marsden, Theor. Comput. Fluid Dyn. 1 (1989) 179;
A. Hübler and E. Lüscher, Naturwissenshaften 76 (1989) 67;
E.A. Jackson, Phys. Lett. A 151 (1990) 478;
B. Huberman and E. Lumer, IEEE Trans. Circuits Syst. 37 (1990) 547;
S. Sinha, R. Ramaswamy and J.S. Rao, Physica D 43 (1990) 118;
B. Peng, V. Petrov and K. Showalter, J. Phys. Chem. 95 (1991) 4957;
T. Tél, Controlling transient chaos, preprint;
A. Azevedo and S.M. Rezende, Phys. Rev. Lett. 66 (1991) 1342;
Y. Braiman and J. Goldhirsch, Phys. Rev. Lett. 66 (1991) 2545.
[12] K. Ogata, Control engineering, Second Ed. (Prentice-Hall, Englewood Cliffs, NJ, 1990), pp. 782–784.
[13] M. Hénon, Commun. Math. Phys. 50 (1976) 69.
[14] C. Grebogi, E. Kostelich, E. Ott and J.A. Yorke, Physica D 25 (1987) 347; Phys. Lett. A 118 (1986) 448; A 120 (1987) 497 (E).
[15] C. Grebogi, E. Ott and J.A. Yorke, Chaotic attractors in crisis, Phys. Rev. Lett. 48 (1982) 1507.
[16] C. Grebogi, E. Ott and J.A. Yorke, Physica D 7 (1983) 181.
[17] C. Grebogi, E. Ott and J.A. Yorke, Phys. Rev. Lett. 57 (1986) 1284.
[18] C. Grebogi, E. Ott, F.J. Romeiras and J.A. Yorke, Phys. Rev. A 36 (1987) 5365.
[19] G. Nitsche and U. Dressler, Controlling chaotic dynamical systems using time delay coordinates, preprint.
[20] G.M. Zaslavsky, Phys. Lett. A 69 (1978) 145.
[21] G. Benettin, L. Galgani, A. Giorgilli and J. Strelcyn, Meccanica 15 (1980) 9.
[22] G. Benettin, L. Galgani, A. Giorgilli and J. Strelcyn, Meccanica 15 (1980) 21.
[23] J.D. Farmer, E. Ott and J.A. Yorke, Physica D 7 (1983) 153.
[24] J. Kaplan and J.A. Yorke, Chaotic behavior of multidimensional difference equations, in: Functional differential equations and the approximation of fixed points, Lecture Notes in Mathematics, eds. H.O. Peitgen and H.O. Walther, Vol. 730 (Springer, Berlin, 1978), p. 228–237.
[25] T. Shinbrot, E. Ott, C. Grebogi and J.A. Yorke, Phys. Rev. Lett. 65 (1990) 3215.
[26] E. Ott, C. Grebogi and J.A. Yorke, Controlling chaotic dynamical systems, in CHAOS/XAOC, Soviet–American perspective on nonlinear science, ed. D. Campbell (American Institute of Physics, New York, 1990), pp. 153–172.
[27] F. Takens, Detecting strange attractors in turbulence, in: Dynamical systems and turbulence, eds. D.A. Rand and L.-S. Young, Lecture Notes in Mathematics, Vol. 898 (Springer, New York, 1980), pp. 366–381.
[28] J.C. Sommerer, W.L. Ditto, C. Grebogi, E. Ott and M. Spano, Phys. Lett. A 153 (1991) 105.
[29] D.P. Lathrop and E.J. Kostelich, Phys. Rev. A 40 (1989) 4028.
[30] G.H. Gunaratne, P.S. Linsay and M.J. Vinson, Phys. Rev. Lett. 63 (1989) 1.
[31] J.C. Sommerer, W.L. Ditto, C. Grebogi, E. Ott and M.L. Spano, Phys. Rev. Lett. 66 (1991) 1947.

Bifurcation Control of Chaotic Dynamical Systems

Hua Wang and Eyad H. Abed
Department of Electrical Engineering
and the Systems Research Center
University of Maryland, College Park, MD 20742 USA

Abstract

A nonlinear system which exhibits bifurcations, transient chaos, and fully developed chaos is considered, with the goal of illustrating the role of two ideas in the control of chaotic dynamical systems. The first of these ideas is the need for *robust control*, in the sense that, even with an uncertain dynamic model of the system, the design ensures stabilization without at the same time changing the underlying equilibrium structure of the system. Secondly, the paper shows how focusing on the control of primary bifurcations in the model can result in the taming of chaos. The latter is an example of the 'bifurcation control' approach. When employed along with a dynamic feedback approach to the equilibrium structure preservation issue noted above, this results in a family of robust feedback controllers by which one can achieve various types of 'stability' for the system.

Keywords: Chaos, dynamical systems, bifurcation, feedback, control.

1. Introduction

Recently, significant attention has been focused on developing techniques for the control of chaotic dynamical systems [1–6]. Of course, at the outset, one must realize that there is no obvious way to define the 'control of chaos' problem. This is in direct contrast to more traditional dynamical system control problems, such as the textbook problem of stabilization of an equilibrium position of a nonlinear system. Although even this textbook problem allows for various interpretations for the achieved margin of stability, decay rate, etc., these can all be viewed within the same basic framework. Chaos, on the other hand, is a rich, global dynamic behavior, and its 'stabilization' can have vastly differing interpretations. For example, references [2,3] employ a small amplitude control law in a restricted region of the state space, thereby stabilizing a pre-existing equilibrium or periodic orbit. Since the control vanishes in most of the state space, closed-loop system trajectories follow erratic paths for some time, until they enter part of the neighborhood in which the control is effective, after which they are attracted to the equilibrium or periodic orbit of interest. Other authors apply nonlocal linear or nonlinear feedback to stabilize nominal equilibrium points [1,5]. Also, some authors are taking a control systems approach to the analysis of chaos, which may prove useful in control design (see [7,8]). This summary of previous work on control of chaos is of necessity very brief, and the reader is referred to the original papers for details.

In general, the techniques for feedback control of chaos presented thus far in the literature have some common features, which we feel are important to briefly summarize. The control is usually designed for parameter values where the system is known to exhibit chaotic motion, and is typically of the form $u = u(x - x_0)$ where x is the system state vector, and x_0 is an unstable equilibrium of interest, which lies on a chaotic attractor. The control function u is not necessarily smooth. Thus, the control consists of direct state (or output) feedback around x_0, a specific equilibrium of interest. Note that x_0 can also be a periodic orbit, as observed in [2,3].

The approach pursued in the present paper is directed toward nonlinear systems which undergo bifurcations, and possibly chaotic motion, as a parameter is quasistatically varied. Such systems naturally possess several, and possibly infinitely many, equilibria and periodic orbits. The approach is of particular relevance to systems for which the model possesses a high degree of uncertainty. Often, an engineering system is designed to perform well, and to be stable, for a large range of parameter values. However, technological demands are pushing systems to the limits of their performance, and many engineering systems are being operated under conditions which may be viewed as 'stressed.' It is this stressed operation which gives rise to nonlinear dynamic phenomena, such as bifurcations leading, in some cases, to chaos. We take an approach which is in mathematical synergy with this description.

We consider nonlinear systems depending, for simplicity, on a single bifurcation parameter. For the 'usual' values of the parameter, the system operates at a stable equilibrium, and perturbations away from this mode of operation tend to be attenuated (stability). As the parameter is varied, the equilibrium loses stability at a bifurcation point, giving rise to new equilibria or periodic orbits, perhaps. If any of the bifurcated solutions is stable, the system may operate at such a solution. For greater variations of the parameter, these bifurcated solutions may also lose stability, and so on. There are several scenarios by which successive bifurcations can result in a chaotic invariant set; these are discussed extensively in the chaos literature. What is important about these scenarios from a control of chaos perspective, however, is that the appearance of chaos depends heavily on various aspects of the succession of bifurcations. Suppose a particular control significantly reduces the amplitude of a bifurcated solution, or significantly enhances its stability, over a nontrivial parameter range. Then, one might expect that the occurrence of chaos might be 'delayed' to even greater variations in the parameter, or might be extinguished completely.

This work differs from previous techniques in another respect, related to nonlinear model uncertainty. Under model uncertainty, a nonlinear static state feedback controller designed relative to a given equilibrium will influence not only the stability, but also the location, of this and other system equilibria. To circumvent this difficulty, we employ a form of dynamic feedback which exactly preserves all system equilibria. This uses washout filters in a way which retains sufficient freedom to stabilize bifurcations, and to delay their occurrence if desired (see [9]). Besides preserving system equilibria, the incorporation of washout filters in the feedback control facilitates the design of a control which does not depend on the bifurcation parameter. This is also important to achieving a control which is effective over a range of parameter values, instead of at one specific parameter value.

In the remainder of the paper, we focus on a system studied in [1], as a vehicle for illustrating our approach. Singer, Wang and Bau [1] study a thermal convection loop using an experimental apparatus, and compare their experimental results with simulations based on an analytical model. Others have also studied chaos in this type of thermal convection loop; see for instance Jackson [10]. The model employed in [1] is a set of three nonlinear autonomous ordinary differential equations, depending on a parameter, the Rayleigh number R, which is proportional to the heating rate of the loop. The parameter R is taken as the bifurcation parameter of the model, and forced fluctuations of the effective heating rate constitute the control signal. In [1] it is observed that for values of R exceeding a threshold value, the system exhibits motion which is in some sense chaotic. It is then shown experimentally and numerically that feedback ('active control') can be used to suppress the chaotic behavior.

The remainder of the paper is organized as follows. Section 2 reviews the thermal convection loop model of [1], and discusses its open loop dynamical behavior. Section 3 employs bifurcation control results from [9,11,12] to determine control laws for suppressing both the transient chaotic and chaotic motion of the thermal convection loop model. Concluding remarks are given in Section 4.

2. Thermal Convection Loop Model

Singer, Wang and Bau [1] study a thermal convection loop using a combination of experimentation, modeling, and simulation. The analytical model used in [1] is given by the third order system

$$\dot{x}_1 = -px_1 + px_2, \tag{1}$$

$$\dot{x}_2 = -x_1x_3 - x_2, \tag{2}$$

$$\dot{x}_3 = x_1x_2 - x_3 - R. \tag{3}$$

where $x_i, i = 1, 2, 3$, are real, and p and R are positive parameters. The experiment studied in [1] involves thermal convection in a toroidal vertical loop heated from below and cooled from above. The variables x_1, x_2, x_3 correspond, respectively, to the cross-sectionally averaged velocity in the loop, the temperature difference along the horizontal direction (side to side), and the temperature difference along the vertical direction (top to bottom). The parameter R is the Rayleigh number, which is proportional to the net heating rate, and p denotes the Prandtl number. It is observed experimentally that, as the heating rate increases, the fluid flow in the loop goes through transitions. For a low heating rate, the fluid is in the no-motion state. As the heating rate increases, a state of steady convection arises (clockwise or counterclockwise). Further increases in the heating rate result in temporally oscillatory, and, eventually, chaotic motion of the fluid.

The transitions above are also reflected by the model (1)-(3). To facilitate discussion of this model, set $p = 4.0$ and view R as the bifurcation parameter. A bifurcation diagram related to this model is given in Figure 1. In this diagram, a solid line represents a stable equilibrium, a dashed line represents an unstable equilibrium, and an open circle represents the maximum amplitude of an unstable periodic orbit of (1)-(3). The bifurcation diagram is obtained by employing the package AUTO [13]. The model (1)-(3) possesses symmetry, in that replacing (x_1, x_2, x_3) with $(-x_1, -x_2, x_3)$ results in the same set of equations. This symmetry is reflected in the bifurcation diagram of Fig. 1.

For $R \leq 1.0$ the system (1)-(3) has a single, glob-

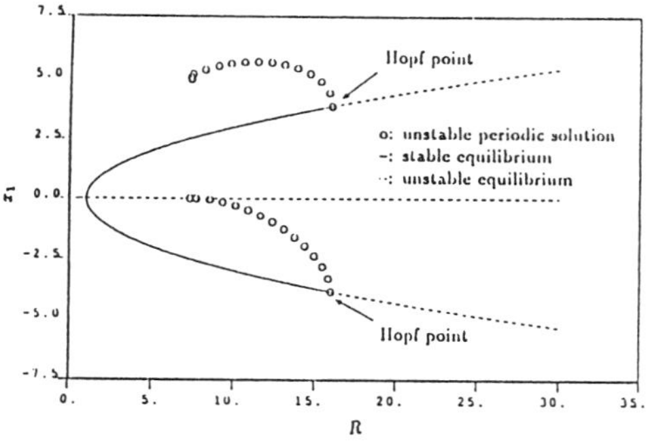

Figure 1. Bifurcation diagram of open loop system

ally attracting, equilibrium point. This equilibrium, given by $x_1 = x_2 = 0, x_3 = -R$, corresponds to the *no-motion state*. At $R = 1.0$ two additional equilibrium points appear through a pitchfork bifurcation. These equilibria, which are present for all $R > 1.0$, are given by $(x_1 = x_2 = \pm\sqrt{R-1}, x_3 = -1.0)$. Denote these equilibria by C_+ and C_-, respectively. These two equilibrium points represent the states of steady convection in the counterclockwise or clockwise directions, respectively. The no-motion equilibrium state $(0, 0, -R)$ loses its stability at the pitchfork bifurcation point, i.e., at $R = 1.0$. The convective equilibria $(\pm\sqrt{R-1}, \pm\sqrt{R-1}, -1.0)$ lose their stability in Hopf bifurcations occurring at $R = 16.0$, as depicted in Fig. 1. The bifurcation diagram of Fig. 1 illustrates that the Hopf bifurcations at the convective equilibria result in *unstable* periodic solutions, i.e., these bifurcations are subcritical. Moreover, Fig. 1 also illustrates the disappearance of the unstable periodic orbit in a blue sky catastrophe [14] at the approximate value $R = 7.3198$. Not discernible from Fig. 1 is the fact that the model (1)-(3) admits erratic behavior for a large range of values of R. This erratic behavior may or may not be chaotic. To be more precise, one observes trajectories which appear chaotic for a long time interval, after which they settle to an equilibrium. One also observes trajectories which are chaotic in the usual sense. The former type of behavior is often referred to as "transient chaos." Transient chaos is observed in simulations of (1)-(3) for parameter values $7.3198 < R < 16$, for *some* initial conditions. (Extensive simulation shows that initial conditions resulting in transient chaos are more common for larger values of R in this interval.) For $R > 16$, typical trajectories of the system (1)-(3) are chaotic.

The foregoing is a necessarily brief description of the qualitative behavior of (1)-(3) and its dependence on the parameter R. There are, however, intricate details associated with the various behaviors and their bifurcations. The resemblance of (1)-(3) to the Lorenz equations suggests that studies such as [15] will provide a reasonable starting point for the further analysis of this model.

3. Bifurcation Control of Convection Dynamics

In this section, we employ bifurcation control results from [9,11,12] to determine control laws for suppressing both the transient chaotic and chaotic motion of the system (1)-(3). In the course of seeking control laws for suppression of chaos, we shall also employ feedback to achieve other, subsidiary goals. For instance, in the next subsection we consider use of feedback to delay to higher values of the Rayleigh number the occurrence of the Hopf bifurcations from the convective equilibria $C_\pm$. This in itself has little to do with suppressing chaos. However, it addresses a question which arises rather naturally in the context of using feedback to modify the phase portrait of system (1)-(3) in useful ways. The control laws developed for achieving this delay in Hopf bifurcation parameter values have a feature which occurs throughout this paper: *they do not result in any change in the set of equilibria, even in the presence of model uncertainty.* This is achieved using dynamic feedback incorporating washout filters, as proposed in [9,12].

3.1. Delaying the Hopf Bifurcations

Recall that the convective equilibria $C_\pm$ lose their stability at Hopf bifurcations occurring at $R = 16$. In this subsection, we give controllers which result in changing this critical value of R to some prescribed value. In practice, the prescribed value would likely be greater than the nominal (open loop) value, so as to result in an increase in the range of parameter values for which the system exhibits stable steady motion.

Linearizing the model (1)-(3) at the upper equilibrium C_+ of Fig. 1, we find that, for $R = 16$, the Jacobian matrix has a pair of imaginary eigenvalues $\pm i\omega_c$ where $\omega_c = 4.47214$ (recall that $p = 4$). Next we present a feedback control scheme which allows one to modify the critical value of R at which the Hopf bifurcations occur, and to do so without modifying the equilibria of (1)-(3). The state variable x_3 is readily observable.

A *linear washout filter aided feedback* with measurement of x_3 is a dynamic feedback described as follows. The closed loop system is given by

$$\dot{x}_1 = -px_1 + px_2, \tag{4}$$
$$\dot{x}_2 = -x_1x_3 - x_2, \tag{5}$$
$$\dot{x}_3 = x_1x_2 - x_3 - R + u, \tag{6}$$
$$\dot{x}_4 = x_3 - dx_4, \tag{7}$$

where x_4 is the washout filter state, and where the

control u is of the form

$$u = -k_l y, \tag{8}$$

with y an output variable, given by

$$y := x_3 - dx_4. \tag{9}$$

Here, k_l is a scalar (linear) feedback gain.

This control preserves the symmetry inherent in the model (1)-(3). Thus, in discussing the effects of the controller above, remarks specific to the upper equilibrium branch C_+ apply also to the lower branch C_-.

The control above is a dynamic feedback control. By adjusting the linear control gain k_l one can delay the Hopf bifurcations to occur at any desired parameter value. The relationship between the critical parameter value R and the control gain k_l can be determined by finding the conditions under which the Jacobian of the overall system (4) - (9) possesses a pair of pure imaginary eigenvalues. This relationship translates to the conditions

$$\begin{aligned}&(Rd - 2p + 2Rp + dp)^2\\&+ (2 + d + k_l + p)^2(-2dp + 2Rdp)\\&- (2 + d + k_l + p)(Rd - 2p + 2Rp\\&+ dp)(R + 2d + k_l + p + dp + k_l p) = 0,\end{aligned} \tag{10}$$

$$k_l + p + d + 2 > 0, \text{ and } R > 1 \tag{11}$$

In the case $p = 4.0$ and $d = 0.5$, these conditions are tantamount to the restriction

$$-1.5 < k_l < 2 \tag{12}$$

on the gain k_l. To *delay* occurrence of the Hopf bifurcations, however, one must further restrict k_l to be positive. Indeed, negative values of k_l in the interval $-1.5 < k_l < 2$ result in moving the Hopf bifurcations to smaller values of R.

The foregoing discussion has resulted in linear, dynamic feedback control laws which can be tuned to result in moving the Hopf bifurcation points to any desired value of $R > 1$. These control laws also ensure asymptotic stability of the convective equilibria for all values of R up to the desired critical value. Despite this positive conclusion, the closed loop system incorporating the control laws given above still exhibits chaotic and transient chaotic behavior. This chaotic behavior is delayed to greater values of R if $0 < k_l < 2$, and moved ahead to lesser values if $-1.5 < k_l < 0$.

In this particular model, there are at least two remedies for the continued presence of chaotic motion after the introduction of linear feedback. First, one can continue to employ the same type of dynamic linear feedback, but with a higher feedback gain. Specifically, for $p = 4$, and taking any $k_l \geq 2$, $d > 0.04533$, it can be seen that both the upper and lower convective equilibria are rendered asymptotically stable, and that the system no longer exhibits chaos or transient chaos. Second, one can employ a combined linear-plus-nonlinear feedback to suppress chaos in the closed loop system. The linear part of the feedback is chosen to delay the Hopf bifurcations to a desired value of R, and the nonlinear part of the feedback is chosen so as to stabilize the Hopf bifurcations occurring in the closed loop system. Both of these alternatives deserve consideration, but it is the authors' opinion that the latter is more versatile and of more general applicability.

Before proceeding to issues of nonlinear control design, we remark that the control introduced in the foregoing does not affect the stability of the nominal equilibrium branch, $(0, 0, -R, -R/d)$. This is easy to prove by examining the associated characteristic polynomial.

3.2. Stabilizing the Hopf Bifurcations

Suppose a dynamic linear feedback has been introduced as in the foregoing subsection, resulting in positioning the Hopf bifurcations to a desired value of R. One result of such a control is to affect the bifurcated periodic solutions which emerge at the two Hopf bifurcations. Recall that these bifurcations are subcritical for the open loop system (see Fig. 1). The subcriticality of the Hopf bifurcations is crucial to the appearance of transient chaos and chaos in the model for various values of R. Thus the question arises as to whether or not the feedback controller of the previous subsection can be modified to result in stabilization of the Hopf bifurcations. Next, we summarize some positive results in this direction.

In [11], the problem of stabilization of Hopf bifurcations via direct state feedback is considered. This was extended to stabilization of Hopf bifurcations using dynamic feedback through washout filters in [9,12]. The stability of a Hopf bifurcation is determined by the sign of a coefficient, often denoted β_2, in the expansion of the Floquet exponent near zero of the bifurcated periodic solutions. See [11,12] for details.

Using techniques from these references, one can show that there is a family of stabilizing, purely cubic nonlinear controllers. We now choose the simplest such stabilizing control law. The closed loop system again takes the form (4)-(9), except that now the controller is

$$u = -k_n y^3. \tag{13}$$

Here, k_n denotes the nonlinear feedback gain.

To illustrate the utility of such a nonlinear control law, we state a simple result obtained using a formula for β_2 [11]. Any choice of control law (13) with

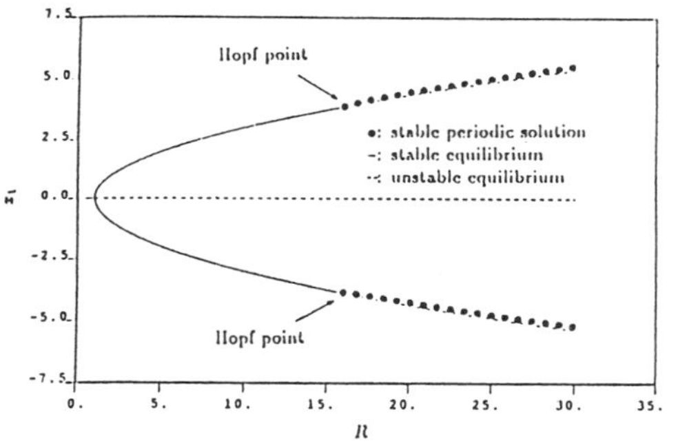

Figure 2. Bifurcation diagram for a nonlinear control $u = -k_n y^3$ with $k_n = 2.5$ ($d = 0.5$)

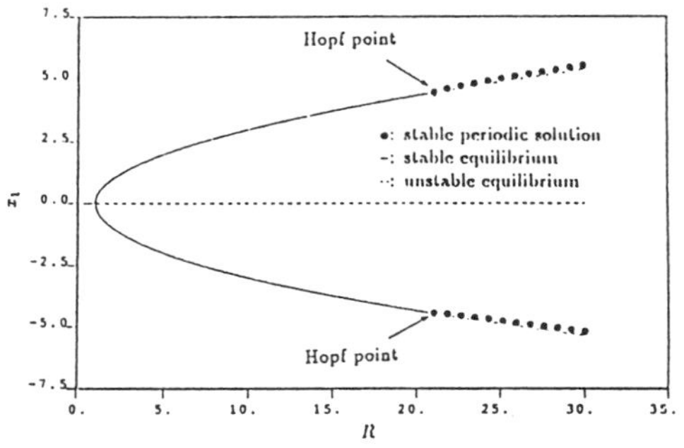

Figure 3. Bifurcation diagram for linear-plus-nonlinear control $u = -k_l y - k_n y^3$ with $k_l = 0.182538$, $k_n = 2.5$ ($d = 0.5$). Note that the Hopf bifurcation is delayed to $R = 21$, as well as stabilized.

$k_n > 0.009$ stabilizes the Hopf bifurcations occurring at $R = 16$. This is a local result. To assess the degree to which this is reflected in the global dynamics of the system, one resorts to extensive computation. Simulations along with application of the bifurcation analysis tool AUTO indicate that larger values of the gain k_n result in an increased stability margin in parameter space. With this control, transient chaos is successfully suppressed, and the previous chaotic trajectories are replaced by small amplitude stable limit cycles near the convective equilibria. This is illustrated in Figure 2, which shows a bifurcation diagram of the closed loop system with this type of control. Solid circles indicate stable limit cycles. The amplitude of a limit cycle is given by the distance of a solid circle from the equilibrium. Figure 3, on the other hand, shows a bifurcation diagram of the closed loop system with a combined linear-plus-nonlinear feedback control

$$u = -k_l y - k_n y^3 \tag{14}$$

Note that the control law (14) effects both a delay in the occurrence of the Hopf bifurcations, as well as stabilization of these bifurcations.

4. Concluding Remarks

Using bifurcation control ideas, control laws have been designed which result in the suppression of both transient chaotic and chaotic motion of a thermal convection system. The control laws exactly preserve all the equilibrium branches of the system, and simultaneously stabilize *both* convective equilibrium branches. This stabilization can take one of two forms. One can literally stabilize the equilibria using linear dynamic feedback. Alternatively, it is possible to re-locate the Hopf bifurcations to occur at higher values of the Rayleigh number R, and then employ nonlinear control to ensure stability of these bifurcations. In this way, a small amplitude stable limit cycle is introduced which surrounds the equilibrium for parameter values at which it is unstable. Simulations show that this control scheme is effective in suppressing chaos for any parameter range. The choice between linear feedback and linear-plus-nonlinear feedback depends on several factors, including degree of confidence in the model and available net gain. However, both types of feedback are related in their structure, especially in their incorporation of washout filters and preservation of model symmetry. The paper has also shown that control directed at stabilizing a primary bifurcation in a succession of bifurcations leading to chaos is a viable technique for suppression of chaos. The results presented here are characteristic of other results obtained by the authors for the system under consideration. For example, another result is that one can "target" a particular equilibrium or periodic orbit of the system. That is, other equilibria or periodic orbits are rendered unstable, while the target equilibrium or periodic orbit is stabilized. Reports of these results are forthcoming.

Acknowledgment

The authors are grateful to Dr. J.-H. Fu for helpful discussions. This research has been supported in part by the National Science Foundation's Engineering Research Centers Program: NSFD CDR-88-03012, by NSF Grant ECS-86-57561, by the Air Force Office of Scientific Research under URI Grant AFOSR-87-0073, and by the TRW Foundation.

References

[1] J. Singer, Y.Z. Wang & H.H. Bau, "Controlling a chaotic system," *Physical Review Letters* 66 (1991), 1123–1125.

[2] E. Ott, C. Grebogi & J.A. Yorke, "Controlling chaos," *Physical Review Letters* 64 (1990), 1196–1199.

[3] F.J. Romeiras, E. Ott, C. Grebogi & W.P. Dayawansa, "Controlling chaotic dynamical systems," *Proc. 1991 American Control Conference*, Boston (1991).

[4] T.B. Fowler, "Application of stochastic control techniques to chaotic nonlinear systems," *IEEE Trans. Automatic Control* AC-34 (1989), 201–205.

[5] T.L. Vincent & J. Yu, "Control of a chaotic system," *Dynamics and Control* 1 (1991), 35–52.

[6] A. Hübler, "Adaptive control of chaotic systems," *Helvetica Physica Acta* 62 (1989), 343–346.

[7] J. Baillieul, R.W. Brockett & R.B. Washburn, "Chaotic motion in nonlinear feedback systems," *IEEE Trans. Circuits and Systems* CAS-27 (1980), 990–997.

[8] R. Genesio & A. Tesi, "Harmonic balance methods for the analysis of chaotic dynamics in nonlinear systems," *Automatica* (to appear).

[9] H.C. Lee & E.H. Abed, "Washout filters in the bifurcation control of high alpha flight dynamics," *Proc. 1991 American Control Conference*, Boston (June 1991).

[10] E.A. Jackson, *Perspectives of Nonlinear Dynamics* #2, Cambridge Univ. Press, Cambridge, 1990.

[11] E.H. Abed & J.H. Fu, "Local feedback stabilization and bifurcation control, I. Hopf bifurcation," *Systems and Control Letters* 7 (1986), 11–17.

[12] H.C. Lee, "Robust Control of Bifurcating Nonlinear Systems with Applications," University of Maryland, College Park, Ph.D. Dissertation, 1991.

[13] E.J. Doedel, "AUTO: A program for the automatic bifurcation analysis of autonomous systems," *Cong. Num.* 30 (1981), 265–284.

[14] J.M.T. Thompson & H.B. Stewart, *Nonlinear Dynamics and Chaos*, Wiley, Chichester, 1986.

[15] C. Sparrow, *The Lorenz Equations: Bifurcations, Chaos, and Strange Attractors*, Springer-Verlag, New York, 1982.

CHAPTER 13

Control: Experimental Stabilization of Unstable Orbits

W. Ditto, S.N. Rauseo, M.L. Spano
Experimental control of chaos
Phys. Rev. Lett. **65**, 3211 (1990).

J. Singer, Y.-Z. Wang, H.H. Bau
Controlling a chaotic system
Phys. Rev. Lett. **66**, 1123 (1991).

A. Garfinkel, M. Spano, W. Ditto, J. Weiss
Controlling cardiac chaos
Science **257**, 1230 (1992).

E.R. Hunt
Stabilizing high-period orbits in a chaotic system: The diode resonator
Phys. Rev. Lett. **67**, 1953 (1991).

Z. Gills, C. Iwata, R. Roy, I. Schwartz, I. Triandaf
Tracking unstable steady states: Extending the stability regime of a multimode laser system
Phys. Rev. Lett. **69**, 3169 (1992).

V. Petrov, V. Gáspár, J. Masere, K. Showalter
Controlling chaos in the Belousov-Zhabotinsky reaction
Nature **361**, 240 (1993).

Editors' Notes

Experiments in many areas of science of engineering have been successful in implementing control of unstable periodic orbits or steady states embedded in a chaotic attractor, as described in the previous chapter. One of the earliest was the work of **Ditto et al.** (1990) on a gravitationally buckling magnetoelastic ribbon. This was rapidly followed by **Singer et al.** on thermally driven fluid convection, **Hunt** (1991) on an electrical circuit, **Garfinkel et al.** (1992) on chaotically oscillating rabbit cardiac tissue, **Gills et al.** (1992) on a laser system, and **Petrov et al.** (1993) on the Belousov-Zhabotinskii chemical reaction. These articles are a representative sample of control of chaos in the laboratory. Many more examples are listed in the bibliography.

Experimental Control of Chaos

W. L. Ditto, S. N. Rauseo, and M. L. Spano
Naval Surface Warfare Center, Silver Spring, Maryland 20903-5000
(Received 27 August 1990)

We have achieved control of chaos in a physical system using the method of Ott, Grebogi, and Yorke [Phys. Rev. Lett. **64**, 1196 (1990)]. The method requires only small time-dependent perturbations of a single-system parameter and does not require that one have model equations for the dynamics. We demonstrate the power of the method by controlling a *chaotic* system around unstable periodic orbits of order 1 and 2, switching between them at will.

PACS numbers: 05.45.+b, 75.80.+q

In a recent Letter, Ott, Grebogi, and Yorke[1] (OGY) demonstrated that one can convert the motion of a chaotic dynamical system to periodic motion by controlling the system about one of the many unstable periodic orbits embedded in the chaotic attractor, through only small time-dependent perturbations in an accessible system parameter. They demonstrated their method numerically by controlling the Hénon map.

Far from being a numerical curiosity that requires experimentally unattainable precision, we believe this method can be widely implemented in a variety of systems including chemical, biological, optical, electronic, and mechanical systems. In this Letter we report the control of chaos in a physical system, a parametrically driven magnetoelastic ribbon, using the method of OGY.

Theoretical background.—The method is based on the observation that unstable periodic orbits are dense in a typical chaotic attractor. Their method assumes only the following four points. First, the dynamics of the system can be represented as arising from an n-dimensional nonlinear map (e.g., by a surface of section or time one return map), the iterates given by $\xi_{n+1} = \mathbf{f}(\xi_n, p)$, where p is some accessible system parameter. Second, there is a specific periodic orbit of the map which lies in the attractor and around which one wishes to stabilize the dynamics. Third, there is maximum perturbation δp_* in the parameter p by which it is acceptable to vary p from the nominal value p_0. Finally, one assumes that the position of the periodic orbit is a function of p, but that the local dynamics about it do not vary much with the allowed small changes in p. Note that while the dynamics is assumed to arise from a map, one needs no model for the global dynamics. These assumptions would seem to allow for the control of any chaotic system for which a faithful Poincaré section can be constructed. The construction of a map from and the location of periodic orbits in[2] experimental data are straightforward processes.

To control chaotic dynamics one only needs to learn the *local* dynamics around the desired periodic orbit by observing iterates of the map near the desired orbit and fitting them to a local linear approximation of the map $\mathbf{f}$.[3] From this, one can find the stable and unstable eigenvalues as well as the local stable and unstable manifolds (given by the eigenvectors). Next, by changing p slightly and observing how the desired orbit changes position, one can estimate the partial derivatives of the orbit location with respect to p.

To control the chaos, one attempts to confine the iterates of the map to a small neighborhood of the desired orbit. When an iterate falls near the desired orbit, we change p from its nominal value p_0 by δp, thereby changing the location of the orbit and its stable manifold, such that the *next* iterate will be forced back toward the stable manifold of the *original* orbit for $p = p_0$. [Figure 1 illustrates this method for the case of a saddle fixed point located at $\xi_F(p_0)$.] That the method of OGY

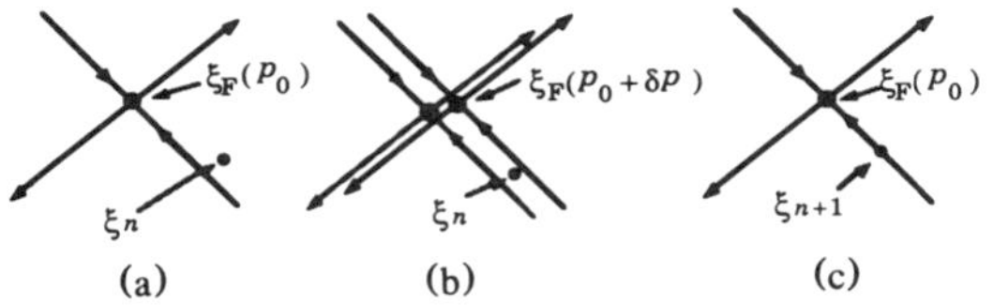

FIG. 1. Schematic of the OGY control algorithm for a saddle fixed point: (a) The nth iterate ξ_n falls near the fixed point $\xi_F(p_0)$. (b) Turn on the perturbation of p to move the fixed point. (c) The next iterate is forced onto the stable manifold of $\xi_F(p_0)$. Turn off the perturbation.

rests on attempting to force the dynamics to stay in the neighborhood of an unstable periodic orbit in the attractor makes it quite different from other previously published methods from removing chaos.[4]

Experimental setup and results.—The experimental system consisted of a gravitationally buckled, amorphous magnetoelastic ribbon. The ribbon material belongs to a new class of amorphous magnetostrictive materials[5] that have been found to exhibit very large reversible changes of Young's modulus $E(H)$ with the application of small magnetic fields.[6,7] The ribbon was clamped at the base to yield a free vertical length greater than the Euler buckling length, thus giving an initially buckled configuration. The ribbon was placed within three mutually orthogonal pairs of Helmholtz coils, which allowed us to compensate for the Earth's magnetic field and to apply an approximately uniform vertical magnetic field along the ribbon. The Young's modulus of the ribbon was varied by applying a vertical magnetic field having the form $H = H_{dc} + H_{ac}\cos(2\pi ft)$. To lowest order, the ribbon was not driven by magnetic forces, but was forced by gravity as $E(H)$ was varied. The magnetic-field amplitudes were typically set in the range 0.1–2.5 Oe. A sensor measured the curvature of the ribbon near its base. Other details of the experiment can be found in Refs. 6 and 7.

The data were time-series voltages $V(t)$ acquired from the output of the sensor. Voltages were sampled at the drive period of the ac field (at times $t_n = n/f$) by triggering a voltmeter off the ac signal.

By considering the sampled voltages as arising from iterates of a map, $\mathbf{X}_n = V(t_n)$, we are able to directly apply the control theory outlined above. We selected H_{dc} to be the parameter to be varied to achieve control (i.e., $p = H_{dc}$). First, we chose a parameter region (H_{ac}, H_{dc}, and f) such that the ribbon was oscillating chaotically. In order to simplify the comparison with the theory, the parameter region chosen was one in which the dynamics of the iterates near the orbits of interest clearly appears to be two dimensional (i.e., the two-dimensional return map, $\mathbf{X}_{n+1}$ vs $\mathbf{X}_n$, is always single valued in the neighborhood of the orbits of interest). The first 2350 iterations (in gray) in Fig. 2(a) are of the uncontrolled time-series data for $H_{ac} = 2.050$ Oe, $H_{dc} = 0.112$ Oe ($= p_0$), and $f = 0.85$ Hz (from 1 to 2350 iterations). In Fig. 2(b), the return map for the uncontrolled system is shown in gray. We estimate the dynamical noise in our system, i.e., the deviation of the motion of the ribbon away from deterministic chaos, to be ± 0.005 V, since any structure on the attractor below this scale is blurred out.

We found the approximate location X_F of an unstable period-1 orbit of the map (i.e., a fixed point) by noting that any fixed point of the dynamics must lie along the $\mathbf{X}_{n+1} = \mathbf{X}_n$ line in the plot of the return map. To stabilize this fixed point we next examined the data series and found all pairs of iterates *both* of which fell within 0.05 V of the approximate fixed point. To these pairs of iterates we fit the approximate local *linear* map $\mathbf{M}$,

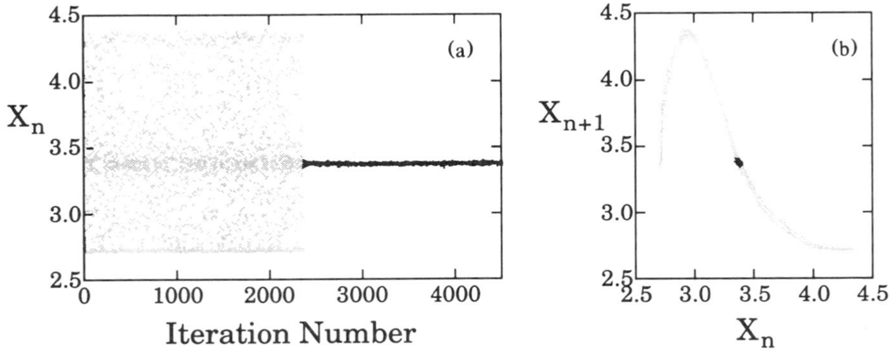

FIG. 2. (a) Time series of $\mathbf{X}_n = V(t_n)$ for H_{dc}(nominal) $= 0.112$ Oe, $H_{ac} = 2.050$ Oe, and $f = 0.85$ Hz. Control was initiated after iteration 2350. (b) The first return map ($\mathbf{X}_{n+1}$ vs $\mathbf{X}_n$) for the controlled system (in black) is superimposed on the map for the uncontrolled system (in gray). The large density of points of low values of $\mathbf{X}_n$ is due to the saturation of the sensor for large excursions of the ribbon away from the sensor.

where

$$\xi_{n+1}-\xi_F=\mathbf{M}(\xi_n-\xi_F),\quad \xi_{n+1}=\begin{pmatrix}X_{n+2}\\X_{n+1}\end{pmatrix},$$

$$\xi_n=\begin{pmatrix}X_{n+1}\\X_n\end{pmatrix},\quad \xi_F=\begin{pmatrix}X_F\\X_F\end{pmatrix}.$$

Knowing $\mathbf{M}$, we could extract the stable and unstable eigenvalues (λ_s,λ_u) and eigenvectors $(\mathbf{e}_s,\mathbf{e}_u)$. We actually only needed λ_u and the unstable *contravariant* eigenvector[1] $\mathbf{f}_u$, given by $\mathbf{f}_u\cdot\mathbf{e}_u=1$ and $\mathbf{f}_u\cdot\mathbf{e}_s=0$.

Next, we changed H_{dc} slightly ($H_{dc}=0.120$ Oe) and collected another set of data. We again found the precise location of the fixed point and calculated $\mathbf{g}=\partial\xi_F/\partial p\approx\delta\xi_F/\delta H_{dc}$.

To control the oscillations of the ribbon, we set $p=p_0$; when $|\xi_n-\xi_F|<\delta\xi_*$, we attempted control. Here, $\delta\xi_*\approx[(\lambda_u-1)/\lambda_u]\delta p_*(\mathbf{g}\cdot\mathbf{f}_u)$ is the maximum distance from the stable manifold of ξ_F for which one can achieve control for a given δp_*. As long as the iterate was within $\delta\xi_*$ of ξ_F, we perturbed p from p_0 by $\delta p=C(\xi_n-\xi_F)\cdot\mathbf{f}_u$, where Ref. 1 gives $C=[\lambda_u/(\lambda_u-1)]/\mathbf{g}\cdot\mathbf{f}_u$. Since noise and errors in determining ξ_F, $\mathbf{f}_u$, $\mathbf{g}$, and λ_u, as well as any inaccuracies due to the linear approximation, prevented us from getting the next iterate exactly on the stable manifold, a new δp was calculated for each iterate. Note that both $\delta\xi_*$ and C can be computed at the start of the run, and that the calculations at each iterate are very simple. We could apply the changes to the applied magnetic field and change the Young's modulus of the ribbon in under 1 ms. Thus, our change in p was effectively instantaneous in relation to the 1.2-s period of the ac drive.

At the values of H_{ac}, H_{dc}, and f mentioned above, we calculated $X_F=3.398\pm0.002$, $\delta X_F/\delta H_{dc}=-337\pm50$, $\mathbf{f}_u=(^{-0.2}_{1.2})\pm(^{0.2}_{0.1})$, and $\lambda_u=-1.2\pm0.2$. These numbers are typical of our data in that the fixed point can be determined with a great deal of accuracy, but the computed values of the eigenvalues and eigenvectors are sensitive to the noise on the attractor. Fortunately, the control is quite insensitive to variations in λ_u and $\mathbf{f}_u$ (e.g., using $\lambda_u=-1.4$ yielded results similar to those using $\lambda_u=-1.2$).

We have been able to control the oscillations of our ribbon for over 200000 iterates (>64 h), with a maximum allowed perturbation of 0.01 Oe. Figure 2(a) (after 2350 iterations) shows the controlled time series (in black) and Fig. 2(b) the return map (superimposed, in black, on the attractor for the uncontrolled system). The control was to ±0.015 V of the desired fixed point, about triple the dynamic noise present on the uncontrolled attractor.

We have also controlled the motion about a period-2 oscillation (again for over 50000 iterates and with $\delta p_*=0.01$ Oe). The same procedure outlined above is followed except using $\xi_n=(^{X_{2n+2}}_{X_{2n}})$. The control was adjusted only at every other data iterate, about the periodic point at $X_F=3.926\pm0.004$. As a demonstration of the versatility of the method, Fig. 3 shows time-series data while the system was switched between no control and control about the fixed point or the period-2 orbit, again with the same values of H_{dc}(nominal), H_{ac}, and f as for Fig. 2.

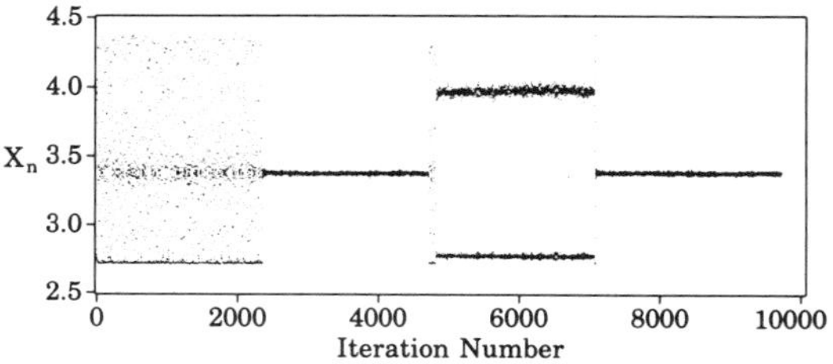

FIG. 3. Time-series data as the system is switched from no control to control about the fixed point (at $n=2360$), to control about the period-2 orbit (at $n=4800$), and back to control around the fixed point (at $n=7100$).

In conclusion, we have demonstrated the first control of chaos in a physical system, using the method of Ott, Grebogi, and Yorke. Some advantages of this method are the following: (1) No model for the dynamics is required; (2) the computations required at each iterate are minimal; (3) the required changes in the parameter can be quite small; (4) different periodic orbits can be stabilized for the *same* system in the *same* parameter range; (5) control can be achieved even with imprecise measurements of the eigenvalues and eigenvectors; and (6) this method is *not* restricted to periodically driven mechanical systems, but extends to any system whose dynamics can be characterized by a nonlinear map.

This work was supported by the Naval Surface Warfare Center Independent Research Program and by ONR through the Navy Dynamics Institute Program. We wish to thank Ed Ott for helpful comments and Heather D. Lynn and Bryan D. Lee for their assistance.

[1]E. Ott, C. Grebogi, and J. A. Yorke, Phys. Rev. Lett. **64**, 1196 (1990); in *Chaos*, edited by D. K. Campbell (American Institute of Physics, New York, 1990), pp. 153–172.

[2]D. P. Lathrop and E. J. Kostelich, Phys. Rev. A **40**, 4028 (1989).

[3]J.-P. Eckman and D. Ruelle, Rev. Mod. Phys. **57**, 617 (1985).

[4]These include T. B. Fowler, IEEE Trans. Autom. Control **34**, 201 (1989); B. A. Huberman and E. Lumer, IEEE Trans. Circuits Syst. **37**, 547 (1990); A. Hübler and E. Lüscher, Naturwissenschaften **76**, 67 (1989); R. Georgii, W. Eberl, E. Lüscher, and A. Hübler, Helv. Phys. Acta **62**, 291 (1989); A. Hübler, Helv. Phys. Acta **62**, 343 (1989). In particular, we

contrast this method with the resonant control method of Hübler and co-workers that has been applied to nonlinear pendula and other oscillators with nonlinear potentials. Their method differs from the method described above in at least three important respects. (1) One must have or construct model equations for the dynamics. (2) One must be able to modify the *driving force* of these equations, and these modifications can be rather large. The method of OGY requires no model equations and the perturbations could be to *any* accessible system parameter. (3) Rather than apply corrections as the dynamics wanders from a given unstable orbit, the resonant control method seeks to modify the underlying dynamical system such that the goal dynamics become *stable* solutions of the system (and thus uses no feedback).

[5]C. Modzelewski, H. T. Savage, L. T. Kabacoff, and A. E. Clark, IEEE Trans. Magn. **17**, 2837 (1981).

[6]W. L. Ditto, S. Rauseo, R. Cawley, C. Grebogi, G.-H. Hsu, E. Kostelich, E. Ott, H. T. Savage, R. Segnan, M. L. Spano, and J. A. Yorke, Phys. Rev. Lett. **63**, 923 (1989).

[7]H. T. Savage and C. Adler, J. Magn. Magn. Mater. **58**, 320 (1986); H. T. Savage and M. L. Spano, J. Appl. Phys. **53**, 8002 (1982); H. T. Savage, W. L. Ditto, P. A. Braza, M. L. Spano, S. N. Rauseo, and W. C. Spring, III, J. Appl. Phys. **67**, 5619 (1990).

Controlling a Chaotic System

J. Singer, Y-Z. Wang, and Haim H. Bau[(a)]

Department of Mechanical Engineering and Applied Mechanics, University of Pennsylvania, Philadelphia, Pennsylvania 19104-6315

(Received 22 October 1990)

Using both experimental and theoretical results, this Letter describes how low-energy, feedback control signals can be successfully utilized to suppress (laminarize) chaotic flow in a thermal convection loop.

PACS numbers: 05.45.+b

Chaotic behavior is abundant both in nature and in man-made devices. On occasion, chaos is a beneficial feature as it enhances mixing and chemical reactions and provides a vigorous mechanism for transporting heat and/or mass. However, in many other situations, chaos is an undesirable phenomenon which may lead to vibrations, irregular operation, and fatigue failure in mechanical systems, temperature oscillations which may exceed safe operational conditions in thermal systems, and increased drag in flow systems. Also, since chaotic behavior cannot be predicted in detail, it may be detrimental to the operation of various devices. Clearly, the ability to control chaos (i.e., promote or eliminate it) is of much practical importance. Although the topic of enhancing chaos has attracted some attention in the scientific literature,[1] there are, indeed, very few theoretical publications[2,3] and even fewer experimental works which address the probably more difficult topic of chaos suppression.

In the first part of this Letter, we describe an experiment conducted with a thermal convection loop, in which for heating rates exceeding a certain threshold value the flow exhibited chaotic behavior. By making small adjustments to the heating rate in response to events detected inside the loop (feedback control), we succeeded in suppressing the chaotic behavior and "laminarizing" the flow. In order to achieve a better understanding as to how our controller operates, we applied a similar control strategy to a simplified mathematical model capable of qualitatively describing the flow in the loop. This theoretical investigation is described in the second part of the Letter. The success of our effort gives hope that it may be possible to suppress chaos in more complicated systems.

The experimental apparatus consists of a pipe of diameter d ($=0.030$ m) bent into a torus of diameter D ($=0.760$ m) containing liquid (i.e., water). The apparatus stands in the vertical plane. The lower half of the apparatus is heated with a uniform-heat-flux resistance heater while the upper half is submerged in a jacket containing flowing coolant so as to approximate a uniform wall temperature (Fig. 1). The apparatus is similar to the one employed by Creveling *et al.*[4] and Gorman, Widmann, and Robins,[5] who have described it in detail. We measured the temperature differences between positions 3 and 9 o'clock and between 6 and 12 o'clock around the loop as functions of time. The heating (cooling) of the lower (upper) half causes temperature gradients within the liquid which under certain conditions may cause fluid motion inside the loop.

For low heating rates ($Q < 190$ W), the flow inside the loop is steady and unidirectional. That is, depending on initial conditions, the fluid flows in either the counterclockwise or the clockwise direction. Above a certain critical heating rate (about 190 W in our experiment), the steady motion loses its stability and the flow becomes chaotic. The chaotic flow appears as irregular oscillations in the flow rate and occasional reversals in the direction of the flow. For example, in Fig. 2, we depict the experimentally obtained temperature difference between positions 3 and 9 o'clock as a function of time for the heating rate of 600 W. The corresponding Rayleigh number is 3.16 times its value at the onset of chaos. Positive (negative) values of the temperature difference in Fig. 2 indicate flow in the counterclockwise (clockwise) direction. Witness the relatively high temperature oscillations associated with the chaotic flow.

Our objective is to suppress the oscillations so as to make the flow approximately steady. That is, we wish to retain the steady unidirectional flow as it existed before the onset of chaos (albeit with higher cross-sectionally

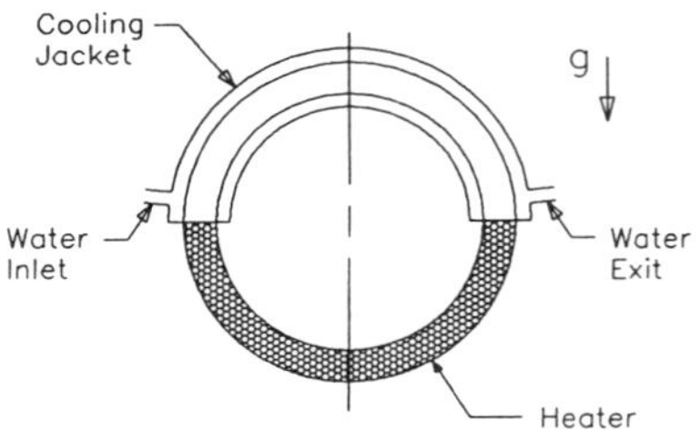

FIG. 1. Schematic description of the thermal convection loop. The lower half of the loop is heated with a uniform-flux resistance heater. The upper half is cooled by passing water through the jacket.

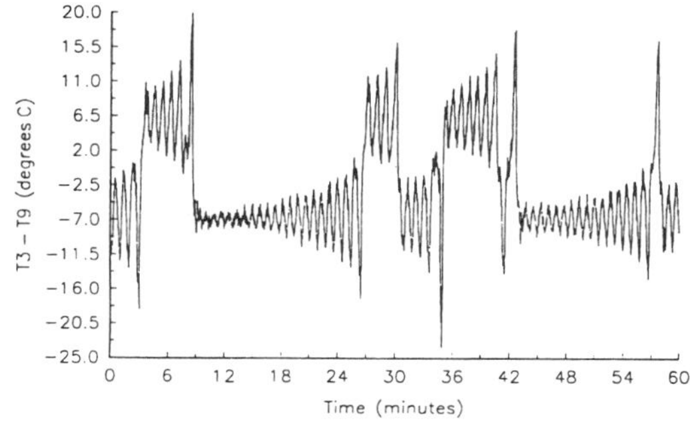

FIG. 2. The experimentally obtained temperature difference between positions 3 and 9 o'clock around the loop as a function of time for a heating rate of 600 W without a controller. The change in sign indicates a change in the direction of the flow.

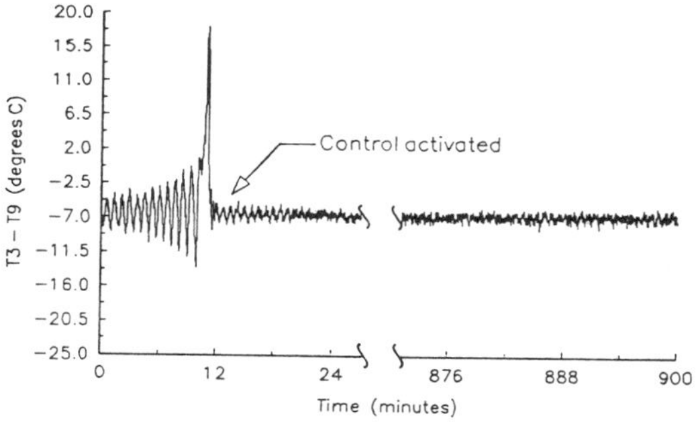

FIG. 3. The experimentally obtained temperature difference between positions 3 and 9 o'clock around the loop as a function of time for a heating rate of 600 W. The controller is turned on 12.5 min into the run, "laminarizing" the flow.

averaged velocity, reflecting the higher heating rate). To accomplish this objective, we adopted a relatively simple control strategy. We change the heating rate by a relatively small increment as a function of the low-pass-filtered temperature difference between positions 6 and 12 o'clock around the loop. When the above temperature difference exceeds or drops below some average value, the heating rate is increased or decreased by a preset increment (i.e., 25 W in Fig. 3) after a time delay of a few seconds. The results of this strategy are depicted in Fig. 3 where we show the temperature difference between positions 3 and 9 o'clock depicted as a function of time. Initially, the flow was uncontrolled and we observed similar oscillations to the ones depicted in Fig. 2. The controller was activated 12.5 min into the run in Fig. 3. The transition from the chaotic flow into a relatively steady, laminar flow is self-evident. We ran the experiment for over 15 h maintaining the type of steady flow shown in Fig. 3. The controller also succeeded in overcoming finite-amplitude disturbances purposely introduced into the loop. It is likely that the magnitude of the control signal could be further reduced by adopting a more sophisticated control strategy than the one reported here.

In order to gain physical insight into how the controller operates, it is useful to briefly describe the mechanism responsible for the chaotic, oscillatory behavior of the flow in the loop.[6] To this end, imagine that a small disturbance causes the flow to slow down below the steady-state flow rate. As a result, the fluid spends more time in the heater (cooler) section, gains (loses) more (less) heat than usual, and emerges from the heater (cooler) with a temperature higher (lower) than usual. This, in turn, causes an increase in the buoyancy force with a corresponding increase in the fluid's velocity. Once the fluid velocity increases, the reverse effect occurs with a subsequent reduction in the fluid velocity. Under appropriate conditions, in the absence of a control mechanism, these oscillations amplify and eventually lead to the chaotic behavior depicted in Fig. 2. The controller detects the appearance of disturbances by monitoring deviations in the temperature difference between top and bottom from the corresponding steady-state value $(z-z_0)$. Once such a deviation is detected, the controller takes action to counteract the effect of this deviation. For instance, if the deviation tends to accelerate (decelerate) the flow, the heating rate is increased (decreased) to counteract this effect. As the controller applies only relatively small perturbations to the input power, it will be able to counteract only small oscillations. Consequently, when the controller is applied to a chaotic flow, it may take some time before the temperature oscillations become small enough for the controller to take effect. This amount of time will decrease as the magnitude of the control signal increases. Ott, Grebogi, and Yorke[3] argue that this length of time is proportional, on the average, to a negative power of the control signal. Once the controller succeeds in laminarizing the flow, it will prevent the oscillations from increasing beyond the controllable magnitude. It should be noted that, due to the presence of noise in the system, it is necessary to maintain the control signal above some minimal value.

To attain further insight into how the controller operates, we examined a simple mathematical model based on the Lorenz equations.[7,8] The solutions of the Lorenz equations provide a good qualitative resemblance to the observed flow in the loop.[9,10] The solutions of these equations, depending on the magnitude of the Rayleigh number (which in our case is proportional to the heating rate), include a no-motion state, two steady flow states (consisting of flows in the counterclockwise and clockwise directions), and chaotic flow of the type depicted in Fig. 2. The model also predicts periodic windows within the chaotic regime, but these have not yet been observed in experiments. The variables (x,y,z) in the equations below correspond, respectively, to the cross-sectionally averaged velocity in the loop, the temperature difference between positions 3 and 9 o'clock, and the temperature difference between positions 12 and 6 o'clock. The Lorenz equations with the on-off controller are

$$\frac{dx}{dt}=p(y-x),\quad \frac{dy}{dt}=-xz-y\ ,$$
$$\frac{dz}{dt}=xy-z-[R+\varepsilon\,\mathrm{sgn}(z-z_0)]\ . \tag{1}$$

As in the experiment, the controller reacts to deviations of z from some preset, average value and modifies the magnitude of the Rayleigh number (R) which is proportional to the heating rate in the experiment. In the above, p is the Prandtl number, ε represents the magnitude of the control signal (in the classical Lorenz equations $\varepsilon=0$), and sgn corresponds to the sign of $z-z_0$. We carried out numerical experiments to observe the effect of the controller on the behavior of the flow. The results of our numerical experiments are depicted in Fig. 4, where we show the controlled (uncontrolled) signals with thick (light) lines. In Fig. 4, the controller has been switched on at a nondimensional time $t=9$. Comparing Figs. 4 and 3, we observe that the physical and simulated controller cause a similar effect.

In order to analyze the controller's action, it is convenient to construct the Lyapunov functional for the controlled system. To this end, we define a new set of dependent variables $\{X,Y,Z\}=\{x-\sqrt{R-1},y-\sqrt{R-1},z+1\}$, where the fixed point $\{X,Y,Z\}=\{0,0,0\}$ corresponds to a steady-state-motion solution of the Lorenz system (1). We focus on R values for which this solution is nonstable. The Lyapunov functional

$$E=\tfrac{1}{2}(X^2+pY^2+pZ^2)\geq 0 \tag{2}$$

satisfies

$$\frac{1}{p}\frac{dE}{dt}=-(X-Y)^2-Z^2 + [\sqrt{R-1}X-\varepsilon\,\mathrm{sgn}(Z)]Z\ . \tag{3}$$

For stability, we require that (3) be negative. This can be satisfied provided that X is sufficiently small. Thus, there is a domain of attraction in phase space $\{X,Y,Z\}$ in the vicinity of the fixed point $\{0,0,0\}$. In other words, once the system enters into this domain of attraction (and eventually it would), it will stay in it as long as externally imposed perturbations are not too large. The Lyapunov functional presented above, although not optimal, demonstrates the effect of the controller on the stability characteristics of the loop.

We note in passing that the chaotic attractor includes, in addition to the nonstable time-independent flow, also an assortment of nonstable periodic and quasiperiodic orbits. Ott, Grebogi, and Yorke[2,3] argue that it might be possible to stabilize any of the aforementioned orbits. In numerical experiments, using a controller, we indeed succeeded in obtaining a stable periodic flow in the nominally chaotic regime.

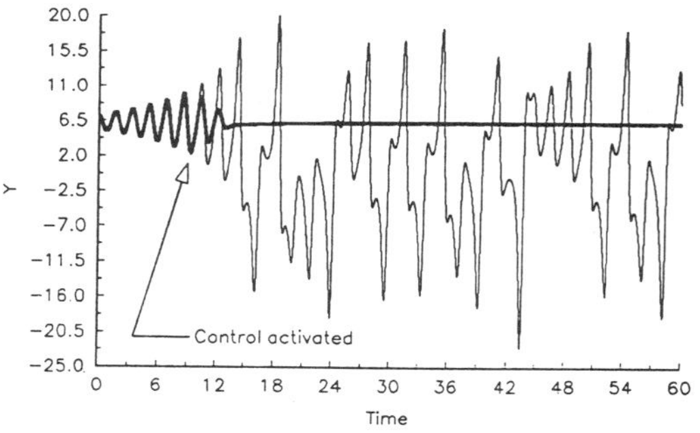

FIG. 4. The numerically generated temperature difference between positions 3 and 9 o'clock around the loop as a function of time. The uncontrolled and controlled signals are shown in light and thick lines, respectively.

In conclusion, we have demonstrated experimentally and theoretically that a simple control strategy can be effectively used to suppress chaos in a simple dynamical system. It is our hope that similar control strategies can be successfully implemented for more complicated situations.

This work was supported, in part, by the National Science Foundation through Grant No. CBT 83-51658.

[a]To whom all corespondence should be addressed.

[1]J. M. Ottino, *The Kinematics of Mixing: Stretching, Chaos, and Transport* (Cambridge Univ. Press, Cambridge, 1989).

[2]E. Ott, C. Grebogi, and J. A. Yorke, Phys. Rev. Lett. **64**, 1196–1199 (1990).

[3]E. Ott, C. Grebogi, and J. A. Yorke, in *CHAOS: Soviet-American Perspectives on Non-Linear Science,* edited by D. K. Campbell (American Institute of Physics, New York, 1990), pp. 153–172.

[4]H. F. Creveling, J. F. De Paz, J. Y. Baladi, and R. J. Schoenhals, J. Fluid Mech. **67**, 65–84 (1975).

[5]M. Gorman, P. J. Widmann, and K. Robins, Phys. Rev. Lett. **52**, 2241–2244 (1984).

[6]P. Welander, J. Fluid Mech. **29**, 17–30 (1967).

[7]E. N. Lorenz, J. Atmos. Sci. **20**, 130–141 (1963).

[8]C. Sparrow, *The Lorenz Equations: Bifurcations, Chaos, and Strange Attractors* (Springer-Verlag, Berlin, 1982).

[9]W. V. R. Malkus, Mem. Soc. Roy. Sci. Liege Collect. 6 **4**, 125–128 (1972).

[10]H. H. Bau and Y.-Z. Wang, in "Annual Reviews in Heat Transfer," edited by C. L. Tien (Hemisphere, Washington, DC, to be published).

SCIENCE

Controlling Cardiac Chaos

Alan Garfinkel, Mark L. Spano, William L. Ditto, James N. Weiss

The extreme sensitivity to initial conditions that chaotic systems display makes them unstable and unpredictable. Yet that same sensitivity also makes them highly susceptible to control, provided that the developing chaos can be analyzed in real time and that analysis is then used to make small control interventions. This strategy has been used here to stabilize cardiac arrhythmias induced by the drug ouabain in rabbit ventricle. By administering electrical stimuli to the heart at irregular times determined by chaos theory, the arrhythmia was converted to periodic beating.

The realization that many apparently random phenomena are actually examples of deterministic chaos offers a better way to understand complex systems. Phenomena that have been shown to be chaotic include the transition to turbulence in fluids (*1*), many mechanical vibrations (*2*), irregular oscillations in chemical reactions (*3*), the rise and fall of epidemics (*4*), and the irregular dripping of a faucet (*5*). Several studies have argued that certain cardiac arrhythmias are instances of chaos (*6, 7*). This is important because the identification of a phenomenon as chaotic may make new therapeutic strategies possible.

Until recently the main strategy for dealing with a system displaying chaos was to develop a model of the system sufficiently detailed to identify the key parameters and then to change those parameters enough to take the system out of the chaotic regime. However this strategy is limited to systems for which a theoretical model is known and that do not display irreversible parametric changes (often the very changes causing the chaos) such as aging.

Recently a strategy has emerged that does not attempt to take the system out of the chaotic regime but uses the chaos to control the system. The key to this approach lies in the fact that chaotic motion includes an infinite number of unstable periodic motions (*8*). A chaotic system never remains long in any of these unstable motions but continually switches from one periodic motion to another, thereby giving the appearance of randomness. Ott, Grebogi, and Yorke (OGY) (*9*) postulated that it should be possible to stabilize a system around one of these periodic motions by using the defining feature of chaos, the extreme sensitivity of chaotic systems to perturbations of their initial conditions.

The OGY theory was first applied experimentally to controlling the chaotic vibrations of a magnetoelastic ribbon (*10*) and subsequently to a diode resonator circuit

A. Garfinkel is in the Department of Physiological Science, University of California, Los Angeles, CA 90024–1527. M. L. Spano is at the Naval Surface Warfare Center, Silver Spring, MD 20903. W. L. Ditto is in the Department of Physics, The College of Wooster, Wooster, OH 44691. J. N. Weiss is in the Department of Medicine (Cardiology), University of California, Los Angeles, CA 90024.

(*11*) and to the chaotic output of lasers (*12*). We have found that it is possible to control a chaotic cardiac arrhythmia using the same basic properties of chaotic systems that were exploited by OGY but that are here employed in a new method of chaos control suitable for use in systems where no systemwide parameters can be readily manipulated as required by the OGY method.

Experimental arrhythmia model. Our cardiac preparation consisted of an isolated well-perfused portion of the interventricular septum from a rabbit heart, arterially perfused through the septal branch of the left coronary artery with a physiologic oxygenated Kreb's solution at 37°C (*13*). The heart was stimulated by passing a 3-ms constant-voltage pulse, typically 10 to 30 V, at twice the threshold between platinum electrodes embedded in the preparation, by means of a Grass SD9 stimulator triggered by computer. Electrical activity was monitored by recording monophasic action potentials with Ag-AgCl wires on the surface of the heart. Monophasic action potentials and a stimulus marker tracing were recorded on a modified videocassette recorder (Model 420, A. R. Vetter, Inc.) and one of the monophasic action potential traces was simultaneously digitized at 2 kHz by a 12-bit A-D converter board (National Instruments model AT-MIO-16). The digitized trace was processed in real time by a computer to detect the activation time of each beat from the maximum of the first derivative of the voltage signal.

Arrhythmias were induced by adding 2 to 5 μM ouabain with or without 2 to 10 μM epinephrine to the arterial perfusate. The mechanism of ouabain-epinephrine–induced arrhythmias is probably a combination of triggered activity and nontriggered automaticity caused by progressive intracellular Ca^{2+} overload from Na^{+} pump inhibition and increased Ca^{2+} current (*14, 15*). We reasoned that this arrhythmia might progress to chaos because in cardiac myocytes intracellular Ca^{2+} is regulated by several interactively coupled processes whose delicate balance is disrupted by ouabain. The resultant oscillations in intracellular Ca^{2+} cause spontaneous beating by activating arrhythmogenic inward currents from electrogenic Na^{+}-Ca^{2+} exchange and Ca^{2+}-activated nonselective cation channels (*14*). Typically the ouabain-epinephrine combination induced spontaneous beating, initially at a constant interbeat interval and then progressing to bigeminy and higher order periodicity before developing a highly irregular aperiodic pattern of spontaneous beating in ~85 percent of the preparations. The duration of the aperiodic phase was variable, lasting up to several minutes before spontaneous electrical activity irreversibly ceased, probably correspond-

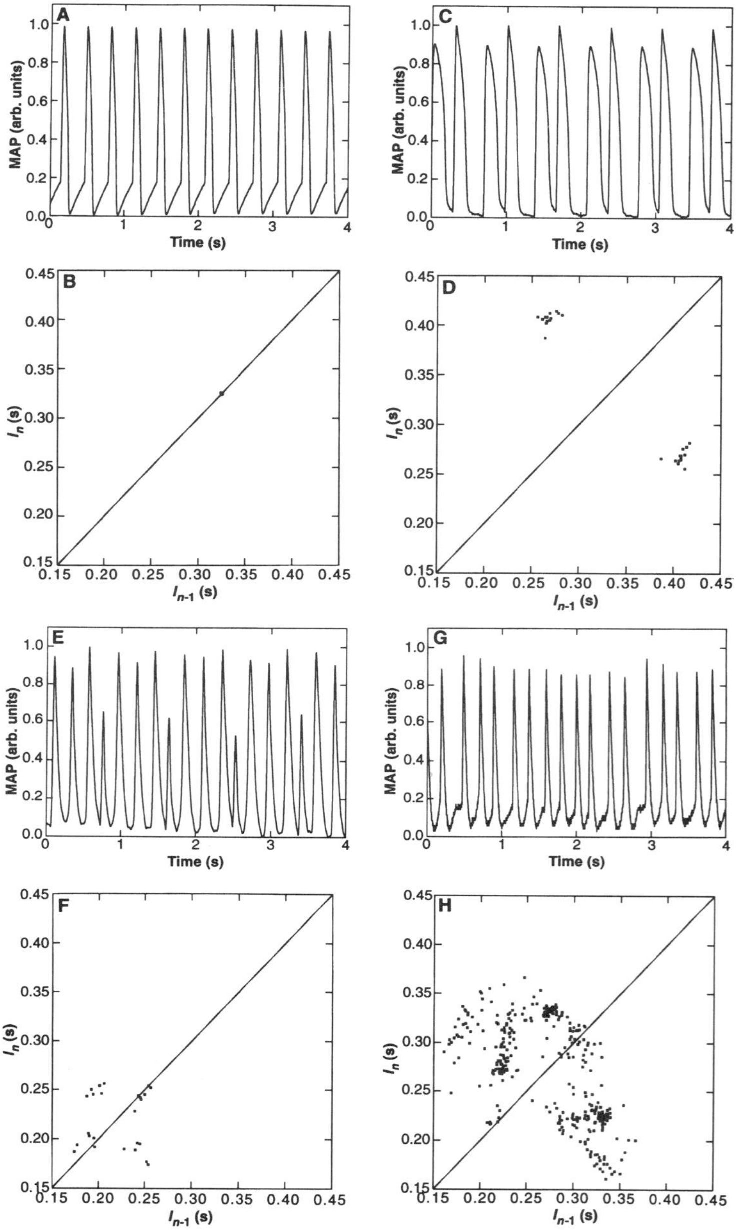

Fig. 1. Recordings of monophasic action potentials (MAPs) (**A**, **C**, **E**, and **G**) and their respective Poincaré maps of interbeat intervals (**B**, **D**, **F**, and **H**) at various stages during arrhythmias induced by ouabain-epinephrine in typical rabbit septa. Typically the arrhythmia was initially characterized by spontaneous periodic beating at a constant interbeat interval (A and B), then developed bigeminal or period 2 patterns (C and D) or higher order periodicities such as a period 4 pattern (E and F), and lastly a completely aperiodic pattern (G and H). Note that in the Poincaré map of the final stage the points form an extended structure that is not point-like or a set of points (that is, not periodic) and is not space-filling (that is, not random). This is a sign of chaos.

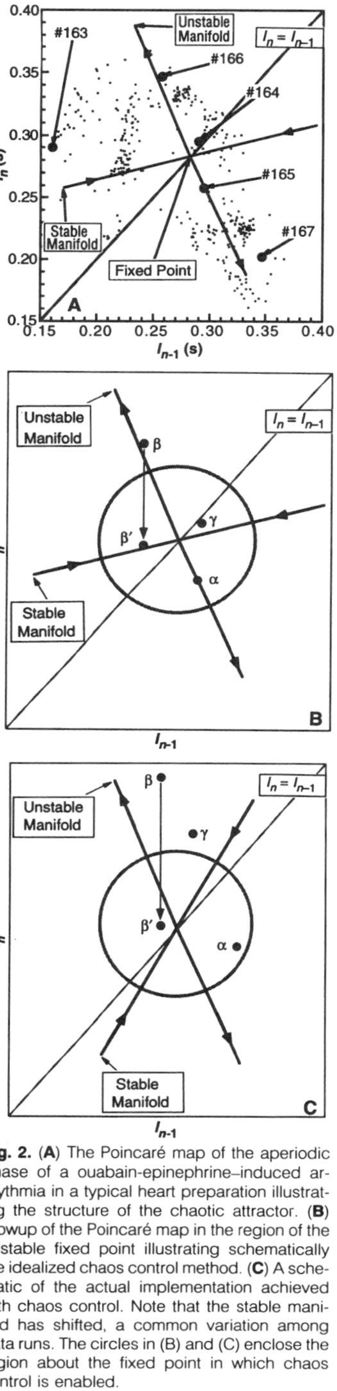

Fig. 2. (**A**) The Poincaré map of the aperiodic phase of a ouabain-epinephrine–induced arrhythmia in a typical heart preparation illustrating the structure of the chaotic attractor. (**B**) Blowup of the Poincaré map in the region of the unstable fixed point illustrating schematically the idealized chaos control method. (**C**) A schematic of the actual implementation achieved with chaos control. Note that the stable manifold has shifted, a common variation among data runs. The circles in (B) and (C) enclose the region about the fixed point in which chaos control is enabled.

ing to progressive severe membrane depolarization from Na^+ pump inhibition. The spontaneous activity induced by ouabain-epinephrine in this preparation showed a number of features symptomatic of chaos. Most important, in progressing from spontaneous beating at a fixed interbeat interval to highly aperiodic behavior, the arrhythmia passed through a series of transient stages that involved higher order periodicities. These features are illustrated in Fig. 1, in which the nth interbeat interval (I_n) has been plotted against the previous interval (I_{n-1}) at various stages during ouabain-epinephrine–induced arrhythmias. Such a Poincaré map (5) allows us to view the dynamics of the system as a sequence of pairs of points (I_{n-1}, I_n), thus converting the dynamics of our system to a map. This map has several special features. The first is that a periodic motion appears as a finite set of points in such a plot (Fig. 1, A to C). When the arrhythmia progressed to the aperiodic stage, truly random aperiodicity would have demonstrated no structure in the Poincaré map. Chaotic aperiodicity however is represented by an extended structure that is not a single point (or finite set of points) and yet is not space-filling. Figure 1D illustrates that the aperiodic phase of the ouabain-epinephrine–induced arrhythmia fell into the latter category, supporting a chaotic rather than a random process. In 11 preparations the Poincaré maps consistently showed patterns similar to Fig. 1D, and these patterns are consistent with chaotic aperiodicity. Figure 1, A to D, as shown are from different preparations. However virtually all preparations exhibited at least one period doubling before the arrhythmia became chaotic (although it is not clear whether the transitions to chaos in our preparations are the result of a period doubling route to chaos).

Method of chaos control. The control of chaos begins with the realization that, where the attractor crosses the line of identity ($I_n = I_{n-1}$), it must contain an unstable periodic motion of period 1. (If it were stable, the attractor would be only a single point lying on the diagonal of Fig. 1). Such a crossing point is called an unstable fixed point and represents constant interbeat intervals (with period 1). These unstable fixed points have associated directions along which the trajectory approaches and diverges from the fixed point. These directions are called the stable and unstable manifolds, respectively. A typical sequence of interbeat intervals during aperiodic beating induced by ouabain-epinephrine in a rabbit septum is indicated in Fig. 2A (points 163 through 167). From point 163 to 164 the state of the system (or state point) moves toward the unstable fixed point. Thus point 163 must lie close to the stable manifold (direction). Points 164 through 167 diverge from the unstable fixed point and hence reveal an unstable manifold (direction). In the chaotic region unstable fixed points on the attractor possess at least one unstable and one stable manifold (*16*). Thus the local geometry around a fixed point in a Poincaré map plot is that of a saddle. In this case the saddle is a flip saddle; that is, while the distances of successive state points from the fixed point increase in an exponential fashion along the unstable manifold (one of the signs of chaos) the state points alternate on opposite sides of the stable manifold (*17*). The flip saddle appears as a short interbeat interval followed by a long interval and vice versa.

Our method of chaos control, which we call proportional perturbation feedback (PPF), consists of delivering a perturbation (near the desired fixed point) that forces the system state point onto the stable manifold of the desired fixed point. Consequently the system will naturally move toward the unstable fixed point rather than away from it. Contrast this with the OGY method, which moves the stable manifold to the current system state point rather than vice versa. Both methods use a linear approximation of the dynamics in the neighborhood of the desired fixed point. OGY then varies a systemwide parameter to move the stable manifold to the system state point; our method perturbs the system state point to move it toward the stable manifold. Thus the magnitude of our perturbation is the same as predicted by the OGY equations, but the sign is opposite. We developed the PPF method when it became obvious that our cardiac preparation possessed no systemwide parameter that could be changed sufficiently quickly to implement classical OGY control.

The procedure begins by determining the location of the unstable fixed point $\boldsymbol{\xi}_F$, as well as its local stable and unstable contravariant eigenvectors, $\mathbf{f}_s$ and $\mathbf{f}_u$ respectively. If $\boldsymbol{\xi}_n$ is the current position of the system on the Poincaré map and p is the predicted timing of the next natural beat, the required advance in timing δp is proportional to the projection of the distance $\boldsymbol{\xi}_n - \boldsymbol{\xi}_F$ onto the unstable manifold (the unstable contravariant eigenvector):

$$\delta p = C(\boldsymbol{\xi}_n - \boldsymbol{\xi}_F)\cdot\mathbf{f}_u \qquad (1)$$

where:

$$C = \frac{\lambda_u}{\lambda_u - 1}\,\frac{1}{\mathbf{g}\cdot\mathbf{f}_u} \qquad (2)$$

The constant of proportionality C depends on the unstable eigenvalue λ_u. This eigenvalue determines the rate of the exponential divergence of the system from the fixed point along the unstable manifold and is easily determined (8) from the sequence of points 164 through 167 in Fig. 2A. Lastly, **g** is the sensitivity of points near the fixed point to an advance in the timing δp, which we approximate with the change in the fixed point with respect to δp:

$$\mathbf{g} \approx \frac{\delta \boldsymbol{\xi}_F}{\delta p} \qquad (3)$$

The constant C is inversely proportional to the projection of this change onto the unstable manifold.

We must stress that Eqs. 1 to 3 are identical to the OGY control equations except that the systemwide parameter p used in the OGY method is replaced in our method by the variable that controls the system perturbation. Thus our system perturbation δp (which replaces the change, δp, in the OGY systemwide parameter) represents the amount of time we must shorten an anticipated natural beat (through the introduction of a stimulus) to force the state point onto the stable manifold.

Our proportional perturbation feedback control consisted of two parts: a learning phase and an intervention phase. In the learning phase the computer monitored the interbeat intervals until it determined the approximate locations of the unstable fixed point and the stable and unstable manifolds (9, 10). The application of Eqs. 1 and 2 during the intervention phase was then straightforward once the geometry of the local map around the fixed point had been determined and the quantity **g** had been found. The interbeat interval was directly manipulated by shortening it with an electrically stimulated beat. The advance in the timing of the next interbeat interval is the perturbation δp. The sensitivity of the state point to changes in p was determined experimentally by noting the change in the interbeat interval in response to a single electrical stimulus when the state point was near the fixed point (Eq. 3).

Chaos control with this approach was complicated by the fact that the intervention was, of necessity, unidirectional; that is, by delivering an electrical stimulus before the next spontaneous beat, the interbeat interval could be shortened but it could not directly be lengthened. This is because a stimulus that elicits a beat from the heart must, by definition, shorten the interbeat interval between the previous spontaneous beat and the beat elicited by the stimulus. Although a stimulus that does not elicit a beat may prolong the interbeat interval by electrotonic effects, in the intact heart this is highly dependent both on the precise location of the pacing site in relation to the site of origin of the arrhythmia and on the history of preceding interbeat intervals. Thus such a stimulus is highly unstable and unpredictable during an aperiodic rhythm and therefore is unsuitable as a cardiac pacing strategy.

The learning phase typically lasted from 5 to 60 seconds, after which the computer waited for the system to make a close approach to the unstable fixed point (indi-

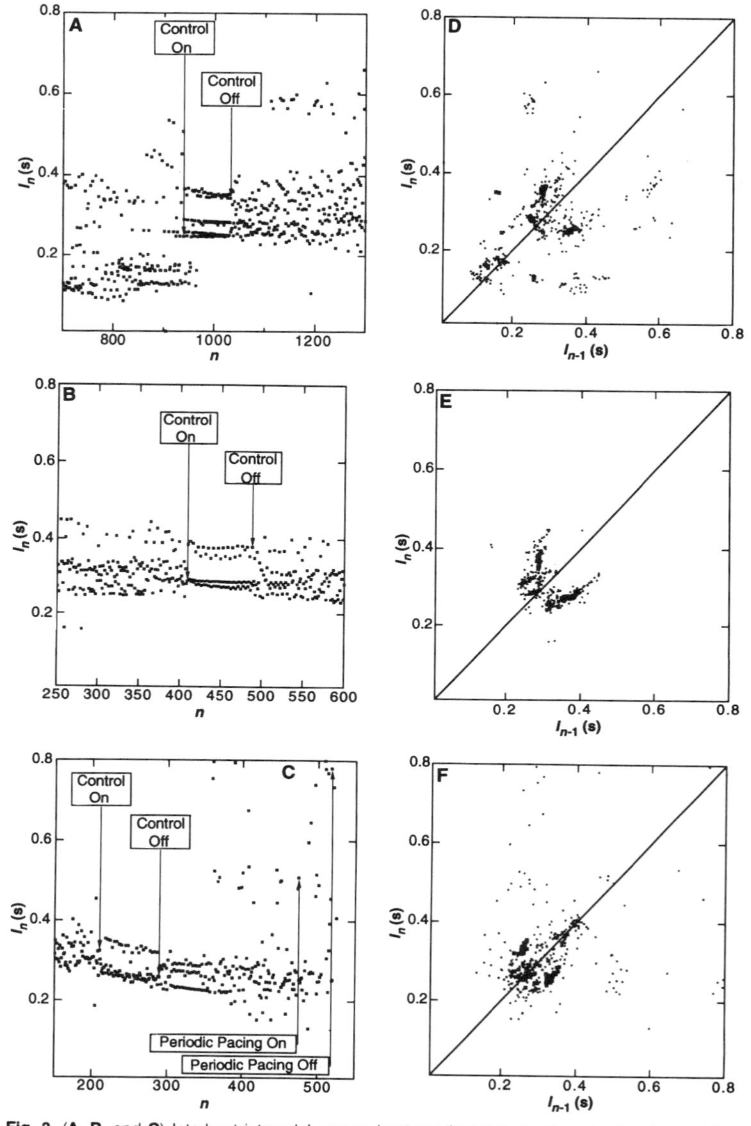

Fig. 3. (**A**, **B**, and **C**) Interbeat interval I_n versus beat number n during the chaotic phase of the ouabain-epinephrine–induced arrhythmia in three typical hearts. The region of chaos control is indicated. In (C) a region during which periodic pacing was applied is also indicated. Note that periodic pacing failed to control the arrhythmia. (**D**, **E**, and **F**) The corresponding Poincaré maps for (A), (B), and (C), respectively. The small points represent interbeat intervals during the uncontrolled arrhythmia, while the larger points represent interbeat intervals during the chaos control. In the (A) and (D) data set, there were several losses of control that occurred early in the control sequence (indicated by the lower points in the early part of the control region). These were always immediately followed by reacquisition of the chaos control.

cated by the point α within the circle in Fig. 2B.) The next point would normally fall further out along the unstable manifold (as well as on the opposite side of the stable manifold) as indicated by point β. However at this point the computer intervened by injecting an electrical stimulus early enough so that this point actually occurred at β', lying directly below β and, by construction, near the stable manifold. Because the system now lay close to the stable manifold, ideally the subsequent beat would tend to move closer to the fixed point along the stable manifold, as indicated by the point γ. Thus the state point would be confined to the region near the unstable fixed point, thereby regularizing the arrhythmia. However, in actual practice this degree of accuracy was not typically obtained. Figure 2C illustrates the usual result. When β' did not fall precisely on the stable manifold, γ often was not extremely close to the fixed point (fell outside the circle) but still lay fairly close to the stable manifold (since β' was close to the stable manifold). Yet γ lay closer to the stable manifold than point α did. Thus the next point fell within the circle in the vicinity of α and restarted the cycle (as a new point α). As these patterns repeated, a period 3 beating resulted. In this manner the chaotic (arrhythmic) beating was made periodic by only intermittent stimuli. The subsequent stabilized period 3 motion represents the combined dynamics of the cardiac tissue, sensor, and computer control. We emphasize that this approach did not require any theoretical model of the heart; all of the quantities needed were calculated in real time from the data. It should also be noted that a control stimulus has only a transient effect on the cardiac preparation (which may influence the timing of several subsequent beats) but otherwise does not affect the preparation in any permanent fashion.

Results of chaos control. Using PPF, control of the chaotic phase of the ouabain-epinephrine–induced arrhythmia was attempted in 11 separate experimental runs and was successful in 8. Figure 3 shows the results of three successful cases. When the arrhythmia became chaotic, the chaos control program was activated. Our criteria for chaos were that the Poincaré map exhibit stable and unstable manifolds with a flip saddle along the linear part of the unstable manifold (*18*). Specifically we looked for the state point to walk toward the unstable fixed point along one direction (thereby determining the stable manifold) and then to walk out from the unstable fixed point along a clearly different direction (the unstable manifold). The program then chose and delivered electrical stimuli as described above. In order to prove that we had achieved and maintained control of the chaos [defined as a clear conversion of a chaotic sequence to a periodic one with a low-integer (2 to 3) period], we turned off the chaos control program and consistently saw a return to chaotic behavior, sometimes preceded by transient complex periodicities, as in Fig. 3C.

Several observations should be made about the pattern of the stimuli delivered by the chaos control program. First, these stimuli did not simply overdrive the heart. Stimuli were delivered sporadically, not on every beat and never more than once in every three beats on average. The pattern of stimuli was initially erratic and aperiodic but soon became approximately periodic as the arrhythmia was controlled into a nearly periodic rhythm. For example, in Fig. 3, the chaos control program rapidly converted the aperiodic behavior of the arrhythmia to a period 3 rhythm. In contrast, periodic pacing, in which stimuli were delivered at a fixed rate, was never effective at restoring a periodic rhythm and often made the aperiodicity more marked. This is illustrated in Fig. 3C, in which the chaos control program converted chaotic behavior to an approximate period 3, whereas periodic pacing at a rate of 75 beats per minute (nearly identical to the time-averaged rate at which stimuli were delivered during chaos control) had no effect. Irregular pacing was similarly ineffective at converting chaotic to periodic behavior, as illustrated in Fig. 3A. In this case chaotic behavior was converted to a period 3 by the chaos control program, but the rhythm quickly became aperiodic again when the parameters of the algorithm were modified by arbitrarily changing C (Eqs. 1 and 2) to eliminate effective chaos control (at the arrow labeled "control off").

In two cases, one of which is shown in Fig. 3A, chaos control had the additional effect of eliminating the shortest interbeat intervals, hence reducing the average rate of the tachycardia. Without an understanding of the chaotic nature of the system, it would seem paradoxical that an intervention that only shortened the interbeat intervals lengthened the average interval. However, because very long interbeat intervals tend to be followed by very short interbeat intervals (a consequence of the properties of a flip saddle), elimination of the very long intervals also tends to eliminate very short intervals. In cases in which very short intervals predominate during the arrhythmia, their elimination during chaos control will tend to lengthen the average interbeat interval.

Future possibilities for chaos control. In the cases where chaos was successfully controlled, the chaotic pattern of the arrhythmia was converted to a low-order periodic pattern. However, as discussed previously for Fig. 2, B and C, we did not observe any period 1 patterns during chaos control. There is no a priori reason why a period 1 pattern cannot be achieved; once period 3 chaos control is established, it is possible that a refinement of the control parameters could "walk" the points along the stable manifold, in essence spiraling in toward the fixed point. This could be implemented in the future either as a second learning phase in which the details of the Poincaré map are learned to higher accuracy than during the first attempts at chaos control or as an adaptive algorithm that responds to the state of the heart after each beat. An adaptive algorithm might also allow chaos control to adapt to changing physiological conditions such as variations in autonomic tone or other extracardiac factors. It may even be possible to use chaos control to walk the fixed point upward along the diagonal (*19*), thereby reducing the rate of the tachycardia.

The relevance of the ouabain-induced arrhythmia model used in our study to clinically important arrhythmias in humans is not established. Although toxicity from cardiac glycosides is a common cause of human arrhythmias, only in lethal doses would it be likely to produce the severe aperiodic arrhythmias observed in our study. However many clinically important rapid cardiac arrhythmias are aperiodic, such as atrial and ventricular fibrillation, polymorphic ventricular tachycardia, and multifocal atrial tachycardia. In cases where aperiodic arrhythmias are examples of deterministic chaos, it is conceivable that a chaos control strategy, perhaps implemented by a "smart" pacemaker, could be used to restore the cardiac rhythm to normal. In this context it is encouraging to note that in several instances in which chaos control was achieved in the ouabain-induced arrhythmia model, the average heart rate decreased as a result of the elimination of very short interbeat intervals (Fig. 3). It remains a challenge to determine whether this chaos control strategy in in vitro cardiac ventricle can be successfully applied to the in vivo heart.

REFERENCES AND NOTES

1. J. P. Gollub and H. S. Swinney, *Phys. Rev. Lett.* **35**, 927 (1975).
2. F. C. Moon, *Chaotic Vibrations* (Wiley, New York, 1987).
3. O. E. Rossler and K. Wegmann, *Nature* **271**, 89 (1978).
4. L. F. Olsen and W. M. Schaffer, *Science* **249**, 499 (1990).
5. J. P. Crutchfield, J. D. Farmer, N. H. Packard, R. S. Shaw, *Sci. Am.* **255**, 46 (December 1986).
6. D. R. Chialvo, R. S. Gilmour, Jr., J. Jalife, *Nature* **343**, 653 (1990); M. R. Guevara, L. Glass, A. Shrier, *Science* **214**, 1350 (1981); A. Garfinkel, D. O. Walter, R. Trelease, R. K. Harper, R. M. Harper, *Life Sci.* **48**, 2189 (1991).
7. M. F. Arnsdorf, *Curr. Opin. Cardiol.* **6**, 3 (1991).

8. C. Grebogi, E. Ott, J. A. Yorke, *Phys. Rev. A* **37**, 1711 (1988); R. Bowen, *Trans. Am. Math. Soc.* **154**, 377 (1971).
9. E. Ott, C. Grebogi, J. A. Yorke, *Phys. Rev. Lett.* **64**, 1196 (1990).
10. W. L. Ditto, S. N. Rauseo, M. L. Spano, *ibid.* **65**, 3211 (1990).
11. E. R. Hunt, *ibid.* **67**, 53 (1991).
12. R. Roy, T. W. Murphy, Jr., T. D. Maier, Z. Gills, E. R. Hunt, *ibid.* **68**, 1259 (1992).
13. J. N. Weiss and K. I. Shine, *Am. J. Physiol.* **243**, H318 (1982). Care and use of all animals was in full accordance with institutional and federal laboratory animal care guidelines. New Zealand White rabbits were killed with an overdose of intravenous pentobarbitol. The composition of the arterial perfusate was (in mM) 120 NaCl, 25 $NaHCO_3$, 4 KCl, 1.5 $CaCl_2$, 1 $MgCl_2$, 0.44 NaH_2PO_4, 5.6 dextrose, and 10 units of insulin per liter of perfusate, pH 7.4 when gassed with a mixture of 95 percent O_2 and 5 percent CO_2.
14. D. Colquhoun, E. Neher, H. Reuter, C. F. Stevens, *Nature* **294**, 752 (1981); D. Fedida, D. Noble, A. C. Rankin, A. J. Spindler, *J. Physiol.* **392**, 523 (1987).
15. J. N. Weiss, in *Cardiac Arrhythmias—Where to Go from Here?*, P. Brugada and H. J. J. Wellens, Eds. (Futura, Mount Kisco, NY, 1987), pp. 83–104.
16. J. Kaplan and J. A. Yorke, in *Functional Differential Equations and Approximation of Fixed Points*, H.-O. Peitgen *et al.*, Eds., "Springer Lecture, Notes in Mathematics" (Springer-Verlag, Berlin, 1979), vol. 730, p. 228.
17. J. Guckenheimer and P. Holmes, *Nonlinear Oscillations, Dynamical Systems, and Bifurcations of Vector Fields* (Springer-Verlag, New York, 1983), p. 105.
18. High-order periodic orbits were specifically disallowed when looking for the chaotic attractor.
19. T. Shinbrot, E. Ott, C. Grebogi, J. A. Yorke, *Phys. Rev. Lett.* **65**, 3215 (1990); T. Shinbrot *et al.*, *ibid.* **68**, 2863 (1992).
20. We thank M. Shlesinger of the Office of Naval Research for his continued support and encouragement, S. Lamp and J. Stuart for technical assistance with the experiments, and J. Middleton for help in preparing this manuscript. Partially supported by the Naval Surface Warfare Center's Independent Research Program and by the Office of Naval Research, Physics Division (M.L.S.) and by NIH grants RO1 HL36729 and RO1 HL44880, Research Career Development Award KO4 HL01890, by the Laubisch Cardiovascular Research Fund, and by the Chizuko Kawata Endowment (J.N.W.).

Stabilizing High-Period Orbits in a Chaotic System: The Diode Resonator

E. R. Hunt

Condensed Matter and Surface Science Program and Department of Physics and Astronomy, Ohio University, Athens, Ohio 45701-2979

(Received 24 June 1991)

The chaotic dynamics found in the diode resonator has been converted into stable orbits with periods up to 23 drive cycles long. The method used is a modification of that of Ott, Grebogi, and Yorke [Phys. Rev. Lett. **64**, 1196 (1990)]. In addition to stabilizing existing low-period orbits, the method allows making small alterations in the attractor permitting previously nonexistent periodic orbits to be stabilized. It is an analog technique and therefore can be very fast, making it applicable to a wide variety of systems.

PACS numbers: 05.45.+b

Recently, there have been considerable theoretical and experimental efforts to control chaos, that is, to convert the chaotic behavior found in many physical systems to a periodic time dependence. Ott, Grebogi, and Yorke [1] (OGY) proposed a general way to achieve this control using a feedback technique in which small, carefully chosen, time-dependent perturbations are made on one of the parameters of the system. Ditto, Rauseo, and Spano [2] implemented the OGY method in a periodically driven physical system, converting its chaotic motion into period-1 and period-2 orbits. Also, laminar flow has been produced in a previously unstable thermal convection loop by a thermostat-type feedback mechanism [3]. Spin-wave instabilities have been suppressed with a nonfeedback technique, the addition of a periodic field [4]. A nonfeedback procedure has also been shown to be feasible for the periodically driven pendulum [5] and the Duffing-Holmes oscillator [6]. None of these systems has been stabilized in a high-period orbit.

We have converted the chaotic dynamics found in a driven diode resonator system to a number of stable orbits with periods as long as 23 drive cycles. The technique used is a modification of the OGY method. Deviations of the chaotic variable within a specified window from a set point are fed back to perturb the controlling drive. In contrast to the OGY method, we allow fairly large perturbations which can change the chaotic system slightly, permitting high-period orbits. Our technique, which is completely analog, allows the system to find stable orbits by itself. In order to find these orbits one varies the set point, window, and amplitude of the feedback; and many different orbits automatically lock in. The technique is fast and could have a wide range of applicability.

It is pointed out in OGY that the presence of chaos, when controllable, can be advantageous. The orbit to be used in a given application can be chosen in order to maximize a system's performance. We have stabilized nineteen different orbits, all initially in the same chaotic regime. Furthermore, we can easily change from one orbit to another. The high-period orbits have the advantage that most regions of the attractor are visited. This is important because the different regions correspond to different physical states of the system, and one may want to sample as many of these as possible.

Ott, Grebogi, and Yorke [1] prescribe a method to transform a system initially in a chaotic state into a controlled periodic one. Their idea is that there exists an infinite number of unstable periodic orbits embedded in the attractor, and that only small, carefully chosen perturbations are necessary to stabilize one of these. Creating new orbits is purposely avoided by requiring that the perturbations be small. To obtain a period-1 orbit, a local map around the desired fixed point, including the stable and unstable directions, is constructed. When the chaotic variable is in the neighborhood of the fixed point, the perturbation is applied to a system parameter so that the next cycle (iterate) will fall on the stable manifold of the point. The chaotic variable will then move toward the fixed point in successive iterations. They verified their technique numerically using the Hénon map, extracting the period-1 orbit, which remained stable in the presence of added noise.

Ditto, Rauseo, and Spano [2] successfully used the OGY method in a physical system comprised of a gravitationally buckled magnetostrictive ribbon. They were able to achieve stable period-1 and period-2 orbits in the chaotic regime by making perturbations, limited to less than 9% in one of the system's available parameters. They found an approximate linear mapping function in the vicinity of the desired fixed point, and used this to calculate how much feedback to apply in order to move the fixed point into the neighborhood of the stable manifold. Because of experimental inaccuracies they could not get the system exactly on the stable manifold, and a new correction was applied each cycle. For period-2 the parameter was adjusted every other cycle.

By the very nature of chaos, high-period orbits are impossible to achieve by making only one correction in the long period. For example, our system has a largest Lyapunov exponent of approximately 0.4, allowing only five cycles on the average before any uncertainty in the chaotic variable increases tenfold. Thus any attempt to obtain a large-N periodic orbit must usually involve multiple corrections. Exceptions to this are possible when the period-N cycle is stable or nearly stable by itself. This situation occurs when one of the iterates falls near the extremum of the first return map.

The system used in this work is the diode resonator, which is composed of a p-n junction rectifier in series with an inductor. When driven with an increasing sinusoidal voltage, the system exhibits the classic period-doubling route to chaos [7]. It is a system well characterized by a two-dimensional map [8]. The peak forward current through the diode provides a convenient chaotic variable. Our resonator utilizes a 1N2858 diode and a 100-mH inductor with 25-Ω dc resistance, and is driven at 53 kHz.

Our method of control uses occasional proportional feedback. The peak current I_n is sampled, and if it is within a given window, the drive voltage is amplitude modulated with a signal proportional to the difference between I_n and the center of the window. If it is not within the window, no modulation signal is applied. The maximum correction is proportional to the size of the window and the system gain, both of which are adjustable. For low-period orbits the method is essentially equivalent to that of OGY. For longer orbits the perturbations can be large enough to alter the attractor slightly, so that a periodic orbit can exist where none did previously.

A block diagram of the system is shown in Fig. 1. The current through the resonator is converted to a voltage by the I/V device. If a peak current signal falls within the adjustable range of the window comparator, which is centered about zero, a trigger is generated for the timing circuit, which in turn generates pulses for the sample/hold (S/H) and the switch. The current peaks may be offset, so that control may be attempted for any amplitude. The deviation of the peak from zero is switched through an amplifier to become the control (feedback) signal, which amplitude modulates the signal generator. The switch can remain closed for a large fraction of a cycle; the remainder of the cycle is used by the S/H. The whole correction process takes less than 20 μs.

Stabilizing low-period orbits is very easy. For period-1, for example, one notes that the fixed point lies on the diagonal of the first return map. The window is opened around the point and the amplitude of the feedback is adjusted until locking occurs. However, it is not essential to know initially where the fixed points are. Scanning the adjustable controls quickly reveals many periodic orbits. Locked into period-1, the control signal is extremely small, corresponding to changes in the drive voltage of less than 0.5%. The technique is quite robust in that the drive voltage may be changed ±10% without losing control.

Figure 2 demonstrates the method for a controlled period-5 orbit. Figure 2(a) is a double exposure of the first return map of the current peaks in the chaotic regime with the five overexposed dots representing the controlled state. Note the small deviations of the dots from

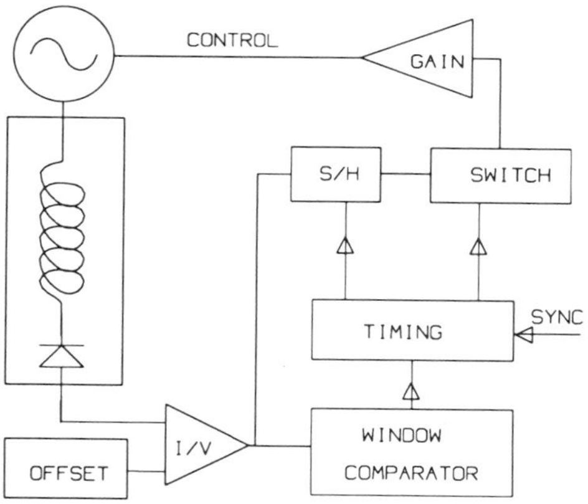

FIG. 1. Block diagram of the system. The diode resonator is boxed. The I/V is a current-to-voltage converter whose peak output is held by the sample/hold (S/H) if the peak is within the window. The held peak is switched and amplified to become the control or feedback signal, which amplitude modulates the signal generator.

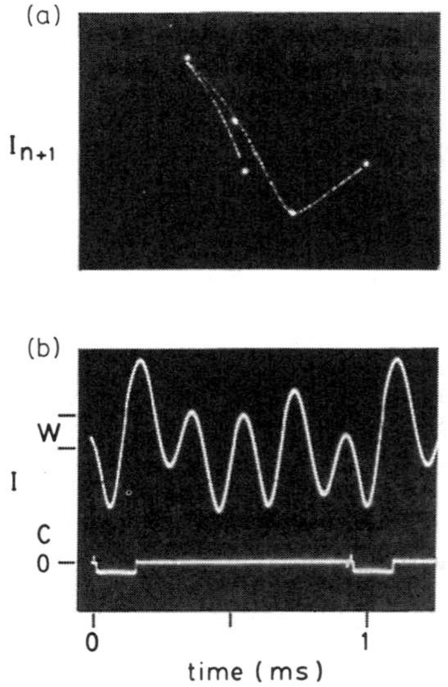

FIG. 2. (a) Double exposure showing the first return map (I_{n+1} vs I_n) and five overexposed dots representing the period-5 stabilized orbit (arbitrary scale). (b) Upper trace: The current through the resonator vs time. Also shown is the window (W). Lower trace: The control signal (C). Only the smallest peak is in the window.

the attractor showing the effect of the control signal. Period-5 is a stable orbit in the uncontrolled system. However, the drive voltage used here is about 5% less than that corresponding to the stable orbit which, therefore, would not exist without the feedback. Figure 2(b) shows the current through the resonator (upper trace) locked to period-5 by the feedback signal (lower trace). The window is designated by W. In this case the feedback signal lasts for only one drive period out of the five. The signal has a negative periodic component, which causes a 3% increase in the drive voltage. The feedback also has a fluctuating component (too small to be seen) which stabilizes the orbit.

A controlled period-21 orbit is shown in Fig. 3. The drive voltage is unchanged and the chaotic attractor is the same as for the period-5 case. As seen in Fig. 3(a), there are some deviations of the dots from the attractor, somewhat larger in this case due to the larger feedback signal. The upper trace of Fig. 3(b) shows the current through the resonator and the window used. The lower trace is the control signal showing six corrections in the 21 cycles. The largest correction shown here corresponds to a 9% modulation of the drive voltage. Some of the correcting perturbations could have no effect or even be opposite to those needed to control, but on the average they must stabilize the orbit or else these orbits would not be found. Note that the different regions of the attractor are represented by a number of dots.

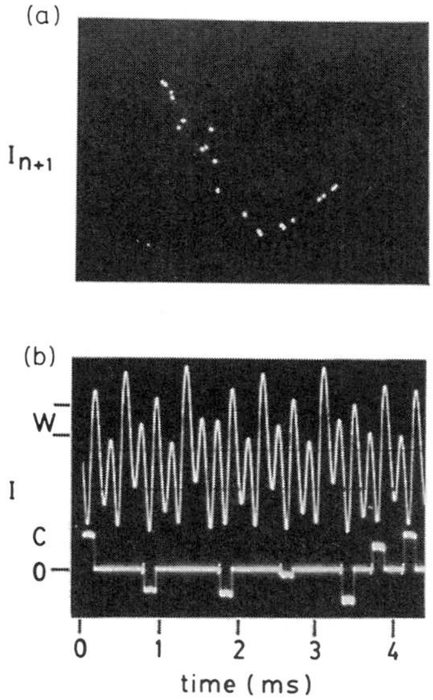

FIG. 3. (a) First return map with dots representing a period-21 orbit. (b) Upper trace: The current and the window used for this orbit. Lower trace: The control signal. The drive voltage is the same as that in Fig. 2.

By changing the level, width of the window, and gain of the feedback signal, we have obtained all periods up to 23 except 13, 14, 15, and 17. Perhaps these would require perturbations larger than we allow. Also, at just one level and gain, we have found periods of 2, 10, 12, 21, and 22 by simply adjusting the width of the window. At present we scan the level, width, and gain to stabilize the different periods. So far we have not found a systematic method to obtain a particular orbit. However, once the conditions are found for particular orbits, they are easily reproduced making it possible to rapidly switch between them.

In conclusion, we have demonstrated that many high-period orbits may be obtained in the chaotic regime of the diode resonator system. The technique involves proportional feedback to amplitude modulate the drive voltage. For low-period orbits only a small control signal is needed, in agreement with that expected from previous work. Larger perturbations, still less than 10% alter the attractor to some extent but can stabilize high-period orbits. These orbits visit most regions of the attractor. We feel that the versatility of the method more than offsets any disadvantage incurred by a small change in the attractor. The method should be applicable to a wide variety of systems. An important advantage is that it can be completely analog and, therefore, works at a fairly high speed.

We wish to acknowledge R. W. Rollins and S. Ulloa for their helpful comments, T. Tigner for contributions to the electronics, and G. Johnson for his assistance in finding stable orbits.

Note added.—After submission of this paper, Peng, Petrov, and Showalter [9] published a paper describing the control of a chaotic system by the same method as described here. Their system, which models a chemical reaction, is described by three coupled equations and a one-dimensional map. They stabilized period-1, -2, and -4.

[1] E. Ott, C. Grebogi, and J. A. Yorke, Phys. Rev. Lett. **64**, 1196 (1990).
[2] W. L. Ditto, S. N. Rauseo, and M. L. Spano, Phys. Rev. Lett. **65**, 3211 (1990).
[3] J. Singer, Y-Z. Wang, and H. H. Bau, Phys. Rev. Lett. **66**, 1123 (1991).
[4] A. Azevedo and S. M. Rezende, Phys. Rev. Lett. **66**, 1342 (1991).
[5] Y. Braiman and I. Goldhirst, Phys. Rev. Lett. **66**, 2545 (1991).
[6] R. Lima and M. Pettini, Phys. Rev. A **41**, 726 (1990).
[7] P. S. Lindsay, Phys. Rev. Lett. **47**, 1349 (1981); J. Testa, J. Perez, and C. Jeffries, *ibid.* **48**, 714 (1982); R. W. Rollins and E. R. Hunt, *ibid.* **49**, 1295 (1982).
[8] See Z. Su, R. W. Rollins, and E. R. Hunt, Phys. Rev. A **40**, 2698 (1989), and references therein.
[9] B. Peng, V. Petrov, and K. Showalter, J. Phys. Chem. **95**, 4957 (1991).

Tracking Unstable Steady States: Extending the Stability Regime of a Multimode Laser System

Zelda Gills, Christina Iwata, and Rajarshi Roy
School of Physics, Georgia Institute of Technology, Atlanta, Georgia 30332

Ira B. Schwartz and Ioana Triandaf
U.S. Naval Research Laboratory, Special Project for Nonlinear Science, Plasma Physics Division, Code 4700.3, Washington D.C. 20375-5000
(Received 23 July 1992)

It is shown that the unstable steady state of a multimode laser system may be stabilized by the occasional proportional feedback technique for dynamical control of chaotic systems. The range of pump excitations over which stabilization can be maintained is extended by more than an order of magnitude through application of a procedure for tracking the unstable steady state as the pump excitation is slowly varied.

PACS numbers: 05.45.+b, 42.50.−p

It was recently demonstrated in several experiments that dynamical control of chaos can be achieved in mechanical, fluid, electronic, and laser systems [1–4]. In these experiments, two related techniques were employed to establish control: the Ott-Grebogi-Yorke (OGY) algorithm [5] and the occasional proportional feedback (OPF) method [3,4,6], both of which consist of applying small, appropriately estimated perturbations to a system parameter. These perturbations stabilize the system on a chosen periodic orbit, or unstable steady state. A crucial question immediately arises: Can we maintain control of a particular orbit or unstable steady state over a wide range of parameter values for the system, including both chaotic and nonchaotic regions?

We study here the particular case of stabilization of an unstable steady state. In the case of the chaotic multimode laser system, it is of great relevance for practical applications that a stabilized steady state be maintained. It is found that when the laser is stabilized by a dynamical control technique, it remains stable only for a very limited range of pump powers, if the parameters of the control circuit are kept fixed. However, it is next demonstrated that through the application of a recently developed procedure for tracking the unstable steady state (through judicious changes in a control parameter), the stabilized steady state can be maintained over a greatly extended range of pump excitations. The tracking procedure was developed [7] particularly for experimental systems where a detailed theoretical model may not be available. For the laser system studied in these experiments, a reasonably accurate model is available [8], but we do not make use of the model for either the control or the tracking procedure. Furthermore, the laser system employed is a higher-dimensional one for which no simple return map is available [4,8], in contrast to the experiments of Refs. [1,3]. Our experiments are thus a striking demonstration that the procedures developed in Ref. [7] may be applied successfully to complex systems with no low-dimensional characterization of the chaotic attractor.

To obtain insight into the tracking technique, let us assume that the chaotic system under study may be controlled through application of a technique that consists of making small perturbations of a system parameter in response to the system output, as measured through detection of a system variable, or suitable combination of variables. Two techniques that have been developed for such control are the OGY method [1,5] and OPF method [3,4,6]. In the OGY algorithm, three basic elements are needed to implement control: a time series, a control point ξ_F about which control is achieved, and an accessible system parameter p. We suppose that there exists an unstable periodic orbit $\xi(p)$ which varies as a function of parameter p. For some initial parameter value p for which the orbit is chaotic, we assume we have determined the eigenvalues and eigenvectors of the orbit, and that the orbit is controlled. The OPF method is essentially a limiting case of the OGY algorithm when the contracting direction is infinite in strength, i.e., the stable eigenvalue λ_s is zero. We review the OGY method here since it is more general and is based on a first-principles attractor reconstruction technique.

To track the orbit, a small change is made in p. Since the orbit location is a function of p, a small error is made if the control point is not changed. To see this, we examine the OGY control algorithm. At a fixed value of p, we choose a small correction $\delta p = C(\xi_{\mathbf{n}} - \xi_{\mathbf{F}})\mathbf{f_u}$, so that the next iterate falls on the local stable manifold. Here, $\xi_{\mathbf{F}}$ is the predicted fixed point, $\xi_{\mathbf{n}}$ is the current iterate of the map, $\mathbf{f_u}$ is the contravariant eigenvector along the unstable direction, and C is a constant $C = \lambda_u/(\lambda_u - 1)\mathbf{g}\cdot\mathbf{f_u}$, where $\mathbf{g} = \partial\xi_{\mathbf{F}}(p)/\partial p$ and λ_u is the unstable eigenvalue. To implement the tracking procedure, we determine the error $\mathbf{e}$ between the predicted fixed point $\xi_{\mathbf{F}}$ and the true fixed point, by examining the mean of the control fluctuations $\langle \delta p \rangle$ about p. If the predicted value of the fixed point is equal to the exact value, it can be shown [7] that the mean of the fluctuations will be close to zero (to within the noise of the experiments). The mean of the fluctuations is given approximately by $\langle \delta p \rangle \approx C\mathbf{e}\cdot\mathbf{f_u}$. Therefore, the error is minimized by varying the estimate of the fixed point at the new parameter value to that which minimizes $|\langle \delta p \rangle|$. This minimization is a correction to the prediction of the fixed point and locates a new point on the branch of orbits, which is now used for further prediction. The connection between the OGY and OPF methods has been addressed by Schwartz, Triandaf, and Roy [9], and it may be shown that the considerations discussed for tracking an orbit are easily extended to the OPF control technique as well.

We have applied the OPF control technique for dynamical stabilization of the unstable steady state of a chaotic multimode Nd:YAG (neodymium doped yttrium aluminum garnet) laser with a nonlinear intracavity KTP (potassium titanyl phosphate) crystal [4,8]. The diode laser pumped solid state laser system displays periodic and chaotic fluctuations of the output intensity for certain operating parameter regimes. To apply the OPF method, an analog electronic feedback system was developed, described in Ref. [4]. The laser output intensity is detected by a photodiode, the signal $V(t)$ from which [corresponding to the intensity $I(t)$] is amplified with a variable gain (proportionality factor "A") and offset with respect to a reference voltage V_{ref} (corresponding to the reference intensity I_{ref}). The signal is sampled periodically with the period determined by an external synchronizing pulse generator. The sampled signal is input to the laser diode driver as a small perturbation on the dc bias level, for a time short compared to the sampling period. A series of minute kicks of fluctuating magnitude and sign are thus applied to the diode drive current. The control parameters are (i) the reference level with respect to which the output intensity is measured, (ii) the proportionality factor A that multiplies $V(t)-V_{\text{ref}}$, (iii) the period T at which the output is sampled, and (iv) the gating period δt over which the correction $A[V(t)-V_{\text{ref}}]$ is applied to the ambient value of the diode drive current. For stabilization of the unstable steady state, I_{ref} is our guess for the average steady-state intensity. If this guess is incorrect, the control signal fluctuations acquire a net dc component. Thus we can zero this dc component by adjustment of I_{ref}, and minimize the control fluctuations by adjustment of the other control parameters.

In Fig. 1, the average value of the Nd:YAG laser output (relative units) is shown as a function of the dc bias applied to the diode laser driver. No control signal is applied. The symbols denote a complicated sequence of stable steady-state, periodic, and chaotic behavior. Note that stable steady-state output is obtained only for a very small range of pump powers, for the given set of laser operating parameters.

Figure 2(a) shows an example of unstable chaotic fluctuations of the output intensity, with no control signal applied. Application of dynamical control results in the stabilization of the steady state, Fig. 2(b). The control signal fluctuations are so small as to be virtually indistinguishable from noise in the digital oscilloscope trace and are significantly less than 1% of the ambient dc bias. The stabilized steady-state intensity has the same average value as that of the chaotic output. The fluctuations of the laser intensity are a few percent of the steady-state dc

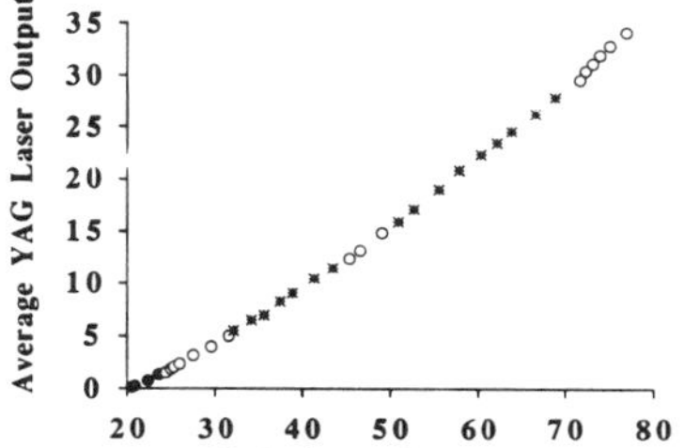

FIG. 1. The average laser output (relative units) at 1.06 μm, without application of the control signal. The symbols designate steady-state (●), periodic (○), and chaotic (∗) behavior of the laser. Stable operation is obtained only very near threshold; for higher pump powers, a complex sequence of periodic and chaotic behavior is found.

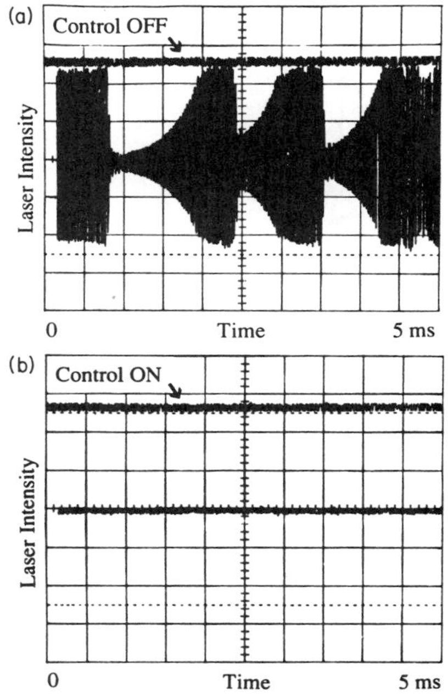

FIG. 2. (a) Digital oscilloscope trace showing the chaotic fluctuations of the laser output intensity vs time. The pump power is about 53 mW, and no control signal is applied (upper trace). (b) The stabilized steady-state intensity vs time with the control signal applied. The upper trace is the control signal; its fluctuations are submerged in the noise level of the digital oscilloscope.

level, reduced from the nearly 100% fluctuations in the chaotic state. These results demonstrate that it is possible to successfully stabilize the unstable steady state for a given pump power (about 50 mW) and control parameter settings.

We display in Fig. 3(a) the consequence of changing the pump power for fixed control parameters. The laser retains stable steady state only for the limited range of pump power indicated by the solid dots, giving rise to complex periodic and chaotic oscillations for small changes of the diode laser power. Figure 3(b) shows that the control signal fluctuations grow on either side of the control points as the pump power is varied, becoming several times larger (~3% to 10% of the dc value) than those at the optimized control point (less than ~1%).

Next, we show the results of application of the tracking procedure. The pump power was steadily increased in small steps from a value close to threshold, where stable steady state was obtained without control. When unstable oscillations of the output intensity were detected, control was switched on, and the reference intensity level I_{ref} was optimized to obtain an essentially zero dc component of the control signal. The standard deviations of the fluctuations of the control signal were also minimized by adjustment of T, δt, and A. It was found that the laser could be stabilized over the entire range of pump power shown in Fig. 4, by tracking the steady state through adjustment of I_{ref} as the pump power was increased, and through minor adjustments in T and δt. The intensity fluctuations about the stabilized steady state are typically more than an order of magnitude smaller than the almost full scale unstable periodic and chaotic fluctuations.

A comparison of Figs. 1 and 4 immediately shows that the tracking technique allows us to obtain about 15 times more output power in a stable steady state for a given set of laser operating parameters. The stabilized steady-state values are very close to the average values of the fluctuating unstable laser output for the same pump power. The control signal fluctuations are a small perturbation and almost no extra pump energy input is required to stabilize the system. Steady-state operation achieved in this manner is extremely stable for long periods of time (many minutes).

In conclusion, we have introduced a new procedure to track an unstable steady state in a multimode laser system as the pump power is varied over a wide range. The results demonstrate over an order of magnitude extension of the stability regime, from a pump power about 20% above threshold to more than 300% above threshold. The control technique requires only small perturbations of the pump power and excellent stability is maintained in the laser output.

A detailed model of the system is not required. However, it is known [8] that the system has a higher-dimensional attractor, with no simple return map. Several globally coupled longitudinal modes oscillate simultaneously in different polarizations. Our results indicate that complex systems of nonlinear oscillators may be stabilized, and the regime of stability extended, through the combination of tracking and control procedures demonstrated here. The technique should be applicable to a wide variety of electronic, fluid, and mechanical systems as well as to chemical and biological processes.

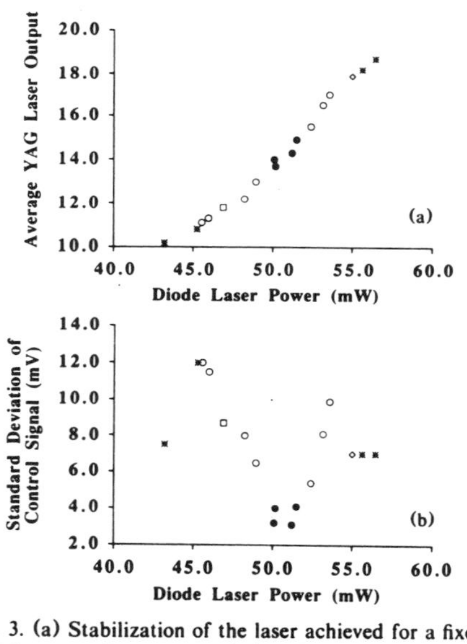

FIG. 3. (a) Stabilization of the laser achieved for a fixed setting of the control parameters is lost for small changes in the pump power. The solid dots indicate the range of stable operation. Period 1 (○), period 4 (□), period 10 (◇), and chaotic (∗) operation were observed as the pump power was changed without any tracking procedure applied to the control parameters. (b) The standard deviation of the control signal fluctuations is shown for the range of pump powers investigated in (a). The fluctuations grow sharply on either side of the optimized control point.

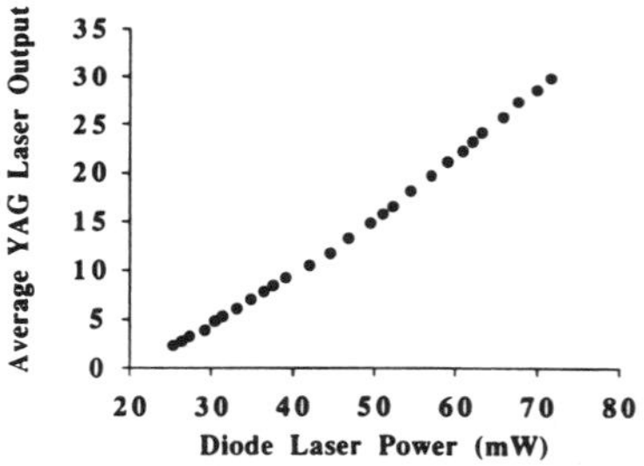

FIG. 4. Stable steady-state output of the laser (same scale as in Fig. 1) vs pump power. Control and tracking were both applied. The regime of stable operation was extended from about 20% above threshold to more than 300% above threshold.

R.R. thanks Pere Colet and Kurt Wiesenfeld for very helpful discussions. He also thanks Jack Hale for an early remark on tracking orbits and a discussion of control strategies for nonlinear systems. Z.G. was supported by an AT&T graduate fellowship. C.I. was supported by an NSF REU program, and R.R. by NSF Grant No. ECS-9114232. We thank Tom Maier for his constant help with electronics instrumentation. I.T. gratefully acknowledges the support of the Office of Naval Technology.

---

[1] W. L. Ditto, S. N. Rauseo, and M. L. Spano, Phys. Rev. Lett. **65**, 3211 (1990).

[2] J. Singer, Y-Z. Wang, and H. H. Bau, Phys. Rev. Lett. **66**, 1123 (1991).

[3] E. R. Hunt, Phys. Rev. Lett. **67**, 1953 (1991).

[4] Rajarshi Roy, T. Murphy, Jr., T. D. Maier, Z. Gills, and E. R. Hunt, Phys. Rev. Lett. **68**, 1259 (1992).

[5] E. Ott, C. Grebogi, and J. A. Yorke, Phys. Rev. Lett. **64**, 1196 (1990).

[6] B. Peng, V. Petrov, and K. Showalter, J. Phys. Chem. **95**, 4957 (1991); V. Petrov, B. Peng, and K. Showalter, J. Chem. Phys. **96**, 7506 (1992).

[7] Ira B. Schwartz and I. Triandaf, Phys. Rev. A (to be published); T. Carroll, I. Triandaf, I. B. Schwartz, and L. Pecora, Phys. Rev. A **46**, 6189 (1992). The tracking procedure is applied to an electronic Duffing oscillator in this paper.

[8] C. Bracikowski and Rajarshi Roy, Chaos **1**, 49 (1991); Rajarshi Roy, C. Bracikowski, and G. E. James, in Proceedings of the International Conference on Quantum Optics, edited by G. S. Agarwal and R. Inguva (Plenum, New York, to be published).

[9] Ira B. Schwartz, Ioana Triandaf, and Rajarshi Roy (to be published).

Controlling chaos in the Belousov–Zhabotinsky reaction

Valery Petrov, Vilmos Gáspár*, Jonathan Masere & Kenneth Showalter*

Department of Chemistry, West Virginia University, Morgantown, West Virginia 26506-6045, USA

DETERMINISTIC chaos is characterized by long-term unpredictability arising from an extreme sensitivity to initial conditions. Such behaviour may be undesirable, particularly for processes dependent on temporal regulation. On the other hand, a chaotic system can be viewed as a virtually unlimited reservoir of periodic behaviour which may be accessed when appropriate feedback is applied to one of the system parameters[1]. Feedback algorithms have now been successfully applied to stabilize periodic oscillations in chaotic laser[2], diode[3], hydrodynamic[4] and magnetoelastic[5] systems, and more recently in myocardial tissue[6]. Here we apply a map-based, proportional-feedback algorithm[7,8] to stabilize periodic

* Permanent address of V. G. is: Department of Physical Chemistry, Kossuth L. University, PO Box 7, 4010 Debrecen, Hungary. Correspondence should be addressed to K.S.

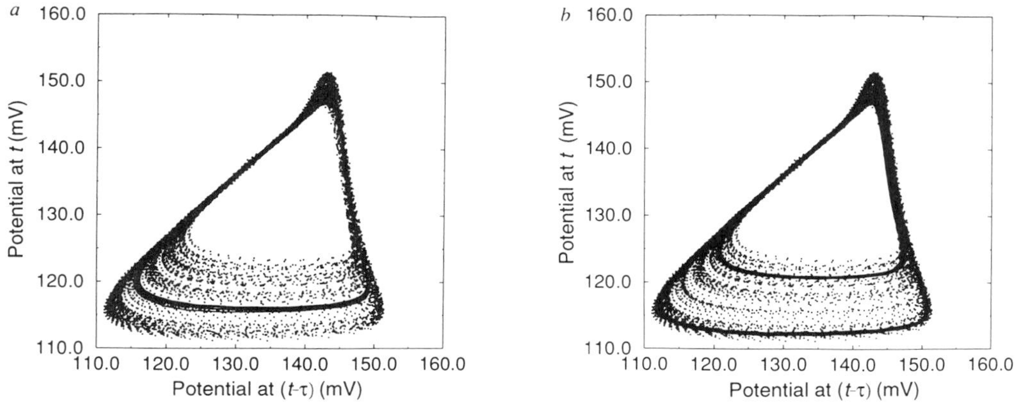

FIG. 1 Stabilized period-1 and period-2 limit cycles embedded in strange attractor of the BZ reaction. Scattered points show chaotic trajectory in time-delay phase space ($\tau = 13$ s) with a reactor residence time of 2.8×10^3 s and concentrations, after mixing, of [malonic acid] $= 2.22 \times 10^{-1}$ M, $[Ce_2(SO_4)_3] = 4.50 \times 10^{-4}$ M, $[NaBrO_3] = 1.02 \times 10^{-1}$ M and $[H_2SO_4] = 0.200$ M. Solid curves show (*a*) period-1 limit cycle stabilized by using $A_s = 33.0$ mV and $g = 18.0$ in equation (1) and (*b*) period-2 limit cycle stabilized by using $A_s = 26.9$ mV and $g = 33.4$. Reaction was carried out in a continuously stirred tank reactor (volume 34.0 ml, stirring rate 1,800 r.p.m.) maintained at 28.0 ± 0.1 °C and fed with separate solutions of malonic acid, cerous sulphate and sodium bromate. Two peristaltic pumps were used, with the malonic acid solution delivered at a fixed flow rate by one, and the cerium and bromate solutions (acidified with sulphuric acid) delivered by the other at a rate regulated by a computer. The potential of a bromide-ion-selective electrode was monitored and collected at a frequency of 10 Hz by a 16-bit data-acquisition board. Moving averages of 10 values were calculated and stored each second, and these values were numerically filtered using a 5-s characteristic period.

behaviour in the chaotic regime of an oscillatory chemical system: the Belousov–Zhabotinsky reaction.

The dynamical behaviour of a chaotic system can be visualized by the strange attractor (an attractor in which the trajectory never exactly repeats itself) traced out by its trajectory in phase space. An infinite number of unstable periodic orbits are embedded in such an attractor, each characterized by a distinct number of oscillations per period. In low-dimensional systems, these orbits are simple saddle-type limit cycles with an attracting stable manifold and a repelling unstable manifold. A general algorithm for stabilizing such orbits, based on targeting the stable manifold of the limit cycle by applying perturbations to a system constraint, has been developed by Ott, Grebogi and Yorke[1] (OGY). The stable manifold is found in a particular Poincaré section in the phase space, and it is targeted in this section each period of oscillation. For systems exhibiting low-dimensional chaos characterized by effectively one-dimensional maps, the OGY method can be reduced to a simple map-based algorithm[7,8]. The simplified method is more convenient in experimental applications because it minimizes the targeting procedures, and it has been used to control chaos in laser[2] and diode[3] systems.

The best-studied example of an oscillatory chemical system is the Belousov–Zhabotinsky (BZ) reaction, in which Ce(IV)/Ce(III) catalyses the oxidation and bromination of $CH_2(COOH)_2$ (malonic acid) by BrO_3^- in H_2SO_4. If the reaction is carried out in a continuous-flow stirred-tank reactor, the flow rate of the reactants into the tank ultimately determines whether the system exhibits steady-state, periodic or chaotic behaviour. Here we use conditions similar to the low-flow-rate Texas experiments[9-11], which ensure that the system is maintained within the chaotic regime (see Fig. 1). An important difference in our experiment is that we apply feedback to the system by perturbing the rate at which the cerium and bromate solutions are fed into the tank (the flow rate of the malonic acid being fixed), permitting the targeting and stabilization of periodic behaviour within the chaotic regime.

Figure 1*a* shows the strange attractor and the stabilized period-1 limit cycle for the BZ reaction. The time-delay phase portrait was reconstructed *in situ* from smoothed values of the potential of a bromide electrode. Except for small positive and negative perturbations to the flow rate of the bromate and cerium reactant stream, the operating conditions were identical for the chaotic and periodic behaviour. Figure 1*b* shows the stabilized period-2 limit cycle embedded in the strange attractor, again obtained with the same average reactant-stream flow rate. The oscillatory behaviour can be switched between period-1, period-2 and chaos by simple adjustments of the proportional feedback. Each time the controlling experiment was repeated (more than 10 times), a bifurcation diagram was first constructed to locate a suitably wide range of chaotic behaviour arising from a period-doubling cascade. The flow rate of the bromate and cerium reactant stream was chosen as the bifurcation parameter, as this choice gave the widest range of period-doubling chaos.

The control algorithm takes advantage of the predictable evolution of a chaotic system in the vicinity of a fixed point in the next-amplitude map, corresponding to a particular unstable periodic orbit. Shown in Fig. 2*a* is the next-amplitude map for the strange attractor and the period-1 orbit shown in Fig. 1*a*. The position of the period-1 fixed point is given by the intersection of the map with the bisectrix, where system state $A_{n+1} = A_n$. As the chaotic system traverses the attractor, the region near the fixed point is eventually visited, and the control algorithm is then activated. Control is achieved by perturbing a system constraint such that the fixed point is targeted on each return.

A shift of the map occurs on varying a system constraint, and this shift can be used to target any particular fixed point. Shown in Fig. 2*b* is a next-amplitude map constructed from the Györgyi–Field[12] model of the BZ reaction with concentrations and residence time similar to the experimental values for Fig. 2*a*. The inset shows a blowup of the map around the period-1 fixed point (right) and the shifted map (left) obtained at a slightly different value of the bifurcation parameter μ (the flow rate of the bromate and cerium reactant stream). For small perturbations, the shift in the linear region around the fixed point is directly proportional to the variation in μ. The map can therefore be shifted to target the fixed point by applying a

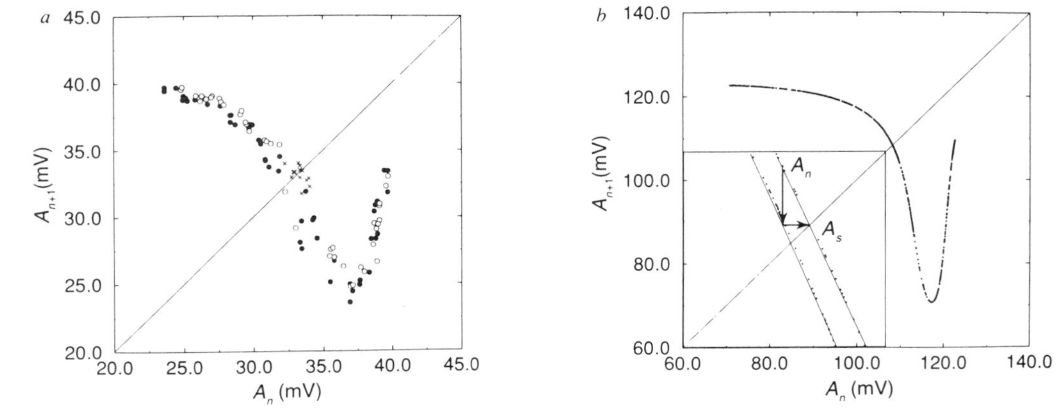

FIG. 2 *a*, Next-amplitude map before (●) and after (○) controlling period-1 and period-2, respectively, as shown in Figs 1 and 3*a*. Also shown are values (×) near fixed point during control corresponding to period-1 limit cycle trajectory in Fig. 1*a*. *b*, Next-amplitude map calculated from three-variable Györgyi–Field model of the BZ reaction with rate constants and residence time (2.17×10^3 s) the same as in ref. 12, and concentrations as in Fig. 1 except [malonic acid] = 0.24 M. Inset shows a 40-fold enlargement of map around fixed point (right) and shifted map (left) resulting from a 0.2% change in the flow rate of the bromate and cerium reactant stream. The system at state A_n is directed to A_s on the next return by varying μ according to equation (1) to shift map[7,8].

perturbation to the bifurcation parameter according to the difference between the system state A_n and the fixed point A_s

$$\Delta\mu = (A_n - A_s)/g \qquad (1)$$

where g is a constant. The current amplitude A_n is measured and with the value of A_s obtained from the map, the value of $\Delta\mu$ necessary for the fixed point to be targeted is calculated. The value of the proportionality constant g can be determined by measuring the horizontal distance between two maps constructed at slightly different values of μ (refs 7, 8), as shown in Fig. 2*b*. Various period-k limit cycles can be similarly stabilized by using the corresponding values of A_s and g (obtained from the appropriate maps of A_{n+k} against A_n) to determine $\Delta\mu$ in equation (1). The control algorithm may be implemented in several variations; for example, the perturbation determined every kth return may be applied for the entire period or for only a fraction of the period. In the stabilization of period-1 and period-2 shown in Fig. 1, the perturbation was applied for 15 s on each return.

Experimental fluctuations in the measured bromide potential result in significant scatter in the next-amplitude map, as shown in Fig. 2*a*. To reduce experimental noise, next-amplitude maps were used in the control algorithm rather than next-return maps; similar results were obtained with slightly more noise using

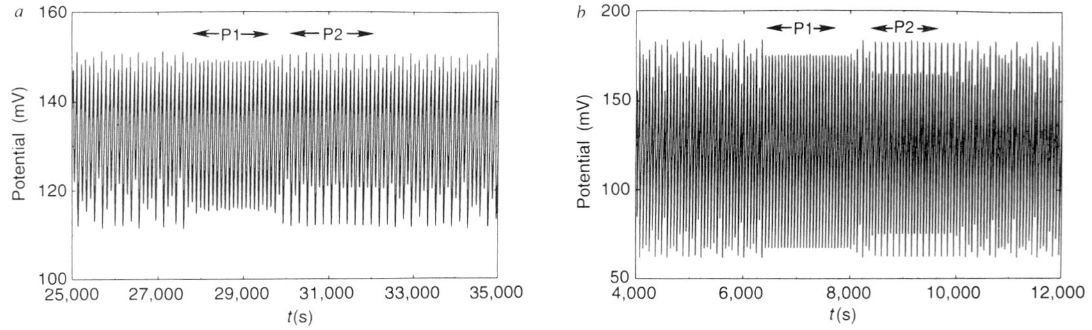

FIG. 3 *a*, Potential of bromide electrode as a function of time in BZ reaction with conditions given in Fig. 1. The control algorithm was switched on to stabilize period-1 at $t = 27{,}800$ s and switched off at $t = 29{,}500$ s, with the parameters corresponding to Fig. 1*a*. The parameters were changed at $t = 30{,}000$ s to stabilize period-2, with values corresponding to Fig. 1*b*, until $t = 32{,}100$ s when control was switched off. The control range was set at ±3.0 mV for both period-1 and period-2. *b*, Oscillations in bromide potential calculated from Györgyi–Field[12] model showing chaos, stabilized period-1 and period-2, and the reappearance of chaos. Potential calculated from the Nernst equation and bromide concentration assuming an Ag/AgCl reference electrode potential of 197 mV. Period-1 limit cycle stabilized by using $A_s = 108.25$ mV and $g = -9.0 \times 10^4$ and period-2 limit cycle stabilized by using $A_s = 90.235$ mV and $g = 7.0 \times 10^4$ in equation (1). Controlling algorithm was applied on each return for 25 s and the control range was set at ±5.0 mV. In both calculation and experiment, the flow rate of the bromate and cerium reactant stream was varied during control; a residence-time decrease of 1.0% therefore gives rise to an increase in these concentrations of 0.5% and a decrease in the malonic acid concentration of 1.0%. The typical residence-time variation during control was ~1.0% in experiment and ~10^{-5}% in calculation.

next-return maps. Although the experimental uncertainty seems to be comparable to that in previously reported maps for chaos in the BZ reaction[9,10], the scatter prevented the direct measurement of the proportionality constant g. The value of g was therefore estimated from the shift of A_s with variation of μ in the bifurcation diagram, and the final value was determined by adjustments around the initial estimate. Once the values for the fixed points and corresponding proportionality constants had been determined, the system could be switched between period-1 and period-2 behaviour and chaos by appropriately changing the values. Figure 3*a* shows the effect of the control algorithm, with the stabilization of the period-1 limit cycle followed by the period-2 limit cycle and the reappearance of chaotic behaviour when control was switched off. Transient aperiodic oscillations appear between period-1 and period-2 as the system leaves the period-1 orbit to arrive eventually at the control range of the period-2 orbit. When control was switched off, the chaotic behaviour was much the same as before the application of the feedback algorithm (Fig. 2*a*). Calculated behaviour using the Györgyi-Field[12] model is shown in Fig. 3*b*, with the stabilization of period-1 followed by period-2 and then a return to chaotic behaviour.

The stabilization of periodic orbits in chaotic systems by imposed feedback was first proposed by Ott, Grebogi and Yorke[1], and implications for biological self-regulation as well as practical uses such as convenient waveform generation have been pointed out[1-6]. The quenching techniques of Sørensen and Hynne[13-15] for targeting the unstable stationary state following a supercritical Hopf bifurcation are closely related to the OGY method. These authors have also successfully targeted the period-1 orbit in the BZ system following the initial period-doubling bifurcation, resulting in the transient appearance of these oscillations[15]. The map-based algorithm is more convenient than the OGY method for some experimental settings. (The OGY method has successfully been applied to stabilize period-1 in our BZ experiment, with results similar to those shown in Fig. 1*a*.) The map-based algorithm is especially useful in controlling high-frequency chaos, such as in laser[2] and diode[3] systems. In these applications, as well as in our present experiments, it is advantageous to apply the controlling perturbation for only a fraction of the oscillatory period. The system is attracted by the stable manifold of the unstable limit cycle during the perturbation-free fraction, thereby reducing the targeting error by ensuring that it resides effectively on the unstable manifold on its next return.

An important feature of both the OGY method and the map-based algorithm is that no knowledge of the underlying dynamics of a system (that is, the governing differential equations) is necessary for stabilizing any particular periodic orbit. This feature can be exploited to characterize experimentally the bifurcation behaviour of a chaotic (or periodic) system by tracking the unstable orbits as a bifurcation parameter is varied. The technique is similar to the computational algorithm AUTO[16] for exploring the bifurcation behaviour of model systems.[17] The tracking algorithm based on stabilizing periodic orbits, however, does not depend on model descriptions. We have made preliminary investigations of the BZ reaction using this technique with promising results; period-1 can easily be tracked after the first period-doubling bifurcation. □

Received 15 October; accepted 11 December 1992.

1. Ott, E., Grebogi, C. & Yorke, J. A. *Phys. Rev. Lett.* **64,** 1196-1199 (1990).
2. Roy, R., Murphy, T. W., Maier, T. D., Gills, Z. & Hunt, E. R. *Phys. Rev. Lett.* **68,** 1259-1262 (1992).
3. Hunt, E. R. *Phys. Rev. Lett.* **67,** 1953-1955 (1991).
4. Singer, J., Wang, Y-Z. & Bau, H. H. *Phys. Rev. Lett.* **66,** 1123-1125 (1991).
5. Ditto, W. L., Rauseo, S. N. & Spano, M. L. *Phys. Rev. Lett.* **65,** 3211-3214 (1990).
6. Garfinkel, A., Spano, M. L., Ditto, W. L. & Weiss, J. N. *Science* **257,** 1230-1235 (1992).
7. Peng, B., Petrov, V. & Showalter, K. *J. phys. Chem.* **95,** 4957-4959 (1991).
8. Petrov, V., Peng, B. & Showalter, K. *J. chem. Phys.* **96,** 7506-7513 (1992).
9. Simoyi, R. H., Wolf, A. & Swinney, H. L. *Phys. Rev. Lett.* **49,** 245-248 (1982).
10. Coffmann, K. G., McCormick, W. D., Noszticzius, Z., Simoyi, R. H. & Swinney, H. L. *J. chem. Phys.* **86,** 119-129 (1987).
11. Noszticzius, Z., McCormick, W. D. & Swinney, H. L. *J. phys. Chem.* **91,** 5129-5134 (1987).
12. Györgyi, L. & Field, R. J. *Nature* **355,** 808-810 (1992).
13. Hynne, F. & Sørensen, P. G. *J. phys. Chem.* **91,** 6573-6575 (1987).
14. Sørensen, P. G. & Hynne, F. *J. phys. Chem.* **93,** 5467-5474 (1989).
15. Sørensen, P. G., Hynne, F. & Nielsen, K. *Reaction Kinet. Catal. Lett.* **42,** 309-315 (1990).
16. Doedel, E. *AUTO: Software for Continuation and Bifurcation Problems in Ordinary Differential Equations* (Applied Mathematics, California Institute of Technology, Pasadena, 1986).
17. Petrov, V., Scott, S. K. & Showalter, K. *J. chem. Phys.* **97,** 6191-6198 (1992).

ACKNOWLEDGEMENTS. We thank the NSF and the Hungarian Academy of Sciences for financial support, and the donors of The Petroleum Research Fund, administered by the ACS, for partial support of this research. We also thank S. K. Scott, Z. Noszticzius and R. H. Simoyi for discussions.

CHAPTER 14

Control: Targeting and Goal Dynamics

T. Shinbrot, W. Ditto, C. Grebogi, E. Ott, M. Spano, J.A. Yorke
Using the sensitive dependence of chaos (the "butterfly effect") to direct trajectories in an experimental chaotic system
Phys. Rev. Lett. **68**, 2863 (1992).

A. Hübler, E. Lüscher
Resonant stimulation and control of nonlinear oscillators
Naturwissenschaften **76**, 67 (1989).

E.A. Jackson
The entrainment and migration controls of multiple-attractor systems
Phys. Lett. A **151**, 478 (1990).

Editors' Notes

The first reprinted paper in this chapter discusses the "targeting" type of control for chaotic systems. The general problem is to *rapidly* direct the orbit from some given initial condition to a small region about some specified point on the chaotic attractor. Because of the inherent exponential sensitivity of chaotic dynamics to perturbations, one expects that this can be accomplished using only small controlling adjustments. A technique for doing this is devised and demonstrated in numerical experiments in Shinbrot et al. (1990). A laboratory experiment on this is reported in **Shinbrot et al.** (1992). Kostelich et al. (1993) give another technique for targeting control which is simpler to program and implement for higher-dimensional chaotic attractors.

The paper of **Hübler and Lüscher** (1989) discusses how a nonlinear system can be driven toward a given goal dynamics. Specifically they consider a system of the form $\dot{\mathbf{x}} = \mathbf{F}(\mathbf{x}) + \mathbf{u}(t)$, where $\mathbf{u}(t)$ represents the driving. If the goal dynamics is denoted $\mathbf{g}(t)$, then the problem they consider is how to choose $\mathbf{u}(t)$ so that $||\mathbf{x}(t) - \mathbf{g}(t)||$ approaches zero as $t \to \infty$. This convergence does not always occur, and **Jackson** (1990) considers the "entrainment regions" such that a given initial condition actually approaches the goal.

Using the Sensitive Dependence of Chaos (the "Butterfly Effect") to Direct Trajectories in an Experimental Chaotic System

Troy Shinbrot,[(1),(a),(b)] William Ditto,[(2)] Celso Grebogi,[(1),(a),(c),(d)] Edward Ott,[(1),(b),(c),(e)] Mark Spano,[(3)] and James A. Yorke[(1),(a),(d)]

[(1)]*University of Maryland, College Park, Maryland 20742*
[(2)]*Department of Physics, The College of Wooster, Wooster, Ohio 44691*
[(3)]*Naval Surface Warfare Center, Silver Spring, Maryland 20902*
(Received 20 November 1991)

In this paper we present the first experimental verification that the sensitivity of a chaotic system to small perturbations (the "butterfly effect") can be used to rapidly direct orbits from an arbitrary initial state to an arbitrary accessible desired state.

PACS numbers: 05.45.+b

Recently, it was demonstrated [1] theoretically and numerically that orbits on a chaotic attractor can be brought rapidly to a desired state by the application of tiny, judiciously chosen perturbations to an available system parameter. In this Letter we describe the first experimental confirmation of this method.

To illustrate the method in the simplest context (i.e., where the attractor dimension is near 1), assume that the dynamics of the system to be controlled are described by a one-dimensional map [2],

$$X_{n+1}=F(p,X_n)\,. \tag{1}$$

We imagine that the parameter p can be varied by some small amount about its nominal value $\bar{p}$, $p=\bar{p}+\delta p$, and we seek a value for the small perturbation δp in some allowed limited range, $-\Delta p \le \delta p \le \Delta p$, which will take us from a current state, X_s, to a desired state, X_t. We observe that the variation in the state after one iterate of our map due to the variation in p is

$$\delta X_1 \cong \left.\frac{\partial F}{\partial p}\right|_{(\bar{p},X_s)} \delta p\,. \tag{2}$$

Since $|\delta p|$ is restricted to be less than or equal to Δp, this defines an interval ΔX_1. This interval will typically grow with each successive iteration of the map until it encompasses the desired point X_t. Once X_t is contained within the interval, we know that some parameter value, p_t, between $p_{\min}=\bar{p}-\Delta p$ and $p_{\max}=\bar{p}+\Delta p$ will lead to X_t. All that remains is to estimate p_t, which can be done by a variety of means.

This procedure will give us a value of p which would, in the absence of noise or modeling errors, lead us along some idealized trajectory directly to the target after some small number of iterations. In a real physical system, noise and modeling errors will cause the actual trajectory to wander off the idealized trajectory, however. Therefore we make periodic corrections by reapplying our targeting algorithm after every iteration. Thus we have a different value of the parameter on each iterate, n, and we denote this value p_n. If the system truly were described by a one-dimensional map, then our replacement of p in (1) by p_n would be valid. For the system we deal with, this turns out to be a useful approximation. In general, however, the validity of such an approximation has to be examined on a case-by-case basis. We discuss this issue at the end of this Letter.

To experimentally evaluate the effectiveness of this method, we used a vertically oriented, magnetoelastic ribbon, which is known [3] to vibrate chaotically in response to an external applied magnetic field of the form $H=H_{dc}+H_{ac}\cos(\omega t)$. The ribbon was clamped at its base but was otherwise free to move. The elastic modulus of the ribbon was nonlinearly dependent on the applied field, so that as the field oscillated, the ribbon alternately buckled and stiffened under the influence of gravity. The position X of the ribbon was measured at a point near its base with an optical sensor.

In order to apply our targeting algorithm, we first constructed a map from the experimental system. We chose [4] to use the dc field H_{dc} as our control parameter p and we constructed a map as follows. First, we selected nominal values of H_{ac}, H_{dc}, and ω, and formed an experimental delay plot of 500 points by sampling the position of the ribbon once per driving period, $2\pi/\omega$. This is shown [5] in Fig. 1. We then fitted these points with a robust spline curve. This gave us a map $F(\bar{p},X_n)$ for a nominal value $\bar{p}$ of the parameter. Next, we changed the map by decreasing the parameter to $\bar{p}-\Delta p$, obtaining 500 new data points which we fitted with a second spline. Finally, we increased the parameter to $\bar{p}+\Delta p$, again obtaining 500 points, and fitted these with a third spline. We used the spline fits to estimate the value of the map function F for the three parameter values at any given value of X. We could then interpolate between these three values to estimate $F(p,X_n)$ for any value of the parameter, p, in our range, $[\bar{p}-\Delta p,\bar{p}+\Delta p]$. Given this model of our map, we could apply our targeting method in a straightforward way to reach a given target point.

To illustrate our procedure, we target the point $X=2.5$. We remark that we also targeted other accessible points. It is interesting to discuss targeting of $X=2.5$,

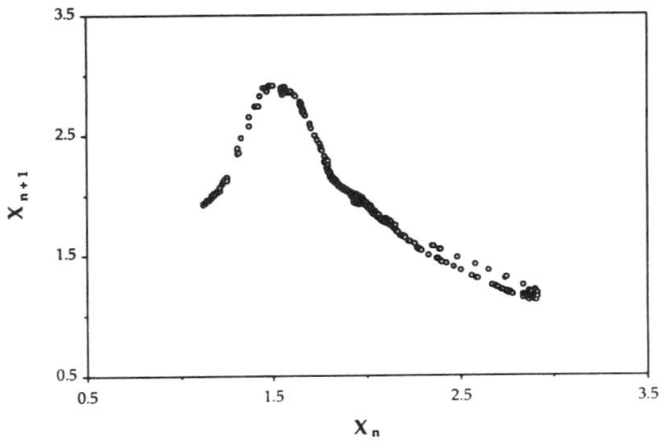

FIG. 1. Experimental delay plot of magnetoelastic ribbon at nominal parameter values.

however, because it is in a region of relatively low measure and therefore its vicinity typically takes a long time to reach in the absence of our targeting procedure. In Fig. 2(a), we show a time series of the ribbon position with fixed, nominal, applied field, which clearly shows that the vicinity of $X=2.5$ is seldom visited. Indeed, for each of our three fixed parameter values, $\bar{p}-\Delta p$, $\bar{p}$, and $\bar{p}+\Delta p$, the ribbon position reaches the neighborhood [2.49,2.51] only about 1 time in 500 iterates.

By contrast, in Fig. 2(b), we show the results of targeting for several representative trajectories. For each trajectory shown, we let the ribbon vibrate chaotically for 100 iterates, and then on the 101st iterate, denoted $n=1$, we initiated the targeting algorithm. The position of the ribbon at $n=1$ is shown as an open circle, and the position when the ribbon reaches the neighborhood, [2.49,2.51], is shown as a solid circle in the figure. Every time that targeting is turned on, the ribbon is brought rapidly to the desired target neighborhood [2.49,2.51]. We reiterate that due to noise and modeling errors, it was necessary to repeat our targeting algorithm at every iteration. We carried out this procedure for 100 randomly chosen initial conditions on the attractor, and rapidly brought the trajectory to the targeted point each and every time. On average, we were able to reach the neighborhood [2.49,2.51] in 20 iterates [6], as compared to 500 iterates that would be required without targeting.

Despite the apparent effectiveness of our targeting algorithm, we encountered two complications which may be of importance in future applications. First, as we mentioned earlier, the technique we used relies on the map (1) being approximately one dimensional. Yet higher-dimensional behavior is manifestly present in our system, as can be seen in Fig. 1 (in particular, note the structure on the right half of the map indicative of fractal behavior). The targeting algorithm is nevertheless successful. It is worthwhile, however, to consider the limits of our algorithm due to our use of a one-dimensional approximation.

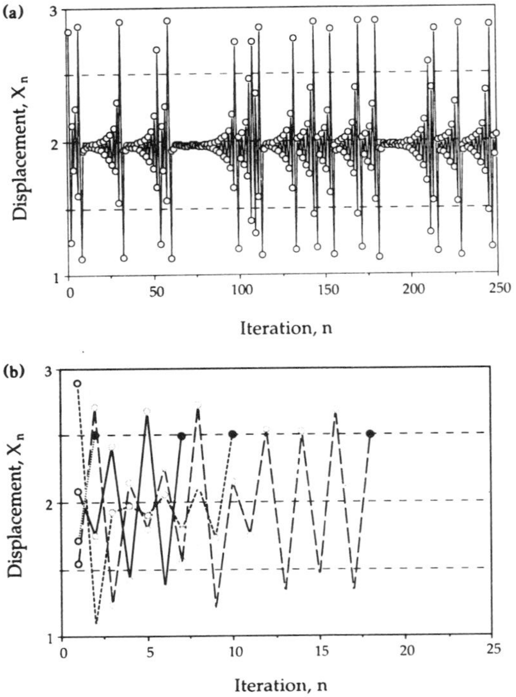

FIG. 2. (a) Ribbon displacement vs iteration for fixed, nominal parameter values. (b) Typical ribbon displacements vs iteration for targeting on. The point $X=2.5$ is targeted starting at iteration $n=1$.

To evaluate the effect of approximating behavior whose dimension is slightly above 1 with a one-dimensional map, we implemented our targeting algorithm numerically on the Hénon system defined by the map

$$X_{n+1}=\tilde{F}(p,X_n,X_{n-1})$$
$$=p+bX_{n-1}-X_n^2\,, \tag{3}$$

for various values of b. We used the parameter p as our control, and varied p by up to $\Delta p=0.05$ about the nominal value $\bar{p}=1.5$. We modeled the system as a one-dimensional map in precisely the same way as in the experimental system. That is, we generated three data sets by iterating Eq. (3) using $p=1.45$, 1.5, and 1.55, and, using only these data, interpolated between quadratic fits of these data sets to estimate $F(p,X_n)$. Using this model, we chose initial points on the attractor at random [7], and targeted a particular point, $X=1.1$, in 25 realizations at each of several values of the parameter b. The results of this process are shown in Fig. 3. We show the rms error of the quadratic fit in the abscissa of the figure [8].

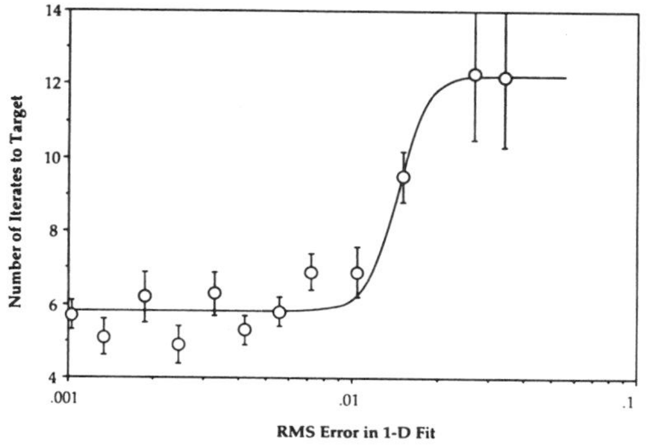

FIG. 3. Effect on targeting of rms error due to approximating higher-dimensional attractor by a one-dimensional map.

It is clear from Fig. 3 that our procedure can tolerate higher dimensionality only until the rms error due to higher dimensionality reaches within an order of magnitude of the change (± 0.05) which we can produce by parametrically varying the position on the attractor. In our physical experiment, the maximum change [9] in the position on the attractor which we produced by varying the parameter amounted on average to 5% of the position itself. By constrast, the rms error in our spline fits was about 1% of the position. According to our simulation of the effect of higher dimensionality, our experiment is apparently operating very near the margin of targeting effectiveness. Indeed, our targeting algorithm was observed in the experiment to be substantially less effective for smaller parameter variations.

The second complication has to do with the fact that in our analysis we have assumed that the system dependence on the parameter is given by $x_{n+1}=F(x_n,p_n)$ with F a one-dimensional map. We note, however, that in this paper we are dealing with a surface of section for a continuous time (t), infinite-dimensional system. Under such circumstances, even if a one-dimensional map applies at each fixed parameter value, and even if contraction to the attractor is very rapid compared to the surface of section map iteration time, Δt, then the switching of p_{n-1} to p_n in principle leads to a dependence of x_{n+1} on *both* p_{n-1} and p_n. To see this, we note that the attractor position in the full infinite-dimensional phase space depends on p. A point with scalar coordinate x_n at time n is on the $p=p_{n-1}$ attractor at time $t=t_n^-$. At a time shortly after the switch from p_{n-1} to p_n it is essentially on the $p=p_n$ attractor (assuming rapid contraction), and in general now has moved to a new location that has an x coordinate, $\hat{x}_n=g(x_n,p_n,p_{n-1})$, that depends on the location of the attractors for $p=p_{n-1}$ and for $p=p_n$. The location of x_{n+1} is determined by iterating the point $\hat{x}_n$ via the map that applies for p fixed for all time at p_n. Thus x_{n+1} depends on both p_n and p_{n-1}. An added dependence of x_{n+1} on p_{n-1} is also implied by the use of delay coordinates [10].

To examine the dependence of x_{n+1} on p_m for $m\leq n-1$, we performed the following experiment. We fixed the parameter at the minimum value, $p_{\min}=\bar{p}-\Delta p$, for 10 iterates. Then, we switched to the maximum value, $p_{\max}=\bar{p}+\Delta p$, for 10 more iterates. At this time, we switched back to $p_{\min}$. We repeated this process for 50 cycles and produced a delay plot of the position of the ribbon, as shown in Fig. 4(a). For comparison, Fig. 4(b) shows plots of x_{n+1} vs x_n for the case where p is kept fixed at $p=p_{\max}$ and for the case where p is kept fixed at $p=p_{\min}$. For the data in Fig. 4(a), for most of the switches, the subsequent x point does indeed land on the appropriate approximate curve in Fig. 4(b). However, occasionally, when p is switched from $p=p_{\min}$ to $p=p_{\max}$, and the $p=p_{\min}$ x coordinate is in the region labeled 0 in Fig. 4(a), then the subsequent x value occurs in the region labeled 1. Region 1 is not on the $p=p_{\max}$ curve [the lower curve of Fig. 4(b)]. Region 1 in turn iterates to region 2 (which also is appreciably far from

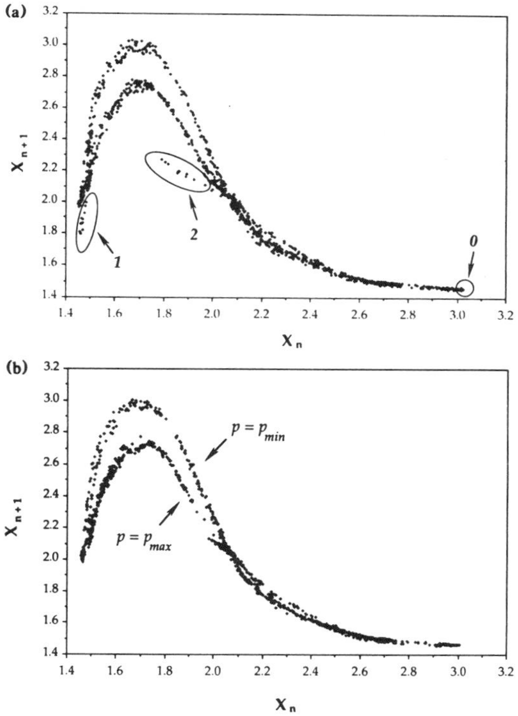

FIG. 4. (a) Delay plot for a parameter alternately switched between $p_{\min}$ and $p_{\max}$. The nominal parameters H_{dc}, H_{ac}, and ω used in this experiment are slightly different from those in Fig. 1. (b) Delay plots for a parameter fixed at $p_{\min}$, and for a parameter fixed at $p_{\max}$.

the appropriate curve) and thereafter rapidly approaches the $p = p_{max}$ curve. While this effect degrades the targeting performance, it occurs relatively rarely. Presumably, a more elaborate procedure could, in principle, correct for this, but we have not found such a procedure to be necessary.

In conclusion, this experiment has confirmed for the first time that it is possible to direct a chaotic system from an arbitrary initial state to a particular desired state using only small perturbations to an accessible system parameter. The method used is particularly suitable for systems which are describable by a one-dimensional map, although some higher dimensionality can be tolerated. We stress that we have achieved control of an experimental system in the absence of an *a priori* theoretical model [11].

The authors wish to acknowledge the assistance and helpful comments of Dr. M. Wun-Fogle of the NSWC. This research was supported by the U.S. Department of Energy (Scientific Computing Staff Office of Energy Research).

[a]Institute for Physical Science and Technology.
[b]Department of Physics.
[c]Laboratory for Plasma Research.
[d]Department of Mathematics.
[e]Department of Electrical Engineering.

[1] T. Shinbrot, E. Ott, C. Grebogi, and J. A. Yorke, Phys. Rev. Lett. **65**, 3215 (1990).

[2] T. Shinbrot, E. Ott, C. Grebogi, and J. A. Yorke, Phys. Rev. A **45**, 4165 (1992). This reference considers chaotic systems which are describable by a one-dimensional map. Reference [1] considers chaotic systems whose fractal dimension is between 1 and 2 and which cannot be approximately described by a one-dimensional map.

[3] W. L. Ditto, S. Rauseo, R. Cawley, C. Grebogi, G.-H. Hsu, E. Kostelich, E. Ott, H. T. Savage, R. Segnan, M. L. Spano, and J. A. Yorke, Phys. Rev. Lett. **63**, 923 (1989).

[4] The other parameters H_{ac} or ω were tried as well. Our method worked about equally well with either ac field H_{ac} or dc field H_{dc}. We wanted to avoid changing the frequency since we sampled the ribbon position once per frequency cycle.

[5] It is worth mentioning that we sampled the position of the ribbon at zero phase of the sinusoidal driving function. Since the experimental system is continuous, we could instead have sampled the position at any other phase. Our delay plot would then have looked different, but our technique would work equally well.

[6] In our 100 realizations, we observed a few outlying cases requiring abnormally long targeting times (~ 50 iterates). Deletion of these few cases from the record results in a significantly shorter average targeting time (~ 10 iterates).

[7] To choose random initial points, we iterated Eq. (3) 25 times using $p = 1.5$, starting from $X_{n-1} = 0$ and using X_n randomly distributed on $[-0.5, 0.5]$.

[8] To aid the eye, we have also shown in the figure a best-fit hyperbolic tangent obtained by nonlinear regression.

[9] This is the average of the difference between robust spline positions from the three curves divided by the nominal position. We normalized the position of the ribbon to a scale from 0 to 1.

[10] U. Dressler and G. Nitsche (to be published).

[11] Experimental targeting for attractors of higher dimensionality (as in our numerical work of Ref. [1]) remains a problem for future study.

Resonant Stimulation and Control of Nonlinear Oscillators

A. Hübler
Institut für Theoretische Physik und Synergetik der Universität, D-7000 Stuttgart

E. Lüscher
Physikdepartment der Technischen Universität München, D-8046 Garching

Generalized dimensions, entropies, Lyapounov exponents [1], and approximations of the flow vector field [2, 3] are used to describe the periodic and chaotic dynamics of nonlinear experimental systems. In addition to the passive observation of a nonlinear oscillator and the description of the measured data using statistical quantities, it is possible to characterize a nonlinear oscillator by an active method, namely by determining its response to specific driving forces [4]. Active methods, like resonance spectroscopy, are superior to passive methods, especially when the experimental system is a set of identical, weakly coupled oscillators, behaving incoherently. If one gets without driving force a compound signal of all oscillators, small and complex due to interference, a strong response emerges at resonance. At resonance every single oscillator is forced to coherent oscillations, sychronized by the driving force.

For example, when a damped nonlinear mechanical pendulum is perturbed by a sinusoidal force, the response is comparatively small in amplitude [5] and does not fulfill any well-defined resonance condition [6], even when the frequency of the driving force coincides with a peak(resonance) in the power spectrum of the dynamics of the unperturbed system [7]. Outside the region of entrainment the response is complicated, in many cases chaotic [8]. In order to obtain a large response, the frequency of the driving force has to be varied in such a way, that it coincides at all amplitudes with a characteristic frequency of the oscillator. The characteristic frequencies of a nonlinear oscillator depend on the amplitude. For example, a weekly damped, mechanical oscillator with an anharmonic potential energy $V(y) = 10^{-12}y^2 + y^4$, where y is the amplitude, has a very small characteristic frequency in the vicinity of the minimum of the potential. The characteristic frequency increases strongly at larger amplitude. If the unperturbed system is in the vicinity of the minimum of the potential, an ideal, resonant driving force has to start with a small frequency. At higher amplitude the driving frequency must be gradually increased to meet the resonance condition [4, 6]. But even when the frequency of the driving force coincides at all amplitudes with a leading resonance (characteristic frequency) of the unperturbed system, the resonance condition is not exactly fulfilled, since all the other peaks [8] of the power spectrum of the unperturbed system have to be taken into account as well. Here we present a method to control nonlinear systems by a special driving force. By this method, it is possible to stimulate damped oscillations in a nonlinear potential resonantly, taking into account all higher-order resonances. In the next section we present a general control theory for nonlinear systems. Afterwards we show that this control theory can be used to calculate resonant perturbations of nonlinear oscillators, and that it can be used to calculate perturbations which force a system with a chaotic motion to perform a periodic motion.

Oscillators with marked nonlinearity and chaotic solutions provide good mathematical models in various fields of science, as in classical mechanics, plasma physics, and celestial mechanics [9], medical physics [10], chemical thermodynamics [11], etc. Haken [12] and others [13, 2, 3] have shown that continuous systems and complex systems can often be well described by a low-dimensional system of ordinary differential equations. Therefore we assume that the investigated real system can be modeled by:

$$\dot{\mathbf{y}} = \mathbf{f}(\mathbf{y},\mathbf{p}_1), \quad (1)$$

where $\mathbf{y}(t)$ is an n-dimensional state vector, $\mathbf{p}_1(t)$ represents a set of m parameters which may depend on time t, and $\mathbf{f}$ is an n-dimensional nonlinear flow vector field. In order to control the experimental system, we integrate an aim equation

$$\dot{\mathbf{x}} = \mathbf{f}(\mathbf{x},\mathbf{p}_2) \quad (2)$$

numerically, where $\mathbf{x}(t)$ is an n-dimensional state vector and $\mathbf{p}_2(t)$ is another set of parameters. After the integration of the aim equation (2), the driving force

$$\mathbf{F}(t) = -\mathbf{f}(\mathbf{x},\mathbf{p}_1) + \mathbf{f}(\mathbf{x},\mathbf{p}_2), \quad (3)$$

where $\mathbf{F}$ is an n-dimensional vector, which is independent of the state $\mathbf{z}$ of the real system. The differential equation of the driven experimental system is:

$$\dot{\mathbf{z}} = \mathbf{f}(\mathbf{z},\mathbf{p}_3) + \mathbf{F}(t), \quad (4)$$

where $\mathbf{z}(t)$ is the n-dimensional state vector. Since $\mathbf{z}(t) = \mathbf{z}_s(t) := \mathbf{x}(t)$ is a special solution of Eq. (4), if the model (1) is correct, i.e., $\mathbf{p}_3(t) = \mathbf{p}_1(t)$, Eq. (2) is called an aim equation. For example, if Eq. (1) describes the reaction rates of chemical substances in a continuously stirred reactor [13], the perturbation $\mathbf{F}(t)$ can be done by a time-dependent addition of the different substances according to Eq. (3). The aim equation (2) can be the solution of a variation principle or it can be defined directly in order to get useful dynamics for special application. Both applications of the control theory will be discussed in the following sections.

In order to investigate resonant driving forces we consider in the following a damped oscillation in a nonlinear potential $V(y)$:

$$\ddot{y} + \eta_1 \dot{y} + \frac{dV(y,t)}{dy} = 0, \quad (5)$$

where η_1 is a friction constant. Equation (5) might be the equation of motion of a nonlinear mechanical pendulum, where $\ddot{y}$ represents inertial forces, $\eta_1 \dot{y}$ friction, and the last term potential forces. In order to calculate resonant driving forces, we integrate, according to [4], the special aim equation

From *Naturwissenschaften* **76** (1989). Reprinted with permission from Springer-Verlag.

$$\ddot{x} + \eta_2 \dot{x} + \frac{dV(x,t)}{dx} = 0, \quad (6)$$

where η_2 is a friction constant. If Eqs. (5) and (6) are transformed into a system of first-order differential equations in the usual way, the driving force

$$F(t) = (\eta_1 - \eta_2)\dot{x} \quad (7)$$

results by Eq. (3), for the driven experimental system

$$\ddot{z} + \eta_1 \dot{z} + \frac{dV(z,t)}{dz} = F(t). \quad (8)$$

For the special solution $z(t) = z_s(t)$: $= x(t)$ of Eq. (8), the driving force $F(t)$ is a resonant perturbation [4, 6], since the driving force and the velocity are in phase, i.e., $\dot{z}(t) \sim F(t)$. For $\eta_2 = -\eta_1$ Eq. (6) results from Eq. (5) by a reflection of time. In this case the driving force (7) is equivalent to the driving force calculated in [4]. In order to investigate the stability of the special solution $z_s(t)$, we consider the case where the experimental system z and x differs by a small value ϵ. To lowest order in ϵ, we obtain

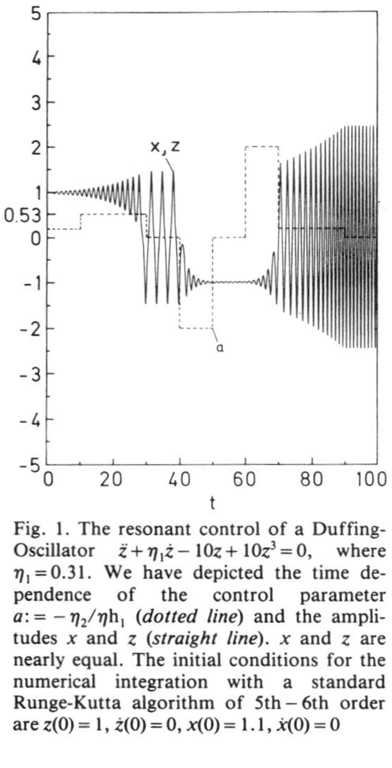

Fig. 1. The resonant control of a Duffing-Oscillator $\ddot{z} + \eta_1\dot{z} - 10z + 10z^3 = 0$, where $\eta_1 = 0.31$. We have depicted the time dependence of the control parameter $a := -\eta_2/\eta h_1$ (*dotted line*) and the amplitudes x and z (*straight line*). x and z are nearly equal. The initial conditions for the numerical integration with a standard Runge-Kutta algorithm of 5th – 6th order are $z(0) = 1$, $\dot{z}(0) = 0$, $x(0) = 1.1$, $\dot{x}(0) = 0$

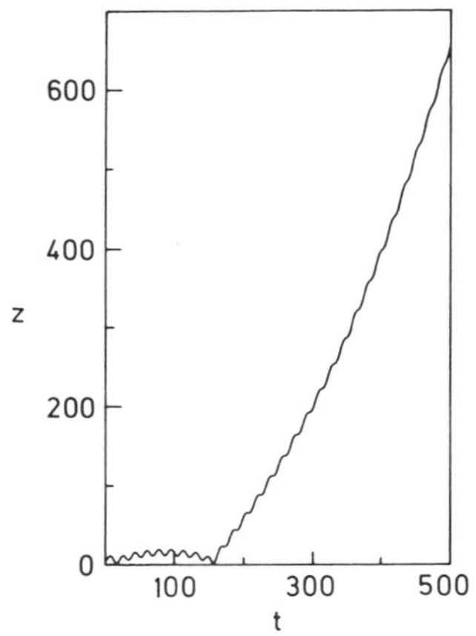

Fig. 2. "Ionization" of a chaotic oscillator in an effective Kepler potential with a resonant driving force [Eqs. (11) – (13)], where $a := -\eta_2/\eta_1 = 0.1$. x and z are plotted versus time for two different time scales. The numerical simulation shows that x and z are nearly equal

$$\ddot{\epsilon} + \eta_1 \dot{\epsilon} + \frac{dV(z,t)}{dz^2} \, |z_s(t) \cdot \epsilon = 0. \quad (9)$$

If V is harmonic, ϵ tends to zero and z_s is stable. For anharmonic potentials Eq. (9) represents a damped, parametrically driven oscillator. If friction is large enough, such oscillators may possess solutions with decaying amplitude. For a more detailed discussion see [14]. The parameter η_2 controls the energy absorption of the experimental oscillator. The energy transferred from the driving force to the oscillator per unit of time is

$$P = F(t) \cdot \dot{z}(t) \overset{z = z_s}{=} \begin{cases} > 0 \text{ for } \eta_2 < \eta_1 \\ = 0 \text{ for } \eta_2 = \eta_1 \\ < 0 \text{ else} \end{cases} \quad (10)$$

For $\eta_2 < \eta_1$, there is no energy transfer from the experiment to the system which produces the driving force, i.e., there is no reflection of energy. For $\eta_2 > \eta_1$ there is no energy transfer from the driving system to the driven system. For all values of η_2 the energy transfer does not change its sign and there is no energy oscillating between the two systems. The corresponding quantity W_0 is zero and has its minimal value if aim equation (6) is used. For all resonant perturbations W_0 is zero. We found numerically that variations of the parameter η_2, which are slow compared to the typical time scale of the oscillator, do not destroy the stability of z_s (Fig. 1). Therefore η_2 can be used for controlling the energy absorption of the experimental system. Since the driving force is resonant, this is called a resonant control of the system.

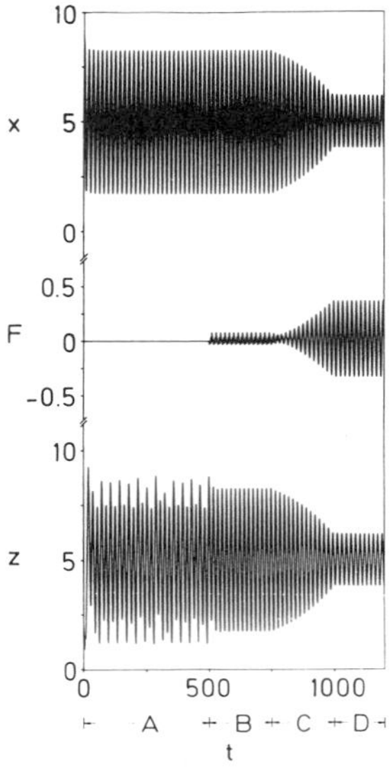

Fig. 3. Control of a chaotic oscillator according to Eqs. (11), (14), and (15). Plotted are the amplitudes of the experimental system z, the driving force F, and the amplitude x of the aim equation. In the area A, x and z are uncoupled ($\eta_2 = \sqrt{8}/10$, $c = 5$), i.e., $F(t)$ is set zero. At time $t = 500$ x and z differ. In the area B the experimental system z is perturbed by an additional force $F(t)$, calculated according to Eq. (15). After a short period of time, z is sychronized with x. When Eqs. (11) and (14) are transformed into a system of first-order differential equations their attractors lie in the same region of the state space and in the surrounding of the attractors, the corresponding flow vector fields are approximately the same. The driving force is the difference between the flow vectors of the corresponding flow vector fields [Eq. (3)]. Therefore the driving force which switches the chaotic dynamics into a periodic one is small. In the area C the aim equation (14) is slowly varied [$\eta_2 = \sqrt{8}/(32.5 - 0.03 \cdot t)$], until finally in the area D, the experimental system z is forced to a sinusoidal oscillation with a small amplitude. In the whole course of the controlling procedure, measurement of the state z is not necessary, i.e., the control is possible without any feedback from the experimental system

If the aim equation differs from Eq. (6), the driving force is not resonant, but still can be used to control the system, for instance to switch chaotic dynamics into a periodic one. In order to illustrate this effect, we consider a damped oscillation in an effective Kepler potential with a sinusoidal driving force

$$\ddot{y} + \eta_1 \dot{y} + 1/y^2 - 1/y^3 + 0.5\cos\omega t = 0, \quad (11)$$

where $\omega = 1/\sqrt{8}$ (Figs. 2, 3). The solution turns out to be chaotic for $\eta_1 = 0.31$ [15]. The aim is to calculate an additional driving force $F(t)$, which forces the controlled experimental system

$$\ddot{z} + \eta_2 \dot{z} + 1/z^2 - 1/z^3 + 0.5\cos\omega t = F(t) \quad (12)$$

to perform a motion which coincides, for example, with the solution of one of the following aim equations:

$$\ddot{x} + \eta_2 \dot{x} + 1/x^2 - 1/x^3 + 0.5\cos\omega t = 0 \quad (13)$$

or

$$\ddot{x} + \eta_2 \dot{x} + 0.01 \cdot (x - c) + 0.5\cos\omega t = 0. \quad (14)$$

The stable solution of Eq. (14) is a harmonic oscillation, centered around c. The corresponding driving force

$$F(t) = (\eta_1 - \eta_2)\dot{x} + 1/x^2 - 1/x^3 - 0.01 \cdot (x - c) \quad (15)$$

is not in phase with the velocity of the experimental system, i.e., $F(t)/\dot{z}(t)$ is not constant for $z = z_s$. Therefore the driving force is not resonant in contrast to the driving force for Eq. (14)

$$F(t) = (\eta_1 - \eta_2)\dot{x}, \quad (16)$$

which is of type (6).

The dynamics of nonlinear oscillators is in many cases sensitive to experimental noise [1]. With imprecise initial conditions, the dynamics of those oscillators can be only roughly predicted and only for a short period of time [3]. With the described method nonlinear oscillators and in particular chaotic oscillators can be properly manipulated by a small perturbation, even without knowing explicit analytical solutions. The dynamics of the manipulated system can be easily predicted, even when the imposed dynamics is chaotic. The generalization of the above techniques to systems of differential equations as well as iterated maps is straightforward [16]. Its application to controlling Navier Stokes flows [17], e.g., the vortex shedding from a circular cylinder, to controlling chaotic trajectories in plasma physics, and to resonantly stimulating Rydberg atoms [18], large molecules and related systems in solid-state physics by aperiodic electromagnetic perturbations may have important consequences.

We would like to thank H. Haken, W. Kroy, O. Wohofsky, W. Schirmacher, P. Deisz, A. Altmann, Ch. Berding, R. Friedrich, R. Blümel and K. Schulten for many discussions. This work was supported by the Messerschmitt-Bölkow-Blohm company.

Received December 5, 1988

1. Eckmann, J. P., Ruelle, D.: Rev. Mod. Phys. *57,* 617 (1985)
2. Cremers, J., Hübler, A.: Z. Naturforsch. *42a,* 797 (1987); Crutchfield, J. P., McNamara, B. S.: Complex Syst. *1,* 417 (1987)
3. Farmer, J. D., Sidorowich, J. J.: Phys. Rev. Lett. *59,* 845 (1987)
4. Reiser, G., Hübler, A., Lüscher, E.: Z. Naturforsch. *42a,* 803 (1987)
5. Nayfeh, A. H., Mook, D. T.: Nonlinear Oscillations, Chapt. 4.1. New York: Wiley 1976; Rothhardt, R., Hübler, A., Lüscher, E.: Helv. Phys. Acta *61,* 220 (1988)
6. Hueter, T. F., Bolt, R. H.: Sonics, p. 20. New York: Wiley 1966
7. Ruelle, D.: Phys. Rev. Lett. *56,* 405 (1986); Parlitz, U., Lauterborn, W.: Phys. Lett. *107A,* 351 (1985)
8. Huberman, B. A., Crutchfield, J. P.: Phys. Rev. Lett. *43,* 1743 (1979); Humieres, D. D., et al.: Phys. Rev. A *26,* 3483 (1982)
9. For a review see Lichtenberg A. J., Liebermann, M. A.: Regular and Stochastic Motion. New York: Springer 1982; Richter, P., Scholz, H. J., in: Ordnung aus dem Chaos (ed. B. Küppers). München: Piper 1987
10. Babloyantz, A., Salazar, J. M., Nicolis, C.: Phys. Lett. *111A,* 152 (1985)
11. Weitz, D. A., et al.: Phys. Rev. Lett. *54,* 1416 (1985)
12. Haken, H.: Synergetics, Chapt. 8. New York: Springer 1983
13. For a review on experiments see Swinney, H. L.: Physica *7D,* 3 (1983)
14. Nayfeh, A. H., Mook, D. T.: Nonlinear Oscillations, Chapt. 5.2. New York: Wiley 1976
15. Wachinger, Ch., et al.: Helv. Phys. Acta *59,* 132 (1986)
16. Merten, J., et al.: ibid. *61,* 88 (1988); Hübler, A.: Ph. D. Thesis, Technische Universität München 1987, Chapt. 2
17. Jensen, M. H., et al: Phys. Rev. Lett. *55,* 2798 (1985); Olinger, D. J., Sreenivasan, K. R.: ibid. *60,* 797 (1988)
18. Blümel, R., Smilansky, U.: ibid. *58,* 2531 (1987)

The entrainment and migration controls of multiple-attractor systems

E. Atlee Jackson

Department of Physics, 1110 West Green Street and Center for Complex Systems Research, Beckman Institute, 405 North Mathews, University of Illinois at Urbana-Champaign, Urbana, IL 61801, USA

Received 10 August 1990; revised manuscript received 9 October 1990; accepted for publication 18 October 1990
Communicated by A.R. Bishop

A method of imposing limited controls on systems which can generally have complex dynamics with many dynamic attractors, is illustrated using several map-dynamics and the Lorenz system. The two types of controls involve: (a) the entrainment of the system, x, to a class of "entrainment goals" $g(t)$, meaning that $\lim_{t\to\infty} |x-g|=0$, and (b) the transfer of the system from one attractor to another, using "migration-goals" in the control.

Many of the more interesting and important complex dynamic systems in nature have more than one dynamic attractor (multiple-attractor systems; MAS). Examples are known in hydrodynamics, mechanics, optics, chemistry, biology, ecology, the dynamics of the heart, neural networks, etc. These attractors generally have topologically distinct forms of dynamics (e.g., stable fixed points, limit cycles, semi-periodicity, intermittency, chaotic, etc.), some of which may be destructive to the system, or which may be responsible for the system's ability to successfully respond to a changing environment (see, e.g., refs. [1–3]). Recently, a general theory has been developed [4], for the limited control of any MAS described by maps or ordinary differential equations, based on only macroscopic information concerning the initial state of the system. In the terminology of standard control theories, this is an "open-loop" or "no-feedback" method of controlling the system.

The present control method is based on the existence of convergent regions, $C_i(k)$, in each basin of attraction, BA_k, of an attractor A_k ($k=1, 2, \ldots$; $i=1, 2, \ldots$). Convergent regions are regions of phase space where nearby solutions converge along all eigendirections for a limited time (or at least for one iteration, in the case of maps). Even systems with strange attractors apparently always have such regions. This has been confirmed for all of the classic maps and flows in one through three dimensions, but a general proof is presently lacking. If the experimental system (E) is governed by either a map or a system of ODEs,

$$x_{n+1}=E(x_n) \quad (x\in\mathbb{R}^n)\,, \tag{1a}$$

$$\dot{x}\equiv \mathrm{d}x/\mathrm{d}t=E(x) \quad (x\in\mathbb{R}^n)\,, \tag{1b}$$

their convergent regions, C_i, are the connected regions related to the local characteristic equations (respectively),

$$C=\{x|\ \|\partial E_i/\partial x_j-\mu(x)\delta_{ij}\|=0 \quad (x\in\mathbb{R}^n;\ i,j=1,\ldots,n),\ |\mu(x)|<1,\ \forall\mu\}\,, \tag{2a}$$

$$C=\{x|\ \|\partial E_i/\partial x_j-\lambda(x)\delta_{ij}\|=0 \quad (x\in\mathbb{R}^n;\ i,j=1,\ldots,n),\ \mathrm{Re}\,\lambda(x)<0,\ \forall\lambda\}\,. \tag{2b}$$

The nature of the control can take a variety of forms, based on two control goals; entrainment and migration. In both cases, a "goal" dynamics, $g(t)$, is selected which one wants the system to behave like. Entrainment-goals (e-goals) involve goals in some convergent region, $g(t)\in C_i$, and the system, $x(t)$, is said to be entrained to $g(t)$, provided that

$$\lim_{n\to\infty} |x_n-g_n|=0, \quad \lim_{t\to\infty} |x(t)-g(t)|=0\,. \tag{3}$$

We wish to obtain such entrainment for e-goals which have any topological character. The condition of $q(t) \in C$, (2), is then required (for any dimension n) to ensure that (3) holds for some initial state, $x(0)$, and all topological forms of goals. This follows simply from a linear stability analysis about a fix-point goal $g(t)=g_0$ (a special case). However, experimental situations require that controls be effective even when $|x(0)-q(0)|$ is not sufficiently small for such linear analyses to be relevant, and when other goals are desired. These essential extensions of the control objectives will be discussed below.

The controlled system is governed according to the action

$$x_{n+1}=E(x_n)+F(g_n, g_{n+1})S_n\,, \qquad (4a)$$

$$\dot{x}=E(x)+F(g, \dot{g})S(t)\,, \qquad (4b)$$

where the on/off switch, $S=0$ for $t<0$, and $0\leqslant S\leqslant 1$ otherwise. Eq. (3) requires that the controlling action is given by

$$F(u, v)=v-E(u)\,. \qquad (5)$$

Note that $F(u, v)$ does not depend on the state of the system, x, so there is no feedback of dynamic information in this control.

This form of control was first proposed by Hübler and Lüscher [5], and a number of pioneering studies of particular examples have now been explored [3,4,6–11]. The fact that $g(t) \in C_i$, (2), ensures that the solution $x(t)=g(t)$ of (4) is stable to infinitesimal perturbations, but not to the finite initial differences, $\epsilon_i=|x(0)-g(0)|$, which are required by experimental limitations. The global region of phase space in which the initial conditions, $x(0)$, yields solutions of (4) satisfying (3), is referred to as the basin of entrainment, BE(g). The control (4) is only initiated when $x(0) \in \mathrm{BE}(g)$ (which defines $t=0$), after which the state of the system does not need to be monitored to ensure control (no-feedback). Specific examples of such basins of entrainment, and topologically varied forms of $x(t)$ and $g(t)$ are given in ref. [4].

Figs. 1a–1d illustrate the entrainment of topologically varied logistic-map dynamics, $x_{k+1}=cx_k(1-x_k)$, to a topological variety of goal dynamics (order→order, chaos→order, chaos→chaos, order→chaos). The figures show x_k and g_k versus k. It has been proved [4] that the basin of entrainment of this system, for general $\{g\} \subset C$, is

$$\mathrm{BE}(\{g\}) \subset C=\{x|(c-1)/2c<x<(c+1)/2c\}$$

(that is, any open region interior to C). The first and third vertical lines indicate when the search for the satisfaction of the condition $x \in C$ is satisfied, and the control is initiated ($S=1$ in (4a)); the second line is when the control is terminated ($S=0$). Two entrainments are illustrated to show the reproducibility. If the control is initiated when x_k is not in BE$(\{g_k\})$, x_k may tend to another attractor ($x_k \to -\infty$) or some undesired period/chaotic response. This aspect of controlling the logistic dynamics was studied in detail by Jackson and Hübler [8].

An example of entrainment for the two-dimensional Hénon map, $x_{n+1}=y_{n+1}-ax_n$, $y_{n+1}=bx_n$ ($a=1.4$; $b=0.2$), is illustrated in fig. 2. The convergent region of this map is $C=\{x, y|\,|2ax|<1-b, |b|<1\}$, indicated by the two vertical lines in the figure (C is independent of y). As noted above, the condition $\{g\} \subset C$ is required for general entrainment goals, in particular for fixed points, g_0. The basin of entrainment for all goal dynamics is not presently known. However, for fixed points, BE(g_0) is known to be some region contained within C, and to become very small if g_{0y} is large. The details of these facts will be reported later. The illustrated example in fig. 2 involves a period-three goal dynamics, circling the origin at radius of 0.05. The control is not initiated until the system (which is on a strange attractor, SA) enters the region C. Numerous numerical examples indicate that the region SA$\cap$C falls within the basin of entrainment for this e-goal. Note that it is important in practice for the basin of entrainment of an e-goal to intersect the attractor, in order to initiate a reliable control. A variety of other e-goals have also been established for this system, including other strange attractors, verifying the topological freedom of this control method.

The second type of control involves goal dynamics which connect two convergent regions in a finite time, so $g(0) \in C_i$ and $g(t=T) \in C_j$ ($i \neq j$). This migration goal set (m-goal) is denoted by M_{ij}. There are a variety of uses for such m-goals. Perhaps the most important is to be able to transfer a system from one attractor to another attractor in a MAS. This may be

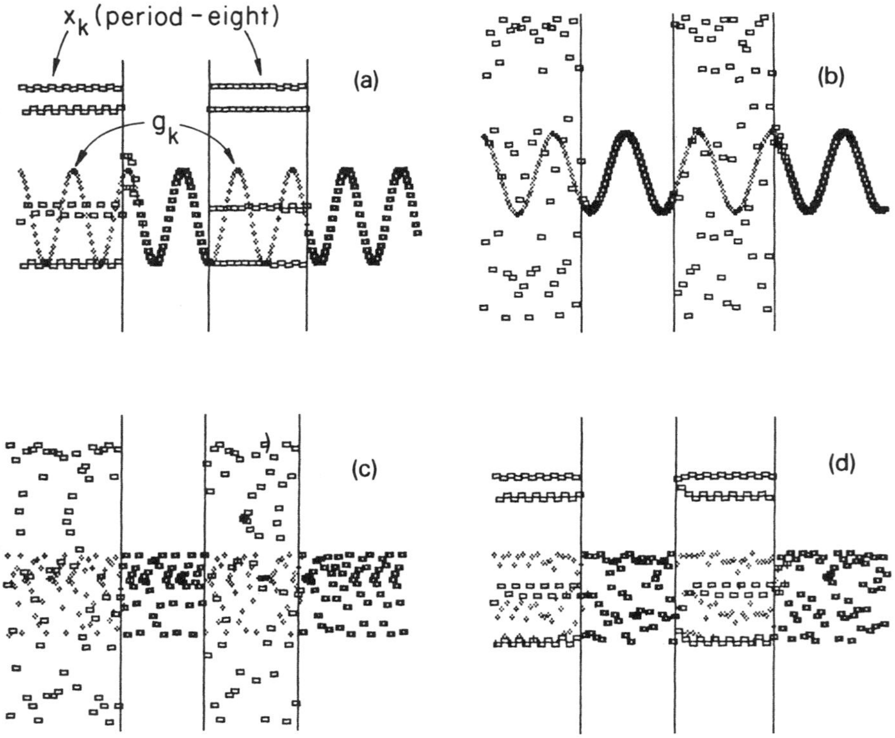

Fig. 1. The topological freedom of entrainment is illustrated by logistic dynamics entrained (twice) to various goal dynamics, confined to the convergent region $(c-1)/2c<g_k<(c+1)/2c$. The goals are either trigonometric functions, or chaotic. (a) order → order, (b) chaos → order, (c) chaos → chaos, (d) order → chaos.

desired in order to "improve" the state of the system; e.g., to make it less self-destructive (chaos → order) or to free its dynamics from a "frozen" state (order → "chaos") as in the case of epileptic seizures [2] or simply to change its periodic state. In some cases, this transfer can be accomplished with e-goals, which have large basins of entrainment [4], but generally m-goals are required. In either case, the transfer between attractors is accomplished by controlling the system only for a finite time. It may also be desirable to transfer between convergement regions in order to determine the accuracy of the dynamic models of the experimental systems, or to retain control in the presence of noise. These aspects are discussed in ref. [4].

As an example of a simple MAS, consider the Gaussian map

$$E(x)=r\alpha x\exp(-2x^2+ax)\quad (x\in\mathbb{R})\,, \tag{6}$$

which has two attractors, $A^{\pm}\subset\mathbb{R}^{\pm}$, that can be topologically distinct (if $a\neq 0$). These attractors are shown in fig. 3, for $2\leqslant r\leqslant 8$, and for $a=0.3$. Each basin of attraction has two convergent regions, $C_i^{\pm}$ $(i=1,2)$, one of which is infinite, $C_{\infty}^{\pm}$, and the other $C_i^{\pm}$ intersects the attractor $A^{\pm}$. Any goal-set $\{g_k\}\subset C_{\infty}^{\pm}$ has an infinite basin of entrainment, $BE=\mathbb{R}$, as proved in ref. [4]. Fig. 4 shows two methods of transferring this system between these attractors (for any value of r in fig. 3). For $r=4.3$, A^+ is a strange attractor and A^- is a period-two attractor

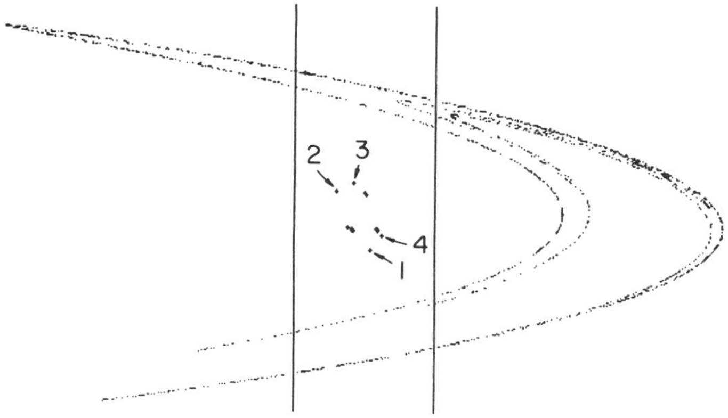

Fig. 2. The entrainment of the Hénon strange attractor ($a=1.4$; $b=0.2$) to a period-three dynamics about the origin with radius 0.05. The convergent region lies between the vertical lines, $-0.25<x<0.25$, where the control is initiated. The controlled points are indicated by +, and require about five iterations to be entrained on the inner triangle.

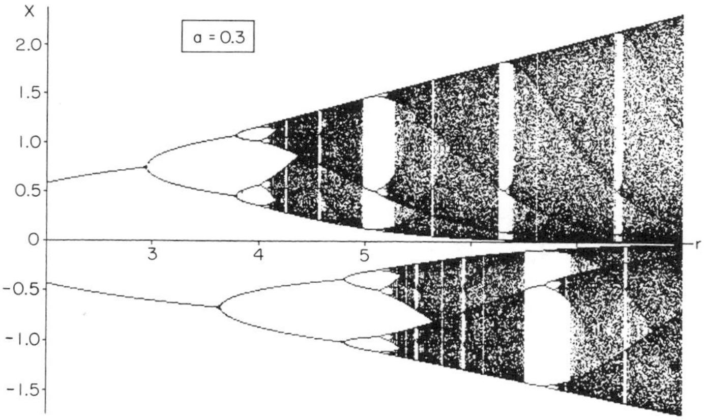

Fig. 3. The attracting sets for the Gaussian map (1a) and (1b) when $a=0.3$, for $2\leqslant r\leqslant 8$. The basins of attraction to these sets are $x>0$ and $x<0$. Transfers can be made between any two sets with the same r.

(fig. 3). In fig. 4, the vertical and horizontal directions correspond to values of (x_n, g_n) versus n, as in fig. 1. The horizontal lines are the boundaries of the convergent regions. The system begins in the attractor A^+, and the e-goal is taken in $C_\infty^\pm$. As soon as the control is initiated ($S=1$), the system rapidly entrains to this goal. Since this goal is in the basin of attraction of A^-, once the control is terminated ($S=0$), the system rapidly tends to A^-. This is one way to transfer $A^+ \to A^-$ (in this system). The second transfer method begins by putting the e-goal in C_2^-, which intersects A^-. This can be used to entrain the system. This is then followed by a migration-goal set (e.g., see (9)), which transfers the system from C_2^- to C_2^+. Once the control is terminated, the system tends to A^+, giving the transfer $A^- \to A^+$. While the first transfer method is simpler, it is not possible in all systems [4].

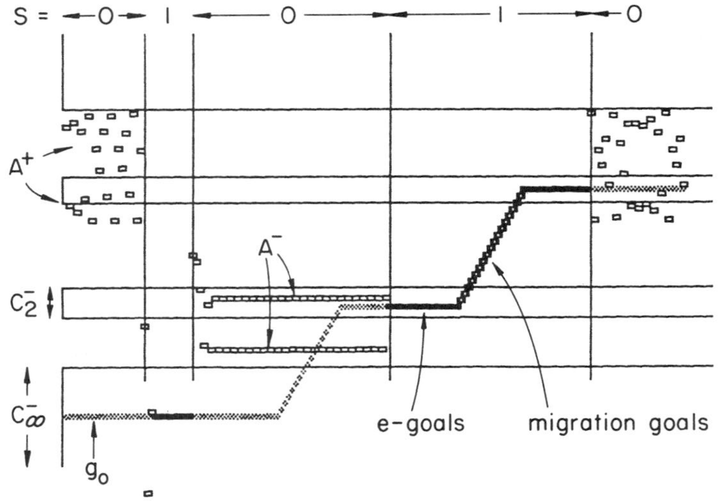

Fig. 4. Two methods to transfer the system (1a) and (1b) ($a=0.3$, $r=4.3$) between attractors; first, from the strange attractor A^+ to the period-two attractor, A^-, using a constant goal in $g_0 \in C_\infty^-$, which has a global basin of entrainment. g_0 is in BA(A^-), so when $S=0$, $x_k \to A^-$; second, entrain to e-goal$\in C_2^-$, then use migration-goal to transfer to A^+.

As an illustration of a transfer from a strange attractor to a stable fixed point, in the case of flows (1b), when $n=3$, the Lorenz system is considered,

$$\dot{x}=\sigma(y-x), \quad \dot{y}=rx-y-xz, \quad \dot{z}=-bz+xy, \qquad (7)$$

when $\sigma=10$, $b=8/3$ and $r=24.2$. For these parameters, (7) has three attractors; two fixed points and a strange attractor. It can be shown that the convergent region of phase space, (2b), is given by the two inequalities

$$(1+\sigma+b)[\sigma(1-r+z)+b(1+\sigma)+x^2] > b\sigma(1-r+z)+\sigma x(x+y)>0, \qquad (8)$$

which is satisfied for large enough positive z [12].

Fig. 5 illustrates the transfer from the strange attractor to the stable fixed point

$$x=y=-[b(r-1)]^{1/2}, \quad z=r-1.$$

First, an entrainment-goal is used which is the fixed point

$$g_0=\{x_0=y_0=+[b(r-1)]^{1/2}, \quad z_0=r+49\}$$

that is in the convergent region C, given by (8). This choice is quite arbitrary, within C, and was selected simply to make the figure clear. The figure shows the autonomous strange attractor dynamics of (7), representing (4) with $S=0$. When the control (5) is switched on at any time ($S=1$ in (4)), using the e-goal, g_0, the system tends toward g_0 as shown. The simplicity of this particular case is due to the large basin of entrainment which these goals have in the Lorenz system. The details of these facts are presented in ref. [12]. Once near-entrainment is obtained,

$$|x(t)-g_0|=\epsilon_e \approx 0.5,$$

the m-goal set

$$\begin{aligned} M=\{g(k)\in\mathbb{R}^3 | g_x &= g_y \\ &= [b(r-1)]^{1/2}[2\exp(-10^{-3}k)-1], \\ g_z &= g_0\exp(-10^{-3}k) \\ &\quad +(r-1)[1-\exp(-10^{-3}k)]; \\ k &= 0, 1, 2, \ldots\} \end{aligned} \qquad (9)$$

is introduced, and the system is transferred from g_0 toward the stable fixed point. Once the system is in the basin of attraction of this attractor, the control can be terminated ($S=0$), completing the transfer

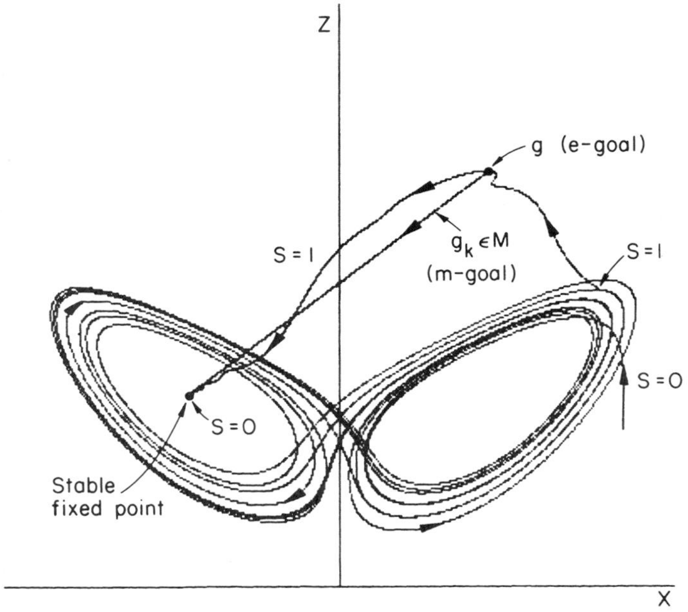

Fig. 5. The transfer of the Lorenz system ($\sigma = 10$, $b = 8/3$, $r = 24.2$) from a strange attractor to a stable fixed point. First the system is entrained to $g_0 \in C$ (eq. (8)), by switching on the control at any time ($S = 1$). After near-entrainment, a migration-goal (eq. (9)) is used to transfer to the basin of attraction of a stable fixed point. In this region the control is terminated ($S = 0$), completing the transfer.

strange attractor → stable fixed point.

There are many aspects of such attractor transfers which need to be clarified in more complicated MAS. Among these are the use of convergent regions to re-entrain systems, because of the inherent instability outside of the convergent regions or when they are subject to environmental noise in the process of making such transfers. There are also a number of potential uses of convergent regions to more precisely characterize physical systems, and thereby obtain more reliable models of the system. Some initial explorations of these issues have been made [4], but much more needs to be known about this potentially important way of characterizing truly complex dynamic systems (MAS).

Finally, we note that another novel method for controlling chaotic attractors has been recently proposed by Ott, Grebogi, and Yorke [13]. This method involves changing the physical system, by making a small change in one of its control parameters, which is assumed to be accessible for such a modification. The possible entrainment which can thereby be achieved is to one of the unstable periodic orbits of the chaotic attractor, which has been made stable by this parameter modification. The basin of entrainment of such a control appears to be unknown, but the entrainment will become effective if the system ergodically wanders into a region sufficiently close to this stabilized periodic orbit at some time. This method of control differs fundamentally from the no-feedback control, (4), which can be applied to any attractor system, and whose goal dynamics can have any topological character within a convergent region, and which have "macroscopic" basins of entrainment. Moreover, the migration aspects within MAS present new opportunities for flexible controls.

References

[1] J.S. Nicholis, in: Chaos in biological systems, eds. H. Degn, A.V. Holden and L.F. Olsen (Plenum, New York, 1987) pp. 221–232.

[2] R. Pool, Science 243 (1989) 604.
[3] E.A. Jackson, Understanding complex systems, appendix in: Perspectives of nonlinear dynamics, Vol. 2 (Cambridge Univ. Press, Cambridge, 1990).
[4] E.A. Jackson, On the control of complex dynamic systems, CCSR Report-90-8, Center for Complex Systems Research, Beckman Institute, University of Illinois (1990).
[5] A. Hübler and E. Lüscher, Naturwissenschaften 76 (1989) 67.
[6] K. Chang, A. Kodogeorgiou, A. Hübler and E.A. Jackson, General resonance spectroscopy, in: SIAM Conf. on Dynamical systems, May 1990, to be published.
[7] R. Georgii, Control of nonlinear continuous systems based on Poincaré maps, in: SIAM Conf. on Dynamical systems, May 1990, to be published.
[8] E.A. Jackson and A. Hübler, Physica D 44 (1990) 407.
[9] A. Kodogeorgiou, Optimal control of catastrophes, in: SIAM Conf. on Dynamical systems, May 1990, to be published.
[10] R. Shermer, Control of the dynamics of shock waves and complicated flows by aperiodic perturbations, in: SIAM Conf. on Dynamical systems, May 1990, to be published.
[11] R. Shermer, A. Hübler and N. Packard, Nonlinear control of Burgers equation, CCSR Report-90-6 (1990).
[12] E.A. Jackson, Controls of dynamic flows with attractors, CCSR Report-90-15, Center for Complex Systems Research, Beckman Institute, University of Illinois (1990).
[13] E. Ott, C. Grebogi and J.A. Yorke, Phys. Rev. Lett. 64 (1990) 1196.

CHAPTER 15

Synchronism and Communication

L.M. Pecora, T.L. Carroll
Synchronization in chaotic systems
Phys. Rev. Lett. **64**, 821 (1990).

K.M. Cuomo, A.V. Oppenheim
Circuit implementation of synchronized chaos with applications to communication
Phy. Rev. Lett. **71**, 65 (1993).

S. Hayes, C. Grebogi, E. Ott
Communicating with chaos
Phys. Rev. Lett. **70**, 3031 (1993).

P. So, E. Ott, W.P. Dayawansa
Observing chaos: Deducing and tracking the state of a chaotic system from limited observation
Phys. Lett. A **176**, 421 (1993).

Editors' Notes

Pecora and Carroll (1990) consider how identical, or almost identical, chaotic systems can be synchronized by a chaotic reference signal so that the two systems follow the same chaotic orbit. He and Vaidya (1992) showed how this synchronization can be understood in many representative cases by the existence of a global Lyapunov function of the difference signals.

One possible use of the ability to synchronize chaotic systems is in secure communications. A sender of the information might add a very large chaotic component to the information-containing signal, thus masking the information in the signal from any third party who intercepts it. Now imagine that the chaotic masking component results from the output of a system synchronized to a chaotic reference signal. If the reference signal is also transmitted, the information can be extracted (decoded) by someone who possesses a replica of the system that is synchronized by the reference signal. This is discussed in the paper of **Cuomo and Oppenheim** (1993).

Another way of using chaos in communication is proposed in the paper of **Hayes et al.** (1993). The idea is to control the dynamics of a chaotic oscillator so that it follows a given sequence in its symbolic dynamics. Since this sequence can be controlled, it can be used to transmit information.

Assuming an accurate mathematical description of a chaotic system of interest is available, the paper of **So et al.** (1993) considers the following question: How can one track the *full* state of the system if the only available observation is the time series of a single observed scalar function of the system state? In control theory this is a standard consideration (the construction of an "observer"). The presence of chaos results in new aspects to this problem.

Synchronization in Chaotic Systems

Louis M. Pecora and Thomas L. Carroll
Code 6341, Naval Research Laboratory, Washington, D.C. 20375
(Received 20 December 1989)

Certain subsystems of nonlinear, chaotic systems can be made to synchronize by linking them with common signals. The criterion for this is the sign of the sub-Lyapunov exponents. We apply these ideas to a real set of synchronizing chaotic circuits.

PACS numbers: 05.45.+b

Chaotic systems would seem to be dynamical systems that defy synchronization.[1] Two identical autonomous chaotic systems started at nearly the same initial points in phase space have trajectories which quickly become uncorrelated, even though each maps out the same attractor in phase space. It is thus a practical impossibility to construct identical, chaotic, synchronized systems in the laboratory.

In this paper we describe the linking of two chaotic systems with a common signal or signals. We show that when the signs of the Lyapunov exponents for the subsystems are all negative the systems will synchronize. By synchronize we mean that the trajectories of one of the systems will converge to the same values as the other and they will remain in step with each other. The synchronization appears to be structurally stable.

We apply these ideas to several well-known systems[2] (e.g., Lorenz and Rössler) as well as the construction of a real set of chaotic synchronizing circuits.

The capability of synchronization is not obvious in nonlinear systems. We derive the results for flows (differential equations), but only a slight variation is needed to use them for iterated maps. Consider an autonomous n-dimensional dynamical system,

$$\dot{u} = f(u) . \tag{1}$$

Divide the system, arbitrarily, into two subsystems $[u = (v, w)]$,

$$\dot{v} = g(v,w), \quad \dot{w} = h(v,w) , \tag{2}$$

where $v = (u_1, \ldots, u_m)$, $g = (f_1(u), \ldots, f_m(u))$, $w = (u_{m+1}, \ldots, u_n)$, and $h = (f_{m+1}(u), \ldots, f_n(u))$.

Now create a new subsystem w' identical to the w system, substitute the set of variables v for the corresponding v' in the function h, and augment Eqs. (2) with this new system, giving

$$\dot{v} = g(v,w) , \quad \dot{w} = h(v,w) , \quad \dot{w}' = h(v,w') . \tag{3}$$

Examine the difference, $\Delta w = w' - w$. The subsystem components w and w' will synchronize only if $\Delta w \to 0$ as $t \to \infty$. In the infinitesimal limit this leads to the variational equations for the subsystem,

$$\dot{\xi} = D_w h(v(t), w(t))\xi , \tag{4}$$

where $D_w h$ is the Jacobian of the w subsystem vector field with respect to w only. The behavior of Eq. (4) or its matrix version[9] depends on the Lyapunov exponents of the w subsystem. We refer to these as sub-Lyapunov exponents. We now have the following theorem: The subsystems w and w' will synchronize only if the sub-Lyapunov exponents are all negative.

The above theorem is a necessary, but not sufficient, condition for synchronization. It says nothing about the set of "initial conditions" in w' which will synchronize with w. We do not mention here any results regarding these sets of points. They are under investigation and will be reported elsewhere.

Taking a broader view, one can think of the $v = (v_1, \ldots, v_m)$ components as being driving variables and the $w' = (w'_{m+1}, \ldots, w'_n)$ as being responding variables. We take just such a view in our application to a chaotic electronic circuit, below.

It is natural to ask how the synchronization is affected by differences in parameters between the w and w' systems which would be found in real applications. Let μ be a vector of the parameters of the y subsystem and μ' of the w' subsystem, so that $h = h(v, w, \mu)$, for example. If the w subsystem were one dimensional, then for small Δw and small $\Delta\mu = \mu' - \mu$,

$$\Delta\dot{w} \approx h_w \Delta w + h_\mu \Delta\mu , \tag{5}$$

where h_w and h_μ are the derivatives of h. Roughly, if h_w and h_μ are nearly constant in time, the solution of this will follow the form

$$\Delta w(t) = \left[\Delta w(0) - \frac{h_\mu}{h_w}\right] e^{h_w t} + \frac{h_\mu}{h_w} . \tag{6}$$

If $h_w < 0$, the difference between w and w' will level off at some constant value. Although this is a simple one-dimensional approximation, it turns out to be the case for all systems we have investigated numerically, even when the differences in parameters are rather large ($\sim$10%–20%).

The phenomena of synchronization is reminiscent of the "slaving principle" of Haken.[10] Haken applied his principle mostly to systems near singularities, like bifurcations, showing that the degrees of freedom of the system for which the eigenvalue of the linear part of the

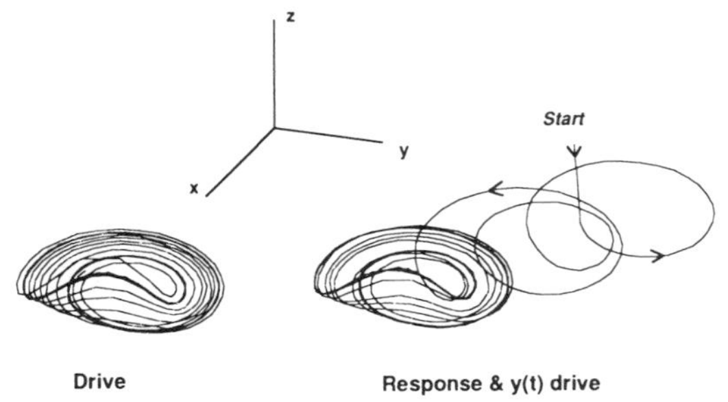

FIG. 1. The attractors for the Rössler drive system and the $(x'\text{-}z')$ response system and $y(t)$ drive variable.

vector field were ≥ 0 determined the behavior of all other variables associated with negative eigenvalues. Just as the Lyapunov exponent is the generalization of the Jacobian for stability studies, our use of the sub-Lyapunov exponents appears to be a generalization of concepts like Haken's slaving.

We have tested these ideas on several models, including several two-dimensional maps. Here we present the results for the Rössler[4] and Lorenz[3] attractors which are typical for all our systems.[2]

We found that in the Rössler system it was possible to use the y component to drive an (x',z') response Rössler system and attain synchronization with the (x,z) components of the driving system. Figure 1 shows three-dimensional views of the drive and response systems for a particular set of parameters in the chaotic regime. One can see that although the response system starts far away from the drive values it soon spirals into the same type of attractor where it remains in synchronization with the drive-system attractor. Table I shows the sub-Lyapunov exponents[11] of various configurations of drive and response for the Rössler system. Note that only the y drive configuration will synchronize.

Table I also shows the sub-Lyapunov exponents for the Lorenz system in the chaotic regime. In this case, synchronization will occur for either x or y driving. Figure 2(a) shows a plot of time versus log of the differences $y'-y$ and $z'-z$ for the Lorenz attractor. The convergences to synchronization are consistent with the values in Table I.

Figure 2(b) shows the results for the same situation, but with a slight change in the parameters of the response system. As expected from the simple one-dimensional argument above, the differences level off. The systems partially synchronize in that y' and z' stay within some neighborhood of y and z as they proceed around the attractor.

We have investigated all the above phenomena in other models[2] and have found similar results.

We used a modified version of an electronic chaotic circuit by Newcomb and Sathyan[6] to test these ideas on a real system. The drive circuit consists of an unstable second-degree oscillator coupled to a hysteritic circuit which continually shifts the center of the unstable focus causing the system to be reinjected into the region near one of two unstable focii. This keeps the motion bounded and chaotic in certain parameter regimes. This is a three-dimensional dynamical system. The response circuit was chosen to be a subcircuit in which the hysteritic circuitry was mostly cut off, so the drive signal came from a point just at the cutoff. The details of the circuits and these experiments will be given elsewhere.

The equations of motion for the model of the drive circuit can be written in terms of the above oscillator-hysterisis description (see Ref. 12 for a description of modeling hysteresis). These must be transformed so that

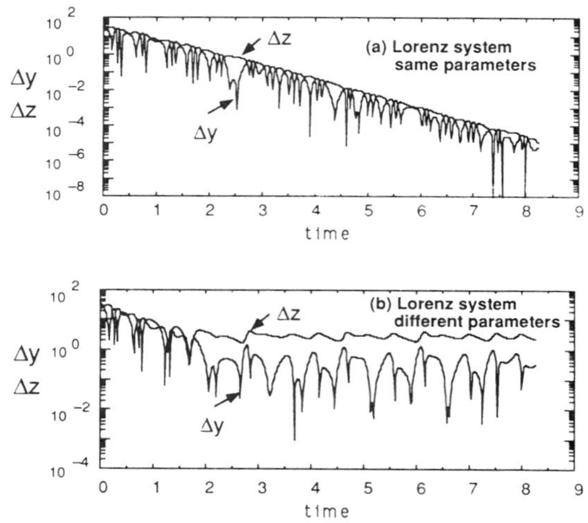

FIG. 2. The differences $y'-y$ and $z'-z$ between the response variables and their drive counterparts for the Lorenz system for (a) when parameters are the same for both systems and (b) when the parameters differ by 5%.

TABLE I. A listing of the various subsystems and driving components for the Lorenz and Rossler systems and their sub-Lyapunov exponents.

System	Drive	Response	Sub-Lyapunov exponents
Rössler	x	(y,z)	$(+0.2, -8.89)$
$a=0.2,\ b=0.2$	y	(x,z)	$(-0.056, -8.81)$
$c=9.0$	z	(x,y)	$(+0.1, +0.1)$
Lorenz	x	(y,z)	$(-1.81, -1.86)$
$\sigma=10,\ b=\frac{8}{3}$	y	(x,z)	$(-2.67, -9.99)$
$r=60.0$	z	(x,y)	$(+0.0108, -11.01)$

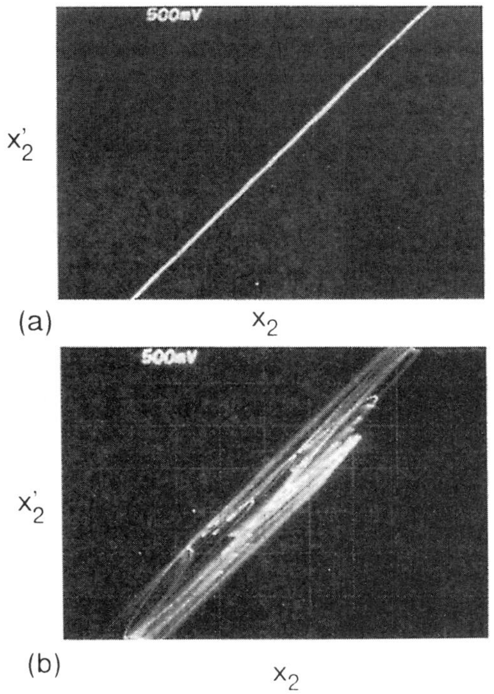

FIG. 3. Oscilloscope traces of the response voltage x_2' vs its drive counterpart voltage x_2 for (a) circuit parameters the same and (b) circuit parameters different by 50%.

the drive signal, x_3, is explicitly shown. This gives

$$\begin{aligned} \dot{x}_1 &= x_2 + \gamma x_1 + c(\alpha x_3 - \beta x_1)\,, \\ \dot{x}_2 &= -\omega_2 x_1 - \delta_2 x_2\,, \\ \epsilon \dot{x}_3 &= \alpha^{-1}\{[1-(\alpha x_3 - \beta x_1)^2](s x_1 - r + \alpha x_3 - \beta x_1) \\ &\quad - \delta_3 \alpha x_3 - \beta x_1 - \beta x_2 - \beta \gamma x_1 - \beta c(\alpha x_3 - \beta x_1)\}\,. \end{aligned} \tag{7}$$

The equations for x_1 and x_2 model the response circuit as well. For the chaotic regime the circuit settings dictate that $\gamma=0.2$, $c=2.2$, $\alpha=6.6$, $\beta=7.9$, $\delta_2=0.01$, $\omega_2=10$, $s=1.667$, and $r=0.0$. The sub-Lyapunov exponents can be calculated directly since the Jacobian for Eqs. (7) is a constant in the x_1 and x_2 variables. The exponents are -16.587 and -0.603, implying synchronization will occur.

The circuit itself runs in the realm of a few kHz. We find that the response synchronizes with the drive within about 2 ms which is consistent with the above sub-Lyapunov exponents whose units are inverse milliseconds. Figure 3 shows oscilloscope traces of the variable x_2 versus its response counterpart x_2' for the synchronizing circuits for two different parameter values. The parameter varied was a resistor in the response circuit which effectively changed α and β. In Fig. 3(b) $\alpha=9.9$ and $\beta=10.4$. The values for the driving circuit remained unchanged. This shows changes ($\sim 50\%$) of the circuit parameters effect synchronization greatly. Even though the sub-Lyapunov exponents in the latter cases both remain negative, synchronization is degraded.

At this point much more remains to be done (theoretically and experimentally) on synchronizing systems. All of the systems studied so far have been low dimensional with one positive Lyapunov exponent. Can synchronization be accomplished in the case of two or more positive exponents, but with only one drive? Can one predict which components will synchronize based on the structure of the center, unstable, and stable manifolds? Despite these and other open questions, we would like to offer some speculations.

The ability to design synchronizing systems in nonlinear and, especially, chaotic systems may open interesting opportunities for applications of chaos to communications, exploiting the unique features of chaotic signals. One now has the capability of having two remote systems with many internal signals behaving chaotically yet still synchronized with each other through the one linking drive signal.

Recent interesting results[13,14] suggest the possibility of extending the synchronization concept to that of a metaphor for some neural processes. Freeman has suggested that one should view the brain response as an attractor. The process of synchronization can be viewed as a response system that "knows" what state (attractor) to go to when driven (stimulated) by a particular signal. It would be interesting to see whether this dynamical view could supplant the more "fixed-point" view of neural nets.[15,16]

We would like to acknowledge useful conversations with R. W. Newcomb and the continued encouragement of A. C. Ehrlich, S. Wolf, M. Melich, and W. Meyers. One of us (T.L.C.) was supported on an Office of Naval Technology Postdoctoral Associateship.

---

[1] Y. S. Tang, A. I. Mees, and L. O. Chua, IEEE Trans. Circuits **30**, 620 (1983).

[2] References to "all systems" in this paper include the Lorenz (Ref. 3), Rössler (Ref. 4), scroll (Ref. 5), Newcomb hysteresis (Ref. 6), three-mode spin system (Ref. 7), and laser emulation (Ref. 8) systems. We hope to report on these results in the future.

[3] J. Gukenheimer and P. Holmes, *Nonlinear Oscillations, Dynamical Systems, and Bifurcations of Vector Fields* (Springer-Verlag, New York, 1983), pp. 92–102.

[4] O. E. Rössler, Phys. Lett. **57A**, 397 (1976).

[5] T. Matsumotot, L. O. Chua, and M. Komuro, IEEE Trans. Circuits Syst. **32**, 798 (1985).

[6]R. W. Newcomb and S. Sathyan, IEEE Trans. Circuits Syst. **30**, 54 (1983).

[7]T. L. Carroll, L. M. Pecora, and F. J. Rachford, Phys. Rev. A **40**, 377 (1989).

[8]F. Mitschke and N. Flüggen, Appl. Phys. B **35**, 59 (1984).

[9]J. Gukenheimer and P. Holmes, *Nonlinear Oscillations, Dynamical Systems, and Bifurcations of Vector Fields* (Springer-Verlag, New York, 1983), p. 25.

[10]H. Haken, *Synergetics* (Springer-Verlag, Berlin, 1977); *Advanced Synergetics* (Springer-Verlag, Berlin, 1983).

[11]Lyanpunov exponents were calculated by using the technique suggested by J.-P. Eckmann and D. Ruell [Rev. Mod. Phys. **57**, 617 (1985)] employing QR decompositions of the fundamental solution matrix of the equation of motion at points along the trajectory.

[12]O. E. Rössler, Z. Naturforsch. **38a**, 788 (1983).

[13]C. Skarda and W. J. Freeman, Behav. Brain Sci. **10**, 161 (1987), and the commentaries following the article.

[14]A. Garfinkel, Am. J. Physiol. **245**, R455 (1983).

[15]*Proceedings of the IEEE First Annual International Conference on Neural Networks, San Diego, 1987*, edited by M. Caudil and C. Butler (IEEE, New York, 1987).

[16]C. Skarda and W. J. Freeman, Behav. Brain Sci. **10**, 170 (1987).

Circuit Implementation of Synchronized Chaos with Applications to Communications

Kevin M. Cuomo and Alan V. Oppenheim

Research Laboratory of Electronics, Massachusetts Institute of Technology, Cambridge, Massachusetts 02139

(Received 21 January 1993)

An analog circuit implementation of the chaotic Lorenz system is described and used to demonstrate two possible approaches to private communications based on synchronized chaotic systems.

PACS numbers: 05.45.+b, 43.72.+q, 84.30.Wp

In 1990 Pecora and Carroll [1] reported that certain chaotic systems possess a self-synchronization property. A chaotic system is self-synchronizing if it can be decomposed into subsystems: a drive system and a stable response subsystem that synchronize when coupled with a common drive signal [1-3]. They showed numerically that synchronization occurs if all of the Lyapunov exponents for the response subsystems are negative. For some synchronizing chaotic systems the ability to synchronize is robust. For example, the Lorenz system is decomposable into two separate response subsystems that will each synchronize to the drive system when started from any initial condition. As discussed in [4-6], the combination of synchronization and unpredictability from purely deterministic systems leads to some potentially interesting communications applications. In this Letter, we focus on the synchronizing properties of the Lorenz system, the implementation of the Lorenz system as an analog circuit, and the potential for utilizing the Lorenz circuit for various communications applications. It should be stressed that the applications indicated are very preliminary and presented primarily to suggest and illustrate possible directions.

The Lorenz system [7] is given by

$$\begin{aligned} \dot{x} &= \sigma(y-x)\,, \\ \dot{y} &= rx - y - xz\,, \\ \dot{z} &= xy - bz\,, \end{aligned} \tag{1}$$

where σ, r, and b are parameters. As shown by Pecora and Carroll an interesting property of (1) is that it is decomposable into two stable subsystems. Specifically, a stable (x_1,z_1) response subsystem can be defined by

$$\begin{aligned} \dot{x}_1 &= \sigma(y-x_1)\,, \\ \dot{z}_1 &= x_1y - bz_1\,, \end{aligned} \tag{2}$$

and a second stable (y_2,z_2) response subsystem by

$$\begin{aligned} \dot{y}_2 &= rx - y_2 - xz_2\,, \\ \dot{z}_2 &= xy_2 - bz_2\,. \end{aligned} \tag{3}$$

Equation (1) can be interpreted as the drive system since its dynamics are independent of the response subsystems. Equations (2) and (3) represent dynamical response systems which are driven by the drive signals $y(t)$ and $x(t)$, respectively. The eigenvalues of the Jacobian matrix for the (x_1,z_1) subsystem are both negative and thus $|x_1-x|$ and $|z_1-z|\to 0$ as $t\to\infty$. Also, it can be shown numerically that the Lyapunov exponents of the (y_2,z_2) subsystem are both negative and thus $|y_2-y|$ and $|z_2-z|\to 0$ as $t\to\infty$.

As we show below, the two response subsystems can be used together to regenerate the full-dimensional dynamics which are evolving at the drive system. Specifically, if the input signal to the (y_2,z_2) subsystem is $x(t)$, then the output $y_2(t)$ can be used to drive the (x_1,z_1) subsystem and subsequently generate a "new" $x(t)$ in addition to having obtained, through synchronization, $y(t)$ and $z(t)$. It is important to recognize that the two response subsystems given by Eqs. (2) and (3) can be combined into a single system having a three-dimensional state space. This produces a full-dimensional response system which is structurally similar to the drive system (1). Further discussion of this result is given below in the context of the circuit implementations.

A direct implementation of Eq. (1) with an electronic circuit presents several difficulties. For example, the state variables in Eq. (1) occupy a wide dynamic range with values that exceed reasonable power supply limits. However, this difficulty can be eliminated by a simple transformation of variables. Specifically, we define new variables by $u=x/10$, $v=y/10$, and $w=z/20$. With this scaling, the Lorenz equations are transformed to

$$\begin{aligned} \dot{u} &= \sigma(v-u)\,, \\ \dot{v} &= ru - v - 20uw\,, \\ \dot{w} &= 5uv - bw\,. \end{aligned} \tag{4}$$

This system, which we refer to as the transmitter, can be more easily implemented with an electronic circuit because the state variables all have similar dynamic range and circuit voltages remain well within the range of typical power supply limits.

An analog circuit implementation of the circuit Eqs. (4) is shown in Fig. 1. The operational amplifiers (1-8) and associated circuitry perform the operations of addition, subtraction, and integration. Analog multipliers implement the nonlinear terms in the circuit equations. We emphasize that our circuit implementation of (4) is exact,

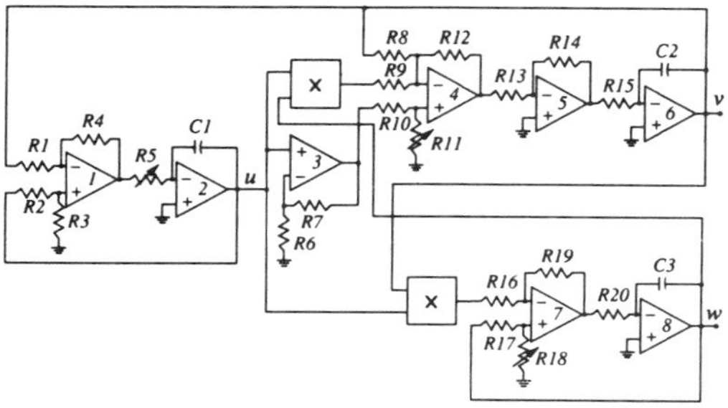

FIG. 1. Lorenz-based chaotic circuit.

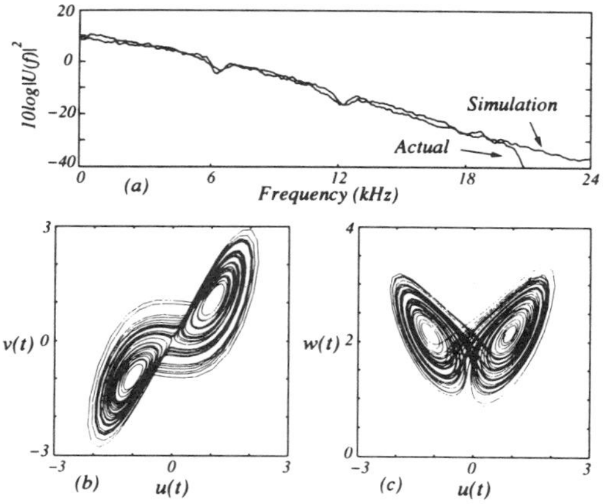

FIG. 2. Circuit data: (a) averaged power spectrum of $u(t)$; (b) chaotic attractor projected onto the uv plane; (c) chaotic attractor projected onto the uw plane.

and that the coefficients σ, r, and b can be independently varied by adjusting the corresponding resistors R_5, R_{11}, and R_{18}. In addition, the circuit time scale can be easily adjusted by changing the values of the three capacitors, C_1, C_2, and C_3, by a common factor. We have chosen component values [Resistors (kΩ): $R_1, R_2, R_3, R_4, R_6, R_7, R_{13}, R_{14}, R_{16}, R_{17}, R_{19} = 100$; $R_5, R_{10} = 49.9$; $R_8 = 200$; $R_9, R_{12} = 10$; $R_{11} = 63.4$; $R_{15} = 40.2$; $R_{18} = 66.5$; $R_{20} = 158$; capacitors (pF): $C_1, C_2, C_3 = 500$ Op-Amps (1-8): LF353 multipliers: AD632AD] which result in the coefficients $\sigma = 16$, $r = 45.6$, and $b = 4$.

To illustrate the chaotic behavior of the transmitter circuit, an analog-to-digital (A/D) data recording system was used to sample the appropriate circuit outputs at a 48-kHz rate and with 16-bit resolution. Figure 2(a) shows the averaged power spectrum of the circuit wave form $u(t)$. The power spectrum is broadband which is typical of a chaotic signal. Figure 2(a) also shows a power spectrum obtained from a numerical simulation of the circuit equations. As we see, the performance of the circuit and the simulation are consistent. Figures 2(b) and 2(c) show the circuit's chaotic attractor projected onto the uv plane and uw plane, respectively. These data were obtained from the circuit using the stereo recording capability of the A/D system to simultaneously sample the x-axis and y-axis signals at a 48-kHz rate and with 16-bit resolution. A more detailed analysis of the transmitter circuit is given in [6].

A full-dimensional response system which will synchronize to the chaotic signals at the transmitter (4) is given by

$$\begin{aligned} \dot{u}_r &= \sigma(v_r - u_r), \\ \dot{v}_r &= ru - v_r - 20uw_r, \\ \dot{w}_r &= 5uv_r - bw_r. \end{aligned} \tag{5}$$

We refer to this system as the receiver in light of some potential communications applications. We denote the transmitter state variables collectively by the vector $\mathbf{d} = (u,v,w)$ and the receiver variables by the vector $\mathbf{r} = (u_r, v_r, w_r)$ when convenient.

By defining the dynamical errors by $\mathbf{e} = \mathbf{d} - \mathbf{r}$, it is straightforward to show that synchronization in the Lorenz system is a result of stable error dynamics between the transmitter and receiver. Assuming that the transmitter and receiver coefficients are identical, a set of equations which govern the error dynamics are given by

$$\begin{aligned} \dot{e}_1 &= \sigma(e_2 - e_1), \\ \dot{e}_2 &= -e_2 - 20u(t)e_3, \\ \dot{e}_3 &= 5u(t)e_2 - be_3. \end{aligned}$$

The error dynamics are globally asymptotically stable at the origin provided that $\sigma, b > 0$. This result follows by considering the three-dimensional Lyapunov function defined by $E(\mathbf{e},t) = \frac{1}{2}(1/\sigma)e_1^2 + e_2^2 + 4e_3^2]$. The time rate of change of $E(\mathbf{e},t)$ along trajectories is given by

$$\begin{aligned} \dot{E}(\mathbf{e},t) &= (1/\sigma)e_1\dot{e}_1 + e_2\dot{e}_2 + 4e_3\dot{e}_3 \\ &= -(e_1 - \tfrac{1}{2}e_2)^2 - \tfrac{3}{4}e_2^2 - 4be_3^2, \end{aligned}$$

which shows that $E(\mathbf{e},t)$ decreases for all $\mathbf{e} \neq 0$. As $E(\mathbf{e},t)$ goes to zero synchronization occurs. Note that the transmitter and receiver need not be operating chaotically for synchronization to occur. In [8], a similar Lyapunov argument is given for the synchronization of the (y,z) subsystem of the Lorenz equations.

A comparison of the receiver equations (5) with the transmitter equations (4) shows that they are nearly identical, except that the drive signal $u(t)$ replaces the receiver signal $u_r(t)$ in the $(\dot{v}_r, \dot{w}_r)$ equations. This similarity allows the transmitter and receiver circuits to be built in an identical way, which helps to achieve perfect synchronization between the transmitter and receiver. In [6] we discuss and illustrate the synchronization performance of the receiver circuit.

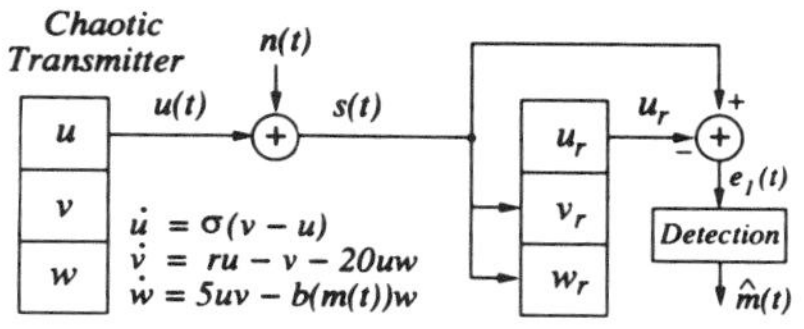

FIG. 3. Chaotic communication system.

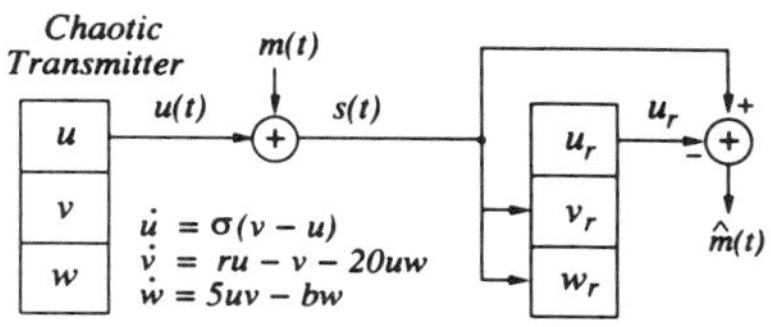

FIG. 5. Chaotic signal masking system.

As one illustration of the potential use of synchronized chaotic systems in communications, we describe a system to transmit and recover binary-valued bit streams [6]. The basic idea is to modulate a transmitter coefficient with the information-bearing wave form and to transmit the chaotic drive signal. At the receiver, the coefficient modulation will produce a synchronization error between the received drive signal and the receiver's regenerated drive signal with an error signal amplitude that depends on the modulation. Using the synchronization error the modulation can be detected.

The modulation/detection process is illustrated in Fig. 3. In this figure, the coefficient b of the transmitter equations (4) is modulated by the information-bearing wave form, $m(t)$. For purposes of demonstrating the technique, we use a square wave for $m(t)$ as illustrated in Fig. 4(a). The square wave produces a variation in the transmitter coefficient b with the zero-bit and one-bit coefficients corresponding to $b(0)=4$ and $b(1)=4.4$, respectively. In [6] we show that the averaged power spectrum of the drive signal with and without the embedded square wave present are very similar. Figure 4(b) shows the synchronization error power, $e_1^2(t)$, at the output of the receiver circuit. The coefficient modulation produces significant synchronization error during a "1" transmission and very little error during a "0" transmission. Figure 4(c) illustrates that the square-wave modulation can be reliably recovered by low pass filtering the synchronization error power wave form and applying a threshold test. This approach has also been shown to work using Chua's circuit [9].

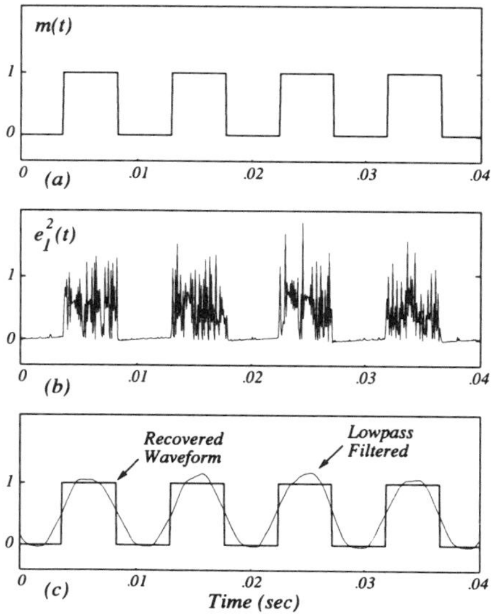

FIG. 4. Circuit data: (a) modulation wave form; (b) synchronization error power; (c) recovered wave form.

Another potential approach to communications applications is based on signal masking and recovery. In signal masking, a noiselike masking signal is added at the transmitter to the information-bearing signal $m(t)$ and at the receiver the masking is removed. In our system, the basic idea is to use the received signal to regenerate the masking signal at the receiver and subtract it from the received signal to recover $m(t)$. This can be done with the synchronizing receiver circuit since the ability to synchronize is robust, i.e., is not highly sensitive to perturbations in the drive signal and thus can be done with the masked signal. It is interesting to note that this idea is not restricted to just the Lorenz circuit but has wider potential; for example, Kocarev *et al.* [10] have also demonstrated our signal masking concept in [4,5] using Chua's circuit. While there are many possible variations, consider, for example, a transmitted signal of the form $s(t) = u(t) + m(t)$. It is assumed that for masking, the power

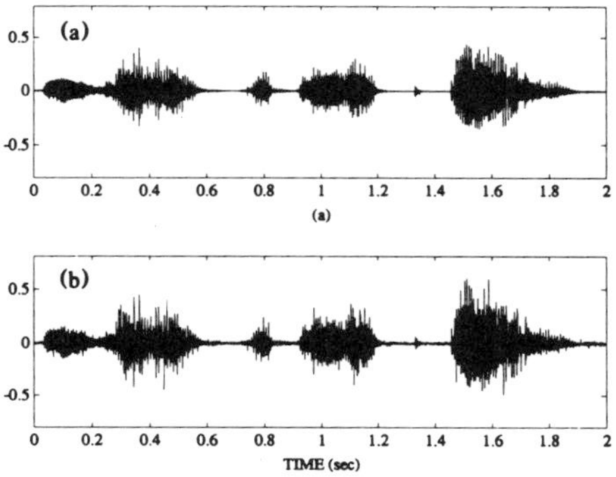

FIG. 6. Circuit data: speech wave forms. (a) Original; (b) recovered.

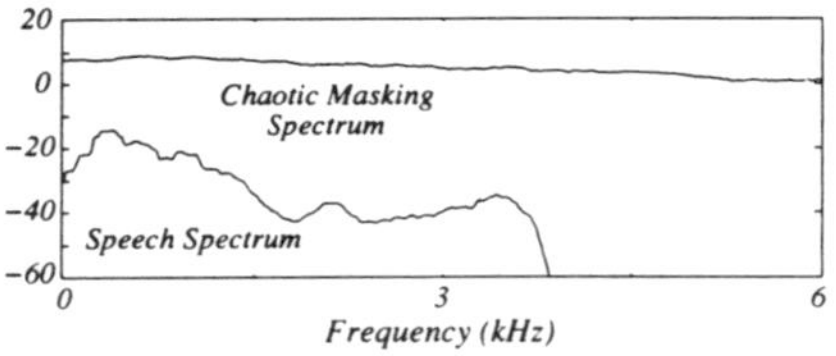

FIG. 7. Circuit data: power spectra of chaotic masking and speech signals.

level of $m(t)$ is significantly lower than that of $u(t)$. The dynamical system implemented at the receiver is

$$\dot{u}_r = 16(v_r - u_r),$$

$$\dot{v}_r = 45.6s(t) - v_r - 20s(t)w_r,$$

$$\dot{w}_r = 5s(t)v_r - 4w_r.$$

If the receiver has synchronized with $s(t)$ as the drive, then $u_r(t) \simeq u(t)$ and consequently $m(t)$ is recovered as $\hat{m}(t) = s(t) - u_r(t)$. Figure 5 illustrates the approach.

Using the transmitter and receiver circuits, we demonstrate the performance of this system in Fig. 6 with a segment of speech from the sentence "He has the bluest eyes." As indicated in Fig. 7 the power spectra of the chaotic masking signal, $u(t)$, and the speech are highly overlapping with an average signal-to-masking ratio of approximately -20 dB. Figures 6(a) and 6(b) show the original speech, $m(t)$, and the recovered speech signal, $\hat{m}(t)$, respectively. Clearly, the speech signal has been recovered and in informal listening tests is of reasonable quality.

We thank S. Isabelle and S. Strogatz for helpful discussions. This work was sponsored in part by the Air Force Office of Scientific Research under Grant No. AFOSR-91-0034-A, in part by a subcontract from Lockheed Sanders, Inc., under ONR Contract No. N00014-91-C-0125, and in part by the Defense Advanced Research Projects Agency monitored by the Office of Naval Research under Grant No. N00014-89-J-1489. K.M.C. is supported in part through the MIT/Lincoln Laboratory Staff Associate Program.

[1] L. M. Pecora and T. L. Carroll, Phys. Rev. Lett. **64**, 821 (1990).
[2] T. L. Carroll and L. M. Pecora, IEEE Trans. Circuits Syst. **38**, 453 (1991).
[3] L. M. Pecora and T. L. Carroll, Phys. Rev. A **44**, 2374 (1991).
[4] K. M. Cuomo, A. V. Oppenheim, and S. H. Isabelle, MIT Research Laboratory of Electronics TR No. 570, 1992 (unpublished).
[5] A. V. Oppenheim, G. W. Wornell, S. H. Isabelle, and K. M. Cuomo, in Proceedings of the IEEE International Conference on Acoustics, Speech, and Signal Processing, 1992 (to be published).
[6] K. M. Cuomo and A. V. Oppenheim, MIT Research Laboratory of Electronics TR No. 575, 1992 (unpublished).
[7] E. N. Lorenz, J. Atmos. Sci. **20**, 130 (1963).
[8] R. He and P. G. Vaidya, Phys. Rev. A **46**, 7387 (1992).
[9] U. Parlitz, L. Chua, Lj. Kocarev, K. Halle, and A. Shang, Int. J. Bif. Chaos **2**, 973 (1992).
[10] Lj. Kocarev, K. Halle, K. Eckert, and L. Chua, Int. J. Bif. Chaos **2**, 709 (1992).

Communicating with Chaos

Scott Hayes[(a)]
U.S. Army Research Laboratory, Adelphi, Maryland 20783

Celso Grebogi[(b),(c)] and Edward Ott[(b),(d),(e)]
University of Maryland, College Park, Maryland 20742
(Received 18 December 1992)

The use of chaos to transmit information is described. Chaotic dynamical systems, such as electrical oscillators with very simple structures, naturally produce complex wave forms. We show that the symbolic dynamics of a chaotic oscillator can be made to follow a desired symbol sequence by using small perturbations, thus allowing us to encode a message in the wave form. We illustrate this using a simple numerical electrical oscillator model.

PACS numbers: 05.45.+b

Much of the fundamental understanding of chaotic dynamics involves concepts from information theory, a field developed primarily in the context of practical communication. Concepts from information theory used in chaos include metric entropy, topological entropy, Markov partitions, and symbolic dynamics [1]. On the other hand, because of their exponential sensitivity, chaotic systems are often said to evolve randomly. This terminology is partially justified if one regards the information obtained by *detailed* observation of the chaotic orbit as being less significant than the statistical properties of the orbits. The object of this Letter is to show that we can use the close connection between the theory of chaotic systems and information theory in a way that is more than purely formal. In particular, we show that the recent realization that chaos can be controlled with *small* perturbations [2] can be utilized to cause the symbolic dynamics of a chaotic system to track a prescribed symbol sequence, thus allowing us to encode any desired message in the signal from a chaotic oscillator. The natural complexity of chaos thus provides a vehicle for information transmission in the usual sense. Furthermore, we argue that this method of communication will often have technological advantages.

Specifically, assume that there is an electrical oscillator producing a large amplitude chaotic signal that one wishes to use for communication. The so-called double scroll electrical oscillator [3] yields a chaotic signal consisting of a seemingly random sequence of positive and negative peaks. If we associate a positive peak with a 1, and a negative peak with a 0, the signal yields a binary sequence. Furthermore, we can use *small* control perturbations to cause the signal to follow an orbit whose binary sequence represents the information we wish to communicate. Hence the chaotic power stage that generates the wave form for transmission can remain simple and efficient (complex chaotic behavior occurs in simple systems), while all the complex electronics controlling the output remains at the low-power microelectronic level.

The basic strategy is as follows. First, examine the free-running (i.e., uncontrolled) oscillator and extract from it a symbolic dynamics that allows one to assign symbol sequences to the orbits on the attractor. Typically, some symbol sequences are never produced by the free-running oscillator. The rules specifying allowed and disallowed sequences are called the *grammar*. Methods for determining the grammar (or an approximation to it) of specific systems have been considered in several theoretical [4] and experimental [5] works. (In the engineering literature, a similar concept exists in the context of constrained communication channels.) The next step is to choose a code whereby any message that can be emitted by the information source can be encoded using symbol sequences that satisfy suitable constraints imposed by the dynamics in the presence of the control. (The construction of codes with such constraints is a standard problem in information theory [6], and will be discussed in the context of communicating with chaos, along with the required generalizations, in a longer paper [7].) The code cannot deviate much from the grammar of the free-running oscillator because we envision using only tiny controls that cannot grossly alter the basic topological structure of the orbits on the attractor. Once the code is selected, the next problem is to specify a control method whereby the orbit can be made to follow the symbol sequence of the information to be transmitted. Finally, the transmitted signal must be detected and decoded.

We now present a simple numerical example illustrating how the preceding strategy is carried out. Figure 1(a) is a schematic diagram of the electrical circuit producing the so-called double scroll chaotic attractor [3]. The nonlinearity comes from a nonlinear negative resistance represented by the voltage v_R in Fig. 1. (Different realizations of the negative resistance are possible; we have constructed one using an operational amplifier circuit, and are designing an experiment using this oscillator to demonstrate information transmission using chaos.) The differential equations describing the double scroll system are

$$C_1\dot{v}_{C_1}=G(v_{C_2}-v_{C_1})-g(v_{C_1}),$$

$$C_2\dot{v}_{C_2}=G(v_{C_1}-v_{C_2})+i_L,$$

$$L\dot{i}_L=-v_{C_2}.$$

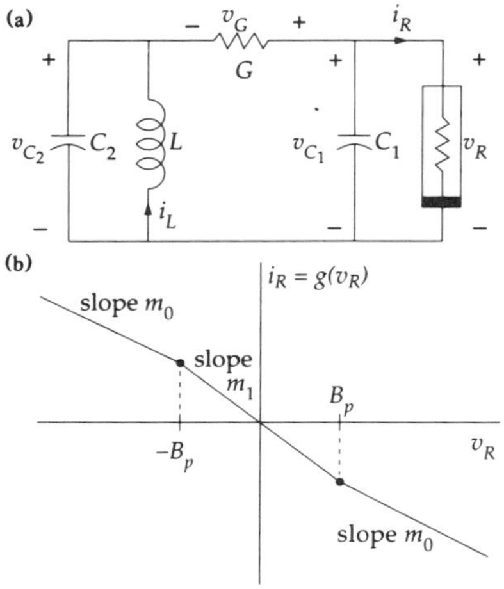

FIG. 1. Double scroll oscillator: (a) electrical schematic and (b) nonlinear negative-resistance i-v characteristic g.

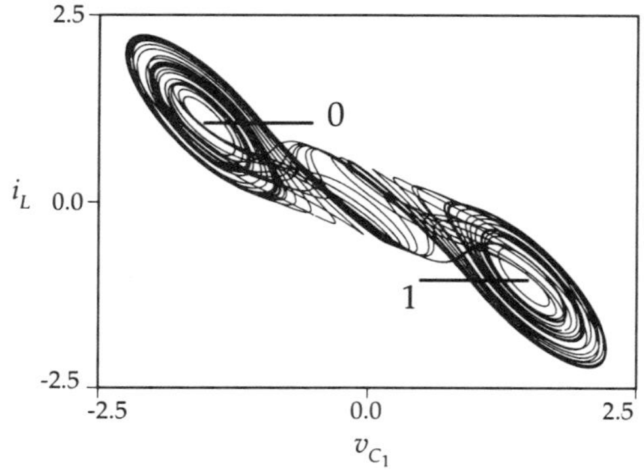

FIG. 2. Double scroll oscillator state-space trajectory projected on the i_L-v_{C_1} plane showing the two branches of the surface of section.

The negative-resistance i-v characteristic g is shown in Fig. 1(b). For our example, we use the normalized parameter values used by Matsumoto [3]: $C_1 = \frac{1}{9}$, $C_2 = 1$, $L = \frac{1}{7}$, $G = 0.7$, $m_0 = -0.5$, $m_1 = -0.8$, and $B_p = 1$. For a Poincaré surface of section (see Fig. 2), we take the surfaces $i_L = \pm GF$, $|v_{C_1}| \leq F$, where $F = B_p(m_0 - m_1)/(G + m_0)$, so that these half planes intersect the attractor with edges at the unstable fixed points at the center of the attractor lobes. Figure 2 shows a trajectory of the double scroll system with the two branches of the surface of section labeled 0 and 1. (The plane surfaces are edge-on in the picture.) The intersection of the strange attractor with the surface of section is approximately a single straight line segment on each of the two branches. Let x denote the distance along this straight line segment from the fixed point at the center of the respective lobe, $x = (F - |v_{C_1}|)\cos\theta + |v_{C_2}|\sin\theta$, where θ is the angle that the line segment makes with the i_L-v_{C_1} plane. Because absolute values are used in defining x, we can use the same x coordinate for both lobes of the attractor.

To construct a description of the symbolic dynamics of the system, we run the computer simulation without control. When the free-running system state point passes through the surface of section, we record the value of the generalized coordinate x (restricted to 1000 discrete bins for the computer simulation), and then record the symbol sequence that is generated by the system after the state point crosses through the surface. Suppose the system generates the binary symbol sequence $b_1b_2b_3\ldots$. We represent this by the real number $0.b_1b_2b_3\ldots$, so that each symbol sequence corresponds to the real number $r = \sum_{n=1}^{\infty} b_n 2^{-n}$, and symbols that occur at earlier times are given greater weight. We refer to the number r, specifying the future symbol sequence, as the *symbolic state* of the system. This defines a function mapping the state-space coordinate x on the surface of section to the symbolic coordinate r. This function $r(x)$ (which we call the *coding function*) is shown in Fig. 3. (The function gives actual symbol sequences when referring to the 0 lobe, and the bitwise complement when referring to the 1 lobe.) Because the oscillator is only approximately described by a binary sequence, multiple values of x lead to the same future symbol sequence. (We only need to track one of them. More sophisticated techniques both for symbol assignment and symbol sequence ordering are discussed in the longer paper [7].) Because the intersection of the attractor with the surface of section is only approximately one dimensional, there is a slight uncertainty in the symbolic state for some values of x; this uncertainty is indicated by the shading in the regions between the upper and lower bounds on the value of r in Fig. 3. Observations of the time wave form produced by the oscillator suggest that the grammar is simple: Any sequence of

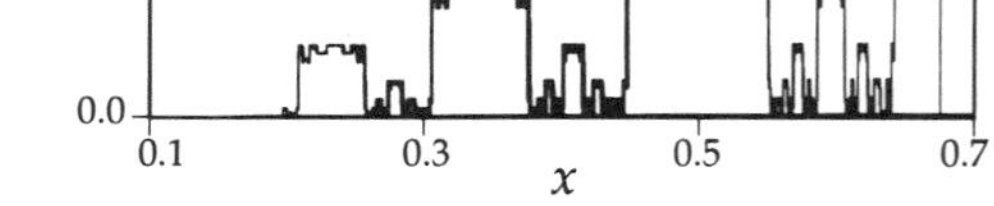

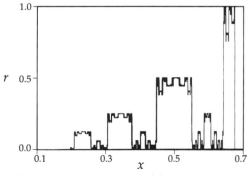

FIG. 3. Binary coding function $r(x)$ for the double scroll system.

binary symbols is allowed, except there can never be less than two oscillations of the same polarity. (We do not discuss the full grammar here, but instead adopt this simple grammar for simplicity of description.) This no-single-oscillation rule leads to a very simple coding: Insert an extra 1 after every block of 1's in the binary stream to be transmitted, and an extra 0 after every block of 0's. This altered data stream now satisfies the constraints of the grammar, and is uniquely decodable: Simply remove a 1 from every block of 1's upon reception, and a 0 from every block of 0's. Thus k oscillations of a given polarity represent $k-1$ information bits.

We now discuss how we control the system to follow a desired binary symbol sequence. Say the system state point passes through branch 0 of the surface of section (shown in Fig. 2) at $x=x_a$, and next crosses the surface of section (on either branch 0 or 1) at $x=x_b$. Because we have previously determined the function $r(x)$, we can use the stored values to find the symbolic state $r(x_a)$. We then convert the number $r(x_a)$ to its corresponding binary sequence truncated at some chosen length N, and store this finite-length symbol sequence in a *code register*. As the system state point travels towards its next encounter with the surface of section at $x=x_b$, we shift the sequence in the code register left, discarding the most significant bit (the leftmost bit), and insert the first desired information code bit in the now empty least significant slot (the rightmost slot) of the code register. We then convert this new symbol sequence to its corresponding symbolic state r_b'. Now, when the system state point crosses the surface of section at $x=x_b$, we use a simple search algorithm to find the nearest value of the coordinate x that corresponds to the desired symbolic state r_b'; call this x_b'. By construction, $|r(x_b)-r(x_b')| \leq 2^{-N}$. [If $r(x)$ is continuous, as in the Lorenz system, for example, this search can be replaced by a more efficient local derivative projection to find the desired value of x.] Now let $\delta x = x_b - x_b'$. Because we have chosen the branches of the surface of section at constant values of the inductor current i_L, the deviation δx in the generalized coordinate corresponds to a deviation in the voltages v_{C_1} and v_{C_2} across the two capacitors in Fig. 1. We thus apply a vector correction parallel to the surface of section (at constant i_L) along the attractor cross section to put the orbit at $x=x_b'$. This small correcting voltage perturbation is given by $\delta v_{C_1} = \pm \delta x \cos(\theta)$, $\delta v_{C_2} = \pm \delta x \sin(\theta)$, where the + signs are used for lobe 1 of the attractor, and the − signs for lobe 0. We plan to do this experimentally with current pulse generators connected in parallel with each capacitor. (Many methods of applying control perturbations are possible, but this one is particularly straightforward.) On each successive pass through the surface of section, a new code bit is shifted into the code register, and we repeat the procedure to correct the state-space coordinates, and thus the symbolic state, of the system. The coded information sequence, because it is shifted through the code register, does not begin to appear in the output wave form until N iterations of the procedure, where N is the length of the code register. If the symbol sequence is coming from a properly coded discrete ergodic information source, the process of shifting the information sequence through the code register can be viewed as locking the symbolic dynamics of the oscillator to the information source. Thus, there is a short transient phase during which the symbolic dynamics of the oscillator is being locked to the information source, and the symbolic dynamics of the oscillator is always N bits behind the information source.

Figure 4 shows an encoded wave form for the double scroll system produced by the described technique. This wave form corresponds to the voltage wave form $v_G(t)$ across the passive conductance G. If the conductance G is replaced by a transmission channel of the same impedance, the signal produced can be transmitted through the channel. We have represented each letter of the Roman alphabet by the five-bit binary number for its location in the alphabet, and added the extra bits to satisfy the no-single-oscillation constraint to encode the word "chaos." We have applied the technique to first bring the system to a periodic orbit about lobe 1 of the attractor, then to execute the writing of the word, and then to bring the system back to a periodic orbit about lobe 0. The trajectory shown in Fig. 2 is actually the encoded trajectory, but this is not apparent in the figure because the controlled trajectory approximates a possible natural trajectory. The root-mean-squared amplitude of the control signal over the writing of the word was of order 10^{-3} in the normalized units. The control probably cannot be made much smaller using this simple technique, primarily because the one-dimensional approximation in the surface of section causes the coding function to be slightly inaccurate. This control amplitude, though already very small compared to the oscillator signal voltages, does not

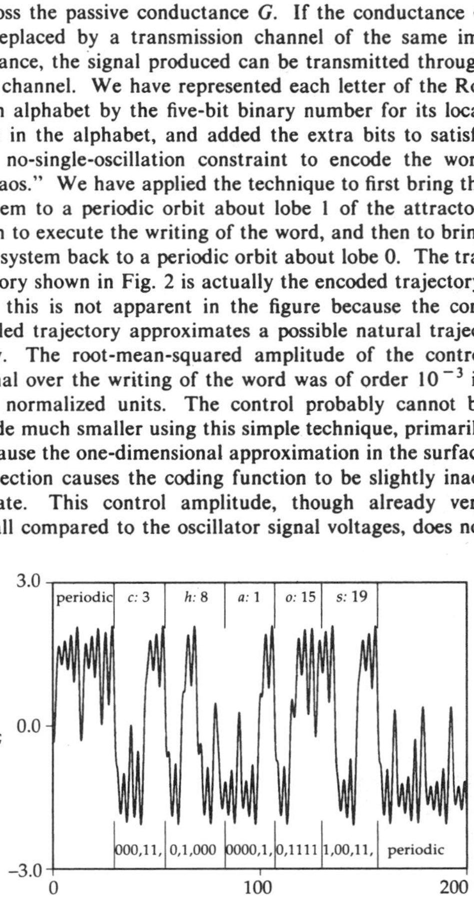

FIG. 4. Controlled $v_G(t)$ signal for the double scroll system encoded with "chaos." Each letter is shown at the top of the figure, along with its numerical position in the alphabet. Shown at the bottom are the corresponding binary code words. Extra bits (indicated by commas) are added to satisfy the constraints imposed by the grammar.

appear to be a fundamental limit, and we are developing control techniques to reduce it.

We conclude with some comments concerning the scope, application, and theoretical significance of our technique.

(1) Since we envision the transmitted signal to be a single scalar, its instantaneous value does not specify the full system state of the chaotic oscillator. If more state information is needed to extract the symbol sequence, time delay embedding [8] can be used. As our example using the double scroll equations shows, however, time delay embedding is not always necessary.

(2) Because our control technique uses only small perturbations [9], the dynamical motion of the system is approximately described by the equations for the uncontrolled system. Knowing the equations of motion greatly simplifies the task of removing noise [10] from a received signal. The basic bipolar nature of the signal in Fig. 4 implies that the message can still be extracted for noise amplitudes that are significant, but not too large compared to the signal. We consider the effects of additive noise on the detection of chaos signals quantitatively in the longer paper [7].

(3) Signals that are generated by chaotic dynamical systems and carry information in their symbolic dynamics have an interesting and possibly useful property: More than one encoded symbol can be extracted from a single sample of the trajectory if time delay embedding is used. This is done by using the state-space partition for a higher order iterate of the return map [7] of the system.

(4) Much of the theory needed to understand information transmission using the symbolic dynamics of chaotic systems already exists [11]. For example, because the topological entropy [12] of a dynamical system is the asymptotic growth exponent of the number of finite symbol sequences that the system can generate (given the best state-space partition), the channel capacity of a chaotic system used for information transmission is given by the topological entropy. The types of channel constraints that arise with a chaotic system will be discussed in a longer paper [7], along with other theoretical considerations.

(5) We emphasize that the particular methods for control and coding used in our double scroll example were chosen for simplicity, and that other more optimal methods are possible. Also, the double scroll oscillator itself was chosen because it is simple, and a large body of research is available about its dynamics. It is not intended as an example of a practical oscillator for communication wave form synthesis. It may be possible to use a higher-dimensional radio-frequency band-limited chaotic system for improved performance (higher information rate and better noise immunity), roughly analogous to the use of complex signaling constellations in classical communication systems. We are now developing more practical high-speed symbolic control techniques that could be used at higher bit rates than an implementation of the straightforward example given here.

(6) There has been much discussion of the role of chaos in biological systems, and we speculate that the control of chaos with tiny perturbations may be important for information transmission in nature.

This research was supported by Harry Diamond Laboratories (now the U.S. Army Research Laboratory), and by the U.S. Department of Energy (Office of Scientific Computing, Office of Energy Research).

---

[a]Also at Department of Physics and Astronomy, University of Maryland, College Park, MD 20742.
[b]Laboratory for Plasma Research.
[c]Institute for Physical Science and Technology and Department of Mathematics.
[d]Department of Physics and Astronomy.
[e]Department of Electrical Engineering.

[1] For a pedagogical discussion emphasizing the role of concepts from information theory in chaos, see R. Shaw, Z. Naturforsch. **36A**, 80 (1981).

[2] (a) E. Ott, C. Grebogi, and J. A. Yorke, Phys. Rev. Lett. **64**, 1196 (1990); (b) T. Shinbrot, E. Ott, C. Grebogi, and J. A. Yorke, Phys. Rev. Lett. **65**, 3215 (1990).

[3] Several articles about this circuit appear in Proc. IEEE **75**, No. 8 (1987): e.g., T. Matsumoto, p. 1033.

[4] P. Cvitanović, G. Gunaratne, and I. Procaccia, Phys. Rev. A **38**, 1503 (1988); P. Grassberger, H. Kantz, and U. Moenig, J. Phys. A **22**, 5217 (1989).

[5] D. P. Lathrop and E. J. Kostelich, Phys. Rev. A **40**, 4028 (1989); L. Flepp, R. Holzner, E. Brun, M. Finardi, and R. Badii, Phys. Rev. Lett. **67**, 2244 (1991).

[6] R. E. Blahut, *Principles and Practice of Information Theory* (Addison-Wesley, Reading, MA, 1988); C. E. Shannon and W. Weaver, *The Mathematical Theory of Communication* (University of Illinois Press, Chicago, 1963).

[7] S. Hayes, C. Grebogi, and E. Ott (to be published).

[8] F. Takens, in *Dynamical Systems and Turbulence*, edited by D. A. Rand, Springer Lecture Notes in Mathematics Vol. 898 (Springer-Verlag, New York, 1981), p. 366.

[9] Our control technique can also be used to target a chaotic system [2(b)] in state space. The relationship between symbolic states and state-space coordinates is established by the coding function $r(x)$.

[10] E. J. Kostelich and J. A. Yorke, Phys. Rev. A **38**, 1649 (1988); S. M. Hammel, Phys. Lett. A **148**, 421 (1990). It is easier to filter noise that is introduced in the communication channel than it is to filter noise that is present in the chaotic oscillator itself.

[11] A very different concept is that of using the synchronization of two chaotic systems [L. M. Pecora and T. L. Carroll, Phys. Rev. Lett. **64**, 821 (1990)] for secure communications. In this case a small information-bearing signal is masked by a large chaotic signal. Several papers discussing this appear in *Proceedings of the International Conference on Acoustics, Speech, and Signal Processing* (IEEE, New York, 1992). This mechanism is not, however, based on the information theoretic formalism of chaos.

[12] R. L. Adler, A. G. Konheim, and M. H. McAndrew, Trans. Am. Math. Soc. **114**, 309 (1965).

Observing chaos: deducing and tracking the state of a chaotic system from limited observation

Paul So [a], Edward Ott [a,b] and W.P. Dayawansa [b,c]

[a] Department of Physics and Laboratory for Plasma Research, University of Maryland, College Park, MD 20742, USA
[b] Department of Electrical Engineering, University of Maryland, College Park, MD 20742, USA
[c] Systems Research Center, University of Maryland, College Park, MD 20742, USA

Received 24 August 1992; revised manuscript received 22 March 1993; accepted for publication 26 March 1993
Communicated by A.P. Fordy

Assuming that an accurate mathematical description of the system is available, a method is proposed whereby the full state vector of a chaotic system can be reconstructed and tracked using only the time series of a single observed scalar function of the system state. Noise effects on the procedure are investigated using as an example a kicked mechanical system which results in a four-dimensional dissipative map.

Consider the situation where there is some experimental system behaving chaotically, and one is able to accurately observe a single scalar measure of the system state. Formally, the system state is given by some vector $\boldsymbol{X}$ in $\mathbb{R}^d$ which is a function of time. The observed scalar can be expressed as some function of the system state, $O=g(\boldsymbol{X})$. The question we ask is the following: Assuming that an accurate mathematical description of the system is available (e.g., for a discrete time system, the map is known), how can we deduce the system state $\boldsymbol{X}$ from measurements of O? As an example, say we have a mechanical system consisting of interconnected levers, gears, springs, etc. and this system is behaving chaotically on an attractor of not too large dimension. Can we deduce the positions of all the parts of the system from observations of the time series of the position of just one of the levers?

One way of addressing this general problem is via the delay coordinate embedding technique. Takens [1] shows that a delay coordinate vector $[O_n, O_{n-1}, ..., O_{n-(N-1)}]$ of sufficiently large N uniquely determines the system state $\boldsymbol{X}_n$. Thus by using a computer to solve the known mathematical description of the system (assumed here to be a discrete time system), one can build up a mapping at each point on the attractor from an N-dimensional delay coordinate vector $[O_n, O_{n-1}, ..., O_{n-(N-1)}]$ to the system state, $\boldsymbol{X}_n$ at time n. A drawback to this procedure is that it will typically require the generation, storing and searching of a large amount of data. Another way to address this problem is by utilizing the so called "extended Kalman filter" [2]. By taking the statistics of noise into consideration, it can be shown that the Kalman filter for a *linear* system is optimal in the sense that the error variance between the actual state and the estimated state is minimal. However, since the implementation of the extended Kalman filter requires, at each iterate, the manipulation of matrix equations which have the same dimension as the full dynamical system, the calculation can get quite cumbersome when the dimension of the system is large. (In addition, when the Kalman filter is extended to a nonlinear system, the sense in which this method is optimal becomes unclear [#1].) Thus, while the embedding method and the extended Kalman filter may be useful for the purpose we address, they have drawbacks that motivate us to investigate another

[#1] We will initially be interested in noiseless situations, and it should be noted that the Kalman filter procedure is still well defined in the zero noise limit.

From *Phys. Lett. A* **176** (19903). Reprinted with permission from Elsevier Science Publishers.

approach [#2]. Here, we propose a more efficient tracking technique for relatively high dimensional chaotic systems but with low dimensional attractors. The stability of our technique to the addition of small noise will also be investigated. Unlike the embedding technique, computer memory requirements for our procedure are minimal and the procedure is comparatively fast. Unlike the Kalman filter, which requires manipulation of matrix equations whose dimension is the full dimensionality of the system, the number of calculations in our method is of the order of the number of expanding directions, which may be much smaller than the system dimensionality.

The general procedure for this tracking method is similar to the design of an observer in *linear* control theory. The observer is built upon a numerical copy of the actual system but with an additional time dependent correction term which compares the actual output of the chaotic system and the estimated output of the observer. Depending on the difference between the actual and the estimated output, the time dependent parameters in the correction term are adjusted so that the difference will exponentially decay to zero with time. In this paper, we shall present a procedure for doing this in the case of *nonlinear chaotic* systems.

In *linear* control theory, it is possible to estimate unmeasured state variables using a "state observer". To be specific, consider a linear time independent d-dimensional system, $X_{n+1} = \mathbf{A}X_n$, $O_n = GX_n$, where X is a d-dimensional column vector, $\mathbf{A}$ is a constant $d \times d$ matrix, and G is a constant d-dimensional row vector. The scalar function O_n is the observed physical output of the system. In order to reconstruct the actual state X_n of the system from a time series of the scalar output O_n, one can employ a state observer, defined by

$$\hat{X}_{n+1} = \mathbf{A}\hat{X}_n + C[O_{n+1} - \hat{O}_{n+1}] ,$$

where $\hat{O}_{n+1} = G\mathbf{A}\hat{X}_n$. The idea of this technique is to choose the control vector C such that the numerically generated state $\hat{X}_n$ will converge to the actual state X_n with increasing n. To derive the necessary condition for this to happen, one can look at the dynamics of the error equation,

$$X_{n+1} - \hat{X}_{n+1} = [\mathbf{A} - CG\mathbf{A}](X_n - \hat{X}_n) .$$

If the control vector C can be chosen so that the magnitudes of the eigenvalues of $[\mathbf{A} - CG\mathbf{A}]$ are all less than one, then the error will exponentially decrease to zero as n approaches infinity. A standard technique exists for choosing the control vector C to do this, and can be found in many control theory textbooks (e.g., see ref. [4]).

The design of our observer for nonlinear chaotic systems is similar in spirit to the linear observer. We assume that the chaotic system that we want to observe is given by the following equations: $X_{n+1} = M(X_n)$, $O_n = g(X_n)$. Here M and g are *nonlinear* functions of the d-dimensional vector X_n. The corresponding state observer is taken to be

$$\hat{X}_{n+1} = M(\hat{X}_n) + C_n[O_{n+1} - \hat{O}_{n+1}] ,$$

where $\hat{O}_{n+1} = g(M(\hat{X}_n))$ and C_n is a *time dependent* d-dimensional control column vector which we need to adjust at each iterate. These equations for X_{n+1} and $\hat{X}_{n+1}$ yield the error equation,

$$X_{n+1} - \hat{X}_{n+1} = M(X_n) - M(\hat{X}_n)$$
$$- C_n[g(M(X_n)) - g(M(\hat{X}_n))] .$$

Linearizing about $\hat{X}_n$ gives

$$\delta X_{n+1} = [DM(\hat{X}_n) - C_n Dg(M(\hat{X}_n))DM(\hat{X}_n)]\delta X_n ,$$

where $DM(\hat{X}_n)$ and $Dg(M(\hat{X}_n))$ are the derivatives of $M(\hat{X}_n)$ and $g(M(\hat{X}_n))$, respectively, and Dg is a d-dimensional row vector. Looking back at our discussion of observers for linear time independent systems, the matrix $[\mathbf{A} - CG\mathbf{A}]$ was a constant, and the long term evolution of the observer error is determined by $[\mathbf{A} - CG\mathbf{A}]^n$, which converges to zero with increasing n if the eigenvalues of $[\mathbf{A} - CG\mathbf{A}]$ have magnitudes less than one. In the chaotic case, however, the long term behavior of the error is governed by the product of matrices of the form

$$[DM(\hat{X}_n) - C_n Dg(M(\hat{X}_n))DM(\hat{X}_n)]$$

[#2] Another common situation arises when the system is acted upon by external time dependent inputs u_n. In that case one seeks X_n given knowledge of u_n and O_n. Now, the use of the embedding technique to obtain X_n becomes inapplicable, but the alternative which we shall discuss still applies.

Our method is closely related to the recent work of Pecora and Caroll [3] on synchronizing chaos, the main difference being that we seek to ensure conditions where synchronism is guaranteed.

which change at each iterate,

$$\delta X_{n+1} = \prod_{m=0}^{n} [DM(\hat{X}_m) - C_m Dg(M(\hat{X}_m))DM(\hat{X}_m)]\delta X_0 .$$

While one can adjust each individual matrix at each iterate to have eigenvalues with magnitudes less than one, that does not guarantee that the product goes to zero as n goes to infinity [#3]. Below we give a procedure which yields convergence of our observer in the chaotic case.

For specificity of the discussion, we will assume the chaotic attractor of our system to be hyperbolic and to have two positive Lyapunov exponents with the rest negative. Thus, the tangent space at each point on the attractor can be decomposed into the sum of a two-dimensional unstable subspace and a $(d-2)$-dimensional stable subspace. Noting that $DM(\hat{X}_n)$ maps the unstable subspace at $\hat{X}_n$ into the unstable subspace at $M(\hat{X}_n)$ and similarly maps the stable subspace at $\hat{X}_n$ into the stable subspace at $M(\hat{X}_n)$, we see that, if C_n is chosen to lie in the unstable subspace at $\hat{X}_n$, then the matrix representation of

$$[DM(\hat{X}_n) - C_n Dg(M(\hat{X}_n))DM(\hat{X}_n)]$$

can be put in the following block form,

$$\begin{pmatrix} \mathbf{U}_n & \mathbf{W}_n \\ \mathbf{0} & \mathbf{S}_n \end{pmatrix},$$

where $\mathbf{U}_n$ is a 2×2 submatrix acting on the unstable subspace, $\mathbf{S}_n$ is a $(d-2)\times(d-2)$ submatrix acting on the stable subspace, and $\mathbf{W}_n$ is a $2\times(d-2)$ submatrix taking vectors from the stable subspace into the unstable subspace. Since the matrix is block-upper-triangular and the space acted on by $\mathbf{S}_n$ is already stable ($\mathbf{S}_n$ is given by $DM(\hat{X}_n)$ restricted to the stable subspace), the stability of the product of these matrices, which is also block-upper-triangular, depends solely on the product of the $\mathbf{U}_n$ [#4]. If we consider lower triangular matrices ($\mathbf{U}$ is lower triangular if $U_{ij}=0$ for $i<j$), then we have the following: (i) the product of two or more such matrices will still be lower triangular; (ii) the eigenvalues are just the diagonal elements; and most importantly, (iii) the eigenvalues of the product of such matrices will be the product of their respective eigenvalues taken from their diagonals. Thus, if we choose C_n so that $\mathbf{U}_n$ is stable (i.e., the magnitudes of its eigenvalues are all less than one) and is lower triangular, then the product of the $\mathbf{U}_n$ will also be stable and lower triangular. The choice of basis for $\mathbf{S}_n$ can be any vectors which span the stable subspace at $\hat{X}_n$ while the choice of basis spanning the unstable subspace at $\hat{X}_n$ (i.e., the basis for $\mathbf{U}_n$) is essential in the design of our observer. We make this choice by defining two numbers, $\lambda_n^{(1)}$, and $\lambda_n^{(2)}$, and two unit basis column vectors, $e_n^{(1)}$ and $e_n^{(2)}$, for the unstable subspace at $\hat{X}_n$ according to the following procedure,

$$\lambda_n^{(1)} e_{n+1}^{(1)} = [DM(\hat{X}_n)]e_n^{(1)} ,$$

$$\lambda_n^{(2)} e_{n+1}^{(2)} = [DM(\hat{X}_n) - \{C_n^{(1)} e_{n+1}^{(1)}\} Dg(M(\hat{X}_n))DM(\hat{X}_n)]e_n^{(2)}. \qquad (1)$$

Thus given a pair of initial values $e_0^{(1)}$ and $e_0^{(2)}$ spanning the unstable subspace at $\hat{X}_0$, eq. (1) yields $e_n^{(1)}, e_n^{(2)}, \lambda_n^{(1)}$, and $\lambda_n^{(2)}$ at all subsequent times $n>0$. In practice, we choose $e_0^{(1)}$ and $e_0^{(2)}$ arbitrarily (not necessarily in the unstable subspace) and then evolve them via eq. (1); as n increases $e_n^{(1)}$ and $e_n^{(2)}$ asymptote to the unstable subspace. The matrix $\mathbf{U}_n$ will then be in lower triangular form with two free parameters, $C_n^{(1)}$ and $C_n^{(2)}$,

$$\mathbf{U}_n = \begin{pmatrix} \lambda_n^{(1)} - C_n^{(1)} Dh_n^{(1)} & 0 \\ -C_n^{(2)} Dh_n^{(1)} & \lambda_n^{(2)} - C_n^{(2)} Dh_n^{(2)} \end{pmatrix}, \qquad (2)$$

[#3] As an example, the product of the following sequence of matrices,

$$\begin{pmatrix} \frac{1}{2} & 2 \\ 0 & \frac{1}{2} \end{pmatrix}\begin{pmatrix} \frac{1}{2} & 0 \\ 2 & \frac{1}{2} \end{pmatrix}\begin{pmatrix} \frac{1}{2} & 2 \\ 0 & \frac{1}{2} \end{pmatrix}\cdots$$

will be infinite while the eigenvalues of each individual matrix are less than one, i.e., $\frac{1}{2}$.

[#4] We can write down the product of n of these matrices in a simple formula,

$$\begin{pmatrix} \mathbf{U}_n\mathbf{U}_{n-1}\ldots\mathbf{U}_1 & \sum_{i=1}^{n}\left(\prod_{j=i+1}^{n}\mathbf{U}_j\mathbf{W}_i\prod_{k=1}^{i-1}\mathbf{S}_k\right) \\ \mathbf{0} & \mathbf{S}_n\mathbf{S}_{n-1}\ldots\mathbf{S}_1 \end{pmatrix}.$$

Because this matrix is in a block-upper-triangular form, its eigenvalues are just the eigenvalues of $\mathbf{U}_n\mathbf{U}_{n-1}\ldots\mathbf{U}_1$ plus the eigenvalues of $\mathbf{S}_n\mathbf{S}_{n-1}\ldots\mathbf{S}_1$. With both $\mathbf{U}_n\mathbf{U}_{n-1}\ldots\mathbf{U}_1$ and $\mathbf{S}_n\mathbf{S}_{n-1}\ldots\mathbf{S}_1$ going to zero as n goes to infinity, it is not hard to see that the whole matrix will go to zero as well.

where $Dh_n^{(i)} = Dg(M(\hat{X}_n))DM(\hat{X}_n)e_n^{(i)}$, $C_n^{(i)} = f_{n+1}^{(i)}C_n$, and the contravariant row vectors $f_n^{(i)}$ are defined by $f_n^{(i)}e_n^{(i)} = \delta_{ij}$. Thus, if we adjust $C_n^{(1)}$ and $C_n^{(2)}$ so that the eigenvalues (i.e. the diagonal elements) of $\mathbf{U}_n$ are less than one, then the product of the matrices

$$[DM(\hat{X}_n) - C_n Dg(M(\hat{X}_n))DM(\hat{X}_n)]$$

will be a stable matrix.

A possible concern with our method, as outlined above, is that, as time n increases, the vectors $e_n^{(1)}$ and $e_n^{(2)}$ might tend to become more and more nearly parallel. (This would invalidate our procedure since we assume that $e_n^{(1)}$ and $e_n^{(2)}$ span the two-dimensional unstable subspace). We note, however, that we have the freedom of choosing the eigenvalues of $\mathbf{U}$, $\Lambda_n^{(i)} = \lambda_n^{(i)} - C_n^{(i)}Dh_n^{(i)}$. If we choose $\Lambda_n^{(1)}$ to be zero, then the collapse of $e_n^{(1)}$ and $e_n^{(2)}$ to a common direction can be prevented (see next paragraph), leaving one degree of freedom in choosing $\Lambda_n^{(2)}$. Setting both eigenvalues to zero yields,

$$C_n = (\lambda_n^{(1)}/Dh_n^{(1)})e_{n+1}^{(1)} + (\lambda_n^{(2)}/Dh_n^{(2)})e_{n+1}^{(2)} . \quad (3)$$

The expression in eq. (3) for the control vector C_n is valid as long as the denominators, $Dh_n^{(i)}$, are not zero. (This is similar to the "observability" condition in linear control theory [4].) In our numerical program, we set a minimum value such that whenever $Dh_n^{(i)}$ falls below that value, we set the control vector to zero. Thus, when the $Dh_n^{(i)}$ are small, we do not attempt to bring X and $\hat{X}$ together. But, if they were already close, they will still be close one iterate later. Hence, little is lost by turning the control off for one iterate, provided that this is done only infrequently.

To see that $e_n^{(1)}$ and $e_n^{(2)}$ do not typically approach a common direction as n increases, we assume that they are nearly parallel at time n, and then demonstrate that eq. (1) and the expression for $C_n^{(1)}$ in eq. (3) imply that they are not nearly parallel at time $n+1$. Setting $e_n^{(2)} = e_n^{(1)} + \delta e_n$ with $|\delta e_n| \ll 1$, eqs. (1) and (3) yield

$$\lambda_n^{(2)}e_{n+1}^{(2)} = DM(\hat{X}_n)\delta e_n$$
$$-\lambda_n^{(1)}e_{n+1}^{(1)}[(Dg(M(\hat{X}_n))DM(\hat{X}_n)\delta e_n)/Dh_n^{(1)}] .$$

Thus $e_{n+1}^{(2)}$ consists of two terms which are both typically of order δe_n, one pointing in the direction $e_{n+1}^{(1)}$ (which is the direction of $DM(\hat{X}_n)e_n^{(1)}$), and the other pointing in the direction $DM(\hat{X}_n)\delta e_n$, which is different from the direction of $e_{n+1}^{(1)}$ because δe_n is approximately perpendicular to $e_n^{(1)}$ by assumption. Thus, the only problem arises in the rare cases when $Dh_n^{(1)}$ is small, but we skip over these "glitches" anyway by setting $C_n \equiv 0$ when this occurs.

Although the above discussion is in the context of a two-dimensional unstable subspace, we note that no essential change is produced in the case where the unstable subspace has an arbitrary dimension d_u. In particular, eq. (1) generalizes in a natural way [5] to d_u equations for the d_u basis vectors $e^{(i)}$. Computationally, since we only need to keep track of the evolutions of d_u basis vectors in eq. (1), the number of calculations required in our method will be a factor of d/d_u less than the number of calculations required in the extended Kalman filter which involves the evolution of a $d \times d$ matrix.

A variant of the above technique can be formulated in the special, but common, case where the observed quantity $g(X_n)$ is simply a projection onto one of the state variables. That is, $g(X_n)$ is one of the d components of $X_n = (X_n^1, ..., X_n^{d-1}, X_n^d)$. Without loss of generality, we can take the observed component to be X_n^d, so that $g(X_n) = X_n^d$. In this case, the observed state variable does not need to be estimated by the observer. It is then convenient to rewrite the state equation $X_{n+1} = M(X_n)$ as $Y_{n+1} = M_Y(Y_n, Z_n)$ and $Z_{n+1} = M_Z(Y_n, Z_n)$, where

$$X = (X^1, ..., X^{d-1}, X^d) \equiv (Y^1, Y^{d-1}, Z) = (Y|Z) .$$

$Z_n = X_n^d$ is the observed state variable and the $(d-1)$dimensional vector Y_n contains the remaining unmeasured state variables. Similarly, $M_Z(\;)$ is the component of the d-dimensional vector function $M(\;)$ associated with Z_n, and the $(d-1)$-dimensional vector function $M_Y(\;)$ is the remaining part of $M(\;)$ associated with Y_n. The "reduced-order" observer for the unmeasured part of the state vector Y_n can then defined as

$$\hat{Y}_{n+1} = M_Y(\hat{Y}_n, Z_n) + C_n[Z_{n+1} - M_Z(\hat{Y}_n, Z_n)] ,$$

where C_n is the $(d-1)$-dimensional control vector corresponding to the reduced order observer. Forming the error equation, we have

$$\delta Y_{n+1} = [DM_Y(Y_n, Z_n) - C_n DM_Z(Y_n, Z_n)]\delta Y_n ,$$

which can be treated using exactly the same techniques as already discussed. A similar expression for the control vector C_n can be derived as before (eq. (3)), but with $Dg(M(\hat{X}_n))DM(\hat{X}_n)$ replaced by $DM_Z(Y_n, Z_n)$. The main obvious advantage achieved by doing this is that the dimensionality of the observer is reduced by one. In addition, in numerical experiments, we have found that the noise performance of this "reduced order" observer is substantially improved as compared with the "full order" observer (i.e., numerical estimate of *all* the state variables as discussed previously). (From a different viewpoint, the reduced order observer and the actual system can be considered as a pair of coupled systems with the actual system providing the driving signal and the observer as the response function. This pair of driven-response systems reduce to the one studied by Pecora and Caroll [3] when the feedback control in our reduced order observer is turned off (i.e., $C_n=0$).)

In order to assess the performance characteristics of our chaotic observer technique, we illustrate its performance without and with noise using a reduced order observer (the results for the full order observer will be given in ref. [5]) for a specific map. The particular map employed describes the time evolution of a mechanical system (fig. 1), called the kicked double rotor [6,7]. As shown in the illustration, fig. 1, the kicked double rotor consists of two massless rods of lengths L_1 and L_2 connected at the pivot P_2 and with the other end of rod 1 connected to a fixed pivot at P_1. Point masses m_1 and $\frac{1}{2}m_2$ are attached at the end of rod 1 and the two ends of rod 2 as shown. At one of the ends of rod 2, an impulse force is applied at times $t=0,\ T,\ 2T,\ \ldots,\ f(t)=f_0\sum_{n=0}^{\infty}\delta(t-nT)\hat{y}$. The kicked double rotor is governed by the following set of equations,

$$X_{n+1}=\begin{pmatrix}\theta_{n+1}\\ \dot{\theta}_{n+1}\end{pmatrix}=\begin{pmatrix}\mathbf{K}\dot{\theta}_n+\theta_n\\ \mathbf{L}\dot{\theta}_n+G(\theta_{n+1})\end{pmatrix}, \qquad (4)$$

where $\theta=(\theta_1,\ \theta_2)^\dagger$, $\dot{\theta}=(\dot{\theta}_1,\ \dot{\theta}_2)^\dagger$, and $G(\theta)=(a_1\sin\theta_1,\ a_2\sin\theta_2)^\dagger$ (here the dagger denotes transpose), θ_1 and θ_2 are angle variables giving the positions of the rotor arms, and $\dot{\theta}_1$ and $\dot{\theta}_2$ are the angular velocities of the rotor arms at the instant immediately after the nth kick. $a_{1,2}=(f_0/I)L_{1,2}$ are constants proportional to the strength of the periodic kick $f(t)$. The moments of inertia about pivots 1 and 2 are chosen to be equal,

$$I=(m_1+m_2)L_1^2=m_2L_2^2 .$$

$\mathbf{K}$ and $\mathbf{L}$ are constant matrices defined by

$$\mathbf{L}=\sum_{i=1}^{2}\mathbf{W}_i\exp(\zeta_iT),\quad \mathbf{K}=\sum_{i=1}^{2}\mathbf{W}_i\frac{\exp(\zeta_iT)-1}{\zeta_i},$$

$$\mathbf{W}_1=\begin{pmatrix}\alpha & \beta\\ \beta & \sigma\end{pmatrix},\quad \mathbf{W}_2=\begin{pmatrix}\sigma & -\beta\\ -\beta & \alpha\end{pmatrix},$$

$$\alpha=\tfrac{1}{2}(1+\nu_1/\Delta),\quad \sigma=\tfrac{1}{2}(1-\nu_1/\Delta),\quad \beta=-\nu_2/\Delta,$$

$$\zeta_{1,2}=-\tfrac{1}{2}(\nu_1+2\nu_2\pm\Delta),\quad \Delta=(\nu_1^2+4\nu_2^2)^{1/2},$$

where ν_1 and ν_2 are the friction coefficients at the pivots (see fig. 1 and refs. [6,7]). In our numerical experiment, we used $g(X)=\theta_2$ and have chosen a particular set of values for the physical parameters [#5]. The resulting chaotic attractor has two positive Lyapunov exponents and its Lyapunov dimension is approximately 2.81. It is interesting to note that, although the attractor has two positive Lyapunov exponents, there exist periodic orbits on the attractor with two-dimensional unstable tangent spaces, and other periodic orbits on the attractor with only one-dimensional unstable tangent spaces [6,8]. Thus, the kicked double rotor map is not globally hyperbolic as assumed in the previous theoretical dis-

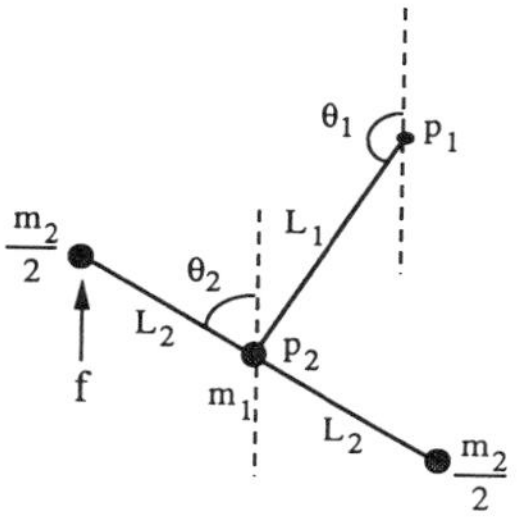

Fig. 1. The double rotor.

[#5] In terms of the notation used in fig. 1 (see also ref. [6], the parameters used in our numerical example are: $L_1=1\sqrt{2}$, $L_2=1$, $m_2=1$, $m_1=1$, $\nu_1=\nu_2=1$, the time interval between successive kicks is $T=1$, the impulse strength of a kick is $f_0=8.7$ (which gives $a_1=8.7/\sqrt{2}$ and $a_2=8.7$). For this case synchronism as defined by Pecora and Carroll [3] does not occur.

cussion. Nevertheless, we find that our method still works if we use two basis vectors as in eqs. (1)–(3).

Typically, the observer orbit $\hat{X}_n$ begins to track the true orbit X_n when $\hat{X}_n$ is close enough to X_n. Furthermore, if we wait long enough, an initially non-tracking observer orbit $\hat{X}_n$ will typically come close to X_n at some future time n. In our numerical experiments, we found that the average transient time before tracking sets in could be quite long. To remedy this problem, we use many observer test points with randomly chosen initial conditions on the attractor, and continuously test each one to see if it has locked onto the true orbit X_n. We do this by calculating $g(X_n) - g(\hat{X}_n)$ and declaring the orbits X_n and $\hat{X}_n$ locked if this quantity is small for several successive iterates. We then take the observer state as $\hat{X}_n$ for such a locked orbit. To determine the average transient time $\langle \tau \rangle$ for our method (eq. (4)), we begin with a large number of randomly chosen observer test orbits. Then, a semi-log plot of the number $\hat{N}$ of orbits which are still not tracking the true orbit after a time interval n is generated. Since the number of such orbits typically decays exponentially with n, i.e., $\hat{N}(n) = N_0 e^{-n/\langle \tau \rangle}$, the inverse of the slope of this graph defines an average time $\langle \tau \rangle$ needed for a single observer orbit to converge to the true orbit. In fig. 2, we used five thousand randomly chosen observer test orbits and estimated $\langle \tau \rangle \sim 4500$.

To look at the effect of noise on our observer, we add a term $\epsilon \boldsymbol{\delta}_n$ to the right hand side of the double rotor map (4), where the four components of $\boldsymbol{\delta}_n$ are uncorrelated random variables with zero mean and uniform distribution, while ϵ is the maximum magnitude of the noise. (We assume that the observer system has no knowledge of the noise). We expect the method to work well when $|\epsilon \boldsymbol{\delta}_n|$ is less than the typical radius of the linear region of the map. For a sufficiently small value of ϵ ($< 10^{-5}$), the observer was able to track the actual state continuously. However, as ϵ increases, the probability of the observer being kicked out of the linear region increases. When this happens, we try to lock back onto the actual orbit again by first going back a few iterates to a point where the observer and the actual state are still close together. Then, we activate a set of N observer test points randomly chosen within a neighborhood centered on that past iterate of the observer. We then evolve the observer separately using all N of the initial test points. When one of these N observer test points begins to lock onto the actual state, we pick that particular observer test orbit as our new observed orbit and drop the rest of the N test points. On the other hand, if none of the N observer test points is able to lock back onto the actual orbit within a given short time limit, we reinitiate the procedure with another set of N randomly chosen observer test points. Figure 3 shows a plot of $|X_n - \hat{X}_n|$ versus n with $N = 4500$ and $\epsilon = 1 \times 10^{-2}$. The norm $|\ |$ is the standard Euclidean norm with each component of $X_n - \hat{X}_n$ normalized by the average values, $\langle X_n^i \rangle$, of the components of X_n. The spikes with negative sign indicate the moments when multiple observer

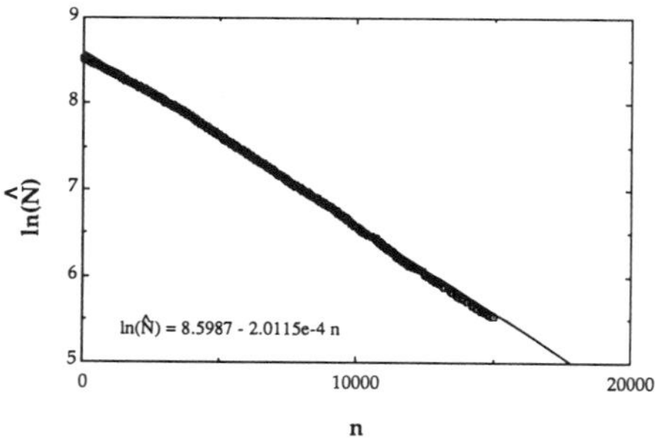

Fig. 2. $\ln(\hat{N})$ versus n with 5000 randomly chosen observer test points. $\hat{N}$ is the number of orbits which are still not tracking the true orbit after a time interval n.

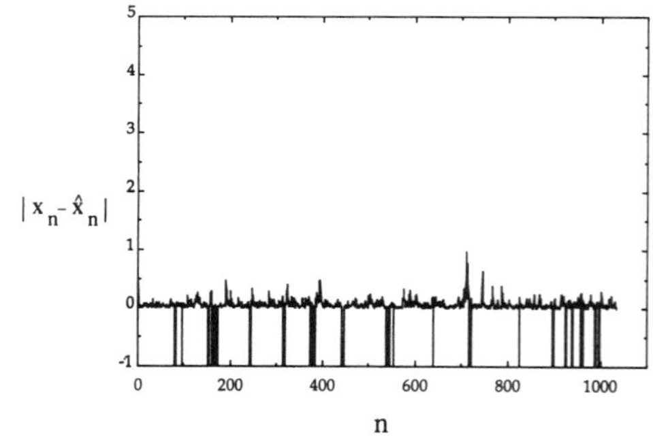

Fig. 3. $|X_n - \hat{X}_n|$ versus n with $N = 4500$ and $\epsilon = 1 \times 10^{-2}$. The diameter of the attractor is approximately five units in this figure.

test points were initiated. We see that the observer successfully tracks $\boldsymbol{X}_n$ even when ϵ is relatively large (signal to noise ratio ≈ 300) provided that N is sufficiently large.

The work of P. So and E. Ott was supported by the Department of Energy (Scientific Computing) and by the Office of Naval Research (Physics), and the work of W.P. Dayawansa was supported by the National Science Foundation (Engineering Research Center Program).

References

[1] F. Takens, in: Dynamical systems and turbulence, eds. D. Rand and L.S. Young (Springer, Berlin, 1981) p. 366.
[2] B.D. Anderson and J.B. Moore, Optimal filtering (Prentice-Hall, Englewood Cliffs, NJ, 1990) pp. 193–211.
[3] L. Pecora and T. Carroll, Phys. Rev. Lett. 64 (1990) 821.
[4] K. Ogata, Modern control engineering (Prentice-Hall, Englewood Cliffs, NJ, 1990) pp. 706–713.
[5] P. So, E. Ott and W. Dayawansa, in preparation.
[6] F. Romeiras, C. Grebogi, E. Ott and W. Dayawansa, Physica D 58 (1992) 165.
[7] C. Grebogi, E. Kostelich, E. Ott and J.A. Yorke, Phys. Lett. A 118 (1986) 448; 120 (1987) 497 (E).
[8] F. Romeiras, E. Ott, C. Grebogi and W. Dayawansa, in: Proc. of the American Control Conference (IEEE Service Center, Piscataway, 1991) pp. 1113–1119.

Bibliography

ABARBANEL, H.D., BROWN, R., KADTKE, J., Predictions and system identification in chaotic nonlinear systems: Time series with broadband spectra. Phys. Lett. A **138**, 401 (1989).

ABARBANEL, H.D., BROWN, R., KADTKE, J., Prediction in chaotic nonlinear systems: Methods for time series with broadband Fourier spectra. Phys. Rev. A **41**, 1742 (1990).

ABARBANEL, H.D., BROWN, R., SIDOROWICH, J., TSIMRING, L., The analysis of observed chaotic data in physical systems. Rev. Mod. Phys. (1993).

ABARBANEL, H.D., CARROLL, T.A., PECORA, L.M., SIDOROWICH, J.J., TSIMRING, L.S., Predicting physical variables in time-delay embedding. Phys. Rev. E **49**, 1840 (1994).

ABARBANEL, H.D., KENNEL, M.B., Lyapunov exponents in chaotic systems: Their importance and their evaluation using observed data. Int. J. Mod. Phys. B **5**, 1347 (1991).

ABARBANEL, H.D., KENNEL, M.B., Local false nearest neighbors and dynamical dimensions from observed chaotic data. Phys. Rev. E **47**, 3057 (1993).

ABRAHAM, N. ET AL., Eds. *Measures of Complexity and Chaos*. Plenum, New York (1989).

AFRAIMOVICH, V.S., EZERSKY, A.B., RABINOVICH, M.I., SHERESHEVSKY, M.A., ZHELEZNYAK, A.L., Dynamical description of spatial disorder. Physica D **58**, 331 (1992).

★ALBANO, A.M., MUENCH, J., SCHWARTZ, C., MEES, A.I., RAPP, P.E., Singular-value decomposition and the Grassberger-Procaccia algorithm. Phys. Rev. A **38**, 3017 (1988).

ALBANO, A.M., PASSAMANTE, A., FARRELL, M.E., Using higher-order correlations to define an embedding window. Physica D **54**, 85 (1991).

ALBANO, A.M., PASSAMANTE, A., HEDIGER, T., FARRELL, M.E., Using neural nets to look for chaos. Physica D **58**, 1 (1992).

ALESIĆ, Z., Estimating the embedding dimension. Physica D **52**, 362 (1991).

ARANSON, I., LEVINE, H., TSIMRING, L., Controlling spatiotemporal chaos. Phys. Rev. Lett. **72**, 2561 (1994).

AUERBACH, D., Controlling extended systems of chaotic elements. Phys. Rev. Lett. **72**, 1184 (1994).

[1] References marked with ★ are reprinted in this volume

BABLOYANTZ, A., Some remarks on nonlinear data analysis of physiological time series. In: *Quantitative Measures of Dynamical Complexity in Nonlinear Systems*, N. Abraham et al., Eds. Plenum (1989).

BADII, R., Measures of complexity and chaos: Some remarks on nonlinear data analysis of physiological time series. In: *Quantitative Measures of Dynamical Complexity in Nonlinear Systems*, N. Abraham et al., Eds. Plenum (1989).

BADII, R., BROGGI, G., DERIGHETTI, B., RAVAIN, M., CILIBERTO, S., POLITI, A., RUBIO, M.A., Dimension increase in filtered signals. Phys. Rev. Lett. **60**, 979 (1988).

BADII, R., POLITI, A., Statistical description of chaotic attractors: The dimension function. J. Stat. Phys. **40**, 720 (1985).

BADII, R., POLITI, A., On the fractal dimension of filtered chaotic signals. In: *Dimensions and Entropies in Chaotic Systems*, ed. G. Mayer-Kress (Springer, Berlin) (1986).

BAUER, M., HENG, H., MARTIENSSEN, W., Characterization of spatiotemporal chaos from time series. Phys. Rev. Lett. **71**, 521 (1993).

BENETTIN, G., GALGANI, L., GIORGILLI, A., STRELCYN, J.-M., Lyapunov characteristic exponents for smooth systems and for Hamiltonian system: A method for computing all of them. Part 2: Numerical Application, Meccanica **15**, 21 (1980).

BEN-MIZRACHI, A., PROCACCIA, I., GRASSBERGER, P., Characterization of experimental (noisy) strange attractors. Phys. Rev. A **29**, 975 (1984).

BENZI, R., CARNEVALE, G.F., A possible measure of local predictability. J. Atmos. Sci. **46**, 3595 (1989).

BEUTER, A., LABRIE, C., VASILAKOS, K., Transient dynamics in motor control of patients with Parkinson's disease. Chaos **1**, 279 (1991).

BIELAWSKI, S., DEROZIER, D., GLORIEUX, P., Controlling unstable periodic orbits by a delayed continuous feedback. Phys. Rev. E **49**, R971 (1994).

BINGHAM, S., KOT, M., Multidimensional trees, random searching, and correlation dimension algorithms of reduced complexity. Phys. Lett. A **140**, 327 (1989).

BRAIMAN, Y., GOLDHIRSCH, I., Taming chaotic dynamics with weak periodic perturbations. Phys. Rev. Lett. **66**, 2545 (1991).

★BRANDSTATER, A., SWINNEY, H.L., Strange attractors in weakly turbulent Couette-Taylor flow. Phys. Rev. A **35**, 2207 (1987).

BREEDEN, J., DINKELACKER, F., HÜBLER, A.W., Using noise in the modeling and control of dynamical systems. Phys. Rev. A **42**, 5827 (1990).

BREEDEN, J.L., HÜBLER, A.W., Reconstructing equations of motion from experimental data with hidden variables. Phys. Rev. A **42**, 5817 (1990).

BREEDEN, J.L., PACKARD, N., Nonlinear analysis of data sampled nonuniformly in time. Physica D **58**, 273 (1992).

BRIGGS, K., Improved methods for the analysis of chaotic time series. Phys. Lett. A **151**, 27 (1990).

BROCK, W.A., Distinguishing random and deterministic systems. J. Econ. Theo. **40**, 168 (1986).

BROCK, W.A., Causality, chaos, explanation and prediction in economics and finance. In: *Beyond Belief: Randomness, Prediction and Explanation in Science*, J. Casti et al., Eds. CRC Press, Boca Raton, FL (1991).

BROCK, W.A., DECHERT, W.D., SCHEINKMAN, J.A., LEBARON, B., A test for independence based on the correlation dimension. University of Wisconsin Press, Madison (1988).

BROCK, W.A., HSIEH, D., LEBARON, B., *A Test of Nonlinear Dynamics, Chaos, and Instability: Statistical theory and economic evidence.* MIT Press, Cambridge, MA (1991).

BROCK, W.A., LAKONISHOK, A.J., LEBARON, B., Simple technical trading rules and the stochastic properties of stock returns. J. Finance **47**, 1731 (1992).

BROCK, W.A., POTTER, S.M., Diagnostic testing for nonlinearity, chaos, and general dependence in time series data. In: *Nonlinear Modeling and Forecasting*, M. Casdagli, S. Eubank, Eds. Addison-Wesley, Reading, MA (1992).

★BROOMHEAD, D.S., KING, G.P., Extracting qualitative dynamics from experimental data. Physica D **20**, 217-236 (1986).

BROOMHEAD, D.S., JONES, R., KING, G.P., Topological dimension and local coordinates from time series data. J. Phys. A **20**, L563 (1987).

BROOMHEAD, D.S., LOWE, D., Multivariable functional interpolation and adaptive networks. Complex Systems **2**, 321 (1988).

BROOMHEAD, D.S., HUKE, J.P., MULDOON, M.R., Linear filters and nonlinear systems. J. Roy. Stat. Soc. B **54**, 373 (1992).

BROWN, R., Calculating Lyapunov exponents for short and/or noisy data sets. Phys. Rev. E **47**, 3962 (1993).

BROWN, R., BRYANT, P., ABARBANEL, H.D., Computing the Lyapunov spectrum of a dynamical system from observed time series. Phys. Rev. A **43**, 27 (1991).

★BRYANT, P., BROWN, R., ABARBANEL, H., Lyapunov exponents from observed time series. Phys. Rev. Lett. **65**, 1523 (1990).

BUZUG, T., PFISTER, G., Optimal delay time and embedding dimension for delay-time coordinates by analysis of the global static and local dynamical behavior of strange attractors. Phys. Rev. A **45**, 7073 (1992).

BUZUG, T., PFISTER, G., Comparison of algorithms calculating optimal embedding parameters for delay time coordinates. Physica D **58**, 127 (1992).

BUZUG, T., REIMERS, T., PFISTER, G., Optimal reconstruction of strange attractors from purely geometrical arguments. Europhys. Lett. **13**, 605 (1990).

CAPUTO, J.G., Practical remarks on the estimation of dimension and entropy from experimental data. In: *Measures of Complexity and Chaos*, N. Abraham et al., Eds. Plenum, New York (1989).

CAPUTO, J.G., MALRAISON, B., ATTEN, P., Determination of attractor dimension and entropy for various flows: An experimentalist's viewpoint. In: *Dimensions and Entropies in Chaotic Systems: Quantification of Complex Behavior*, G. Mayer-Kress, Ed. Springer-Verlag, Berlin (1986).

CARROLL, T.L., HEAGY, J., PECORA, L.M., Synchronization and desynchronization in pulse coupled relaxation oscillators. Phys. Lett. A **186**, 225 (1994).

CARROLL, T.L., PECORA, L.M., Synchronizing chaotic circuits. IEEE Trans. Cir. Sys. **38**, 453 (1991).

CARROLL, T.L., PECORA, L.M., Cascading synchronized chaotic systems. Physica D **67**, 126 (1993).

★CASDAGLI, M., Nonlinear prediction of chaotic time series. Physica D **35**, 335 (1989).

CASDAGLI, M., Chaos and deterministic versus stochastic nonlinear modeling. J. Roy. Stat. Soc. B **54**, 303 (1991).

CASDAGLI, M., DES JARDINS, D., EUBANK, S., FARMER, J.D., GIBSON, J., THEILER, J., HUNTER, J., Nonlinear modeling of chaotic time series: Theory and applications. In: Applied Chaos, J.H. Kim, J. Stringer, Eds. Wiley, New York (1992).

CASDAGLI, M., EUBANK, S., FARMER, J.D., GIBSON, J., State space reconstruction in the presence of noise. Physica D **51**, 52 (1991).

CASDAGLI, M., EUBANK, S., *Nonlinear Modeling and Forecasting*, Santa Fe Institute Studies in the Science of Complexity XII. Addison-Wesley, Reading MA (1992).

CAWLEY, R., HSU, G.-H., A local geometric projection method for noise reduction in maps and flows. Phys. Rev. A **46**, 3057 (1992).

CAWLEY, R., HSU, G.-H., SNR performance of a noise reduction algorithm applied to coarsely sampled chaotic data. Phys. Lett. A **166**, 188 (1992).

CENYS, A., LASIENE, G., PYRAGAS, K., Estimation of interrelations between chaotic observables. Physica D **52**, 332 (1991).

CENYS, A., PYRAGAS, K., Estimation of the number of degrees of freedom from chaotic time series. Phys. Lett. A **129**, 227 (1988).

CHEN, G., DONG, X., On feedback control of chaotic dynamical systems. Int. J. Bif. Chaos **2**, 407 (1992).

CHEN, G., DONG, X., From chaos to order – perspectives and methodologies in controlling chaotic nonlinear dynamical systems. Int. J. Bif. Chaos **3**, 1363 (1993).

CHRISTIANSEN, F., PALADIN, G., RUGH, H.H., Determination of correlation spectra in chaotic systems. Phys. Rev. Lett. **65**, 2087 (1990).

CILIBERTO, S., Fractal dimension and metric entropy in extended systems. Europhys. Lett. **4**, 685 (1987).

COHEN, A., PROCACCIA, I., Computing the Kolmogorov entropy from time signals of dissipative and conservative dynamical systems. Phys. Rev. A **31**, 1872 (1985).

CREMERS, J., HUBLER, A., Construction of differential equations from experimental data. Z. Naturforsch **42**a, 797 (1987).

CRUTCHFIELD, J.P., MCNAMARA, B.S., Equations of motion from a data series. Complex Systems **1**, 417 (1987).

CRUTCHFIELD, J.P., YOUNG, K., Inferring statistical complexity. Phys. Rev. Lett. **63**, 105 (1989).

★CUOMO, K.M., OPPENHEIM, A.V., Circuit implementation of synchronized chaos with applications to communication. Phy. Rev. Lett. **71**, 65 (1993).

CUTLER, C.D., Some results on the behavior and estimation of the fractal dimensions of distributions on attractors. J. Stat. Phys. **62**, 651 (1991).

CVITANOVIĆ, P., Ed., *Universality in Chaos*. Adam Hilger, Bristol (1984).

DÄMMIG, M., MITSCHKE, F., Estimation of Lyapunov exponents from time series: The stochastic case. Phys. Lett. A **178**, 385 (1993).

DEGN, H., HOLDEN, A.V., OLSEN, J.F., Eds. *Chaos in Biological Systems*. Plenum, New York (1986).

DE SOUSA VIEIRA, M., LICHTENBERG, A.J., LIEBERMAN, M.A., Nonlinear dynamics of self-synchronizing systems. Int. J. of Bif. Chaos **1**, 691 (1991).

DEISSLER, R.J., FARMER, J.D., Deterministic noise amplifiers. Physica D **55**, 155 (1992).

DIEBOLD, F.X., NASON, J.M., Nonparametric exchange rate forecasting? J. Intl. Econ. **28**, 315 (1990).

★DING, M., GREBOGI, C., OTT, E., SAUER, T., YORKE, J.A., Plateau onset for correlation dimension: When does it occur? Phys. Rev. Lett. **70**, 3872 (1993).

DING, M., OTT, E., Enhancing synchronism of chaotic systems. Phys. Rev. E **49**, R945 (1994).

★DITTO, W., RAUSEO, S.N., SPANO, M.L., Experimental control of chaos. Phys. Rev. Lett. **65**, 3211 (1990).

DRAZIN, P.G., KING, G.P., Eds. *Interpretation of Time Series from Nonlinear Systems*. (Special issue of Physica D **58**, nos. 1–4.)

★DRESSLER, U., NITSCHE, G., Controlling chaos using time delay coordinates. Phys. Rev. Lett. **68**, 1 (1992).

DVORAK, I., Takens versus multichannel reconstruction in EEG correlation exponent estimates. Phys. Lett. A **151**, 225 (1990).

ECKMANN, J.-P., KAMPHORST, S., RUELLE, D., Recurrence plots of dynamical systems. Europhys. Lett. **4**, 973 (1987).

★ECKMANN, J.-P., KAMPHORST, S.O., RUELLE, D., CILIBERTO, S., Liapunov exponents from time series. Phys. Rev. A **34**, 4971 (1986).

ECKMANN, J.-P., RUELLE, D., Ergodic theory of chaos and strange attractors. Rev. Mod. Phys. **57**, 617 (1985).

★ECKMANN, J.-P., RUELLE, D., Fundamental limitations for estimating dimensions and Lyapunov exponents in dynamical systems. Physica D **56**, 185 (1992).

ELLNER, S., Estimating attractor dimensions from limited data: A new method, with error estimates. Phys. Lett. A **133**, 128 (1988).

ELLNER, S., Detecting low-dimensional chaos in population dynamics data: A critical review. In: *Chaos and Insect Ecology*, J.A. Logan, F.P. Hain, Eds. University of Virginia Press, Blacksburg, VA (1991).

ELLNER, S., GALLANT, A.R., MCCAFFERY, D., NYCHKA, D., Convergence rates and data requirements for Jacobian-based estimates of Lyapunov exponents from data. Phys. Lett. A **153**, 357 (1991).

ELSNER, J.B., TSONIS, A.A., Nonlinear prediction, chaos and noise. Bull. Am. Met. Soc. **73**, 49 (1992).

ENGE, N., BUZUG, T., PFISTER, G., Noise reduction on chaotic attractors. Phys. Lett. A **175**, 402 (1993).

ESSEX, C., LOOKMAN, T., NERENBERG, M.A.H., The climate attractor over short time scales. Nature **326**, 64 (1987).

ESSEX, C., NERENBERG, M.A.H., Comments on 'Deterministic chaos: the science and the fiction' by D. Ruelle. Proc. Roy. Soc. Lond. A **435**, 287 (1991).

★FARMER, J.D., SIDOROWICH, J.J., Predicting chaotic time series. Phys. Rev. Lett. **59**, 845 (1987).

FARMER, J.D., SIDOROWICH, J.J., Exploiting chaos to predict the future and reduce noise. In: *Evolution, Learning and Cognition*, Y.C. Lee, Ed. World Scientific (1988).

FARMER, J.D., SIDOROWICH, J.J., Optimal shadowing and noise reduction. Physica D **47**, 373 (1991).

★FLEPP, L., HOLZNER, R., BRUN, E., FINARDI, M., BADII, R., Model identification by periodic-orbit analysis for NMR-laser chaos. Phys. Rev. Lett. **67**, 2244 (1991).

FOWLER, A.C., KEMBER, G., Delay recognition in chaotic time series. Phys. Lett. A **175**, 402 (1993).

FOWLER, T.B., Application of stochastic control techniques to chaotic nonlinear systems. IEEE Trans. Auto. Control **34**, 201 (1989).

FRAEDRICH, K., Estimating the dimension of weather and climate attractors. J. Atmos. Sci. **43**, 419 (1986).

FRAEDRICH, K., WANG, R., Estimating the correlation dimension of an attracator from noisy and small data sets based on re-embedding. Physica D **65**, 373 (1993).

FRANK, G.W., LOOKMAN, T., NERENBERG, M.A.H., ESSEX, C., Chaotic time series analysis of epileptic seizures. Physica D **34**, 391 (1989).

FRASER, A., Phase space reconstructions from time series. Physica D **34**, 391 (1989).

FRASER, A., SWINNEY, H., Independent coordinates for strange attractors from mutual information. Phys. Rev. A **33**, 1134-1140 (1986).

FRONZONI, L., GIOCONDO, M., Experimental evidence of suppression of chaos by resonant parametric perturbations. Phys. Rev. A **43**, 6483 (1991).

GANG, H., KAIFEN, H., Controlling chaos in systems described by partial differential equations. Phys. Rev. Lett. **71**, 3794 (1993).

GANG, H., ZHILIN, Q., Controlling spatiotemporal chaos in coupled map lattice systems. Phys. Rev. Lett. **72**, 68 (1994).

GANTERT, C., HONERKAMP, J., TIMMER, J., Analyzing the dynamics of hand tremor time series. Biol. Cyber. **66**, 479 (1992).

★GARFINKEL, A., SPANO, M., DITTO, W., WEISS, J., Controlling cardiac chaos. Science **257**, 1230 (1992).

GENCAY, R., Nonlinear prediction of noisy time series with feedforward networks. Phys. Lett. A **187**, 397 (1994).

GERR, N.L., ALLEN, J.C., Stochastic versions of chaotic time series: Generalized logistic and Hénon time series models. Physica D **68**, 232 (1993).

GERSHENFELD, N.A., An experimentalist's introduction to the observation of dynamical systems. In: *Directions in Chaos*, Vol. 2, H. Bai-lin, Ed. World Scientific (1988).

GERSHENFELD, N.A., Dimension measurement on high-dimensional systems. Physica D **55**, 135 (1992).

GIBSON, J., FARMER, J.D., CASDAGLI, M., EUBANK, S., An analytic approach to practical state space reconstruction. Physica D **57**, 1 (1992).

★GILLS, Z., IWATA, C., ROY, R., SCHWARTZ, I., TRIANDAF, I., Tracking unstable steady states: Extending the stability regime of a multimode laser system. Phys. Rev. Lett. **69**, 3169 (1992).

GIONA, M., LENTINI, F., CIMAGALLI, V., Functional reconstruction and local prediction of chaotic time series. Phys. Rev. A **44**, 3496 (1991).

GLASS, L., MACKEY, M.C., Pathological conditions resulting from instabilities in physiological control systems. Ann. N.Y. Acad. Sci. **316**, 214 (1979).

GLASS, L., MACKEY, M.C., *From Clocks to Chaos: The Rhythms of Life*. Princeton Univ. Press (1988).

GLASS, L., SHRIER, A., BÉLAIR, J., Chaotic cardiac rhythms. In: *Chaos*, A.V. Holden, Ed. Princeton Univ. Press (1986).

GOLDBERGER, A.L., RIGNEY, D.R., WEST, B.J., Chaos and fractals in human physiology. Sci. Amer. **263**, 43 (1990).

GOUESBET, G., MAQUET, J., Construction of phenomenological models from numerical scalar time series. Physica D **58**, 202 (1992).

GRASSBERGER, P., Do climatic attractors exist? Nature **323**, 609; **326**, 524 (1986).

GRASSBERGER, P., Finite sample corrections to entropy and dimension estimates. Phys. Lett. **128**, 369 (1988).

GRASSBERGER, P., An optimized box-assisted algorithm for fractal dimensions. Phys. Lett. A **148**, 63 (1990).

★GRASSBERGER, P., HEGGER, R., KANTZ, H., SCHAFFRATH, C., SCHREIBER, T., On noise reduction methods for chaotic data. Chaos **3**, 127 (1993).

GRASSBERGER, P., PROCACCIA, I., Measuring the strangeness of strange attractors. Physica D **9**, 189 (1983).

GRASSBERGER, P., PROCACCIA, I., Characterization of strange attractors. Phys. Rev. Lett. **50**, 346 (1983).

GRASSBERGER, P., PROCACCIA, I., Estimation of the Kolmogorov entropy from a chaotic signal. Phys. Rev. A **28**, 2591 (1983).

GRASSBERGER, P., SCHREIBER, T., SCHAFFRATH, C., Nonlinear time sequence analysis. Int. J. Bif. Chaos **1**, 512 (1991).

GREEN, M.L., SAVIT, R., Dependent variables in broadband continuous time series. Physica D **50**, 521 (1991).

★GUCKENHEIMER, J., BUZYNA, G., Dimension measurements for geostrophic turbulence. Phys. Rev. Lett. **51**, 1438 (1983).

GUPTE, N., AMRITKAR, R.E., Synchronization of chaotic orbits: The influence of unstable periodic orbits. Phys. Rev. E **48**, 1620 (1993).

★HAMMEL, S., A noise reduction method for chaotic systems. Phys. Lett. A **148**, 421 (1990).

HAVSTAD. J.W., EHLERS, C.L., Attractor dimension of nonstationary dynamical systems from small data sets. Phys. Rev. A **39**, 845 (1989).

★HAYES, S., GREBOGI, C., OTT, E., Communicating with chaos. Phys. Rev. Lett. **70**, 3031 (1993).

HE, R., VAIDYA, P.G., Analysis and synthesis of synchronous periodic and chaotic systems. Phys. Rev. A **46**, 7387 (1992).

HEALEY, J.J., Identifying finite dimensional behavior from broadband spectra. Phys. Lett. A **187**, 59 (1994).

HENNEQUIN, D., GLORIEUX, P., Symbolic dynamics in a passive Q-switching laser. Europhys. Lett. **14**, 237 (1991).

HENTSCHEL, H.G.E., JIANG, Z., Learning to control dynamical behavior. Physica D **67**, 141 (1993).

HENTSCHEL, H.G.E., JIANG, Z., Prediction using unsupervised learning. Physica D **67**, 151 (1993).

HUERTA, R., SANTA CRUZ, C., DORRONSORO, J.R., LOPEZ, V., State-space reconstruction using averaged scalar products of the dynamical system flow vectors. Phys. Rev. E **49**, 1962 (1994).

HOGG, T., HUBERMAN, B.A., Controlling chaos in distributed systems. IEEE Trans. Sys. Man. Cyber. **21**, 1325 (1991).

HOLZFUSS, J., PARLITZ, U., Lyapunov exponents from time series. In: *Proceedings of the Conference "Lyapunov Exponents, Oberwolfach 1990,"* L. Arnold, H. Crauel, J.-P. Eckmann, Eds. Lecture Notes in Mathematics. Springer-Verlag (1991).

HSIEH, D., Testing for nonlinear dependence in daily foreign exchange rates. J. Business **62**, 339 (1989).

HSIEH, D., Chaos and nonlinear dynamics: Applications to financial markets. J. Finance **46**, 1839 (1991).

HÜBLER, A., Adaptive control of chaotic systems. Helv. Phys. Acta **62**, 343 (1989).

★HÜBLER, A., LÜSCHER, E., Resonant stimulation and control of nonlinear oscillators. Naturwissenschaften **76**, 67 (1989).

★HUNT, E.R., Stabilizing high-period orbits in a chaotic system: The diode resonator. Phys. Rev. Lett. **67**, 1953 (1991).

★JACKSON, E.A., The entrainment and migration controls of multiple-attractor systems. Phys. Lett. A **151**, 478 (1990).

JACKSON, E.A., On the control of complex dynamical systems. Physica D **50**, 341 (1991).

JACKSON, E.A., Controls of dynamic flows with attractors. Phys. Rev. A **44**, 4839 (1991).

JACKSON, E.A., HÜBLER, A.W., Periodic entrainment of chaotic logistic map dynamics. Physica D **44**, 407 (1990).

JACKSON, E.A., KODOGEORGIOU, A., Entrainment and migration controls of two dimensional maps. Physica D **54**, 253 (1992).

JADITZ, T., SAYERS, C., Is chaos generic in economic data? Int. J. Bif. Chaos **3**, 745 (1993).

JÁNOSI, I.M., TÉL, T., Time series analysis of transient chaos. Phys. Rev. E **49**, 2756 (1994).

JENSEN, M.H., KADANOFF, L.P., LIBCHABER, A., Global universality at the onset of chaos: Results of a forced Rayleigh-Bénard experiment. Phys. Rev. Lett. **55**, 2798 (1985).

JUDD, K., An improved estimator of dimension and some comments on providing confidence intervals. Physica D **56**, 216 (1992).

KADTKE, J.B., BRUSH, J., HOLZFUSS, J., Global dynamical equations and Lyapunov exponents from noisy chaotic time series. Int. J. Bif. Chaos **3**, 607 (1993).

KAPLAN, D.T., COHEN, R.J., Is fibrillation chaos? Circ. Res. **67**, 886 (1990).

★KAPLAN, D.T., GLASS, L., Direct test for determinism in a time series. Phys. Rev. Lett. **68**, 427-430 (1992).

KAPLAN, D.T., GLASS, L., Coarse-grained embeddings of time-series: Random walks, Gaussian random processes, and deterministic chaos. Physica D **64**, 431 (1993).

KAPLAN, D.T., TALAJIC, M., Dynamics of heart rate. Chaos **1**, 251 (1991).

KARLSSON, M., YAKOWITZ, S., Nearest-neighbor methods for nonparametric rainfall-runoff forecasting. Water Resources Res. **23**, 1300 (1987).

KANTZ, H., A robust method to estimate the maximal Lyapunov exponent of a time series. Phys. Lett. A **185**, 77 (1994).

KANTZ, H., SCHREIBER, T., HOFFMAN, I., BUZUG, T., PFISTER, G., FLEPP, L.G., SIMONET, J., BADII, R., BRUN, E., Nonlinear noise reduction: A case study on experimental data. Phys. Rev. E **48**, 1529 (1993).

KEMBER, G., FOWLER, A.C., A correlation function for choosing time delays in phase portrait reconstructions. Phys. Lett. A **179**, 72 (1993).

KENNEL, M.B., ISABELLE, S., Method to distinguish possible chaos from colored noise and to determine embedding parameters. Phys. Rev. A **46**, 3111 (1992).

★KENNEL, M., BROWN, R., ABARBANEL, H., Determining embedding dimension for phase-space reconstruction using a geometrical construction. Phys. Rev. A **45**, 3403-3411 (1992).

KING, G.P., STEWART, I., Phase space reconstruction for symmetric dynamical systems. Physica D **58**, 216 (1992).

KIRBY, M., MIRANDA, R., Nonlinear reduction of high-dimensional dynamical systems via neural networks. Phys. Rev. Lett. **72**, 1822 (1994).

KIVSHAR, Y.S., RÖDELSPERGER, F., BENNER, H., Suppression of chaos by nonresonant parametric perturbations. Phys. Rev. E **49**, 319 (1994).

KOSTELICH, E.J., Problems in estimating dynamics from data. Physica D **58**, 138 (1992).

KOSTELICH, E.J., GREBOGI, C., OTT, E., YORKE, J.A., Higher dimensional targeting. Phys. Rev. E **47**, 305 (1993).

KOSTELICH, E.J., SCHREIBER, T., Noise reduction in chaotic time-series data: A survey of common methods. Phys. Rev. E **48**, 1752 (1993).

KOSTELICH, E.J., SWINNEY, H.L., Practical considerations in estimating dimension from time series. Phys. Scr. **40**, 436 (1989).

★KOSTELICH, E.J., YORKE, J.A., Noise reduction in dynamical systems. Phys. Rev. A **38**, 1649 (1988).

KOSTELICH, E.J., YORKE, J.A., Noise reduction: Finding the simplest dynamical system consistent with the data. Physica D **41**, 183 (1990).

KRONENBERG, F., Menopausal hot flashes: Randomness or rhythmicity. Chaos **1**, 271 (1991).

KURTHS, J., HERZEL, H., An attractor in a solar time series. Physica D **25**, 165 (1987).

LAI, Y.-C., GREBOGI, C., Synchronization of chaotic trajectories using control. Phys. Rev. E **47**, 2357 (1993).

LAI, Y.-C., TÉL, T., GREBOGI, C., Stabilizing chaotic–scattering trajectories using control. Phys. Rev. E **48**, 709 (1993).

LANDA, P., ROSENBLUM, M., Time series analysis for system identification and diagnosis. Physica D **48**, 232 (1991).

LAPEDES, A., FARBER, R., Nonlinear signal processing using neural networks: Prediction and system modeling. Los Alamos technical report LA-UR-87-2662 (1987).

LASKAR, J., FROESCHLÉ, C., CELLETTI, A., The measure of chaos by the numerical analysis of the fundamental frequencies: Applications to the standard mapping. Physica D **56**, 253 (1992).

★LATHROP, D.P., KOSTELICH, E.J., Characterization of an experimental strange attractor by periodic orbits. Phys. Rev. A **40**, 4028 (1989).

LAYNE, S.P., MAYER-KRESS, G., HOLZFUSS, J., Problems associated with dimensional analysis of EEG data. In: *Dimensions and Entropies in Chaotic Systems*, G. Mayer-Kress, Ed. Springer (1986).

LEE, T.-H., WHITE, H., GRANGER, C., Testing for neglected nonlinearity in time series models: A comparison of neural network methods and alternative tests. J. Econometrics **56**, 269 (1993).

LEFEBVRE, J.H., GOODINGS, D.A., KAMATH, M.V., FALLEN, E.L., Predictability of normal heart rhythms and deterministic chaos. Chaos **3**, 267 (1993).

LIEBERT, W., PAWELZIK, K., SCHUSTER, H.G., Optimal embeddings of chaotic attractors from topological considerations. Europhys. Lett. **14**, 521 (1991).

LIEBERT, W., SCHUSTER, H.G., Proper choice of time delay for the analysis of chaotic time series. Phys. Lett. A **142**, 107 (1989).

LIEBOVITCH, L.S., TOTH, T., A fast algorithm to determine fractal dimension by box counting. Phys. Lett. A **141**, 386 (1989).

LIMA, R., PETTINI, M., Suppression of chaos by resonant parametric perturbations. Phys. Rev. A **41**, 726 (1990).

LINSAY, P., An efficient method of forecasting chaotic time series using linear interpolation. Phys. Lett. A **153**, 353 (1991).

LIU, Y., RIOS LEITE, J.R., Control of Lorenz chaos. Phys. Lett. A **185**, 35 (1994).

LONGTIN, A., Nonlinear forecasting of spike trains from sensory neurons. Int. J. Bif. Chaos **3**, 651 (1993).

LORENZ, E., Atmospheric predictability as revealed by naturally occurring analogies. J. Atmos. Sci. **26**, 636 (1969).

MACKEY, M., GLASS, L., Oscillation and chaos in physiological control systems. Science **197**, 287 (1977).

MALINETSKII, G.G., POTAPOV, A.B., RAKHMANOV, A.I., Limitations of delay reconstruction for chaotic dynamical systems. Phys. Rev. E **48**, 904 (1993).

MANDELL, A.J., SELZ, K.A., Brain stem neuronal noise and neocortical "resonance". J. Stat. Phys. **70**, 355 (1993).

MAÑÉ, R., On the dimension of the compact invariant sets of certain nonlinear maps. Lecture Notes in Math. **898**, Springer-Verlag (1981).

MARITAN, A., BANAVAR, J.R., Chaos, noise, and synchronization. Phys. Rev. Lett. **72**, 1451 (1994).

MARTEAU, P.-F., ABARBANEL, H.D., Noise reduction in chaotic time series using scaled probabilistic methods. J. Nonlin. Sci. **1**, 313 (1991).

MATÍAS, M.A., GÜÉMEZ, J., Stabilization of chaos by proportional pulses in the system variables. Phys. Rev. Lett. **72**, 1455 (1994).

MAYER-KRESS, G., Ed., *Dimensions and Entropies in Chaotic Systems: Quantification of Complex Behavior*. Springer-Verlag, Berlin (1986).

MEES, A.I., Dynamical systems and tesselations detecting determinism in data. Int. J. Bif. Chaos **1**, 777 (1991).

MEES, A.I., Parsimonious dynamical reconstruction. Int. J. Bif. Chaos **3**, 669 (1993).

MEES, A.I., AIHARA, K., ADACHI, M., JUDD, K., IKEGUCHI, T., MATSUMOTO, G., Deterministic prediction and chaos in squid axon response. Phys. Lett. A **169**, 41 (1992).

MEES, A.I., JUDD, K. Dangers of geometric filtering. Physica D **68**, 427 (1993).

MEES, A.I., RAPP, P., JENNINGS, L., Singular-value decomposition and embedding dimension. Phys. Rev. A **36**, 340 (1987).

MEHTA, N.J., HENDERSON, R.M., Controlling chaos to generate aperiodic orbits. Phys. Rev. A **44**, 4861 (1991).

MEUCCI, R., GADOMSKI, W., CIOFINI, M., ARECCHI, F.T., Experimental control of chaos by means of weak parametric perturbations. Phys. Rev. E **49**, R2528 (1994).

MEYER, T.P., PACKARD, N.H., Local forecasting of high-dimensional chaotic dynamics. In: *Nonlinear Modeling and Forecasting*, Santa Fe Institute Studies in the Science of Complexity XII, M. Casdagli and S. Eubank, Eds. Addison-Wesley, Reading, MA (1992).

MINDLIN, G.B., HOU, X.-J., SOLARI, G., NATIELLO, M.A., GILMORE, R., TUFILLARO, N.B., Classification of strange attractors by integers. Phys. Rev. Lett. **64**, 2350 (1990).

MITSCHKE, F., Acausal filters for chaotic signals. Phys. Rev. A **41**, 1169 (1990).

MITSCHKE, F., MÖLLER, M., LANGE, W., Measuring filtered chaotic signals. Phys. Rev. A **37**, 4518 (1988).

MOLINARI, L., A note on bias correction for correlation dimension. Phys. Lett. A **187**, 163 (1994).

MÖLLER, M., LANGE, W., MITSCHKE, F., ABRAHAM, N.B., HÜBNER, U., Errors from digitizing and noise in estimating attractor dimensions. Phys. Lett. A **138**, 176 (1989).

MULDOON, M.R., MACKAY, R.S., HUKE, J.P., BROOMHEAD, D.S., Topology from a time series. Physica D **65**, 1 (1993).

MUNDT, M.D., MAGUIRE, W.B., CHASE, R.R.P., Chaos in the sunspot cycle: Analysis and prediction. J. Geophys. Res. **96**, 1705 (1991).

MURALI, K., LAKSHMANAN, R.E., Transmission of signals by synchronization in a chaotic Van der Pol – Duffing oscillator. Phys. Rev. E **48**, 1624 (1993).

MURRAY, D.B., Forecasting a chaotic time series using an improved metric for embedding space. Physica D **68**, 318 (1993).

NERENBERG, M.A., ESSEX, C., Correlation dimension and systematic geometric effects. Phys. Rev. A **42**, 7065 (1990).

NESE, J.M., Quantifying local predictability in phase space. Physica D **35**, 237 (1989).

NEWELL, T.C., ALSING, P.M., GAVRIELIDES, A., KOVANIS, V., Synchronization of chaotic diode resonators by occasional proportional feedback. Phys. Rev. Lett. **72**, 1647 (1994).

NEWELL, T.C., ALSING, P.M., GAVRIELIDES, A., KOVANIS, V., Synchronization of chaos using proportional feedback. Phys. Rev. E **49**, 313 (1994).

NICOLIS, C., NICOLIS, G., Is there a climatic attractor? Nature **311**, 529 (1984).

NYCHKA, D., ELLNER, S., MCCAFFREY, D., GALLANT, A.R., Finding chaos in noisy systems. J. Roy. Stat. Soc. B **54(2)**, 399 (1992).

OLSEN, L.F., SCHAFFER, W.M., Chaos versus noisy periodicity: Alternative hypotheses for childhood epidemics. Science **249**, 499 (1990).

OSBORNE, A.R., KIRWAN, A.D., PROVENZALE, A., BERGAMASCO, L., A search for chaotic behavior in large and mesoscale motions in the Pacific Ocean. Physica D **23**, 75 (1986).

OSBORNE, A.R., PROVENZALE, A., Finite correlation dimension for stochastic systems with power-law spectra. Physica D **35**, 357 (1989).

★OTT, E., GREBOGI, C., YORKE, J.A., Controlling chaos. Phys. Rev. Lett. **64**, 1196 (1990).

★PACKARD, N., CRUTCHFIELD, J., FARMER, D., SHAW, R., Geometry from a time series. Phys. Rev. Lett. **45**, 712 (1980).

PALUŠ, M., DVOŘÁK, I., Singular-value decomposition in attractor reconstruction: Pitfalls and precautions. Physica D **55**, 221 (1992).

★PAPOFF, F., FIORETTI, A., ARIMONDO, E., MINDLIN, G.B., SOLARI, H., GILMORE, R., Structure of chaos in the laser with saturable absorber. Phys. Rev. Lett. **68**, 1128 (1992).

PARKER, T.S., CHUA, L.O., *Practical Numerical Algorithms for Chaotic Systems*. Springer-Verlag, New York (1990).

★PARLITZ, U., Identification of true and spurious Lyapunov exponents from time series. Int. J. Bif. Chaos **2**, 155 (1992).

PARLITZ, U., ERGEZINGER, S., Robust communication based on chaotic spreading sequences. Phys. Lett. A **188**, 146 (1994).

PARMANANDA, P., SHERARD, P., ROLLINS, R.W., DEWALD, H.D., Control of chaos in an electrochemical cell. Phys. Rev. E **47**, 3003 (1993).

PAWELZIK, K., SCHUSTER, H.G., Generalized dimensions and entropies from a measured time series. Phys. Rev. A **35**, 481 (1987).

★PECORA, L.M., CARROLL, T.L., Synchronization in chaotic systems. Phys. Rev. Lett. **64**, 821 (1990).

PECORA, L.M., CARROLL, T.L., Driving systems with chaotic signals. Phys. Rev. A **44**, 2374 (1991).

PENG, B., PETROV, V., SHOWALTER, K., Controlling chemical chaos. J. Phys. Chem. **95**, 4957 (1991).

★PETROV, V., GÁSPÁR, V., MASERE, J., SHOWALTER, K., Controlling chaos in the Belousov-Zhabotinsky reaction. Nature **361**, 240 (1993).

PFISTER, G., BUZUG, T., ENGE, N., Characterization of experimental time series from Taylor-Couette flow. Physica D **58**, 441 (1992).

PRICE, C.P., PRICHARD, D., On the embedding statistic. Phys. Lett. A **184**, 83 (1993).

PROVENZALE, A., OSBORNE, A.R., SOJ, R., Convergence of the K_2 entropy for random noises with power law spectra. Physica D **47**, 361 (1991).

PROVENZALE, A., SMITH, L.A., VIO, R., MURANTE, G., Distinguishing between low-dimensional dynamics and randomness in measured time series. Physica D **58**, 31 (1992).

PYRAGAS, K. Predictable chaos in slightly perturbed unpredictable chaotic systems. Phys. Lett. A **181**, 203 (1993).

QIN, F., WOLF, E.E., CHANG, H.-C., Controlling spatiotemporal patterns on a catalytic wafer. Phys. Rev. Lett. **72**, 1459 (1994).

QU, Z., HU, G., MA, B., Controlling chaos via continuous feedback. Phys. Lett. A **178**, 265 (1993).

RAJASEKAR, S., LAKSHMANAN, M., Controlling of chaos in Bonhoeffer-Van der Pol oscillator. Int. J. Bif. Chaos **2**, 201 (1992).

RAPP, P.E., ALBANO, A.M., SCHMAH, T.I., FARWELL, L.A., Filtered noise can mimic low-dimensional chaotic attractors. Phys. Rev. E **47**, 2289 (1993).

REYL, C., FLEPP, L., BADII, R., BRUN, E., Control of NMR–laser chaos in high-dimensional embedding space. Phys. Rev. E **47**, 267 (1993).

ROLLINS, R.W., PARMANANDA, P., SHERARD, P., Controlling chaos in highly dissipative systems: A simple recursive algorithm. Phys. Rev. E **47**, 780 (1993).

★ROMEIRAS, F., GREBOGI, C., OTT, E., DAYAWANSA, W.P., Controlling chaotic dynamical systems. Physica D **58**, 165 (1992).

ROSENSTEIN, M.T., COLLINS, J.J., DELUCA, C.J., A practical method for calculating largest Lyapunov exponents from small data sets. Physica D **65**, 117 (1993).

ROUX, J.C., ROSSI, A., BACHELART, S., VIDAL, C., Representation of a strange attractor from an experimental study of chemical turbulence. Phys. Lett. A**77**, 391 (1980).

ROUX, J.C., SWINNEY, H., Topology of chaos in a chemical reaction. In: Nonlinear Phenomena in Chemical Dynamics, C. Vidal and A. Pacault, Eds. (Springer, Berlin, 1981).

ROY, R., THORNBURG, K.S., Experimental synchronization of chaotic lasers. Phys. Rev. Lett. **72**, 2009 (1994).

RUELLE, D., Deterministic chaos: The science and the fiction. Proc. Roy. Soc. Lond. A **427**, 241 (1990).

SANO, M., SAWADA, Y., Measurement of the Lyapunov spectrum from chaotic time series. Phys. Rev. Lett. **55**, 1082 (1985).

SATO, S., SANO, M., SAWADA, Y., Practical methods of measuring the generalized dimension and the largest Lyapunov exponent in high dimensional chaotic systems. Prog. Theor. Phys. **77**, 1 (1987).

SAUER, T., A noise reduction method for signals from nonlinear systems. Physica D **58**, 193 (1992).

★SAUER, T., Time series prediction using delay coordinate embedding. In: *Time Series Prediction: Forecasting the Future and Understanding the Past*, Santa Fe Institute Studies in the Science of Complexity XV, A.S. Weigend and N.A. Gershenfeld, Eds. Addison-Wesley, Reading, MA (1993).

SAUER, T., YORKE, J.A., How many delay coordinates do you need? Int. J. Bif. Chaos **3**, 737 (1993).

SAUER, T., YORKE, J.A., CASDAGLI, M., Embedology. J. Stat. Phys. **65**, 579 (1991).

SCARGLE, J.D., An introduction to chaotic and random time series analysis. Int. J. Imag. Sys. Tech. **1**, 243 (1989).

SCHIFF, S.J., CHANG, T., Differentiation of linearly correlated noise from chaos in a biologic system using surrogate data. Biol. Cyb. **67**, 387 (1992).

SCHREIBER, T., An extremely simple nonlinear noise reduction method. Phys. Rev. E **47**, 2401 (1993).

SCHREIBER, T., Determination of the noise level of chaotic time series. Phys. Rev. E **48**, 13 (1993).

SCHREIBER, T., GRASSBERGER, P., A simple noise-reduction method for real data. Phys. Lett. A **160**, 411 (1991).

SEPULCHRE, J.A., BABLOYANTZ, A., Controlling chaos in a network of oscillators. Phys. Rev. E **48**, 945 (1993).

SHARIFI, M.B., GEORGAKAKOS, K.P., RODRIGUEZ-ITURBE, I., Evidence of deterministic chaos in the pulse of storm rainfall. J. Atmos. Sci. **47**, 888 (1990).

SCHEINKMAN., J, LEBARON, B., Nonlinear dynamics and stock returns. J. Business **62**, 311 (1989).

★SHINBROT, T., DITTO, W., GREBOGI, C., OTT, E., SPANO, M., YORKE, J.A., Using the sensitive dependence of chaos (the "butterfly effect") to direct trajectories in an experimental chaotic system. Phys. Rev. Lett. **68**, 2863 (1992).

SHINBROT, T., OTT, E., GREBOGI, C., YORKE, J.A., Using chaos to direct orbits to targets in systems described by a one-dimensional map. Phys. Rev. A **45**, 4165 (1992).

SHINBROT, T., OTT, E., GREBOGI, C., YORKE, J.A., Using chaos to direct trajectories to targets. Phys. Rev. Lett. **65**, 3215 (1990).

SHINBROT, T., OTT, E., GREBOGI, C., YORKE, J.A., Using chaos to target stationary states of flows. Phys. Lett. A **169**, 349 (1992).

Shinbrot, T., Grebogi, C., Ott, E., Yorke, J.A., Using small perturbations to control chaos. Nature **363**, 411 (1993).

Simoyi, R., Wolf, A., Swinney, H.L., One-dimensional dynamics in a multicomponent chemical reaction. Phys. Rev. Lett. **49**, 245 (1982).

★Singer, J., Wang, Y.-Z., Bau, H.H., Controlling a chaotic system. Phys. Rev. Lett. **66**, 1123 (1991).

Sinha, S., Ramaswamy, R., Rao, J.S., Adaptive control in nonlinear dynamics. Physica D **43**, 118 (1990).

Smith, L.A., Intrinsic limits on dimension calculations. Phys. Lett. A **133**, 283 (1988).

Smith, L.A., Identification and prediction of low dimensional dynamics. Physica D **58**, 50 (1992).

Smith, R.L., Optimal estimation of fractal dimension. In: *Nonlinear Modeling and Forecasting*, Santa Fe Institute Studies in the Science of Complexity XII, M. Casdagli and S. Eubank, Eds. Addison-Wesley, Reading, MA (1992).

Smith, R.L., Estimating dimension in noisy chaotic time series. J. Roy. Stat. Soc. B **54**, 329 (1992).

★So, P., Ott, E., Dayawansa, W.P., Observing chaos: Deducing< and tracking the state of a chaotic system from limited observation. Phys. Lett. A **176**, 421 (1993).

★Sommerer, J.C., Ditto, W.L., Grebogi, C., Ott, E., Spano, M., Experimental confirmation of the theory for critical exponents of crises. Phys. Lett. A **153**, 105 (1991).

Stone, L., Coloured noise or low-dimensional chaos? Proc. Roy. Soc. London B **250**, 77 (1992).

Stoop, R., Parisi, J., Calculation of Lyapunov exponents avoiding spurious elements. Physica D **50**, 89 (1991).

Sugihara, G., Grenfell, B., May, R., Distinguishing error from chaos in ecological time series. Phil. Trans. Roy. Soc. London B **330**, 235 (1990).

★Sugihara, G., May, R.M., Nonlinear forecasting as a way of distinguishing chaos from measurement error in time series. Nature **344**, 734 (1990).

Szpiro, G.G., Measuring dynamical noise in dynamical systems. Physica D **65**, 289 (1993).

Takens, F., Detecting strange attractors in turbulence. Lecture Notes in Math. **898**, Springer-Verlag (1981).

Takens, F., Detecting nonlinearities in stationary time series. Int. J. Bif. Chaos **3**, 241 (1993).

Tarroja, M.F.H., Sicam, V.A., The effect of smoothing on the dynamical behavior of noisy chaotic signals. Int. J. Bif. Chaos **3**, 1591 (1993).

THEILER, J., Spurious dimension from correlation algorithms applied to limited time series data. Phys. Rev. A **34**, 2427 (1986).

THEILER, J., Efficient algorithm for estimating the correlation dimension from a set of discrete points. Phys. Rev. A **36**, 4456 (1987).

THEILER, J., Lacunarity in a best estimator of fractal dimension. Phys. Lett. A **133**, 195 (1988).

THEILER, J., Some comments on the correlation dimension of $1/f^\alpha$ noise. Phys. Lett. A **155**, 480 (1989).

THEILER, J., Statistical precision of dimension estimators. Phys. Rev. A **41**, 3038 (1990).

THEILER, J., Estimating fractal dimensions. J. Opt. Soc. Am. **A7**, 1055 (1990).

THEILER, J., EUBANK, S., Don't bleach chaotic data. Chaos **3**, 771 (1993).

★THEILER, J., EUBANK, S., LONGTIN, A., GALDRAKIAN, B., FARMER, J.D., Testing for nonlinearity in time series: The method of surrogate data. Physica D **58**, 77 (1992).

THEILER, J., GALDRAKIAN, B., LONGTIN, A., EUBANK, S., FARMER, J.D., Using surrogate data to detect nonlinearity in time series. In: *Nonlinear Modeling and Forecasting*, M. Casdagli, S. Eubank, Eds. Addison-Wesley, Reading, MA (1992).

THEILER, J., LINSAY, P.S., RUBIN, D.M., Detecting nonlinearity in data with long coherence times. In: *Time Series Prediction: Forecasting the Future and Understanding the Past*, Santa Fe Institute Studies in the Science of Complexity XV, A.S. Weigend, N.A. Gershenfeld, Eds. Addison-Wesley, Reading, MA (1993).

TIMMER, J., GANTERT, C., DEUSCHL, G., HONERKAMP, J., Characteristics of hand tremor time series. Biol. Cybern. **70**, 75 (1993).

TONG, H., *Nonlinear Time Series Analysis: A Dynamical Systems Approach*. Oxford University Press (1990).

TOWNSHEND, B., Nonlinear prediction of speech signals. In: *Nonlinear Modeling and Forecasting*, M. Casdagli, S. Eubank, Eds. Addison-Wesley, Reading, MA (1992).

TRIANDAF, I., SCHWARTZ, I., Stochastic tracking in nonlinear dynamical systems. Phys. Rev. E **48**, 718 (1993).

TSONIS. A.A., *Chaos: From Theory to Applications*. Plenum, New York (1992).

TSONIS, A.A., ELSNER, J.B., The weather attractor on very short time scales. Nature **333**, 545 (1988).

TSONIS, A.A., ELSNER, J.B., Comments on dimension analysis of climatic data. J. Climate **3**, 1502 (1990).

TUFILLARO, N.B., HOLZNER, R., FLEPP, L., BRUN, E., FINARDI, M., BADII, R., Template analysis for a chaotic NMR laser. Phys. Rev. A **44**, 4786 (1991).

TUFILLARO, N.B., SOLARI, H.G., GILMORE, R., Relative rotation rates – fingerprints for strange attractors. Phys. Rev. A **41**, 5717 (1990).

VAUTARD, R., YIOU, P., GHIL, M., Singular spectrum analysis: A toolkit for short, noisy signals. Physica D **58**, 95 (1992).

VOHRA, S., SPANO, M., SHLESINGER, M., PECORA, L., DITTO, W., *Proceedings of the First Experimental Chaos Conference*. World Scientific, Singapore (1992).

WALES, D.J., Calculating the rate of loss of information from chaotic time series by forecasting. Nature **350**, 485 (1991).

★WANG, H., ABED, E.H., Bifurcation control of chaotic dynamical systems. Proceedings of IFAC Nonlinear Control Systems Design Symposium, Bordeaux (1992).

WAYLAND, R., BROMLEY, D., PICKETT, D., PASSAMANTE, A., Recognizing determinism in a time series. Phys. Rev. Lett. **70**, 580-582 (1993).

WEIGEND, A., GERSHENFELD, N.A., eds., *Time Series Prediction: Forecasting the Future and Understanding the Past*, Santa Fe Institute Studies in the Science of Complexity XV, Addison-Wesley, Reading, MA (1993).

WEIGEND, A., HUBERMAN, B., RUMELHART, D., Predicting the future: A connectionist approach. Int. J. Neur. Sys. **7**, 403 (1990).

WIENER, N., Nonlinear prediction and dynamics. In: Proceedings of the Third Berkeley Symposium. J. Neyman, Ed. University of California Press, Berkeley (1956).

WOLF, A., SWIFT, J.B., SWINNEY, H., VASTANO, J.A., Determining Lyapunov exponents from a time series. Physica D **16**, 285 (1985).

YAMAMOTO, Y., HUGHSON, R.L., Extracting fractal components from time series. Physica D **68**, 250 (1993).

YULE, G., On a method of investigating periodicity in disturbed series with special reference to Wolfer's sunspot numbers. Phil. Trans. Roy. Soc. London A **226**, 267 (1927).

ZENG, X., EYKHOLT, R., PIELKE, R.A., Estimating the Lyapunov exponent spectrum from short time series of low precision. Phys. Rev. Lett. **66**, 3229 (1991).

Index